International Conference on Water Resource and Environment 2016 (WRE 2016)

IOP Conference Series: Earth and Environmental Science
Volume 39

Shanghai, China
23 – 26 July 2016

Editor:

Miklas Scholz

ISBN: 978-1-5108-2910-7
ISSN: 1755-1307

Printed from e-media with permission by:

Curran Associates, Inc.
57 Morehouse Lane
Red Hook, NY 12571

Some format issues inherent in the e-media version may also appear in this print version.

Printed by Curran Associates, Inc. (2016)

For permission requests, please contact the Institute of Physics at the address below.

Institute of Physics
Dirac House, Temple Back
Bristol BS1 6BE UK

Phone: 44 1 17 929 7481
Fax: 44 1 17 920 0979

techtracking@iop.org

Additional copies of this publication are available from:

Curran Associates, Inc.
57 Morehouse Lane
Red Hook, NY 12571 USA
Phone: 845-758-0400
Fax: 845-758-2633
Email: curran@proceedings.com
Web: www.proceedings.com

Table of contents

Volume 39

International Conference on Water Resource and Environment 2016 (WRE2016)
23–26 July 2016, Shanghai, China

Accepted papers received: 29 July 2016
Published online: 1 September 2016

Preface

Papers

IOP Conf. Series: Earth and Environmental Science **39** (2016) 011001 doi:10.1088/1755-1315/39/1/011001

Preface

The *2016 International Conference on Water Resource and Environment (WRE 2016)* was held from July 23rd to July 26th, 2016 in Shanghai, the biggest city in China and one of the largest urban areas in the world. The technical program comprised eleven international keynote speakers, four interesting oral presentation sessions and a stimulating poster session.

The key focus of the conference was on the areas "Wastewater and Water Treatment", "Soil, Water and Environment" and "Water Resource and Hydrology". A double presentation session was necessary for the latter area.

This volume is based on the proceedings of the WRE 2016 and contains 73 specially selected manuscripts submitted to the WRE 2016 conference. The electronic submission and handling of manuscripts via the conference website, including the selection of reviewers and evaluation of manuscripts, were comparable to the procedures usually applied to manuscripts submitted as regular contributions to high-ranking journals.

The organization of this conference and the preparation of this volume would have been impossible without the tremendous efforts and dedication of many individuals as well as a large team of reviewers. We express our sincere thanks to all authors and presenters for their valuable contributions.

The 2017 International Conference on Water Resources and Environment (WRE 2017) will be held in Qingdao, which is one of the most beautiful tourism cities in China. Everybody is welcome to submit high quality papers to WRE 2017!

Editor

Prof. Miklas Scholz

Lund University, Sweden

Peer review statement

All papers published in this volume of *IOP Conference Series: Earth and Environmental Science* have been peer reviewed through processes administered by the proceedings Editors. Reviews were conducted by expert referees to the professional and scientific standards expected of a proceedings journal published by IOP Publishing.

IOP Conf. Series: Earth and Environmental Science **39** (2016) 012001 doi:10.1088/1755-1315/39/1/012001

Application of chaotic prediction model based on wavelet transform on water quality prediction

L Zhang[1], Z H Zou[1, 2] and Y F Zhao[1]

[1]School of Economics and Management, Beihang University, Beijing 100191, China

E-mail: zouzhihong@buaa.edu.cn

Abstract. Dissolved oxygen (DO) is closely related to water self-purification capacity. In order to better forecast its concentration, the chaotic prediction model, based on the wavelet transform, is proposed and applied to a certain monitoring section of the Mentougou area of the Haihe River Basin. The result is compared with the simple application of the chaotic prediction model. The study indicates that the new model aligns better with the real data and has a higher accuracy. Therefore, it will provide significant decision support for water protection and water environment treatment.

1. Introduction

The first paragraph after a heading is not indented (Bodytext style). Along with economic development and improvements in people's standard of living, bodies of water are facing more and more serious pollution and destruction. However, polluted water has certain self-purification abilities. It can reduce the concentration of pollutants to normal levels by physical, chemical and biological functions. Dissolved oxygen (DO) is the basis for studying water bodies' self-purification abilities. Consequently, research concerning trends of DO in water and predicting its concentration has substantial theoretical and realistic significance. Because of the complexity of the water environment and the limitation of single methods, combined forecasting methods have been widely used, such as the radial basis function (RBF) water quality prediction model based on particle swarm optimization, the BP neural network prediction model based on ant-colony optimization and the gray and neural network combination model [1-3]. These methods led to accurate predictions of water quality trends and concentrations of pollutants.

Water environment systems are quite complex. The DO value could be affected by plenty of factors. As a signal processing technology, the wavelet transform has been widely used recently. The decomposition and reconstruction of original signals can lead to better extraction of the feature information of the original signal [4]. The chaotic prediction model based on the wavelet transform is established by combining the wavelet transform and the chaotic prediction model, which has simple calculations and high precision. Comparing it with the chaotic prediction model proves the validity of the new model.

2. The chaotic prediction model based on wavelet transform

2.1. Preparing the new file with

The first paragraph after a heading is not indented (Bodytext style). Chaos comes from nonlinear dynamic systems. It is a kind of pseudo randomness showed by deterministic systems. Thus,

IOP Conf. Series: Earth and Environmental Science **39** (2016) 012001 doi:10.1088/1755-1315/39/1/012001

seemingly random phenomena produced in chaotic systems are predictable. Chaos is a kind of resilience of periodic motion produced by deterministic systems. This motion is very sensitive to initial conditions, which makes it suitable for short-term forecasting [5]. Wavelet analysis has developed in the past 20 years. As an analytic method of wavelet analysis, the wavelet transform makes up for the defects of the discrete Fourier transform with a fixed window and no discrete orthogonal basis. The multi-scale discrete wavelet transform can decompose the original water quality time series on different temporal resolutions [6] and gain high frequency and low frequency sequences with more single frequencies. This can strengthen the chaotic features of sequences and make them more predictable.

The steps of the chaotic prediction model based on the wavelet transform are as follows:

- Perform wavelet decomposition of the original time series data based on the à Trous algorithm.

Many algorithms of the wavelet transform require down sampling of the original sequences. The resulted approximation and detail sequences contain only half of the points of the original sequences, which decreases prediction accuracy. With its easy calculations and translational invariance, the à Trous algorithm is conducive for extracting the feature information of the original sequences. The choice of the wavelet function is an important factor affecting the accuracy of the prediction. After repeated trial calculations, db2 of the Daubechies wavelet group was picked for this model. The algorithm decomposition process is as follows:

Let the original sequence be c0(t), t=0,1,2, …,n-1. N is the length of the sequence. Then

$$c_j(t)=\sum_l h(l)c_{j-1}(t+2^{j-1}l), j=1,2,\cdots k \quad (1)$$

$$w_j(t)=c_{j-1}(t)-c_j(t), j=1,2,\cdots k \quad (2)$$

where k is the decomposition scale, $h(l)$ is the low-pass filter and $cj(t)$ and $wj(t)$ are the scale coefficients and wavelet coefficients under scale j, respectively.

- Refactor the decomposed scale coefficients and wavelet coefficients.

The scale coefficients and wavelet coefficients can be refactored to get original time sequences, namely:

$$c_0(t)=\sum_{j=1}^{k} w_j(t)+c_k(t) \quad (3)$$

where ck is an approximate sequence, and $\sum wj$ is a detail sequence, denoted by dk. With a single frequ ency component and simpler sequence structure, these two sequences are more predictable.

- Test the chaos characteristics for ck and dk . Establish the chaotic model to predict sequences with chaos characteristics. Use the neural network method to predict sequences without chaos characteristics. Add the two predicted outcomes together to get the final prediction result.

The chaos characteristics of the chaotic time sequences have to be tested before prediction, which mainly includes phase-space reconstruction and the calculation of the maximum Lyapunov exponent [7]. The chaotic forecasting model can predict the time series with chaos characteristics. The chaos weighted first-order local prediction method believes that the prediction center in the phase space has approximately a first-order linear relationship with its next point. This relationship is approximately the same as the relationship between point F, which is nearest to the prediction center by Euclidean distance and its next point, namely: Y(q+1)=a+bY(q). The specific steps are as follows:

a. Reconstruct the phase space and determine the embedding delay time[8] and embedding dimension [9].

b. Calculate the F points nearest the final point and their weights.

$$p_j=\frac{\exp(-(d_j-d_{min}))}{\sum_{j=1}^{n}\exp(-(d_j-d_{min}))} \quad (4)$$

where j=1,2,⋯,n. dj is the Euclidean distance between the first j point and the final point. $dmin$ is the minimum among di。

c. Use the least squares method to fit the first-order linear relationship between point F and its next point, and then obtain the corresponding coefficient fitting coefficient, namely:

$$y(N_j+1) = a_j e + b_j y(N_j) \qquad (5)$$

where $y(Nj+1)$ is the next point of $y(Nj)$ and aj、bj are fitting coefficients. eT=(1,1,...,1)F.

d. To predict the point, weight the gained fitting coefficients and substitute to the prediction center and points waiting for prediction.

e. Separate the prediction of the original sequence from the point to be predicted.

3. Case study

3.1. Date source

The research object is the average weekly monitoring value of DO for 258 weeks between 2010 and 2014 of a monitoring section in the Mentougou section of the Haihe river Basin. All data comes from China's environmental monitoring network.

3.2. Establishment of prediction model

The model is established based on the DO concentrations of the former 256 weeks to predict concentrations in the last two weeks. The DO concentration curve of the former 256 weeks in this district is shown in Figure 1.

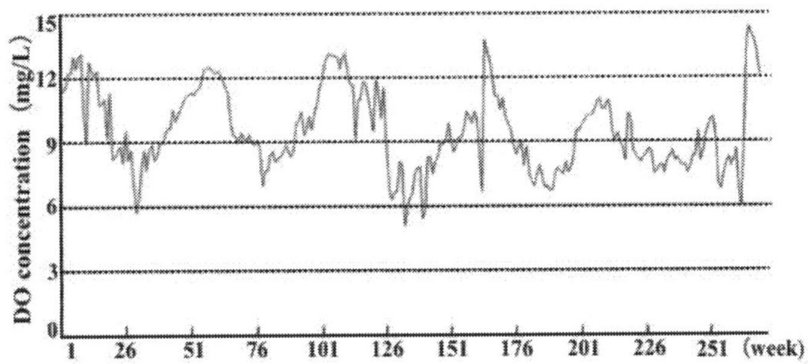

Figure 1. DO concentration for 256 weeks of a monitoring section of Haihe.

When using the auto-correlation method to calculate the embedding time delay τ of DO concentration sequence, we calculate the auto-correlation coefficient under different time delays. When the auto-correlation coefficient becomes negative for the first time, the corresponding time delay is the result that we desire. According to Table 1, τ equals 13.

Table 1. Auto-correlation coefficient changing with the embedding time delay.

T	1	2	3	4	5	6	7	8	9	10	11	12	13
R	0.85	0.76	0.71	0.66	0.62	0.60	0.66	0.47	0.36	0.26	0.17	0.09	-0.02

The cao algorithm is applied to calculate changes of $E1(m)$ and $E2(m)$ along with the embedding dimensions (see Figure 2) and $m \geqslant 3$.

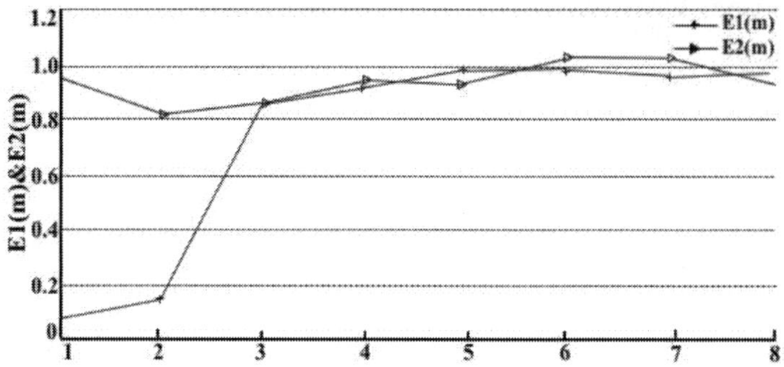

Figure 2. Changes of E1(m) and E2(m) along with embedding dimensions.

The DO concentration sequence has an obvious annually periodic feature, namely the average orbital period $\tau p \approx 52$weeks, which equals the result of FFT. This provides important additional information for reasonably determining time delay and dimensions. Embedding dimensions separately determined can hardly consider the combination of both. Given the principle of $\tau p \approx \tau w = \tau(m-1)$, we chose $\tau=13$, $m=5$, so $\tau w=52$. This was plugged into the minimum date account method to calculate the maximum Lyapunov exponent of the DO concentration sequence. The result was 0.0025; the sequence was a chaotic sequence. Meantime, *E2(m)* changed along with m, which also shows the chaotic characters of the sequence.

The appropriate decomposition scale should be selected for wavelet decomposition. The standard for choosing the decomposition level is minimizing the prediction error. Normally, the decomposition level is 3~5. To be prudent, a decomposition level of 1~5 was calculated, and the results showed that 3 is the best choice.

Next, we refactored the decomposed scale coefficients and the wavelet coefficients to get an approximating sequence and a detail sequence (see Figure 3).

Then, we selected parameters for the approximating sequence and the detail sequence and calculated the maximum Lyapunov exponent (see Table 2). Results showed that the maximum Lyapunov exponent for both sequences were positive, which indicates chaotic features. Thus, the chaotic prediction model can be applied to these two sequences.

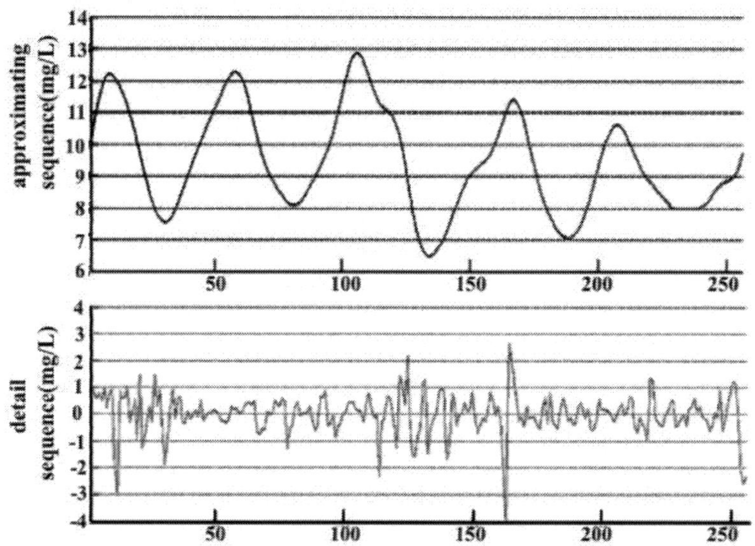

Figure 3. Approximate sequence and detail sequence after wavelet decomposition and refactoring.

Table 2. Parameters of approximating and detail sequences based on maximum Lyapunov exponent.

Research Object	τ	m	Maximum Lyapunov exponent
Approximating Sequence *ck*	13	5	0.0030
Detail Sequence *dk*	2	4	0.0017

3.3. Model fitting and prediction

3.3.1. Model fitting. The combined prediction model was applied to fit the data of the previous 256 weeks and the predicted DO concentration of the upcoming 2 weeks. The two results were added together to get the fitting result of the combined model (see Figure 4). According to the figure, the chaotic prediction model based on the wavelet transform was the best fit to the trend.

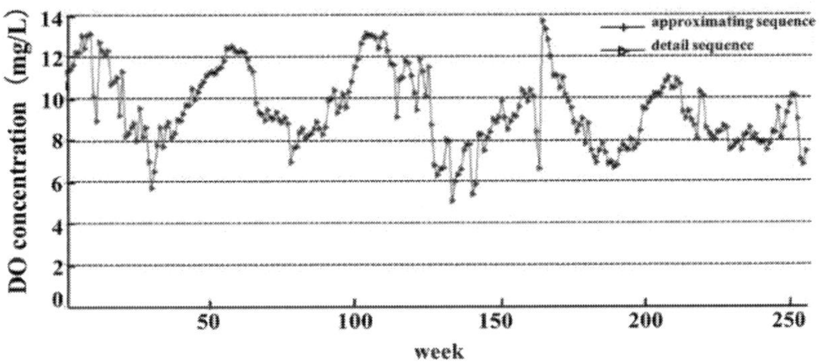

Figure 4. Model fitting figure.

3.3.2. Model prediction. The nonlinear system is very sensitive to initial conditions, which makes it suitable for short-term forecasting. The chaotic prediction model (model 1), the artificial neural network prediction model (model 2) and the chaotic prediction model based on the wavelet transform (model 3) were applied to forecast DO concentration in the next two weeks. The MRE (mean relative error) and RMSE (root mean square error) were used to evaluate the models (see Table 3). The RMSE is very sensitive to the maximum and minimum error, which enables it effectively reflect the accuracy of the prediction results.

Table 3. Comparison of two prediction models.

Model	Model 1 (Week 257)	Model 1 Week 258)	Model 2 (Week257)	Model 2 (Week258)	Model 3 (Week 257)	Model 3 (Week 258)
Predictor	7.670	7.684	8.196	8.873	8.049	8.829
Real value	8.02	8.27	8.02	8.27	8.02	8.27
Errors	0.349	0.585	0.176	0.603	0.029	0.559
MRE	0.043	0.071	0.022	0.073	0.004	0.068
Average MRE	0.0572		0.0474		0.0356	
RMSE	0.4824		0.0444		0.3956	

According to Table 3, the average MRE and RMSE of the combined prediction model is lower than Model 1 and Model 2, implying that the combined model has better accuracy. With the same data amount, approximating and detail sequence can easily find points near to the center by Euclidean distance, which makes them better predictors. A more accurate prediction of DO concentration can provide better reference for water self-purification abilities and water conservation work in this district.

Conclusion

The water environment is very complicated, and water indicators could show different sequence features. The chaotic prediction model based on the wavelet transform can be applied to the short-term prediction of the fluctuation index with large-scale samples. Based on the volatility of the DO concentration and making use of the good stripping effect of the wavelet transform, the low frequency and high frequency data sequences were gained. The chaotic prediction model based on the wavelet transform was established by combining the wavelet transform and the chaotic prediction model, which involves simple calculations and high precision. Through the comparison of average MRE, the combined model has higher accuracy and can provide a more solid basis for water protection and water environment treatment.

Acknowledgments

This work was supported by the National Natural Science Foundation of China (Nos. 51478025).

References

[1] Bahaa M K, Ayman G A, Hussein K and Ashraf E.-S 2012 Application of artificial neural networks for the prediction of water quality variables in the nile delta *J. Water Resour.* Protection
[2] Cao L Y 1997 Practical method for determining the minimum embedding dimension of a scalar time series *Physical D* **110**: 43-50
[3] Chen J J, Mao H X, Li P C, et al. 2014 A wavelet autoregressive model and its application to water quality forecast *Hydro-Science and Engineering* **2**: 95-99
[4] Chen M and Ye X Z 2008 Research on the chaotic time series decision method *Information Technology* **2008**(6): 23-26
[5] Cheng S F, Cheng X H, Yang L 2014 Application of wavelet neural network with improved particle swarm optimization algorithm in power transformer fault diagnosis *Power. Syst. Prot. Control* 42
[6] Ding J, Wang W S and Zhao Y L 2003 Characteristics of daily flow variation in the Yangtze River ,1, optimum determination of delay time for reconstruction of a phase space *Advances In Water Science* **14**(4): 407-11
[7] Ding J, Wang W S and Zhao Y L 2003 Characteristics of daily flow variation in the Yangtze River ,2, optimum determination of embedding dimension for reconstruction of a phase space *Advances In Water Science* **14**(4):412-16
[8] Dogan E, Sengorur B and Koklu R 2009 Modeling biological oxygen demand of the Melen River in Turkey using an artificial neural network technique *J. Environ. Manage.* **90** 1229–35
[9] Gazzaz N M, Yusoff M K, Aris A Z, Juahir H and Ramli M F 2012 Artificial neural network modeling of the water quality index for Kinta River (Malaysia) using water quality variables as predictors *Mar. Pollut. Bull.* **64** 2409–20
[10] Govindaraju R S 2000 Artificial neural networks in hydrology. II: hydrologic applications *J. Hydrol. Eng.* **5** 124–37
[11] Liu D J and Zou Z H 2011 Applications of Gray Forecast Model combined with Artificial Neural Networks Model to water quality forecast *Systems Engineering* **29(**9): 105–09
[12] Mohanmad J A and Mohanmad R K 2015 Development of wavelet-ANN models to predict water quality parameters in Hilo Bay, Pacific Ocean *Mar. Pollut. Bull.* **98**: 171-78

[13] Musavi S and Golabi M 2008 Application of artificial neural networks in the river water quality modeling: Karoon River, Iran *J. Appl. Sci.* **8(12)** : 2324–28

[14] Nian H, Gu S M and Li X 2014 Back propagation neural network based on ant colony optimization algorithm for the prediction of water quality *J. Minnan Normal University(Nat. Sci.)* **83**: 62-67

[15] Okkan U 2012 Wavelet neural network model for reservoir inflow prediction *Sci. Iranica* **19**: 1445–55

[16] Palani S, Tkalich P, Balasubramanian R and Palanichamy J 2011 ANN application for prediction of atmospheric nitrogen deposition to aquatic ecosystems *Mar. Pollut. Bull.* **62**: 1198–206

[17] Qi J X and Chen J 2014 Radial basis function (RBF) water quality prediction model based on particle swarm optimization *Gansu Water Resour. Hydropower Technol.* **50**(5): 1-4

[18] Xu L Q and Liu S Y 2013 Study of short-term water quality prediction model based on wavelet neural network *Math. Comput. Model.* **58**(3): 807-13

[19] Yao S P, Hu C Z and Zheng L 2006 Chaotic time sequence prediction based on wavelet decomposition and refactoring *J. Projectiles, Rockets, Missiles and Guidance* **26**(2): 1133-35

[20] Zhang F, Dai H and Tang D 2014 A conjunction method of wavelet transform- particle swarm optimization-support vector machine for stream flow forecasting *J. Appl. Math.* **10**: 1-10

[21] Zou Z H and Feng W 2008 Add-weighted one-rank local-region method based on phase space reconstruction and application to dissolved oxygen prediction *Math. Practice Theory* **38**(13): 70-75

Assessing the hydrologic alteration of the Yangtze River using the histogram matching approach

F Huang[1, 2, 5], N Zhang[3], L D Guo[4] and Z Q Xia[1, 2]

[1] State Key Laboratory of Hydrology-Water Resources and Hydraulic Engineering, Hohai University, Nanjing, 210098, China
[2] College of Hydrology and Water Resources, Hohai University, Nanjing, 210098, China
[3] Yellow River Institute of Hydraulic Research, Zhengzhou, 450003, China
[4] Business School, Hohai University, Nanjing, 210098, China

E-mail: hfeng0216@163.com

Abstract. Hydrologic changes of the Yangtze River, an important river with abundant water resources in China, were investigated using the Histogram Matching Approach. Daily streamflow data spanning the time interval from 1955 to 2013 was collected from Yichang and Datong stations, which monitor the hydrologic processes of the upper and lower reach of the Yangtze River, respectively. The Gezhouba Dam, the first dam constructed at the main stream of the Yangtze River, started operations in 1981. 1981 was used to differentiate the pre-dam (1955-1980) and post-dam (1981-2013) hydrologic regimes. The hydrologic regime was quantified by the Indicators of Hydrologic Alteration. The overall alteration degree of the upper Yangtze River was 31% and the alteration degree of every hydrologic indicator ranged from 10% to 81%. Only 1, 5 and 26 hydrologic indicators were altered at high, moderate and low degrees, respectively. The overall alteration degree of the lower Yangtze River was 30%, and the alteration degree of every hydrologic indicator ranged from 8% to 49%. No high alteration degree was detected at the Datong station. Ten hydrologic indicators were altered at moderate degrees and 22 hydrologic indicators were altered at low degrees. Significant increases could be observed for the low-flow relevant indicators, including the monthly flow from January-March, the annual minimum 1, 3, 7, 30 and 90-day flows, and the base flow index.

1. Introduction

The natural hydrological regime is a basic driver of water ecosystems and vitally important for keeping the health of river, lake and wetland ecosystems. It is affected by both anthropogenic activities and the climate, both of which have been the subject of many studies investigating worldwide hydrologic changes [1-5].

The Yangtze River plays an irreplaceable role in the ecological environmental protection and socio-economic construction of China, with its abundant water, hydropower, and biological resources. Over 50,000 dams/reservoirs have been built in the basin [6]. The total storage capacity of these dams is almost 180 billion m^3, which amounts to approximately 20% of the annual runoff of the Yangtze River [7]. The impacts of these reservoirs on the hydrologic regime and its related ecological environmental implications have attracted considerable attention in recent years [8-12]. The present study seeks to evaluate the hydrologic changes of the Yangtze River using the Histogram Matching Approach (HMA). The results of this research will provide reference for water resource utilization and for the protection of riverine ecosystems.

2. Study site and data

The Yangtze River originates in the Qinghai-Tibet Plateau in western China (figure 1). Its length is about 6300 km and its catchment area is approximately 1.8 million km^2. The basin lies in the monsoon region of East Asia, which is a subtropical climate zone. The monsoon-driven precipitation is unevenly distributed throughout the year, which causes seasonal variability in river runoff. The rainy season starts in April and determines the hydrologic year. The Yangtze River's hydrologic year begins from April and goes until March of the next calendar year.

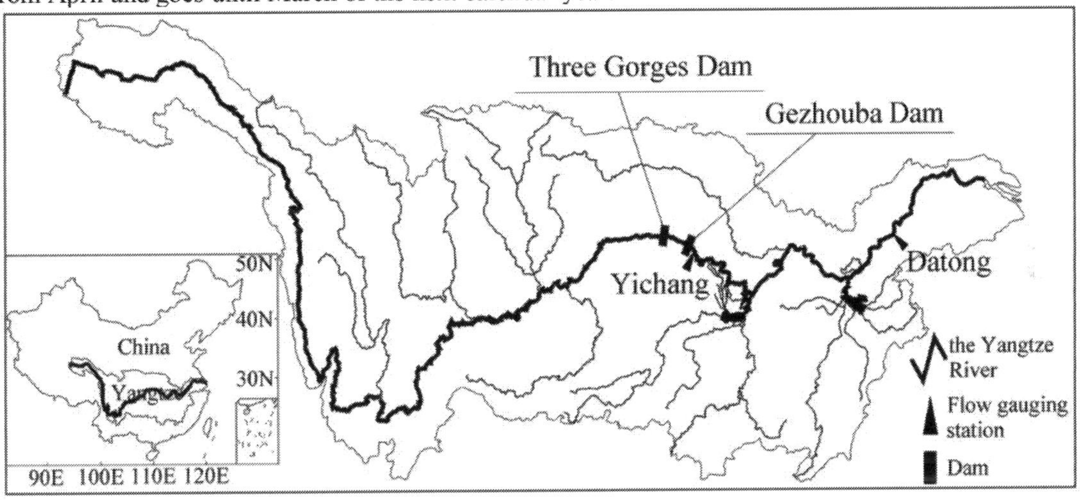

Figure 1. Locations of the Yangtze River and flow gauging stations.

The Yichang and Datong flow gauging stations were selected for assessing hydrologic alteration. The Yichang station is located at the outlet of the upper reach, approximately 6 km below the Gezhouba Dam and 44 km below the Three Gorges Dam. It monitors the hydrologic regime of the upper Yangtze River. The Datong station distributes at the lower reach where the annual mean runoff is 896.4 billion m3 and the annual mean sediment load is 0.39 billion t.

The Changjiang Water Resources Commission (China) provided daily streamflow data spanning from 1955 to 2013. The Gezhouba Dam, which was the first dam constructed at the main stream of the Yangtze River, started operations in 1981. 1981 was used as the dividing point between the pre-dam (1955-1980) and post-dam (1981-2013) hydrologic regimes. The hydrologic regime during 1955-1980 may not be wholly unaltered because some dams were constructed at the tributaries of the upper Yangtze River prior to 1981 [12,13]. This assessment aims to reveal the degree of alteration of the hydrologic regime from 1981 to 2013 compared with the hydrologic regime during the pre-dam era.

3. Methods

3.1. Indicators of Hydrologic Alteration (IHA)

The IHA method is widely applied to characterize the various components of hydrologic regimes [14]. It contains 33 hydrologic parameters that are grouped into five categories of hydrologic features. These are, group 1: magnitude of monthly water conditions, group 2: magnitude and duration of annual extreme water conditions, group 3: timing of annual extreme water conditions, group 4: frequency and duration of high and low pulses, and group 5: rate and frequency of water condition changes. The parameter "number of zero-flow days" was not applied because no zero-flow events occurred during the study period. Each IHA parameter was calculated for every hydrologic year, and the results were used to assess hydrologic alteration.

3.2. Histogram Matching Approach

The Range of Variability Approach (RVA) has been widely applied in worldwide assessments of hydrologic alteration and environmental flow design [15-18]. Shiau and Wu [19] have noted that the RVA method has two main limitations: (1) it only considers the frequency of the hydrologic parameters that fall within the target range, and ignores variation; and (2) it ignores the frequency and variation of the hydrologic parameters that fall above the upper-target or below the lower-target. These limitations might potentially lead to a false assessment of hydrologic alteration. The HMA method has been proposed to address the limitations of the RVA method. Its detailed procedure can be found in the literature of Shiau and Wu [19].

The hydrologic alterations were assessed by applying the HMA method. Then, following the class evaluation system [20], the degrees of hydrologic alteration for individual hydrologic indicators were classified into the categories "little alteration" (ranging from 0 to 33%, denoted by the letter "L"), "moderate alteration" (ranging from 34 to 67%, indicated by the letter "M"), and "high alteration" (ranging from 68 to 100%, indicated by the letter "H").

4. Results and discussion

The pre-dam and post-dam hydrologic regimes of the Yichang and Datong stations were compared in the histograms in figure 2 and figure 3, respectively, and the results of the hydrological alteration for every hydrologic indicator are shown in figure 4.

At the Yichang station, the overall degree of hydrologic alteration was 31%, which is classified as "little alteration." The deviations of the median flow in August, January, and March ranged from 37% to 43%, and this range is within the classification of "moderate alteration." The median flow deviations of the other months ranged from 10% to 31%, categorized as "little alteration." Figure 2 shows the change characteristics of the post-dam histograms as compared to the pre-dam histograms. The histograms of the median flow of January and March displayed similar change patterns. The post-dam histogram moved to the right of the pre-dam histogram, indicating an increased flow in January and March. The alteration degree of magnitude and duration of annual extreme flow ranged from 10% to 33%, classified as "little alteration." Similar change patterns were observed in the annual minimum flow and the base flow index. The relative frequencies of the post-dam histogram were distributed to the right of the relative frequencies of the pre-dam histogram, indicating an increased flow in the dry season. The alteration degree of the Julian calendar date of the annual 1-day minimum flow was 35% ("moderate alteration") and the alteration degree of the Julian calendar date of the annual 1-day maximum flow was 13% ("little alteration"). The annual 1-day minimum flow usually occurred in January, February or March in the pre-dam years, but after 1981, it sometimes occurred in December of the previous calendar year (figure 2 (24)). The alteration degree of frequency and the duration of high and low pulses fell about 20%, which is classified as "little alteration." The rise rate was found to have "moderate alteration" and the fall rate had "little alteration." The number of reversals showed "high alteration," and this was the only hydrologic indicator that exceeded 67% deviation. The number of reversals increased drastically after 1981.

At the Datong station, the overall degree of hydrologic alteration was 30%, which was very close to the alteration degree at the Yichang station. According to the data, no hydrologic indicators experienced "high alteration" at the Datong station. The median flows in January, February and March showed "moderate alteration", and during the other months the median flows showed "little alteration." The alteration degree of the annual minimum flow's relevant indicators ranged from 36% to 49% ("moderate alteration"). The annual maximum flow's relevant indicators ranged from 19% to 32% ("little alteration"). An increased flow in the dry season was also observed at the Datong station (figure 3), which was similar to the change features at the Yichang station. All of the hydrologic indicators in Group 3 and 4 had "little alteration." The rise and fall rates showed "little alteration," but the number of reversals was found to have "moderate alteration." Different features of hydrologic alterations at the lower reach, as compared with hydrologic alterations at the upper reach, might have been influenced by the runoff from tributaries such as the Hanjiang River, as well as hydraulic regulation of the river channels and lakes between the Yichang and Datong stations, e.g. the Dongting and Poyang Lakes.

The hydrologic regime of the Yangtze River represents a catchment response to climate input and anthropogenic activities. Precipitation may be a dominant factor in hydrologic changes. The spatial distribution of the precipitation regime can lead to spatial patterns of alterations in runoff [21]. Reservoir regulation is another important factor that affects the hydrologic regime. The significant increase of the low-flow relevant indicators may be caused by the water supply function of the reservoirs in the dry season [22-24]. The assessment results of the HMA method showed that the alteration degree of the hydrologic regime was affected by both climate change and anthropogenic activities, but this model was unable to definitively differentiate the individual roles of the two factors.

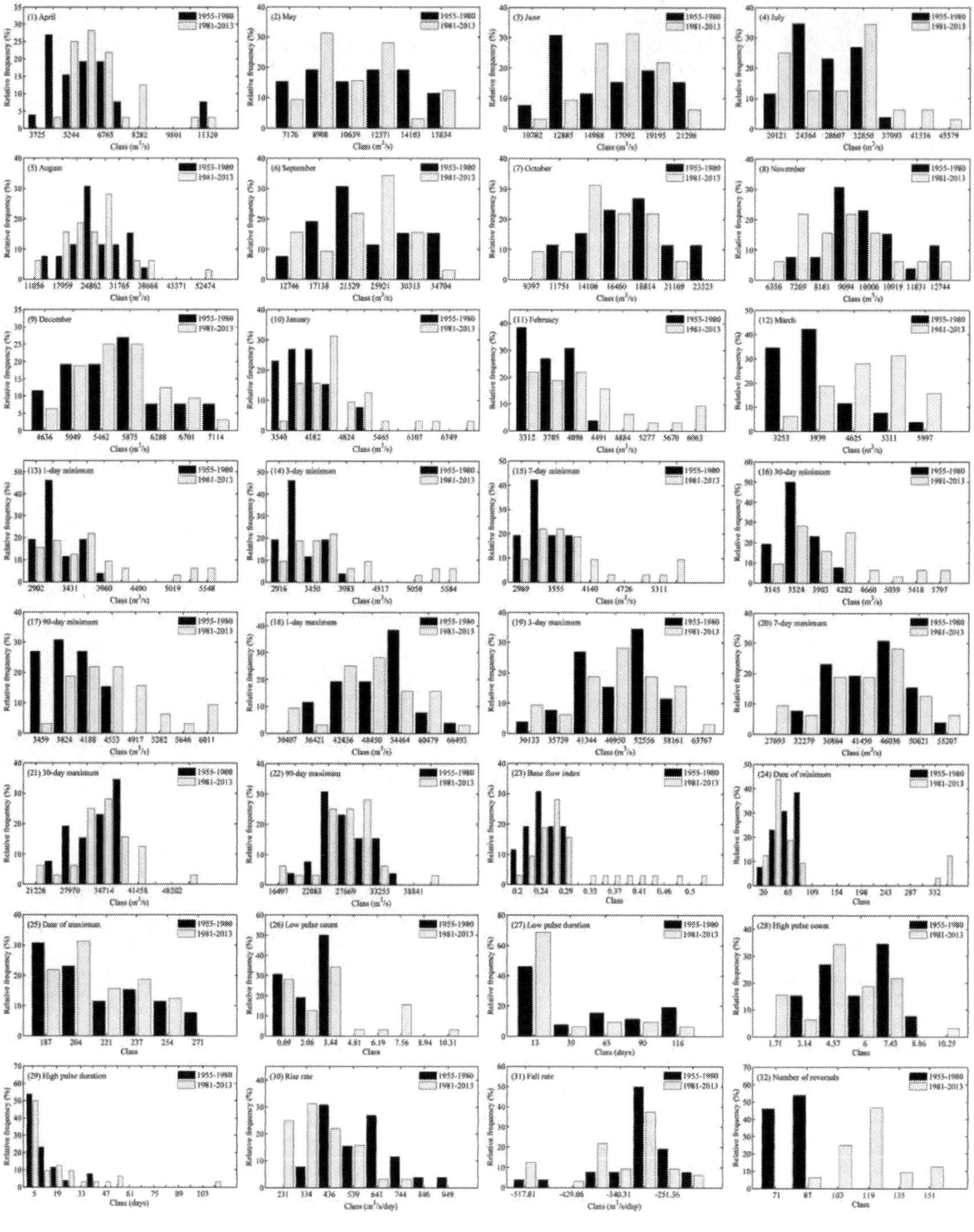

Figure 2. Histogram comparisons of the hydrologic regimes at the Yichang station.

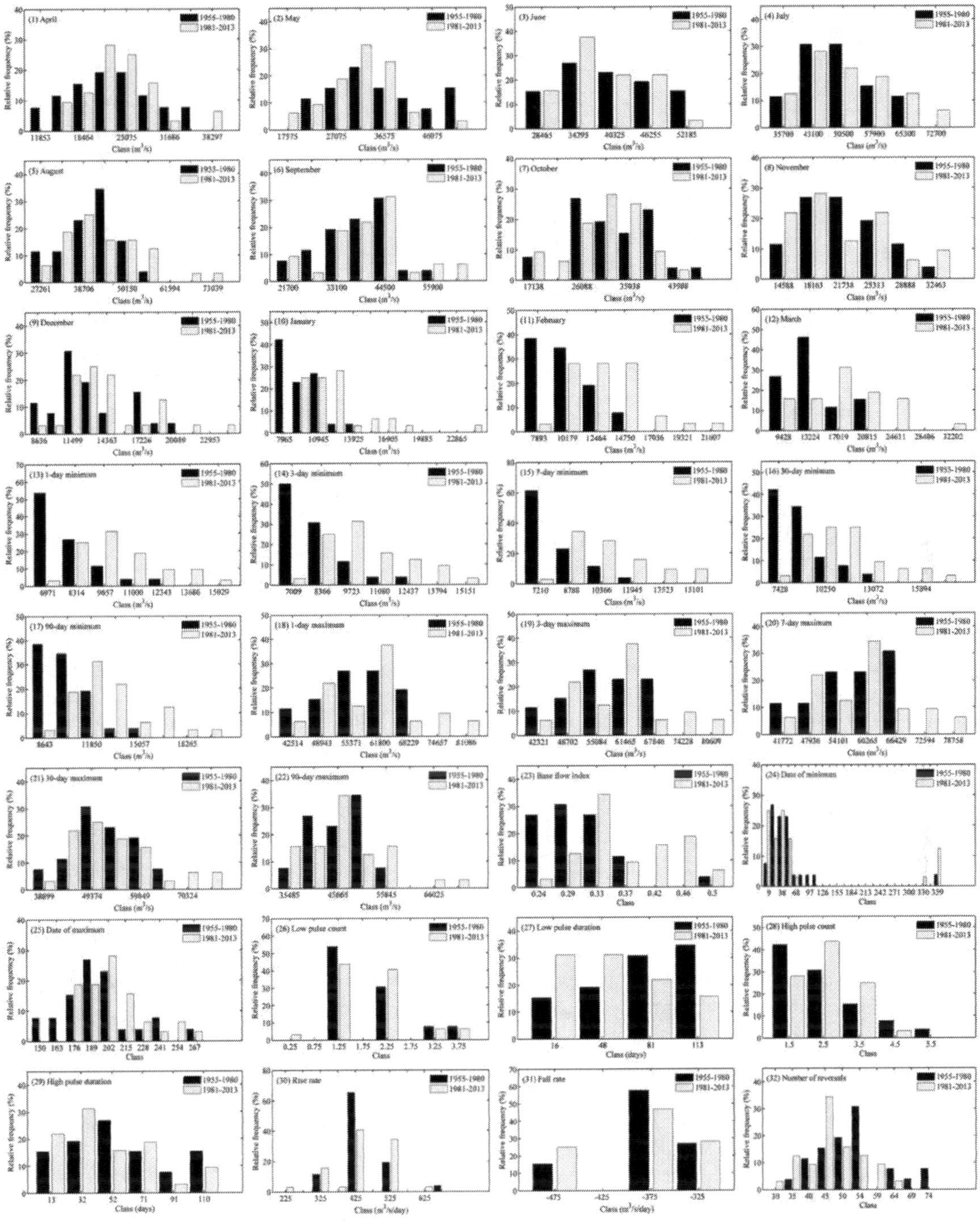

Figure 3. Histogram comparisons of the hydrologic regimes at the Datong station.

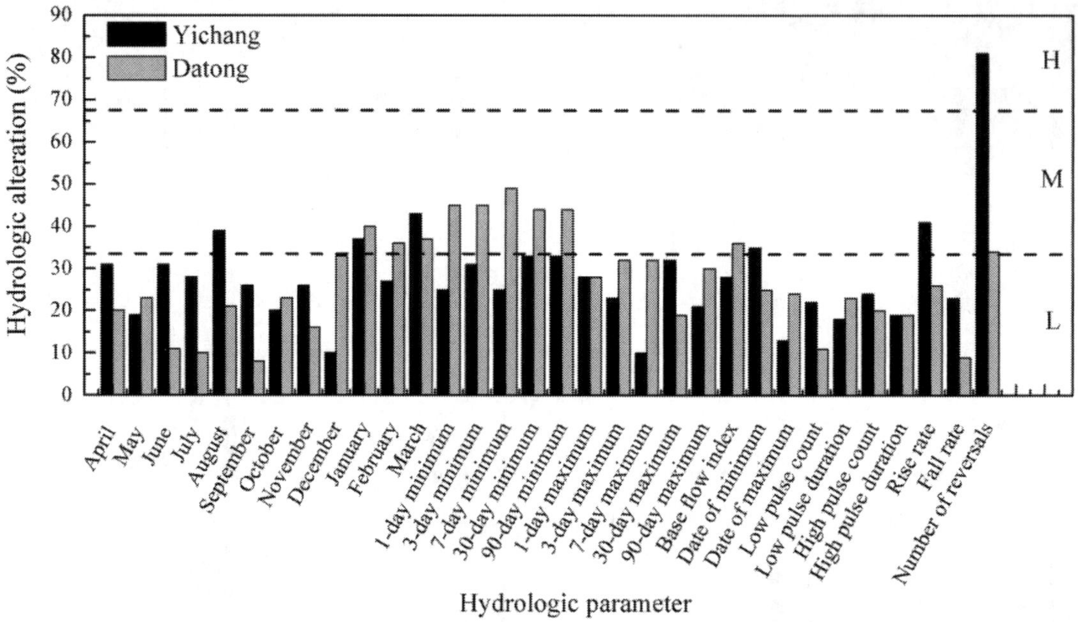

Figure 4. Hydrologic alterations of the Yangtze River evaluated using the HMA method.

5. Conclusions

The hydrologic alteration degree of the Yangtze River was assessed by applying the HMA method to long-term daily streamflow data from the Yichang and Datong stations. The overall alteration degree was approximately 30%, which is in the range of "little alteration." The alteration degree of the hydrologic indicators of the Yichang station ranged from 10% to 81%. 1, 5 and 26 hydrologic indicators were altered with high, moderate and low degree, respectively. The hydrologic indicator with the highest level of alteration was the number of reversals, which increased drastically after 1981. The alteration degree of the hydrologic indicators from the Datong station ranged from 8% to 49%. No hydrologic indicators showed high alteration. Ten hydrologic indicators exhibited "moderate alteration" and 22 hydrologic indicators showed "little alteration." Significant increases were observed for the low-flow relevant indicators, such as the monthly flow from January-March, the annual minimum 1-, 3-, 7-, 30- and 90-day flow, and the base flow index. The increased flow in the dry season in the post-dam period has helped to preserve the environmental and ecological water requirements, and also to protect the riverine ecosystem.

References

[1] Liechti T C , Matos J P , Boillat J L , et al. 2015 Influence of hydropower development on flow regime in the Zambezi River Basin for different scenarios of environmental flows *Water Resources Management* **29**(3) :731-47

[2] Chen Q H , Zhang X , Chen Y Y , et al. 2015 Downstream effects of a hydropeaking dam on ecohydrological conditions at subdaily to monthly time scales *Ecological Engineering* **77**: 40-50

[3] Dyer F, ElSawah S, Croke B, et al. 2014 The effects of climate change on ecologically-relevant flow regime and water quality attributes *Stochastic Environmental Research and Risk Assessment* **28**(1) : 67-82

[4] Gain A K and Giupponi C 2014 Impact of the Farakka Dam on thresholds of the hydrologic flow regime in the lower ganges river basin (Bangladesh) *Water* **6**(8) : 2501-18

[5] Souter N J and Schultz M 2014 Using simulated data and the range of variability approach to assess changes to the flow regime of the river Murray at the South Australian border *Transactions of the Royal Society of South Australia* **138**(2) : 277-92

[6] Yang S L, Milliman J D, Li P, et al. 2011 50,000 dams later: Erosion of the Yangtze River and its delta *Global and Planetary Change* **75**(1-2) : 14-20

[7] Wei W, Chang Y and Dai Z 2014 Streamflow changes of the Changjiang (Yangtze) River in the recent 60 years: Impacts of the East Asian summer monsoon, ENSO, and human activities *Quaternary International* **336**(SI) : 98-107

[8] Huang F, Xia ZQ, Li F, et al. 2013 Assessing sediment regime alteration of the upper Yangtze River *Environmental Earth Sciences* **70**(5) : 2349-57

[9] Li Q F, Yu M X, Zhao J H, et al. 2012 Impact of the Three Gorges reservoir operation on downstream ecological water requirements *Hydrology Research* **43**(1-2) : 48-53

[10] Zhang N, He H M, Zhang S F, et al. 2012 Influence of reservoir operation in the upper reaches of the Yangtze River (China) on the inflow and outflow regime of the TGR-based on the mproved SWAT model *Water Resources Management* **26**(3) : 691-705

[11] Li Q F, Yu M X, Lu G B, et al. 2011 Impacts of the Gezhouba and Three Gorges reservoirs on the sediment regime in the Yangtze River, China *Journal of Hydrology* **403**(3-4) : 224-33

[12] Huang F, Xia Z Q, Zhang N, et al. 2011 Flow-complexity analysis of the upper reaches of the Yangtze River, China *Journal of Hydrologic Engineering* **16**(11) : 914-19

[13] Gao B, Yang D W, Zhao T T G, et al. 2012 Changes in the eco-flow metrics of the upper Yangtze River from 1961 to 2008 *Journal of Hydrology* **448** : 30-38

[14] Richter B D, Baumgartner J V, Powell J, et al. 1996 A method for assessing hydrologic alteration within ecosystems *Conservation Biology* **10**(4) : 1163-74

[15] Richter B D, Baumgartner J V, Wigington R, et al. 1997 How much water does a river need? *Freshwater Biology* **37**(1) : 231-49

[16] Yang T, Zhang Q, Chen Y D, et al. 2008 A spatial assessment of hydrologic alteration caused by dam construction in the middle and lower Yellow River, China *Hydrological Processes* **22**(18) : 3829-43

[17] Babel M S, Chien N D, Mullick M R A, et al. 2012 Operation of a hydropower system considering environmental flow requirements: a case study in La Nga river basin, Vietnam *Journal of Hydro-Environment Research* **6**(1) : 63-73

[18] Lian Y Q, You J Y, Sparks R, et al. 2012 Impact of human activities to hydrologic alterations on the Illinois River [J] *Journal of Hydrologic Engineering* **17**(4) : 537-46

[19] Shiau J T and Wu F C 2008 A histogram matching approach for assessment of flow regime alteration: application to environmental flow optimization *River Research and Applications* **24**(7) : 914-28

[20] Richter B D, Baumgartner J V , Braun D P, et al. 1998 A spatial assessment of hydrologic alteration within a river network *Regulated Rivers-Research & Management* **14**(4) : 329-40

[21] Zhang Q, Singh V P, Xu C Y, et al. 2013 Abrupt behaviours of streamflow and sediment load variations of the Yangtze River basin, China *Hydrological Processes* **27**(3) : 444-52

[22] Yu M X, Li Q F, Lu G B, et al. 2013 Investigation into the impacts of the Gezhouba and the Three Gorges reservoirs on the flow regime of the Yangtze River *Journal of Hydrologic Engineering* **18**(9) : 1098-106

[23] Wang Y K, Wang D and Wu J C 2015 Assessing the impact of Danjiangkou reservoir on ecohydrological conditions in Hanjiang river, China *Ecological Engineering* **81** : 41-52

[24] Jiang L Z, Ban X, Wang X L, et al. 2014 Assessment of hydrologic alterations caused by the Three Gorges Dam in the middle and lower reaches of Yangtze River, China *Water* **6**(5) : 1419-34

Acknowledgements
This research is supported financially by the National Natural Science Foundation Projects of China (No.41401011, No.51309107 and No.41401010), the Fundamental Research Funds for the Central Universities of China (No. 2013/B13020312), and the Major Program of National Social Science Foundation of China (No.11&ZD168).

Study on Treatment of acidic and highly concentrated fluoride waste water using calcium oxide-calcium chloride

T Ren [1], X R Gao[1], T Zheng [1, 3] and P Wang [1, 2]

[1]School of Municipal and Environmental Engineering, Harbin Institute of Technology, Harbin, 150090, China
[2]State Key Laboratory of Urban Water Resource and Environment, Harbin Institute of Technology, Harbin, 150090, China

E-mail: zhengtong@hit.edu.com

Abstract. There are problems with treating acidic waste water containing high concentration fluorine by chemical precipitation, including the low sludge setting velocity and the high difficulty of reaching the criterion. In Heilongjiang province, a graphite factory producing high-purity graphite generates acidic waste water with a high concentration of fluorine. In this paper, the effect of removals on the concentration of fluoride with the combined treatment of calcium oxide and calcium chloride were discussed with regard to acid waste water. The study improved the sludge characteristics by using polyacrylamide (PAM) and polymeric aluminum chloride (PAC). The effect of different coagulants on sludge was evaluated by the sludge settlement ratio (SV), sludge volume index (SVI) and sludge moisture content. The results showed that the optimal combination for 100 ml waste water was calcium oxide addition amount of 14 g, a calcium chloride addition amount of 2.5 g, a PAM addition amount of 350 mg/L, and the effluent fluoride concentration was below 6 mg/L. PAM significantly improved the sludge settling velocity. The sludge settlement ratio reduced from 87.6% to 60%. The process for wastewater treatment was easily operated and involved low expenditure.

1. Introduction

Fluorine in the environment can be accumulated into the human body, creating a serious threat to people's health and life. The hydrofluoric acid method is widely applied in many domestic enterprises that produce high-purity graphite at present. After this treatment, the carbon content in the scaly graphite is increased to 99.94% [1]. The method that is used commonly has some advantages, such as its low production cost and the high quality of the product. However, acidic waste water containing a high concentration of fluorine is produced during production of high-purity graphite.

At present, waste water defluoridation is accomplished by several methods, including chemical precipitation, adsorption, chemical deposition, nanofiltration and other methods. Li Li applied lime-fly ash spectrometry technology in the treatment of 100mg/L fluoride wastewater [2]. The chemical deposition method is characterized by simple equipment, easy controls and low operation cost. It often chooses quicklime and hydrated lime as precipitants. However, in practical application, chemical precipitation still faces many problems, such as its difficulty keeping fluoride concentrations below 10mg/L [3] and its low sludge setting velocity.

To solve these problems, in this paper, acidic waste water containing a high concentration of fluorine was treated by the combined treatment of calcium oxide and calcium chloride. Calcium oxide reacted with both acids and fluorine ions to form calcium fluoride, which is insoluble in water.

Because of the common-ion effect, calcium chloride effectively reduced the concentration of fluoride. PAM and PAC were used to increase the settling rate and shorten the time required for settling. The study of treatment technologies of fluorine-containing waste water is favorable to the popularization of the hydrofluoric acid method, and it promotes the development of the graphite industry.

2. Materials and methods

2.1. Main instruments and reagents

Instruments: HJ-4magnetic stirrer, fluoride ion-selective electrode with lanthanum fluoride single crystal membrane, 217calomel electrode.

Reagents: AR grade calcium oxide made in Tianjin bodi chemical co., LTD., AR grade calcium chloride made in Tianjin bodi chemical co., LTD., AR grade non-ionic PAM, AR grade PAC.

The fluoride aqueous solution used in this study came from a graphite factory in Heilongjiang province. The concentration of fluoride was 3×104mg/L, with a pH around 1. The mud used in this study came from the treatment of a fluoride aqueous solution using calcium oxide-calcium chloride.

2.2. Experimental methods

2.2.1. Defluoridation by calcium oxide. A designated amount of calcium oxide was added to 100ml of fluoride solutions. The solutions were vigorously stirred for 40min on at a rate of 300 rpm. There were two ways to add calcium oxide: dry powder, or emulsion (emulsion concentration was 20%). The dosage of calcium oxide had an influence on the treatment. In the experiment, the effect of dosing calcium oxide on pH, concentration of fluoride and sludge production was studied, and the treatment effects of dry powder and emulsion were compared. Fluoride ions were measured in the solution after 24h using a selective electrode for fluoride ions.

2.2.2. Defluoridation by calcium chloride. At 20℃, in order to increase the pH of the reaction system to neutral (pH=7), 14g calcium oxide was added to 100ml of fluoride solutions. The concentration of fluoride decreased to 12mg/L. The solutions were vigorously stirred for 40min at a rate of 300rpm. Then, designation calcium chloride was added to the solutions. The solutions were vigorously stirred for 10min at a rate of 300 rpm. The concentration of fluoride in the supernatant was measured after 24h.

2.2.3. Flocculation experiment. The PAC solution concentration was 4%, PAM 0.1%. The flocculant was added to 200ml of mud. The mixture was vigorously stirred at a rate of 120 rpm for 1min, then stirred at a rate of 40 rpm for 10min. The sludge volume was recorded at 5min, 10min, 20min, 30min, 60min and 120min. In the paper, the sludge settlement ratio (SV), sludge volume index (SVI) and sludge moisture content were used to evaluate the sludge settleability.

3. Results and discussion

3.1. Defluoridation by calcium oxide

The waste water has strong acidity. When calcium oxide dissolves in water, the liquid will have alkalinity, and emit heat. Calcium oxide reacts with acid to form a salt. It increases the pH value to make the effect better. In the waste water, F^- reacts with Ca^{2+}. Then, the fluoride ion will be reduced by forming CaF2, which is hardly dissolved by water. The reaction equation is: $CaO + 2H^+ = Ca^{2+} + H_2O$, $Ca^{2+} + 2F^- = CaF_2\downarrow$. The results are shown in figure 1 and figure 2.

The results showed that with the increasing of calcium oxide, the concentration of fluoride in the waste water decreased sharply. When n(Ca2+/F-) was more than 0.8, the concentration of fluoride tended to be constant at about 15mg/L. Whether calcium oxide was added as dry powder or emulsion, the concentration of fluoride was not lower than 10mg/L. At the same time, from figure 2, the higher the calcium oxide dosage was, the higher the pH of the solution was. When n(Ca2+/F-) was 0.8, the solution was neutral, with a pH value of 7. The proper treatment condition obtained from the experiment results was: n(Ca2+/F-) of 0.8.

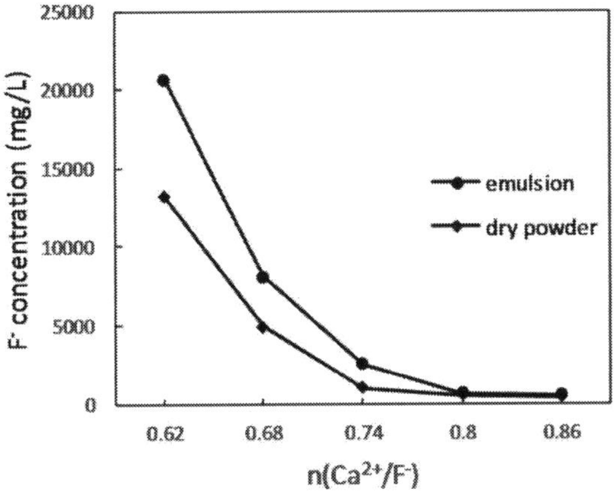

Figure 1. Influence of calcium oxide dosage on F- removal efficiency.

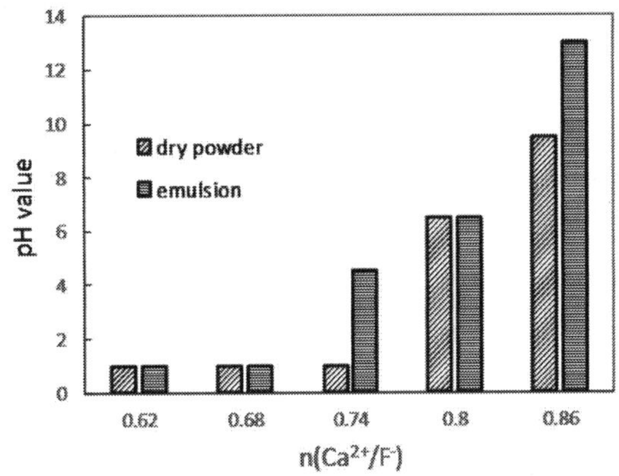

Figure 2. Influence of calcium oxide dosage on pH.

3.2. *Defluoridation by calcium chloride*

Calcium fluoride is difficult to dissolve in water. In theory, because its solubility product (Ksp) is 4.0×10-11 at 25 degrees Celsius, the concentration of fluoride is 8.17mg/L. Waste water from the factory is characterized by complicated components and many salts, so the solubility of calcium fluoride is improved. If calcium oxide alone is added to waste water, the concentration of fluoride has difficulty reaching the criterion. According to the common ion effect, when calcium chloride is added to a saturated calcium fluoride solution, the calcium ion concentration is increased. The solubility equilibrium of precipitation to calcium fluoride can occur with an increase in the calcium ion concentration. The concentration of fluoride is therefore reduced [4].

The best dosing quantity was determined by comparing the treatment effects. The results are shown in figure 3 and figure 4.

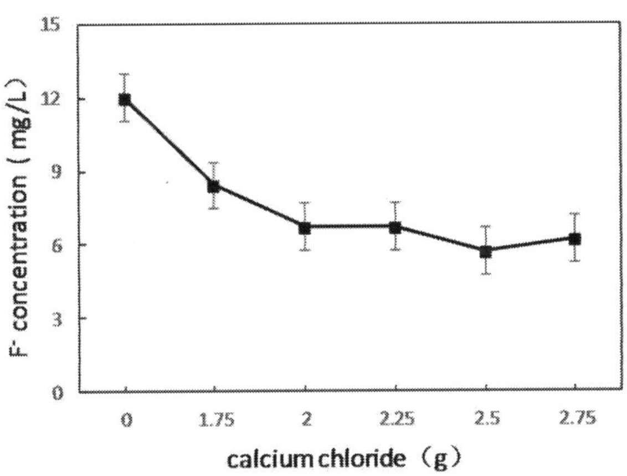

Figure 3. Influence of calcium chloride dosage on F- removal efficiency.

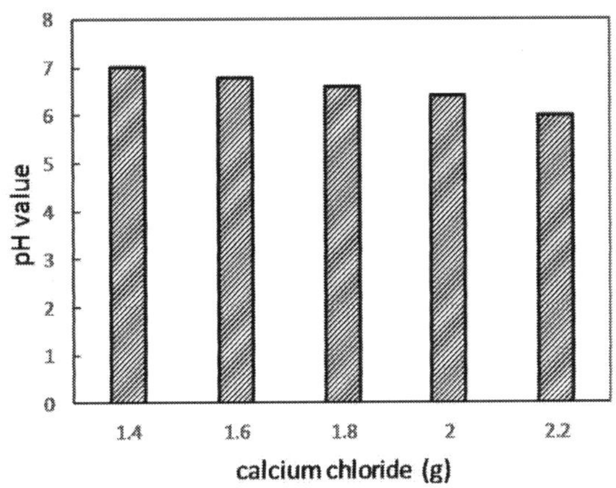

Figure 4. Influence of calcium chloride dosage on pH.

The results showed that with the increasing of calcium chloride, the concentration of fluoride in the waste water was significantly reduced. When the added calcium chloride was 2.5g, the concentration of fluoride was reduced to 5.7mg/L. When the added calcium chloride was more than 2.5g, the concentration of fluoride tended to be constant at about 6mg/L. The reasons could be attributed to the following. Overmuch calcium ion caused the formation of a complex, such as CaCl+ and CaF+ [5], which made it unable to further reduce the concentration of the fluoride in the solution.

3.3. Flocculation settling experiment

Sludge was a by-product of the wastewater-treatment system. In the combined treatment system of calcium oxide and calcium chloride, a large amount of complicated ingredient sludge was formed. It mainly contained calcium hydroxide and calcium fluoride. If the sludge was not being disposed of properly, the factory would be affected. During the experimental process, the sludge settling velocity was slow. In order to improve the sludge behaviour of the precipitation, flocculant was used in this paper. The flocculation mechanisms include the interaction of hydroxides, the compression of double charged and so on [6]. Flocculation was applied to add to the instability and aggregation of suspended particles. The effect of the flocculant type and agent dosage on the settling speed was studied in the flocculation settling experiment.

3.3.1. Settleability improved by PAC. PAC is an inorganic cationic polymeric coagulant, which has the very good characteristics of coagulation and precipitation [7]. PAC can flocculate the suspended particles in the solution by the function of gathering, adsorbing and unwrapping. The results were shown in figure 5 and table 1.

Table 1. Effect of PAC dosages on the characteristics of sludge.

PAC (mg/L)	SV (%)	SVI (ml/g)	Moisture content(%)
0	90.8	9.5	89
200	88.4	9.2	88
1000	83.2	8.7	85

The dose of PAC was 200mg/L and 1000mg/L, which was 1.3‰ and 6.7‰ of the dried sludge. From figure 4, when no PAC was added in the sludge, the sludge volume reduced to 4/5 after 2h. After PAC was added, the curve showed an obvious decreasing trend. This showed that the sludge sedimentation velocity increased with the growth of the dose. When 1000mg/L PAC was added, the sludge settlement ratio decreased to 83.2%, the sludge volume index to 8.7ml/g and the sludge moisture content to 85%. The sludge volume reduced to 3/5 after 2h.

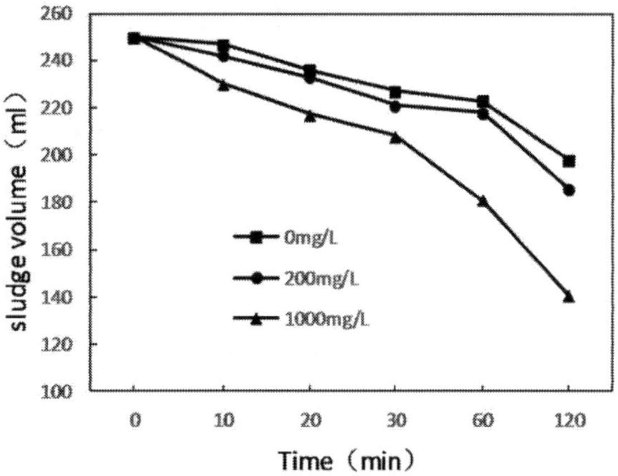

Figure 5. Settlement curve of different PAC dosage.

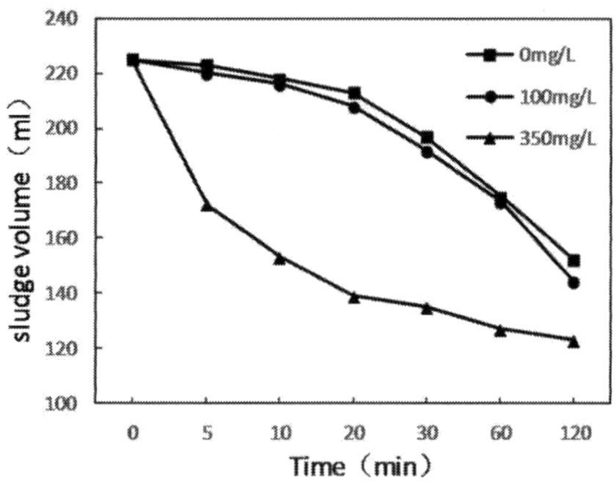

Figure 6. Settlement curve of different PAM dosage.

3.3.2. Settleability improved by PAM. PAM, as a linear high polymer, flocculates the suspended particles though charge neutralization and bridging action. In the paper, sludge settlement ratio (SV), sludge volume index (SVI) and sludge moisture content were used to evaluate the sludge settleability. The results are shown in figure 6 and table 2.

Table 2. Effect of PAM dosages on the characteristics of sludge.

PAC (mg/L)	SV (%)	SVI (ml/g)	Moisture content(%)
0	87.6	8.2	86
100	85.3	8	85
350	60	5.6	83

The dose of PAM was 100 mg/L and 350 mg/L, which was 0.7‰ and 2.3‰ of the dried sludge. The results indicate that when 100mg/L PAM was added in the sludge, there was no obvious change in the sludge settleability. When 350mg/L PAM was added, the curve showed an obvious decreasing trend. The sludge settlement ratio decreased to 60%, the sludge volume index to 5.6ml/g and the sludge moisture content to 83%. The sludge volume reduced to 1/2 after 2h. This showed that the sludge sedimentation velocity increased with the growth of the dose.

4. Conclusion

- In the treatment of acidic and highly concentrated fluoride waste water using calcium oxide-calcium chloride, the optimum dosage for the 100 ml waste water was a calcium oxide addition amount of 14 g, calcium chloride addition amount of 2.5g, PAM addition amount of 350mg/L. The effluent fluoride concentration is 5.7mg/L, which met the integrated wastewater discharge standard (GB8978-1996).
- Both PAC and PAM could accelerate the settlement velocity, but the treatment effects were better with PAM. When 350 mg/L PAM was added, the sludge settlement ratio decreased from 87.6% to 60%, the sludge volume index from 8.2 ml/g to 5.6ml/g and the sludge moisture content from 86% to 83%. The sludge volume reduced to 1/2 after 2h.
- The process featured simple operation, low cost of operation and easy maintenance. Its effects were satisfactory. PAM was used to solve the low settlement velocity.

Acknowledgements
Thanks for the financial support of the key science and technology project subject to tender in Heilongjiang province (GA14A102).

References

[1] Zhang R and Yu L X Study on 2007 Preparation of high-purity graphite by stepwise purification with sulfuric acid/hydrofluoric acid *Non-Metallic Mines* **30**(3): 42-44
[2] Li L, Ding L Q, Ma Y F, et al. 2012 Study on the treatment of high concentration containing fluoride wastewater by the processing of lime-fly ash *Guangzhou Chemical Industry* **40**(17): 51-56
[3] Liu H J, Li P, He J S, et al. 2012 Brief analysis on industrial security & protection system *Copper Engineering* **118**: 81-84
[4] Du M, Ning J G, Yang D W, et al. 2015 Treatment of acidic and highly concentrated fluoride wastewater using calcium chloride-poly aluminum chloride *Chinese Journal of Environmental Engineering* **9**(4): 1837-41
[5] Jiang K, Zhou K G and Li C W 2011 Effect of particle size on precipitation-dissolution equilibrium of CaF2 *The Chinese Journal of Nonferrous Metals* **21**(12): 3195-201
[6] Liu C G, Zhang P Y, Zeng G M, et al. 2010 Sewage sludge conditioning by bioleaching-dual

IOP Conf. Series: Earth and Environmental Science **39** (2016) 012003 doi:10.1088/1755-1315/39/1/012003

PAC and PAM addition *Environmental Science* **31**(9): 2124-29
[7] Jin J H and Miao Z J 2006 Experiment on sludge dewatering by using PAC and PAM *Jiangsu Environmental Science and Technology* **19**(2): 4-7

IOP Conf. Series: Earth and Environmental Science **39** (2016) 012004 doi:10.1088/1755-1315/39/1/012004

Calculation of ecological compensation for water sources for water diversion projects

H B Su [1,3], T M Zhang [2], C Y Hu [1] and L Y Long [1]

[1]Institute of Water Conservancy, Yunnan Agricultural University, Kunming 650201, China
[2]Water Conservancy Bureau of Yunnan Province, Kunming 650021, China

E-mail: 2568097575@qq.com

Abstract. This study considers the compensation of water diversion projects for the values of the terrestrial biological resources, water environment, and aquatic biological resources in water sources. An analysis of capital dynamics was conducted, and the economic development coefficient was used to correct the current method for calculating ecological compensation. A model was constructed to calculate_the ecological compensation for the water sources for water diversion projects. This model was used to calculate the ecological compensation for the Niulanjiang River provided by the Niulanjiang River to the Dianchi Lake water diversion project, which was calculated to be 136,799,400 RMB. As long as we know the occupying area of the project, the change of the river net flow after diversion and the local average GDP, the ecological compensation for water sources could be calculated by the model. The proposed model for calculating the ecological compensation for water sources is simple and incorporates the compensation provided by water diversion projects for the various environmental effects on water sources. It provides a guarantee for the capital to be used for the environmental protection of water sources and facilitates the sustainable development of the ecological environments of water sources.

1. Introduction

Water diversion projects use water resources to satisfy needs [1]. These projects often destroy ecological integrity because their implementation occupies a great deal of farmland, destroys forest vegetation, and affects the animal habitats around the water sources. In particular, the decrease in the amount of available water resulting from water diversion has affected the water volume, quality of rivers, and aquatic organisms in rivers, which has given rise to eco-environmental problems [2]. The smooth implementation of water diversion projects requires ecological compensation for the ecological losses experienced by the water sources. This compensation not only resolves the conflict between engineering construction and environmental protection, but also eases the shortage of capital required for environmental protection. The majority of cost-benefit analyses of environmental protection measures of water conservancy projects fail to consider the loss and gain of the functional value of the life support system provided to human beings from the ecosystem. This lack of consideration leads to incomplete calculations of the investment costs of projects. Domestic scholars have conducted an exploratory study on ecological compensation for basins and reservoirs [3] [4], and some methods have been proposed for proper calculation [5]. However, the ecological compensation for water sources in water diversion projects involves the quality of rivers and the aquatic organisms in rivers, and this is also calculated using a different method. This paper establishes a model for the calculation of the ecological compensation for water resources for water diversion projects that then also calculates ecological compensation for the water resources of the Niulanjiang River, provided by the Niulanjiang River to Dianchi Lake water diversion project. This calculation avoids the one-sidedness of the conventional ecological compensation calculation of water diversion projects and ensures proper implementation of water diversion projects.

2. Model for calculation of ecological compensation for water sources for water diversion projects

The expenses spent on protective measures reduce the adverse effects of a project on the environment are used to reflect the loss incurred by the project due to its environmental effects, the calculated value of which is the ecological compensation for the environment. The water sources for water diversion projects are primarily affected by the land occupation of the projects, which leads to decreased terrestrial biological resources, decreased water in rivers due to diversion, and poorer water quality. Dams are often built as well to block the activity of aquatic organisms. Thus, the ecological compensation for a water resource consists of the compensation for the values of its terrestrial biological resources, water environment, and aquatic biological resources.

2.1. Calculation of compensation for the value of terrestrial biological resources

The corrected method for calculating ecological value was used. Xie gaodi listed the value of the ecological services provided per unit area of China's terrestrial ecosystems [6], as shown in Table 1.

Table 1. Value of the ecological services provided per unit area of China's terrestrial ecosystems.

Unit: RMB/ hm^2.a.

Ecosystem services and functions	Forest	Grassland	Farmland	Wetland	Water	Desert
Gas regulation	3097.0	707.9	442.4	1592.7	0	0
Climate regulation	2389.1	796.4	787.5	15130.9	407.0	0
Water conservation	2831.5	707.9	5 30.9	13715.2	18033.2	26.5
Soil formation and protection	3450.9	1725.9	1291.9	1513.1	8.8	17.7
Waste disposal	1159.2	1159.2	1451.2	16086.6	16086.6	8.8
Biodiversity protection	2884.6	964.5	628.2	2212.2	2203.3	300.8
Food production	88.5	265.5	884.9	265.5	88.5	8.8
Raw material	2300.6	44.2	88.5	61.9	8.8	0
Entertainment	1132.6	35.4	8.8	4910.9	3840.2	8.8
Total	16237.0	5698.6	5672.6	53896.3	40676.4	371.4

Table 1 shows the unit price of the average value of the ecological services provided by China's ecosystems. The unit prices of the service values of the ecosystems in different regions depend on local and national economic development. Moreover, the value natural resources constantly changes. With the development of the social economy, the increase in average income, and the decrease in natural resources over time, the value of nature will increase, resulting in a general increase in willingness to pay for these resources. Therefore, an area more economically developed requires higher ecological compensation. The unit price of the service value of an ecosystem in an area must be multiplied by the local economic development factor. The economic development factor can be expressed by the ratio of the local gross domestic product (GDP) to the national average GDP. Additionally, the land occupied by a project has a time value. Land not only has current biological value, but it will also constantly produce ecological value in the future. Therefore, the compensation for the ecological value of the occupied land must be calculated through an analysis of the capital dynamics by converting the ecological value that the land cannot gain after being occupied into the current ecological value at a discount rate. The corrected unit price of the service value of an ecosystem and the ecological compensation for terrestrial resources are defined as [7]:

$$w_i = a_i[\frac{1}{1+r} + \frac{1}{(1+r)^2} + \cdots + \frac{1}{(1+r)^n}] = \frac{a_i}{r}[1 - \frac{1}{(1+r)^n}] \tag{1}$$

$$D_i = e\ w_i = e\frac{a_i}{r}[1 - \frac{1}{(1+r)^n}]$$

A project occupies land permanently, meaning $n \to \infty$. The above equation can thus be corrected to:

$$D_i = \lim_{n\to\infty} e\frac{a_i}{r}[1 - \frac{1}{(1+r)^n}] = e.\frac{a_i}{r} = \frac{g}{G}\cdot\frac{a_i}{r} \tag{2}$$

$$F_l = \sum_{i=1}^{6} D_i A_i \tag{3}$$

where wi represents the present value of the ecological services provided by unit area of an ecosystem (RMB); a_i represents the unit price of the service value (RMB/hm2.a) of the different ecosystems listed in Table 1; Di represents the corrected unit price (RMB/hm2.a) of the service value of the ecosystems; e represents the economic development factor; g represents the local GDP per capita (RMB/person); G represents the national GDP per capita (RMB/person); r represents the discount rate; n represents the number of years of land use (a); Ai represents the area of land (hm2); Fl represents the ecological compensation for terrestrial resources (RMB); and i represents the six ecosystem types shown in Table 1.

2.2. Calculation of compensation for the value of water environment

Utilization of more than 20% of runoff after water diversion is generally believed to result in significant effects on the environment. Water diversion can lead to a decrease in both the percentage of a river's flow in a non-flood season and the river's self-purification capacity, which results in an increase in the expenses required for further processing of the wastewater along the river's bank. Rivers exhibit low self-purification capacities primarily in non-flood seasons. The dilution ratio is the ratio of the amount of wastewater discharged to the amount of water contained in the water body that contains the wastewater. The compensation for the water environment of a river consists of the pollution abatement expenses required for a river from which water is diverted in a non-flood season to achieve a dilution ratio that is the same as the dilution ratio prior to water being diverted. This calculation is performed according to the following steps [8]: (1) calculate the dilution ratio of a river in a non-flood period before and after water diversion; (2) determine the amount of wastewater that must be reduced for the dilution ratio of the river after water diversion to be equal to the dilution ratio of the river before water diversion; (3) use the shadow engineering method to construct a wastewater treatment plant for treating the amount of wastewater that must be reduced. The incurred expenses are therefore the compensation for the water environment of the river. The calculation of compensation for the water environment of the river is defined as:

Then, $Q_{yw} = \frac{Q_w}{Q_j} \cdot Q_{yj}$ can be obtained from $b = \frac{Q_w}{Q_j} = \frac{Q_{yw}}{Q_{yj}}$.

$$\Delta Q_w = Q_w - Q_{yw} = \frac{Q_w}{Q_j} \cdot \Delta Q_j \tag{4}$$

$$F_S = C \times \Delta Q_w \times N \tag{5}$$

where b represents the dilution ratio; Q_w represents the amount of wastewater discharged from the river before water diversion (m^3/a); Q_j represents the river runoff before water diversion (m^3/a); Q_{yj} represents the river runoff after water diversion (m^3/a); Q_{yw} represents the amount of wastewater discharged from the river after water diversion (m^3/a); ΔQ_w represents the amount of water required to be processed (m^3/a); ΔQ_j represents the decrease in the river runoff (m^3/a); Fs represents the compensation for the water environment of the river (RMB); C represents the wastewater treatment plant's cost for processing the wastewater (RMB/m^3); and N represents the life of the wastewater treatment plant (a).

2.3. *Calculation of compensation for the value of aquatic biological resources*

The decrease in the amount of water caused by a water diversion can result in a decrease in the quantity of fish. Building a dam on a river for a reservoir can also result in the division of the habitat for fish, which prevents fish species from genetic exchange and leads to decreased genetic diversity and increased probability of species extinction. The raised water level resulting from dam construction can lead to the disappearance of the natural fish spawning area. The cost of protecting rare fish from extinction is the ecological value of fish. The effects and damage to natural fish caused by a project are irreversible. As the quantity of rare fish continues to decrease, people will become more willing to pay for them, and their value will increase. Therefore, the protection expenses must be discounted indefinitely and multiplied by the local economic development factor. Based on Equation (2), the compensation for the ecological value of rare fish can be calculated as follows:

$$F_y = e \cdot \frac{b}{r} = \frac{g}{G} \cdot \frac{b}{r} \tag{6}$$

where Fy represents the ecological compensation for rare fish (RMB) and b represents the cost of protecting rare fish (RMB/a).

3. Calculation of ecological compensation for Niulanjiang River provided by the Niulanjiang River to Dianchi Lake water diversion project

3.1. *Overview of the Niulanjiang River to Dianchi Lake water diversion project*

The Niulanjiang River to Dianchi Lake water diversion project is a recent key project that diverts water to the central Yunnan Province. It is designed to fill Dianchi Lake with ecological water, improve the water environment of Dianchi Lake, offer water to Kunming City for urban, domestic and industrial uses in case of a water supply crisis in 2020, and supply water to Qujing City for production and living in 2030. This project diverts water from Niulanjiang River, which is a tributary of the Jinsha River, originates in Xundian County of Kunming City in the northeast of Yunnan Province, and has a mouth located within the borders of Zhaotong City. Niulanjiang River has a mainstream length of 440 m and an average runoff of 4.35 billion m3 per year. This project consists of the Deze reservoir, the Ganhe river pumping station in Deze, and the line for delivery of water from the Ganhe river pumping station to Kunming.

3.2. *Assessment of the ecological compensation for the terrestrial biological resources in Niulanjiang River*

The land in Niulanjiang River occupied by the project is shown in Table 2 according to the 2010 feasibility study report on the Niulanjiang River to Dianchi Lake water diversion project. The GDP in Niulanjiang River in 2007 was 3,469 RMB per capital. Based on the fact that the national GDP in 2007 was 18,934 RMB per capital [9], the economic development factor of Niulanjiang River was calculated to be 0.18. The discount rate, r, was assigned a value of 2.79%, the one-year term deposit rate.

Table 2. Land in Niulanjiang River occupied by the project.

Farmland (hm^2)	Woodland (hm^2)	Grassland (hm^2)	Waters (hm^2)	Bare land (hm^2)	Total (hm^2)
218.51	1034.50	468.17	103.87	91.76	1916.81

According to the above data and Equations (2) and (3), the ecological compensation in Niulanjiang River granted for the Niulanjiang to Dianchi Lake water diversion project is 136,460,000 RMB for the terrestrial biological resources, 7,990,000 RMB for farmland, 108,340,000 RMB for woodland, 17,200,000 RMB for grassland, 2,720,000 RMB for waters, and 210,000 RMB for bare land. The ecological compensation for woodland is the greatest.

3.3. Assessment of the ecological compensation for the water environment of Niulanjiang River

According to the Plan for Protection of the Water Environment of Niulanjiang River (2009), it costs 2000,000 RMB to build a wastewater treatment plant with a capacity of 500 m^3/a. The life of a wastewater treatment plant is generally thirty years. The cost of treating wastewater by building a wastewater treatment plant is expressed as:

$$C = \frac{T}{G \times N} \tag{7}$$

where C represents the cost of treating wastewater by building a wastewater treatment plant (RMB/m^3); G represents the amount of wastewater to be treated (m^3/a); T represents the cost of building a wastewater treatment plant (RMB); and N represents the lifespan of a wastewater treatment plant.

According to Equation (7), the cost of treating wastewater by building a wastewater treatment plant in Niulanjiang River is 133.33 RMB/m3. According to the 2010 feasibility study report on the Niulanjiang River to Dianchi Lake water diversion project, the average annual runoff of Niulanjiang River is 59.5 m3/s in a dry season and 36.5 m3/s after water diversion, with a reduction of 23 m3/s. Niulanjiang River discharges wastewater at a rate of 239.58 m3/a. According to Equations (4) and (5), the ecological compensation for the value of the water environment of Niulanjiang River is 37,000 RMB.

3.4. Assessment of the ecological compensation for the rare fish in Niulanjiang River

Deze Reservoir, having a dam height of 142 m, was built for the Niulanjiang River to Dianchi Lake water diversion project. The dam and runoff regulation have resulted in a change in the downstream hydrology. The fish affected are primarily the rare fish that spawn in torrents, including Sinocyclocheilus graham, Sinogastromyzon dezeensis, Jinshaia niulanjiangensis, and Paraprotomyzon niulanjiangenssis. It is necessary to build a fish proliferation station in the Niulanjiang River basin for long-term artificial population propagation in order to protect these rare fish. In order to compensate for the effects caused by water conservancy projects to rare fish, American fish ecologists have studied the protection of Oncorhynchus nerka, a rare species of fish and built a fish spawning area at a cost of 7,440.62 USD/a[10], or 46,875.90 RMB/a (at a USD to RMB exchange rate of 6.3). The discount rate r was assigned a value of 2.79%, and the economic development factor of Niulanjiang River was assigned a value of 0.18. Based on Equation (6), the ecological compensation for the rare fish in Niulanjiang River provided by the water diversion project is 302,400 RMB.

3.5. Assessment of the total ecological compensation for Niulanjiang River

The ecological compensation for Niulanjiang River is equal to the sum of the ecological compensation for its terrestrial biological resources, water environment, and rare fish:

$$F = F_l + F_s + F_y = 136{,}799{,}400\,\text{RMB} \tag{8}$$

4. Conclusions

The Niulanjiang River to Dianchi Lake Water Diversion Project provided 136,799,400 RMB for Niulanjiang River water sources as ecological compensation. Among all of the ecological compensation, the most expensive compensation is for woodland, which is 108,340,000 RMB. In order to minimize the water diversion project's adverse effects on the ecological environment of water sources, during the construction, it is necessary to protect some remaining native species that survive in the steep environment. Since Niulanjiang River water sources are located on the plateau of the central region of Yunnan Province, they have sufficient water content and appropriate heat quantities to endow vegetation with strong self-repairing capabilities. Thus, the remaining native species could be effectively used for vegetation restoration after the construction.

The proposed method for calculating the ecological compensation for water sources for water diversion projects considers the compensation for the various environmental effects on water sources provided by water diversion projects and local social development. This method enables the ecological compensation to cover the environmental protection and ecological restoration in water sources while facilitating the implementation of water diversion projects.

References

[1] Heng L 2011 *Key technical issue study of risk management of diverting water from south to North* Beijing: Science Press pp 1-20

[2] Lu C X 2013 *Development of eco-friendly effects and ecological investigations research of China Drainage Basin Reservoir Projects* Beijing: Science Press pp 1-15

[3] Zhou D J, Dong W J and Sun L Y 2005 Research on ecological compensation of river basin water management *Academic Journal of Beijing Normal University*: Social Science Press **4**: 131-35

[4] Liu Y L 2007 *Ecological compensation and co-construction and sharing of basin ecology* Beijing: China Water & Power Press pp 19-31

[5] Wan B T and Zou S M 2008 *Heading for ecological compensation practice: case analysis and exploration* China Environmental Science Press pp 83-115

[6] Chen K L 2013 *Ecological functions and eco-compensation in Qinghai Lake Basin* Beijing: Science Press pp 48-54

[7] Zhang R S and Shi G Q 2008 Study on Evaluation of Farmland Resources Submersed by Water Conservancy and Hydroelectric Engineering *China Rural Water and Hydropower* (3): pp 99-102

[8] Zhang Z Z and Qiao P S 2012 *Production mechanism and calculation method for reservoir compensation benefits* Beijing: China Water & Power Press pp 105-25

[9] *Consumption Per Capita and GDP Per Capita in China for years 1978 through 2007* Baidu Wenku

[10] Liu N F 2005 *Ecological capacity and evaluation of environmental value loss* Beijing: Chemical Industry Press pp 81-191

Removal of Pb, Cd, and Cr in a water purification system using modified mineral waste materials and activated carbon derived from waste materials

H R Lu[1, 2], L C Su[1] and H D Ruan[1]

[1]28 Jinfeng Road, Tang jia wan, Zhuhai, Guangdong, China

[2]E-mail:h230400019@mail.uic.edu.hk

Abstract. This study attempts to find out and optimize the removal efficiency of heavy metals in a water purification unit using a low-cost waste material and modified mineral waste materials (MMWM) accompanied with activated carbon (AC) derived from waste materials. The factors of the inner diameter of the purification unit (2.6-5cm), the height of the packing materials (5-20cm), the size of AC (200-20mesh), the size of MMWM (1-0.045mm), and the ratio between AC and MMWM in the packing materials (1:0 - 0:1) were examined based on a L18 (5) 3 orthogonal array design. In order to achieve an optimally maximum removal efficiency, the factors of the inner diameter of the purification unit (2.6-7.5cm), the height of the packing materials (10-30cm), and the ratio between AC and MMWM in the packing materials (1:4-4:1) were examined based on a L16 (4) 3 orthogonal array design. A height of 25cm, inner diameter of 5cm, ratio between AC and MMWM of 3:2 with size of 60-40mesh and 0.075-0.045mm, respectively, were the best conditions determined by the ICP-OES analysis to perform the adsorption of heavy metals in this study.

1. Introduction

In the past two decades there has been an increasing public awareness of the hazards that exist from the contamination of the environment by toxic substances. Heavy metal pollutions are one of the most significant hazards in our daily life. The term "heavy metal" refers to any metallic element with a high density and toxicity.

Heavy metals exist in the environment naturally, or they can be emitted into an environment through the discharge of domestic and industrial wastes. Some human activities such as mining operations, electronics manufacturing and recycling, are the major causes of heavy metal emission [1]. Heavy metal pollution of surface and underground water sources result in considerable soil pollution. Another cause to soil pollution is when mined ores are dumped on the ground surface for manual dressing [2].

Heavy metals are one of the most significant concerns in water treatment. Heavy metals (e.g, lead (Pb), cadmium (Cd), chromium (Cr)) are of great environmental concern because of their toxicity [3]. A mail challenge remains in dealing with the water safety for local citizens. To date, there are a number of technologies available for the removal of heavy metals from wastewater. For example, chemical precipitation by adding flocculants including those containing OH^{-}, CO_3^{2-} and S^{2-}, are commonly employed in industrial water treatment. However, some compounds among them are amphoteric with a point of low solubility [4]. Other more advanced technologies such ion exchange, ultra-filtration, and reverse osmosis can achieve better removal efficiency of heavy metals, but the high operating costs make them uncompetitive.

Adsorption has served as an economical choice in terms of removing trace metals from water [5]. A lot of research has demonstrated that AC with its high surface area has a high ability of metal ion adsorption due to its micro porous character and chemical nature; therefore it can be utilized to remove the heavy metals from industrial wastewater [6]. In addition to AC, modified mineral waste materials (MMWM), another green and environmentally friendly adsorption material, is proposed and tested in this study. This kind of material is recycled from construction waste materials through physical and chemical treatments. MMWM has been applied not only for the removals of heavy metals and toxic organic compounds in water [7], but has also been using for the remediation of contaminated soils through neutralizing the acidic soil to an appropriate pH due to high pH value of MMWM relative to the contaminated soil, which has been verified in our current experiments. It is a good practice that pollutions can be treated using waste materials without introducing new wastes. Moreover, a possible application of this material in wastewater purification enables the use of mineral wastes to efficiently alleviate the burden of landfill.

It is hoped that MMWM could be promoted to be a synergic adsorption material with AC. This study aims at obtaining the most effective purification unit by applying the adsorption mechanism to remove heavy metals in water. It also aims to promote the use of MMWM. A purification unit is designed with adsorption materials of AC and MMWM, and its efficiency is expectedly improved to perform optimally. Besides the adsorption materials, optimum conditions of removal efficiency including the most suitable size ration of activated carbon and MMWM, the height of adsorption materials, as well as the inner diameter of purification unit are optimized. This research focuses on removal of three heavy metals in water, namely the Pb, Cd, and Cr.

2. Materials and methods

2.1. Materials

2.1.1. Solutions of contaminated water.

The influents used for purification treatment by different testing columns were designed. Three types of heavy metals were added for stage I and stage II: Cd (0.1ppm), Cr (0.5ppm), Pb (1.0ppm). These solutions were prepared in the laboratory by diluting the standard stocks of 1000ppm of concentrated heavy metals with deionized water. The concentrations of heavy metals are based on the standard of GB 8978-1996

2.1.2. Packing materials

One of the packing materials—Coconut derived granular AC (200, 60-40, and 40-20mesh in size) with an iodine value of around 800mg/g was purchased from YAOU Company.

The other packing material is MMWM (1-0.3mm, 0.3-0.075mm, and 0.074-0.045mm in size). The characteristics, including the SEM images, XRD and the mechanism of binding heavy metals, of MMWM were previously studied and provided in the other published paper [7]. Both AC and MMWM were directly packed into the column without any pre-treatment.

2.1.3. Instruments

Peristaltic pump—BT100-4 (Shanghai Yue Ming scientific instrument co., LTD) was employed to control the flow rate of inlets. Chromatographic columns (Shanghai mansion biochemical science and technology development co., LTD) served as purification columns loading with adsorbents. ICP-OES—Optima_2100DV (PerkinElmer, Inc.,) was required in ICP-OES analysis to obtain results of concentrations of heavy metals.

2.2. Methods

2.2.1. Experimental setup

The experimental setup of purification unit is presented in Figure 1.

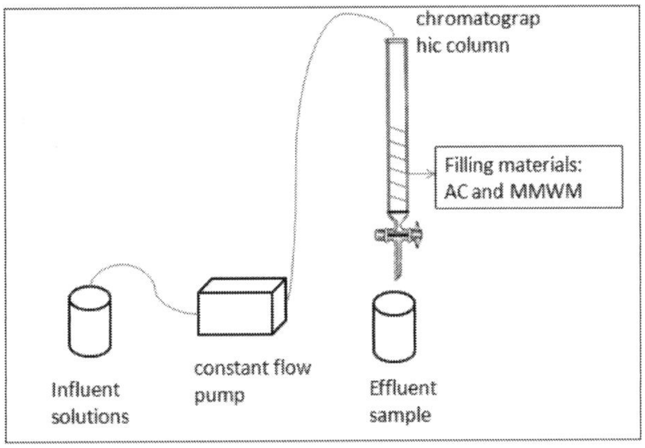

Figure 1. Experimental setup of purification unit.

Generally, based on the purification unit setup, the influents with known concentrations of heavy metals were pumped into the testing column by a constant flow pump. After the solutions going through the column, the effluents were collected with containers and stored in a fridge at 4°C.

2.2.2. Orthogonal experimental design

The orthogonal experiments were designed and demonstrated separately in stages.

Five factors of the experimental design were involved in stage I: the height of packing materials (H), the inner diameter of purification unit (D), the sizes of activated carbon (SAV) and MMWM (SMMWM), and the ratio between MMWM and AC on the composition of purification materials (MMWM/AC). In this experiment, orthogonal experimental design method was used to analyze the influence degree of H, D, SAV, SMMWM, and MMWM/AC on purification efficiency of purification units, then the best overall performance setup was selected. It is assumed that any two factors are independent of each other. The orthogonal experimental design table L18 (3) 5 employed in the test program is presented in Table1.

Table 1. Orthogonal table L18 (3)5 of stage I.

No. of test	H(cm)	D(cm)	SAC (meshes)	SMMWM (mm)	MMWM/AC
1	5	5	200	1-0.3	0:1
2	5	3.5	60-40	0.3-0.075	1:1
3	5	2.6	40-20	0.075-0.045	1:0
4	10	5	200	0.3-0.075	1:1
5	10	3.5	60-40	0.075-0.045	1:0
6	10	2.6	40-20	1-0.3	0:1
7	20	5	60-40	1-0.3	1:0
8	20	3.5	40-20	0.3-0.075	0:1
9	20	2.6	200	0.075-0.045	1:1
10	5	5	40-20	0.075-0.045	1:1
11	5	3.5	200	1-0.3	1:0
12	5	2.6	60-40	0.3-0.075	0:1
13	10	5	60-40	0.075-0.045	0:1
14	10	3.5	40-20	1-0.3	1:1
15	10	2.6	200	0.3-0.075	1:0

16	20	5	40-20	0.3-0.075	1:0
17	20	3.5	200	0.075-0.045	0:1
18	20	2.6	60-40	1-0.3	1:1

Three factors of the experimental design were involved in stage II: the height of packing materials (H), inner diameter of purification unit (D), and the ratio between AC and MMWM on the composition of purification materials (MMWM/AC). As opposed to stage I, this stage consists of 4 levels of each factor instead of 3 levels. The orthogonal experimental design table L16 (4) 3 employed in the test program is presented in Table 2.

Table 2. Orthogonal table L16 (4)3 of stage II.

No. of test	H(cm)	D(cm)	MMWM/AC
1	10	2.6	1:4
2	10	3.5	2:3
3	10	5	3:2
4	10	7.5	4:1
5	20	2.6	2:3
6	20	3.5	1:4
7	20	5	4:1
8	20	7.5	3:2
9	25	2.6	3:2
10	25	3.5	4:1
11	25	5	1:4
12	25	7.5	2:3
13	30	2.6	4:1
14	30	3.5	3:5
15	30	5	2:3
16	30	7.5	1:4

2.2.2.1. Analytical method and data handling
The concentrations of heavy metals in both influents and effluents were determined by ICP-OES analysis. The original concentrations of influents and effluent samples can be obtained by timing the dilution factor.

$$C_x = 2.5 \times C_{rx} \quad (1)$$

Cx is the actual concentration of metal x
2.5 is the dilution factor
Crx is the raw concentration of metal x given out by ICP analysis

The removal efficiency in percentage for each heavy metal was calculated through equation 2.

$$\mathrm{E}_x = \frac{1 - C_{xf}}{C_{xi}} \times 100\% \quad (2)$$

Ex is removal efficiency in percentage of heavy metals x (x can be any one of the 3-type heavy metals)
Cxf is the actual concentration of heavy metal x in effluent from ICP analysis
Cxi is the actual concentration of heavy metal x in influent from ICP analysis

The average Ex(average) of three heavy metals in each test was calculated.

$$\mathrm{E}_{x(average)} = \frac{E_{x1}+E_{x2}+E_{x3}}{3} \quad (3)$$

Ex1 is the removal efficiency in percentage of Cd
Ex2 is the removal efficiency in percentage of Cr
Ex3 is the removal efficiency in percentage of Pb

The removal efficiencies in percentage for all the tests were handled with SPSS for two statistical analysis methods: range analysis and variance analysis.

3. Results and analysis

3.1. *Orthogonal test*

Results of heavy metals removal are presented in stages. The analysis methods on the results including range analysis and variance analysis are basing on the book *Design and Modeling of Experiments [8]*.

3.1.1. *Stage I*

3.1.1.1. Range analysis. The results are shown in Table 3 and Fig 2. The sequence of factors that affected the removal efficiency was: ratio between AC and MMWM > size of AC > size of MMWM > inner diameter > height. The optimal conditions were obtained by using the orthogonal experimental design to be as follows: ratio 1:1 of AC to MMWM, 60-40mesh of AC in size, 0.3-0.075mm of MMWM in size, inner diameter 3.5cm, and height 20cm. The removal efficiency under these conditions was measured individually, and the result was 93.52% regarded as the optimal removal efficiency.

Table 3. Parameters of purification test results and intuitive analysis (Stage I).

	A	B	C	D	E	
Experimental number	H(cm)	D(cm)	SAC (meshes)	SMMWM (mm)	MMWM/AC	Removal efficiency
1	5	5	200	1-0.3	0:1	72.34%
2	5	3.5	60-40	0.3-0.075	1:2	92.31%
3	5	2.6	40-20	0.075-0.045	1:0	66.32%
4	10	5	200	0.3-0.075	1:2	91.23%
5	10	3.5	60-40	0.075-0.045	1:0	82.34%
6	10	2.6	40-20	1-0.3	0:1	72.32%
7	20	5	60-40	1-0.3	1:0	84.20%
8	20	3.5	40-20	0.3-0.075	0:1	78.32%
9	20	2.6	200	0.075-0.045	1:2	82.34%
10	5	5	40-20	0.075-0.045	1:2	81.54%
11	5	3.5	200	1-0.3	1:0	72.30%
12	5	2.6	60-40	0.3-0.075	0:1	78.43%
13	10	5	60-40	0.075-0.045	0:1	74.29%
14	10	3.5	40-20	1-0.3	1:2	78.53%
15	10	2.6	200	0.3-0.075	1:0	75.02%
16	20	5	40-20	0.3-0.075	1:0	74.21%
17	20	3.5	200	0.075-0.045	0:1	76.13%
18	20	2.6	60-40	1-0.3	1:2	83.24%
m1	77.20%	79.60%	78.20%	77.20%	75.30%	78.63%
m2	79.00%	80.00%	82.50%	81.60%	84.90%	(Average)
m3	79.70%	76.30%	75.20%	77.20%	75.70%	
Range	0.025	0.037	0.073	0.044	0.096	

m: the average removal efficiency in percentage of every factor (5 factors in total) under the same level (3 levels in total). For example, m1 of factor A is the average removal efficiency of factor A when the level is 5cm. m1= ((72.34+92.31+66.32+81.54+72.30+78.43)/6) %= 77.20%

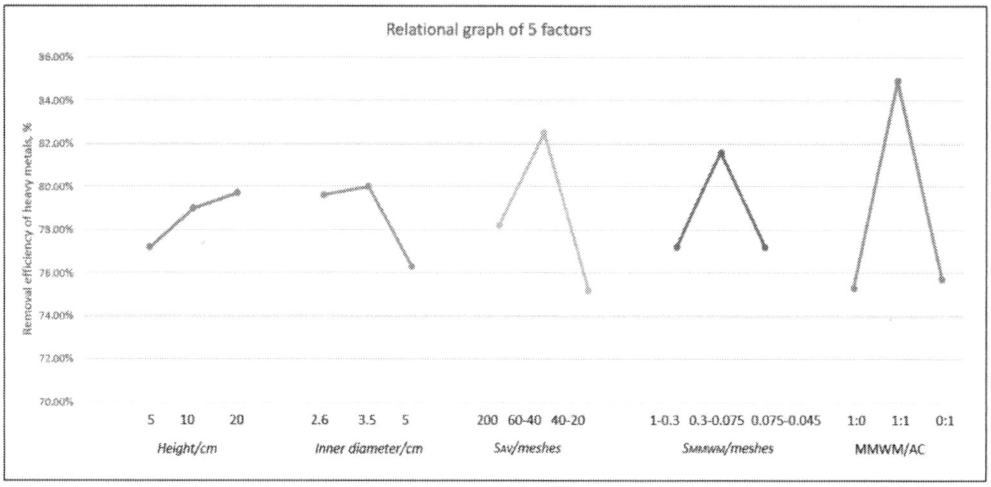

Figure 2. Relational figure between removal efficiencies and five factors (stage I).

3.1.1.2. Analysis of variance. The orthogonal experimental and analysis of variance results (Table 4) indicated that: there was a significant difference in factor E—the ratio between AC and MMWM with the F value of 7.010 which had the most significant effect on improving the removal efficiency of the purification unit, while there was no significant difference in the sizes of AC and MMWM, height of packing materials and inner diameter of testing column. Overall, the order from high to low significance level of factors was: E>C>D>B>A, which was consistent with the range analysis.

Table 4. Variance analysis table of tests of removal efficiencies (stage I).

Factors	sum of squares of deviations	Degree of freedom	Mean Square	F	Fcritical, (95%)
A	0.002	2	0.001	0.396	3.74
B	0.0049	2	0.0025	0.970	3.74
C	0.0162	2	0.0081	3.208	3.74
D	0.0077	2	0.0039	1.525	3.74
E	0.0354	2	0.0177	7.010	3.74 *
Error (blanks)	0.0101	4	0.0025		
Total	0.0763	14			

* A sign indicates that there is a significant difference on the factor with it.

3.1.2. Stage II

3.1.2.1. Range analysis. Similar to analysis in stage I, the results are shown in Table 5and Figure3. The order from high to low significance level of the factors that affected the removal efficiency was: ratio between AC and MMWM> height>inner diameter. The optimal conditions were obtained by using the orthogonal experimental design shown as the follows: ratio 3:2 of AC to MMWM, height 25cm, and inner diameter 5cm. The removal efficiency under these conditions was 98.52%.

Table 5. Purification test results and intuitive analysis table (Stage II).

	A	B	C	
Experimental number	H(cm)	D(cm)	MMWM/AC	Removal efficiency
1	10	2.6	1:4	89.02%
2	10	3.5	2:3	93.08%
3	10	5	3:2	97.30%
4	10	7.5	4:1	91.57%
5	20	2.6	2:3	93.32%
6	20	3.5	1:4	93.26%
7	20	5	4:1	94.31%
8	20	7.5	3:2	98.28%
9	25	2.6	3:2	98.12%
10	25	3.5	4:1	93.64%
11	25	5	1:4	97.30%
12	25	7.5	2:3	93.49%
13	30	2.6	4:1	92.90%
14	30	3.5	3:5	96.79%
15	30	5	2:3	92.80%
16	30	7.5	1:4	93.28%
ml	92.70%	93.30%	93.20%	94.23%
m2	94.80%	94.20%	93.20%	(Average)
m3	95.60%	95.40%	97.60%	
m4	93.90%	94.20%	93.10%	
Range	0.029	0.021	0.045	

m: the average removal efficiency in percentage of every factor (3 factors in total) under the same level (4 levels in total). For example, ml of factor A is the average removal efficiency of factor A when the level is 10cm.
ml= ((89.02+93.08+97.30+91.57)/4)%= 92.70%

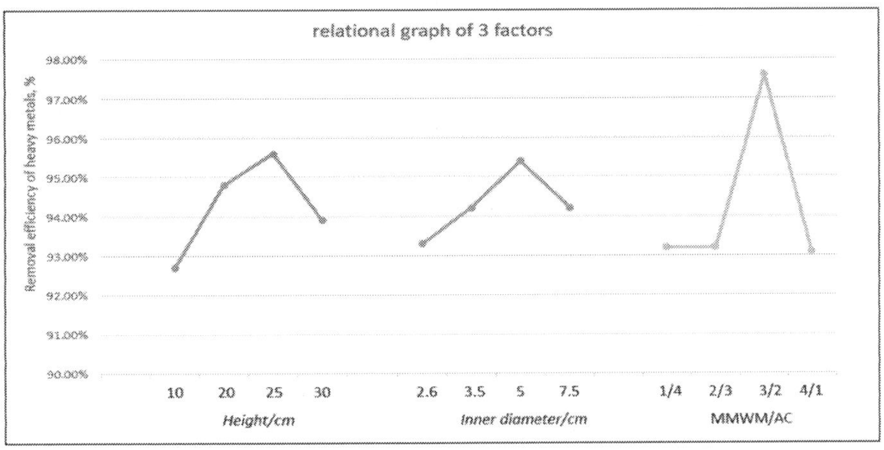

Figure 3. Relational figure between removal efficiencies and three factors (stage II).

3.1.2.2. Analysis of variance. The orthogonal experimental and analysis of variance results (Table 6) indicated that: there was a significant difference in factor C—the ratio between AC and MMWM with the F value of 5.488 which had the most significant effect on improving the removal efficiency of the purification unit; there were no significant difference in the height of packing materials and inner diameter of testing column, which were consistent with the results of stage I. Overall, the order from high to low significance level was: C>A>B. This consisted with the range analysis but with statistical support.

Table 6. Variance analysis table of tests of removal efficiencies (stage II).

Factors	sum of squares of deviations	Degree of freedom	Mean Square	F	Fcritical, (95%)	
A	0.001804	3	0.000601	1.853	3.287	
B	0.000889	3	0.000296	0.913	3.287	
C	0.005343	3	0.001781	5.488	3.287	*
Error (blanks)	0.001947	6	0.000325			
Total	0.009983	15				

* A sign indicates that there is a significant difference on the factor with it.

4. Discussion

The purpose of the study was to find out and optimize the conditions of the purification unit with AC and MMWM as packing materials while increasing the ratio of the use of MMWM, which is the predominant factor to improve the efficiency of purification.

Referring to Figure 2, there were no obvious peaks of the effects of height and the inner diameter of purification unit; in other words, the results of height and inner diameter were not significant. However, Figure 3 shows sharp peaks for the effects of height and inner diameter of purification. Hence, the final optimum conditions of height and inner diameter of the purification unit were 25cm and 5cm, respectively. The significant effects of height and inner diameter on improving the removal efficiency were different between two analyses. This discrepancy further demonstrated that the effects of height and inner diameter were small enough to be neglected.

Among these factors which had sharp peaks, the sizes of AC (60-40mesh) and MMWM (0.075-0.035mm) were determined in stage I, while the ratio between AC and MMWM (3:2) had the best effect on improving the purification efficiency determined by a collaboration testing of both stage I and II. Overall, the best process condition of purification column is ratio 3:2 of AC to MMWM, 60-40mesh of AC in size, 0.3-0.075mm of MMWM in size, inner diameter 5cm, and height 25cm. The removal efficiency was up to 98.5% under these conditions.

Our results are similar to the result of a study on the adsorption of AC for heavy metal ions showed that heavy ion removals were above 98% under optimal conditions of pH, temperature, amount of used AC and the adsorption time [9]. However, according to the study, the adsorption time is 2 hours and other conditions are well designed and strictly controlled while in this research, the adsorption materials including AC and MMWM were not in optimal conditions of pH and temperature and the average adsorption time is around 10 minutes for each purification column. In other words, AC cooperating with MMWM, without strict control of pH and temperature, can achieve a similar effect with pure use of AC under strict conditions. Hence, it can be deduced that with a collaboration of MMWM, AC maximizes its removal efficiency.

5. Conclusion

The process conditions of purification columns removed heavy metals by AC and MMWM were investigated through orthogonal tests in this paper. The best process conditions of the purification column were a ratio of 3:2 of AC to MMWM, 60-40mesh of AC in size, 0.3-0.075mm of MMWM in size, inner diameter 5cm, and height 25cm. The removal efficiency was up to 98.5% under the best conditions. The sequence of factors that affect the removal efficiency: the ratio between AC and MMWM > the size of AC > the size of MMWM (noted that the effects of height and inner diameter were excluded), was acquired by means of range analysis and variance analysis. There were significantly differences in ratio between AC and MMWM.

Acknowledgements
This research was supported financially by Beijing Normal University-Hong Kong Baptist University United International College (UIC) (Project Code: R201404). I am grateful to Jiasui Li and Redon Xharja who have made valuable comments on this study.

References

[1] Hutton M and Symon C 1986 The quantities of cadmium, lead, mercury and arsenic entering the UK environment from human activities *Science of the total environment* **57** 129-50
[2] Garbarino J R, Hayes H, Roth D, Antweider R, Brinton T I. and Taylor H 1995 Contaminants in the Mississippi river *US Geological Survey Circular* **1133**
[3] Bishop P 2000 *Pollution prevention: Fundamentals and practice* Boston: McGraw-Hill
[4] Eckenfelder W 1966 *Industrial water pollution control* New York: McGraw-Hill
[5] Kadirvelu K, Faur-Brasquet C and Cloirec P L 2000 Removal of Cu (II), Pb (II), and Ni (II) by adsorption onto activated carbon cloths *Langmuir* **16**(22) 8404-09
[6] Kobya M, Demirbas E, Senturk E and Ince M 2005 Adsorption of heavy metal ions from aqueous solutions by activated carbon prepared from apricot stone *Bioresource Technology* **96**(13) 1518-21
[7] Jiang Y, Ruan H D, Lai S Y, Lee C H, Yu C F, Wu Z, Chen X, and He S 2013 Recycling of solid waste material in Hong Kong: I. properties of modified clay mineral waste material and its application for removal of cadmium in water *Earth Sciences* 40-40
[8] Fang K., Liu, Q and Zhou Y 2011 Orthogonal experiments design *Design and Modeling of Experiments* Higher Education Press.
[9] Hong H, Chen H, Ji H, and Fan X 2013 The adsorption of activated carbon for heavy metal ions *Tianjin Chemical Industry* **27**(2)

IOP Conf. Series: Earth and Environmental Science **39** (2016) 012006 doi:10.1088/1755-1315/39/1/012006

Improvement of activated sludge bacteria growth by low intensity ultrasound

Y X Yan[1, 2], J Y Ding[1] and J L Gao[1]

[1]School of Water Conservancy and Environment, Zhengzhou University, Zhengzhou, Henan, China

E-mail: yxyan@zzu.edu.cn

Abstract. Influence of low intensity ultrasound (US) on growth rate of bacteria separated from aerobic activated sludge was studied. In order to reveal the optimal ultrasonic conditions,specific oxygen uptake rate (SOUR) of activated sludge was first detected and results showed that the maximum SOUR was obtained (increased by 40%) at US intensity of 0.3 Wcm-2 and irradiation time of 10min. Under the optimal conditions, 2 species of bacteria isolated from activated sludge were sonicated and then cultivated for 36h, and increment of 6% and 10% of growth rate were detected for the 2 species of bacteria, respectively, indicating US irradiation of suitable parameters effectively improved activated sludge bacteria growth.

1. Introduction

As a new technology in the wastewater biological treatment field, low intensity ultrasound (US) can enhance the wastewater purification effect with simple equipment, low cost and convenient combination with various biological treatment processes. Yan et al. [1] applied low intensity US to the Sequencing Batch Reactor (SBR) process, the specific oxygen uptake rate (SOUR) increased by 14%, and the effluent COD decreased by more than 40%. In sewage anaerobic treatment and removal of nitrogen and phosphorus processes, the low intensity US showed significant improving effects [2-4]. Duan et al. [5] and Yu et al. [6] enhanced the Anammox process using low intensity US, and a maximum of 2.6 times increment of TN (total nitrogen) removal effect was obtained.

Many studies in biological fields have indicated that the stable cavitation effect induced by low intensity US increases the permeability of cell membranes, and the consequent acceleration of the mass transfer thus increases the bacteria growth [7-11]. In order to study the influence of low intensity US on activated sludge bacteria, the appropriate ultrasonic irradiation conditions were first explored by SOUR detection, and then the growth rate of activated sludge bacteria irradiated under these conditions was detected and compared with the control.

2. Materials and Methods

2.1. Materials

Aerobic activated sludge was from a pump room of returning sludge of a wastewater treatment plant (WWTP) in Zhengzhou, China. The wastewater used was from a primary settling tank in the same WWTP with the water quality, as shown in table 1.

2.2. Ultrasonic irradiation apparatus

An ultrasonic cleaning bath (DL-60D, Shanghai ZhiSun Instruments Co., Ltd., Shanghai, China) with a fixed frequency of 35 kHz and variable powers from 0~80 W was used to insonate the activated sludge in a beaker. During exposure, the beaker containing activated sludge was immersed in the sonic bath and fixed at the center of the support, and the activated sludge sample was stirred gently.

Table1. Water quality (mgL^{-1}).

Contaminant	COD_{cr}	BOD_5	NH_3-N	TP	SS
Concentration	409.5	282.3	43.1	7.6	70.1

2.3. *Experimental processes*

*2.3.1. Optimal ultrasonic irradiation conditions.*An optimization test of ultrasonic irradiation conditions was conducted with the sludge activity index of SOUR. The activated sludge samples werefirst sonicated with different ultrasonic intensities for 20 min and then detected for SOUR. Then, the USintensity inducing the highest SOUR was adopted to irradiate the sludge samples for different times to discern the optimal ultrasonic irradiation time length.

*2.3.2. Influence of low intensity US on growth rate of activated sludge bacteria.*Single colonies of bacteria samples separated from activated sludge were sonicated under the optimal ultrasonic irradiation conditions. Their growth curves were then ascertained according to thephotoelectric turbidimetry results, which showed the influence of US on bacterial growth rate by comparison with the unsonicated control.

2.4. *Analytical methods*

The mixed liquor suspended solids (MLSS), single colony separating and purification, gram stain and photoelectric turbidimetry were determined using Standard Methods [12,13]. SOUR was measured by a self-made respirometer, as shown in figure 1.

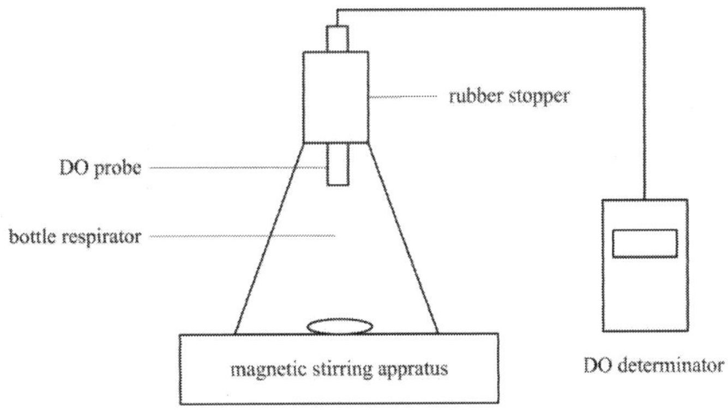

Figure1. Experimental setup for measuring SOUR.

The activated sludge sample was washed with distilled water three times to ensure all organic compounds were removed, and then saturated with oxygen by feeding of fully-aerated wastewater. During the measurement,the conical flask was carefully closedwith no air bubbles left, and the activated sludge sample was mixed by a magnetic stirrer (Shanghai Sile Appliance Factory, Shanghai, China) at 150 rpm. The change of dissolved oxygen (DO) was detected by an oxygen electrode (HI9146, HANNA) to calculate SOUR (mgO2(gMLSS.min)-1)according to the slope of the linear parts of the recorded DO profiles(Equation(1)).

$$SOUR= (DO_1—DO_2) / MLSS\cdot (t_2—t_1) \quad (1)$$

All the tests were conducted in triplicate at a temperature of 25-26℃.

3. Results and Discussion

3.1. Optimal Ultrasonic irradiation conditions

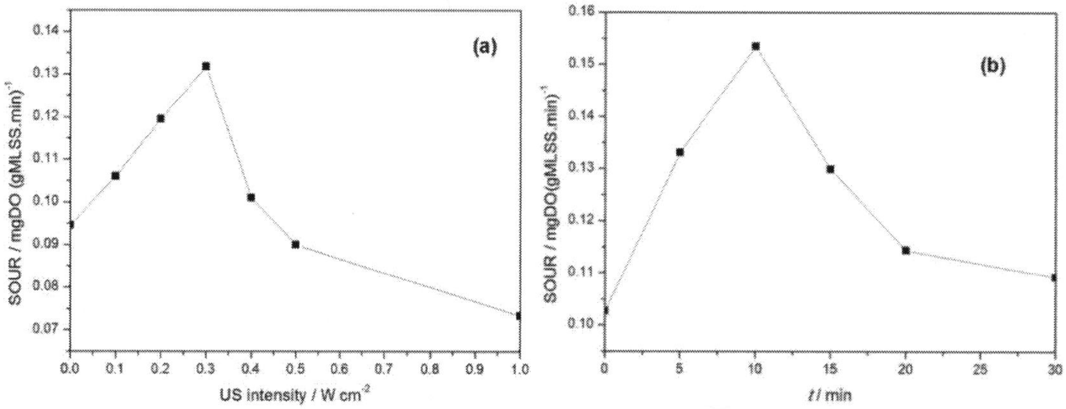

Figure 2. Effects of US irradiation conditions on sludge activity (a:US intensity; b: irradiation time).

The influence of US intensity on sludge activity isshown in figure 2(a). Initially, sludge activity progressively increased with the increment of US intensity before the maximum of SOUR was achieved at 0.3 Wcm^{-2}. Then, continuous increments of US intensity caused a declineof SOUR, and the SOUR was even inferior to the control when the intensityexceeded 0.5wcm^{-2}.

On this basis, sludge samples were irradiated for 0-30 min with the US intensity of 0.3 Wcm^{-2} to obtain the optimal irradiation time length. As shown in figure 2(b), the activated sludge achieved the maximum SOUR (increased by 40% compared with the control) when irradiated for 10 min, and longer irradiation time caused a downward trend. Therefore, the optimal ultrasonic irradiation conditions were 0.3 Wcm^{-2} of US intensity and 10 min of irradiation time length, and a higher intensityor longer irradiation time would bring adverse influences on the sludge activity.

US is characterized by the cavitation effect. Stable cavitation thatis gentle and regular occurs when US intensity is low enough, thus inducing oscillation of cavitation bubbles and consequent acoustic microstreaming in the medium, which in turn causes shear stress on cells around cavitation bubbles [14]. Besides, the vibration energy of medium particles increases due to the mechanical vibration induced by US. In combination with thestable cavitation and mechanical vibration effects, the substrate entering active sites of enzymes or cells and the product entering the liquid medium are both improved by low intensity US through thinning the mass transfer boundary layer and accelerating solute particle movement [15,16]. Therefore, enzymatic activity and cell metabolism are enhanced when irradiated by low intensity US. However, when US intensity exceeds a limit, transient cavitation occurs and induces cavitation bubbles to shrink until they collapse, and the produced high temperature and high pressure in the bubbles causes the formation of radical and powerful shockwave and jet flow. Thus, the resultant cell structure damage and enzyme inactivation decrease the cell activity. Therefore, there is always an optimal US intensity, below which no significant enhancing effect would be achieved, and above which the cell activity would be inhibited or even destroyed.

Activated sludge is an ecological population of microbes, and the individual differences to US induce aninconformity in adaptive intensity for different microorganisms. However, most of the microorganisms adapted to ultrasonic irradiation of low intensity and showed notable increments of sludge activitym as shown in figure 2(a), and with increasing intensity the sludge activity was also enhanced. When the intensity increased beyond the optimal value (0.3 Wcm^{-2} in this test), the amount of unadaptive microbes increased and the sludge activity declined to even lower than the control, indicating that excessive US intensity (beyond 0.5 Wcm^{-2} in this test) induced severe destruction of the microbial population.

Furthermore, US-induced shear stress may inflict minor wounds on the cell surface. When irradiation time is short enough, the smaller wounds may be bearablefor most cells, and the cell activity is improved with the produced increments of cell permeability and extra and intracellular material exchange. However, an excessive irradiation time may deteriorate the wound and thus cause a declineof sludge activity. Therefore, US intensity of 0.3 Wcm^{-2} and irradiation time of 10min were usedto conduct ultrasonic irradiation on activated sludge bacteria.

3.2. Influences of ultrasonic irradiation on cell growth rate

*3.2.1. Separation and purification of activated sludge bacteria.*Activated sludge samples were separated and purified to obtain 2 species of bacteria prior to gram staining,and their morphological characteristics were as shown in table 2.

Table 2. Morphological characteristics of 2 bacteria separated from activated sludge.

Bacteria	Community form	Gram stain	Cell shape
W	white, hyaline, round, regular edge	G-	rod
Y	yellow, hyaline, round, regular edge	G+	spherical

3.2.2. Influences of ultrasonic irradiation on cell growth rate. Bacteria W and bacteria Y were sonicated with a US intensity of 0.3 Wcm-2 for 10min. A comparison of the growth rates with the control over the course of36h is shown in figure 3.

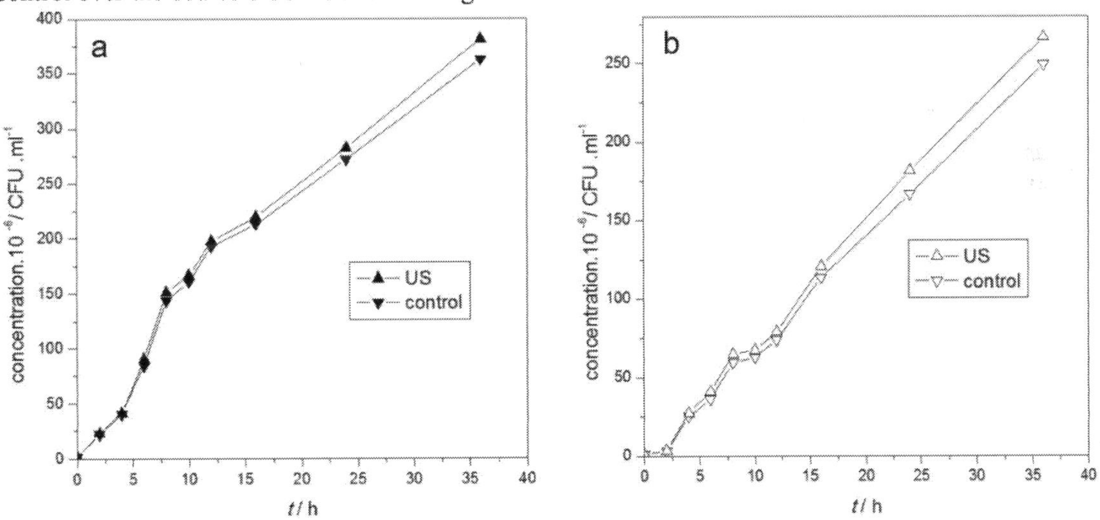

Figure 3. Influences of ultrasonic treatment on the growth of bacteria (a: bacteria W; b: bacteria Y).

As shown in figure 3, the growth rate of bacteria W and bacteria Y increased by 6% and 10%, respectively, after being sonicated, and the t-test analysis indicated that the results had a significant difference.

Considerable research on cell growth accelerated by low intensity US has been conducted in the medical, food and pharmaceutical fields. Gao et al. [15] investigated the effects of low intensity US on the growth of Saccharomyces cerevisiae cells, and notable promotion of growth rate was achieved. Pitt and Ross [16] sonicated S.epidermidis, E.coli and P.aeruginosa with low intensity US, and obvious acceleration was observed in the growth ofall the three species of bacteria. Similarly, activated sludge bacteria showed a significant improvement in growth rate after being sonicated by low intensity US. In combination with the above SOUR test results, it canbe speculated that the enhancement of sludge activity induced its growth acceleration.

In addition, more significant improvement of growth rate was observed for bacteria Y than bacteria W, which wereG+ and G- respectively. Likewise, Pitt and Ross [16] observed different degree improvement effects of US on different species of bacteria, among which the improvement effects followed the descending order: S. epidermidis (G+)>E.Coli (G-)>P. aeruginosa (G-). Therefore, the difference of the cell wall structures of gram-positive bacterium and gram-negative bacterium may lead to adifference in theimprovement effects of low intensity US on bacteria growth.

4. Conclusions

- The optimal irradiation conditions of low intensity US were0.3 Wcm^{-2} of US intensity and 10min of irradiation time length, with which a 40% improvementof SOUR could be achieved over the control.
- Bacteria W and bacteria Y were separated from the activated sludge, and their growth rates accelerated by 6% and by 10%, respectively, after being sonicated under optimal irradiation conditions.

Acknowledgments

The authors thank the Henan province education department for financial support (15A610022) and thank the reviewers for helpful suggestions.

References

[1] Yan Y X, Liu H, Zhang S L and Xie B Z 2006 *Env. Sci.***27**(8) 1596-602
[2] Xie B Z, Wang L and Liu H 2008 *Ultrason. Sonochem.* **15** (5) 775-81
[3] Xie B Z, Liu H and Yan Y X 2009*J. Environ. Manage.***90** (1) 260-64
[4] Xie B Z and Liu H 2010 *Water, Air, Soil Pollut.***211** (1-4) 157-63
[5] Duan X M, Zhou J T, Qiao S and Wei H 2011 *Bioresour. Technol.***102** 4290-93.
[6] Yu J J, Chen H, Zhang J, Ji Y X, Liu Q Z and Jin R C 2013*Bioresour. Technol.***142** 693-96
[7] Keris-Sen U D, Sen U, Soydemir G and Gurol MD 2014 *Bioresour. Technol.***152** 407-13
[8] Liu YY, Takatsuki H, Yoshikoshi A, Wang B and Sakanishi A 2003 *Colloids Surf. B* **32** 105-16
[9] Wu J Y and Lin L D 2002 *Biotechnol. Progr.***18** 862-66
[10] Wang B C, Yoshikoshi A and Sakanishi A 1998 *Colloids Surf.B***12** 89-95
[11] Su H, Jiang H X, Zheng H M, Li D R and Du Y H 2015 *Chinese J. Ultrasound Med.***31**(11):1038-40
[12] American Public Health Association (APHA) 2001*Standard Methods for the Examination of Water and Wastewater*(Washington DC: American Public Health Association)
[13] Shen P, Fan X Rand and Li G W 1999 *Microbiology Experiment* (Beiing: Higher Education Press)
[14] Nyborg W L 1982 *Br. J. Cancer Supl.***45** 156-60
[15] Gao D W, Gao W H, Lei D Z, Yu S J and Zhang Z H 1999 *J. South China Univ. Technol.***27**(12) 34-7
[16] Pitt W G and Ross S A2003 *Biotechnol. Progr.***19**(3) 1038-44

Distribution and assessment of residual mercury from gold mining in Changbai Mountain Range Northeastern China

D Meng[1], N Wang[2, 5], J C Ai[3], G Zhang[2] and X J Liu[4]

[1] College of Resources Environment and Tourism, Capital Normal University, Beijing 100048, China
[2] College of Urban and Environmental Science, Northeast Normal University, Changchun, 130117, China
[3] Tianjin No.1 High School, Tianjin, 300051, China
[4] Middle school of the 61th, Daqing, 163714, China

E-mail: nwang@nenu.edu.cn

Abstract. Gold mining was first initiated in Jiapigou area, Huadian city of Northeastern China about 200 years ago. Before 2006, the mercury amalgamation technique was used in the gold mining process, which led to severe mercury contamination. The aim of this paper is to explore the influences of residual mercury on the environment media after eliminating the amalgamation process to extract gold. The mercury concentrations of the atmosphere and the soil were determined in autumn of 2011 and spring of 2012. The soil environmental quality was assessed by the index of geoaccumulation. The results indicated that the maximum value of gaseous mercury was 25ng•m-3 in autumn and 19.5ng•m-3 in spring; the maximum value of mercury in the soil was 2.06mg•kg-1 in autumn and 2.51mg•kg-1in spring. It can be seen that the peak concentrations of the gaseous mercury happened at the gold mine area and tailings, while the peak mercury concentrations in the soil were located at the places near the mining sites and the residential area in the valley. Furthermore, the regression analysis of the total mercury contents between the atmosphere and the soil showed a significant correlation, which indicated that there was certain circulation of the mercury between the regional atmosphere and soil. In general, after the elimination of the amalgamation technique in gold extraction, the distance to the mercury source, the special conditions of hilly weather and landforms and the mercury exchange flux are the main factors of mercury contamination.

1. Introduction

The accumulation, migration and circulation of mercury in the environment could contaminate the air, water and soil, and cause problems for human health. After the Minamata disease event of Japan, poison accidents of mercury and methyl mercury to organisms including humans have recurred frequently, which has become the focus of research [1-3]. The mercury released in the gold extraction process by the amalgamation technique was the man-made source of mercury pollution in the gold mining area [4]. The mercury amalgamation technique is relatively inexpensive and readily available, so the overwhelming majority of artisanal and small-scale gold miners utilize mercury to separate gold from the unwanted gangue materials [5, 6].

Miners separate the gold from the amalgam by "burning" the amalgam, volatilizing the elemental mercury and dispersing it throughout mining camps. Then, the mercury continually enters into the soil, water and other environmental media near the mines, which becomes a threat to the local animal, plant and human communities. Some countries ban mercury use in artisanal and small-scale gold mining (ASGM) outright [7, 8], but some less developed areas still keep using this technology. One good example is that the privately owned Clean Tech Mine in the Manica Province in Mozambique has eliminated the use of mercury at their mine. Instead, gold is isolated by centrifugation and magnetic removal of gangue materials [9]. There are still many small-scale/artisanal gold mines that use the mercury amalgamation technique. However, the mercury contamination in the ASGM area and its surrounding area caused by long-term gold mining activity should not be neglected [10, 11]. As a result, the remaining mercury pollution of the gold mine area has gradually aroused great attention in recent years [12-14]. The gold mining has caused severe pollution in the local environment and a big threat to the human health of the local mine area in South America [15, 16], Africa [17] and Asia [18].

Mercury releases from Artisanal and small scale gold mining (ASGM) were estimated to be about 1400 tonnes/year, making it the largest global demand sector for mercury. This sector produces about 12-15% of the world's gold. An estimated 10-15 million miners, including 4-5 million women and children, are involved in the sector [7]. The study of Lecce et al. [10] on historic gold mining in North Carolina, USA showed that the mercury concentrations in the channel and floodplain sediment downstream from the Gold Hill mining district ranged from 0.01 to 2.2mg•kg-1, with a maximum concentration more than 35 times background levels. Balzino et al. [19] suggested that contrary to common belief, Hg was recovered from the bottom of the river instead of being released into it. The total Hg-mass balance is actually closed with a positive recovery and with an emission factor. Despite this result, Hg continues to be volatilized into the atmosphere in the form of vapors. The investigation by Saiki et al. [20] showed that the mercury and methyl mercury concentration of the samples from water, soil and sediment in the downstream area of the Sierra Nevada foothill Reservoir are much higher than they are upstream [20], so gold mine areas often became important pollution sources in these areas. Egler et al. [21] studied the total mercury concentration in soil and vegetables and the mercury concentration relationship between soil and vegetables in two small gold mine areas, São Chico and Creporizinho of Amazon, Brazil. They found that the total mercury concentrations of the soil samples in the two areas were all above the local background value, and the total mercury of the edible parts of the vegetables in São Chico was close to the standard of Food and Agricultural Organization/World Health Organization/JECFA [21].

Lacerda [22] argued that most mercury was discharged into the atmosphere due to the uncontrolled burning of the amalgam and bullion smelting in small-scale gold mining. China is rich in gold resources, and gold mining has a long history, but in the past, the mercury amalgamation technique was widely used in the extraction of the primary gold. The reports on gold mine mercury contamination were mainly concentrated on the Shanxi areas. The study of Feng et al. [23] on the gold mine area in Tongguan, Shanxi, showed that the total gaseous mercury (TGM) in the environmental atmosphere was severely higher than the background concentration, and its average TGM concentration reached 18,000 ng·m-3. The largest mercury concentration in the soil was 19.50mg•kg-1, but the mercury concentration in the background area was only 0.25mg•kg-1. The total mercury concentration range in vegetables and wheat was from 42mg•kg-1 to 640mg•kg-1, which were all beyond the upper limit of the standard value of Chinese vegetables (0.01mg•kg-1) and food fish (0.02mg•kg-1) [23].

There has been a long history of gold mining in Northeastern China, and most gold mines were located in the mountainous areas, such as the Jiapigou gold mining area, which is located in the upper reaches of the Songhua River, Changbai Mountain Range. In this area, mercury contamination was serious in ambient air, water and sediment. Some research of the mercury pollution investigations had already been done in this area [24] before the ending of amalgamation use, but there is no deeper discussion of the contamination at the present situation and the control measures after the elimination of the mercury amalgamation technique in the gold mine area. Based on the regional gold mining features and the terrain and weather conditions, this article studied the current distribution features of mercury concentrations in the atmosphere and soil in this region. Furthermore, the reasons of the formation of the mercury contamination in the air/soil of the area and the mechanism of mercury migration and circulation in the environment were discussed. Finally, the geoaccumulation index was used to assess the local soil environmental quality. The study provides the scientific basis to understand more about the characteristics of gold mine mercury contamination, as well as methods to control this type of regional mercury contamination.

2. Study area and methods

2.1. Study area

The Jiapigou gold mine area in Jiapigou Town, Huadian City, Jilin Province, Northeastern China was selected as the study area, which is located in the northwest of Changbai Mountain (42°41′–43°00′N and 127°15′–127°30′E). As a typical hilly area, the topography of this area is high in the southeast and low in the northwest. There are many valleys and channels in the area, and most of the landforms are two mountains with one valley in the middle. There are more than 20 mountains that are above 450 m in the region, and the highest mountain is 557 m. It has a typical mid-latitude zone continental monsoon climate, which is cold and dry in winter and wet and rainy in summer. The average annual temperature is 1.9–4.4℃, and the average temperature is 9.5-12℃ in spring and 13-20℃ in autumn. The average annual rainfall is 650–850 mm, the predominant wind direction is southwest in summer and northwest in winter, and the average wind speed is 2.2 m/s. It is rainy in summer and snowy in winter, so the weather conditions are not conducive to sampling and research. The average temperature in autumn is higher than that in spring, so the authors sampled in both spring and autumn.

Jiapigou gold mine area is situated in the northeast of the Zhongchao ancient land plate and the southeast of the Huifa River deep fracture. The northwest west Jiapigou-Dashilizi tectonic zone controls the distribution of the goldfields and deposits. This gold mine belt is 50 km long and 1–3 km wide, and mainly includes the Jiapigou gold mine, the Laoniugou gold mine, the Erdaodianzi gold mine, the Daxiangou gold mine, the Banmiaozi gold mine, the Sandaogou gold mine, the Huanan gold mine and the Laojinchang gold mine, as well as other small-scale gold mines. Placer gold was exploited in this gold mine area in 1821, and rock gold was exploited in 1845. The Jiapigou gold mine has been exploited for nearly 190 years. First, indigenous alchemy was adopted for gold extraction after 1821, and then the mercury amalgamation has been used for the gold extraction since 1940 until 2006. Considering the amalgamation of serious pollution to the environment, the international community does not advocate the use of amalgamation, and China's government officially prohibited the mercury amalgamation technique in 1996 [8], so the miners in this area have used cyanidation instead of amalgamation for gold extraction since 2006. According to investigations, 2000 t ore can annually be exploited in Jiapigou alone, approximately 20 kg•yr-1 of mercury were used in the amalgamation, and 50%–60% of the mercury could enter into the water, soil and atmospheric environment through point and non-point source ways [24].

In order to further understand the effects of the past and current gold mining activities on the environment, three operating gold mines (Laoniugou gold mine, Jiapigou gold mine and Sandaocha little gold mine) and one abandoned gold mine (Laojinchang gold mine) as well as 240 km2 of its surrounding area in Jiapigou gold mine area were chosen as the study area.

2.2. Sampling and analytical methods

In order to explore more about the effects of mercury contamination on environmental quality in the gold mine and the surrounding areas, the sampling sites should contain all of the current types of mercury discharge (gold mine site, mineral processing plant and tailings) and residential area; in addition, resident coal-fired cooking time and busy traffic times of the study area should be avoided when sampling.

In October 2011 and April 2012, considering the landform characteristics of Jiapigou Town and the position of the gold mines, the sampling sites of the atmosphere and soil were set along the main roads and country roads that led to the gold mines. Laojinchang village, the center of Jiapigou gold mine area, was selected as the starting site, and one sample site every 3 km was set along the roads that led

IOP Conf. Series: Earth and Environmental Science **39** (2016) 012007 doi:10.1088/1755-1315/39/1/012007

in the Wudaogou (Laoniugou gold mine) direction, Jiapigou gold mine direction and Erdaogou direction respectively. There were 23 sampling sites in total. The locations and description of the natural conditions of the sampling sites are shown in figure 1 and table 1.

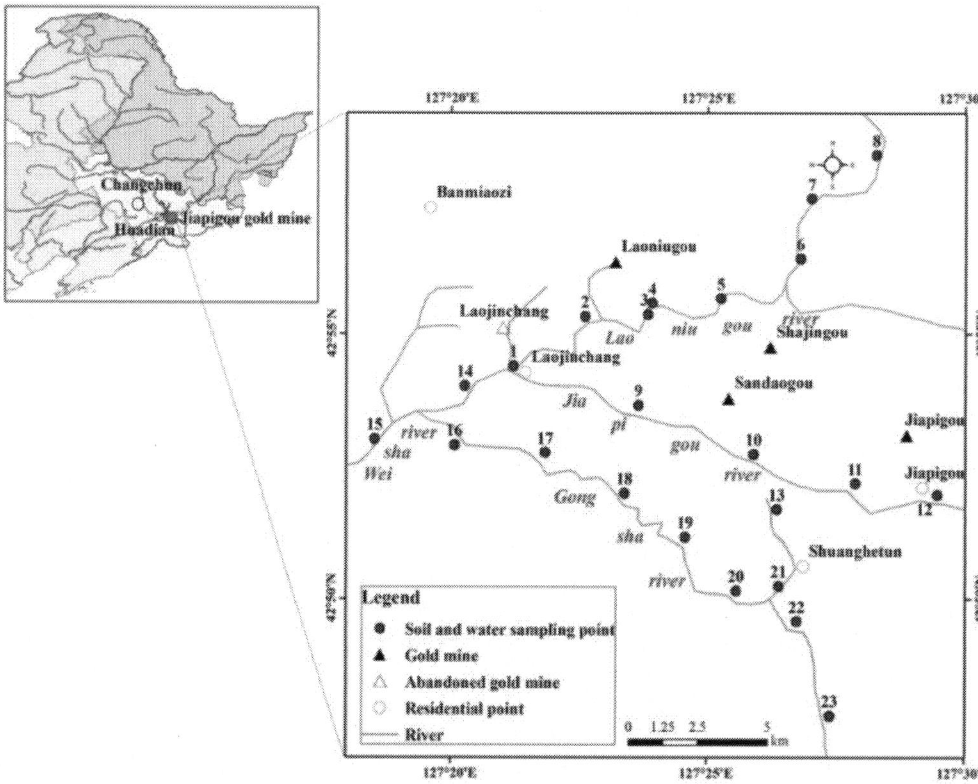

Figure 1. Locations of sampling sites in study area.

Table 1. Natural conditions of sampling sites.

Sample No.(#)	Name	Natural conditions	Sample No.(#)	Name	Natural conditions
1	Laojinchang	Abandoned gold mine	13	Xingan village	Residential area, country road
2	Laoniugou	Residential area	14	Erdaogoukou	Road, river confluence
3	Laoniugou gold mine	Gold mine	15	Banchang	Road
4	Gaolitun	Mineral processing plant	16	Xinshengtun	Country road
5	Shajingou	Tributary converge site, Gold mine	17	Erdaogou	Country road
6	Sidaogou	Country road	18	Lingqian	Country road
7	Wudaogou	Country road	19	Daheyan	Country road
8	Hamagou	*Rana*	20	Xiaoyinggo	Country road

		chensinensis(Chinese brown frog) cultivation farm		u	
9	Toudaocha	Regional road	21	Shuanghetu n	Residential area
10	Sandaocha	Small-scale gold mine, tailing pond, mineral processing plant, residential area	22	Majiadian	Wood farm, country road
11	Miaolingtun	Residential area, tailing pond	23	Majiadian′	*Rana chensinensis*(Chinese brown frog) cultivation farm
12	Jiapigou	Gold mine, residential area, road			

As figure 1 and table 1 show, Laoniugou gold mine (3#), Sandaocha (10#) and Jiapigou (12#) are near the sites of the gold mines that are still currently operating; Laojinchang (1#) is near the site of abandoned gold mine; there are small mineral processing plants around Gaolitun (4#) and Sandaocha (10#); gold mine tailing ponds are in Sandaocha (10#) and Miaolingtun (11#); residential areas are in Laoniugou (2#), Sandaocha (10#), Miaolingtun (11#), Jiapigou (12#), and Shuanghetun (21#); Erdaogoukou (14#) and Banchang (15#) have the roads that lead to Huadian City; Toudaocha (9#) and Jiapigou (12#) have the roads that lead to Jiapigou Town; others are country roads; and Laojinchang (1#), Shajingou (5#) and Erdaogoukou (14#) are the places where rivers converged. The sampling sites have covered all the conditions of study area.

The atmosphere and soil control samples were collected at Longwan Nature Reserve, Huinan County, Jilin Province, which was set as the clean area, 56 km southwest of the study area. The natural conditions of the control site including the topography, hydrometeorological and ecological environment are similar to those in the polluted area, but without large-scale gold mines or mine area.

The concentration of gaseous mercury in the air was determined by automatic mercury vapor analyzer RA-915+ produced by Lumex Company, Russia. Three consecutive values of gaseous mercury were read at each sampling site, and nine instantaneous values were obtained in all. Meanwhile, the latitude and longitude coordinates and altitude of the sampling sites were measured by GPS.

The mixed surface soil samples of multi-sampling sites were collected, and each of the mixed soil samples was about 1 kg and the surplus soil was discarded by the quarter method. The samples were kept in a PVC ziplock bag and brought back to the laboratory. The concentration of mercury in the soil was determined by the national standard method. The soil samples were wet digested by HNO3-H2SO4-V2O5 and a constant volume of the solutions was determined by cold atomic fluorescence spectrometry AFS-2202. Parallel samples were taken from each of the samples, and the standard reference sample of the soil was used for validation.

2.3. Calculation method

2.3.1. Kriging interpolation. Interpolation predicts values for cells in a raster from a limited number of sample data points. It can be used to predict unknown values for any geographic point of data. The Kriging interpolation method was used to analyze the spatial distribution of the mercury concentration, and ArcGIS software was used to build the model of Kriging interpolation in the study.

It is not required to have a normal distribution of the original points data for prediction. However, when the data were normally distributed, Kriging intepolation will be the best method of unbiased estimation. In the study, only the gaseous mercury data set in autumn were characterized by normal distribution, so logarithmic transformation for all of data have been done before the Kriging interpolation.

2.3.2. Geoaccumulation index. The geoaccumulation index (Igeo) was used to analyze and evaluate the mercury contamination level in the soil of Jiapigou gold mine area, and this could provide solid information for the management monitoring of the soil environment in this area.

The geoaccumulation index method was proposed by Muller [25] to evaluate the heavy metal contamination level in sediment. Due to the common characteristics of soil and sediment, the method was also widely used in assessing the heavy metal contamination level in soil. The computational formula is:

$$I_{geo}=\mathrm{Log}_2[C_n/(k\times B_n)] \quad (1)$$

where *Cn* is the measured concentration in the sediment for metal n; *Bn* is the geochemical background value for metal n in the local rock; *k* is the coefficient of the change of the background value of the metal elements caused by the difference of local rock (usually 1.5). Generally, the geoaccumulation index includes seven grades, and the corresponding contamination levels are shown in table 2 [25].

Table 2. Ranks of geoaccumulation index and corresponding contamination level.

Item	Contamination level						
	Uncontaminated	Uncontaminated/ moderately contaminated	Moderately contaminated	Moderately/strongly contaminated	Strongly contaminated	Strongly/extremely contaminated	Extremely contaminated
Igeo	≤0	0-1	1-2	2-3	3-4	4-5	>5
Level	0	1	2	3	4	5	6

Because of industrial pollution and human activity, the mercury background value in the soil was difficult to determine, so the geochemical background value (*Bn*) of the mercury in the local rock that the above formula mentioned was usually the mercury value in less polluted areas. Kabata-Pendias [26] estimated the background mercury values of all different types of soils in the world, which were all between 0.05-0.30 mg·kg-1. The study of Wang *et al.* [27] on the soil environmental element chemistry of China indicated that the average mercury background value in soil of China was 0.038 mg·kg-1, and the average mercury background value in soil in Jilin Province (0.027 mg·kg-1) was lower than that of the whole nation, which was related to the geochemical characteristics of mercury being the dispersed element; moreover, the mercury background values were different in different types of soils [27]. The average mercury concentrations for dark brown forest soil and dark brown forest soil were 0.044 -0.086 mg·kg-1, so we selected 0.063 mg·kg-1 as the background value in this research.

3. Results and Discussion

3.1. Results of mercury concentration

3.1.1. Total mercury concentrations in atmosphere and soil. The characteristic values of the total mercury concentrations in the regional atmosphere and soil of the study area are shown in table 3 and Table 4 respectively, which are calculated in Excel. All of the mean values between spring and autumn were significant by the one-way ANOVA test.

IOP Conf. Series: Earth and Environmental Science **39** (2016) 012007 doi:10.1088/1755-1315/39/1/012007

Table 3. Characteristic values of total Hg concentrations in the atmosphere of Jiapigou gold mine area in spring and autumn.

Time	Maximum /ng·m-3	Minimum /ng·m-3	Arithmetic mean /ng·m-3	Standard deviation /ng·m-3	Coefficient of variation (C.V.) /%	Number of samples (n)	Arithmetic mean±Standard deviation in Control site /ng·m-3
Spring	19.5	1.33	5.27	6.56	5.81	207	5.0±0.51
Autumn	25.0	1.00	11.45	8.16	7.11	138	8.0±0.23

Table 4. Characteristic values of total Hg concentration in the soil of Jiapigou gold mine area in spring and autumn.

Time	Maximum /mg·kg-1	Minimum /mg·kg-1	Arithmetic mean /mg·kg-1	Standard deviation/mg·kg-1	Coefficient of variation (C.V.) /%	Number of samples (n)	Arithmetic mean±Standard deviation in Control site /mg·kg-1
Spring	2.51	0.09	0.58	0.62	109.54	46	0.39±0.16
Autumn	2.06	0.06	0.44	0.52	117.33	46	0.43±0.14

The results from table 3 and table 4 showed that the mean value of the mercury concentration in the atmosphere was 5.27 ng·m-3 in spring, and11.45 ng·m-3 in autumn in the study area, and the mean mercury value in the soil was 0.58 mg·kg-1 in spring and 0.44 mg·kg-1 in autumn. All of the mean values were more than those in the control sites.

The Kriging interpolation method in ArcGIS was used to analyze the spatial distribution of the total mercury concentrations of the atmosphere and soil in spring and autumn in Jiapigou gold mine area, and the results are shown in figure 2 and figure 3.

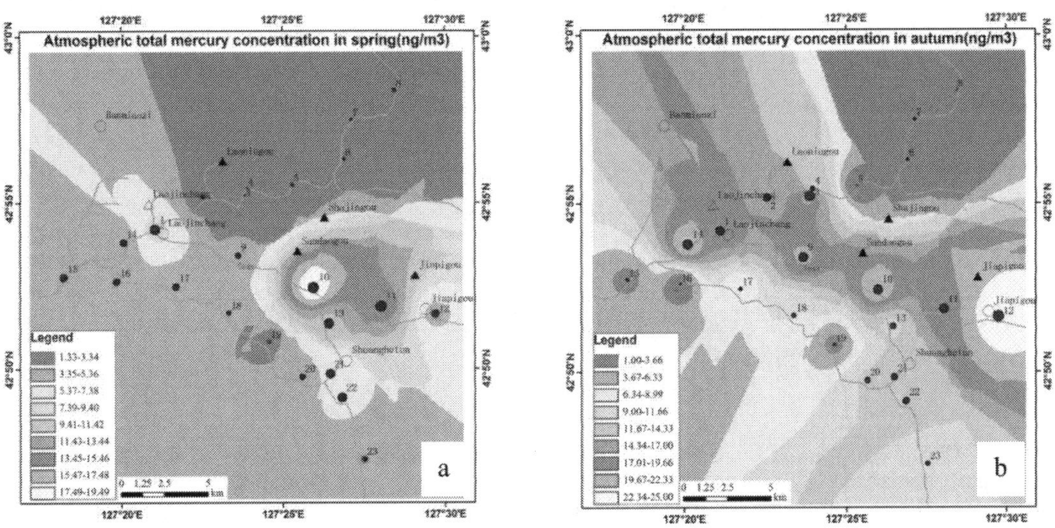

Figure 2. Spatial distribution of the total mercury concentration in the atmosphere in spring (2a) and in autumn (2b).

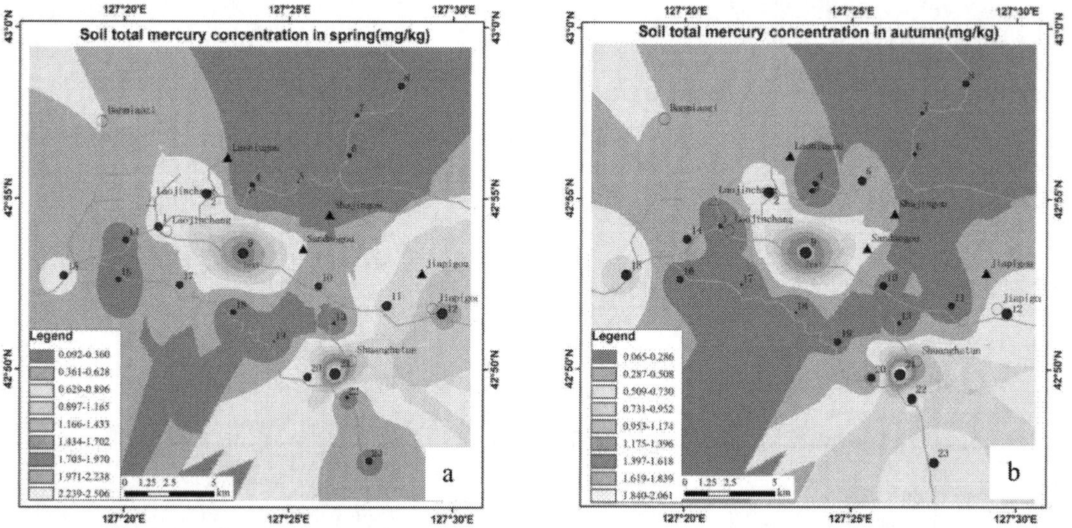

Figure 3. Spatial distribution of the total mercury concentration in the soil in spring (3a) and in autumn (3b).

It was indicated that the concentration of mercury in atmosphere gradually decreased from the centers, Sandaocha (10#), Miaolingtun (11#) and Xingancun (13#), to their surrounding areas in spring; and from the center, Jiapigou (12#), Laoniugou gold mine (3#), Toudaocha (9#) and Sandaocha (10#), to its surrounding areas in autumn. Meanwhile, the mercury concentrations in the soil were similar in spring and autumn, that is, the concentration of mercury gradually decreased from the centers, Shuanghetun (21#), Toudaocha (9#), Jiapigou (12#) and Laoniugou (2#), to their surrounding areas (table 2 and figure 2 to figure 5).

3.1.2. Correlation of mercury concentrations in atmosphere and soil. Correlation analysis was done after the elimination of several extreme outliers of mercury contents in atmosphere and soil, and it should be within the allowable range. The correlation analysis was calculated in Excel. R2 is the regression determination coefficient and r is the Pearson correlation coefficient.

The regression equation of the total mercury concentrations in atmosphere and soil in spring was $y = 3.0086x + 2.6007$, $R2 = 0.6505$ and $r = 0.8065$ ($P < 0.001$), and in autumn, $y = 15.191x + 2.3336$, $R2 = 0.5802$ and $r = 0.7617$ ($P < 0.001$). Therefore, the mercury concentrations in the atmosphere and soil have a significant correlativity.

3.2. Distribution characteristics of mercury concentrations

3.2.1. Distribution characteristics of mercury peak values. The spatial distributions of gaseous mercury concentrations in spring and autumn are shown in figure 2. It can be seen that the concentrations of gaseous mercury substantially decreased from the center (gold mine area) to the surrounding area in general. In autumn, the peak values of gaseous mercury concentration emerged at Jiapigou gold mine, Laoniugou gold mine and Sandaocha gold mine, which showed that the mercury concentration in the atmosphere had a close relationship with gold mining activity; in spring, the peak value of gaseous mercury concentration emerged only at Sandaocha gold mine and the mercury concentrations were not high at the other two gold mines. The reason of the mercury peak values emerging at Sandaocha both in spring and autumn was that there was a small-scale gold mine, a tailing pond and a mineral processing plant near Sandaocha site. The concentration of gaseous mercury tends to decrease, which was related to the elimination of the mercury amalgamation technique [28, 29]. In the autumn of 2008, the maximum total gaseous mercury concentration was 119.0 ng·m-3 in Jiapigou [29], while the concentration was 19.5 ng·m-3 in spring of 2012 and 25.0 ng·m-3 in autumn of 2011 in our study. Excluding weather-related conditions, it can be mainly related to prohibiting the use of the plate amalgamation technique for a long time.

The spatial distributions of the soil mercury concentration in spring and in autumn are shown in Figure 3. It was indicated that the mercury peak values were not at the gold mining sites at all, but at the places near mining sites and residential areas.

3.2.2. Correlation of mercury concentrations in atmosphere and soil. The characteristics of mercury determine that its contamination scope has regional limitations. The circulation and transportation of mercury in the atmosphere and soil was the main mode of movement for regional mercury contamination, and mercury could migrate far away through the media of water and atmosphere. The regression analysis of the total mercury contents between the atmosphere and the soil showed a significant correlation, which indicated that there was certain circulation of the mercury between the regional atmosphere and soil. The residual mercury in the soil can be released into the atmosphere, and the gaseous mercury attached to the particulate matter can come into the soil through dry and wet deposition in certain weather conditions [19].

3.2.3. Causes for mercury contamination in atmosphere and soil. Mason et al. [30] estimated the contribution of gaseous mercury to terrestrial mercury at different latitudes in the northern hemisphere. At 10°N~30°N and 30°N~70°N, the mercury settlement values were 19.8μg·m-2·yr-1 and 15.8μg·m-2·yr-1 respectively. Schlüter [31] summarized the mechanism and related reports about mercury evaporation at present, which showed that in the topsoil, the concentrations of elemental mercury and methyl mercury formed from biotic and abiotic ways were significantly affected by the evaporation rate from the soil to the atmosphere. Furthermore, the evaporation rate of mercury was restricted by climate change and the migration of soil volatile matter in mercury-enriched mine areas. So there are many factors affecting the transfer and circulation of mercury between the atmosphere and soil.

For the study area, the mercury contamination in the atmosphere and soil might be caused by the following factors.

- Gold mine production. It is assumed that the distribution of mercury in the soil and atmosphere was related to long-term production of gold in the Jiapigou gold mine area. Jiapigou gold mine has been exploited since 1921, and the plate amalgamation technique has been used in gold extraction in the area from 1940 to 2006. In the mining process, mercury metal ions were heated into elemental mercury and then released into the atmosphere. The gaseous mercury migrates with the air movement in the atmosphere, and it can enter into soil through the dry and wet deposition around the mining area. Mercury was absorbed by clay particles in the soil, so the mercury concentration of the surface soil increased.

In 2006, the mercury amalgamation technique was replaced by cyanidation for the gold extraction in Jiapigou gold mine, which reduced and even terminated the release of mercury. However, the residual mercury in the soil was constantly released into atmosphere and became the mercury source of the study area.

- The mechanism of soil absorption. The mercury concentration in the soil correlated significantly with the gaseous mercury concentrations in Chongqing region, China, and the correlation coefficient was 0.741 [32]. The authors concluded that the soil organic matter and clay had a strong ability of adsorption to mercury, so the wet and dry mercury deposition into the soil is mostly fixed, which can increase the mercury concentration in the soil [32]. Jiapigou gold mine is located in the northeast of China. The soil type in this area was dark brown forest soil with high organic matter (5.0%~5.72%). The soil texture is mostly loam or clay loam; in addition, the climate is humid for these are mountainous and forest areas, therefore the gaseous mercury showed strong adsorption. As so many similar conditions above, it was suggested that the soil in this area had a similar absorption mechanism with Wang *et al.*'s [32] study.

- Mercury exchange flux. Zhang *et al.* [28] suggested the mechanism of mercury exchange flux between atmosphere and soil at the gold mining site and its surrounding area. The soil intensely released mercury into the atmosphere in autumn, but in spring, the mercury in the atmosphere could be settled into the soil by wet or dry depositon, so the mercury concentration in the soil in spring was generally slightly higher than that in autumn. The results of the paper were consistent with the finding published by Zhang et al. [28].
- Landform characteristics. The landform characteristics of the study area was of two mountains with a valley in the middle, which made the closed valley landforms of Shuanghetun and Toudaocha, and became obstructive areas for the diffusion of atmospheric flow and peak values areas for soil mercury.

In summary, the source of soil mercury contamination is complex, not only through mining release into the atmosphere and affects the surrounding areas, but also due to mercury dry and wet deposition into the soil through the atmosphere released from coal fired by local people. In addition, it also includes the soil's natural mercury release. This research avoided sampling at the time of coal firing and traffic peak hours, and it can be partially excluded from these sources, but it can also be further proof that the mercury in the soil is the cumulative result of a variety of sources for many years, which are sinks and sources of mercury contamination in gold mining district, and the result is similar with other scholars' results [33].

3.3. Assessment of mercury contamination in soil. Considering the determined results of mercury concentrations in the study area, the geo-accumulation indexes of 23 sampling sites were calculated (table 5).

Table 5. Geo-accumulation index of mercury contamination in Jiapigou gold mine area.

Sampling site	Spring		Autumn		Sampling site	Spring		Autumn	
	*I*geo	Level	*I*geo	Level		*I*geo	Level	*I*geo	Level
1# Laojinchang	2.942	3	0.112	1	13# Xingancun	0.596	1	0.099	1
2# Laoniugou	3.266	4	3.057	4	14# Erdaogoukou	1.560	2	2.337	3
3# Laoniugou gold mine	0.268	1	0.532	1	15# Banchang	2.580	3	-0.542	0
4# Gaolitun	1.082	2	0.322	1	16# Xinshengtun	1.214	2	0.752	1
5# Shajingou	0.090	1	2.357	3	17# Erdaogou	2.863	3	3.128	4
6# Sidaogou	0.842	1	-0.189	0	18# Lingqian	1.209	2	-0.400	0
7# Wudaogou	1.004	2	-0.069	0	19# Daheyan	-0.047	0	1.307	2
8# Hamagou	1.7	2	0.78	1	20#	2.78	3	1.20	2

	95		2		Xiaoyinggou	2		5	
9# Toudaocha	4.301	5	4.180	5	21# Shuanghetun	4.731	5	4.449	5
10# Sandaocha	2.422	3	1.331	2	22# Majiadian	0.493	1	2.681	3
11# Miaolingtun	3.322	4	0.785	1	23# Majiadian	1.768	2	2.556	3
12# Jiapigou	3.963	4	2.881	3					

Kriging interpolation was used to analyze the spatial differentiation of mercury contamination levels (figure 4). As indicated in the figures, the sampling sites of Shuanghetun (21#) and Toudaocha (9#) in the study area were strongly/extremely contaminated with Hg (in level 5) both in autumn and spring, and the sampling site of Laoniugou (2#) was strongly contaminated with Hg (in level 4) both in autumn and spring; all these places were not far from the gold mining site and in the valley of low topography, where the surrounding mountains blocked the atmospheric flow with mercury and the mercury was deposited in this place. Laoniugou gold mine (3#) and Xingancun (13#) were uncontaminated/moderately contaminated with Hg (in level 1) both in autumn and spring. (table 2) In addition, the contamination levels of the other sampling sites were different in spring and autumn. Details can be found in table 5.

The study indicated that, for soil mercury contamination, strongly/extremely contaminated (level 5) and strongly contaminated (level 4) area accounted for 21.74% of Jiapigou gold mine area, and moderately/strongly contaminated (level 3) area accounted for 21.74% and uncontaminated (level 0) area only accounted for 4.35% in spring; and the corresponding values accounted for 17.4%, 21.74% and 17.35% respectively in autumn. In conclusion, the mercury contamination in soil was severe in Jiapigou gold mine area, especially in the mining area and its surrounding area; they were all beyond the moderately/strongly contaminated level.

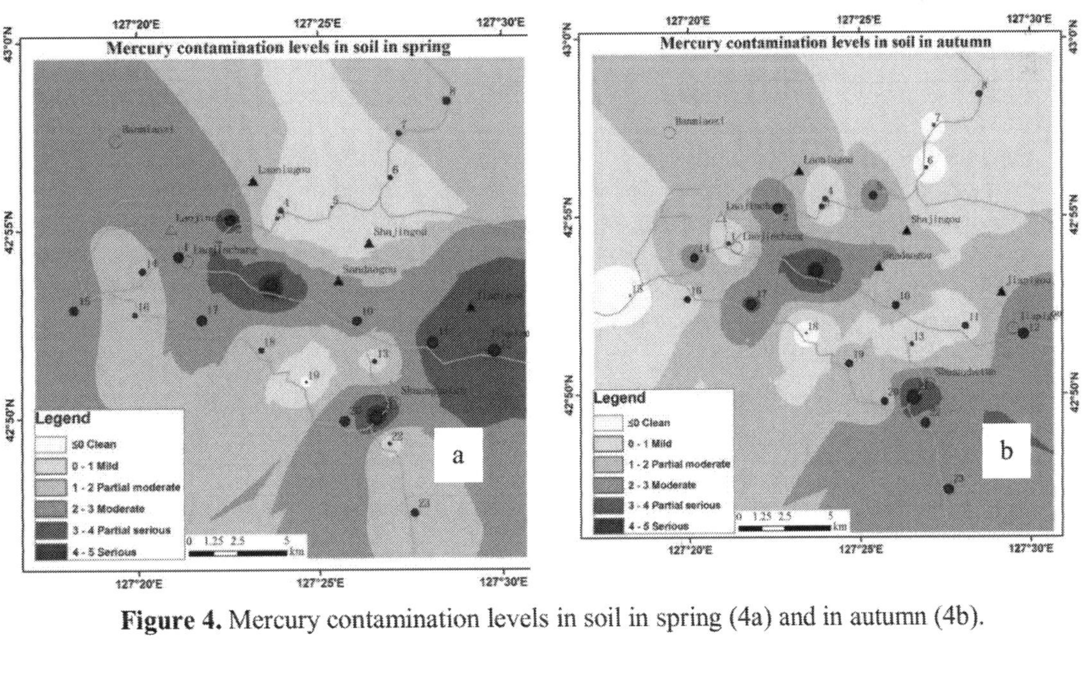

Figure 4. Mercury contamination levels in soil in spring (4a) and in autumn (4b).

4. Conclusions

- The Jiapigou gold mine has been exploited for nearly 190 years; the highest atmosphere mercury concentrations were 3.9 and 3.13 times those at the control site in spring and autumn respectively. The distribution of TGM in the study area was significantly related with the original mercury sources. The peak of TGM is in the existing gold mine area, mineral processing plant and tailings, decreasing gradually from the center to the surrounding area. Overall, the mercury contamination was attributed to long-term use of backward technology for gold extraction, and it gradually reduced after elimination of the technology.
- The highest soil mercury concentrations were 6.44 and 4.79 times those at the control site in spring and autumn respectively. The ranking of mercury concentrations in soil from high to low were the places near mining sites, the gold mining sites and then the others. The mercury concentrations in soil were significantly positively correlated with those in the atmosphere, which indicated that mercury can be transported between the atmosphere and the soil. The data from the study strongly supported the conclusion that the mercury circulation consisted of mercury release from soil into atmosphere and gaseous mercury settlement into the soil through dry and wet deposition in certain weather conditions.
- The mercury peak value of the soil emerged at the valley with low topography that was close to the gold mining site. The combination of strongly/extremely contaminated (level 5) and strongly contaminated (level 4) area accounted for 21.74% and 17.4% in spring and autumn respectively. Based on the local meteorological conditions, it was predicted that the soil mercury contamination was affected not only by long-term amalgamation of gold mine and tailings, but also by the local impact of the closure of the valley topography and climate conditions.

The study presents the mercury contamination situation in the Jiapigou gold mining area after termination of the use of amalgamation technique for gold extraction. These investigations are reasonable for better understanding the potential risk of metal-contaminated areas for human health in these areas.

Acknowledgement

This work was financially supported by the National Natural Science Foundation of China (No.40673059).

References

[1] Ikingura J R, Akagi H, Mujumba J, et al. 2006 Environmental assessment of mercury dispersion, transformation and bioavailability in the Lake Victoria Goldfields, Tanzania *Journal of Environmental Management* **81**(2): 167-73

[2] Zhang H, Feng X B, Larssen T, et al. 2010 Bioaccumulation of Methylmercury versus Inorganic Mercury in Rice (Oryza sativa L.) Grain *Environmental Science & Technology* **44**:4499-504

[3] Nartey V K, Klake R K, Hayford E K, et al. 2011 Assessment of mercury in rivers and streams around artisanal gold mining areas of the Birim north district of Ghana *Journal of Environmental Protection* **2**:1227-39

[4] Hilson G 2006 Abatement of mercury pollution in the small-scale gold mining industry: Restructuring the policy and research agendas *Science of the Total Environment* **362**(1-3):1-14

[5] Telmer K H and Veiga M M 2009 World emissions of mercury from artisanal and small scale gold mining *Mercury Fate and Transport in the Global Atmosphere* New York: Springer pp 131-72

[6] Li X S 1989 Investigation on the technology of vacuum distillation for gold amalgam *Gold* **10**(8):43-46

[7] Schmidt C W 2012 Quick silver & gold: mercury pollution from artisanal and small-scale gold mining *Environmental Health Perspectives* **120**(11): A424-A429

[8] Dai Q J and Feng X B 2004 Study on progress of mercury contamination to the environment associated with gold extraction by amalgamation *Techniques and Equipment for Environmental Pollution Control* **5**(7):13-17

[9] Drace K, Kiefer A M, Veiga M M, et al. 2012 Mercury-free, small-scale artisanal gold mining in Mozambique: utilization of magnets to isolate gold at clean tech mine *Journal of Cleaner Production* **32**: 88-95

[10] Lecce S, Pavlowsky R and Schlomer G 2008 Mercury contamination of active channel sediment and floodplain deposits from historic gold mining at Gold Hill, North Carolina, USA *Environmental Geology* **55**(1):113-21

[11] Strode S, Jaeglé L and Selin N E 2009 Impact of mercury emissions from historic gold and silver mining: global modeling *Atmospheric Environment* **43**(12): 2012-17

[12] Diringer S E, Feingold B J, Ortiz E J, et al. 2015 River transport of mercury from artisanal and small-scale gold mining and risks for dietary mercury exposure in Madre de Dios, Peru *Environmental science-Processes & impacts* **17**(2): 478-87

[13] Pacyna E G, Pacyna J M, Steenhuisen F, et al. 2006 Global anthropogenic mercury emission inventory for 2000 *Atmospheric Environment* **40**(22):4048-63

[14] Loredo J, Soto J, Alvarez R, et al. 2007 Atmospheric monitoring at abandoned mercury mine sites in Asturias (NW Spain) *Environmental Monitoring and Assessment* **130**(1-3):201-14

[15] Marrugo-Negrete J, Benitez L N and Olivero-Verbel J 2008 Distribution of mercury in several environmental compartments in an aquatic ecosystem impacted by gold mining in northern Colombia *Archives of Environmental Contamination and Toxicology* **55**(2):305-16

[16] Cordy P, Veiga M M, Salih I, et al. 2011 Mercury contamination from artisanal gold mining in Antioquia, Colombia: The world's highest per capita mercury pollution *Science of the Total Environment* **410–411**:154-60

[17] Taylor H, Appleton J D, Lister R, et al. 2005 Environmental assessment of mercury contamination from the Rwamagasa artisanal gold mining centre, Geita District, Tanzania *Science of the Total Environment* **343**(1-3):111-33

[18] Castilhos Z C, Rodrigues-Filho S, Rodrigues A P, et al. 2006 Mercury contamination in fish from gold mining areas in Indonesia and human health risk assessment *Science of the Total Environment* **368**(1): 320-25

[19] Balzino M, Seccatore J, Marin T, et al. 2015 Gold losses and mercury recovery in artisanal gold mining on the Madeira River, Brazil *Journal of Cleaner Production* **102**: 370-77

[20] Saiki M K, Martin B A, May T W, et al. 2010 Mercury concentrations in fish from a Sierra Nevada foothill reservoir located downstream from historic gold-mining operations *Environmental Monitoring and Assessment* **163**(1-4):313-26

[21] Egler S G, Rodrigues-Filho S, Villas-Boas R C, et al. 2006 Evaluation of mercury pollution in cultivated and wild plants from two small communities of the Tapajos gold mining reserve, Para State, Brazil *Science of the total environment* **368**(1):424-33

[22] Lacerda L D 2003 Updating global Hg emissions from small-scale gold mining and assessing its environmental impacts *Environmental Geology* **43**(3):308-14

[23] Feng X B, Dai Q Q, Qiu G L, et al. 2006 Gold mining related mercury contamination in Tongguan, Shaanxi Province, PR China *Applied Geochemistry* **21**(11):1955-68

[24] Wang N, Zhu Y M, Sheng L X, et al. 2005 Mercury pollution in rana chensinensis in Weisha River reach, in the upstream region of Songhua river *Chinese Science Bulletin* **50**(19): 2166-70

[25] Müller G 1969 Index of geo-accumulation in sediments of the Rhine River *Geo. Journal* 2(3):108-18

[26] Kabata-Pendias A 1992 *Trace Elements in Soils and Plants(2nd Edition)* Florida Boca Raton: CRC Press p 149

[27] Wang Y and Wei F S 1995 *Element Chemistry of Soil Environment* Beijing: China Environmental Science Press p 10

[28] Zhang G, Wang N, Wang Y, et al. 2012 Characteristics of mercury pollution in soil and atmosphere in Songhua River upstream Jia-pi-gou gold mining area of Jia-pi-gou at the upper Songhua River *Environmental Science* **33**(9):29-35

[29] Zou T T, Wang N, Zhang G, et al. 2010 Songhua River upsteam gold mining area, the environmental spatial distribution of mercury pollution *Environmental Science* **31**(9):2228-33

[30] Mason R P, Fitzgerald W F and Morel F M M 1994 The biogeochemical cycling of elemental mercury: Anthropogenic influences *Geochimica et Cosmochimica Acta* **58**(15):3191-98

[31] Schlüter K 2000 Review: evaporation of mercury from soils. An integration and synthesis of current knowledge *Environmental Geology* **39**(3-4):249-71

[32] Wang D Y, Shi X H and Yang X C 1998 Transformation of atmospheric mercury in soil and its influence on the accumulation of mercury in soil *Chongqing Environmental Science* **20**(5):22-25

[33] Oliveira S M B, Melfi A J, Fostier A H, et al. 2001 Soils as an Important Sink for Mercury in the Amazon *Water, Air, & Soil Pollution* **126**(3-4):321-37

Rational allocation of water resources based on ecological groundwater levels:a case study in Jinghui Irrigation District in China

H Li [1,2], W B Zhou [1,2,3], Q G Dong [1,2], B Y Liu [1,2] and C Ma [1,2]

[1] College of Environmental Science and Engineering, Chang'an University, Xi'an 710054, China

[2] Key Laboratory of Subsurface Hydrology and Ecological Effect in Arid Region of Ministry of Education, Chang'an University, Xi'an 710054, China

Email: zwbzyz823@163.com

Abstract. Aimed at the hydrogeological environmental problems caused by over-exploitation and unreasonable utilization of water resources in Jinghui Irrigation District, this paper discusses the ecological groundwater level of the study area and establishes a three-layer optimal allocation model of water resources based on the theory of large scale systems. Then, the genetic algorithm method was employed to optimize the model and obtain the optimal allocation of crop irrigation schedule and water resources under the condition of a 75% assurance rate. Finally, the numerical simulation model of the groundwater was applied to analyze the balance of the groundwater on the basis of the optimal allocation scheme. The results show that the upper limitation of the ecological groundwater in Jinghui Irrigation District ranged from 1.8m to 4.2m, while the lower limitation level ranged from 8m to 28m. By 2020, the condition of the groundwater imbalance that results from adopting the optimal allocation scheme will be much better than that caused by current water utilization scheme. With the exception of only a few areas, the groundwater level in most parts of Jinghui Irrigation District will not exceed the lower limitation of ecological groundwater level.

1. Introduction

Water resources are of great importance to the development of economies human society and ecosystems. Due to the increasing demand for water resources caused by urbanization, industrial growth, climate change and human activity,China is experiencing severe water shortages. Therefore, how to allocate water resources becomes an important issue.

So far in the literature, ample researches have been conducted on water allocation, and most researchers have taken the three principles of equity, efficiency and sustainability into consideration when planning water allocation schemes. For example, a multilayer model of water resource allocation was built by E Romjin (1983)[1].This model considered the relationship between the utilization and benefit of water resources, as well as reflecting the facts of the multi-objective and multi-bending in the water resource system. Carlos et al.(1997) took maximum synthetical economic benefit as the objective function and created a multi-resources management model of surface water, sewage and groundwater[2]. Rosegrant et al.(2000) optimized the allocation of water resources on the basis of combining a hydrological model with economic benefits in order to improve water utilization efficiency[3]. Reca et al.(2001) established an economic optimization model for water resource management in a deficit irrigation system[4]. Mohammed et al. (2007) established a coupled hydrologic-economic spreadsheet water allocation model that could be applied to both agricultural and environmental sectors with various policy scenarios[5]. Dedi Liu (2010) presented a model that was comprised of four modules for the optimal allocation of water resources in a saltwater intrusion area[6].

However, as groundwater plays an important role in northwestern irrigation areas in China, the impact of groundwater must be taken into account when studying optimal allocation of water resources. Aimed at such an issue, scholars have produced many researches and achievements. For example, Yang (2003) proposed a scheme of groundwater utilization in Hetao irrigation district in Inner Mongolia on the basis of analyzing groundwater recharge and reasonable exploitation[7]; Zhou (2006) used the multivariate nonlinear method to forecast groundwater levels and analyzed the reasonable irrigation water ratio of wells to canals[8]; Yue (2011) established an optimal allocation model of water resources by adopting groundwater level as a constraint condition[9]; He (2011) developed a groundwater stochastic-determination model and optimized the allocation of water resources in Jinghui Irrigation District[10]; Dai and Cai (2012) set up 10 kinds of water-saving irrigation simulation scenarios to simulate the change of the groundwater level according to the actual situation of agricultural irrigation, and studied a suitable irrigation water ratio of well to canal through groundwater equilibrium analysis[11].

In conclusion, most of the current studies have redistributed water resources on a macro level and the optimal allocation of crop water requirements was sketchy. While some other studies took the groundwater level, irrigation consumption and economic benefits into account when optimizing the allocation of water resources, researches that discuss the groundwater equilibrium on the basis of water resource optimal allocation have been much less common.

Jinghui Irrigation District is located in the semi-arid region in northwest of China. In the past few years, because of over-exploitation,the groundwater level in Jinghui irrigation district has been declining, which has caused a series of hydrogeological problems. Against the issue, this paper first discusses the upper and lower limitations of ecological groundwater level based on observation data from 1981-2012, then optimizes the allocation of crop irrigation schedule and water resources on the basis of large scale system theory. Finally, changes to the groundwater budget based on the water utilization scheme were simulated by the groundwater numerical simulation model. The results may provide a useful basis for the sustainable development of groundwater in Jinghui Irrigation District.

2. Study area

Jinghui Irrigation District is located in the middle of Guanzhong Plain of Shaanxi Province, and its west, south and east sides are surrounded by the Jinghe River, Weihe River, and Shichuanhe River. The total area of Jinghui Irrigation District which governs four counties and two administrative districts is 1180km^2. It is a typical well-canal combined irrigation district that owns a well-functional irrigation system (figure 1). In Jinghui Irrigation District, about 50%~60% of groundwater recharge is from canal leakage and irrigation infiltration, and the average annual precipitation and evaporation are 540mm and 1212mm, which presents a character of high evaporation and precipitation in summer, as well as a character of low temperature and precipitation in winter. Jinghe River and groundwater in the area are the main irrigation water resources. Since the 1980s, due to the increasing demand of water resources and decreased runoff of Jinghe river, groundwater in the Jinghui Irrigation District was over-exploited and the irrigation water ratio of well to canal increased from 0.3 in 1981 to 1.3 in 2012. As a result, the groundwater level sharply dropped and formed regional groundwater depressions (figure 2).

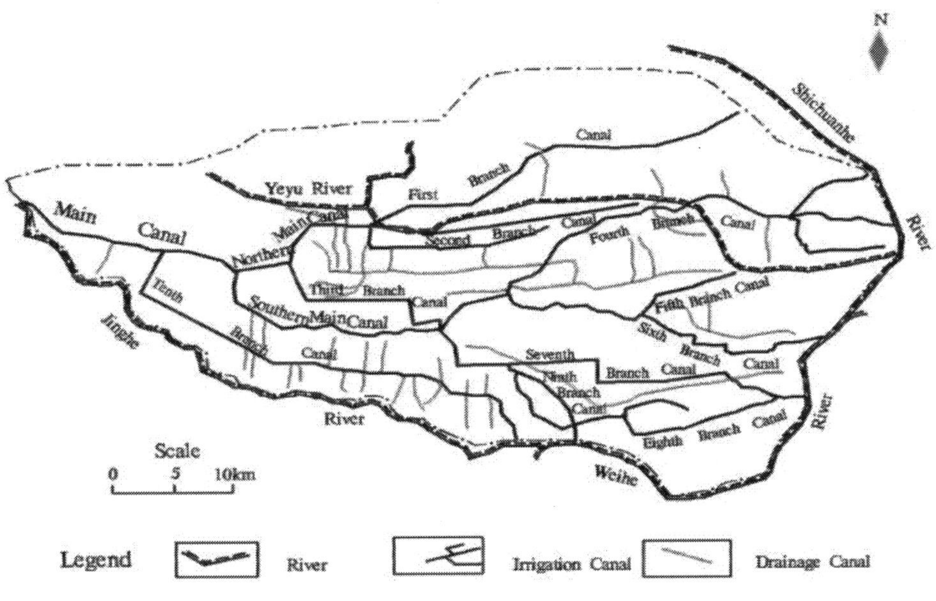

Figure 1. Distribution map of canal system in Jinghui Irrigation District.

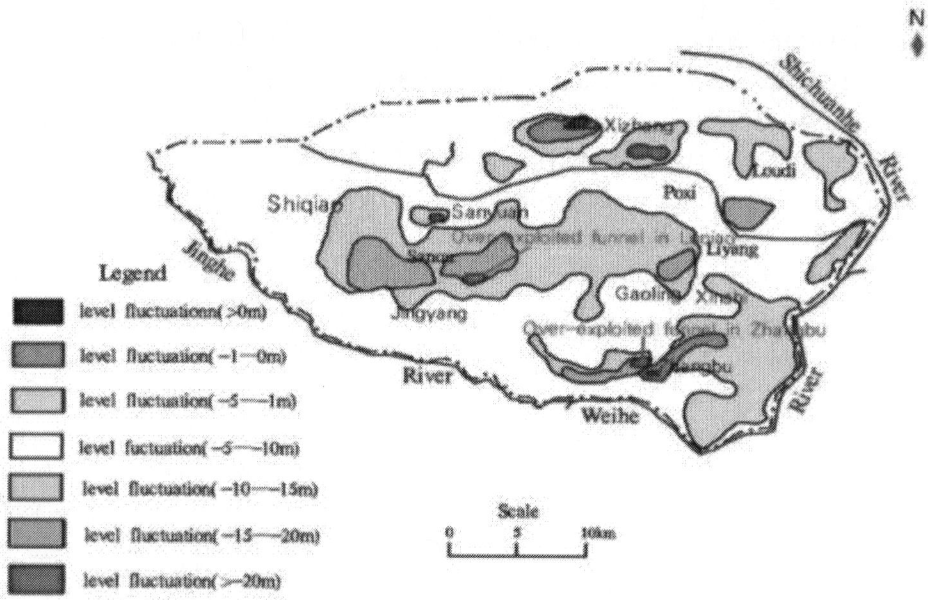

Figure 2. Depressions of groundwater in Jinghui Irrigation District.

3. Discussion on ecological groundwater level

The ecological groundwater level is defined as a level interval that not only can meet the need of ecological environment, but also will not lead to ecological environment exacerbation. It plays an important role in maintaining the sustainable development of the groundwater ecological environment.

3.1. Upper limitation of groundwater level

Increasing the groundwater level, which is caused by consuming a large amount of surface water irrigation, may lead to secondary salinization problems in irrigation districts. Therefore, the upper limitation of groundwater level is supposed to be the minimum groundwater level that does not cause salt accumulation in soil, which could affect crop growth. This paper discusses the upper limitation of groundwater level with a vital index of preventing soil salinization, height of capillary water rise and mainly active layer thickness of crop roots. The calculation formula is as follows:

$$h = h_p + Z \tag{1}$$

where h represents the upper limitation of the groundwater ecological level; h_p represents the height of capillary water rise; and Z represents the mainly active layer thickness of crop roots.

The height of the capillary water rise varies with soil texture, and the mainly active layer thickness of crop roots varies with the type of crops. The soil texture types in the study area are similar to part of the soil texture types in the study area of reference [12].Thus, the height of the capillary water rise can be determined by the experimental results of literature [12].The mainly active layer thickness of crop roots can be read in reference [13]. Then, the upper limitation of ecological groundwater level was calculated by formula 1 and the Kriging spatial interpolation method was employed to obtain the contour map of the upper limitation of the ecological groundwater level (figure 3). As can be seen

IOP Conf. Series: Earth and Environmental Science **39** (2016) 012008 doi:10.1088/1755-1315/39/1/012008

from figure 3, the upper limitation of the ecological groundwater level in Jinghui Irrigation District ranged from 1.8m to 4.2m, and it increased from south to north, which presented a good similarity with the distribution of aquifer media. The reason why this makes a difference is that the aquifer media in the southern part of the study area is mainly composed of medium-coarse sands, so the heights of the capillary water rise in such regions are low and the mainly active layer thickness of crop roots are thin. Thee aquifer media in the southern part of the study area is composed of silty clays and clays, so the heights of the capillary water rise in this region are high and the mainly active layer thickness of crop roots are thick.

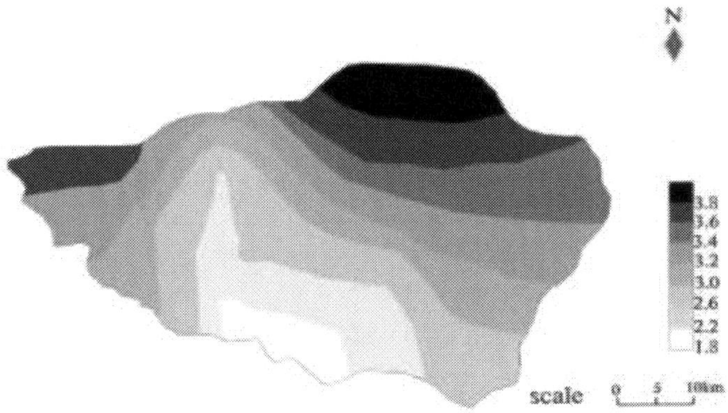

Figure 3. Upper limitation of ecological groundwater level.

3.2. Lower limitation of groundwater level

As described above, Jinghui Irrigation District is located in a semi-arid region in the northwest of China,where precipitation is concentrated in summer and groundwater level changes greatly in the wet season and dry season. In the wet season, the water table is high and the evaporation loss is great. However, in the dry season, the water table is low. When the water table is lower than the limited level of groundwater evaporation, the groundwater ecological environment may be destroyed. Therefore, in this paper, it is considered that the lower limitation of groundwater level should be the limited level of groundwater evaporation, that is, the level where the evaporation of groundwater is zero[14-15].

The limited level of groundwater evaporation was calculated by the dynamic data correlation methods based on data from 49 observation wells. The result of the calculation has shown that the limited level of groundwater evaporation in Jinghui Irrigation District ranged from 7.8m to 21.4m. However, as mentioned above, due to long term over-exploitation, the groundwater level in the study area has sharply decreased,and two obvious depressions were formed,as shown in figure 1. Therefore, it is unrealistic to recover a healthy environment of groundwater in a short period. Based on this, in order to gradually improve the conditions of this groundwater imbalance, this paper modified the lower limitation of groundwater on the basis of the limited level of groundwater evaporation and the practical situation of groundwater level in the study area. The modified lower limitation of ecological groundwater level is shown in figure 4.

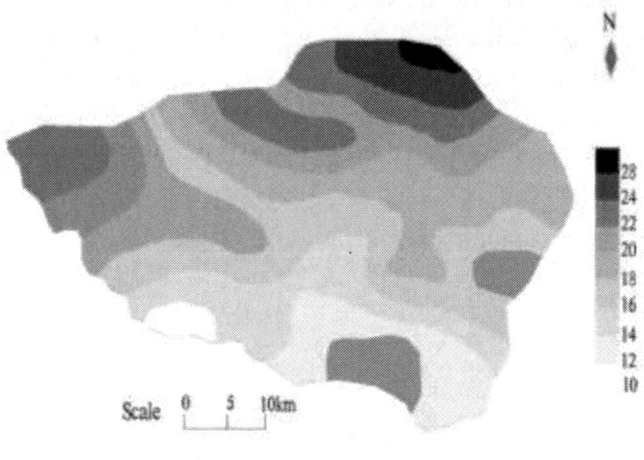

Figure 4. Lower limitation of ecological groundwater level.

4. Optimal allocation of water resources

With the development of society, many complex issues have become difficult to describe by simple mathematical models. Sometimes, even though a model could be established, the result was very difficult to obtain. Therefore, under the conditions of many constraints and effect factors, the optimization theory of the large scale system] was appeared as a burgeoning theory that had more advantages to solve such problems. The large scale system contains several subsystems; the upper systems play a role of coordination while the lower systems are used to decompose the upper systems. The main decomposition purpose of the lower systems is to make the complex issues simple and find out an optimized solution for each simple problem. Meanwhile,the main purpose of the coordination of upper systems is to link up all the subsystems in order to ensure their correlations and find out the optimal solution of the whole system[16-18].

As described earlier, Jinghui Irrigation District is a large irrigation area that consists of six administrative regions, and the main supplying water resources are surface water and groundwater. Therefore, the water utilization system in Jinghui Irrigation District is complex. Under these circumstances, this paper adopted the theory of large scale systems to establish a three-layer structure model of optimal allocation of water resources. Then, the Genetic Algorithm Method was employed to obtain the solution of the model. The final results give the optimal irrigation schedule of crops and optimal allocation of different water resources under the condition of a 75% assurance rate（P=75%）, common drought year, because Jinghui Irrigation District is located in semi-arid zone).

The estimation of crop water requirements was calculated by multiplying the estimated hectare by the net water requirement (m^3/ha) of each crop. The required meteorological data (P=75%) were obtained from the local weather station, while other required data for calculation were collected from the Administration Bureau of Jinghui Irrigation District. The six administrative regions were considered to be the subareas, and wheat, corn and cotton were considered as the main crops. Thus, the structure of the three-layer model based on the theory of large scale can be built (figure 5).

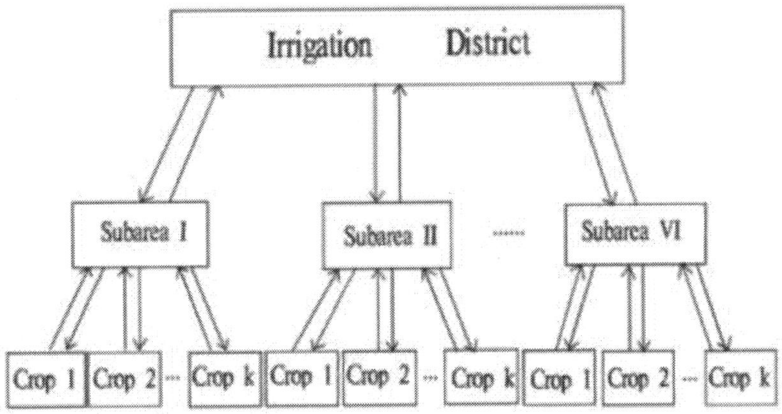

Figure 5. The structure of the optimal allocation model of water resources.

4.1. Water Optimal Allocation Model in Jinghui Irrigation District

4.1.1. Water Optimal Allocation Model of Single Crop in Different Growth Stages

- Objective Function

The Jensen model is a multiplicative model that takes the ratio of the real yield per unit area(Y_a) to the maximum yield per unit area (Y_m) as the goal and assumes that the crop will not yield if the crop evaporation at certain stages is zero. This model is widely used in arid and semi-arid areas because it takes interactions of different growing stages into account and has a high sensitivity to gross product. Therefore, the objective function was established by the Jensen model:

$$f = \max\left(\frac{Y_a}{Y_m}\right) = \max\prod_{i=1}^{n}\left(\frac{ET_a}{ET_m}\right)_i^{\lambda_i} \tag{2}$$

where ET_a represents the real evaporation of the crop; ET_m represents the maximum evaporation of the crop; λ_i represents the sensitivity coefficient of the crop; and n represents the growth period of the crop.

- Constraints

a. Constraints of Irrigation Water in the Growth Period

The water that is allocated to the crop in its whole growth period cannot exceed the maximum water that is supplied by the water resources:

$$\sum_{n=1}^{N} m_i \le Q;\ 0 \le m_i \le q_i \tag{3}$$

where m_i represents the irrigation water in each stage of the growth period; Q represents the distributable irrigation water per unit area in the whole growth period of the crop; and q_i represents the distributable irrigation water at the beginning of each stage.

b. Constraints of evaporation in the growth period

The evaporation of the crop in each stage of the growth period cannot exceed the maximum evaporation of the crop:

$$\left(ET_{a}\right)_{i} \leq \left(ET_{\max}\right)_{i} \tag{4}$$

where $(ET_a)_i$ represents the evaporation of the crop in i stage of the growth period; $(ET_{max})_i$ represents the maximum evaporation of the crop in i stage of the growth period.

4.1.2. Water optimal allocation between different crops

There are several crops in the irrigation district, and the sensitivity degree of reduction of each crop in its growth period is different. Once the cultivated area of each crop and the amount of supply water are certain, the benefits change with the allocation of irrigation water between different crops. Therefore, how to allocate the irrigation water reasonably between different crops in order to reach maximum economic benefits is the goal of the optimal calculation of this layer.

- Objective function

Considering the maximum economic benefits as the objective function:

$$\max \sum_{i}^{N} z_{i} = \sum_{j}^{M} f_{ij}\left(Q_{ij}\right) \bullet S_{ij} \bullet C_{j} \tag{5}$$

where i represents the number of subsystems; j represents the number of crops; f_{ij} represents the yield per unit area of j crop in i subsystem; Q_{ij} represents the irrigation water consumption per unit area of j crop in i subsystem; S_{ij} the cultivated area of j crop in i subsystem; and C_j represents the unit price of j crop.

- Constraints

The sum of crop wager consumptions in the subsystem should be less than or equal to the amount of water that is distributed by the irrigation district:

$$\sum_{j}^{M} Q_{ij} \bullet S_{ij} \leq V_{ij}\eta_{s} + V_{gt} \bullet \eta_{g} \tag{6}$$

where V_{si} represents the amount of surface water allocated by i subsystem; V_{gi} represents the amount of groundwater allocated by i subsystem; η_s represents the utilization coefficient of surface water; and η_g represents the utilization coefficient of groundwater. The means of Q_{ij} and S_{ij} are the same as the means of the variables in equation[5].

4.1.3. Water optimal allocation between different subsystems

In this layer, the available supply water of the whole area was distributed to the subsystems of the second layer.In each subsystem, the optimal allocation of irrigation water between different crops and the economic benefits of each subsystem were determined by the optimal water allocation model between different crops. Then, the formerly calculated results should be fed back to the upper layer. Finally, the optimal allocation scheme of water resources that could reach the maximum economic benefits was obtained through the coordination of each layer.

- Objective function

Considering the maximum economic benefits of the whole irrigation districts as the objective:

$$\max Z = \sum_{i=1}^{6} z_i(w_i) - y_g \sum_{i=1}^{6} Q_{gt} - y_s \sum_{i=1}^{6} Q_{si} \tag{7}$$

where Z represents the economic benefits of the whole area; z_i represents the economic benefits of i subsystem; w_i represents the amount of irritation water distributed by the irrigation district; y_g represents the price per cubic meter of groundwater; Q_{gi} represents the amount of groundwater irrigation of i subsystem; y_s represents the price of per cubic meter of surface water; and Q_{si} represents the amount of surface water irrigation.

- Constraints

a. The sum of the supply water in the canal system of each subsystem should be less than or equal to the amount of water in the whole irrigation canal system:

$$\sum_{i=1}^{6} W_{si} \le W_{st} \bullet \eta_{st} \tag{8}$$

where W_{st} represents the amount of water in whole irrigation canal system and η_{st} represents the water utilization coefficient of canal system.

b.The sum of the groundwater consumption in each subsystem should be less than or equal to the allowable groundwater exploitation:

$$\sum_{i}^{6} Q_{gi} \le U_t \eta_{gt} \tag{9}$$

where U_t represents the allowable groundwater exploitation and η_{gt} represents the groundwater utilization coefficient of the canal system.

c. The ratio of canal-well water utilization should be in a reasonable range.

$$1 \le Q_{st} : Q_{gi} \le 2 \tag{10}$$

The means of Q_{st} and Q_{gi} are the same as the means of the variables in equation (7).

4.2. Optimization technique

This paper employed the Genetic Algorithm method (GA) to work out the three-layer model. The procedures of the solution are as follows. First, distribute the total surface water and groundwater to the second layer at a certain proportion,which is taken as an initial value. Secondly, taking the initial value into each subsystem and allocating the irrigation water to the crops in each subsystem and optimizing the allocation of irrigation water of a single crop in each growth stage by the Genetic Algorithm method (GA)[19-20]. Thirdly, calculating the economic benefit of the whole district according to the output of crops, which was obtained by the optimal allocation scheme of water resources. Finally, judging whether the solution is the optimal solution;if not, returning to the upper layers in turn. The calculation will not stop until the optimal solution is received or the maximum number of iterations is reached. The optimal processes of the GA method are shown in figure 6a and 6b, and the results of the optimal allocation are shown in table 1 and table 2.

a)

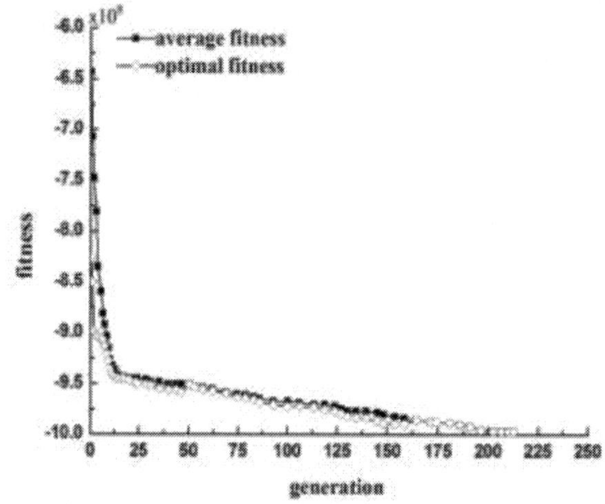

Figure 6a. Changing process of fitness by GA.

b)

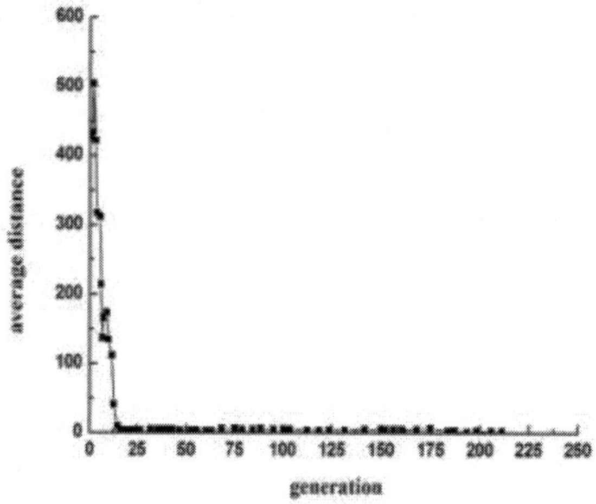

Figure 6b. Changing process of individual average distance by GA.

Table 1. Optimal allocation of crop irrigation schedule in m^3/ha.

crop	growth stage	Jingyang	Sanyuan	Gaoling	Lintong	Yanliang	Fuping
wheat	winter irrigation	1006.4	1000.2	1031.6	1012	998.3	942.4
	elongation	864.6	760.1	839.7	841	847.3	838.2
	earing	745.1	740.2	761	783.8	768.8	775.4
corn	elongation	727.1	735.3	719.7	658.3	688.3	705.1
	earing	706.8	700.4	721.8	711.9	689.6	658.8
	flowering	603.1	615.3	593	600.5	610.2	633.6
cotton	pre-sowing	734	719.3	706.6	723.7	747.2	739.2
	squaring	606	641.4	603.2	619.6	614.3	626.6
	florescence	536	562.4	531.9	526.8	544.3	515.8

Table 2. Optimal allocation of different water resources in $/10^4m^3$.

District	water resource	wheat			corn			cotton		
		Winter irrigation	elongation	earing	elongation	earing	flowering	Pre-sowing	squaring	florescence
Jingyang	canal	963.91	699.25	778.12	689.45	923.52	795.13	357.54	327.51	343.45
	well	858.23	582.98	653.64	519.82	798.58	699.81	249.32	234.22	236.87
Sanyuan	canal	598.17	493.49	577.31	449.87	540.54	461.12	290.03	273.47	255.23
	well	509.58	401.45	465.89	321.56	437.12	377.65	182.86	181.49	179.58
Gaoling	canal	639.39	572.56	575.37	462.56	553.19	491.59	283.29	278.24	264.56
	well	549.71	458.91	460.63	358.91	468.56	407.94	195.77	181.45	183.64
Lintong	canal	556.15	484.56	491.34	385.85	405.78	355.68	297.89	283.56	181.35
	well	479.36	407.28	405.92	273.46	318.95	263.45	206.91	199.68	209.23
Yanliang	canal	459.12	365.17	375.15	263.56	358.51	297.85	224.33	181.63	178.78
	well	338.79	247.65	264.79	204.77	277.84	223.56	131.67	122.75	121.57
Fuping	canal	209.64	187.45	201.89	186.12	198.43	176.14	158.96	142.75	146.36
	well	144.97	126.77	135.47	130.52	131.92	121.78	92.25	95.13	93.45

5. Numerical simulation of groundwater

In order to study the impact of the optimal allocation scheme of water resources on groundwater balance,a numerical simulation of groundwater was developed.

5.1. Modeling

As the aquifer in Jinghui irrigation district is of large thickness and the hydraulic gradient is relatively slow, the groundwater in this area can be generalized as a two-dimensional unsteady flow,while the hydrogeological model is isotropic and heterogeneous in type. The northern boundary is considered to be the second type of recharge boundary because it receives lateral recharge from the Loess Platform region. As mentioned earlier, the west, south and east sides of Jinghui irrigation district are surrounded by Jinghe River, Weihe River, and Shichuanhe River, they are the main discharge area of groundwater, they only recharge groundwater in the flood season, and the amount of recharge is small[21]. Therefore , they are set as the river boundary. Meanwhile, the canal seepage is set as the line source, the irrigation water and precipitation are set as the area source, and the evaporation is set as the area sink[22]. Because there are about 20,000 wells in the study area, the exploited wells are also set as the area sink. According to the above conditions, a mathematical model is established as follows:

$$\begin{cases} \frac{\partial}{\partial x}\left[K(h-z)\frac{\partial h}{\partial x}\right]+\frac{\partial}{\partial y}\left[K(h-z)\frac{\partial h}{\partial y}\right]+W=\mu\frac{\partial h}{\partial t}, & (x,y)\in\Omega, t>0 \\ h(x,y,t)=h_0(x,y), & (x,y)\in\Omega, t=0 \\ K(h-z)\left.\frac{\partial h}{\partial n}\right|_{\Gamma_2}=q(x,y,t), & (x,y)\in\Gamma_2, t>0 \end{cases}$$

where h represents the groundwater level (m); K represents the permeability coefficient of the aquifer(m/d), z represents the lower bed level of the aquifer (m); μ represents the specific yield; h_0 represents the initial groundwater level (m); q represents the inflow or outflow per unit area of the second type of recharge boundary($m^3/d\cdot m^2$); W represents the source or sink term (m^3/d);Γ_2 represents the second type of recharge boundary; Ω represents the calculation area; x, y represent the x and y coordinate axis respectively; and t represents time.

5.2. Model identification and verification

According to the spatial structure and parameter features, the aquifer of Jinghui irrigation district was divided into a series of grids in 300*150 and also into seven hydrogeological subdivisions (figure 7). The initial values of the parameters were obtained by pumping tests. The groundwater level of January, 2009 was regarded as the initial flow field, and the observation levels of groundwater from January 2009 to December 2010 were used to identify the model, while the observation levels of groundwater from January 2011 to December 2012 were used to verify the model. The relative error of the calculation results ranged from 3% to7% by comparing the calculated values and the measured values (figure 8). Thus, the results showed that the established model was reliable, the simulation results could reflect the characteristics of the groundwater flow field of Jinghui Irrigation District, and the model can also be used for numerical analysis. The final hydrogeological parameters of the model are shown in table 3.

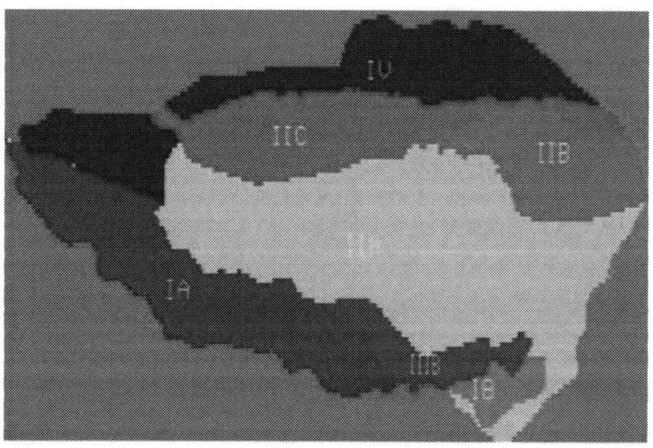

Figure 7. Map of hydrological subdivisions of the study area.

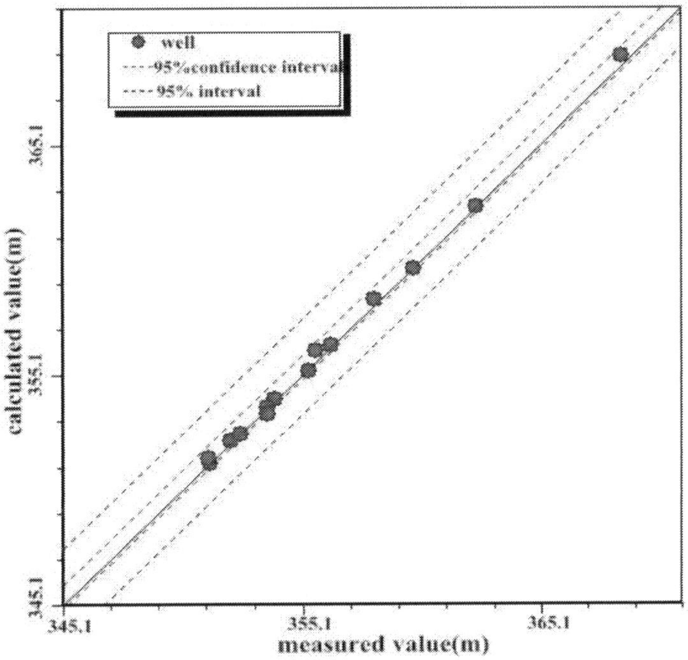

Figure 8. Fitting accuracy of groundwater level by numerical simulation model.

Table 3. Model parameters of each hydrogeological subdivision.

Hydrogeological Area	Permeability Coefficient K (m/s)	Specific Yield μ
IA	2.90×10^{-6}	0.04
IB	1.00×10^{-4}	0.1
IIA	1.74×10^{-4}	0.07
IIB	1.16×10^{-5}	0.06
IIC	7.9×10^{-6}	0.11
IIIB	1.01×10^{-5}	0.01
IV	4.4×10^{-6}	0.045

5.3. Result analysis of numerical simulation

The irrigation infiltration and groundwater exploitation were calculated on the basis of optimal allocation of the crop irrigation schedule and water resources. Then,the results were substituted into the established model in order to forecast the groundwater level and water balance in 2020. Finally, the forecasted water balance and level that resulted from adopting the optimal allocation scheme were compared with the level and balance condition caused by the current water utilization scheme. Also the forecasted level should also be compared with the ecological groundwater level that was discussed previously.The results are shown in table 4, figure 9a and figure 9b.

Table 4. Comparison of water balance under different scenarios in the year 2020 10^4 in m^3/a.

scheme	precipitation	Canal leakage	Irrigation infiltration	Well irrigation regression	Lateral inflow	exploitation	evaporation	Lateral outflow	budget
present	6498.8	6342.5	1788.6	3852.5	1987.2	22589	2365	1893.1	-6377.5
optimization	6589.4	7189.3	2978.9	3218.7	1895.7	19000	2658	1893.8	-1679.8

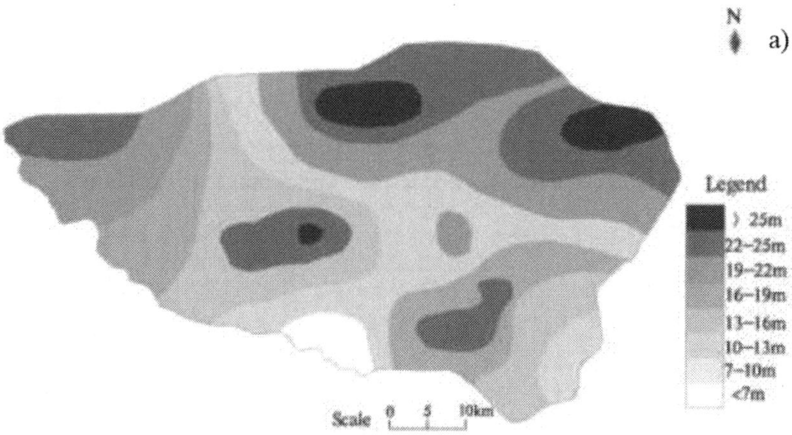

Figure 9a. Difference between forecasted level and upper limitation level.

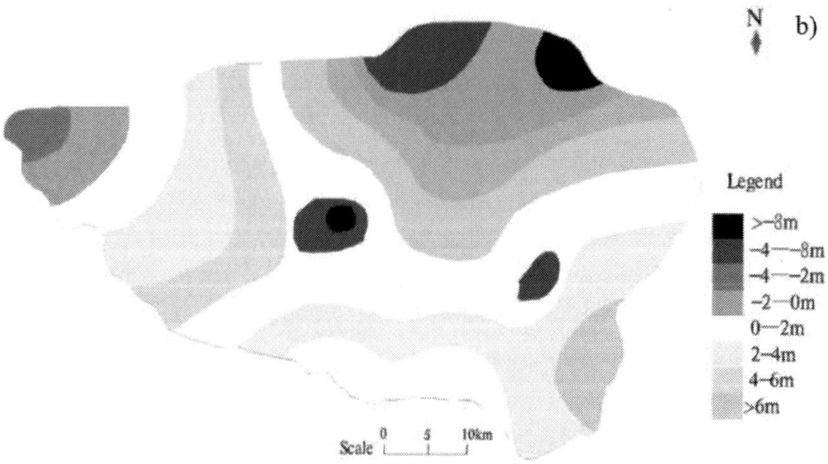

Figure 9b. Difference between forecasted level and lower limitation level.

As shown in table 4, the situation of the groundwater imbalance was greatly improved by the optimal allocation scheme of water resources. By the year 2020, the groundwater level will be lower than the upper limitation of the groundwater ecological level, that is, the phenomenon of soil salinization will not appear in Jinghui Irrigation District,which is as shown in figure 9a. Additionally, according to figure 9b, the groundwater level in 2020 will not exceed the lower limitation of the groundwater ecological level in most areas in Jinghui Irrigation District, only in a few areas will the level be deeper than that value.

6. Conclusions

The ecological groundwater level was obtained according to the limited level of groundwater evaporation, the height of capillary water rise and the mainly active layer thickness of crop roots. The results have shown that the upper limitation of the ecological groundwater level in Jinghui irrigation district ranged from 1.8m to 4.2m, and the lower limitation of the ecological groundwater level ranged from 8m to 28m.

A three-layer optimal allocation model of water resources on the basis of the theory of large scale systems was established. The model applied economic benefits as the objective function and took the ratio of canal-well water utilization, the supply capacity of different water resources, the water requirements and other conditions as constraints. Then, the optimal allocation of the crop irrigation schedules and water resources under the condition of a 75% assurance rate was calculated through the GA method. The results of optimal allocation indicate that the amount of canal irrigation water is 2.17 $\times 10^8 m^3$ while the amount of well irrigation water is $1.69 \times 10^8 m^3$, and the ratio of canal-well water utilization is 1.28:1.

The groundwater balance based on the optimal configuration of the crop irrigation schedule and water resources was simulated by a numerical simulation model. The simulation results indicate that the imbalance condition of the groundwater will be greatly improved by adopting the optimal allocation scheme. By 2020, the groundwater level will be lower than the upper limitation of the ecological groundwater level and will not exceed the lower limitation ecological groundwater level in most areas of Jinhui Irrigation District,although the level in some areas will be deeper than the lower limitation level. The results show that the optimizing allocation scheme of water resources is able to realize the sustainable utilization of water resources and gain good integral benefits.

Acknowledgements
This research was supported by the National Key Technology Support Program of China. The authors sincerely thank editors and reviewers for reviewing the manuscript and providing important comments and inputs to help improve this paper.

References

[1] Salewics K A and Iducks D P 1989 Interactive simulation for planning, managing, and negotiating *Oxford shire: Proceeding of the IAHS Symposium IAHS Pub* 181-83
[2] Carlos P, Gideon O and Abraham M 1997 Optimal operation of regional system with diverse water quality sources *Journal of Water Resources Planning and Management* **123**(2) 105-15
[3] Rosegrant M W, Ringle R C, Mckinny D C, *et al.* 2000 Integrated economic-hydrologic water modeling at the Basin Scale: The Maipo River Basin *Agricultural Economics* **24**(1) 33-46
[4] Reca J, Roldan J, Alcaide M and Camacho E 2001 Optimization model for water allocation in

deficit irrigation system: I. Description of the model *Agric Water Manage* **48**(2) 103-16

[5] Mohammed M, Mac Kirby M and Qureshi E 2007 Integrated hydrologic-economic modeling for analyzing water acquisition strategies in the Murray River Basin *Agric Water Management* **93**(1) 123-35

[6] Liu D D, Chen X H and Lou Z H 2010 A model for the optimal allocation of water resources in a saltwater intrusion area: A case study in Pear River Delta in China *Water Resource Management* 63-81

[7] Yang L H, Shen R K and Cao X L 2003 Scheme of groundwater use in Hetao irrigation district in Inner Mongolia *Transactions of the Chinese Society of Agricultural Engineering* **19**(5) 56-9

[8] Zhou W B and Zeng F C 2006 Groundwater level forecast and the appropriate channel and well water analysis of irrigation area combined with well and canal *Journal of Irrigation and Drainage* **25**(1) 6-9

[9] Yue W F, Yang J Z and Zhan C S 2011 Coupled model for conjunctive use of water resources in the Yellow River irrigation district *Transactions of the Chinese Society of Agricultural Engineering* **27**(4) 35-40

[10] He Y and Zhu T D W 2011 The water use optimization allocation of the irrigation district combined with the channel and well based on the random determined model *Arid Zone Research* **28**(3) 421-26

[11] Dai F G, Cai H J, Liu X M, Liu X, *et al.* 2012 Analysis of suitable irrigation water ratio of well to channel based on groundwater model *Transactions of the Chinese Society of Agricultural Engineering* **28**(15) 45-51

[12] Zhao H Q 2012 Study on ecological groundwater level and the amount of regulation in plain region of western Jilin Province *Beijing: China University of Geosciences*

[13] Wang S 1995 Overview of crop cultivation (northward vision) *Beijing:China Agriculture Press*

[14] Fan Z L, Chen Y N, Li H P, et al.2008 Determination of suitable ecological groundwater depth in arid area in northwest part of China *Journal of Arid Land Resources and Environment* **22**(2) 1-5

[15] Zhang C C, Shao J L, Li C J, *et al.* 2003 A study on the ecological groundwater table in the North China Plain *Journal of Jilin University(Earth Science Edition)* (03)323-26+ 330

[16] Jamshidi M 1983 Large-scale system modeling and control *Amsterdam NewYork*

[17] Dong Q G 2014 Research of health allocation of water resources and numerical simulation of groundwater in Jinghui Irrigation District *Xi'an: Chang'an University*

[18] Rosegrant M W, Ringle R C, Mckinny D C, *et al.* 2000 Integrated economic-hydrologic water modeling at the Basin Scale: The Maipo River Basin *Agricultural Economics* **24**(1) 33-46

[19] Yang G 2003 Optimal scheduling study on water resources of Yellow-river Irrigation District in Ningxia Province *Xi'an: Xi'an University of Technology*

[20] Zhou L 2002 Study on optimal allocation of regional water resources based on Genetic Algorithms *Zhengzhou: Zhengzhou University*

[21] Morshed J and Kaluarachchi J J 2000 Enhancements to genetic algorithm for optimal groundwater management *Journal of Hydraul Engineer ASCE* **5**(1):67-73

[22] Li H, Zhou W B, Ma C, *et al.* 2015 Research on regulation capacity of groundwater in Jinghui Irrigation District *Arid Zone Research* **32**(4):828-33

IOP Conf. Series: Earth and Environmental Science **39** (2016) 012009 doi:10.1088/1755-1315/39/1/012009

Evaluation model coupling exploitable groundwater resources and land subsidence control in regional loose sediments

Z J Luo [1,3], S J Zhao [1], W Z Jin [1], Q S Ma [1] and X H Wu [2]

[1] School of earth science and engineering, Hohai University, Nanjing, China
[2] Institue for Computational and Mathematical Engineering, Stanford University, California94305, USA

Email: luozujiang@sina.com

Abstract. The loose sediments in the Yangtze River Delta, the North China Plain, the plain of Northern Jiangsu and other districts in China are of great thickness, complex in structure and abundant in groundwater. Groundwater overexploitation easily results in geological disasters of land subsidence. Aiming at the issues, assessment models coupling exploitable groundwater resources and land subsidence control in regional loose sediments were brought up in this paper. The two models were: (1) a three dimensional groundwater seepage model with land subsidence based on the one dimensional Terzaghi consolidation theory; (2) a three dimensional full coupling model on groundwater seepage and land subsidence based on the Biot consolidation theory to simulate and calculate. It can be used to simulate and calculate the problems in real situations. Thus, the groundwater seepage and land subsidence were coupled together in the model to evaluate the amount of exploitable groundwater under the specific requirements of land subsidence control. The full coupling model, which considers the non-linear characteristics of soil mass and the dynamic changes of soil permeability with stress state based on the Biot consolidation theory, is more coincident with the variation characteristics of the hydraulic and mechanical properties of soil mass during the pumping process, making the evaluation results more scientific and reasonable.

1. Introduction

The Quaternary and Tertiary loose sediments in the Yangtze River Delta, the North China Plain, the Northern Jiangsu Plain and some other districts in China are of great thickness, which preserves abundant groundwater resources. The sediments consist of Holocene pore-phreatic aquifers (group), late Pleistocene No. Ⅰ confined aquifers (group), mid-Pleistocene No. Ⅱ confined aquifers (group) and early Pleistocene No. Ⅲ confined aquifers (group) from top to bottom. The middle-upper Pliocene No. Ⅳ confined aquifer (group) is also found in some individual areas (figure 1). The aquifers hydraulically connect to each other through clay aquitards, and some of the aquifers are also interconnected with each other (in local sections). With the development of the national economy since the late 1980s, the groundwater exploitation is increasing sharply. It is under threat from problems that affect both the quantity and quality of water that aquifers provide. Quantity problems may arise due to extensive aquifer dewatering and impacts on local groundwater drainage, flow and discharge. The groundwater level decline dramatically, which causes changes in pore water pressure and effective stress in aquifers and clay aquitards (T J Burbey 2001), and land subsidence of geological hazards is also triggered seriously as a result (Shen et al. 2004). It not only brings about immeasurable loss to the development of national economy, but also great changes in the natural hydrogeological environment of the groundwater system. The regulation and storage capacity of the groundwater system is weakened, which also affects the sustainable exploitation and utilization of groundwater resources. Quality problems include contaminant leaching due to acid mine drainage and impacts on physical hydrogeology (Raghavendra N S, Deka P C 2015). For this purpose, new evaluation models of exploitable groundwater resources under the requirements of land subsidence control must be brought up (Hao Zhi-fu, Kang Shao-zhong 2006), which is of utmost urgency.

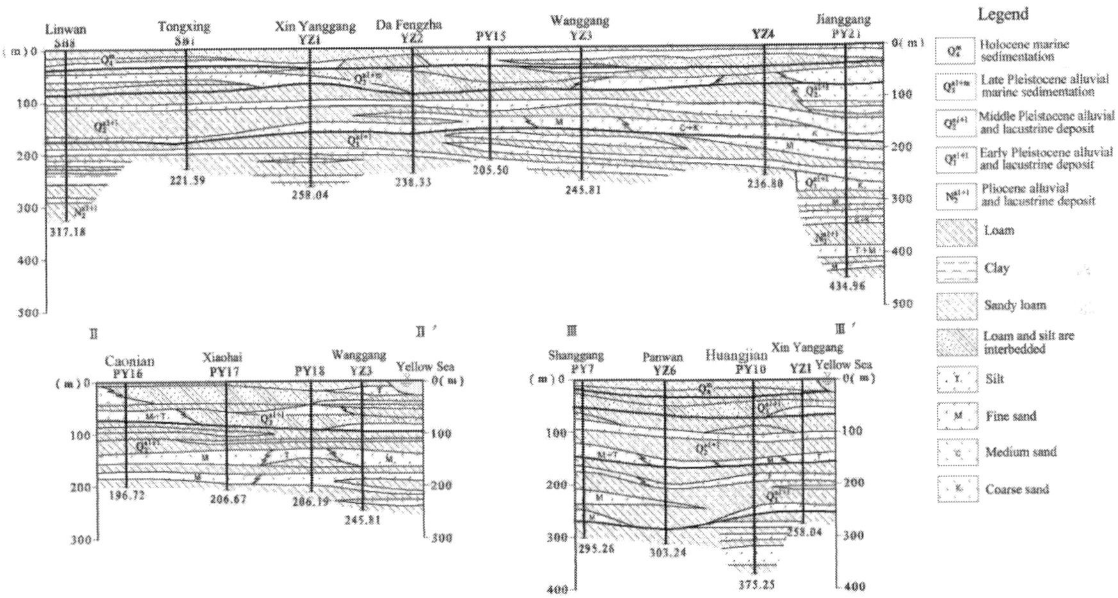

Figure 1. Hydrogeological section map of Quaternary and Tertiary of a city in the Northern Jiangsu Plain.

In this study, based on groundwater seepage, the one dimensional Terzaghi consolidation theory and the Biot consolidation theory, a three dimensional groundwater seepage model coupling land subsidence and a three dimensional full coupling model on groundwater seepage and land subsidence were brought up. The groundwater seepage and land subsidence were coupled together in the model to evaluate the exploitable groundwater resources under the requirements of land subsidence control. The full coupling model considered the non-linear characteristics of the soil mass and the dynamic changes of soil permeability with the stress state based on the Biot consolidation theory, making the evaluation results more scientific and reasonable.

2. Coupling Model of Groundwater Seepage and Land Subsidence

2.1. Three dimensional Model Coupling Groundwater Seepage and Land Subsidence Based on the One Dimensional Terzaghi Consolidation Theory

2.1.1. Mathematical Model of Groundwater Movement (Luo Zu-jiang et al. 2005, Ma Xiu-yuan et al. 2009).

$$\frac{\partial}{\partial x}(k_{xx}\frac{\partial H}{\partial x})+\frac{\partial}{\partial y}(k_{yy}\frac{\partial H}{\partial y})+\frac{\partial}{\partial z}(k_{zz}\frac{\partial H}{\partial z})+W=S_S\frac{\partial H}{\partial t} \qquad (x,y,z)\in\Omega \tag{1}$$

$$H(x,y,z,t)\big|_{t=t_0}=H_0(x,y,z,t_0) \qquad (x,y,z)\in\Omega \tag{2}$$

$$H(x,y,z,t)\big|_{\Gamma_1}=H_1(x,y,z,t) \qquad (x,y,z)\in\Gamma_1 \tag{3}$$

$$k_{xx}\frac{\partial H}{\partial x}\cos(n,x)+k_{yy}\frac{\partial H}{\partial y}\cos(n,y)+k_{zz}\frac{\partial H}{\partial z}\cos(n,z)\Big|_{\Gamma_2}=q(x,y,z,t) \qquad (x,y,z)\in\Gamma_2 \tag{4}$$

$$\begin{cases} H(x,y,z,t)=z \\ k_{xx}\left(\frac{\partial H}{\partial x}\right)^2+k_{yy}\left(\frac{\partial H}{\partial y}\right)^2+k_{zz}\left(\frac{\partial H}{\partial z}\right)^2-\frac{\partial H}{\partial z}(k_{zz}+q_w)+q_w\Big|_{\Gamma_3}=\mu\frac{\partial H}{\partial t} \end{cases} \qquad (x,y,z)\in\Gamma_3 \tag{5}$$

Where:

- S_s is the specific storativity (1/m);
- k_{xx}, k_{yy}, k_{zz} are the anisotropic filtration coefficients of the aquifer in the principal direction (m/d);
- h is the head of point (x, y, z) at t moment (m);
- W is the source and sink (1/d);
- t is time (d);
- Ω is the calculation area.
- $h_0(x,y,z,t_0)$ is the original water table of the point (x, y, z) (m);
- $q(x,y,z,t)$ is the recharge of the unit area at the second boundary (m/d);
- $h_1(x,y,z,t)$ is the head at the first boundary;
- $\cos(n,x)$, $\cos(n,y)$ and $\cos(n,z)$ are the cosine of the exonormal of the discharge flow rate boundary's and the coordinate axis's included angle;
- μ is the saturation deficiency (an increase of free surface) or specific yield (free surface is down), which indicates the water yield that unit section area of aquifer absorbs or discharges when free surface changes a unit height ;
- q_w is the recharge of atmospheric precipitation on the unit area of the free surface (m/d);

- Γ_1, Γ_2 and Γ_3 are the first boundary, the second boundary and the free surface boundary.

2.1.2. Mathematical Model of Land Subsidence (Luo Zu-jiang and Zhang Yue-ping 2007). The calculation model of the deformation extent of the aquifer caused by the decline of the groundwater level is as follows:

The elastic deformation of the confined aquifer is

$$\Delta b = -\Delta h \cdot S_{ske} \cdot b_0 = -\Delta h \cdot S_{fe} \tag{6}$$

The inelastic deformation of the confined aquifer is

$$\Delta b^* = -\Delta h \cdot S_{skv} \cdot b_0 = -\Delta h \cdot S_{fv} \tag{7}$$

The elastic deformation of the phreatic aquifer is

$$\Delta b = -\Delta h \cdot (1 - n + n_w) \cdot S_{ske} \cdot b_0 = -\Delta h \cdot S_{fe} \tag{8}$$

The inelastic deformation of the phreatic aquifer is

$$\Delta b^* = -\Delta h \cdot (1 - n + n_w) \cdot S_{skv} \cdot b_0 = -\Delta h \cdot S_{fv} \tag{9}$$

There into:

$$S_{ske} = \rho_w g \left[\frac{3(1-2v)}{2G(1+v)} \right] \tag{10}$$

$$S_{skv} = \frac{0.434\ C_c \rho_w}{p'(1+e_0)} \tag{11}$$

Where:

- Δb is the elastic deformation of the aquifer (m), positive sign indicates compression and negative sign indicates expansion;
- Δb^* is the inelastic deformation of the aquifer (m), positive sign indicates compression and negative sign indicates expansion;
- Δh is the change of the head (m);
- S_{fe} is the elastic storage factor of the aquifer skeleton (no dimension);
- S_{ske} is the elastic storage rate of the aquifer skeleton (1/m);
- S_{fv} is the inelastic storage factor of the aquifer skeleton (no dimension);
- S_{skv} is the inelastic storage rate of the aquifer skeleton (1/m);
- b_0 is the thickness of the compressible aquifer (m);
- n is the porosity;
- n_w is the moisture capacity above the water table, which is a part of the total volume of the porous medium;
- G is the shear modulus;
- v is the Poisson ratio;
- C_c is the compression index;
- p' is the initial effective stress;
- e_0 is the initial void ratio;
- ρ_w is the water density.

In the subsidence model, the aquifers and the aquitards inside were treated as a whole. The whole skeleton of the aquifer was assumed to be compressed and potentially resulted in land subsidence. The parameters S_{fv} and S_{fv} were equivalent across the whole scale of the aquifer.

The total amount of the land subsidence is calculated only after the amount of compression in every aquifer is obtained.

There were two steps for the coupling of the flow model, land subsidence model and the water release quantity produced by the deformation of soil layers. First, the water level should be worked out using the flow model and take it as a boundary condition of the subsidence model. The deformation amount of soil layers and the resulting water release quantity can be calculated and substituted into the flow model. Finally, compute them with repeated iterations to finish the coupling of water flow and land subsidence.

By coupling the hydraulic heads in the two models together, the coupling simulation model of groundwater exploitation and land subsidence was established.

2.2. The Three Dimensional Full Coupling Model on Groundwater Seepage and Land Subsidence

The change of the groundwater seepage field caused by groundwater exploitation is actually a process of interaction and mutual effects between the seepage field and the stress field. Based on the Biot consolidation theory, the three dimensional full coupling model on groundwater seepage and land subsidence has considered the non-linear characteristics of soil mass and the dynamic changes of soil permeability with stress state, making the results closer to reality and more precise.

2.2.1. Biot Consolidation Theory. The three dimensional Biot consolidation equation is based on the assumptions that (1) the deformation of soil skeleton is linear elastic; (2) the micro-strain is slight; (3) the seepage conforms to the Darcy law; and (4) the water cannot be compressed or micro-compressed in saturated soil. It is as follows (Luo Zu-jang et al. 2008):

$$\begin{cases} -G\nabla^2 w_x - \dfrac{G}{1-2\nu}\dfrac{\partial}{\partial x}\left(\dfrac{\partial w_x}{\partial x}+\dfrac{\partial w_y}{\partial y}+\dfrac{\partial w_z}{\partial x}\right)+\dfrac{\partial u}{\partial x}=0 \\ -G\nabla^2 w_y - \dfrac{G}{1-2\nu}\dfrac{\partial}{\partial y}\left(\dfrac{\partial w_x}{\partial x}+\dfrac{\partial w_y}{\partial y}+\dfrac{\partial w_z}{\partial x}\right)+\dfrac{\partial u}{\partial y}=0 \\ -G\nabla^2 w_z - \dfrac{G}{1-2\nu}\dfrac{\partial}{\partial z}\left(\dfrac{\partial w_x}{\partial x}+\dfrac{\partial w_y}{\partial y}+\dfrac{\partial w_z}{\partial x}\right)+\dfrac{\partial u}{\partial z}=-\gamma \end{cases} \tag{12}$$

$$\begin{aligned} &\frac{\partial}{\partial t}\left(\frac{\partial w_x}{\partial x}+\frac{\partial w_y}{\partial y}+\frac{\partial w_z}{\partial z}\right)-\frac{1}{\gamma_w}\left[\frac{\partial}{\partial x}\left(k_x\frac{\partial u}{\partial x}\right)\right. \\ &\left.+\frac{\partial}{\partial y}\left(k_y\frac{\partial u}{\partial y}\right)+\frac{\partial}{\partial z}\left(k_z(\frac{\partial u}{\partial z}+\gamma_w)\right)\right]=0 \end{aligned} \tag{13}$$

Where:

- G is the shear modulus;
- ν is the Poisson's ratio;
- w_x, w_y and w_z are the displacement components of directions x , y and z;
- u is the pore water pressure;

- k_x, k_y and k_z are the permeability coefficients of directions *x*, *y* and *z*;
- γ is the density of earth's gravity;
- γ_w is the density of water's gravity.

Combined with fixed solution conditions, such as the initial conditions and boundary conditions, the equation above can be solved by the numerical method.

2.2.2. Fixed Solution Conditions.

- The Initial Conditions

A. The Initial Condition of Ground Stress

Using the dead-weight stress of soil to estimate its initial stress,

$$\begin{cases} \sigma_z = \gamma z \\ \sigma_x = K_0 \gamma z \end{cases}$$

Where:

- Z is the depth of the calculation point;
- K_0 is the static lateral compression coefficient, $K_0 = \begin{cases} 1 - \sin\phi' & sandy\ \ layer \\ 0.95 - \sin\phi' & clay\ \ layer \end{cases}$
- φ' is the effective internal friction angle.

B. The Initial Condition of Displacement

$$w(x,y,z,t)\big|_{t=0} = w_0(x,y,z)$$

C. The Initial Condition of Pore Water Pressure

$$u(x,y,z,t)\big|_{t=0} = u_0(x,y,z)\frac{n!}{r!(n-r)!}$$

- Boundary Conditions

A. The Boundary Condition of Pore Water Pressure

$$u(x,y,z,t)\big|_{\Gamma_1} = u_1(x,y,z,t)$$

Where:

- $u_1(x,y,z,t)$ is the known pore water pressure at boundary Γ_1.

B. The Boundary Condition of Flow Rate

$$K\frac{\partial H}{\partial n}\Big|_{\Gamma_2} = q_{\overline{L}}$$

Where:

- $q_{\overline{L}}$ is the known discharge flow rate at the boundary.

C. The Boundary Condition of Free Surface

$$u = 0; \quad q = -\mu\frac{\partial u}{\partial t}\cos\theta$$

Where

- μ is the specific yield of the soil;
- θ is the included angle of the exonormal direction of the free surface and the vertical line.

D. The Boundary Condition of Displacement

$$w(x,y,z,t)|_S = w_s(x,y,z,t)$$

Where:

- $w_s(x,y,z,t)$ is the measure of the displacement at boundary *S*.

2.2.3. Disposal of Correlative Parameters and Physical Quantities in the Model.

- Duncan-Chang Nonlinear Model

By using the Duncan-Chang nonlinear model, the constitutive relation of soil is extended to nonlinearity. Thus, the elastic constants *E* and *v* in matrix [*D*] of the constitutive relation $\{\triangle\sigma\}=[D]\{\triangle\varepsilon\}$ are no longer treated as constants but change with the stress state. The expressions of the tangent elasticity modulus and tangent Poisson's ratio are as follows:

$$E_t = \left[1 - R_f \frac{(1-\sin\varphi)(\sigma_1-\sigma_3)}{2c\cos\varphi + 2\sigma_3\sin\varphi}\right]^2 kp_a\left(\frac{\sigma_3}{p_a}\right)^n \tag{14}$$

$$v_t = \frac{G - F\lg\left(\frac{\sigma_3}{p_a}\right)}{\left(1 - \frac{D(\sigma_1-\sigma_3)}{kp_a\left(\frac{\sigma_3}{p_a}\right)^n\left[1 - R_f\frac{(1-\sin\varphi)(\sigma_1-\sigma_3)}{2c\cos\varphi+2\sigma_3\sin\varphi}\right]}\right)^2} \tag{15}$$

Where:

- *C* and φ are intensity-indexes, and *k*, *n*, R_f, *G*, *F* and *D* are test parameters achieved by the conventional triaxial compression test of soil.

- Dynamic Model of Permeability *k*

The fluid-solid coupling is actually the stress redistribution of the soil skeleton caused by dissipation of pore water stress, which leads to the consolidation deformation on the macro level and changes of porosity factor *n* in soil. Therefore, the permeability of the soil is changed and the seepage is also consequently affected. At present, many models have been proposed in papers about fluid-solid coupling ignore the interrelationship between porosity and permeability and the changes of them. Under the premise of the assumptions of the Biot consolidation theory, the dynamic expression of porosity φ and permeability *k* was built as follows:

$$\varphi = \frac{\varphi_0 + \varepsilon_V}{1+\varepsilon_V} \tag{16}$$

$$k = \frac{k_0}{1+\varepsilon_v} \bullet \left[1+\frac{\varepsilon_v}{\varphi_0}\right]^3 \tag{17}$$

In which:

- φ_0 is the initial porosity;
- k_0 is the initial intrinsic permeability;
- ε_V is the volumetric strain.

The model above couples groundwater seepage with soil deformation in its mechanism. The model involves a large amount of parameters, and the solving of the model is consequently complex. Therefore, at present it is still rare to use the model in regional evaluation of groundwater resources.

3. Applications

3.1. *Instance One*

3.1.1. General Situation, Generalization and Discretization of the Study Zone. Jiangyin City is located in the Yangtze River Delta economic zone, on the south of the Yangtze River, facing Jingjiang across the river, on the north of the Shanghai-Nanjing expressway, on the west of Changshu and Zhangjiagang and on the east of Changzhou and Wujin. It is at the geometric centre of the "Golden Triangle" of Suzhou, Wuxi and Changzhou. The pore water-bearing system in Quaternary shallow loose rock within 60 meters contains pore-phreatic aquifer, micro-confined pore aquifer, clay aquitard and No. I confined aquifer from top to bottom. The internal structure of the system was generalized as inhomogeneous and anisotropic in three dimensional space. The Yangtze River was treated as a constant water head boundary, and the rest of the boundaries around the system were treated as general water head boundaries. The top of the system was treated as a free surface boundary (Bridget R Scanlon et al. 2002), and the bottom of the system was treated as an impervious boundary. The groundwater exploitation took towns as units and was treated as big wells.

The three dimensional model above coupling groundwater seepage and land subsidence based on the one dimensional Terzaghi consolidation theory was solved by the finite difference method (Ma Xiu-yuan et al. 2006, Wang Kun et al. 2009), and the algebraic equations were solved by the strong implicit method (SIP) simultaneously and iteratively. Ignoring the bedrock area, the total area of the study zone was 747.37km^2. The model was divided into rectangular isometric meshes as 200×150 in the plane, four layers on the section which contained phreatic aquifer, micro-confined aquifer, clay aquitard and No. I confined aquifer from top to bottom. The unit number was 120000, among which were 46284 effective calculation cells (figure 2).

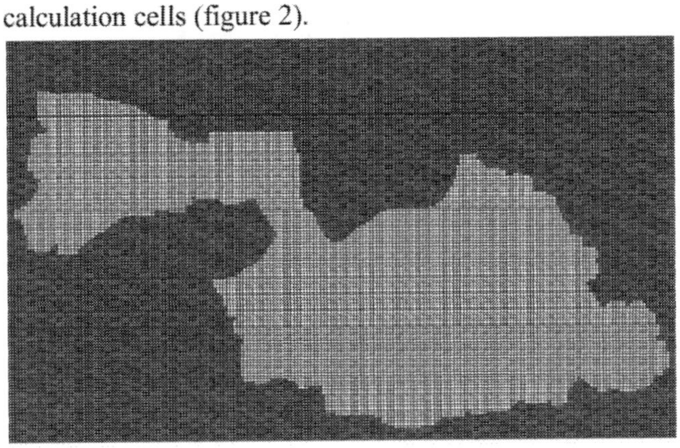

(a) Grid subdivision graph in plane.

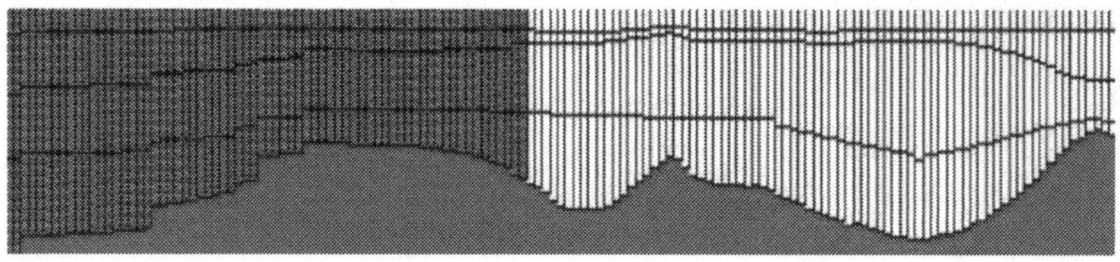

(b) Grid subdivision graph in section at row 111.

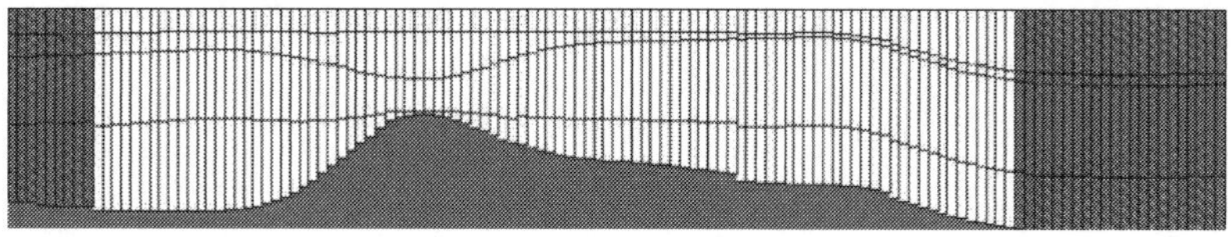

(c) Grid subdivision graph in section at Column 140.

Ineffective calculation cell   Effective calculation cell

Figure 2. Grid subdivision graph.

3.1.2. Identification and Validation of Parameters. Using the model above, the test lasted for 5.25 days according to the data of the observation wells. The whole time, the segment was discretized into 15 stress periods and each stress period was divided into 10 steps. The initial flow field of each aquifer was measured in the test, and the initial displacement was set as 0. The pumpage of each pumping well was obtained by actual statistics. The hydraulic conductivity on the boundaries and the initial parameters of the subareas in each aquifer were all based on previous data, field tests and indoor experiments, and the water level of observation wells were fitted with pumping test data. Then, the inverse calculation method was used to compute the hydro-geological parameters. The parameters involved in coupling analysis included: initial horizontal permeability coefficient K_{0x} and K_{0y}, initial vertical permeability coefficient K_{0z}, specific yield μ, internal friction angle φ, cohesion force c, initial elastic modulus E_0, initial poisson ratio v_0 and unit weight γ, among which the soil mechanics parameters were obtained by indoor experiments. Figure 3 and table 1 illustrate the relevant parameter division in the 4th layer of the model and the parameter values in each subarea. Figure 4 and figure 5 illustrate the fitting precision of the model of groundwater level and land subsidence, from which we can see that the fitting precision of the groundwater level is relatively high and the land subsidence is fitted to the subsidence trend. Thus, the model can be used to simulate the change of water level and land subsidence caused by groundwater exploitation in the study zone.

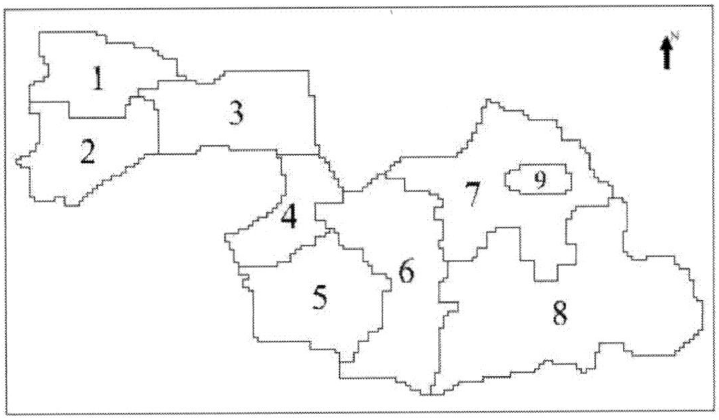

Figure 3. Parameter division of the 4th confined aquifer.

Table 1. Parameters of the 4th confined aquifer.

Parameters Division Number		K_{0x} /m•d^{-1}	K_{0y} /m•d^{-1}	K_{0z} /m•d^{-1}	μ /m^{-1}	E_0 /MPa	v_0	c /kPa	ϕ /(°)	γ /kN•m^{-3}
	1	10	10	1.0	6E-5	943	0.47	5.0	26	19.3
	2	8	8	0.8	5E-5	936	0.47	6.1	28	19.4
	3	8.5	8.5	0.85	5.8E-5	938	0.47	6.2	26	19.3
The 4th	4	0.95	0.95	0.095	1.1E-7	943	0.47	7.1	28	19.4
Confined	5	0.9	0.9	0.09	1E-7	940	0.47	5.1	26	19.3
Aquifer	6	0.89	0.89	0.089	3E-8	943	0.47	7.3	28	19.4
	7	0.8	0.8	0.08	1E-7	956	0.47	7.2	26	19.3
	8	0.88	0.88	0.088	5E-7	967	0.47	7.4	28	19.4
	33	1 E-6	1 E-6	1 E-7	1.0E-10	943	0.47	7.1	26	19.3

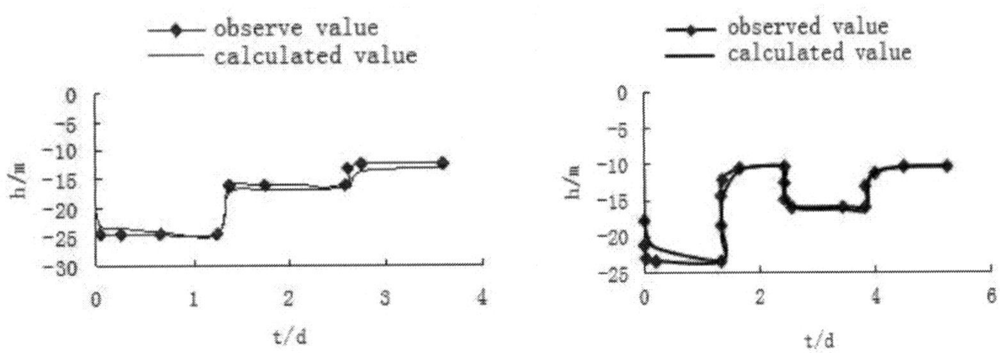

(a) (b)

Figure 4. Fittings of groundwater level with the data from the simulation and observation.

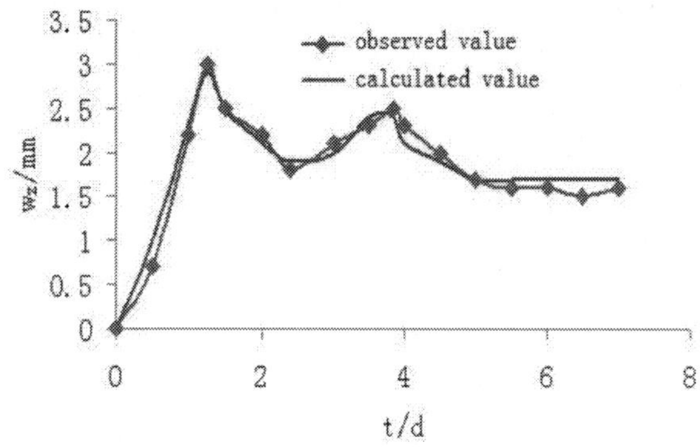

(a)

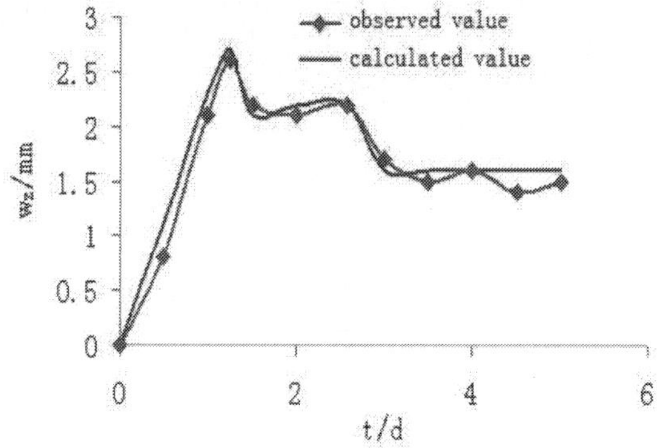

(b)

Figure 5. Fittings of land subsidence with the data from the simulation and observation.

3.1.3. Prediction of Groundwater Level and Land Subsidence. The model regarded the end of April 2007 as the starting point of the calculation, and the buried depth of the groundwater level in the No. Ⅰ confined aquifer, which would not be lower than half the depth of the roof of the aquifer in 10 years, and the quantity of the land subsidence, which would not exceed 50mm in 10 years, as constrained conditions. After the identification of the model, the groundwater recoverable resources of towns in Jiangyin and the mining well number were obtained by running the model, and the results are listed in table 2. Figure 6 and figure 7 are the predictive groundwater level contour maps of the micro-confined aquifer and the 1st confined aquifer at the end of April 2017, and figure 8 is the predictive land subsidence contour map at the end of April 2017. The result shows that the shallow groundwater exploitation in Jiangyin is of great potential as long as the exploitation layout of the shallow groundwater is planned rationally. The total amount of exploitable resources in the micro-confined pore aquifer is 1492.16×10^4 m^3/a, the total amount of exploitable resource in No. Ⅰ confined aquifer is 2052.47×10^4 m^3/a and the maximum quantity of the land subsidence will not exceed 50mm in 10 years.

Table 2. Annual recoverable mining groundwater resources of each town in Jiangyin (10^4 m^3/a).

Towns	Micro-confined aquifer		The 1st confined aquifer	
	Suitable Well	Recoverable	Suitable Well	Recoverable
Huangtu	72	316.71	25	377.78
Ligang	65	364.53	33	698.61
Shengang	41	239.95	14	227.40
Xiagang	47	327.73	10	182.50
Yuecheng	21	9.20	8	49.64
Qingyang	50	25.55	8	26.57
Xiake	23	4.20	72	118.26
Huashi	16	5.84	40	24.82
Zhouzhuang	15	2.19	49	44.71
Xinqiao	6	5.26	13	45.08
Changjing	37	24.31	33	84.32

Gushan	30	17.52	10	38.33
Zhutang	50	140.16	36	118.26
Nanzha	22	8.83	2	4.16
Yuntin	1	0.18	15	12.05

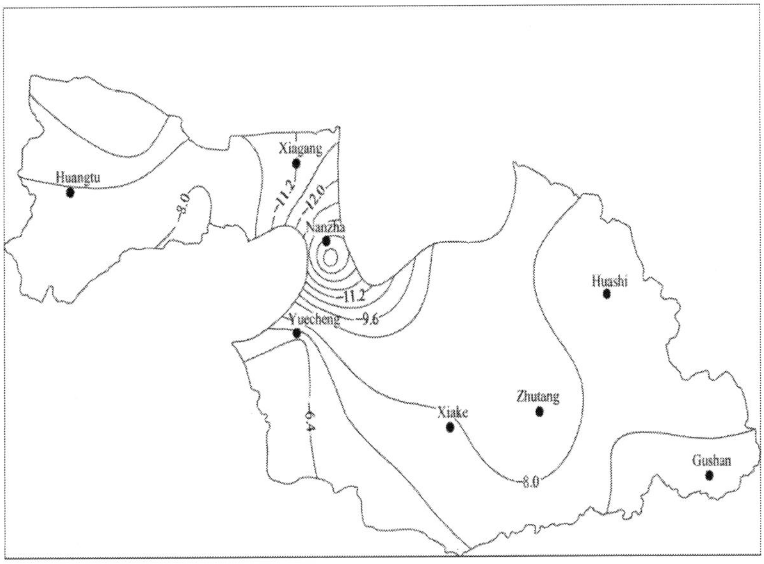

Figure 6. Predictive groundwater level contour map of the micro -confinedaquifer at the end of April 2017 (unit: m).

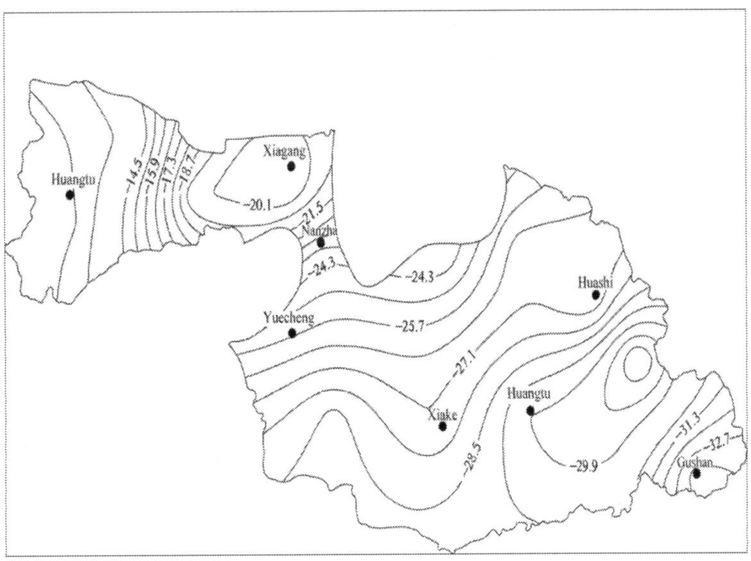

Figure 7. Predictive groundwater level contour map of the1st confined aquiferat the end of April 2017 (unit: m).

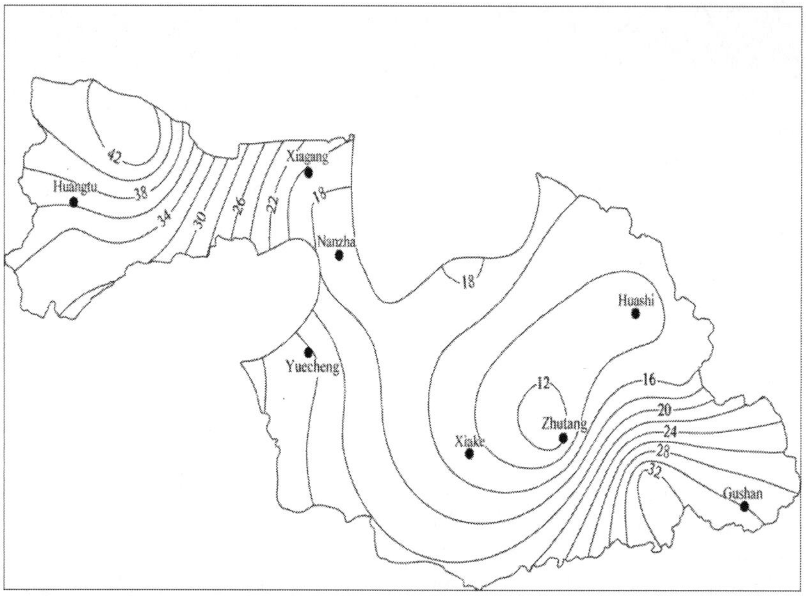

Figure 8. Predictive land subsidence contour map at the end of April 2017 (unit: m).

3.2. *Instance Two*

3.2.1. General Situation, Generalization and Discretization of the Study Zone. Wujiang City lies in the alluvial plain of the Yangtze River Delta, and it has deposited a huge thickness of Quaternary loose sediment. The pore water-bearing system in Quaternary shallow loose rock buried within 60 meters contains phreatic aquifer, clay aquitard and micro-confined aquifer from top to bottom. The internal structure of the system can be generalized as inhomogeneous and anisotropic in three dimensional space. The top of the system was treated as a free surface boundary, and the bottom of the system was treated as an impervious boundary. Considering the exploitation of shallow groundwater in the surrounding cities, the study zone surroundings were generalized as impervious boundaries. The initial water level and initial displacement were given by statistical material. The groundwater exploitation took rural towns as units and was treated as big wells.

The three dimensional full coupling model of the groundwater seepage and land subsidence was solved by the finite element method (Luo Zu-jiang et al. 2008, Li Wen-hu et al. 2005), and the algebraic equations were iteratively solved by the preconditioned conjugate gradient method (Li Yi-min and Zhou Feng-yan 2004, Mansfield L and Damped Jacobi 1991). The total area of the study zone is 111.54 km^2 and the model was divided into 39204 cells and 40000 nodes (figure 9).

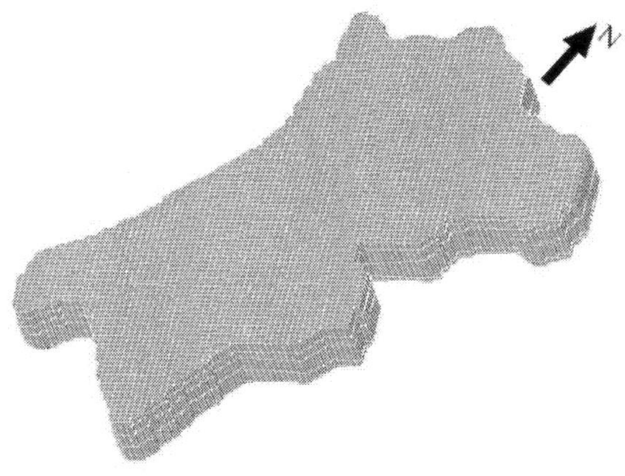

Figure 9. Three dimensional grid subdivision graph.

3.2.2. Identification and Validation of Parameters. The model above was simulated using the Groundwater Seepage and Land Subsidence Software (GWS) that was developed by the research group. According to the data of the observation wells, the test lasted for 1.25 days. The whole time segment was discretized into 10 stress periods, and each stress period was divided into 10 steps. The initial flow field of each aquifer was measured in the test, and the initial displacement was set as 0. The pumpage of each pumping well was obtained by actual statistics. The hydraulic conductivity on the boundaries and the initial parameters of the subareas in each aquifer were all based on previous data, field tests and indoor experiment, and the water level of the observation wells were fitted with pumping test data. Then, the inverse calculation method was used to compute the hydro-geological parameters. The parameters involved in the coupling analysis included: initial horizontal permeability coefficient K_{0x} and K_{0y}, initial vertical permeability coefficient K_{0z}, specific yield μ, internal friction angle φ, cohesion force c, initial elastic modulus E_0, initial poisson ratio v_0 and unit weight γ, among which the soil mechanics parameters were obtained by indoor experiments. Figure 10 and table 3 illustrated the relevant parameter division in the 1st aquifer of the model and the parameter values in each subarea. Figure 11 illustrated the fitting precision of the model of the groundwater level and land subsidence, from which we can see the fitting precision of the groundwater level is relatively high. Thus, the model can be used to simulate the change of the water level and land subsidence caused by groundwater exploitation in the study zone.

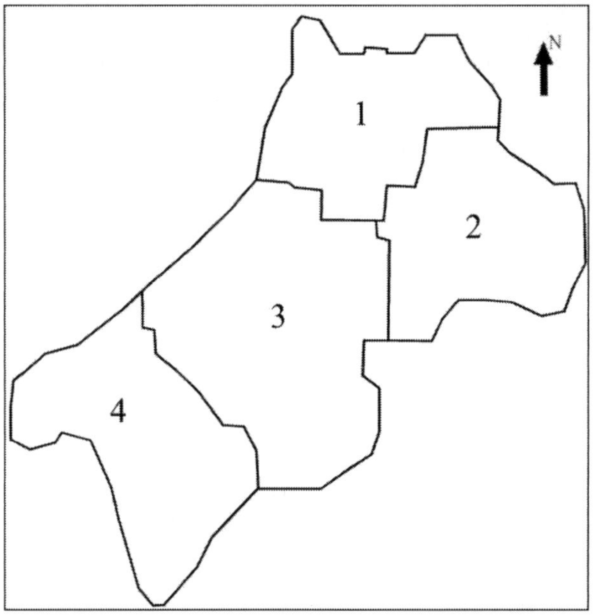

Figure 10. Parameters of the first confined aquifer.

Table 3. Parameters of the first confined aquifer.

Division Number	K_{0x} /m•d^{-1}	K_{0y} /m•d^{-1}	K_{0z} /m•d^{-1}	μ /m^{-1}	E_0 /MPa	V_0	c /kPa	ϕ /(°)	γ /kN•m^{-3}
1	0.17	0.17	0.022	2.5E-8	8.20	0.47	12	12	17.3
2	3	3	0.22	3.5E-7	8.18.	0.47	11	12	17.2
3	0.2	0.2	0.02	1.5E-8	8.25	0.48	12	12	17.9
4	0.2	0.2	0.03	3.5E-8	8.28	0.47	11	12	17.5

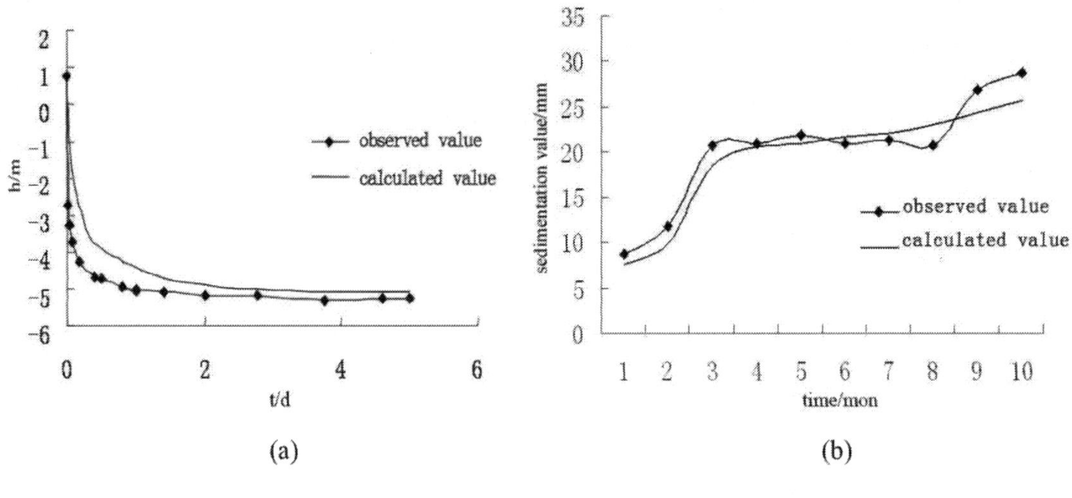

Figure 11. Fittings of groundwater level and subsidence with data from the simulation and observation.

IOP Conf. Series: Earth and Environmental Science **39** (2016) 012009 doi:10.1088/1755-1315/39/1/012009

3.2.3. Prediction of Groundwater Level and Land Subsidence. The model regarded the end of April 2007 as the starting point of the calculations, and the buried depth of the groundwater level in the micro-confined aquifer, which would not be lower than half the depth of the roof of the aquifer in 10 years, and the quantity of the land subsidence, which would not exceed 50mm in 10 years, as constrained conditions. After the identification of the model, the groundwater recoverable resources of towns in Wujiang and the mining well number were obtained by running the model, which are listed in table 4. Figure 12 and figure 13 are the predictive groundwater level contour maps of the phreatic aquifer and the micro-confined aquifer at the end of April 2017, and figure 14 is the predictive land subsidence contour map at the end of April 2017. The result shows that the shallow groundwater exploitation in Wujiang has great potential as long as the exploitation layout of the shallow groundwater is planned rationally. The total amount of exploitable resource in the micro-confined pore aquifer is 821.50×10^4 m^3/a, the total amount of exploitable resource in the phreatic aquifer is 382.09 $\times 10^4$ m^3/a and the maximum quantity of land subsidence will not exceed 50mm in 10 years.

Table 4. Annual recoverable mining groundwater resources of each town in Wujiang (10^4 m^3/a).

Towns	Micro-confined aquifer		Phreatic aquifer	
	Suitable Well	Recoverable	Suitable Well	Recoverable
Tongli	10	127.96	7	54.84
Zhenze	22	40.13	15	27.39
Luxu	30	53.65	22	39.34
Taoyuan	35	35.07	24	24.05
Pingwang	27	47.01	12	20.89
Shengze	40	90.08	13	29.27
Qidu	31	51.05	12	19.76
Hengshan	19	31.78	5	8.36
Songling	27	270.89	14	140.46
Lili	50	73.88	12	17.73

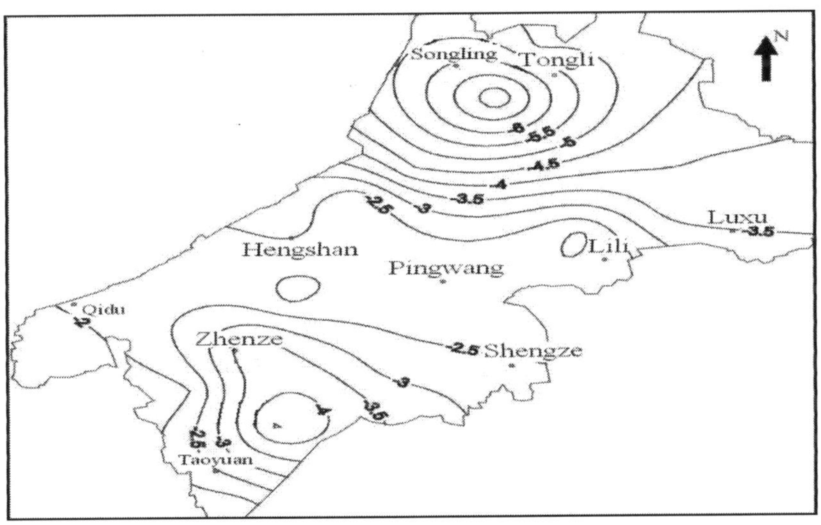

Figure 12. Predictive groundwater level contour map of phreatic aquifer at the end of April 2017 (unit: m).

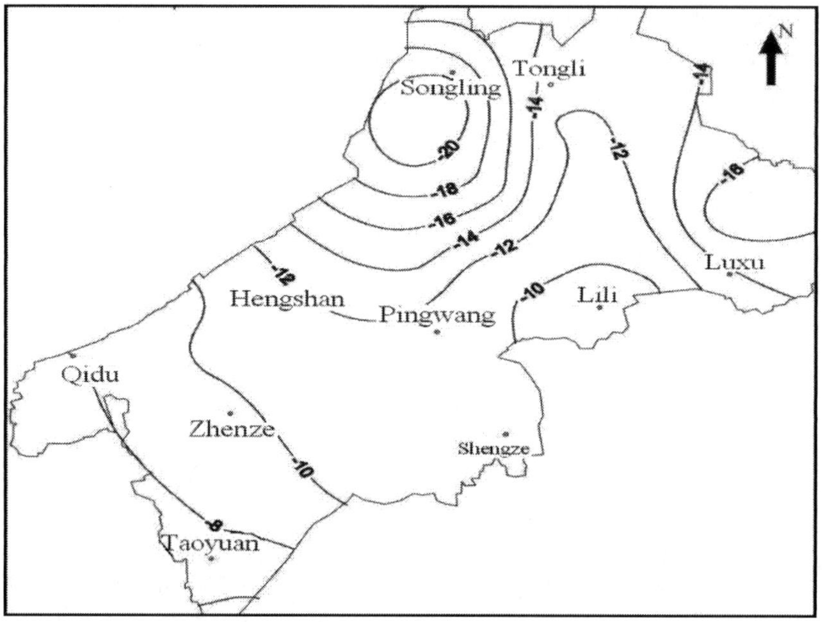

Figure 13. Predictive groundwater level contour map of micro-confined aquifer at the end of April 2017 (unit: m).

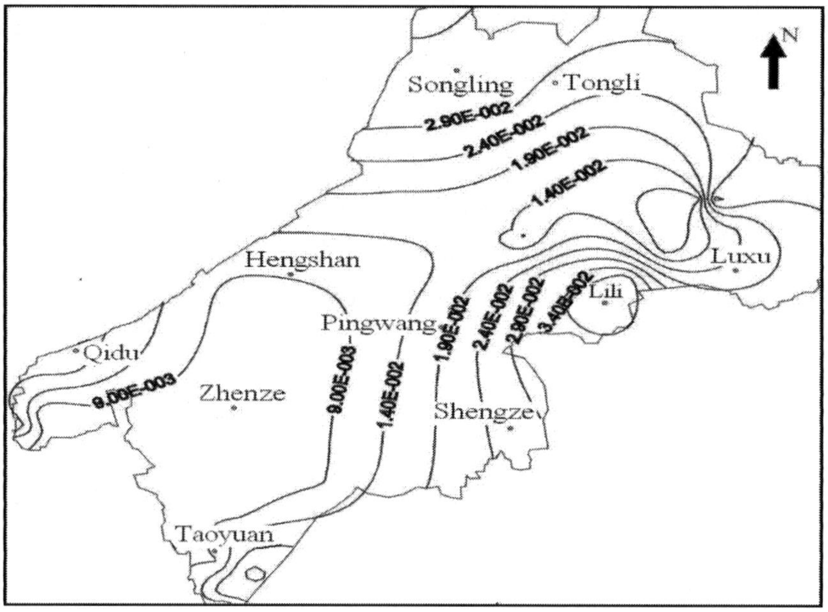

Figure 14. Predictive land subsidence contour map of April 2017 (unit: m).

4. Conclusions

The evaluation model coupling exploitable groundwater resources and land subsidence control can calculate the exploitable groundwater resources based on the land subsidence control in regional loose sediments. The model couples the exploitable groundwater resources with the land subsidence control, thus unifying and integrating the groundwater exploitation and land subsidence control effectively.

Based on the Biot consolidation theory, the full coupling model considered the non-linear characteristics of the soil and the dynamic changes of soil permeability with the stress state. It depicted the variation characteristics of the hydraulic and mechanical properties of the soil mass during the pumping process more correctly than the three dimensional model coupling groundwater seepage and land subsidence based on the one dimensional Terzaghi consolidation theory, and the result of the evaluation is more scientific and reasonable. However, the model is not yet mature enough to be widely used in the evaluation of the exploitable groundwater resources and land subsidence control in regional loose sediments at present because it has too many parameters and the equation is difficult to solve. With the continuing development of calculation technology and the accumulation of the data of the groundwater regime and the stress-strain of the soil with the water, it is inevitable that the theory and method will be widely used in the future.

References:

[1] T J Burbey 2001 Stress-strain analyses for aquifer-system characterization *Ground Water* **39**(1) 128-36

[2] Shen S L, Tohno I N and Ishigaki M, et al. 2004 Land subsidence due to withdrawal of deep-groundwater *Lowl. Technol Int.* IALT **6**(1) 128

[3] Raghavendra N S and Deka P C 2015 Sustainable Development and Management of Groundwater Resources in Mining Affected Areas: A review *Procedia Earth Planet. Sci.* **11** 598 – 604

[4] Hao Z F and Kang S Z 2006 Current situation and development trend of numerical simulation of groundwater system *Advances in Science and Technology of Water Resources* **26**(1) 77-81

[5] Luo Z J, Yao B K and Wang X M 2005 Three dimension numerical model of groundwater resource evaluation of complex aquifer system *Journal of Jilin University (Earth Science Edition)* **35**(2) 188-94

[6] Ma X Y, Li S G and Zhu W S 2009 A new method in groundwater flow modeling *Journal of Hydrodynamic.* B **21**(2) 245-54

[7] Luo Z J, Zhang Y P 2007 Three dimensional numerical model for groundwater withdrawal and land subsidence in development zone along the Yangzte River in Jiangsu Province *Journal of Earth Science and Environment* **29**(3) 280-4

[8] Luo Z J, Liu J B and Li L 2008 Three dimensional full-coupling numerical simulation of groundwater dewatering and land-subsidence in Quaternary loose sediments *Chinese Journal of Geotechnical Engineering* **30**(2) 193-8

[9] Bridget R S, Healy R W and Cook P G 2002 Choosing appropriate techniques for quantifying groundwater recharge *Hydrogeol J.* **10**(1) 18-39

[10] Ma X Y, Yu F and Li J L 2006 Numerical simulation of local groundwater *Rock and Soil Mechanics* **27**(suppl) 131-6

[11] Wang K, Jin S and Liu G 2009 Numerical modeling of free-surface flows with bottom and surface-layer pressure treatment *J Journal of Hydrodynamics* B **21**(3) 352-8

[12] Luo Z J, Zhang Y Y and Wu Y X 2008 Finite element numerical simulation of three

dimensional seepage control for deep foundation pit dewatering *Journal of Hydrodynamics* B **20**(5) 596-602

[13] Li W H, Li T C and Wang X L 2005 Finite element analysis of elastic foundation plate concerning horizontal deformation of foundation *Journal of Jiangsu University*(Natural Science Edition) **26**(1) 89-92

[14] Li Y M and Zhou F Y 2004 Solution to a class of boundary value problem of three dimensional partial differential equation *Journal of Jiangsu University* (Natural Science Edition) **25**(4) 328-31

[15] Mansfield L and Damped J 1991 Preconditioning and coarse grid deflation for conjugate gradient iteration on parallel computers *SIAM J. Sci. Statist. Comput.* **12**(6) 1314-23

IOP Conf. Series: Earth and Environmental Science **39** (2016) 012010 doi:10.1088/1755-1315/39/1/012010

Analysis of the impact of water level fluctuations on macrophytes in Miyun Reservoir after receiving water transferred by the South-to-North Water Diversion Project

L Yu[1, 2], H Gu[1], C H Lou[1], L Zhang[1] and Q Y Meng[1]

[1]Beijing Water science and Technology Institute, Beijing, China

E-mail: yl@bwsti.com

Abastact. As the main primary producers in aquatic ecosystems, macrophytes affect the structure and function of aquatic ecosystems, and their distribution is controlled by water depth. Miyun Reservoir in Beijing will have to experience substantial changes in water level and surface area as it begins to receive water transferred by the South-to-North Water Diversion Project, which will have an adverse impact on the macrophytes growing there. In this study, a hydrodynamic model was constructed with MIKE21 and then used in a simulation in three scenarios: dry year, normal year and wet year. The results suggest that during water diversion, the annual and interannual water level fluctuations will be too significant for them to adapt and as a result, the original macrophytes in the reservoir tend to die and disappear completely. The area of the zone suitable for macrophyte growth, or suitable growth zone (SGZ), fluctuated. Restricted by the main dam and auxiliary dam to its south, the overall suitable growth zone moved toward the northeast and northwest of the reservoir, with a northeastward movement of its centroid. The distance and path of movement varied between scenarios. After the water diversion was completed, the suitable growth zone shrunk in the three scenarios. It is predicted that the macrophyte species diversity and richness of the reservoir can recover to the levels recorded before water diversion only in dry year. These results suggest that manual interventions should be implemented after water diversion to speed up the natural recovery of aquatic plant communities in Miyun Reservoir and thereby maintain the stability of the aquatic ecosystem.

1. Introduction

Miyun Reservoir is located in the mountainous region in northern Miyun District, Beijing (figure 1). It was first constructed in 1958 as a large reservoir for multiple purposes such as water supply, flood control, irrigation, and power generation. The reservoir has a capacity of 4.375 billion m^3, dead storage of 0.419 billion m^3, and its normal water level, dead storage level, and flood control level are 157.5 m, 126.0 m, and 152.0 m, respectively. The inflows of Miyun Reservoir include Bai River, Chao River, Baimaguan River, Duijia River, Andamu River, Qingshui River and so on.

Since 1999, the reservoir had seen declining water level and shrinking water surface as a result of a continued lack of precipitation in the reservoir region and increased domestic water supply to urban Beijing. After considering the capacity of the reservoir, Beijing imposed limitations on the water extraction from Miyun Reservoir in 2004. Since then, the reservoir level has been maintained at around 135 m and the reservoir has been operating at low levels. As the only surface source of drinking water in Beijing, Miyun Reservoir supplies water to Beijing via two primary supply lines: the No. 9 water plant pipeline and Jingmi canal.

According to the Miyun Reservoir Ecosystem Survey Report (2012), there was a total of 15 macrophyte species living in the reservoir, including 8 emergent, 6 submerged and 1 floating-leaved species. Aquatic plants were distributed throughout the reservoir bank, especially concentrated in the northern shallow near-shore regions, such as Yanggezhuang Town, Yanluo Village, Bulaotun Town, and Gaoling Town. Besides, aquatic plants were also found in inner lake, while they scarcely existed in the southern areas due to the restrictive water depth.

Miyun Reservoir Regulation and Storage Project is an important part of South-to-North Water Diversion Project (SNWDP). The project plans to transfer water at a rate of about 0.2 billion m^3 per year into Miyun Reservoir by 9-stage pumping stations and through an 81 km long section of the existing Jingmi Canal and the 22 km long newly built underground pipeline (figure 2). The water inflow will replenish the reservoir so that the reservoir will experience no reduction in storage after supplying water to urban Beijing. Five years of continuous water storage, will raise the reservoir water level and expand the water surface, thus altering the distribution area of local aquatic plants and further affecting their growth and reproduction. In particular, macrophytes are likely to die due to increased water depth and violent water level fluctuations, resulting in changes in their distribution. Therefore, it is necessary to carry out relevant research, and quantitatively analyze the impact of water level fluctuations on macrophytes in different scenarios.

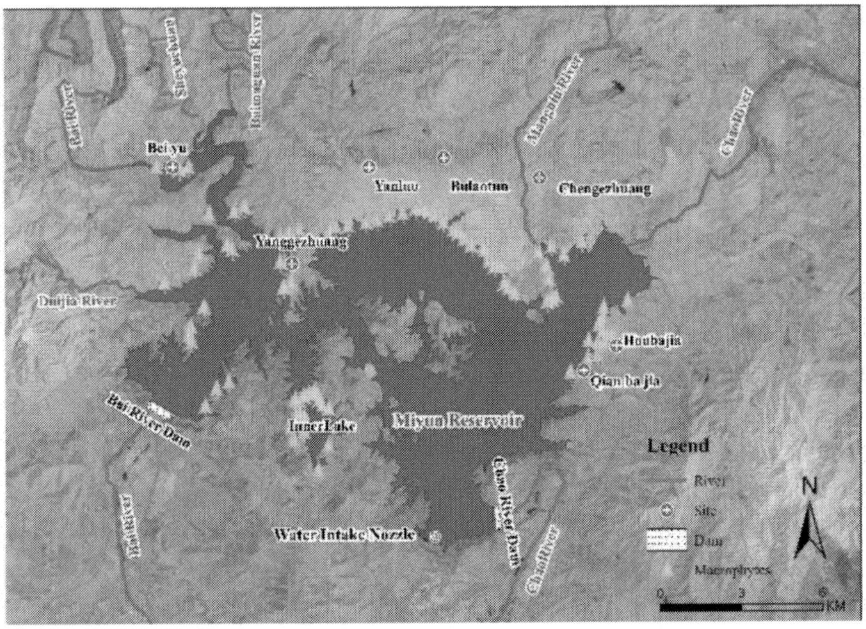

Figure 1. Inflows and macrophytes distribution of Miyun reservoir.

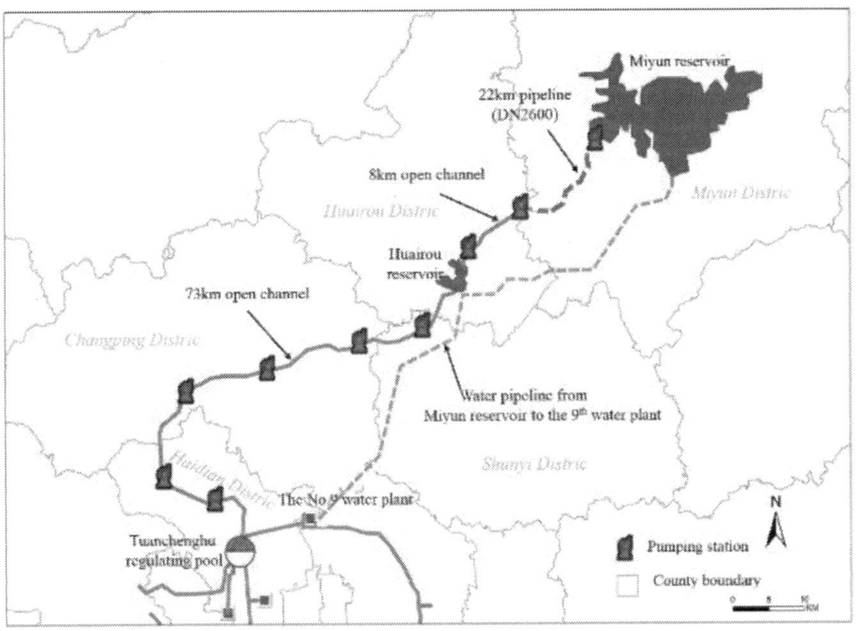

Figure 2. Layout of the Miyun Reservoir Regulation and Storage Project.

2. Method

2.1. *Restrictive factors on macrophytes growth*

As the main primary producers in aquatic ecosystems, macrophytes provide both food and diversified habitats to fish, zooplankton and other animals [1, 2]. Their distribution is primarily determined by water depth. Moreover, water depth can also regulate the composition and richness of aquatic plant species. Slow-flowing water is suitable for growth of macrophytes [3, 4]. From the reservoir bank to deep water, the dominant macrophytes are emergent, floating-leaved, and submerged macrophytes, respectively [5]. Emergent macrophytes live in about 1 m deep water and can tolerate water deeper than 3 m for a short period of time. Floating-leaved macrophytes are distributed in sublittoral zone at depths shallower than 5 m. Submerged macrophytes, the most adaptable macrophytes to water depth, normally live in water no deeper than 6 m [6]. Water depth is a dynamic factor that varies over time and space. Water level fluctuations can exert great influence on aquatic plants [7, 8]. Relevant research has suggested that, in order to maintain the plant species diversity and richness of a lakeside area, it is necessary to keep at least the following two variables within proper limits [9]. (1) Annual water level fluctuation: the most suitable annual water level fluctuation and fluctuation duration are 1 m and 1 month, respectively; the fluctuation duration must not exceed 2 months. (2) Interannual water level fluctuation: the interannual water level fluctuation over 10 years should be limited to between 1.0 and 4.0 m, or a wider range for larger lakes.

Therefore, the key to carry out this research is to find out how the water depth and the water level change, and the hydrodynamic numerical model is the best tool to solve the problem.

2.2. Model Construction

In this study, the MIKE21 software is used to build the two-dimensional hydrodynamic model of the study area.

2.2.1. Creation of terrain file. In order to guarantee the requirements of computational precision and operation speed, terrain grid shape was rectangular and its size was 300 m × 300 m. So the resolution of DEM data (provided by Geospatial Data Cloud site, Computer Network Information Center, Chinese Academy of Sciences. http://www.gscloud.cn), was reduced from 30m to 300 m by Resample, a built-in tool in ArcGIS. After that, the terrain file for Miyun Reservoir was generated using Bathymetries, a built-in tool in MIKE21 tool for creating terrains.

The accuracy of the terrain file was evaluated based on existing water level – storage curve. The results revealed that the storage error was less than 4% and surface area error was below 8%, as shown in figure 3, indicating that the accuracy of the terrain file met the research criteria.

2.2.2. Model setting. Because Bai and Chao Rivers contribute more than 90% of the total inflow to the reservoir, the simulation focused on Bai and Chao Rivers while the other tributaries were considered collectively based on the actual situation, and two inflow boundaries were applied to the model. The water diverted by the SNWDP entered Miyun Reservoir through an inlet at the Bai River dam, which was set as a source point. The outflows of the reservoir include the Bai River dam, the Bai River power plant, the No. 9 water plant at the source, and the Chao River dam. According to the project implementation plan, there would be no outflow except water extraction from the No. 9 water plant at the source. Therefore, only one sink point was set in the model (see figure 4). Figure 5 shows the process of water flow through the source and sink points. Precipitation and evaporation were included in the model. The simulated time period was set between January 1, 2015 and January 1, 2020, i.e. 1/1/2015-1/1/ 2020.

2.3. Scenario setting

A simulation was performed to quantitatively analyze the impact of project implementation and upstream and downstream inflows on the water level of Miyun Reservoir. Three scenarios (wet year, normal year and dry year, see table 1) were set in the simulation. The upstream inflows in different hydrological years (figure 6) were determined based on the *Environmental Impact Assessment of Miyun Reservoir Regulation and Storage Project – A Special Report on the Project's Impact on Miyun Reservoir Aquatic Environment.* The initial water level of the reservoir was set at 134.2 m in light of the latest monitoring data.

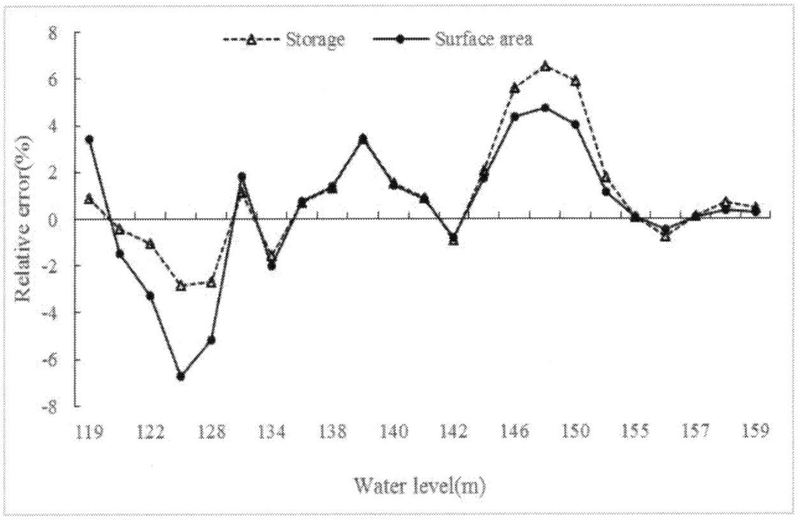

Figure 3. Terrain error.

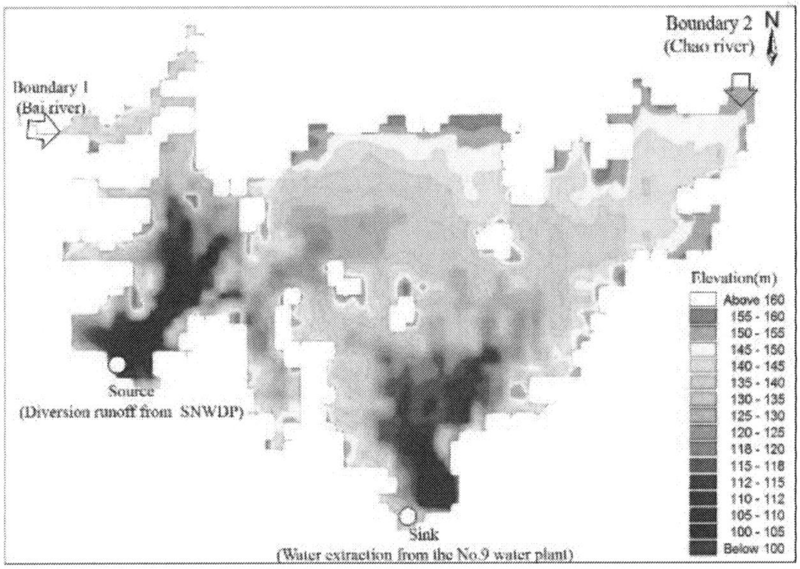

Figure 4. Boundary, source and sink points.

Table 1. Simulation scenarios.

Number	Scenario name	Upstream inflow (billion m^3/year)	Precipitation (mm/year)	Evaporation (mm/year)
1	Dry year	0.174	310	1355
2	Normal year	0.354	469	1144
3	Wet year	0.665	546	1230

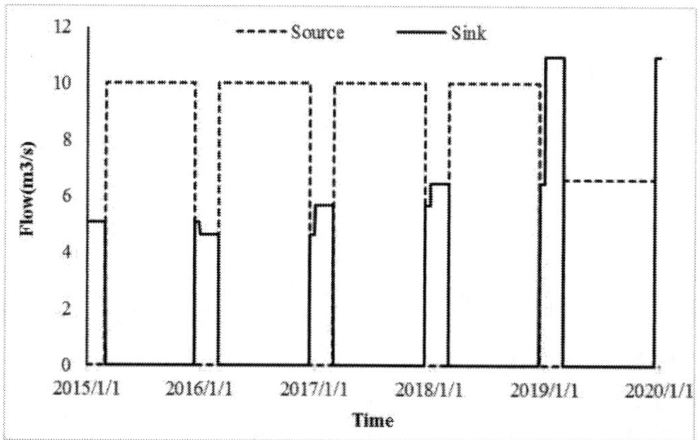

Figure 5. Process of water flow through the source and sink points.

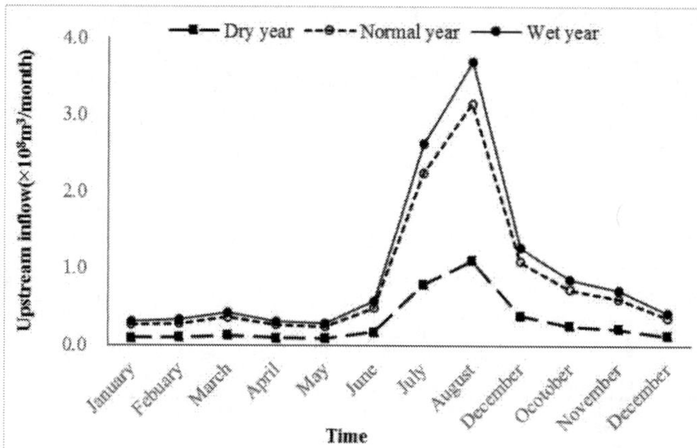

Figure 6. Upstream inflow process.

3. Results and analysis

3.1. Changes in water level and surface area

After water diversion, water level in dry, wet, and normal years had increased by 14.8, 17.8 and 24.5 m, respectively; the corresponding increases in water surface area were 64.27, 79.59 and 105.87 km^2, respectively. Compared to the no diversion mode, the water level changes were between 11% and 18%, and surface area changes were in the range of 81% to 133% (table 2).

Table 2. Simulation results.

Scenario	Water level				Surface area			
	Initial (m)	Final (m)	Difference (m)	Change extent %	Initial (km^2)	Final (km^2)	Difference (km^2)	Change extent (%)
Dry year	134.2	149.0	14.8	11	79.54	143.81	64.27	81
Normal year	134.2	152.0	17.8	13	79.54	159.13	79.59	100
Wet year	134.2	158.7	24.5	18	79.54	185.41	105.87	133

3.2. *Annual and interannual water level fluctuations*

In this study, annual water level fluctuation (AWLF) is the difference between the water levels at the beginning and the end of a year; interannual water level fluctuation (IAWLF) is the water level fluctuation over years. The results of simulation (figure 7) show that during water diversion, both AWLFs and IAWLFs decreased over time. The AWLFs ranged from 1.4 to 8.0 m, and IAWLFs were between 1.8 and 7.0 m. The wet year saw the strongest fluctuations, followed by the normal year, while the dry year saw smaller fluctuations than the other two scenarios.

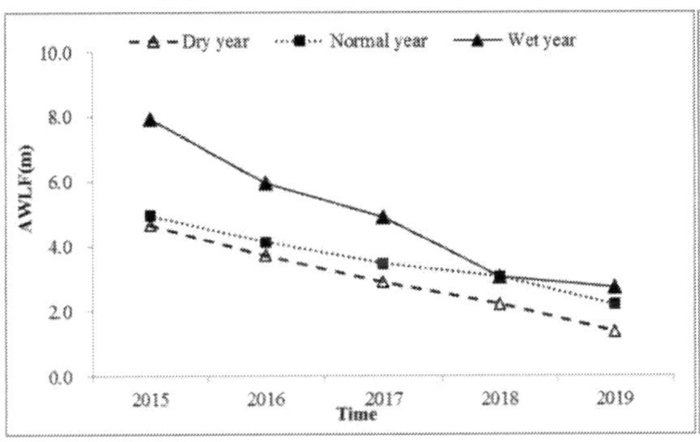

(a) AWLFs

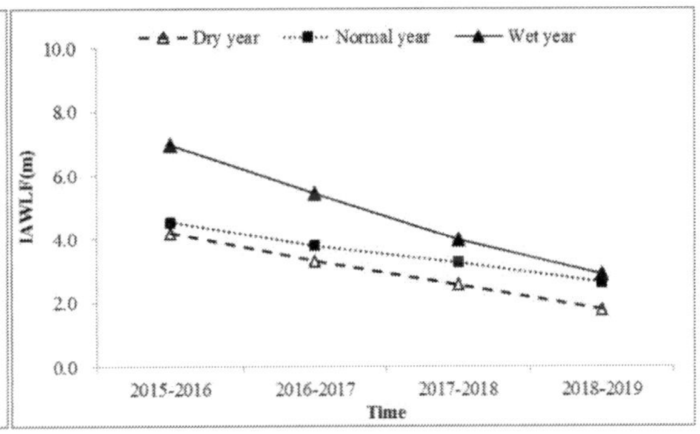

(b) IAWLFs

Figure 7. Water level fluctuations during water diversion.

3.3. *Change in area of suitable growth zone*

Previous analysis suggests that the region at depths shallower than 6 m is suitable for growth of aquatic plants. As shown in figure 8, the area of SGZ fluctuated; in dry, normal, and wet years, it peaked at 33.21, 30.60, and 33.03 km^2 on 2019/1/1, 2018/1/1, and 2017/1/1, respectively. After water diversion was completed, the area of SGZ increased 3.69 km^2 in dry year and decreased 2.43 km^2 and 4.32 km^2 in normal and wet years, respectively, compared to 27.54 km^2 before diversion.

3.4. *Spatial variation in suitable growth zone*

Restricted by the main dam and auxiliary dam to its south, the SGZ moved towards the northeast and northwest. The spatial variation in suitable growth zone was quantitatively studied through centroid analysis, a method commonly used to analyze spatial variation in land use [10]. A centroid, usually defined as the geometric center of a polygon or a plane, is an important index used to describe spatial distribution of geographic phenomena and it can be simplified as the equilibrium point at which the geographic target retains uniform distribution [11]. Centroid analysis is often employed to trace the distribution of geographic targets. The centroid of a SGZ can be calculated using the following formula:

$$X_t = \frac{\sum_{i=1}^{n}(C_{ti} \times X_{ti})}{\sum_{i=1}^{n} C_{ti}} \qquad Y_t = \frac{\sum_{i=1}^{n}(C_{ti} \times Y_{ti})}{\sum_{i=1}^{n} C_{ti}}$$

where X_t and Y_t represent the longitude and latitude coordinates of the centroid of SGZ in t year; C_i is the area of i SGZ patch in t year; X_i and Y_i are the longitude and latitude coordinates of the centroid of i SGZ patch in t year.

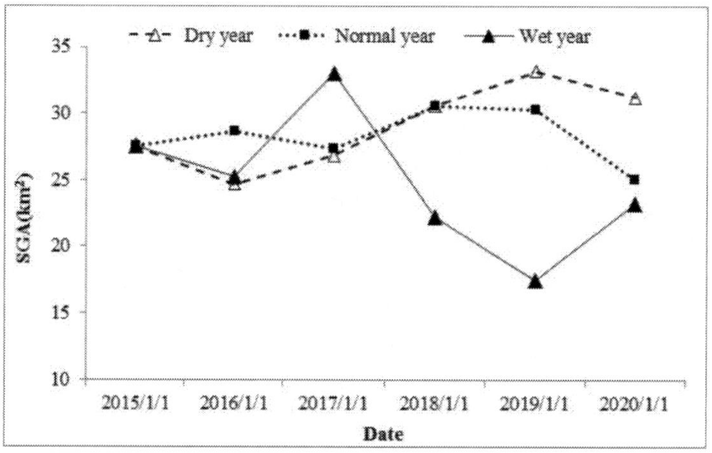

Figure 8. SGZ area change during water diversion.

Figure 9 shows the position and coordinates of the centroid (shown with black dot) in the first year (t=1), third year (t=3), and fifth year (t=5) since the beginning of water transfer (2015/1/1). In order to display its movement more clearly, the centroid was plotted in a two dimensional coordinate system, where the X-axis points towards the east of the reservoir and Y-axis points to the north. It is clear from figure 10 that the centroid followed different trajectories in different scenarios. In dry year, the centroid moved eastward, then northward, and eastward again. In normal year, it moved eastward, then northeastward, and finally northward. In wet year, it moved northeastward in the first two periods, followed by a northwestward shift. After water diversion was completed, the centroid of SGZ had moved 4613 m in dry year, 4838 m in normal year, and 5689 m in wet year, toward the northeast.

3.5. *Analysis of impact on macrophytes*

During water diversion, both AWLFs and IAWLFs will be so significant that macrophytes in the reservoir will fail to adapt and tend to the die. As a result, the reservoir will see declines in both biomass and richness of species.

After water diversion finished, the SGZ area has increased in dry year and decreased in normal and wet years when compared to that before diversion. It can be predicted from the species– area relationship [12, 13] and island biogeography [14] that after water diversion, the macrophytes species diversity and richness could recover to the levels recorded before water diversion only in dry year.

During water diversion, the SGZ moved toward the shallow and bay zones in the northeastern and northwestern parts of the reservoir as water depth increased, with a northeastward movement of its centroid. In dry, normal, and wet years, the centroid of SGZ had moved 4613 m, 4838 m, and 5689 m, respectively, toward the northeast after water diversion.

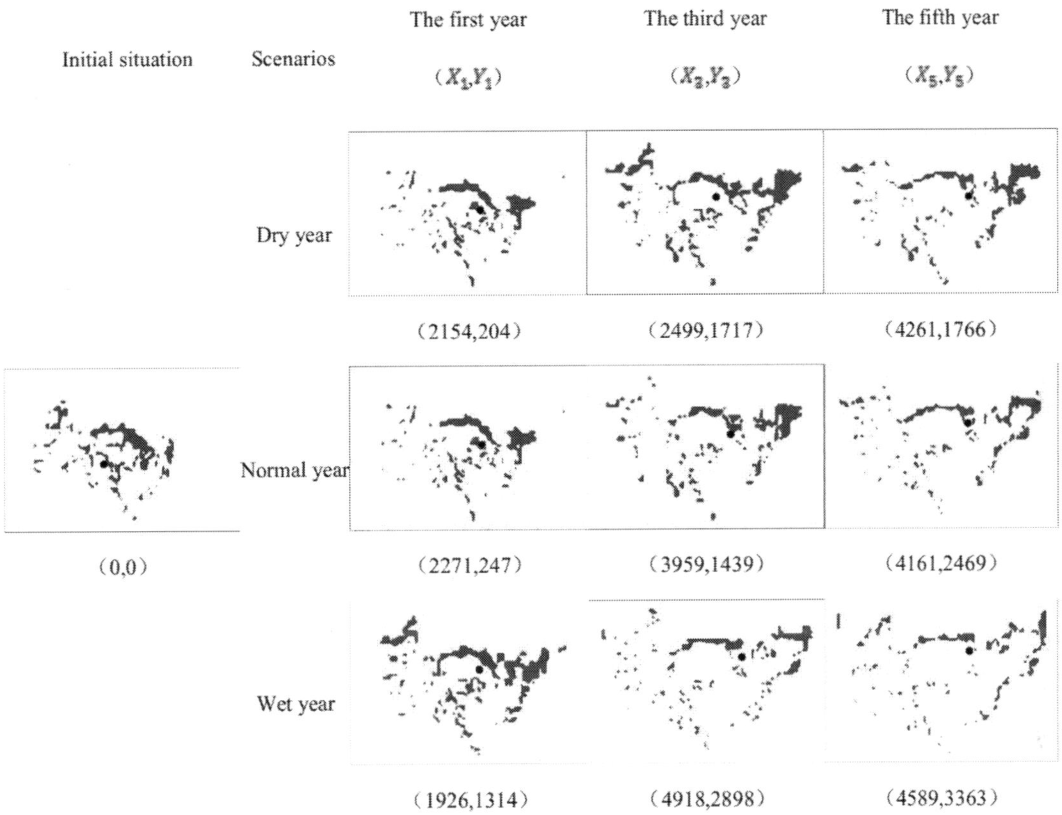

Figure 9. SGZ (blue area) and their centroids (black point).

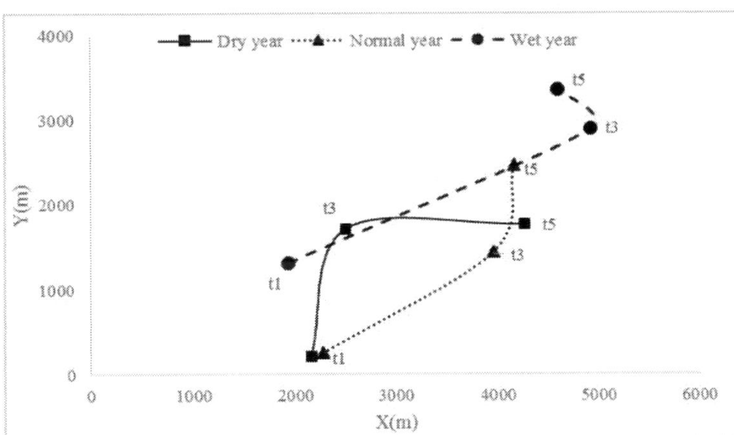

Figure 10. SGZ centroids moving trajectory in different scenarios.

4. Conclusions and Discussion

4.1. Conclusions

The impacts of water level fluctuations and water depth variation on the growth and distribution of macrophytes during and after water diversion are summarized below:

In dry, wet, and normal years, the water level climbed 14.8, 17.8, and 24.5 m, respectively, and the water surface area increased 64.27, 79.59, and 105.87 km2, respectively. The water level changes were between 11% and 18% and water surface area changes varied between 81% and 133%.

During water diversion, the annual and interannual water level fluctuations gradually decreased, but were still too significant for macrophytes to adapt. The original aquatic plants in the reservoir tend to die and disappear completely and the distribution of macrophytes tends to change. The distribution area of macrophytes moved toward the northeast and northwest overall, with a northeastward movement of its centroid. The movement trajectory and distance of the centroid varied between scenarios.

After water diversion, macrophytes living in the reservoir will experience significant changes in distribution area, species biomass, and species richness. In dry, normal, and wet years, centroid of SGZ had moved 4613 m, 4838 m, and 5689 m, respectively, toward the northeast. In dry year, the area of SGZ increased compared to that before water diversion, indicating that the macrophytes species diversity and richness can recover to the levels recorded before water diversion.

4.2. Discussion

If no manual interventions are implemented, it will take a long time for aquatic plant communities to recover naturally after being damaged. Given the importance of aquatic plants in an aquatic ecosystem and the slowness of their natural recovery, manual interventions should be implemented after water diversion to accelerate the recovery of the community structure of Miyun Reservoir and thereby maintain the stability of the aquatic ecosystem. Moreover, special attention should be paid to the problems arising from mass death of macrophytes, such as the drop in water quality caused by decomposition of dead plants and reduction in food supply for herbivorous fish.

In addition to water depth, growth of aquatic plants is also affected by other environmental factors (e.g. wind speed, air temperature, and light), water quality, feeding of herbivorous fish and so on. This study was based on the assumption that the zones at depths shallower than 6 m are suitable growth zones for aquatic plants. Further actual measurements are needed to verify the data about area of suitable growth zone, distribution area of macrophytes, and distribution of macrophyte growth area presented in the conclusions.

Acknowledgements

The authors thank the Beijing Municipal Science &Technology Commission (grant nos. Z151100002115055 and Z121100000312097) and Beijing Municipal Natural Science Foundation of China (grant nos. 8142021) for financial support.

References

[1] Rooney N and Kalff J 2000 Inter-annual variation in submerged macrophyte community

biomass and distribution: the influence of temperature and lake morphometry *Aquatic Botany* **68** 321-35

[2] Beklioglu M and Moss B 1996 Mesocosm experiments on the interaction of sediment influence, fish predation and aquatic plants with the structure of phytoplankton and zooplankton communities *Freshwater Biol.* **36** 315-325

[3] Spence D H N 1982 The zonation of plants in freshwater lake *Adv. Ecol. Res.* **12** 37-125

[4] Wang H Y, Chen J K and Zhou J 1999 Influence of water level gradient on plant growth reproduction and biomass allocation of wetland plant species *Acta Phytoecologica Sinica* **23** 269-74

[5] Andrea J H 2003 *Implementation of a GIS to assess the effects of water level fluctuations on the wetland complex at Long Point Ontario* Waterloo: University of Waterloo

[6] Gu X S , Hu H Y, Wen X H, Wang H and Zhong Y X 2011 *Biology for water and wastewater Treatment* 4nd ed (Beijing: China Architecture & Building Press) Chapter 9 pp 178-80

[7] Tenna R and Ian H 2002 Relationships between water level fluctuations and vegetation diversity in shallow water of New Zealand lakes *Aquatic Botany* **74** 133-48

[8] Coops H, Beklioglu M and Crisman T L 2003 The role of water-level fluctuations in shallow lake ecosystems – workshop conclusions *Hydrobiologia* **506** 23-27

[9] Liu Y, Gu H C, Wang F, Wang Z and Huang K 2006 Role of water level fluctuation on aquatic vegetation in lakes *Acta Ecologia Sinica* **26** 3117-226

[10] Wang X L and Bao Y H 1999 Study on the methods of land use dynamic change research *Progress in Geography* **18** 81-87

[11] Verburg P H, Schot P P, Dijst M J and Veldkamp A 2004 Land use change modelling, current practice and research priorities *Geo. Journal* **61** 309-24

[12] Arrhenius O 1921 Species and area *Journal of Ecology* **9** 95-99

[13] Tjørve E 2003 Shapes and functions of species–area curves: a review of possible models *Journal of Biogeography* **30** 827-35

[14] MacArthur R H and Wilson E O 1967 *Island Biogeography* Princeton University Press Princeton N J

IOP Conf. Series: Earth and Environmental Science **39** (2016) 012011 doi:10.1088/1755-1315/39/1/012011

Research on sudden environmental pollution public service platform construction based on WebGIS

T P Bi[1, 3], D Y Gao[1] and X Y Zhong[2]

[1]School of Management, Shenyang Jianzhu University, Shenyang 110168, China
[2]Urban-rurai planning and design institute of Guizhou, Guiyang550001, China

Email: btp23956710@163.com

Abstract. In order to actualize the social sharing and service of the emergency-response information for sudden pollution accidents, the public can share the risk source information service, dangerous goods control technology service and so on, The SQL Server and ArcSDE software are used to establish a spatial database to restore all kinds of information including risk sources, hazardous chemicals and handling methods in case of accidents. Combined with Chinese atmospheric environmental assessment standards, the SCREEN3 atmospheric dispersion model and one-dimensional liquid diffusion model are established to realize the query of related information and the display of the diffusion effect under B/S structure. Based on the WebGIS technology, C#.Net language is used to develop the sudden environmental pollution public service platform. As a result, the public service platform can make risk assessments and provide the best emergency processing services.

1. Introduction

In order to deal with accidental pollution, many countries have set up relevant public service websites, such as the chemical information net CSIN system developed by The United States National Environmental Protection Agency. For major accident consequence analysis, at present, simulation technologies are widely used, using mathematical models to analyse the results. Based on domestic and foreign advanced technologies, the author proposes research on sudden environmental pollution public service platform construction [1-2].

2. Technical route

Under the guidance of the idea and method of software engineering, we establish a public service platform based on WebGIS for accidental pollution. The research of WebGIS will combine with environmental risk source investigation, database, modeling and simulation technology. The technical route is shown in Figure 1. WebGIS is commonly used in domestic and overseas for performance of environmental information, and it is proven that the combination between them can complement each other.

Taking the accidental pollution actual location as standard, a mathematical model of simulation and indication is used to combine the GIS with the mathematical model solutions to perform analysis of pollution diffusion from two aspects of space and time and calculate the diffusion area of pollution in each period and analyze the pollution degree of different pollutants and the mass concentration of the pollutant itself to obtain some important information for accidental pollution, such as surrounding units, population and so on. Thereby, it can obtain influence and range on the surrounding environment of the accidental pollution and provide the corresponding emergency measures.

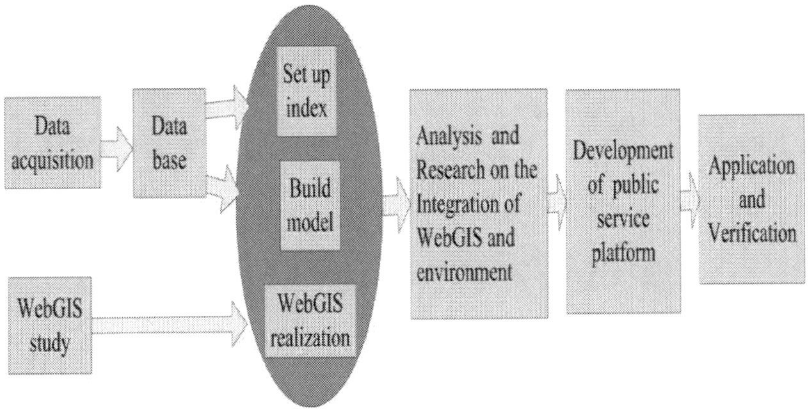

Figure 1. Technology roadmap.

3. System design

In combination with the actual work needs, the overall design of the system must meet the needs of users: easy to use, manage and maintain, and must become a powerful tool for them. On the basis of practicality, it also needs to have some advances that are the guarantee of future utility, advanced technology, method, equipment, etc, and it should be used as much as possible to improve the technical level of the system. The development of a public service platform is to realize social sharing and service of the information of sudden pollution accident emergency measures, and abide by the principles of opening and sharing. As a large number of industries and departments are involved in the implementation of the application of large-scale, critical information system applications, so safety and reliability is also essential. The realization of the system should be based on the function and performance index, and then reduce the cost as much as possible. At the same time, we should pay attention to the integrity of the system structure and technical standard, making it easy to carry out a wide range of promotion [3].

3.1. Architecture design

In order to minimize the load on the user terminals, the map processing and computing work should be done as a service, the system uses a B/S structure, the end-user access to the Web server in the browser is used to operate the Silverlight applications.

From the point of construction of the server, the system is distributed on three different logical servers, as shown in Figure 2. Completing a forecast requires three servers to participate: at first, the IIS providing the user interface, receiving the user's query and the forecast parameters of this stimulation, query control module and then tidy data to transfer the request to the computing service; secondly, computing service after the execution algorithm will return the result to the IIS service, but this result cannot be directly displayed on the interface; finally, the IIS service will find the result then deliver it to GIS service processing request, using the results of the calculation service layer, the IIS service will be used to display [4].

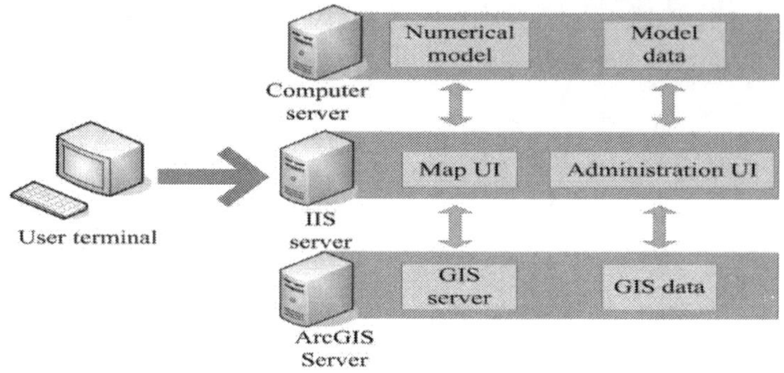

Figure 2. Information system service distribution.

3.2. Database design

The database includes four parts: atmospheric elements, surface water elements, ecological elements and others. Based on existing standards and requirements to organize and classify four types of elements of the relevant data and record it in table form. Atmospheric elements related to data include: pollution source information, air quality monitoring station, atmospheric environment function zone; surface water data elements include: surface water pollution source data, surface water monitoring sections data; data related to ecological factors include: schematic diagrams of the typical ecological protection measures layout, distribution of ecological monitoring, spatial distribution of the ecologically sensitive area, land used maps, project plan, project location map of the area, and the vegetation type map [5].

4. Mathematical model and effect

4.1. Liquid diffusion model

The model of the surface water uses the recommended model of the "technical guidelines for environmental impact assessment of water environment" (HJ/T2.3-1993) to simulate the distribution of pollutants. In the case of general rivers, the length, depth and width are smaller than real size, the pollutant is discharged into the river, and the water quality of the river is considered to be one of the one-dimensional water quality, which can be considered as one-dimensional water quality [6].

When the rivers flow slowly and the pollutant can move across the river fast, it only be used in the one-dimensional water quality model. According to the principle of the conservation of mass, the basic model of the one-dimensional water quality of rivers is derived form equation (1).

$$\frac{\partial c}{\partial t} + \mu_x \frac{\partial c}{\partial x} = D_X \frac{\partial^2 C}{\partial x^2} - K_1 C \qquad (1)$$

In the practical application of the model, according to different pollution sources, it can be divided into a stable release source and an instantaneous release source [7]. The initial conditions and boundary conditions are different, so the calculation model will have some changes. The output on the map is shown in figure 3.

4.2. Atmospheric dispersion model

According to the relevant provisions of the "environmental impact assessment technology guide" (HJ2.2-2008), the EPA provides a partial dispersion model for the standard of atmospheric environmental assessment in China. It recommends model SCREEN 3 for the simulation. The SCREEN 3 model can calculate the maximum ground concentration of the surface source pollution source, and the maximum ground concentration under spatial conditions such as buildings, smoke, and so on, which are embedded in a variety of meteorological combination conditions, including some of the most unfavorable weather conditions, while such weather conditions in a certain area are not necessary. The calculation of the surface source emissions of pollutants near the ground level can be carried out by equation(2).

$$x = \frac{Q_A K}{2\pi u_s} \int_x \frac{V \cdot D}{\sigma_x \sigma_y} \left(\int_y e^{-0.5\left(\frac{y}{\sigma_y}\right)^2} dy \right) dx \qquad (2)$$

The result of the combination of model SCREEN3 and SLAB are packaged to provide a graphical user's interface that facilitates the operator to input the relevant parameters, invoking the encapsulated model to stimulus after acquiring the users' parameter input. Finally, the system makes a visual output on the map, which is about the pollution source, pollutant information, weather conditions, terrain conditions, and so on, as shown in figure 4.

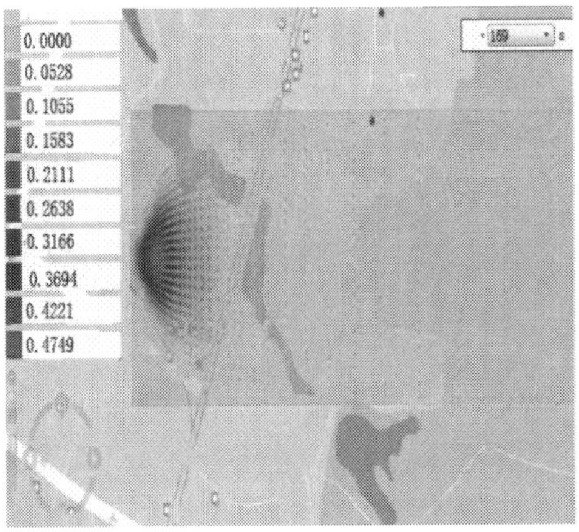

Figure 3. Effect of SCREEN3 diffusion.

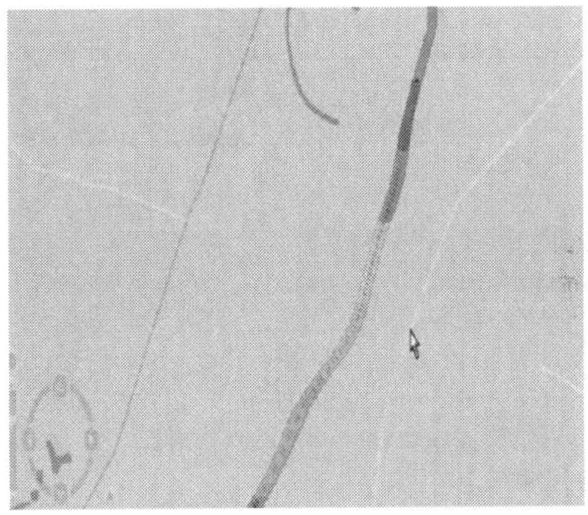

Figure 4. Effect of one-dimensional liquid diffusion.

5. Main functions and effect

To establish the space database of the risk source and dangerous goods, and the method database of the chemicals and emergency disposal, the risk source information of the space database and the more detailed items will be better. Particularly, the space distribution of the risk source should cover the entire study area. Regular storage and preservation of urban images, the risk sources and their related data storage and updating of dangerous goods is much better than ordinary file management, with its characteristics of timeliness, accuracy, and visibility, among others. Provide free information inquiry services to the public will improve the people's ability to deal with emergencies, and enhance the awareness of self-prevention.

The main function is to develop a public service platform for sudden environmental pollution on the Internet that can provide the public with environmental pollution sources, emergency disposal of dangerous goods, evacuation routes, relevant expert inquiry, air pollution diffusion simulations, water pollution diffusion simulations, pollution assessment and other environmental information integrated services, as shown in figure 5.

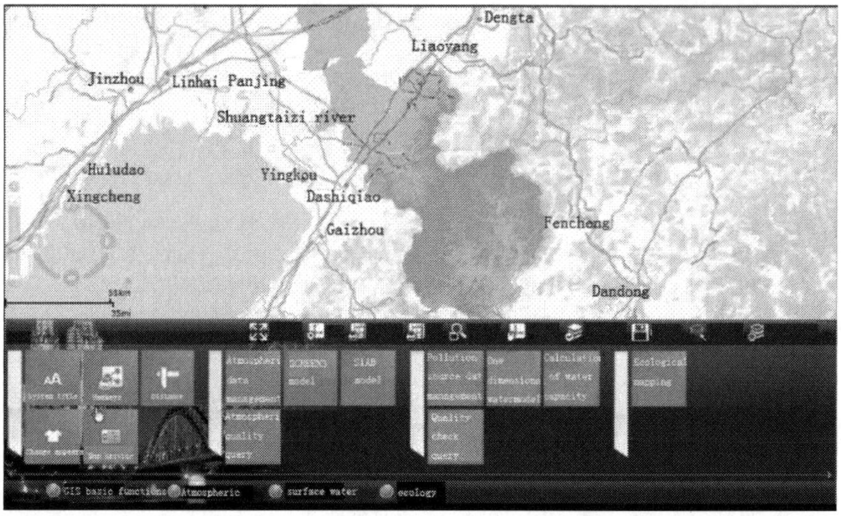

Figure 5. System main interface.

6. Conclusions

The public service platform is the "people's livelihood project", and the public can conveniently obtain the risk source's distribution and characteristics involving their own safety, and can carry out risk assessment simulation analysis. Therefore, it has strong social and environmental benefits. At the same time, the research results can be used in all sorts of cities, which has great economic value and application prospect.

Acknowledgements

The authors would like to acknowledge the support of the Shenyang Jianzhu University of China. This research was supported by the Liaoning Science public welfare research foundation under grant number (2014004010).

References

[1] Yin F 2005 *Theoretical and experimental study on the diffusion of air pollutants* (Shandong: Ocean university of China)
[2] Bi T P, Jin C Z, Zhong S J and Shang J H 2010 Northeastern University (Natural Science) **11** 56
[3] Ma Y J 2007 *ArcObjects based spatial data management platform* (Lanzhou:Lanzhou University)
[4] Erl T 2005 *Service-oriented architecture (SOA)* (London: University College London)
[5] Zhang W, Guo Y G and Wang J Z 2014 *Northern Environment* B **7** 156
[6] Wang Z F, Zhang S Y, Zhang H C, Zhao H and Ji Y Q 2009 *Environmental Engineering* B **6** 100
[7] Zhang Q, Xu Y X, Li H R and Ying S L 2012 *Yellow River* A **3** 100

Enhanced cometabolic degradation of methyl tert-butyl ether by a Pseudomonas sp. strain grown on n-pentane

S S Li[1, 2], S Wang[1] and W Yan[1]

[1]Department of Environmental Science & Engineering, Xi'an Jiaotong University, Xi'an, Shaanxi 710049, China

E-mail: shanshan0320@xjtu.edu.cn

Abstract. When methyl tert-butyl ether (MTBE) is added as oxygenates it increases the octane number and decreases the release of nitric oxide from the incomplete combustion of reformulated gasoline. The extensive use of MTBE allowed it to be detectable as a pollutant in both ground-level and underground water worldwide. The present study focuses on the isolation and characterization of MTB-degrading microorganisms by cometabolism based on the results of growth on different carbon sources. It also focuses on the kinetic analysis and the continuous degradation of MTBE. A bacterial strain WL1 that can grow on both n-alkanes (C5-C8) and aromatics was isolated and named *Pseudomonas sp.* WL1 according to the 16S rDNA sequencing analysis. Strain WL1 could cometabolically degrade MTBE in the presence of n-alkanes with a desirable degradation rate. Diverse n-alkanes with different lengths of carbon chains showed significant influence on the degradation rate of MTBE and accumulation of tert-butyl alcohol (TBA). When strain WL1 cometabolically degraded MTBE in the presence of n-pentane, higher MTBE-degrading rate and lower TBA-accumulation were observed (Vmax = 38.1 nmol/min/mgprotei, Ks = 6.8 mmol/L). In the continuous degrading experiment, the removal efficiency of MTBE by *Pseudomonas sp.* WL1 did not show any obvious decrease after five subsequent additions.

1. Introduction

Cometabolism is defined as the capability of microorganisms to transform the non-growth substrate in the presence of a growth substrate or another biodegradable substrate [1]. It helps amplify the range of biodegrading substrates for microorganisms. Until now, our knowledge about the complete mechanism of cometabolism has been limited. Some studies suggest that microorganisms utilize the primary substrate (growth-supporting substrate) for cell growth, and that they also induce enzymes for the transformation of the growth substrate. In turn, this can also catalyze the conversion of the non-growth substrate [2]. Cometabolism has been confirmed in the elimination of alkane, aromatics, and chlorinated compounds in nature [3, 4], including MTBE.

MTBE is the most frequently used gasoline additive because of its favorable characteristics, such as low production cost, a high octane rating, and the ability to mix with other gas components at any ratio. With the widespread increase in reports of MTBE contamination in underground water, great concern has been raised about its elimination [5]. Due to the presence of a stable ether bond and sterically hindered tertiary carbon in its molecular structure, MTBE is relatively recalcitrant to microbial degradation [6]. It has been reported that both direct-metabolism and cometabolism could potentially degrade MTBE. Only a few microorganisms can grow with MTBE as the sole carbon source, including *Methylibium petrol EIP hilum* PM1 [7], *Aquincola tertiaricarbonis* L108 [8], and

Hydrogenophaga flava ENV735 [9]. Most MTBE-degrading microorganisms transform MTBE via cometabolism. Various compounds present in the gasoline are able to use the cometabolic growth substrates for MTBE degradation (some of these are normal and branched alkanes [6, 10, 11], and aromatics [12]). Smith et al. reported that *Pseudomonas mendocina* KR-1 cometabolically degraded MTBE with an average degradation rate of 61.1 $nmol/min/mg_{protein}$, when grown on *n*-alkanes (C_5-C_8). Although MTBE was consumed, stoichiometric accumulation of *tert*-butyl alcohol (TBA) was observed [13]. Two MTBE-degrading strains, *Pseudomonas aeruginosa* BM-B-450 and *Pseudomonas citronellolis* BM-B-447, were isolated from *n*-pentane-adapted consortium with different kinetic parameters for MTBE, which suggested the cooperative degradation in the consortium [6].

One accepted hypothesis of metabolic pathway is that MTBE integrates with one oxygen atom with the help of monooxygenase and produces tert-butyl formate (TBF), which is subsequently hydrolyzed to TBA. Non-specific monooxygenase enzymes are responsible for the initial oxidation of MTBE, including cytochrome P450 monooxygenases [14, 15] and alkane hydroxylase [16, 17]. Furthermore, alkane hydroxylases and their close homologs appear to be widely distributed, especially among gram-negative organisms that utilize n-alkanes as growth substrates [18]. This may help interpret why alkanes are the preferred cometabolic substrates for the degradation of MTBE. Smith et al. found that an alkane hydroxylase expressed by n-octane grown P. putida GPo1 was responsible for MTBE oxidation [16].

The growing reports of the worldwide contamination of aquifer by MTBE have raised concerns regarding the development of effective elimination techniques of MTBE. Among these, biological treatments are recognized as cost-effective and environmentally friendly options. Therefore, the present study's objective was to isolate the single strain that could degrade MTBE cometabolically from the consortium grown on n-octane. From there, we studied the cometabolic characterizations of this strain, including the range of the cometabolic substrate, the kinetic characterization, and the continuous degradation of MTBE. Though there were a few reports on the cometabolism of MTBE with single Pseudomonas strains grown on either n-alkanes or aromatics, the single strain isolated in this study could grow on both n-alkanes (C5-C8) and aromatics (toluene and xylene). The amplified range of growth substrates of the isolated strain will undoubtedly increase biomass production and improve its adoption in a further application. These results provide a basis for further efforts to eliminate MTBE from natural environments.

2. Materials and methods

2.1. *Strains and media*

Microbial consortium RS was cultivated in salty mineral medium (MSM) with n-octane and MTBE [19]. The MSM [20] for MTBE degradation was designed as follows (g/L): KH_2PO_4 0.9, $Na_2HPO_4 \cdot 12H_2O$ 6.5, (NH4)2SO₄ 0.4, $MgSO_4 \cdot 7H_2O$ 0.2, $CaCl_2 \cdot 2H_2O$ 0.01, $FeSO_4 \cdot 7H_2O$ 0.001, and 1mL original stock solution containing the following trace elements in 1 L of deionized water: 0.1 g of H_3BO_3, 0.4 g of $CoCl \cdot 6H_2O$, 0.25 g of $ZnCl_2$, 1 g of $MnSO_4 \cdot H_2O$, 0.25 g of $Na_2MoO_4 \cdot 2H_2O$, 0.1 g of $NiSO_4 \cdot 6H_2O$, and 0.25 g of $CuCl_2 \cdot 2H_2O$. The pH of the medium was adjusted to 7.0 and the medium was then autoclaved at 115 °C for 15 min. The single strain used in this study was isolated from mixed culture RS [19]. The whole-grown mixed RS culture was diluted and smeared on MSM (n-octane + MTBE) agar plates for screening single colonies. Various individual colonies were first cultivated in LB broth overnight and then cultivated for a second analysis. Cell growth was measured by monitoring the optical density of the culture at 600 nm (OD600) in a Model 680 Microplate Reader (BioRad, CA, USA).

2.2. *Chemicals*

Xi'an Standard Gas Station supplied all gaseous alkanes (methane, ethane, propane, and n-butane). N-pentane (99% purity), n-hexane (99% purity), n-heptane (99% purity), n-octane (99% purity), MTBE

(99% purity) and TBA (99% purity) were obtained from Sinopharm Chemical Reagent Co., Ltd (Shanghai, China).

2.3. Identification of the isolated strain

Genomic DNA of a single, isolated strain was extracted with the Bacterial Genomic DNA Extraction Kit (Generay Biotech, Shanghai, China) per the manufacturer's instructions. Fragments of 16S rDNA were amplified from the DNA extraction with universal primers 27F (5'-AGAGTTTGATCCTGGCTCAG-3') and 1492R (5'-GGTTACCTTGTTACGACTT-3'). The amplification reaction mixtures (50 μL) contained 1μL of template DNA, 1.5 U of Takara TaqTM polymerase (Takara Biotech, Dalian, China), 5 μL of 10× PCR buffer, 2μL each of 10 μM forward and reverse primer, and 5μL of 2.5mM stock dNTP mixture. Amplifications were carried out on Long Gene® Thermal Cycler A300 (Long Gene, Hangzhou, China) under the following conditions: 94 °C for 5 min, then 40 cycles of 94 °C for 1 min, 55 °C for 1 min, and 72 °C for 1.5 min, followed by a final extension at 72 °C for 7 min. The PCR product after purification was cloned onto a TA cloning vector pMD-18T (Takara Biotech, Dalian, China), and sent to Sangon (Sangon Biotech, Shanghai, China) for Sanger sequencing.

Phylogenetic analysis of the 16S rDNA sequence was performed on MEGA 5.1 software [21]. Alignments with the Gen Bank database were generated based on the Clustal method [22]. A phylogenetic tree was constructed via a neighbor-joining method.

2.4. The cometabolic culture conditions of the microorganisms

All the cometabolic MTBE-degrading experiments were conducted in triplicate in 125 mL serum bottles sealed with Teflon Mininert valves with 20 mL of MSM. The carbon source was added to the bottles with syringes at a concentration of 50 mg/L in the presence or absence of MTBE (10 mg/L). The bottles were incubated at 30°C with shaking at 150 rpm. Cell growth was measured by monitoring the optical density of the culture at 600 nm (OD600).

The continuous degrading experiments were performed as described above. Strain WL1 was cultivated in MSM with n-pentane (50 mg/L) and MTBE (10 mg/L) as the substrates. The n-pentane and MTBE were spiked at the time their concentrations were below 5% of initial levels. This addition was conducted four subsequent times.

2.5. Kinetic determination

For the kinetic experiments of the isolated strain, new cells were harvested from the culture with various initial concentrations of MTBE. Cell samples were heated to 65 °C in 3 M sodium hydroxide for 30 minutes. Then, the insoluble material was centrifuged in a Thermo Scientific Sorvall ST 40 Centrifuge before determining its protein content by Folin phenol method. Bovine serum albumin (BSA) was used as a standard. The rate of MTBE oxidation was derived from the final TBA concentration, assuming that the rate of oxidation of MTBE remained constant for the first 30 minutes and that no further oxidation of TBA occurred during the reaction period. The kinetic constants were derived via computer-fitting of the data by nonlinear regression to a single substrate-binding model $y = Vmax \cdot [x/(Ks + x)]$ with Origin version 6.1.

2.6. Analytical method

MTBE and TBA concentrations were analyzed by a headspace solid-phase dynamic extraction-gas chromatography-mass spectrometry (HS-SPDE-GC/MS) method. Liquid culture was collected immediately for HS-SPDE analysis on a CTC CombiPAL-xt auto sampler (Chromtech, Idstein, Germany) equipped with a PDMS/AC syringe (90% polydimethylsiloxane and 10% activated carbon coated) as the aforementioned method [19]. The sample absorbed on the syringe was automatically injected into a Trace GC ULTRA (Thermo Finnigan, Milano, Italy) gas chromatograph fitted with an HP-5MS capillary column (30 m, 0.32 mm ID, 1 μm film) and a TraceISQ (Thermo Finnigan, TX, USA) mass spectrometric detector. The injection was pulsed split less for 1 minute. The oven temperature was maintained at 40°C for 3 minutes and ramped to 120°C at 10°C/min. The flow rate of carrier gas (Helium 5.0) was 1.0 mL/min. The interface and ion source temperatures remained at 280 °C and 230 °C, respectively. The mass spectrometry was operated in electron impact mode at 70 eV in selected ion monitoring (SIM) mode at m/z 73 and 53 for MTBE and TBA, respectively. The MTBE and TBA were quantified using standards of known concentrations [19].

3. Results and discussion

3.1. The isolation and identification of MTBE-cometabolic-degrading strain

One of the key factors that constrain the development of biodegradation of MTBE is the extremely low biomass of microorganisms grown on MTBE. The potential alternative is cometabolism, confirmed during the process of MTBE elimination. Cometabolism describes the transformation of a non-growth substrate when the microbes are cultivated with another compound as a growth substrate. Four single colonies were isolated from a mixed culture with n-octane as the sole carbon source. To determine the single strain with higher biomass production on optimum growth substrate, n-alkanes, and aromatics were tested for their capability to support the growth of the isolated strains. As shown in Table 1, all four strains could grow on the aromatic compounds tested. However, all the isolated strains were unable to use any gaseous n-alkanes (C1 to C4) but grew well on the liquid n-alkanes tested (C5 to C8) with a much higher biomass. Among the four single strains, strain WL1 showed the maximum biomass concentration cultivated with n-alkanes. Furthermore, the final biomass production of WL1 cultivated with different n-alkanes decreased with the length of carbon chain of n-alkanes. MTBE or TBA could not be utilized as the growth substrate of *Pseudomonas sp*. WL1.

Table 1. The cell growth of the four different single strains on potential growth substrates.

Potential growth subatrates[a]	OD_{600} for 5 days cultivation[b]			
	WL1	WL3	WL4	WL6
n-Alkanes				
Methane	<0.01	<0.01	<0.01	<0.01
Ethane	<0.01	<0.01	<0.01	<0.01
Propane	<0.01	<0.01	<0.01	<0.01
n-Butane	<0.01	<0.01	<0.01	<0.01
n-Pentane	0.285 ± 0.013	0.132 ± 0.019	0.192 ± 0.018	0.086 ± 0.003
n-Hexane	0.238 ± 0.010	0.196 ± 0.033	0.227 ± 0.024	0.124 ± 0.007
n-Heptane	0.211 ± 0.013	0.123 ± 0.006	0.175 ± 0.016	0.136 ± 0.018
n-Octane	0.175 ± 0.026	0.145 ± 0.013	0.184 ± 0.026	0.179 ± 0.021
Aromatics				
Benzene	0.022 ± 0.008	0.045 ± 0.004	0.062 ± 0.006	0.086 ± 0.011
Toluene	0.074 ± 0.013	0.031 ± 0.008	0.095 ± 0.004	0.075 ± 0.017
o-Xylene	0.094 ± 0.006	0.051 ± 0.006	0.033 ± 0.007	0.096 ± 0.006
p-Xylene	0.088 ± 0.018	0.062 ± 0.016	0.038 ± 0.009	0.114 ± 0.021
m-Xylene	0.082 ± 0.003	0.057 ± 0.012	0.045 ± 0.003	0.089 ± 0.014
MTBE and its metabolites				
MTBE	<0.01	0.012 ± 0.005	<0.01	<0.01
TBA	<0.01	<0.01	<0.01	<0.01

[a] Triplicate cultures of mixed ERS were grown for 5 days in the presence of each substrate at an initial substrate concentration of 0.02% vol/vol (the initial value of OD_{600} was 0.01).
[b] All optical densities reported are the means ± SD.

Finally, a pure strain *Pseudomonas sp*. WL1, characterized as being gram-positive, short, and rod-shaped, was selected for further study. The 16S rDNA analysis revealed that strain WL1 belongs to the Pseudomonas genus. *Pseudomonas plecoglossicida* NBRC 103162 was found to be phylogenetically closest to the isolated strain WL1, showing 99% similarity between the nucleotides of the 16S rDNA genes (figure 1). The 16S rDNA sequence of the isolated WL1 strain was stored in NCBI GenBank under the accession number of KC217655.

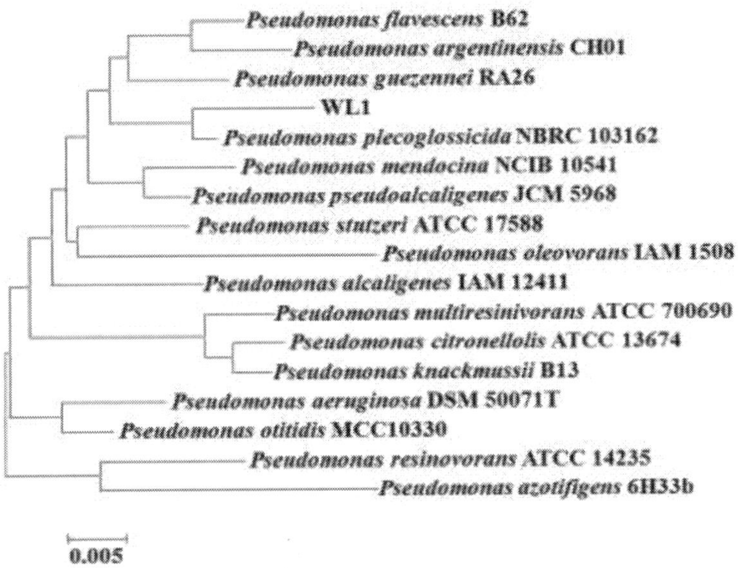

Figure 1. Phylogenetic tree of *Pseudomonas* sp. WL1 based on 16S rDNA sequences.

The percentage of replicate trees in which the associated taxa clustered together in the bootstrap test (1000 replicates) is shown next to the branches. The tree is drawn to scale, with branch lengths in the same units as those of the evolutionary distances used to infer the phylogenetic tree. All positions containing gaps and/or missing data were eliminated from the dataset (complete deletion option).

Based on the reports of previous studies, Pseudomonas genus was found in numerous environmental niches and characterized as capable of degrading various pollutants, such as polycyclic aromatic hydrocarbon (PAH) [23, 24], benzene [25], and alcohols [26, 27]. Additionally, the higher the biomass production, the lower the nutritional demand of Pseudomonas genus strains to make them meet the requirements of industrial biotechnology [28]. Furthermore, lots of typical strains from the Pseudomonas genus have been sequenced for whole-genome, including P. aeruginosin PAO1 [29], P. putida KT2440 [30] and P. mendocina NBRC 14162 [31]. The thick substrates scope and easy accessibility of genetic bioinformations make Pseudomonas genus strains ideal for further use in biotechnology and bioremediation.

3.2. Cometabolic degradation of MTBE with different n-alkanes as growth substrate

Cometabolism is defined as a microorganism's capability to transform any kind of non-growth substrate in the obligatory presence of a growth substrate [2]. Cometabolism has been confirmed in the degrading processes of many pollutants, such as chlorinated compounds [32, 33], aromatic hydrocarbons [34, 35], and gasoline additives [36], including MTBE [10, 14].

Various studies have identified that microorganisms can cometabolically degrade MTBE efficiently when cultured with a variety of growth substrates, such as alkanes, aromatics, and alicyclics [10, 37]. Though strain WL1 was isolated from a mixed culture inoculated with n-octane as the sole carbon source, other n-alkanes and aromatics were tested for their capability to support the cometabolical degradation of MTBE (table 2). For the aromatic compounds tested, *Pseudomonas sp.* WL1 was able to grow on toluene or xylene (low biomass yield), accompanied by the slow degradation of MTBE. All the n-alkanes tested in this study (C5-C8) could support the growth of *Pseudomonas sp.* WL1 with a much higher degradation rate than that of aromatic compounds. Further, as the carbon number of the n-alkanes increased, both the degradation rate of n-alkanes and MTBE (from 178.4

mgMTBE/gprotein/h to 59.3 mgMTBE/gprotein/h) decreased. After two days' cultivation, low levels of TBA accumulation (less than 7%) were observed for all the n-alkanes tested. The simplest interpretation of this phenomenon is that strain WL1 could synthesize the enzyme that can catalyze the oxidation of TBA. This result is similar to the result reported by Morales et al. [6]. Pseudomonas aeruginosin showed no TBA accumulation when the initial ratio of MTBE and n-pentane was below 0.7, indicating complete MTBE mineralization. In the reported MTBE-degrading-microorganisms by cometabolism, the oxidation of MTBE and TBA could be catalyzed by the same monooxygenase [38] or different ones [16]. These results were similar to the reports for MTBE cometabolism by Garnier et al. [39]. All n-alkanes (n-pentane, n-hexane, and n-heptane) tested for cometabolic activity did support the degradation of MTBE. Among them, n-pentane was the most efficient with a deterioration rate of (200 μg/d). In another study, the cometabolic degradation rate of MTBE increased with the carbon number of the n-alkanes used as the growth substrate by P. mendocina KR-1 [13]. Although MTBE was transformed at a high cometabolic degrading rate by P. mendocina KR-1 (61.1 nmol/min/mgprotein), the corresponding accumulation of TBA was observed for all n-alkanes tested (C5 to C8). This indicated the absence of the enzyme catalyzed in the transformation of TBA in strain KR-1.

A cometabolic coefficient (CC) was used to represent cometabolic efficiency. The CC value is defined as the ratio of degraded MTBE to the consumed cometabolic substrate [6]. The lowest and highest CC values of 0.03 and 0.72 for *Pseudomonas sp.* WL1 were obtained from the growth substrate of toluene and n-pentane, respectively (table 2). A lower CC value for aromatic substrates was compatible with a lower degradation rate of MTBE with different alkanes. When cultivated with n-alkanes, the CC value by *Pseudomonas sp.* WL1 decreased in correlation to an increase in the growth substrate's carbon number. The shorter-chain alkane (more soluble than the longer one) could be transferred to the microorganisms more quickly; this may be the potential reason why the CC value obtained from shorter-chain alkane was higher than that from longer-chain alkane.

Table 2. Specific degradation rates and products by *Pseudomonas* sp. WL1 on diverse substrates.

Carbon source[a]	Biomass ($mg_{protein}$ /d)	Degradation rate of substrates[b] ($mg_{substrate}/g_{protein}/h$)	Degradation rate of MTBE[b] ($mg_{MTBE}/g_{protein}/h$)	CC ($mg_{MTBE}/mg_{substrate}$)	Residual TBA[c] (mg/L)
Aromatic substrate					
Benzene	0.3 ± 0.1	3.6 ± 0.2	0	-	-
Toluene	8.3 ± 0.7	247.3 ± 7.9	4.6 ± 0.1	0.03	0.12
Xylene	9.5 ± 0.5	435.8 ± 10.5	12.3 ± 0.9	0.11	1.34
n-Alkanes					
n-Pentane	15.2 ± 1.1	1393.2 ± 78.1	178.4 ± 8.4	0.72 ± 0.2	0.14 ± 0.1
n-Hexane	14.6 ± 1.4	1096.7 ± 56.4	130.2 ± 9.2	0.56 ± 0.1	0.56 ± 0.1
n-Heptane	12.2 ± 0.4	841.7 ± 53.9	72.3 ± 7.1	0.44 ± 0.1	0.67 ± 0.1
n-Octane	10.9 ± 0.8	681.6 ± 45.8	59.3 ± 6.9	0.26 ± 0.1	0.26 ± 0.1
Others					
MTBE	-	-	-	-	9.5 ± 0.2
TBA	-	-	-	-	9.8 ± 0.3

[a] Microbial consortium ERS were cultivated in glass serum vials (125 mL) contained mineral salt medium (20 mL) on diverse carbon source (0.02%, w/v%) in the presence (20 μmol) of MTBE. After 2 days' growth, the concentrations of biomass were examined. All the cultures on diverse carbon sources were performed in triplicate. All the values were the means of triplicate ± SD.

[b] After 2 days cultivation, the amounts of residual carbon source and MTBE in each culture were determined by SPDE-GC-MS. The degradation rates of growth substrate and MTBE were calculated, respectively.

[c] After 2 days cultivation, the amount of TBA produced in each culture was determined by SPDE-GC-MS.

3.3. The kinetic characteristics of cometabolism of MTBE by Pseudomonas sp. WL1 on n-alkanes

The oxidation of TBA, the first stable metabolite detected in MTBE degradation, is the limited step of the mineralization of MTBE. The apparent TBA accumulation during the first period of MTBE oxidation by n-alkane-grown cells indicated that the kinetics of MTBE oxidation could be detected directly by determining the rates of accumulation of TBA. Values of Ks and Vmax for MTBE oxidation were determined for pure culture WL1 grown on n-pentane and n-octane, respectively (figure 2). When cultivated with the equivalent concentration of MTBE and n-alkane with different carbon numbers (C5 and C8) under the same conditions, the pure culture of *Pseudomonas sp*. WL1 showed different kinetic parameters. For example, n-pentane-oxidizing WL1 cells exhibited a Vmax value of 38.1 nmol/min/mgprotein and Ks value of 6.8 mmol/L for MTBE, whereas n-octane-grown WL1 cells had a Vmax value of 19.6 nmol/min/mgprotein and a Ks value of 17.2 mmol/L. The potential interpretation for the variable kinetic parameters of WL1 culture for MTBE on different n-alkanes would be the water solubility of each substrate (C5>C8). The Henry's Law constant of n-pentane and n-octane was 8.1×10-4 M/atm and 3.14×10-4 M/atm (at 25 °C), respectively [40]. When incubated with the same initial concentration of n-pentane or n-octane, the higher dissolved quantity of n-pentane facilitated its capture by the WL1 cells and then induced the transcription of alkane monooxygenase, improving the degradation of n-pentane and MTBE.

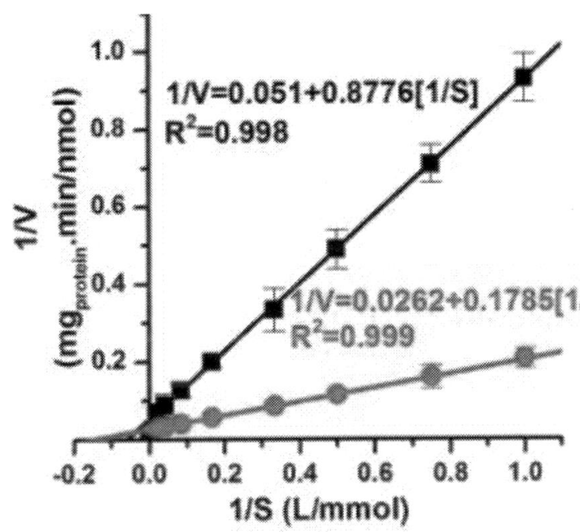

Figure 2. Kinetics of MTBE oxidation by pure strain.
WL1 with *n*-pentane and *n*-octane as a substrate cometabolic.

Values represent means of triplicate samples, and error bars represent the standard deviation of the average. The figure shows the average rate of MTBE oxidation by strain WL1 on n-pentane (filled circles) or n-octane (filled squares).

Another of the study's key findings was that the Ks value for MTBE of strain WL1 was much higher than the values of cometabolically MTBE-degrading microorganisms on the growth substrate of n-alkanes. For instance, the Ks value for MTBE obtained from Arthrobacter sp. ATCC27778 [41] on butane was 24 μmol/L, and another n-alkanes-oxidizing P. mendocina KR-1 had a Ks value of 12.95 μmol/L for MTBE [13]. The clearest interpretation of this result is that variable alkane monooxygenases responsible for the transformation of MTBE are involved in these strains. Particular affinity (a°MTBE = Vmax/Ks) is generally used as an index reflecting substrate specificity [6]. The value for Pseudomonas sp. WL1 (a°MTBE = 0.34 L/gprotein/h) on n-pentane was almost four times higher than the value on n-octane (a°MTBE = 0.07 L/gprotein/h), mainly due to the significant difference in Ks values. Similar a°MTBE values were obtained from P. mendocina KR-1 (0.28 L/gprotein/h), which also has higher Ks values [13]. Though higher a°MTBE values were calculated for Xanthobacter sp. (1.27 L/gprotein/h) [42] and M. vaccae JOB5 (1.08 L/gprotein/h) [18], TBA accumulation was observed as occurring hand in hand with the MTBE degradation process.

3.4. Degradation of MTBE by Pseudomonas sp. WL1 grown on n-pentane in continuous culture

Further research on the cometabolic degradation of MTBE with *n*-pentane as the growth substrate was conducted to determine the continuous degrading ability of strain WL1. MTBE (10 mg/L) and *n*-pentane (50 mg/L) were added when the residual concentrations of both MTBE and *n*-pentane in the cultures fell below 5% of their initial concentrations. Figure 3 below shows the degradation curves of MTBE and *n*-pentane after five subsequent feeding and degrading cycles. Pure strain WL1 can cometabolically degrade MTBE (10 mg/L) at a rate of 45.3-52.5 $mg_{MTBE}/g_{protein}/h$ with *n*-pentane (50 mg/L) as the growth substrate. A distinct lag phase was observed in the MTBE-degradation curve (in comparison with the *n*-pentane-degradation curve), where *n*-pentane transformed itself immediately once it was added. The first degradation cycle of MTBE and *n*-pentane was completed in two days. After additional, *n*-pentane degraded continuously, the removal of MTBE followed. We observed that the degradation rate of MTBE did not decrease after the fifth (and final) addition.

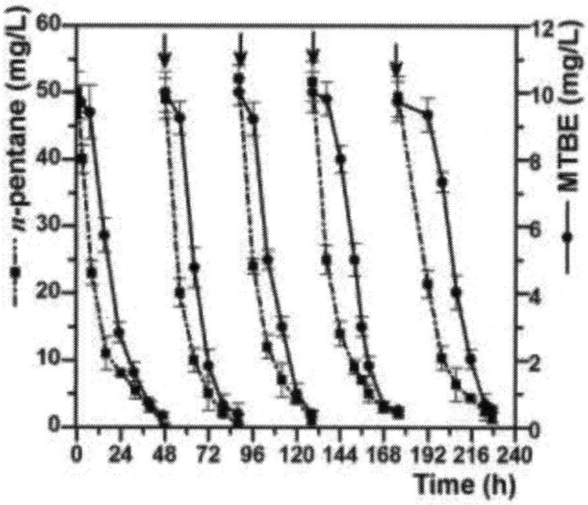

Figure 3. Cometabolic degradation of MTBE with n-pentane as growth substrate. Vertical arrows indicate the additions of MTBE and n-pentane. Values represent means of triplicate samples, and error bars represent standard deviation of the mean.

These results demonstrate that *Pseudomonas sp.* WL1 had the cometabolic capability for the continuous degradation of MTBE when cultivated with n-pentane as the growth substrate. As reported by Fortin et al., a mixed culture F-consortium can degrade MTBE in the presence or after incubation of diethyl ether (DiEE) after two respikes. MTBE would only be transformed after the complete degradation of the cometabolic carbon source, indicating that MTBE biodegradation undergoes some inhibition [43]. Volpe *et al.* inoculated a microbial consortium in a batch reactor wby repeatedly feeding it with MTBE and oxygen. After several instances of feeding and depletion, a significant decrease of the timeframe of complete degradation of MTBE was observed [37].

MTBE has most often been detected in fields of gasoline-contaminated soil and/or in groundwater. This detection is often accompanied with the detection of other gasoline components, such as alkanes. This confirms the growth substrates of MTBE-degrading microorganisms. Various studies have been conducted to solve the problem of MTBE elimination. Both n-alkanes (C5-C8) and aromatics (toluene and xylene) could be used as the growth substrates for MTBE degradation by the *Pseudomonas sp.* WL1 that was isolated in this study. This finding broadens the range of cometabolites for MTBE in a single strain. No visible TBA accumulation was observed during the process of MTBE degradation in the presence of n-pentane by WL1, suggesting complete MTBE mineralization. In the continuous degrading experiment, the removal efficiency of MTBE by *Pseudomonas sp.* WL1 did not show a noticeable decrease after five respikes. The present work seeks to provide a microbial source and theoretical groundwork for further bioremediation of MTBE-contaminated aquifers, a topic of real environmental significance.

Acknowledgements
This work received the support of the National Natural Science Foundation of China (Grant NO. 31300438). In addition, our project was funded by the China Postdoctoral Science Foundation (NO. 2014M552456) and the Shaanxi Province Postdoctoral Science Foundation. We extend our gratitude to these three organizations for their generosity.

References

[1] Tran N H, Urase T, Ngo H H, Hu J Y and Ong S L 2013 Insight into metabolic and cometabolic activities of autotrophic and heterotrophic microorganisms in the biodegradation of emerging trace organic contaminants *Bioresource Technol* **146** 721-31

[2] Arp D J, Yeager C M and Hyman M R 2001 Molecular and cellular fundamentals of aerobic cometabolism of trichloroethylene *Biodegradation* **12** 81-103

[3] Criddle C S 1993 The kinetics of cometabolism *Biotechnology and bioengineering* **41** 1048-56

[4] Horvath R S 1972 Microbial co-metabolism and the degradation of organic compounds in nature *Bacteriological Reviews* **36** 146-55

[5] Johnson R, Pankow J, Bender D, Price C and Zogorski J 2000 Peer reviewed: MTBE-To what extent will past releases contaminate community water supply wells? *Environmental Science & Technology* **34** 210A-7A

[6] Morales M, Nava V, Velasquez E, Razo-Flores E and Revah S 2009 Mineralization of methyl tert-butyl ether and other gasoline oxygenates by Pseudomonads using short n-alkanes as growth source *Biodegradation* **20** 271-80

[7] Kane S R, Chakicherla A Y, Chain P S, Schmidt R, Shin M W, Legler T C, Scow K M, Larimer F W, Lucas S M, Richardson P M and Hristova K R 2007 Whole-genome analysis of the methyl tert-butyl ether-degrading beta-proteobacterium Methylibium petroleiphilum PM1 *J. Bacteriol* **189** 1931-45

[8] Rosell M, Barcelo D, Rohwerder T, Breuer U, Gehre M and Richnow H H 2007 Variations in 13C/12C and D/H enrichment factors of aerobic bacterial fuel oxygenate degradation *Environmental Science & Technology* **41** 2036-43

[9] Streger S H, Vainberg S, Dong H and Hatzinger P B 2002 Enhancing transport of hydrogenophaga flava ENV735 for bioaugmentation of aquifers contaminated with methyl tert-butyl ether *Appl. Environ. Microbiol* **68** 5571-9

[10] Salazar M, Morales M and Revah S 2012 Biodegradation of methyl tert-butyl ether by cometabolism with hexane in biofilters inoculated with Pseudomonas aeruginosa *Journal of Environmental Science and Health. Part A, Toxic/hazardous Substances & Environmental Engineering* **47** 1017-26

[11] Steffan R J V S, Condee C W, McClay K and Hatzinger P B 2000 *Biotreatment of MTBE with a new bacterial isolate*

[12] Steffan R J, McClay K, Vainberg S, Condee C W and Zhang D L 1997 Biodegradation of the gasoline oxygenates methyl tert-butyl ether, ethyl tert-butyl ether, and tert-amyl methyl ether by propane-oxidizing bacteria *Appl. Environ. Microb.* **63** 4216-22

[13] Smith C A, O'Reilly K T and Hyman M R 2003 Cometabolism of methyl tertiary butyl ether and gaseous n-alkanes by Pseudomonas mendocina KR-1 grown on C5 to C8 n-alkanes *Appl. Environ. Microbiol.* **69** 7385-94

[14] Malandain C, Fayolle-Guichard F and Vogel T M 2010 Cytochromes P450-mediated degradation of fuel oxygenates by environmental isolates *FEMS Microbiology Ecology* **72** 289-96

[15] Chauvaux S, Chevalier F, Le Dantec C, Fayolle F, Miras I, Kunst F and Beguin P 2001 Cloning of a genetically unstable cytochrome P-450 gene cluster involved in degradation of the

pollutant ethyl tert-butyl ether by Rhodococcus ruber *J. Bacteriol* **183** 6551-57

[16] Smith C A and Hyman M R 2004 Oxidation of methyl tert-butyl ether by alkane hydroxylase in dicyclopropylketone-induced and n-octane-grown Pseudomonas putida GPo1 *Appl. Environ. Microbiol* **70** 4544-50

[17] Johnson E L and Hyman M R 2006 Propane and n-butane oxidation by Pseudomonas putida GPo1 *Appl. Environ. Microb.* **72** 950-52

[18] Nava V, Morales M and Revah S 2007 Cometabolism of methyl tert-butyl ether (MTBE) with alkanes *Reviews in Environmental Science and Biotechnology* **6** 339-52

[19] Li S, Li D and Yan W 2015 Cometabolism of methyl tert-butyl ether by a new microbial consortium ERS *Environmental Science and Pollution Research International* **22** 10196-205

[20] Li S, Zhao H, Li Y, Niu S and Cai B 2013 Complete nucleotide sequence of plasmid pND6-2 from Pseudomonas putida ND6 and characterization of conjugative genes *Gene* **512** 148-56

[21] Kumar S, Nei M, Dudley J and Tamura K 2008 MEGA: a biologist-centric software for evolutionary analysis of DNA and protein sequences *Briefings in Bioinformatics* **9** 299-306

[22] Larkin M A, Blackshields G, Brown N P, Chenna R, McGettigan P A, McWilliam H, Valentin F, Wallace I M, Wilm A, Lopez R, Thompson J D, Gibson T J and Higgins D G 2007 Clustal W and Clustal X version 2.0 *Bioinformatics* **23** 2947-48

[23] Li S, Zhao H, Li Y, Niu S and Cai B 2012 Complete genome sequence of the naphthalene-degrading Pseudomonas putida strain ND6 *J. Bacteriol* **194** 5154-55

[24] Ferreira L, Rosales E, Sanroman M A and Pazos M 2015 Preliminary testing and design of permeable bioreactive barrier for phenanthrene degradation by Pseudomonas stutzeri CECT 930 immobilized in hydrogel matrices *J. Chem. Technol. Biot.* **90** 500-06

[25] Hasan S A and Jabeen S 2015 Degradation kinetics and pathway of phenol by Pseudomonas and Bacillus species *Biotechnol Biotec. Eq.* **29** 45-53

[26] Vallon T, Simon O, Rendgen-Heugle B, Frana S, Muckschel B, Broicher A, Siemann-Herzberg M, Pfannenstiel J, Hauer B, Huber A, Breuer M and Takors R 2015 Applying systems biology tools to study n-butanol degradation in Pseudomonas putida KT2440 *Eng. Life Sci.* **15** 760-71

[27] Wang X Q, Wu C, Liu N, Li S J, Li W, Chen J M and Chen D Z 2015 Degradation of ethyl mercaptan and its major intermediate diethyl disulfide by Pseudomonas sp strain WL2 *Appl. Microbiol Biot.* **99** 3211-20

[28] Poblete-Castro I, Becker J, Dohnt K, dos Santos V M and Wittmann C 2012 Industrial biotechnology of Pseudomonas putida and related species *Applied microbiology and biotechnology* **93** 2279-90

[29] Winsor G L, Van Rossum T, Lo R, Khaira B, Whiteside M D, Hancock R E and Brinkman F S 2009 Pseudomonas Genome Database: facilitating user-friendly, comprehensive comparisons of microbial genomes *Nucleic Acids Research* **37** D483-8

[30] Nelson K E, *et al.* 2002 Complete genome sequence and comparative analysis of the metabolically versatile Pseudomonas putida KT2440 *Environmental Microbiology* **4** 799-808

[31] Roach D J, Burton J N, Lee C, Stackhouse B, Butler-Wu S M, Cookson B T, Shendure J and Salipante S J 2015 A Year of Infection in the Intensive Care Unit: Prospective whole genome sequencing of bacterial clinical isolates reveals cryptic transmissions and novel microbiota *PLoS Genetics* **11** e1005413

[32] Broholm K, Ludvigsen L, Jensen T F and Ostergaard H 2005 Aerobic biodegradation of vinyl chloride and cis-1,2-dichloroethylene in aquifer sediments *Chemosphere* **60** 1555-64

[33] Elango V, Kurtz H D and Freedman D L 2011 Aerobic cometabolism of trichloroethene and cis-dichloroethene with benzene and chlorinated benzenes as growth substrates *Chemosphere* **84** 247-53

[34] Jones M D, Rodgers-Vieira E A, Hu J and Aitken M D 2014 Association of Growth Substrates and Bacterial Genera with Benzo[a]pyrene Mineralization in Contaminated Soil *Environ Eng.*

Sci. **31** 689-97

[35] Hadibarata T, Kristanti R A, Fulazzaky M A and Nugroho A E 2012 Characterization of pyrene biodegradation by white-rot fungus Polyporus sp S133 *Biotechnol Appl. Bioc.* **59** 465-70

[36] Fayolle F, Vandecasteele J P and Monot F 2001 Microbial degradation and fate in the environment of methyl tert-butyl ether and related fuel oxygenates *Appl. Microbiol Biot.* **56** 339-49

[37] Volpe A, Del Moro G, Rossetti S, Tandoi V and Lopez A 2009 Enhanced bioremediation of methyl tert-butyl ether (MTBE) by microbial consortia obtained from contaminated aquifer material *Chemosphere* **75** 149-55

[38] Smith C A, O'Reilly K T and Hyman M R 2003 Characterization of the initial reactions during the cometabolic oxidation of methyl tert-butyl ether by propane-grown Mycobacterium vaccae JOB5 *Appl. Environ. Microbiol* **69** 796-804

[39] Garnier P M and Auria R 2000 Cometabolic biodegradation of methyl tert-butyl ether by a soil consortium: Effect of components present in gasoline *The Journal of general and applied microbiology* **46** 79-84

[40] Sander R 2015 Compilation of Henry's law constants (version 4.0) for water as solvent *Atmos. Chem. Phys.* **15** 4399-981

[41] Liu C Y, Speitel G E and Georgiou G 2001 Kinetics of methyl t-butyl ether cometabolism at low concentrations by pure cultures of butane-degrading bacteria *Appl. Environ. Microb.* **67** 2197-201

[42] Hyman M, Kwon P, Williamson K and O'Reilly K 1998 Cometabolism of MTBE by alkane-utilizing microorganisms *First International Conference on Remediation of Chlorinated and Recalcitrant Compounds* **3** 321-6

[43] Fortin N Y, Morales M, Nakagawa Y, Focht D D and Deshusses M A 2001 Methyl tert-butyl ether (MTBE) degradation by a microbial consortium *Environ Microbiol* **3** 407-16

IOP Conf. Series: Earth and Environmental Science **39** (2016) 012013 doi:10.1088/1755-1315/39/1/012013

Drawdown changes affected by flow rate and location of pumping wells near a river

S N Chen[1, 2] and L C Shu[1]

[1]State Key Laboratory of Hydrology-Water Resources and Hydraulic Engineering, Hohai University, Nanjing, 210098, China

E-mail: 821121202@qq.com

Abstract. Riverside groundwater exploitation is one of the main means of utilizing groundwater resources. This paper describes the impact of drawdown conditions with changes in the river flow and pumping well locations, and analyzes the interaction between the groundwater and surface water. Based on long-term hydrological gauging data, abundant hydrogeological test data, and the numerical simulation results for a typical well field of the Qinbei Power Plant, this paper presents three different cases of pumping well locations and four different cases of river flow rates. Finally, the abovementioned cases are integrated into 12 extraction groups and the drawdown conditions are calculated for each group. The results show that, for a given set of flow rate conditions, a location set in a recharge zone exhibited the maximum drawdown, while a location in a transition zone had the second-largest drawdown, and a location in a discharge zone had the minimum drawdown. In addition, assuming the same locations for the pumping wells, the drawdown change from small to large corresponded to 100%, 75%, 50%, and 25% of the original flow. This paper provides a foundation for future study of the calculation of riverside groundwater exploration with changes in the flow rate and well locations.

1. Introduction

Riverside groundwater sources offer various advantages such as good supply conditions, minimal depth, and ease of exploitation and management. Exploitation near a river is beneficial to the development and utilization of water resources. Given the purity of water from the unsaturated zone and aquifer, the quality of the riverside groundwater is generally much better than that of the surface water. Besides, due to the interaction between the aquifer and river water, riverside groundwater provides a high-capacity source of water. Therefore, great importance is currently being attached to the development of riverside groundwater resources.

Although the extraction of riverside groundwater would appear to be the exploitation of groundwater, the main supply source is actually the surface water. Groundwater and surface water always influence and restrict each other, so that there will be a wide variety of uncertainties when performing riverside groundwater exploitation. If there is any irrational exploitation of the groundwater, serious problems could result. Therefore, many researchers have been addressing the issue of riverside groundwater exploitation.

Given the special location of riverside groundwater sources, there is always a degree of interaction between the groundwater and the surface water, which results in a complex internal relationship. Many researchers have attached great importance to this interaction, and have made many valuable discoveries. For example, in 1941, Theis calculated the degree of attenuation on river flow of a riverside groundwater source [1]. Chen and Yin [2] also used numerical models to analyze the stream flow depletion in stream-aquifer systems with a baseflow. Shu and Chen [3] used numerical models to calculate the water exchange between the river and groundwater in a river-aquifer system. Wang [4] developed a simple, inexpensive instrument that was used to measure streambed parameters under the condition of a stream being disconnected from the groundwater. Wallace [5] proposed problems of stream depletion resulting from the cyclic pumping of wells and developed equations by applying superposition principles to analytical solutions for steady, continuous pumping. Park [6] evaluated the interference and ratio of river water at riverbank filtration pumping wells and claimed that well interference is caused not only by groundwater drawdown but also by pipe headloss, which depends on the flow rate. Wang [7] set up a three-dimensional mathematical model and analyzed the effects of river width, partial penetration, and the permeability of riverbed sediments on groundwater recharges. The results showed that riverside pumping may cause groundwater to flow beneath the partially penetrating river, and that the river width, penetration, and riverbed permeability obviously influence flows from the partially penetrating river and constant-head boundaries.

Based on the above research, this study set out to analyze the hydrogeological conditions of a typical riverside groundwater resource area, as a basis for establishing a valid numerical groundwater model. Careful analysis of the groundwater dynamics was performed under different flow conditions and a variety of well locations. The drawdown changes were calculated under different exploited cases. This paper provides the necessary theoretical basis and a specific means of calculating the scientific development of a riverside groundwater exploitation program.

2. Method

2.1. Description of study area

The study area was the Qinbei Power Plant in Jiyuan City, Henan Province, China (Figure 1). The general trend of the local terrain is from north to south, west to east. Taihang Mountain is in the north, and the alluvial sectors of the Qin River are in the south. There is abundant rainfall, and according to data gathered between 1956 and 2000, the average annual precipitation is 644.2 mm. The maximum recorded precipitation was 1027.3 mm in 1964, while the minimum recorded amount was 311.5 mm in 1997.

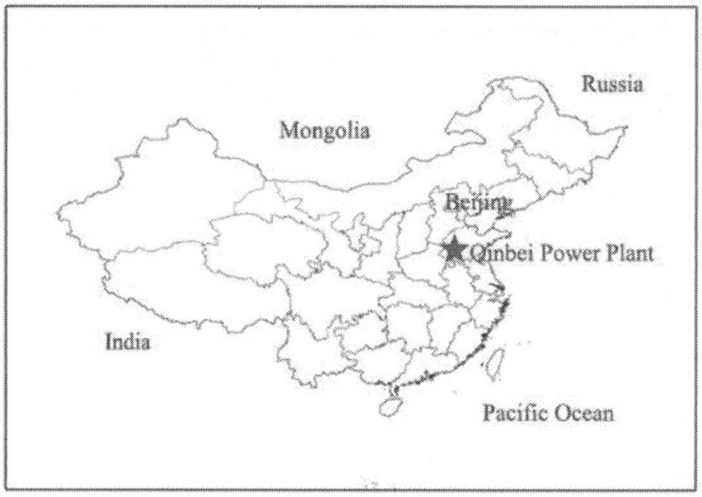

Location of Study area: Qinbei Power Plant, China

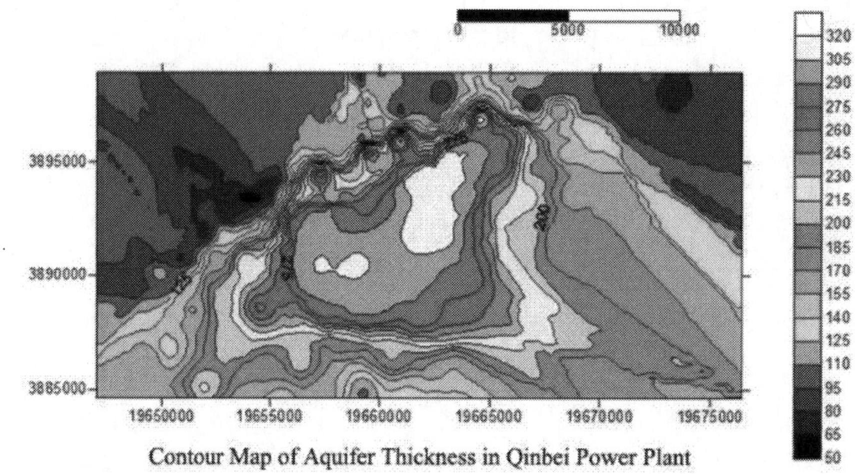

Contour Map of Aquifer Thickness in Qinbei Power Plant

Figure 1. Location and contour map of aquifer thicknessin study area: Qinbei Power Plant, China.

The main river in the study area, the Qin River, has its source in Shanxi Province, and flows from north to south and then though Henan Province, finally emptying into the Yellow River. In this study area, the interaction between the groundwater and surface water changes.

In the area from Wulongkou to Shagou, the groundwater flows in the same direction as the Qin River. Because the river level is higher than that of the groundwater, the groundwater starts to receive seepage from surface water. In the area after Shagou, the riverbed elevation becomes lower and the slope increases. Therefore, the groundwater overflows and drains into the river. The relationship between the groundwater and surface water thus changes. In addition, some of the groundwater will continue flow eastwards of the downstream direction.

The extracted groundwater is mainly used for seasonal irrigation. The total irrigation area is 90,950 acres, of which 50,250 acres are on the left bank. According to the results of previous surveys, the amount of industrial water consumption in the study area is 8,261,480 m3/a, the amount of water consumed as drinking water is 3,586,100 m3/a, and the amount being extracted for irrigation water is

27,058,050 m3/a (a total of 38,905,630 m3/a). Groundwater pollution is not serious. The irrigation wells and flow direction are shown in Figure 2.

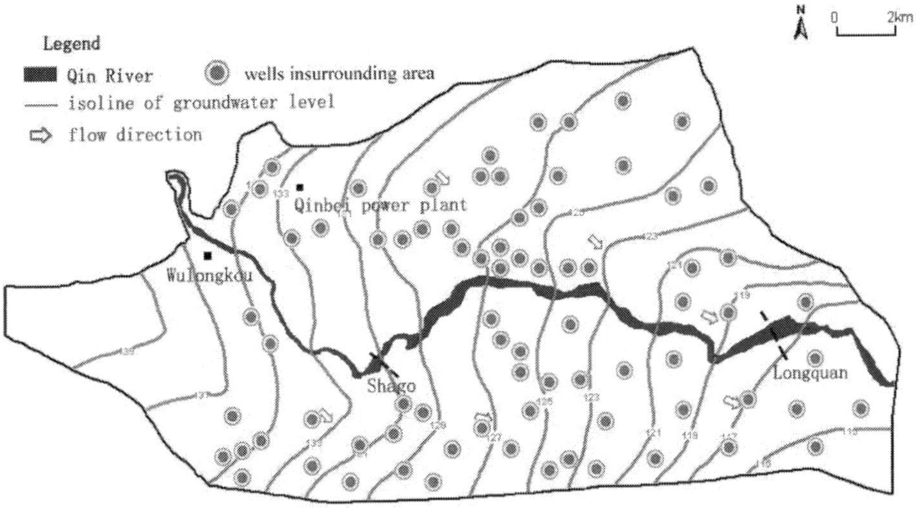

Figure 2. Distribution of surrounding wells and flow direction.

In this study, a numerical model was established by using Visual Modflow software.Modflow, a finite difference method solver developed by USGS, was selected for the groundwater flow model [8]. The solution was based on the finite difference method, and the boundaries were generalized into GHB. The main sources of recharge are lateral inflows, rainfall infiltration, and water seepage. The main discharges are exploitation, evaporation, and the recharging of rivers. As we can see from the groundwater dynamic data, the groundwater level in the area fluctuates in a non-steady state. The internal structure of the study area is non-homogeneous and isotropic, while it can be regarded as being homogeneous in a same parameter partition. The groundwater flow obeys Darcy's law.

According to the hydrogeological conceptual model mentioned above, the corresponding mathematical model can be established as follows:

$$\begin{cases} \frac{\partial}{\partial x}\left[K(H-B)\frac{\partial H}{\partial x}\right]+\frac{\partial}{\partial y}\left[K(H-B)\frac{\partial H}{\partial y}\right]+W-Q_E-Q_V=\mu\frac{\partial H}{\partial t}, & (x,y)\epsilon D \\ H(x,y,t)|_{\Gamma_1}=H_1(x,y), & (x,y)\in\Gamma_1 \end{cases} \tag{1}$$

$$H(x,y)|_{t=0}=H_0(x,y),\quad (x,y)\in D \tag{2}$$

Where H denotes groundwater level; B denotes elevation of confining bed; K denotes hydraulic conductivity; μ denotes specific yield; W denotes the value of recharge; Q_V denotes the value of exploitation; H_0 denotes initial head; H_1 denotes boundary water level of the calculation period; D denotes the total area; $\Gamma_1$denotes the 1st head boundary.

Based on the vertical distribution information collected in the study area, the aquifer system can be generalized into a single layer and partitioned into 146 rows and 184 columns, the length of each cell is 198 meters and its width is 144 meters. The area of this model is a total of 28,494 m^2. To accurately simulate the groundwater flow in the study area, part of the mesh was refined. Lithologic characteristics and a large number of pumping test data were taken into account. The study area was divided into 15 parameter zones, and the initial values were determined by the pumping test data and experience values. For ease of calculation and comparison, in this paper the value of the extraction was determined to be 1 m^3/s.

2.2. Cases of the flow rate

To illustrate the impact of changes in the drawdown conditions with the river flow, in this paper, four different flow rate are presented, namely, 25% of the original flow, 50% of the original flow, 75% of the original flow, and 100% of the original flow, respectively. Each case would be incorporated into the model, with 10 simulation years being examined to determine the maximum drawdown in the aquifer.

According to the data for the river stage and the runoff between 1983 and 1989 (the minimal-frequency monthly data were omitted), the relationship between the stage and discharge for the Qin River was obtained.

$$h = 6.1\times 10^{-6}Q^3 - 0.0008Q^2 + 141.63 \quad (3)$$

Where h denotes the depth of the water in the river and Q is the runoff.

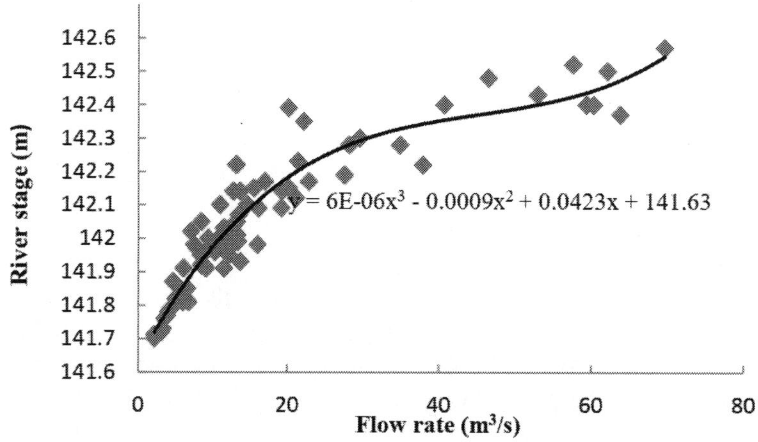

Figure 3. Stage-discharge curve for Qin River.

Given the relationship between the water depth and the discharge, the water depth in the Qin River could be determined for any flow rate provided the flow rate is known.

2.3. Pumping well locations

The locations of the existing wells around the Qinbei Power Plant is haphazard. Any further exploitation of the groundwater in the future will require the development of a scientific exploitation plan regarding the locations of the wells. To calculate the impact of the drawdown conditions with changes in the well locations, three cases were presented, with pumping wells located in the recharge zone, transition zone, and discharge zone, respectively. The extraction rate was set to 1 m^3/s. The simulation time was 10 years. Based on the above, we examined the changes in the drawdown at the end of the simulation. The wells were separated by about 8 km in each case (Figure 4). Observation wells were set up in the middle of the exploited wells. Finally, any changes in the drawdown conditions were recorded in each case.

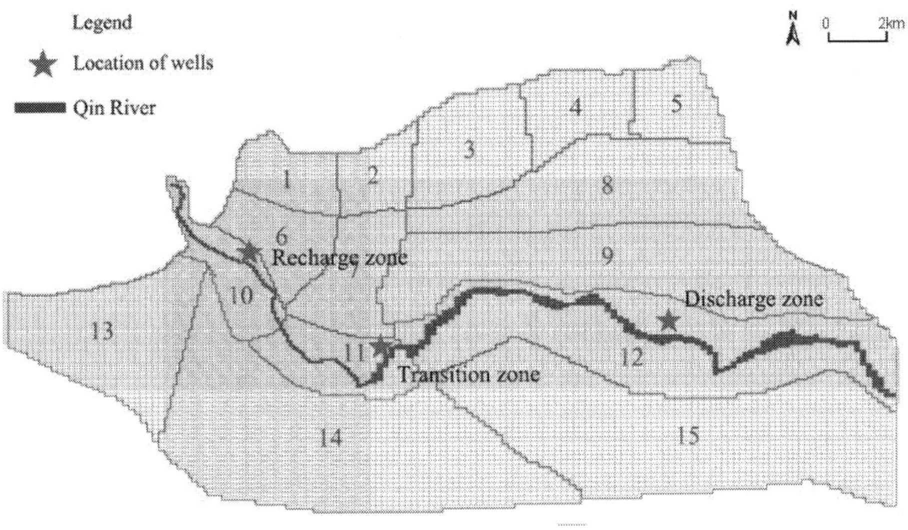

Figure 4. Locations of pumping wells and parameter partitions.

3. Results and discussion

3.1. Output of the model

In this study, a comprehensive analysis of the hydrogeological conditions around the Qinbei Power Plant was undertaken, and a numerical groundwater model was constructed. The model was calibrated and verified by using the data for 1980 to 1982. Wherein: at the end of 1980, the average error between the observed head and the calculated head was 0.868 m, the root mean square was 1.09 m, and the standardized RMS was 7.14%. At the end of 1981, the average error was 0.657 m, the root mean square was 0.812 m, and the standardized RMS was 5.68%. At the end of 1982, the average error was 0.518 m, while the root mean square and standardized RMS were 0.629 m and 4.43%, respectively. These results indicate that the model had been established correctly. Using the numerical model, we did not establish any Qinbei pumping wells, and instead only considered the exploitation of the surrounding irrigation wells. After 10 simulation years, the drawdown condition were as shown in Figure 5.

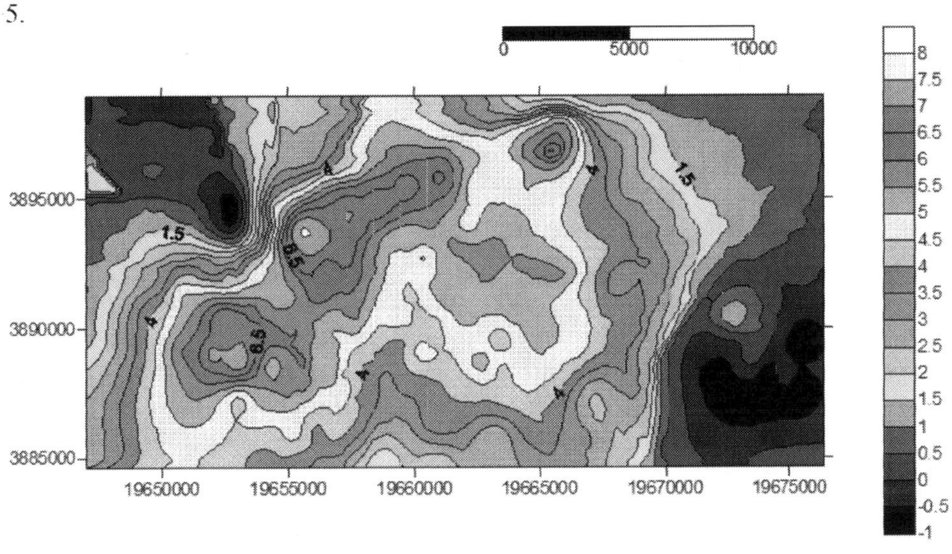

Figure 5. Drawdown conditions after 10 simulation years.

The results of the drawdown showed that the drawdowns in the recharge zone and transition zone were larger, while the drawdown in the discharge zone was smaller. However, to determine whether there is a close relationship between drawdown and the location, the pumping wells should be set in a different location and the simulation performed carefully. In addition, considering the uncertainty of the river flow in the accrual work, observation of the relationship between the river flow rate and groundwater drawdown is necessary.

3.2. *Results for different cases*

On several occasions, the numerical simulation gives consolidated results for different cases, as shown in Figure 6 and listed in Table 1 (both flow rate and pumping well location).

Table 1. Results of drawdown conditions for different cases.

Percentage of original flow	Pumping well locations		
	Recharge zone	Transition zone	Discharge zone
25%	31.11858	21.69895	19.96265
50%	30.36571	21.2009	19.23537
75%	29.13726	20.0006	18.501
100%	28.93994	19.5062	17.38377

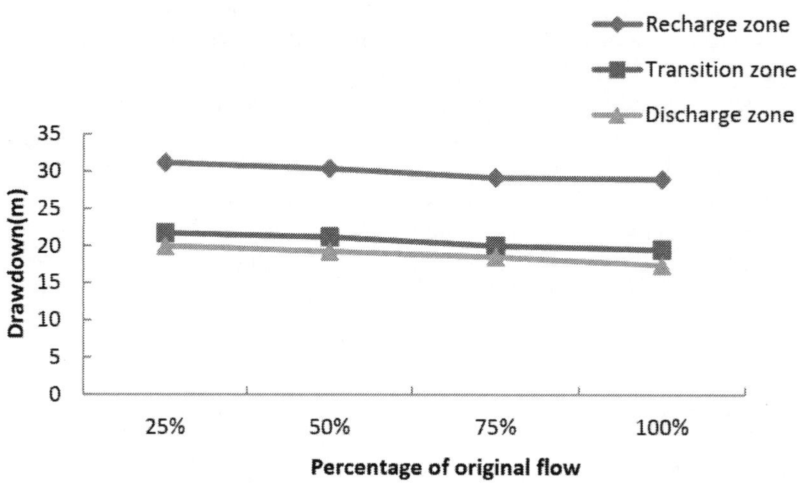

Figure 6. Results obtained for different cases.

Table 1 shows that, when the extraction volume is set to 1 m^3/s, under the same flow rate conditions, the location in the recharge zone has the maximum drawdown, with the value ranging from 28 m to 31 m. The transition zone exhibits the second-largest drawdown, ranging from 19 m to 21 m. The discharge zone is the minimum drawdown area, ranging from only 17 m to 19 m.

As can be seen from the results above, when the pumping wells are located in the recharge zone of the study area, even though the river water level is higher than the groundwater level, the groundwater itself cannot be completely recharged by the river, such that the changes in the drawdown increase. Conversely, when the pumping wells are located in the discharge zone, even though groundwater also supplies the river, the low elevation of the zone means that the groundwater in the discharge zone can be supplied with groundwater from the west, such that changes in the drawdown are minimal. These results tell us that we cannot completely rely on the supply from the river when exploiting riverside groundwater. If the terrain is higher, and there is considerable lateral outflow, it may not be possible to supply water from the surface quickly enough, causing the drawdown to increase, possibly even causing environmental and geological problems.

In addition, when the extraction volume is set to 1 m^3/s, for pumping wells in the same locations, those corresponding to 25% of the original flow rate have the maximum drawdown, with a value ranging from 19 m to 31 m. For the case corresponding to 50% of the original flow rate, the drawdown is the second-largest, ranging from 19 m to 30 m. Then, the drawdown for the case corresponding to 75% of the original flow rate ranges from 17 m to 29 m. Finally, the drawdown for the case equal to 100% of the original flow rate is the smallest, ranging from 17 m to 28 m.

We can see that, for different flow rates, the results are in line with people's perceptions, that is: if the river flow rate is larger, the change in the drawdown becomes smaller when extracting groundwater. The reason for this is: as the river flow increases, more water is available to recharge the groundwater, thus limiting the increasing trend in the drawdown. However, it can be seen from the results that, when the river flow rate is increasing (or decreasing), changes in the drawdown are not that obvious. In the first part of the study area, the surface water recharges the groundwater; in the latter part, however, the relationship between the groundwater and surface water changed, in that the groundwater started to recharge the surface water. The complex interaction between the groundwater and surface water contributes to the fact that a change in the river flow does not result is a significant change in the drawdown.

4. Conclusions

In this study, a comprehensive assessment of the Qinbei Power Plant groundwater resources was carried out, and a numerical groundwater model was established. Careful analysis of the dynamics of the groundwater for different river flows and different well locations allowed us to determine the changes in the drawdown for each case. The following is a summary of the study and our findings:

- Based on the numerical model, without considering the establishment of any Qinbei pumping wells, and only considering the exploitation of surrounding irrigation wells, the drawdown was calculated for 10 simulation years. The results showed that the drawdown in the recharge zone and transition zone were larger, while the drawdown in the discharge zone was smaller.
- Four different cases, specifically, (1) 25% of the original flow rate, (2) 50% of the original flow rate, (3) 75% of the original flow rate, and (4) 100% of the original flow rate were considered. Each case was input into the numerical model to calculate the drawdown conditions.
- Three pumping well location cases have been presented. Pumping wells were located in the recharge zone, transition zone, and discharge zone. The extraction amount was set to 1 m^3/s. Each case was input into the numerical model to calculate the drawdown conditions.
- Based on the numerical simulations, different river flow cases, and different well locations, twelve groups of drawdown conditions were analyzed. The results show that, for the same flow rate, wells in the recharge zone exhibit the maximum drawdown, those in the transition zone have the second-largest drawdown, while the wells in the discharge zone exhibit the minimum drawdown. In addition, for pumping wells in the same locations, the drawdown changes from small to large for flow rates equal to 100%, 75%, 50%, and 25% of the original flow.

References

[1] Theis C V 1941 The effect of a well on the flow of a nearby stream *J. American Geophysical Union Transactions* **22**(3) 734-38
[2] Chen X H and Yin Y 2001 Streamflow depletion: modeling of reduced baseflow and induced stream infiltration from seasonally pumped wells *J. Journal of American Water Resources Association* **37**(1) 185-95

[3] Shu L C and Chen X H 2002 Simulation of water quantity exchange between groundwater and the Platte River water, central Nebraska *J. Journal of Central South University* **9**(3) 212-15
[4] Wang B et al. 2004 Quantitative analysis of groundwater flow near a partially penetrating river under riverside pumping *J. Acta Geologica Sinica* **78**(3) 820-24
[5] Wallace R B, Darama Y and Annable M D 1990 Stream depletion by cyclic pumping of wells *J. Water Resources Research* **26**(6) 1263-70
[6] Park N 2014 Analytical evaluation of interference and ratio of river water at riverbank filtration pumping wells *J. Journal of Korea Water Resources Association* **47**(8) 685-91
[7] Wang W K, Li J T, Wang W M, Chen X H, et al. 2014 Estimating streambed parameters for a disconnected river *J. Hydrological Processes* **28** 3627-41
[8] Michael G M et al. 2003 The history of Modflow *J. Ground Water* **41**(2) 280-83

Acknowledgments
The research work was supported by innovation projects of universities in Jiangsu Province, China, 2014 (No: 2014B35614).

IOP Conf. Series: Earth and Environmental Science **39** (2016) 012014 doi:10.1088/1755-1315/39/1/012014

Environmental impact analysis of mine tailing reservoir

J Z Gong[1, 2]

[1]Hebei Institute of Geophysical Exploration, Langfang 065000, China

E-Mail: gjz212@tom.com

Abstract. Under certain conditions landscape topography which utilizes mine tailing reservoir construction using is likely to increase lateral recharge source regions, resulting in dramatic changes to the local hydrological dynamic field and recharge of downstream areas initiated by runoff, excretion state, elevated groundwater depth, shallow groundwater, rainfall direct communication, and thinning of the vadose zone. Corrosive leaching of topsoil over many years of exposure to chemical fertilizers and pesticides may result in their dissolution into the groundwater system, which may lead to excessive amounts of many harmful chemicals, therby affecting the physical and mental health of human residents and increase environmental vulnerability and risk associated with the water and soil. According to field survey data from Yujiakan, Qian'an City, and Hebei provinces, this paper analyzes the hydrogeological environmental mechanisms of areas adjacent to mine tailing reservoirs and establishes a conceptual model of the local groundwater system and the concentration-response function between NO_3^- content in groundwater and the incidence of cancer in local residents.

1. Introduction

Water pollution is exacerbating the groundwater crisis in China. Experts from the China Geological Survey Bureau determined that 90% of groundwater in China suffers from various degrees of contamination, 60% of which suffers from serious pollution. According to information published by the China Geological Environmental Monitoring Institute China currently exhibits a groundwater trend that indicates worsening pollution levels from point to the surface, from shallow to deep, and from urban to rural [1].

Simultaneously, cancer rates are rising dramatically, and the relationship between cancer and water pollution has gradually become the focus of attention.

Comparring the elemental content of soil, water, agricultural areas, human hair and other medium of high cancer incidence and low risk areas, certain eco-geochemical survey data reported by Li Zhong-Hui, et al., (2012) has identified key geochemical factors which lead to high incidence of cancer in Sichuan Yanting, and predicted the incidence of cancer within the Chengdu Economic Zone [2].

Based on investigation and monitoring, Yang Gonghuan, et al., (2013) established a database of environmental and health factors in Huaihe River Basin. Through spatial analysis, an association model between environmental and health issues was proposed to lay the foundation for the implementation of integrated monitoring[3].

Recently, concurrent with the vigorous development of mineral resources and the construction of the tailing reservoir, the impact on the quality of the surrounding groundwater has become an increasingly prominent and serious threat to human health. In this paper, iron ore from Qian'an City, Hebei Province, and the Xixiakou-Yujiakan region are used as an example in specific case studies.

2. Epidemiological survey of cancer villages

Drinking the water that comes into contact with untreated sewage upstream as well as other sources of air, soil and other environmental contamination can cause serious damage to internal body mechanisms, resulting in massive cancer incidence in local villages.

In 2009, Sun Yuefei confirmed the presence of over 247 cancer villages in China, with two specific geographical distribution characteristics. First, a multi-centered concentralization and regular intersection of azonality and zonality were observed, indicating that the villages are not uniformly distributed. Second, in basic agreement with differences in rainfall, population distribution and regional economy, the number of cancer villages decreases from the eastern portion to the western portion in a gradual gradation. In regard to various causes, cancer villages can be divided into three categories. First are the environmentally polluted cancer vilalges which result from industrial, agricultural and domestic pollution. Second are lifestyle-based cancer villages that reflect vocation, as well as diet and hygiene habits. Last are original environmental cancers which are related to local geographic environment such as geology and hydrology[4].

Experts estimate the number of cancer villages in mainland China to be approximately 459, and gradually to the central and western diffusion. According to statistics representing the years 2009-2015, 37 people in Yujiakan village were identified as cancer patients, indicating an incidence of 1.37% (2012, China cancer 0.2478%) , representing 21 males and 16 females with cancer onset from 1 year to 77 years, indicating an average of 52.6 years and an average age at death of 62.9 years. Digestive system cancers account for 12 people, the number is the most. Some couples or fathers and sons simultaneously cancer.

2007-2015, 117 patients with cardiovascular disease, disease incidence 4.33%, among them, cerebral thrombosis, cerebral hemorrhage 84, heart disease 33. The average pathogenesis age 62.15 years, age at death 71.61 years old. Some couples even father and sons while sick. 2009--2014 year show a gradual increasing trend.

Table 1. Cancer and cardiovascular diseases distribution circumstances of Qian'an City, Yujiakan.

Cancer onset age	1	5	46-50	52-60	63-70	72-77		/ year old	
Patients number	1	1	6	8	4	4			
Cancer types	Lung	colon	renal	breast	leukemia	gastric	liver	esophageal	other
Patients number	8	4	3	3	3	3	2	3	8
Cardiovascular diseases	25-40	41-50	51-60	61-70	71-80	82-93		/ year old	
Patients number	6	15	28	33	27	7			
Fall ill age	2007	2008	2009	2010	2011	2012	2013	2014	2015
Patients number	6	6	2	18	15	17	22	23	8

The village is expected to average life expectancy of 69.88 years, low 4.12 years than the national average life expectancy of 74 years.

The villagers were shocked by the harsh realities of the immediate. In puzzling apart, targeting the village drinking groundwater quality. Villagers found, whenever the northwest monsoon winds, gray tailings flying in the sky. Wells water gradually worse taste, slightly sour, used to fish difficult to survive.

3. Groundwater quality impact of iron ore tailing reservoir

There is a medium-sized sedimentary-metamorphic magnetite bed in Xixiakou village north, start mined from 1980 Generation, waste water generated during the mining, through a pipe discharged into the tailing reservoir in Yujiakan North Hill. 2008, Xixiakou iron ore reservoir dam heightening three meters, become a suspended lake on the ground, daily water surface above 5-20 meters in the south side Yujiakan village .2009-2014, followed by a series of hydro-geological effects.

The area belongs to the new Archean era miniature folding platform structure, Yanshan south erosion - denuded hills and mountains landscape, slightly inclined, northwest was up about 30 meters than southeast. Quaternary genetic types are the late Pleistocene alluvial fan (skirt) accumulation, in the piedmont hydrogeological zoning of Luanhe River system, full fresh water area, mainly diving, partial confined water. Alluvial fan shaft unit inflow more than 40t / (h • m), alluvial fan edge unit inflow 10 ~ 40t / (h • m). The lower part of the aquifer without continuous aquitard, good vertical hydraulic connection, often the lower part of the aquifer mining and dewatering, confined aquifer and the lower aqueous layer aquifer group consisting uniform [5].

The aqueous layer exhibits the dual structure, wherein within the range of 0.5 to 10 meters containing gravel clay, loam, the lower part is weathered bedrock fissure of the Mid Archean Qianxi rock group (Ar_2Q) websterite hornblende plagioclase granulite, New Archean tonalite rocks(Ar_3Togn) and light-colored granite ($Ar_3\gamma$). Aquifers rich aqueous up 5000m^3 / d or more, permeability coefficient> 150m^3 / d, rainfall infiltration coefficient of 25-30%[6].

Under normal conditions, the shallow groundwater recharge main source of atmospheric precipitation. During the summer heavy rainfall, flood into the southeast of Qinglonghe River. In winter and spring non-season rainfall, the performance area is a typical village dry state, water resources is very scarce.

Heightened tailing reservoir, representing increased a huge source of lateral recharge zone, resulting in the region a dramatic change of hydrodynamic field and recharge - runoff - excretion state, specific performance in the following several aspects:

(1) Groundwater depth from the original 10 meters rise to about 2 meters, local area during the rainy season more shallow; (2) In fractured bedrock at the village, many water seepage drain point decline springs, forming a trickle; (3) Shallow groundwater with rainfall to direct communication, vadose zone thinned dramatically or even disappear, in dissolving, leaching, previous years accumulated chemical fertilizers and pesticides in topsoil, integrate mixed into the groundwater system, leading to a variety of harmful ingredients exceeded.

Since the 1980s, application of groundwater system research methods, in terms of establishing a conceptual model of groundwater, Chinese scholars have made significant progress, the Groundwater System Block Diagram of Northeast Songnen Plain, representative in the northern region. Hydrogeological conceptual models mainly reflecting spatial distribution of the whole structure composed of boundary conditions, recharge and discharge, hydrodynamic conditions, water chemical characteristics and its corresponding parameters[7].

According to the spatial distribution of the survey area heterogeneous aqueous medium, recharge and excretion evolution with time, mineral dissolution material transport and changes in vadose zone waters, Author try to set up a local hydrogeological conceptual model (figure 1).

IOP Conf. Series: Earth and Environmental Science **39** (2016) 012014 doi:10.1088/1755-1315/39/1/012014

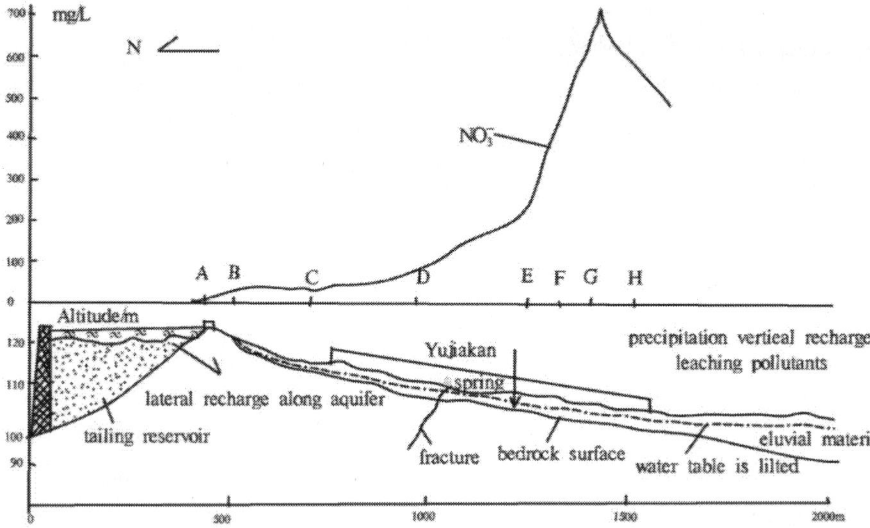

Figure 1. Schematic diagram of tailings reservoir to groundwater effect in YuJiakan Qian'an City.

a.Tailing reservoir above the village become a suspended lake.

b.The water level is raised leading to village low-lying formation of perennial streams

c. Shallow aquifer structure-containing gravel clay, fertilizer and pesticide infiltration properties.

d. Weathering crust aquifer.

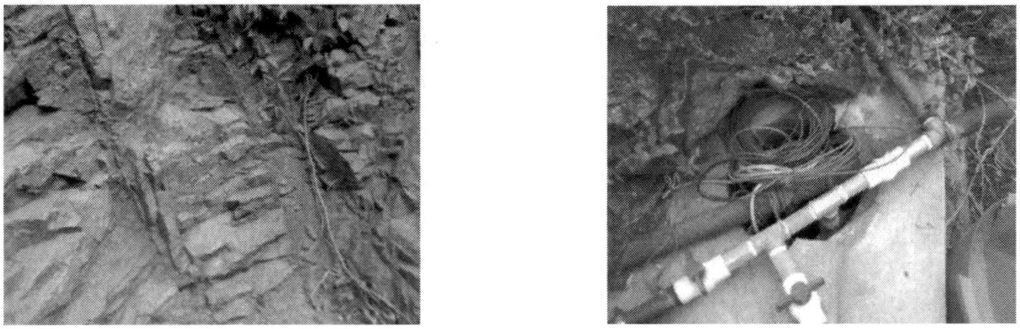

e. Deep aquifer structure, Qianxi rock group granulite.

f. The villager drinking wells.

Figure 2. Hydrogeological photos in YuJiakan Qian'an City.

4. Groundwater quality evaluation

According to the analysis results of the experiment in Hebei Province Water Environment Monitoring Center, We can find: (1) according to the national drinking water quality standards GB/T14848-93 and drinking water health standards GB5749-85, the area most water samples NO_3^- exceeded several times to dozens ten times, Secondly, few water samples hardness, salinity, SO_4^{2-} and Cl^- exceeding 1-2 times; (2) Spatial distribution of all village water samples NO_3^-, gradually increased from the north higher ground to the lower south content doubled by 10–20- 50-100-300-500-700mg/L, entire villages all belong to the scope of excessive, the most serious In village southeast, content contours form closed ring shape distribution; (3) Overall quality of Ⅱ, Ⅲ class water, evaluation from the good, poor to very poor. Compared to 1983[5], by 2015, in addition to decreased pH outside, remaining harmful components have increased.

The village cancer patients mainly in the low-lying area elevation 100-90m, groundwater NO_3^-50-765 mg / L polluted area, show some spatial correlation between the two(table 2, figure 3).

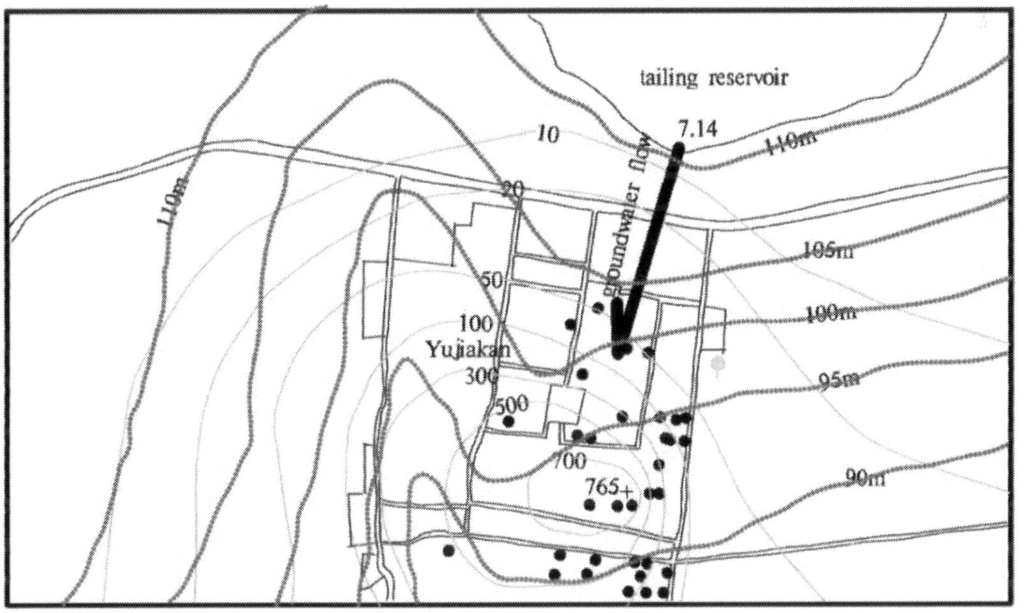

Figure 3. Shallow groundwater NO_3^- content (mg / L) contour map of YuJiakan Qian'an City.

Note: dashed line is the elevation, solid line NO_3^- content, black dots for cancer patients address distribution.

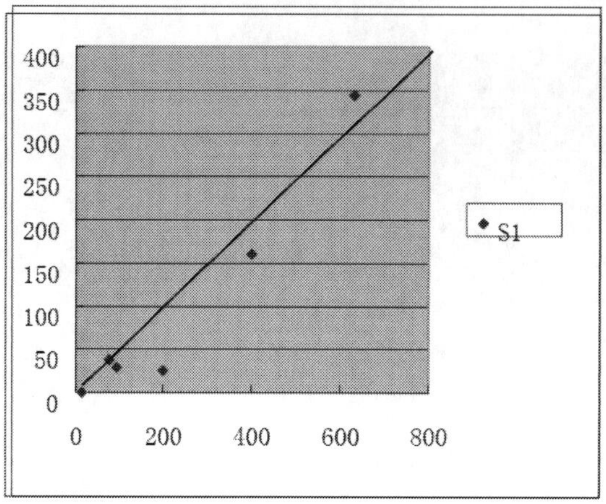

Figure 4. Relationship diagram between x-ground water NO_3^-content and y-patients density in YuJiakan Qian'an City.

Table 2. Interrelation between groundwater NO_3^- content and the number of cancer patients in YuJiakan Qian'an City.

Groundwater NO_3^- content /mg/L	10-20	20-50	50-100	100-300	300-500	500-765
Distribution area / km^2	0.704	0.805	0.135	0.233	0.050	0.032
Number of cancer patients	0	2	5	6	8	11
Patients density people / km^2	0	2.48	37.0	25.8	160	344

Between groundwater NO_3^- content(x) and the patients density(y), We can create a system of linear equations:

Y=0.5363x-26.42 γ=0.9680 $\gamma_{0.05}$=0.7067 n=6

5. Causes of groundwater environmental pollution and health hazards

On the Hebei plain surface soil nitrogen element geochemical maps, the district topsoil N in 1121 -1269mg/L, the province's average of 1048mg/L, This area 1.5 m deep soil N content of 825mg/L or more, compared with the region's average of 581.It is positive abnormal distribution, become groundwater nitrate supply source region.

North China Plain diving around occurred the different levels of nitrite (NO_2-N) pollution, Hebei, Shandong, Henan part of the area, groundwater nitrate content excessive mainly in agricultural areas, Nitrogen fertilizer is caused by infiltration of groundwater[8].

Liu Guangdong (2004) has establish a dose - response relationship between Nitrate contamination and health damage, Predicted the increased deaths number of groundwater pollution exceeded region, Constructed a total evaluation function of human capital loss[9].

Modern medical research shows, Cancer and genetic factors, environment, lifestyle, etc., have shown some correlation. Yanting County, Sichuan Province, a high incidence of cancer, mainly related soil and food, low levels of selenium molybdenum, food and drinking high levels of nitrates[2]。

A team of experts from Chinese Center for Disease Control has studied lasted eight years, the results show, in Huaihe River Basin, irrigation polluted areas compared with the control areas, gastrointestinal cancer prevalence 5 times, child cancer mortality higher than the national average of 68.81 / 100,000, fully confirmed the high incidence of cancer is directly related to the water pollution [3].

6. Conclusions and recommendations

Since the construction of the mine tailing reservoir, increased peripheral lateral groundwater recharge source, resulting in significantly improved downstream water depth, combined with precipitation on surface soil leaching dissolution. As a result, the quality of groundwater beyond national standards, Thereby affecting the residents physical and mental health. In order to do this, we proposed increase water depth, shield shallow groundwater, to mitigate the hazards of environmental pollution.

References

[1] Chen C 2013 *China groundwater pollution serious, Sewage treatment long way to go* http: //news. hefei.cc /2013 /0327 021565420 .shtml

[2] Li Z H, Jin L X, Xu Z, *et al.* 2012 Impact of the Geochemical Enviroment on the cancer Incidence in the Yanting Area, Sichuan Province *Advances in Earth Science* **27** 373-83

[3] Yang G H and Zhuang D F 2013 *Water environment and gastrointestinal cancer mortality portfolio in Huaihe River Valley* Beijing: China Cartographic Publishing House pp1-10

[4] Sun Y F 2009 The geographic distribution of China's cancer villages *Central China Normal University*

[5] Zhang Z H, Shao S Y, Chen Y, *et al.* 1999 *Trend forecast of since late pleistocene epoch geologic environment evolution and future live environment changes in North China* Beijing: Geological Publishing House p175

[6] Hebei Bureau of Geology and Mineral Resources 2005 *Geology Mineral Resources and environment in Hebei Province* Beijing: Geological Publishing House pp576-630

[7] Chen M X 2002 *China groundwater resources and environment* Beijing: Seismological Press pp385-404

[8] Hou C T and Liu X D 2010 *North China plain water and soil geological environment portfolio* Beijing: Geological Publishing House pp22-91

[9] Liu G D, *et al.* 2004 Human capital approach to assess environmental value of agricultural pollution of groundwater loss *China Environmental Science* **24**(3) 372-75

Preliminary study on the influence of coal mining on river runoff

N Zhang[1, 3], H Jiang[2], X W Guo[1], H J Yin[1] and X H Su[1]

[1] Yellow River Institute of Hydraulic Research, Henan, Zhengzhou, 450003, China

[2] Yellow River Basin Water Resources Protection Bureau, HeNan, Zhengzhou, 450004, China

E-mail: zhangnan19810202@126.com

Abstract. Coal mining is a major cause of decreases in river runoff, which can increase the disparity between the supply and demand of water resources. This paper presents runoff variation data from Tou-Tong to analyze the impacts of rainfall and human activity on runoff. This data showed that the annual runoff of Gushanchuan, Kuye and Tuwei river decreased significantly in the middle and latter half of the 1990s. Reservoirs, key dams, terraced fields and vegetation variation are all important factors that affect decreases in runoff. The effect of three types of coal mining on the water cycle is put forward and quantitatively calculated.

1. Introduction

Yellow River is an important energy base, and it is also China's major dryland[1] Scholars such as Wang Hao and Liu Changming have found that 52% of changes in annual runoff can be attributed to human activity[2~6].

The development and utilization of coal resources in the basin of Yellow River has a long history and has flourished, which makes it an important player in our economy. The coal industry and the associated processes are major consumers of water, which can cause a crisis of water resources. In addition, these activities cause serious ecological and environmental questions; the quality of the water environment is suffering serious and increasingly prominent damage, such as groundwater level recession, spring and river water decline, sometimes to the point of drying, and water pollution. Among these consequences, river runoff reduction is especially severe, and it intensifies the mismatch between supply and demand of regional water resources. Information on the influence of coal mining on river runoff and sustainable utilization of water resource management is extremely urgent in the face of river runoff reduction.

2. The runoff variation analysis of Toutong Region

The study region is Toutong in the middle reaches of Yellow River (figure 1). In recent 10 years, the runoff reduced greatly in Toutong region, the average annual rainfall is 501.1 mm between 1952 and 2000, natural runoff is about 17.07 billion m3, average annual rainfall is 502 mm between 2003 and 2010, natural runoff is about 10.09 billion m3, rainfall did not change, natural runoff reduced about 6.98 billion m3.

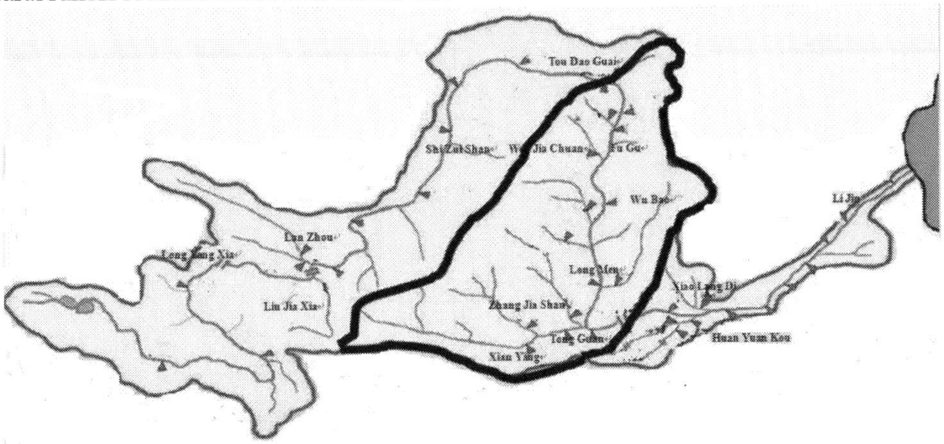

Figure 1. The study region is Toutong in the middle reaches of Yellow River.

2.1. The rainfall

Rainfall and human activity are the main factors that influence the changing of runoff. Changes to the Toutongjian region between June and September are shown in figure 2. From the figure, it can be seen that the change of rainfall is not obvious in the flood season after 2000. Therefore, changes to rainfall alone cannot explain the sharp decline of runoff during this period.

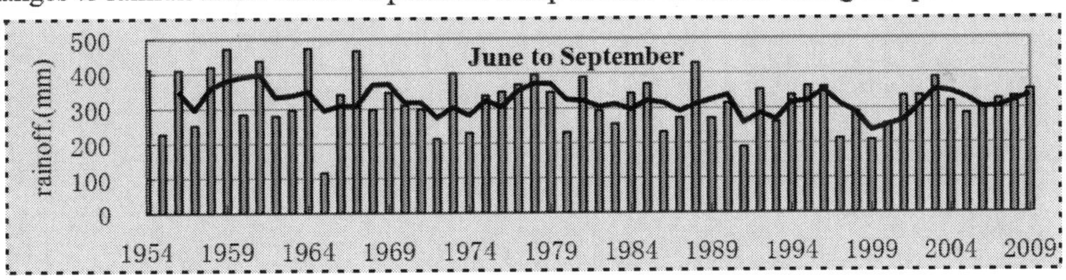

Figure 2. Changes of Toutongjian region from June and September.

2.2. The change of the underlying surface

Table 1 shows big, small and medium-sized reservoirs in the study basins.

As a rough estimate, the evaporation of tributaries in the Toutong region is only about 130 million m3, and the water reducing effect is limited.

During the 1980s, dams that were built were generally planned to be the backbones for dams that have been completed more recently. Through the survey, it was found that there is almost no sedimentation in areas where the vegetation and terraces are in good condition in recent years, and the impoundment was reduced. The new backbones of dams in the Shanbei branch are rare. As a rough estimate, the new evaporation was 150 million m3 at its maximum.

The survey also indicated that the Longtou region does not have obvious increments. A large amount of observation proved that the water reducing effect of the terrace is up to

80~90%; a large number of other studies have shown that the water reducing effect of terrace can be reduced by 80~90%.

The observation in Yanan and Lishi shows that in the loess plateau, as the vegetation cover increased, surface runoff and the total amount of runoff reduced significantly. For example, when coverage increased from 20% to 70%, the runoff could reduce about 1/4.

Table 1. Big small and medium-sized reservoirs in study basins.

Scale		Huang Pu Chuang basin	Gu Shan Chuang basin	Ku Ye He basin	Tu Wei He basin	Jia Lu He basin	Wu Ding He basin	Qing Jian He basin	Yan He basin	Hun He basin	Qiu Shui He basin	San Chuan He basin	Bei Luo He basin
large	number						2		1	1			
	10^8 m^3						3		2.03	1			
middle	number			2	2	1	24	3	1	6	1	3	
	10^8 m^3			1.1	0.834	0.176	9.31	0.678	0.148	1.8	0.241	1.184	
small	number	11	1	10	12	1	63	2	2	2	6	2	2
	10^8 m^3						2		1	1			

3. The influence of coal mining on river runoff

3.1. Annual runoff with the cumulative curves of raw coal production

Annual runoff with the cumulative curves of raw coal production is showed in figures 3-5, include Gushanchuan, Kuyehe and Tuweihe basins in the north Shanxi, which is a region with a large amount of coal production. It can be seen that the raw coal of the three river basins showed an increasing trend after 2000, the runoff accumulation curve in the 1990s showed a decreasing trend, and mine inflow and raw coal production had synchronous growth from 1991 to 2011 year by year. This indicates that coal mining is a major influencing factor on river runoff.

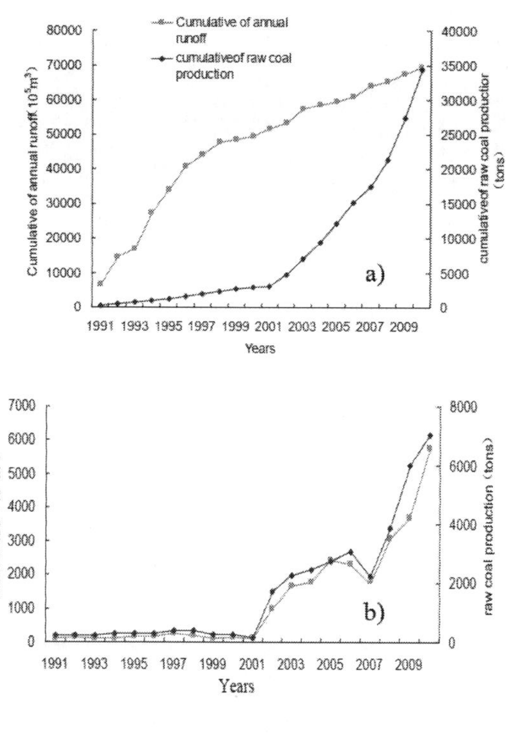

Figure 3. Annual runoff with the cumulative curves of raw coal production and Water inflow with raw coal production in Gu Shan Chuan basin

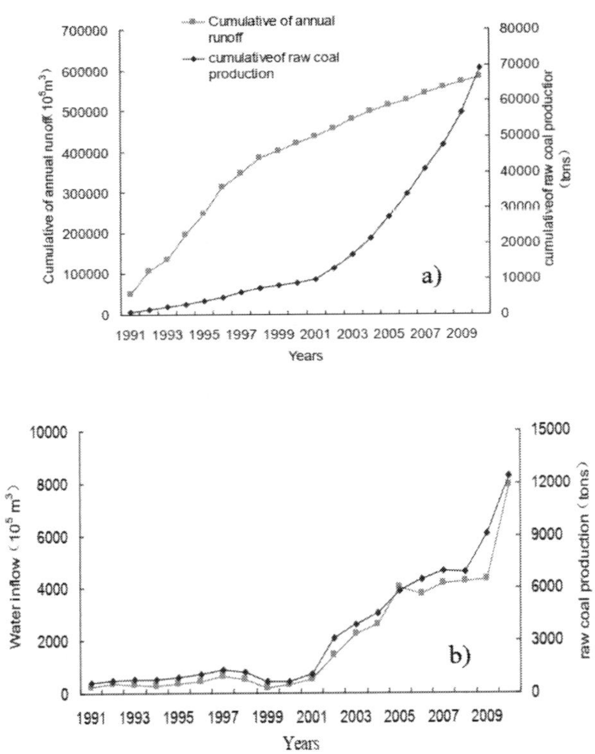

Figure 4. Annual runoff with the cumulative curves of raw coal production and Water inflow with raw coal production in Ku Ye He basin

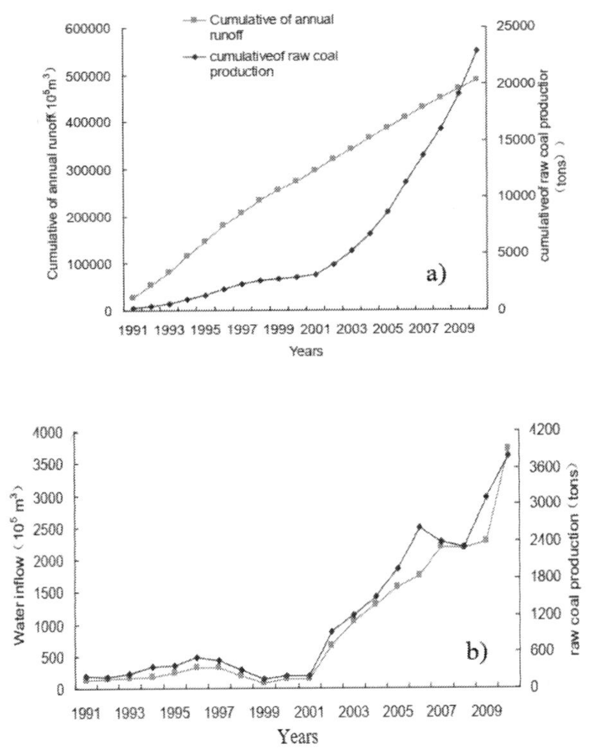

Figure 5. Annual runoff with the cumulative curves of raw coal production and Water inflow with raw coal production in Tu Wei He basin

4. Research method

4.1. Diversion fissure zone and caving development height determination

At present, the research of multiple sets of coal mining under the conditions of water flowing fractured zone is still in its primary stage. In this paper, the building, water, railway, main shaft, coal pillar, and press coal mining regulations (hereafter referred to as "three under" rules) listed in the empirical formula were used to calculate and determine the coal seam mining height of water flowing fractured zones and the development of caving zones.

4.2. Ground subsidence calculation and analysis

The formation of the ground subsidence area and water area were calculated with the coal mining subsidence model. The corresponding simulation program was adjusted based on the future forecast of the ground subsidence deformation under different minable seam mining situations, while the calculation of evaporation changes analyzes the effect of coal mining under the scheme of surface water.

4.3. Calculation of deep and shallow water

In the distributed time-variant gain hydrological model, the hydrological cycle and the hydrologic method are combined with the theory of non-linear systems in order to simulate the hydrological process. The core of the model is the TVGM runoff module. The slope, flow, flow path, river grid, basin boundary and other information are determined by GIS. With consideration of the physical process of the hydrological system, a general pattern was adopted to construct a distributed hydrological model. DTVGM (raster units) in the spatial distribution of hydrological unit with the application of the theory of nonlinear systems with certain mechanisms of the hydrological cycle model were used to calculate the surface of the aquatic product flow, the application linear flow of free water reservoir was used to calculate soil water and the underground aquatic product flow. The actual evaporation glover model and the snowmelt model use the live factor model.

MODFLOW (Modular Three-dimensional Finite-difference Ground-water Flow Model) is a distributed physical model. This model can reflect the spatial characteristics of the groundwater and its movement. The limitations of performance in the model depend upon the specific conditions of input, including tributaries, water supplies, and evaporation data. The model lacks a specified in the form of surface and soil profile hydrological process method of these conditions.

When DTVGM is coupled with MODFLOW, the combined model can retain the advantages of the respective models to simulate surface water and groundwater while performing appropriate quantitative analysis of the hydrological processes in the watershed. The minimum grid and the grid size are the man differences between DTVGM and MODFLOW. First, conversion between the minimum grid and the grid space must be achieved; then, the DTVGM and MODFLOW models can be successfully coupled.

5. Conclusions

- After 2000, the rainfall in Toutongjian from June to September did not obviously chang in flood season. Therefore, the rainfall change alone cannot explain the phenomenon of the drop off of runoff in the same period.

- The water reducing effects of reservoirs and backbones of dams are limited, while terrace (article) fields and vegetation change were found to have a distinct effect in reducing water.
- Through the mechanism combined with the model, three kinds of modes that impact the way coal mining interacts with the water cycle are proposed: water fracture zones without conduction bedrock, water fracture zones with conduction bedrock, the ordovician occurrence patterns of coal seams and water fracture zones. The development of caving zone height is also given, along with a method for the quantitative calculation of deep and shallow water.

Acknowledgements
This study was supported by research of National Natural Science Foundation of China(NSFC) No.51309107 and National Technology Support Project No. 2012BAB02B04. We also thankful to anonymous reviewer for his constructive comments in the manuscript review.

References
[1] Zhang J Y, Wang G Q, He R M and Liu S X 2009 Variation trends of runoffs in the Middle Yellow River basin and its response to climate change *Advances in Water Science* **20**(2): 153-58
[2] Liu C M and Zhang X C 2004 Causal analysis on actual water flow reduction in the mainstream of the Yellow River *Acta Geogaphica Sinca* **59**(3):323-30
[3] Wang H, Chou Y Q and Jia Y W 2010 Development course and tendency of water resources assessment *Journal of Beijing Normal University* (Natural Science) **46**(3): 274-77
[4] Zhang X P, Zhang L, Wang Y and Mu X M 2009 Tempo- spatially responses of the annual streamflow to LUCC in the middle reaches of Yellow River, China *Science of Soil and Water Conservation* **7**(1):19-26
[5] Yellow River Conservancy Commission 2011 *The Yellow River Basin Comprehensive Planning*
[6] Wang H, Yan D H, Jia Y W, Hu D L and Wang L H 2010 Subject system of modern hydrology and water resources and research frontiers and hot issues *Advances in Water Science* **21**(4): 479-89

IOP Conf. Series: Earth and Environmental Science **39** (2016) 012016 doi:10.1088/1755-1315/39/1/012016

Spatio-temporal patterns of redox potential diffusive boundarys around Potamogeton crispus leaves and stems

B Dong[1, 2]

[1]College of Resource and Environment, Linyi University, Linyi 276000, P. R. China

E-mail: dongbin@lyu.edu.cn

Abstract. To understand functions of a diffusive boundary layer around submerged macrophytes, the redox potential (ORP) around the leaves and stems of Potamogeton crispus were measured using a microsensor. Periphyton attached on macrophytes was analysed with the standard method. Results showed that significant spatio-temporal variations of ORP existed around leaves of P. crispus. In a vertical direction (within 2 mm), ORP significantly decreased with reducing distance from the leaf /stem surfaces and reached a minimum (428.51 mv) at the leaf /stem surface. The ORP microprofile was steepest around the mature leaf at the middle shoot and that of the senescent leaves at the basal shoot was relatively flat. At the temporal scale, the ORP microprofile was steepest at the stable growing stage, and those of the seedling stage and declining stage were relatively flat. Periphyton attached on P. crispus became dense gradually upon macrophyte entering growth stages. The results indicated that characteristics of ORP microprofiles were mainly affected by macrophytic physiological characteristics and periphyton synergistically.

1. Introduction

In eutrophic shallow-water ecosysystems, submerged macrophytes potentially play an important role in improving the water environment, and they often provide most of the accessible surface areas for attached periphyton affecting nutrient removal, partly due to the physical trapping of particles, plant uptake, and/or by modifying biogeochemistry [1]. The interface between the submerged macrophyte surfaces and water forms a unique diffusive boundary layer (DBL), because submerged macrophytes are not neutral substrata for periphyton. The microenvironment in DBL is heterogeneous due to the complex composition of periphyton, and to the host plant with biological activity. Understanding the environmental characteristics of DBL is the prerequisite for revealing various ecological functions of DBL. Previous researches suggest that dissolved oxygen (DO) concentrations and pH in DBL around submerged macrophytes Potamogeton crispus and Potamogeton malaianus showed significant spatio-temporal variations [2-4]. Preliminary studies seldom investigate the redox potential (ORP) in DBL around submerged macrophytes, hence little is known about the spatial and temporal patterns of ORP in DBL, which hinders understanding the full functions of DBL. So, in this study, a worldwide species Potamogeton crispus in eutrophic waters was selected. Using a microprofiling system equipped with microsensors, we analyzed the spatio-temporal patterns of ORP at a micro-scale (μm-mm).

2. Materials and methods

Submerged macrophytes *Potamogeton crispus* were collected from the trophic water of the Yihe river (E 118.32°, N 35.11°), in China. There are plenty of inorganic nutrients in the Linyi urban section of the Yihe River. Mature leaves were covered by thick periphyton layers due to high nutrient availability and suspended solids. Ten intact, healthy plants were collected for measurement of ORP microprofiles and periphyton at the seedling stage, rapid growing stage, stable growing stage, and declining stage, respectively, from March to June in 2015.

An intact Potamogeton crispus was placed in a aquarium of 5000 mL river water. The plant was fixed firmly on an agar plate (4%, w/w), which was then placed in a thermostatted water tank (20 °C). Experiment were carrried out under controlled quantum flux densities 200 μE m-2 ·s-l, provided with a halogen lamp through fibre optics [2]. The ORP was determined by a microelectrode with 10 μm in tip diameter, which was combined with a Ag-AgCl reference macromicroelectrode (Unisense, Denmark). The microelectrode was calibrated linearly by redox buffers under the laboratory condition [2], and it was moved vertically by a motor-driven micromanipulator, which could be read at 1 μm precision. To avoid possible disturbance by the moving electrode, only when the electrode was moved down were microprofiles measured.

Periphyton on the P. crispus surfaces was collected on GF/C filters. The filters were dried (105 °C, 24 h) after which the dry weight (DW) was measured, and the filters were then heated at 550°C (4 h) in a muffle furnace (SX3-4-16), after which the ash weight (AW) was determined. The ash free dry weight (AFDW) was the difference between the DW and the AW [2]. The chlorophyll-a (Chl-a) level was measured using the standard method (China EPA).

Univariate data analysis was performed using SPSS for Windows version 17.0. The periphyton ash weight, ash free dry weight, dry weight, chl-a of, ORP at various growing stages, and that of different parts of macrophytes were analyzed using ANOVA. When the difference was significant, the Tukey's honestly significant difference test (Tukey's HSD) is used ($p<0.05$).

3. Results

3.1. Characteristics of ORP microprofiles at different temporal and spatial scales

In a vertical direction from the leaf surface within 2 mm, ORP reduced significantly with decreasing distance from leaf surfaces. On a temporal scale, ORP in the microprofiles changed significantly with highly diverse patterns among the various growing stages in the entire life cycle of macrophyte (figure 1). The variation of ORP was flat at seedling stages, and its amplitude increased as the plants grew. The ORP gradient was the steepest with decreased minimum (428.51 mv) at stable growth stages. Whereas, ORP gradients had spatially significant differences at declining stages due to thick periphyton.

On a spatial scale, ORP microprofiles showed highly diverse patterns among different parts of individual mature plants (figure 2). ORP microprofiles around apical leaves (younger leaf) at the shoot apex varied slightly, and those of the leaves (mature leaf) at the middle section of shoot were steepest. ORP microprofiles around stems at shoot middle section and basal leaves (senescent leaf) at the shoot decreased slightly when entering the dense periphyton layer, and were minimum at the surface of stems and leaves. And that, the periphyton exerted a significant effect on the ORP distribution (figure 2). The ORP amplitude of fluctuations obviously decreased fllowing periphyton removed.

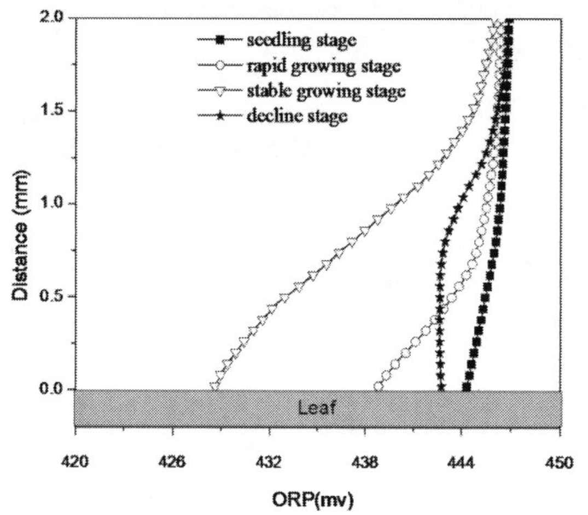

Figure 1. ORP microprofiles around *Potamogeton crispus* leaves during different growth periods. Three profiles were measured at different points on *P. Crispus*. A distance of 0.0 indicates the leaf surface.

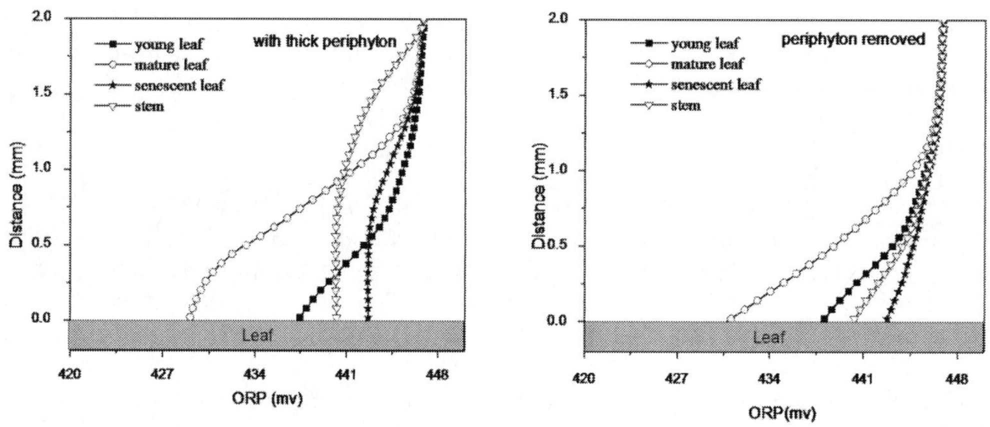

Figure 2. ORP microprofiles around different parts of *Potamogeton crispus*. A distance 0.0 indicates the macrophyte leaf surface.

3.2. Characteristics of periphyton at different temporal and spatial scales

The periphyton attached to P. Crispus leaf surfaces showed obvious dynamics during the life cycle (Figure 3a). Dry weight (DW), ash-free dry weight (AFDW), ash weight (AW), total organic cabon (TOC), total nitrogen (TN), total phosphorus (TP), and thickness of periphyton all increased significantly with the growth of P. Crispus. Periphyton increased significantly in the order of: apical leaves, stems, middle leaves, and basal leaves.

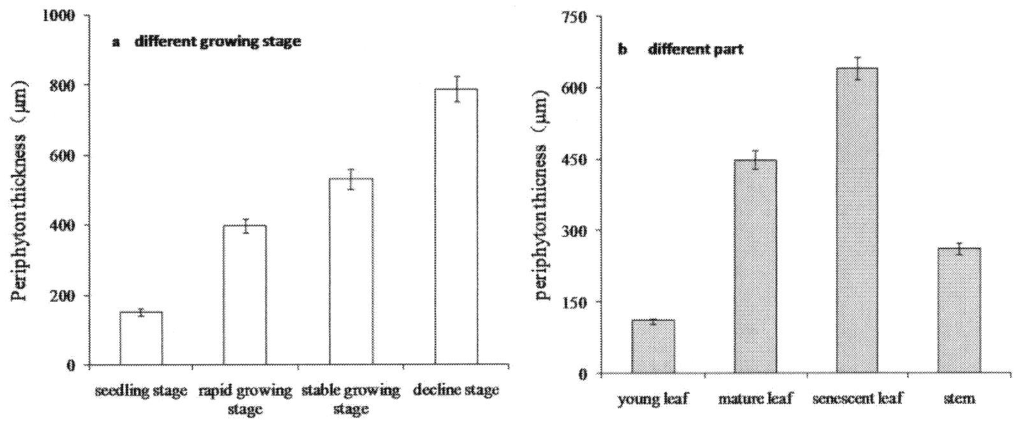

Figure 3. Thickness of periphyton attached on *Potamogeton crispus* during different growth periods and on different parts of *Potamogeton crispus*.

4. Discussion

4.1. Correlations between ORP and periphyton in a diffusive boundary layer

ORP indicates the tendency to receive or donate electrons. It is an important index of the natural aquatic environment affected by many biotic and abiotic factors [2, 5]. A correlation analysis was conducted between ORP and periphyton in the diffusive boundary layer around P. Crispus, in order to study the mechanism of ORP change (Figure 4). Data in the P. crispus life cycle demonstrated a negative correlation between ORP at leaf surface and periphyton indices such as DW, AFDW, AW, total organic carbon (TOC), total nitrogen (TN), total phosphorus (TP), and thickness of periphyton. However, the relatively low relevant coefficient (in the range 0.415 to 0.835) could be caused by a number of factors, as ORP was a combination of various reduction and oxidation [2]. The ORP at leaf surface increased and ORP gradients flattened following the removal of periphyton, possibly because of the diffusion resistance of oxygen decrease, and weakening micro-environment oxidation. Thus, dense periphyton increased the heterogeneity in DBL around leaves and stems.

4.2. The implications of ORP spatio-temporal patterns in DBL for aquatic ecosystems

There are usually only a few millimeters in DBL thickness, with biologically complex periphyton. The physiological activity of host plants made the environment more heterogeneous, which creates many difficuties for relevant research. At present, characteristics of the microenvironment in DBL, especially ORP, are not well understood. The growth of submerged macrophytes associated with the improvement of water quality is perhaps another reason for the significant variations of ORP microprofiles around P. crispus. In addition, the morphological and physiological complexity of P. crispus can provide a greater diversity of microhabitats for microorganisms to live in. Our previous study showed that oxygen produced via macrophyte photosynthesis could diffuse into bulk water. Therefore, the aerobic oxygen microregion formed at the surface of stems and leaves, and the anoxic oxygen microregion formed within periphyton due to aerobic organic matter decomposition [2,6,4], could be the main driving forces of ORP variation. The heterogeneous oxidation-deoxidation microenvironment may help to transform nutrients, and to regulate the nutrient cycle in aquatic ecosystems, and thus to alleviate lake eutrophication [1].

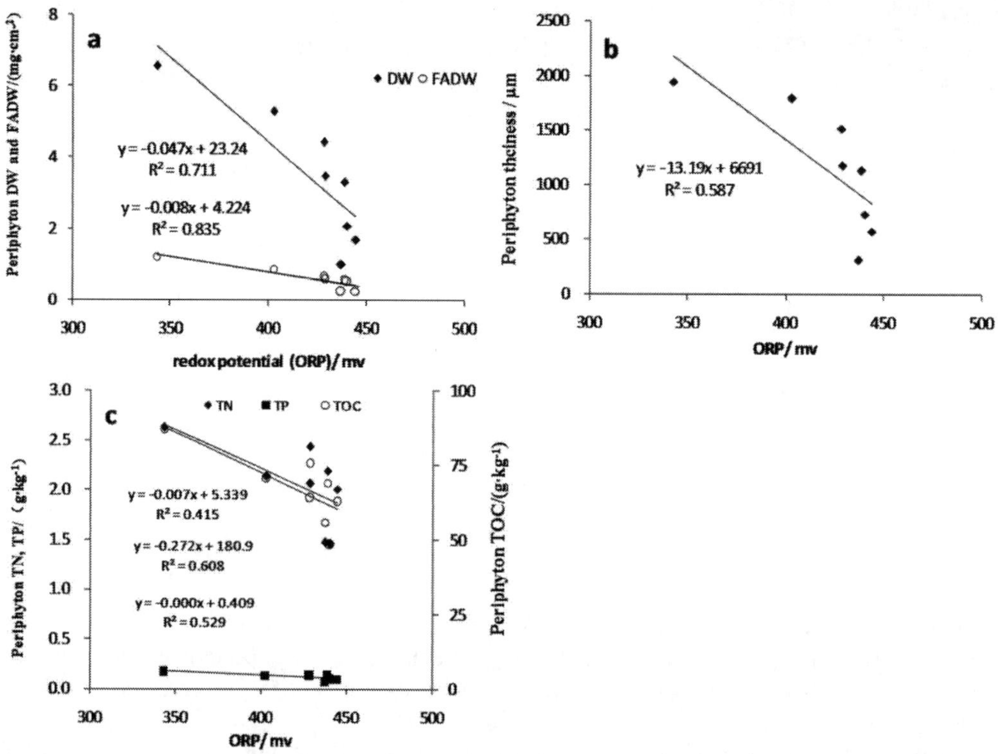

Figure 4. Correlations between the redox potential (ORP) distribution and the periphyton indices on *Potamogeton crispus.*

In the study, we analyzed firstly the spatio-temporal patterns of ORP microprofiles, revealing a heterogeneous oxidation-reduction microenvironment in DBL at the micron scale. The spatial patterns of ORP microprofiles were revealed via studying different parts of a whole plant, and the temporal patterns of ORP microprofiles were revealed via studying different growing stages within macrophyte life cycles. The results indicate that the growth status of submerged macrophytes have a significant impact on ORP microprofiles. Our results provide new information for exploring DBL functions of submerged macrophytes in eutrophic waters. We advocate more studies, focusing on micro-scale distributions of soluble substances (e.g. NH4+-N, NO3—-N, NO2—-N, PO43—, and CO2) and microorganism, because relevant results may help to understand the regulation of DBL on nutrient cycling in eutrophic ecosystems.

5. Conclusion

ORP microprofiles showed significant variations across different growing stages, with the steepest gradient at the stable growing stage affected synthetically by plant photosynthesis capacity, periphyton, light, and temperature. The steepest gradient of ORP occured in the DBL of mature leaves on a whole macrophyte, and was affected by plant physiological activity and periphyton. The negative correlation between ORP level at leaf surface and periphyton indices provided new insights for understanding the function of a DBL.

References

[1] Barko J W and James W F 1998 Effects of submerged aquatic macrophytes on nutrient dynamics, sedimentation, and resuspension The structuring role of submerged macrophytes in lakes *Springer* **131** 197-214

[2] Dong B, Han R, Wang G and Cao X 2014 O_2, pH, and redox potential microprofiles around

Potamogeton malaianus measured using microsensors *PLoS ONE* **9**(7): e101825

[3] Sand-Jensen K, Revsbech N P and Barker Jörgensen B 1985 Microprofiles of oxygen in epiphyte communities on submerged macrophytes *Marine Biology* **89**(1): 55-62

[4] Dong B, Han R M, Wang G X, et al. 2015 Micro-optode analysis of the spatio-temporal dynamics of O_2 in the micro-boundary layer around Potamogeton crispus leaves and stems *Acta Ecologica Sinica* **35**(24): 8202-10

[5] Boros G, Søndergaard M, Takács P, Vári Á and Tátrai I 2011 Influence of submerged macrophytes, temperature and nutrient loading on the development of redox potential around the sediment–water interface in lakes *Hydrobiologia* **665**(1): 117-27

[6] Pietro K C, Chimney M J and Steinman A D 2006 Phosphorus removal by the Ceratophyllum periphyton complex in a south Florida (USA) freshwater marsh *Ecological Engineering* **27**(4): 290-300

Optimization of urban wastewater treatment plants process with low C/N ratio

L Zheng[1, 2], G M Xu[1], J Chen[1], B Chen[1], Z Lv[1] and Y A Yang[1]

[1]Changzhou Drainage Administration, Changzhou 213017, China

Email:39949060@qq.com

Abstract. In southern China, the inflow of water to wastewater treatment plants has a lower concentration of organic matter. This causes treatment plants to face issues in the denitrification and phosphorus removal processes such as deficient carbon sources, high energy consumption, and unstable nitrogen removal. To resolve these issues, we propose the reconstruction of the internal reflux port, improvement of the internal reflux ratio to 200%, the addition of carbon source to anoxic zone, and the addition of phosphorus removal agents in secondary settling tank. The results of study show significantly improved efficiency of nitrogen and phosphorus removal, which ensures the stability of subsequent supply of reused water.

1. Introduction

The concentration of organic matter of influent water is oftenlow in many wastewater treatment plants in southern China, and therefore the actual water quality of outlet is far lower than the designed value [1]. Anaerobic-anoxic-oxic (AAO) is a commonly used process for the biological removal of nitrogen and phosphorus [2, 3]. Inlet with a low C/N ratio could result in several problems, including insufficient carbon source for denitrification, high energy consumption of nitrification and denitrification, and unstable nitrogen removal [4, 5]. This paper illustrates the problems encountered during the actual operation process of a sewage treatment plant in Changzhou, Jiangsu, China. Through partial reconstruction and optimization of operations,the plant improved its nitrogen and phosphorus removal performance, ensuring the stability in subsequent supply of reused water.Furthermore, this reconstruction and optimization can providea reference for the operation management of other sewage treatment plants.

2. Project overview

The designed total treatment capacity of the sewage treatment plant was 50,000m3/d (construction occurred in two phases, the scale of every independent system, excluding the adjoining inlet water pump, each phase was 25,000 m3/dfor each phase). The biochemical unitsused the traditional AAO process. Figure 1 shows the flow of process. The total design coefficient of variation (diurnal coefficient of variation * temporal coefficient of variation) Kzwas1.38. After chlorine dioxide disinfection, part of the effluent water was reused as cooling water in the steel industry. Other effluent water was discharged after passing through the ozone contact unit. The effluent water quality was assessed in accordance with the Class A standard of the 'Discharge standard of pollutants for municipal wastewater treatment plant' (GB18918-2002). The design and actual water quality of influent water and design quality of outletare shown in Table 1.

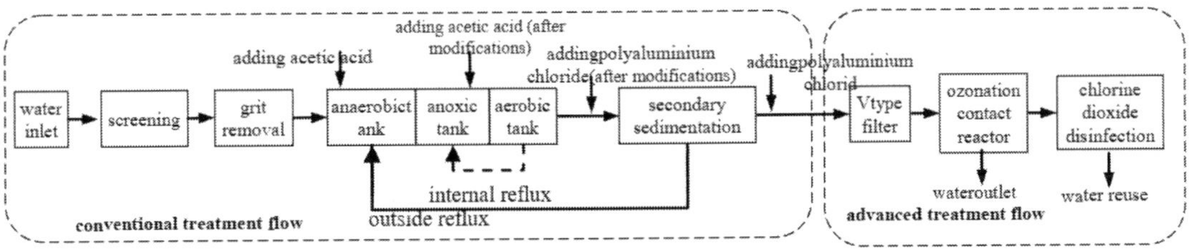

Figure 1. Theflowsheetof wastewater treatment process.

Table 1. The design and actual water quality of influent water and design quality of outlet(notes: actual inlet water quality is the annual average).

Water quality index	Design inletquality	Actualinlet quality(minimum-maximum)	Design outletquality
COD_{cr} (mg/L)	400	206(164-305)	≤50
BOD_5 (mg/L)	180	70.6(60.4-78.8)	≤10
SS(mg/L)	250	80(76-80)	≤10
TN (mg/L)	45	40.7(28.8-46.9)	≤15
NH_3-N(mg/L)	35	33.5(24.7-43.7)	≤5
TP(mg/L)	4	4.15(2.52-9.27)	≤0.5

3. Operational conditions and issues

The flap valve of the internal reflux port is located at the entrance of the anoxic zone, which leads to the occurrence of a vortex on the water surface (Figure 2).In addition, the Oxidation-Reduction Potential (ORP) theoretical control value of the anaerobic tank is -200to-400mv, but the actual values measured by an ORP meter were all around -112mv.

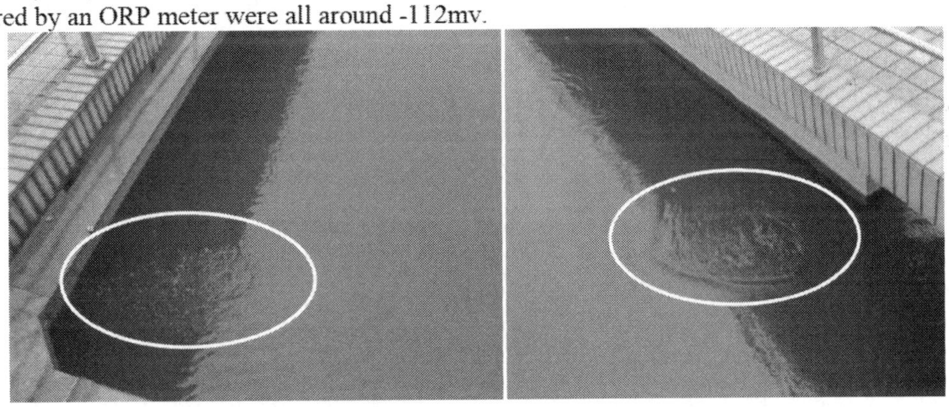

Figure 2. Vortex at the entrance of anoxic zone.

Additionally, the water quality indicators of water samples from differentuntis were also continuously determined (from date1 to date6, 2015.5.4-2015.5.9). The results are shown in Figure 3. Figure 3 shows that since the internal reflux port cannot keep the anaerobic environment of the anaerobic zone, the AAO process turns into the AO process.Due to the lack of anaerobic phosphorus release, the total phosphorus in the secondary sedimentation tankis high, approximately1.0mg/L, which increases the energy consumption of chemical phosphorus removal in the subsequent advanced treatment process. We also observed that concentration of NO_3^--N in the anoxic zone increased, indicating nitrification. Therefore, the anoxic results are not fully achieved, which elevates the outlet NO_3^--N. This posescertain risks for thestable compliance and stability of subsequent supply of reusedwater.

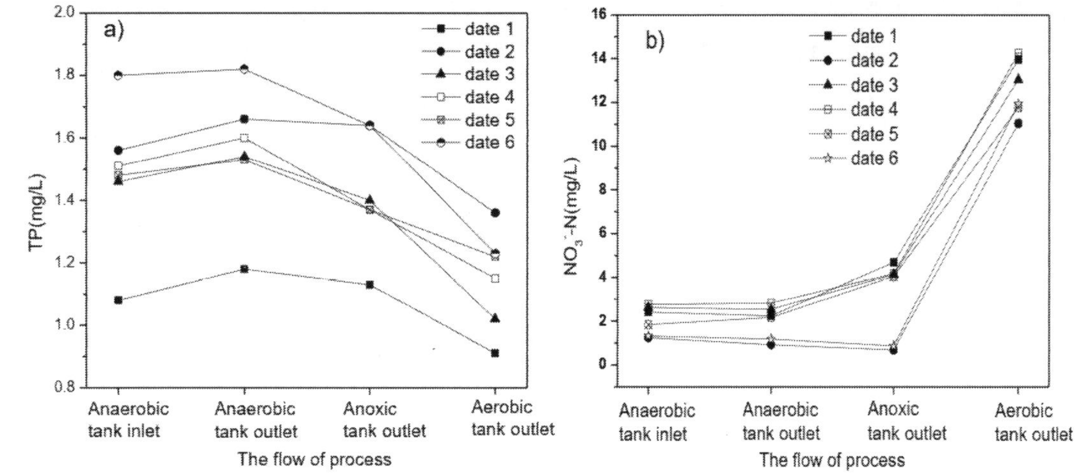

Figure 3. Total phosphorus and nitrate nitrogen indicators of each unit before reconstruction (a. phosphorus indicators, b. nitrate nitrogen indicators).

In order to ensure thestable compliance and the stability of reused water supply, the following measures were adopted.

4. Adopted measures and results

4.1. Reconstruction of internal reflux port

According to phosphorus removal theory, to achieve higher phosphorus removal, phosphorus release must be sufficient [6]. At the same time, phosphorus accumulating bacteria can only release a large amount of phosphorus under strict anaerobic conditions, remaining in a hungry state, which will create a condition for massive phosphorus absorption in the aerobic zone [7]. Since the flap valve was used at the internal reflux port, part of internal reflux nitrification liquid flowed into the anaerobic zone. The ORP measured in that zone was -112mv (theoretical control vale -200to-400mv) due to the NO_3^--N it carried, and the anoxic zone was -90mv (theoretical control vale -50to-110mv), which was unfavorable for phosphorus release.

Therefore, the internal reflux port was reconstructed in a manner in which the position of the original inner reflux port was extended to a certain distance beyondthe anoxic zone to prevent water from returning to the anaerobic zone and destroying the anaerobic environment. The internal reflux ports before and after the reconstructions are shown in Figure 4.

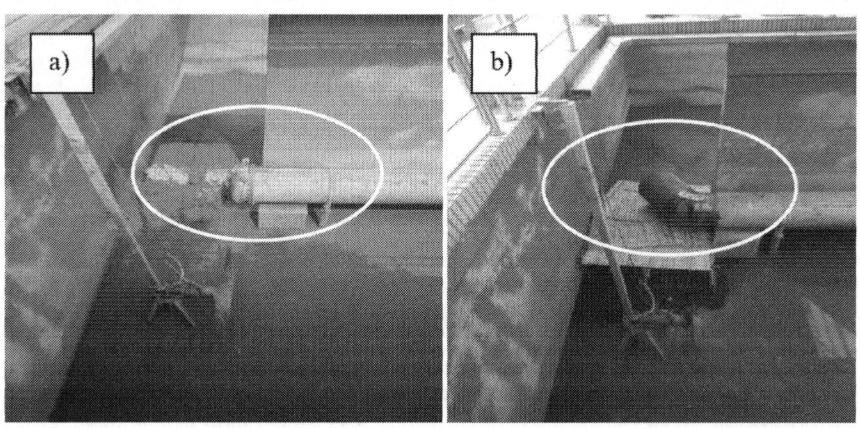

Figure 4. The internal reflux ports before and after the reconstruction.
(a. beforereconstruction b. afterreconstruction).

After the reconstruction of the anaerobic inner reflux pipelinein phase I, the anaerobic conditionsof phaseIsignificantly are improved. ORP was -380mvand -60mv in anaerobic tank and the anoxic tank. After the reconstruction, a strict anaerobic environment was ensured, which provided a good environment for subsequent regulation.

4.2. Improvement of internal reflux ratio

Through the internal reflux system, from aerobic zone, and then to the anoxic zone, nitrate nitrogen is provided for denitrification. When the nitrate load in the anoxic zone exceeds the load of ordinary heterotrophic denitrification, denitrification is stimulated to remove phosphorus [8]. From this perspective, a higher reflux ratio is better for phosphorus removal by denitrification. Therefore, the reflux ratio is closely related to the nitrogen removal by denitrification [9]. There exists the relationship of thedenitrification rateη and reflux ratio R, as shown in equation 1, equation 2 and equation3:

$$\eta=\frac{TN_{inlet}-TN_{outlet}}{TN_{outlet}} \tag{1}$$

$$R=\frac{V_{return}}{V_{inelt}} \tag{2}$$

$$\eta=\frac{R}{1+R}\times 100\% \tag{3}$$

WhereTNinlet TN value of inlet water, TNoutletTN value of outlet water,Vreturnamount of water returned from the aerobic zone to the anoxic zone, Vinelt amount of inlet water flow.

The relationship reveals that a larger internal reflux ratio R could lead to better nitrogen removal by denitrification.However, the effect of theinternal reflux ratio R is not always better as R increases. When R is too high, the power consumption and dynamic costs will significantly increase as well. Furthermore, if the reflux ratio is too high, it shortens the hydraulic retention time in the anoxic zone, affecting the performance of phosphorus removal by denitrification. Moreover, when R is higher, more DO is brought to the anoxic section, which will destroy the anoxic environment to a certain extent. The dissolved oxygen brought to the anoxic section will replace the NO3--N and consume the carbon source in the water, resulting in decreasing total nitrogen removal. However, if the reflux ratio is too low, a large amount of nitrification liquid cannot return to the anoxic pond. Excessive nitrification liquid not only affects the removal rate of total nitrogen but also might cause nitrate nitrogen to undergo denitrification in the secondary sedimentation, which could destroy the sedimentation process of the sludge and affect the performance of discharge [10].

Originally, a 100% internal reflux ratio was used, but the nitrogen and phosphorus removal performance was poor. Under the premise of assured anaerobic environment through reconstruction, the internal reflux ratio increased to 200%.

4.3. Carbon dosing point

The carbon dosing in the anaerobic section showed poor nitrogen and phosphorus removal performance. This may be caused by the added acetic acid being first excessively absorbed by phosphorus accumulating bacteria prior to entering the anoxic section for denitrification, resulting in limited addition of carbon source for the denitrifying bacteria in the anoxic section. Therefore, adjustments were made to add acetic acid in the anoxic section. After running the adjustment measures, i.e., the aforementioned reconstruction of internal reflux port and increased internal reflux ratio, 35mg/L of acetic acid was added to the anoxic and anaerobic zones (anoxic and anaerobic zones in phase I; only the anaerobic zone in phase II) to compare its performance upon nitrogen and phosphorus removal in phase I and phase II.

The monitoring results of the whole process are shown in Figure 5(1-3 were at a different date). The fluctuation of thenitrogen indicator shows that when the carbon source was added to the anaerobic section, the concentration of NO3--N of the anaerobic section decreased due to the absorption of phosphorus by denitrification. The anoxic section continued to absorb phosphorus by denitrification until the absorption of phosphorus and nitrification in the aerobic section occurred. Because priority was given to removal of phosphorus indicator, the NO3--N indicator of that process in the aerobic section was instantaneously increased to 14mg/L, and thus the pressure of the outlet waterto comply with the standard was greater. After a high-quality carbon source was added to the anoxic section, nitrification took place first instead of absorption of phosphorus via denitrification. The concentration of nitrates even decreased to approximately 0.14mg/L. After the nitrification in the aerobic section, the nitrate nitrogen indicator of the effluent water was approximately 1mg/L lower compared to the anaerobic zone with the added carbon source, while the compliance safety of thenitrogen indicator increased.

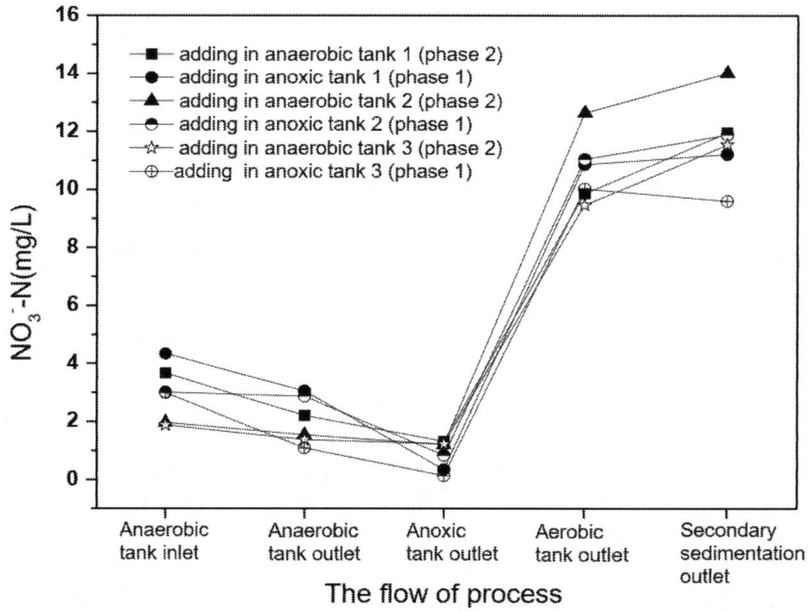

Figure 5. The NO_3^--N values of the entire process when thecarbon source was added to the anaerobic section and anoxic section.

After several days of comparison of the outlet water (Figure 6), the results showed that when the carbon source was added to the anaerobic section, with the addition of the carbon source to the inlet water,anaerobic phosphorus release and absorption of phosphorus by denitrification occurred and the nitrogen and phosphorus indicators lowered. After the carbon dosing point changed, priority was given to the removal of nitrogen. Although PO43- increased, the nitrogen indicator significantly lowered. This showed that the changes in the carbon adding point not only provided the necessary carbon for denitrification, but also effectively reduced the nitrates in the effluent water. On the other hand, the addition of the carbon source reduced the impact of anaerobic phosphorus release by the nitrates in the external influx, which can effectively improved the TN removal without affecting the TP of effluent.

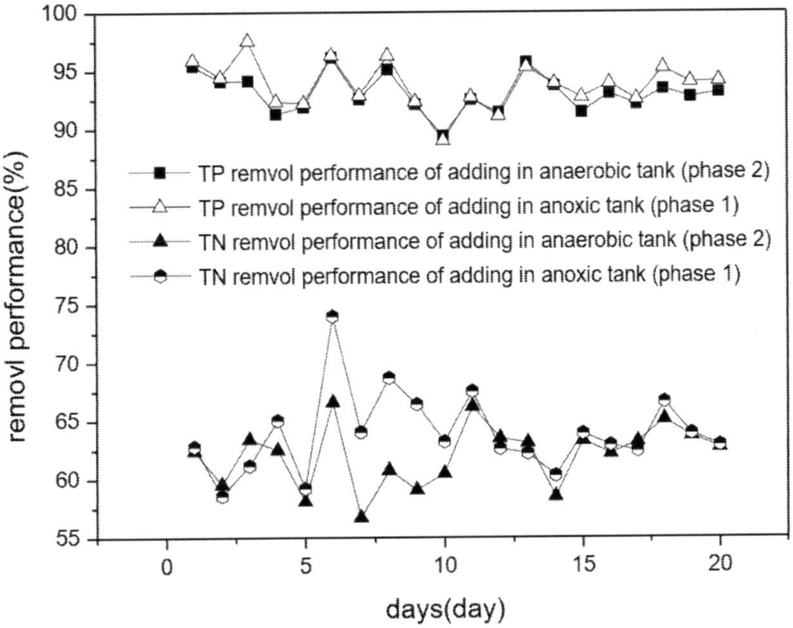

Figure 6. The performance of nitrogen and phosphorus removal when the carbon source was added in the anaerobic section and anoxic section.

4.4. Changing chemical phosphorus removal from V type filter dosingto the multi-point dosing

In the process of phosphorus removal by chemical precipitation, according to the relative position of dosing point of chemical agents and the biological processing unit, the dosing stages can be divided into pre-dosing, central dosing, post-dosing, dosing before filter chamber and multi-point dosing. Originally, a chemical phosphorus removal agent was added before the V type filter. As a result of thehigh fluctuation of TP in the influent water, if the maximum TP of the influent water reached9.27mg/L, then chemical phosphorus removal could not ensure the quality of the effluent water. Therefore, in order to ensure the quality of effluent, multi-point dosing method was adopted where chemical agents were added in the influent of secondary sedimentation and the influent before V type filter respectively. The partial dosing before the secondary sedimentation causes the precipitate formed in the secondary sedimentation to be discharged with the sludge. Comparing the processes before and after the adjustment, when the phosphorous removal agent was added to the inlet water before the V type filter, the TP of the outlet water of the secondary settling tank fluctuated to 1.0mg/L. However, when adding the agent to the secondary settling tank inlet water and before the V type filer, the TP of the outlet water of the secondary settling tank decreased to less than 0.5mg/L, which had a minimal impact on the removal of nitrogen in the biological section. This dosing method can also strengthen emergency phosphorus removal by adding phosphorus removal agents in the secondary sedimentation when abnormal inflow is discovered.

5. Conclusions

This study focused on a sewage treatment process in Changzhou, Jiangsu, China. Through reconstruction and optimization of treatment operations, the results revealed:

- Before the reconstruction, the actual ORP measurement of the treatment plant was around-112mv. In addition, the integrity of the anaerobic environment of the anaerobic zone could not be guaranteed.By extending the original internal reflux port to a certain distance beyond the anoxic zone, the phase I processing after the reconstruction of the anaerobic reflux pipeline significantly improved the anaerobic conditions. ORP was -380mv in anaerobic tank, and -60mv in the anoxic tank. After the reconstruction, a strict anaerobic environment was ensured, which provided a good environment for subsequent regulation.

- The system experienced the following benefits by utilizing biological nutrient removal theory targeting phosphorus and nitrogen: increased internal reflux ratio, carbon dosing in the anoxic section and multi-point dosing of chemical phosphorus removal agents. The results showed that both nitrogen and phosphorus removal efficiency improved. These results ensure the stability of the subsequent supply of reclaimed water.

References

[1] Hou H, Wang H and Zhou Q 2005 Effect of influent COD concentration and C/N ratio on denitrification *China Water and Wastewater-China* **21** 19-23

[2] Lu Q, Wu H, Li H and Yang D 2015 Enhanced biological nutrient removal in modified carbon source division anaerobic anoxic oxic process with return activated sludge pre-concentration *Chinese Journal of Chemical Engineering* **23** 1027-34

[3] Zhang Z, Feng Y, Su H, Xiang L, Zou Q, Gao P and Zhan P 2015 Influence of operating parameters on the fate and removal of three estrogens in a laboratory-scale AAO system *Water Sci. Technol.* **71** 1701-08

[4] Lee Y, Kim Y, Chang N, Lee J and Lee B 2007 Development of sequencing batch reactor with step feed and recycle *Water Sci. Technol.***55** 477-84

[5] Jin P, Wang X, Zhang Q, Wang X, Ngo H and Yang L 2016 A new activated primary tank developed for recovering carbon source and its application *Bioresour Technol.* **200** 722-30

[6] Li J, Jin Y, Guo Y and He J 2013 Enhancement of phosphorus removal in a low temperature A(2)/O process by anaerobic phosphorus release of activated sludge *Water Sci. Technol.* **67** 2437-43

[7] Zeng W, Li B, Wang X, Bai X and Peng Y 2016 Influence of nitrite accumulation on Candidatus Accumulibacter population structure and enhanced biological phosphorus removal from municipal wastewater *Chemosphere* **144** 1018-25

[8] Zeng R, Lemaire R, Yuan Z and Keller J 2003Simultaneous nitrification, denitrification, and phosphorus removal in a lab-scale sequencing batch reactor *Biotechnol Bioeng.* **84** 170-78

[9] Bernat K, Wojnowska-Baryla I and Dobrzynska A 2008 Denitrification with endogenous carbon source at low C/N and its effect on P(3HB) accumulation *Bioresour Technol.* **99** 2410-18

[10] Tan T and Ng H 2008 Influence of mixed liquor recycle ratio and dissolved oxygen on performance of pre-denitrification submerged membrane bioreactors *Water Res.* **42** 1122-32

Acknowledgments

This work was supported by the Key Special Program on the S&T for the Pollution Control and Treatment of Water Bodies of China (No. 2012ZX07301-001), National Natural Science Foundation of China (No. 51308078), Natural Science Foundation of Jiangsu Province, China (No. BK20130252).

Isolation and characterization of an antibacterium against *Vibrio harveyi* 11593 from a mixed pond with *penaeus japonicus* bate, *portunus trituberculatus* and *ruditapes philippinarum* in China

X H An [1, 2, 3], Y J Xu[4], F H Li[1, 2], G Z Ma[1] and L T Li [1, 3, 5]

[1] HuaiHai Institute of Technology, LianYunGang JiangSu, China
[2] Jiangsu Marine Resources Development Research Institute, LianYunGang JiangSu, China
[3] Co-Innovation Center of Jiangsu Marine Bio-industry Technology, LianYunGang JiangSu, China
[4] Lianyungang Normal college, LianYunGang JiangSu, China

Email: lilt@hhit.edu.cn

Abstract. Aquaculture diseases, particularly Vibriosis, are becoming a pressing concern due to incurred aquacultural loss in China. Commercial antibiotics may be used to control it, but its increased antibiotic resistance to prominent pathogenic bacteria has become a prevalent problem nationwide, and a global threat to public health. Probiotics are recommended instead because they are healthy, environment-friendly, and capable of maintaining productivity. An antibacterium against *Vibrio harveyi* 11593 was isolated from a mariculture pond with shrimp, crabs, and shellfish in China. The bacterium, E14, has an inhibitory zone diameter (DIZ) of 24.5 ± 0.5 mm. The strain was identified as *Bacillus pumilus* based on morphological observation, conventional biochemical tests, and 16S rDNA sequence analysis. The gram-positiveand motile bacterium is around 1.10–1.32 μm × 0.67–0.83 μm in size. Optimized conditions for antimicrobial substance production of *B. pumilus* E14 require that it be cultured for 26 h at 28 °C, with an initial pH of 7.0 in 100 mL/500 mL LB with 3% NaCl. The *B. pumilus* E14 cultures were confirmed to be safe and efficacious and actually worked to protect the host animal in shrimp larvae (*Penaeus chinensis*) culture. The *B. pumilus* E14 obtained in this study strengthened the strain's defense against aquaculture disease and made a good candidate for an alternative probiotics and benefit to sustainability of aquaculture.

1. Introduction

Today, the aquaculture is honoured as one of the fastest growing food-producing sectors in the world, posting an average annual growth rate of 8.9% since 1970 [1]. According to FAO statistics in 2013, world food fish aquaculture production attained another all-time high of 70.5 million tonnes. As the largest aquaculture producer in the world, China has produced 43.5 million tonnes of food fish in 2013 [2].

However, aquaculture is currently suffering from grave economic loss due to infectious diseases of microbial origin, which result in high mortality rates and lesions [3]. Vibriosis is one of the biggest causes of aquaculture loss in China [4]. It is caused by bacteria, mainly of the genus *Vibrio*, such as *Listonella anguillarum* (formerly *Vibrio anguillarum*), *Vibrio ordalii*, *Vibrio harveyi*, *Vibrio vulnificus*, *Vibrio parahaemolyticus*, *Vibrio alginolyticus*, and *Vibrio salmonicida* [5-7]. These bacteria have been identified as major reasons for most common disease outbreaks in fish and shellfish. Among them, *V. harveyi* is considered an acute pathogen of ocean fish and invertebrates, particularly penaeid shrimp [8].

Thus, controlling Vibriosis is vital to maintaining aquatic productivity and meeting ever-growing human needs. The disease can be controlled using various methods, notably commercial antibiotics. However, commercial antibiotics used for disease treatment produces ill-natured effects, including toxicity to reared organisms and the release of chemical residues into the environment, which cause risks to animal and human health [9].

These factors prompted the development of new dietary supplementation strategies, such as using health- and growth-promoting compounds, including probiotics, prebiotics, synbiotics, phytobiotics, and other functional dietary supplements[10]. Application of probiotics in aquaculture seems relatively recent, but there remains an increasing enthusiasm in environment-friendly treatments. Probiotics have become quite common, and their treatment may broaden a spectrum for greater nonspecific disease protection [11, 12].

At present, there exists an increasing need to identify new effective antibacteria with few undesirable side effects for shrimp and other sea farming organisms. Hence, aquaculture disease control research has steered toward the discovery of probiotics for prevention and treatment of microbial infections.

The co-culture mode in a mixed pond with *Penaeus japonicus* Bate, *Portunus trituberculatus* and *Ruditapes philippinarum* polycultures is one of the most common and fastest-growing mariculture modes in the east coast of China. A comparison of the bacterial community structure between the co-culture and control pond has been reported, demonstrating the richer diversity of the bacterial communities in co-culture ponds [13]. Samples collected from this common model can serve as a no-rejection-effect candidate for the protection of animals in an aquaculture system. In this study, we aimed to isolate and screen antibacteria against a well-known fish pathogen, such as *V. harveyi*. We then identified the antimicrobial strain and optimized various parameters for maximum antibacterial ability.

2. Material and methods

2.1. Sample collections

The sediments sample was collected from a seawater pond with *Penaeus japonicus* Bate, *Portunus trituberculatus Ruditapes philippinarum* by the scoop with the long handle at 25~30℃ in mid-July, 2012. The pond was one of ten artificial ponds, each pond is about 200×50×1.0 m^3 and located in Ganyu County, Lianyungang city, Jiangsu province, eastern China (34o58'23"W-119o11'36"E). *2.2 Medium*

2216E: Seawater 1000 mL with tryptone 5g, yeast extract 1g, FePO4 0.1g, and Agar 20g (optional). pH 7.0-7.2.

LB with 3% NaCl: Distilled water 1000 mL with tryptone 10 g, yeast extract 5g, NaCl 30 g and Agar 20 g (optional). pH 7.0-7.2.

YPG: Distilled water 1000 mL with tryptone 5g/L, glucose 5g,yeast extract 5g, pH 7.0-7.2.

PPM: Distilled water 1000 mL with tryptone 20g, glucose 18 g, KNO3 5 g, NaCl 1g, pH 7.0-7.2.

NYPD: Distilled water 1000 mL with glucose 10g, beef extract 8g, yeast extract 5g, pH 7.0-7.2.

NA: Distilled water 1000 mL with tryptone 20g, beef extract 1g, yeast extract 2g, NaCl 5 g, pH 7.0-7.2.

CM: Distilled water 1000 mL with tryptone 20 g, beef extract 3g, pH 7.0-7.2.

Above reagents were from commercial sources: tryptone, yeast extract, beef extract, Oxoid; NaCl, FePO4, KNO3, Sinopharm Chemical Reagent Co., Ltd; Agar, Biosharp.

All medium was made suitable for sterilization in the autoclave at 121℃ for 30 min.

2.2. Bacterial isolate

The sediments sample 5 g with seawater sterilized 45 ml were cultured on 120 rpm shaker at 28℃ for 2 h. Then microorganisms were isolated by dilution plate technique described by Ping Shen [14]. The culture was 10-fold serial diluted to 10^{-1}~10^{-7} with seawater sterilized and then inoculated into 2216E agar plates to culture at 28 °C for 3 days and to identify different strains. Single colonies were picked based on its morphology, color, size, and streaked onto 2216E agar plates several times until the morphology of colonies was uniform.

2.3. Selection of antimicrobial strains to inhibit in vitro growth of V. harveyi 11593

The ability of isolated bacteria to inhibit the growth of *V. harveyi* 11593 was detected by agar well diffusion assay [15]. Agar plates were previously coated with 100 μL of *V. harveyi 11593*, which corresponded to 5 x 10^7 cfu/mL. , 6-mm diameter wells were made in the agar plates by a sterilized borer, then 50 μL E14 cutured solution was poured into the wells and incubated at 28℃ for 24 h, and diameter of inhibitory zone (DIZ) was measured for the antimicrobial activity. This test was done in triplicate. A species with the largest DIZ was finally selected for further study.

2.4. Determination for appropriate dose-volume of antimicrobial strain against V. harveyi 11593

The Oxford cup method was used in this test. The broth of antagonistic bacterial strains and *V. harveyi 11593* were 10-fold serial diluted to 5.0 × 10^7 CFU/mL with sterile seawater respectively. Then two 2216E agar plates were previously spread with 100 μL dilution of *V. harveyi 11593*. Five oxford cups (inside diameter, 6 mm; outer diameter, 8 mm; height, 10 mm) were put vertically on every plate. The center cup was poured into 100 μL sterile 2216E as control, and four cups around the center were poured into defferent volume (20, 40, 60 or 80 μL in one plate and100, 120, 140 or160 μL in the other) of antagonistic bacterial strain dilution. All the plates were incubated at 28℃ for 24 h and The DIZ was measured for the antimicrobial activity. The test was repeated three times.

2.5. Identification of antimicrobial strain

The identification of the isolated strain was assessed by morphological observation, conventional biochemical tests and 16S rRNA gene analysis. The biochemical tests were determined with bacterial biochemical tube trace identification kit (Hangzhou Tianhe Microoganism Reagent Co., Ltd.) according to the manufacturer's instructions. Morphological observation was carried out using the Gram-staining and Nikon eclipse 90i microscope (Japan). Two oligonucleotides, based on the report of Dunbar et al. [16], were used to amplicate 16S rDNA of the isolate: 27F -AGAGTTTGATCCTGGCTCAG and 1492R-TACCTTGTTACGACTT (synthesised by Shanghai Sangon Biotech engineering co., LTD). Polymerase chain reaction (PCR) was performed with a T100 thermal cycler (Bio-Rad) using a 50 μL containing: 2.5 U Taq DNA polymelase, 1 × PCR buffer, 200 μM each dNTP, 20 pmol each primer, and 1 μl template DNA extract. The PCR conditions used were: initial denaturation at 95℃ for 5 min, followed by 30-cycles: 94℃ for 1 min, at 52℃ for 1 min, at 72℃for 2 min and with a final elongation step at 72℃ for 5 min. The reagents in this part were purchased from Takara. The 16S rDNA amplified by PCR was sequenced by Shanghai Sangon Biotech engineering co., LTD. The analysis of nucleotide sequences was performed at http://rdp.cme.msu.edu/seqmatch. Sequences were aligned using the program CLUSTAL neighbor-joining and a phylogenetic tree was made by using the MEGA 5.10 program.

2.6. Optimization of conditions for producing the maximum active substanse by B. pumilus E14

The effect of medium (pH 7.0 -7.2) on antibacterial activity of *B. pumilus* E14 was determined by incubation at 160 rpm in different medium (LB with 3% NaCl, PPM, NYPD, YPG, NA and CM) with triangular flask. The effect of temperature on antibacterial production by *B. pumilus* E14 in optimal medium was tested by incubation at different temperatures (18℃, 22℃, 28℃, 32℃ and 40℃). The effect of pH on antimicrobial production was detected by inoculating *B. pumilus* E14 in medium adjusted to different pH values (5.0, 6.0, 7.0, 8.0 and 9.0), and cultured at optimal temperature. The effect of medium volume in 500 mL triangular flask on antibacterial activity of B. pumilus was determined by incubation at medium volumes (50 mL, 75 mL, 100 mL, 125 mL and 150 mL), and cultured at optimal temperature and optimal pH value. Antibacterial activity was analysed using samples from the 24 h culture. The test was repeated three times.

2.7. The growth curve and the antibacterial activity by B. pumilus E14 at growing stage

The growth curve and the antibacterial activity by B. pumilus E14 at growing stage was studied in LB with 3% NaCl with optimized pH and temperature. Samples were fetched out every 2h, from 0 to 36h, followed by measurement of the optical density at 600 nm and the antibacterial activity. The test was repeated three times.

IOP Conf. Series: Earth and Environmental Science **39** (2016) 012018 doi:10.1088/1755-1315/39/1/012018

2.8. Feeding test of shrimp larvae (Penaeus chinensis) with strain E14

To demonstrate that the strain E14 cultures are safe and effective to be used in aquaculture, we do the following work based on Fu [17] and modified. E14 and *V. harveyi* 11593 were cultured in LB incubated at 28℃ for 24h and diluted to 5.0 × 107 CFU/mL with sterile seawater. The shrimp larvae (*Penaeus chinensis*) were obtained from a private shrimp farmer in Lianyungang, China and acclimatized in the laboratory for 3 days in PVC tanks before experimentation. Then 240 vibrant shrimp larvae with 10.0 ± 2.0 mm length were selected and divided randomly into 4 groups marked A,B,C and D, and each group was sub-grouped into 3 sub-groups named I, II, III. Group A, 50 mL sterile saline and 10 mL *B. pumilus* E14; Group B, 50 mL sterile saline and 10 mL *V. harveyi* 11593; Group C, 40 mL sterile saline and 10 mL *B. pumilus* E14 and 1 mL *V. harveyi* 11593; Group D, 60 mL sterile saline (control). The survived shrimp larvae were counted every 8 hours in the following 48 hours, during which no other feed was added.

2.9. Statistic analysis

SPSS statistical 17.0 was use in the data analyzing , and Duncan's multiple test was applied to distinguished a significant difference among samplesat the level of $p<0.05$.

3. Results

3.1. Bacterial isolates and screening

Sixteen bacterial isolates were obtained from sediments samples based on color, morphology, size, edge uniformity, and clone transparency. Eleven strains named E1-2, E2-1, E3-1, E3-2, E3-3, E5-2, E7, E11, E12, E13-2-2 and E14, respectively, were verified by agar well diffusion assay to inhibit the growth of *V. harveyi* 11593. Their DIZs were listed, and their significant differences marked in table 1. Strains E14 and E13-2-2 exhibited the top two maximum antimicrobial activities, which did not show any significant difference, but were significantly different from those of other strains. Strain E14 showed a clearer and more visible DIZ (data not shown) than E13-2-2. Strain E14 was thus selected for further research.

Table 1. The diameter of inhibitory zone from eleven isolated strains against *V. harveyi* 11593.

Strian number	Diameter Of Inhibitory Zone /mm				Strian number	Diameter Of Inhibitory Zone /mm			
	I	II	III	Mean		I	II	III	Mean
E13-2-2	18.0	17.0	16.0	17.0^{a}	E7	11.0	10.0	12.0	11.0^{bc}
E14	16.0	16.5	17.0	16.5^{a}	E3-3	10.0	10.5	10.5	10.3^{cd}
E12	12.0	12.0	12.5	12.2^{b}	E5-2	11.0	11.0	9.0	10.3^{cd}
E2-1	11.5	12.0	12.5	12.0^{b}	E3-2	10.5	8.0	9.0	9.2^{d}
E1-2	11.0	12.5	12.0	11.8^{b}	E3-1	7.0	6.5	7.5	7.0^{e}
E11	12.0	11.5	11.0	11.5^{bc}					

a, b, c, d, e: significant differences at the 5% level.

3.2. Determination for appropriate dose-volume of the antagonistic bacterial strain E14 against V. harveyi 11593

The appropriate dose-volume of the antagonistic bacterial strain E14 against *V. harveyi* 11593 was determined using the Oxford cup method (figure 1). The DIZ increased with the E14 dose-volume. The maximum DIZ (23.7 ± 0.5 mm) appeared at 80 μL (6.0 × 107 CFU/mL) and had no significant differences with those at 100, 120, 140, and 160 μL. Hence, 80 μL was was defined as the optimized dose-volume.

3.3. Classical and phylogenetic identification of antimicrobial strain E14

To know more detail about this strain, morphological, biochemical, and genetic analyses were carried out as described in the methodology. Gram-staining showed that strain E14 was a gram-positive rod approximately 1.10–1.32 μm × 0.67–0.83 μm in size (figure 2a). The flagella (figure 2b) in strain E14 was also reported with alveolate. For the genetic characterization of strain E14, PCR was performed to amplify 16S rRNA gene as described above. The 16S rDNA sequence (1512 bp) of strain E14 was aligned for comparison with available sequences from GenBank, and the nucleotide sequence was deposited (accession number JX987716). Sequences of strain E14 shared the greatest similarity with those of *Bacillus pumilus* (identical to the *B. pumilus* MCCC 1A08208 and *B. pumilus* MCCC 1A06858). A phylogenetic tree based on the bacterial 16S rDNA sequences showed the close relationship of E14 with *B. pumilus* (figure 3).

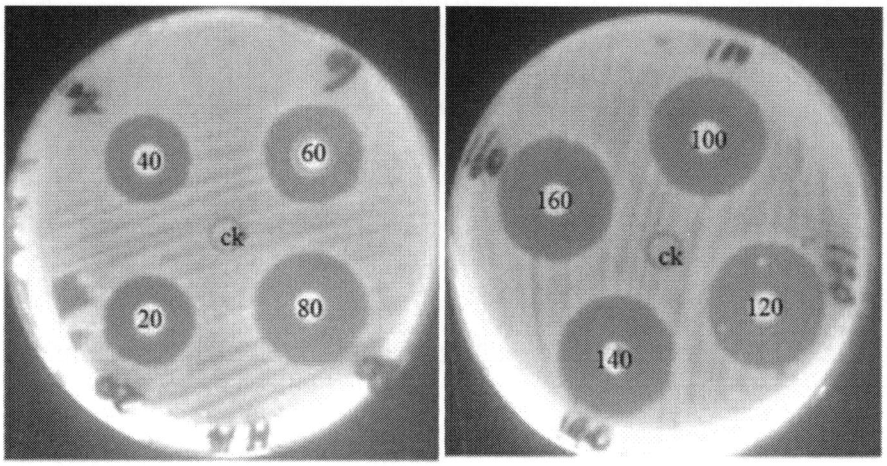

Figure 1. The effect of volume of antagonistic bacterial E14 to *V. harveyi* 11593 . The number in the plates means the volume ofE14 (μL); ck, sterile 2216E medium.

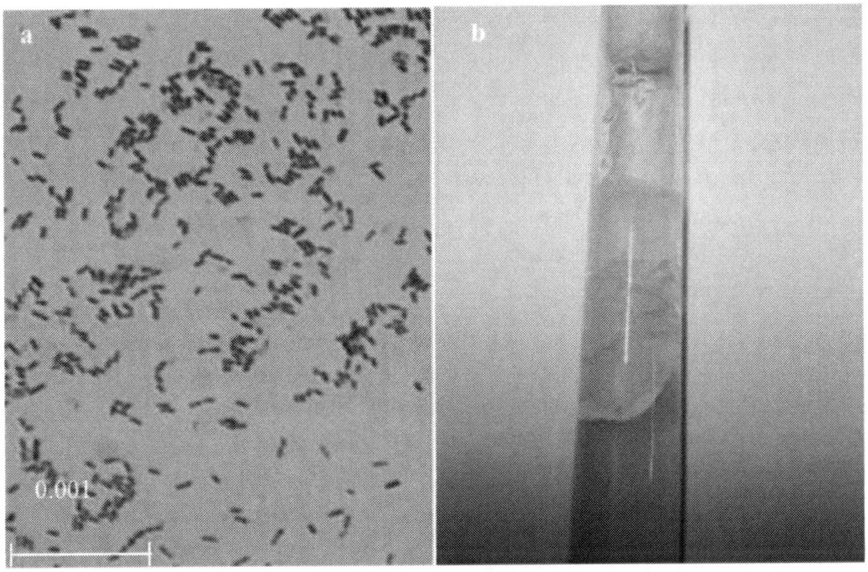

Figure 2. The morphological characteristics of strain E14. a, Gram staining; b, flagella motility.

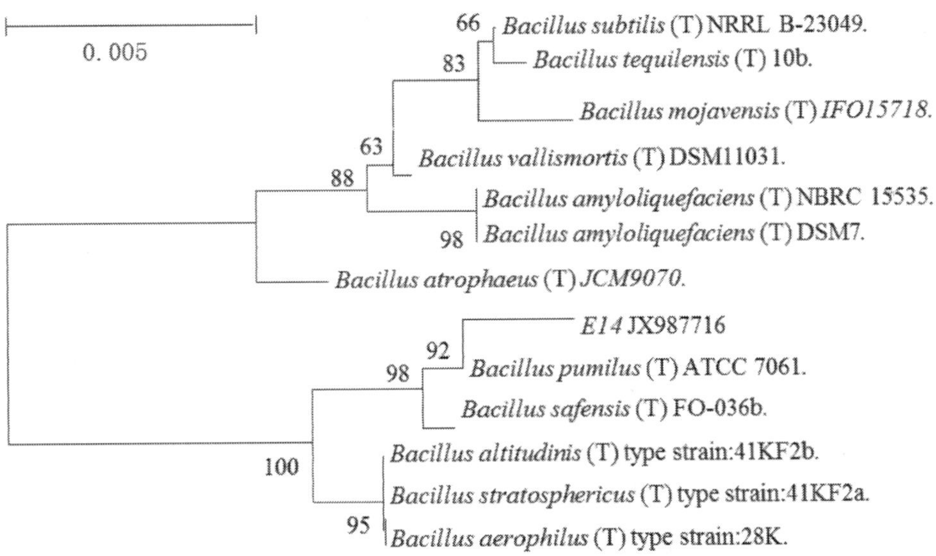

Figure 3. A phylogenetic tree analysis based on the bacterial 16S rDNA sequences of E14 and related members of the genus *Bacillu* . Numbers at nodes are levels of bootstrap support based on neighbor-joining analyses of 1000 replications. The scale bar represents 0.005 substitutions per nucleotide position.

The physiological characteristics of strain E14 are shown in table 2. Strain E14 exhibited physiological characteristics very similar to those of *B. pumilus*. Considering the results of morphological, biochemical, and genetic analyses, the isolated strain E14 was identified and named as *B. pumilus* E14.

3.4. Optimization of conditions for producing the maximum active substanse by B. pumilus E14

The effects of culture mediums, temperature, initial pH of medium, and medium volumes rate were designed to determine conditions that produce the maximum antibacterial ability of *B. pumilus* E14 (figure 4). All seven media were available for *B. pumilus* E14 (figure 4a), but antimicrobial activities with DIZ 24.5 ± 0.5 mm in LB, 2216E, and PPM showed significant differences with those of NYPD, YPG, CM, and NA. No significant difference was available in the antimicrobial activities of LB, 2216E, and PPM. LB is a widely used medium; thus, LB with 3% NaCl was adopted as the optimal medium for *B. pumilus* E14 cultivation.

Table 2. Phenotypic characteristics of strain E14 in conventional biochemical tests.

Item	E14	Item	E14	Item	E14	Item	E14
Gelatin liquefaction	+	amylolysis	-	lipid hydrolysis	-	Proteolysis （Casein）	+
Glucose	+	Cellobiose	-	Mannitol	-	Sulfuretted hydrogen	-
Lactose	-	Galactose	+	Sobitol	-	Propylene hydrochloric acid	-
Maltose	-	Inositol	-	Phenylalanine	-	Editpotassium cyanide test	+
Fructose	+	Arbaitol	-	Urea	-	Editpotassium cyanide control	+
Mannose	+	Rhamnose	-	Tartrate	+	Arginine dihydrolase	-
Xylose	-	Raffinose	-	Acetamide	+	Arginine dihydrolase control	-

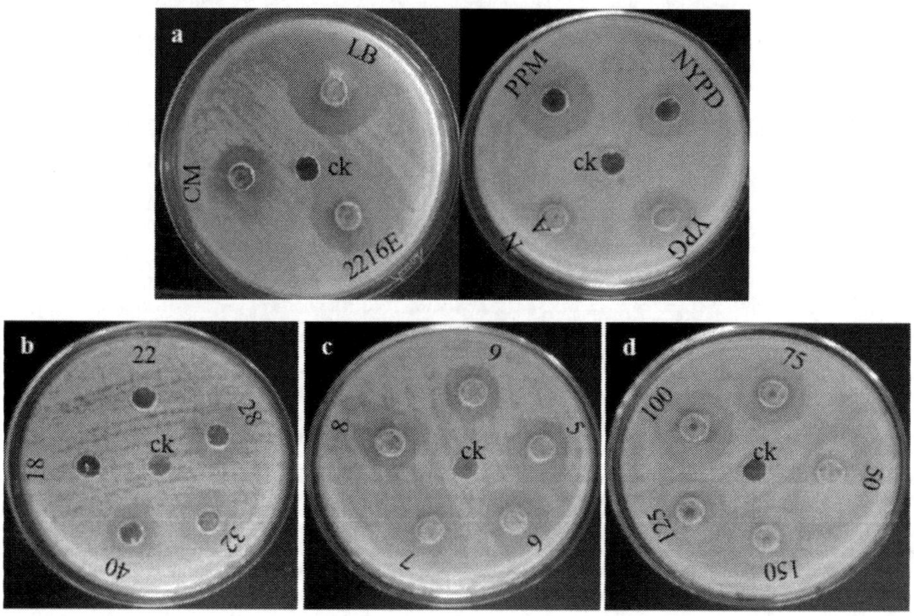

Figure 4. Optimization of various culture parameters for maximum antimicrobial production. a, different medium; b, tempreature/℃; c, pH value; d, medium volume in 500 mL triangular flask/mL.

The optimal temperature for antimicrobial production ranged from 28–40 °C, with no significant difference in antimicrobial activity reported throughout the range. No antimicrobial activity was observed at 18 °C and 22 °C, thus 28–40 °C was recommended as the optimal temperature (figure 4b).

No significant difference in antimicrobial production was observed in the initial pH range of 5.0–9.0 (figure 4c). Thus, the initial pH of 7.0 was chosen because maximum activity was detected there.

Maximum antimicrobial activity was observed at 100 mL/500 mL, which differed significantly from other dose-volumes. Hence, it was chosen as the optimized volume rate for *B. pumilus* E14 cultivation (figure 4d).

3.5. The growth curve and time course of the antibacterial activity by B. pumilus E14

The growth curve and time course of the antimicrobial production of *B. pumilus* E14 is shown in figure 5. Lag phase began before 2h, logarithmic phase at 2–24h and stable phase at 24h, but no decline phase appeared until 36h. During antimicrobial production, inhibitory activity appeared at 8h and increased significantly during cultivation up to 28h; then, it decreased until no inhibitory activity was detected at 36h. Maximum antimicrobial activity was observed at fermentation times ranging from 26–28h (no significant difference was tested within this period). Thus, *B. pumilus* E14 must be cultured for 26 h to achieve maximum production.

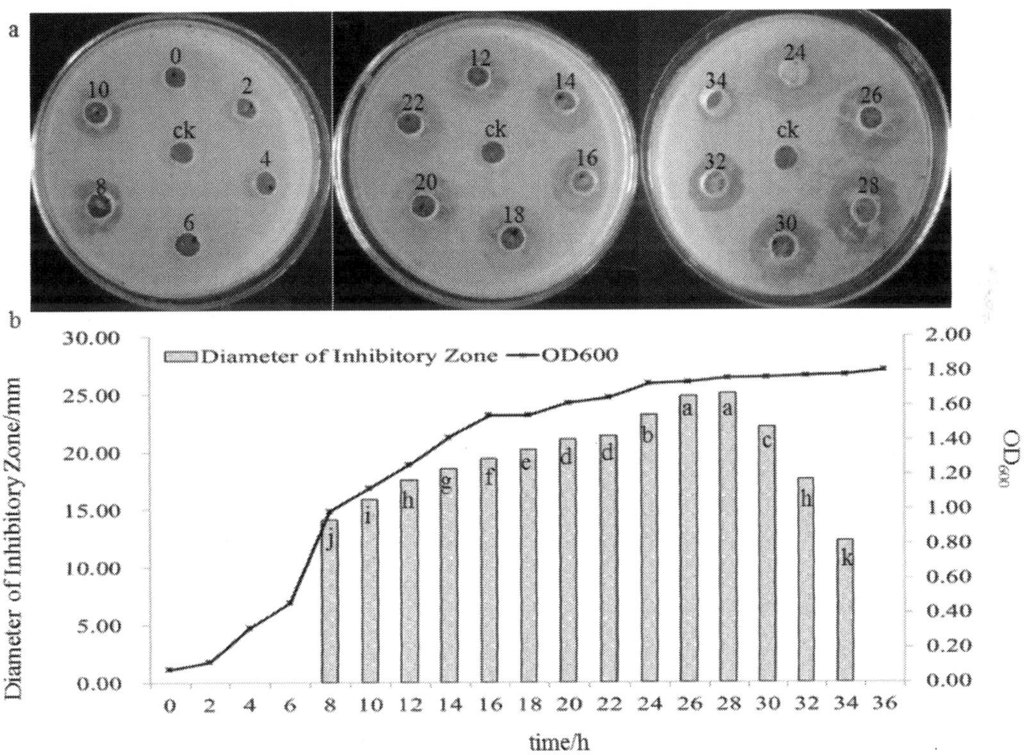

Figure 5. The growth curve and the time course of antibacterial activity by *B. pumilus* E14.
a, Antibacterial activity at the time course; b, Growth curve and Diameter of Inhibitory Zone; Different letters within a column are significantly different ($p<0.05$) as determined by Duncan's multiple range test.

In summary, the optimized conditions for the maximum production of antimicrobial substance by *B. pumilus* E14 require it to be cultured for 26 h, with an initial pH of 7.0 at 28 °C in 100 mL/500 mL LB with 3% NaCl. The DIZ 24.5 ± 0.5 mm under optimized conditions had no significant difference with the DIZ under non-optimized conditions.

3.6. The preliminary effect of strain E14 on shrimp larvae (Penaeus chinensis) culture

The counts of survived shrimp larvae are shown in table 3. The results demonstrated significant differences in larval survival of shrimp across the 4 groups. The count of larval survival shrimp was A >D >C >B. This means survivability in the 4 groups was *B. pumilus* E14>Control> *B. pumilus* E14 mixed *V. harveyi* 11593>*V. harveyi* 11593.

Almost all larval shrimps survived in the E14 group, but 25.00% died in the *V. harveyi* 11593 group 8 h after cultured. The E14 mixed *V. harveyi* 11593 group also presented significant differences in larval shrimp survival compared to the *V. harveyi* group in the 8~24 h timeframe. The survival rate was 20.00% and 5.00% in the *B. pumilus* E14 mixed *V. harveyi* 11593 group at 24h and 32 h after culture, but 6.67 % and 0.00 % in the *V. harveyi* 11593 group at 24 h and 32 h respectively. At 48 h after culture, survived larval shrimps were only found in the *B. pumilus* E14 group. The significant difference was also observed between the *B. pumilus* E14 group and the control in the 24~48 h period.

Table 3. Comparison of larval survival of shrimp in different treating groups at different times.

Group	A				B				C				D			
Time/h	I	II	III	Mean	I	II	III	Mean	I	II	III	Mean	I	II	III	Mean
0	20	20	20	20.0	20	20	20	20.0	20	20	20	20.0	20	20	20	20.0
8	20	20	19	19. 7^{a}	15	14	16	15.0^{c}	18	16	19	17.7^{b}	19	19	18	18.7^{ab}
16	17	17	16	16.7^{a}	8	6	6	6.7^{c}	10	8	10	9.3^{b}	16	15	13	14.7^{a}
24	12	10	12	11.3^{a}	2	0	2	1.3^{d}	5	4	3	4.0^{c}	8	6	7	7.0^{b}
32	6	5	5	5.3^{a}	0	0	0	0^{c}	2	1	0	1.0^{c}	3	3	2	2.7^{b}
40	2	3	3	2.7^{a}	0	0	0	0^{b}	0	1	0	0.3^{b}	1	0	1	0.7^{b}
48	0	0	1	0.3^{a}	0	0	0	0^{a}	0	0	0	0^{a}	0	0	0	0^{a}

a, b, c: significant differences at the 5% level.

All the results showed that the strain B. pumilus E14 inhibited the growth of V. harveyi 11593 and lengthened the lives of the larval shrimps which was fed only with bacteria.

4. Discussion

The genus Bacillus is known as one of the major probiotic resources for the production of bacterial pathogens [18-20]. Bacillus species produce biologically active lipopeptides [21], and exhibit anti-HIV [22] and other antibacterial activities [23]. B. subtilis is a common species with excellent antibacterial capabilities [23-28]. An antibacteria against V. harveyi was isolated from a mariculture pond with shrimp, crabs, and shellfish in China. The strain was identified as B. pumilus based on morphological observation, conventional biochemical tests, and 16S rDNA sequence analysis. The inhibitory zone diameter against V. harveyi of the bacterium, named E14, was 24.5 ± 0.5 mm, thus making it a good candidate for an alternative probiotics agent against V. harveyi. 11593 Interestingly, B. pumilus E14 maybe add a new member to the genus Bacillus for probiotics. More research is needed on this topic.

Strains of B. pumilusare used for enzyme production in an industrial scale and many other fields [29]. B. pumilus is another well-known species with an excellent capability of producing multifarious biological activities [30-34]. B. pumilus E14, which is gram-positive and motile, produces antibacterial substances that can hydrolyze gelatin and casein. The strain E14 tested positive for glucose, mannose, fructose, galactose , tartrate, acetamide and potassium cyanide; whereas it tested negative for lactose, maltose, xylose, cellobiose, inositol, arbaitol, rhamnose, raffinose, mannitol, sobitol, phenylalanine, urea, sulfuretted hydrogen, propylene hydrochloric acid, and arginine dihydrolase. Most of its characteristics were shared with other B.pumilus strains [35], except no amylase and lipid hydrolysis were detected in B. pumilus E14.

The antimicrobial substances produced by strains of genus Bacillus is attributed to mazy genetic regulation, and different culture conditions may induce the different antimicrobial substances. All seven mediums were available for B. pumilus E14, and significant differences among them were reported. Antimicrobial production could be achieved from 28–40 °C. This result implies that B. pumilus E14 is mesophilic, i.e., it is a heat-resistant vegetative cell. B. pumilus E14 is also a neutrophilic species given that maximum activity had been detected at an initial pH of 7.0. Maximum antimicrobial activity was tested at 26–28 h. These results are shared with a previous study, which showed that most Bacillus antimicrobial compounds are produced in the late log phase, possibly due to sporulation [36].

The safety and effectiveness test of E14 used in shrimp larvae (Penaeus chinensis) culture was very limited in this paper. Further evidence is need in the future, for example, expanded culturing volume and sample size of tested shrimp. Additionally, it will also be helpful to reduce observing intervals to every 2 h, especially after 24 h after culture.

The co-culture pond is a complex ecosystem, such that a number of factors (e.g., the introduction of non-native microbial species, input and circulation of nutrients, and human interference) can mediate microbial communities. Competition among microbes for nutrients and space in marine environments is a powerful selection tool for marine microorganisms employed in the production of marine natural products for medical and industrial purposes [37]. The discovery of antimicrobial agents against V. harveyi from a pond with shrimp, crabs, and shellfish in China is dependent on various substances exhibiting different antimicrobial activities. These have been found among certain bacteria due to the specialized role they play in their respective hosts [38]. Bacterium-bacterium antagonism in the environmen may influence its species richness and diversity , thus those antagonisms may become important variables in the ecology of pelagic bacteria, and in bacterium-mediated carbon cycling in the ocean [39]. B. pumilus E14 comes from the co-culture pond; thus, its use may prevent complementarity-mutual exclusiveness and turn it into the dominant bacterial community in the pond.

This strain may also solve problems due to increasing antibiotics, benefit the environment, and ultimately ensure sustainable aquaculture development.

5. Conclusion

The strain E14 antibacterium against Vibrio harveyi 11593 has been obtained from a mixed pond with Penaeus japonicus Bate, Portunus trituberculatus and Ruditapes philippinarum in China. The strain belongs to Bacillus pumilus and has a 24.5 ± 0.5 mm inhibitory zone diameter. The gram-positive, encapsulated, and motile bacterium is around 1.10–1.32 μm × 0.67–0.83 μm in size. B. pumilus E14 also shows protease activity. Optimized conditions for antimicrobial substance production of B. pumilus E14 require that it be cultured for 26 h at 28 °C, with an initial pH of 7.0 in 100 mL/500 mL LB with 3% NaCl.

Acknowledgments

The authors wish to thank Jiangsu provincial science and technology department policy guidance class plan -- The industrial-academic-research cooperation (BY2016056-04), two projects Funded by Lianyungang Science and Technology Bureau (CXY1421, CXY1510), a natural science foundation of Jiangsu province，China (BK20141248), a government science and technology plan projects funded by Entry-Exit Inspection and Quarantine Bureau of Jiangsu province (2015KJ01), a project Funded by the Priority Academic Program Development of Jiangsu Higher Education Institutions (PAPD), a project Funded by Co-Innovation Center of Jiangsu Marine Bio-industry Technology. The funders had no role in study design, data collection and analysis, decision to publish, or preparation of the manuscript.

References

[1] Pandiyan P, Balaraman D, Thirunavukkarasu R., George E G J, Subaramaniyan K, Manikkam S and Sadayappan B 2013 Probiotics in aquaculture *Drug Invention Today* **5**: 55–59

[2] Fao 2014 *The state of world fisheries and aquaculture (opportunities and challenges)* In: FAO Rome, Italy

[3] Kim M S, Cho J Y and Choi H S 2014 Identification of Vibrio harveyi, Vibrio ichthyoenteri, and Photobacterium damselae isolated from olive flounder Paralichthys olivaceus in Korea by multiplex PCR developed using the *rpoB* gene *Fish Sci.* **80**: 333–39

[4] Dong B, Xiang J H, Yang M, Wang J Y and Liu G 2003 The basic study of the disease and in marine aquaculture *Chin Bas. Sci.* **6**: 19–24

[5] Ponprateep S, Somboonwiwat K and Tassanakajon A 2009 Recombinant anti-lipopolysaccharide factor isoform 3 and the prevention of vibriosis in the black tiger shrimp *Penaeus monodon*. *Aquaculture* **289**: 219–24

[6] Radjasa O K S, Sabdono A., Prayitno S B and Hutabarat S 2009 Phylogenetic diversity of the causative agents of vibriosis associated with groupers fish from Karimunjawa Islands, Indonesia *Curr. Res. Bacteriol.* **2**:14–21

[7] Toranzo A E, Magariños B and Romalde J L 2005 A review of the main bacterial fish diseases in mariculture systems *Aquaculture* **246**: 37–61

[8] Austin B and Zhang X H 2006 Vibrio harveyi: A significant pathogen of marine vertebrates and

invertebrates *Lett. Appl. Microbiol.* **43**: 119–24

[9] Cabello F C 2006 Heavy use of prophylactic antibiotics in aquaculture: A growing problem for human and animal health and for the environment *Environ. Microbiol.* **8**: 1137–44

[10] Denev S A 2008 Ecological Alternatives of Antibiotic Growth Promoters in the Animal Husbandry and Aquaculture *DSc. Thesis* Stara Zagora Bulgaria: Department of Biochemistry Microbiology Trakia University 294

[11] Paningrahi A, Azad I S 2007 Microbial intervention for better fish health in aquaculture: The Indian scenario *Fish Physiol. Biochem.* **33**: 429-40

[12] Rengpipat S, Rukpratanporn S, Piyatiratitivorakul S and Menasaveta P 2000 Immunity enhancement in black tiger shrimp (Penaeus monodon) by probiotic bacterium (Bacillus S11) *Aquaculture* **191**: 271–88

[13] Li L T, Yan B L, Li S H, Xu J T and An X H 2016 A comparison of bacterial community structure in seawater pond with shrimp, crab, and shellfish cultures and in non-cultured pond in Ganyu, Eastern China *Ann. Microbiol.* **66**: 317–28

[14] Shen P, Fan X R and Li G W 1999 *Experiment of Microbiology* 3rd Higher Education Press, Beijing pp 101–2

[15] Muhammad S, Ahmad S and Hameed A 2009 Report: Antibiotic production by thermophilic Bacillus specie SAT-4 Pak *J. Pharm. Sci.* **22**: 339–45

[16] Dunbar J, Ticknor L O and Kuske C R 2000 Assessment of microbial diversity in four Southwestern United States soils by 16S rRNA gene terminal restriction fragment analysis *Appl. Environ. Microbiol.* **66**: 2943–50

[17] Fu S Z, Song B B, Liu Y and Liu Z P 2009 Screening of probiotic strains and their protection against vibriosis in juvenile shrimp (Litopenaeus vannamei) *China Environmental Science* **29**: 867-72

[18] Radjasa O K, Khoeri M M, Darussalam C C, Trimasanto H and Sudoyo H 2013 Bacterial symbionts of reef invertebrates: screening for antipathogenic bacteria activity *Biodiversity Conservancy International* Taylor & Francis London, United Kingdom 108

[19] Rengpipat S, Phianphak W, Piyatiratitvorakul S and Menasveta P 1998 Effects of probiotic bacterium on black tiger shrimp *Penaeus Monodon Survival and Growth Aguaculture* **167**: 301–13

[20] Stein T 2005 Bacillus subtilis Antivbiotics: Structures, Syntheses and Specific Functions *Mol. Microbiol.* **56**: 845–57

[21] Trischman J A, Jensen P R and Fenical W 1994 Halobacilin: a cytotoxic cyclic acylpeptide of the iturin class produced by a marine *Bacillus Tetrahedron Lett* **35**: 5571–74

[22] Itokawa H, Miyashita T, Morita H, Takeya K, Hirano T, Homma M and Oka K 1994 Structural and conformational studies of [Ile] and [Leu] surfactins from Bacillus subtilis natto *Chem. Pharm. Bull.* **42**: 604–7

[23] Wu W J, Park S M and Ahn B Y 2013 Isolation and Characterization of an Antimicrobial Substance from Bacillus subtilis BY08 Antagonistic to Bacillus cereus and Listeria monocytogenes *Food Sci. Biotechnol.* **22**: 433–40

[24] Chen H, Wang L, Su C X, Gong G H, Wang P and Yu Z L 2008 Isolation and characterization

of lipopeptide antibiotics produced by Bacillus subtilis *Lett. Appl. Microbiol.* **47**: 180–86

[25] Kim Y, Cho J Y, Kuk J H, Moon J H, Cho J I, Kim Y C and Park K H 2004 Identification and antimicrobial activity of phenylacetic acid produced by Bacillus licheniformis isolated from fermented soybean, cheonggukjang *Curr. Microbiol.* **48**: 312–17

[26] Kugler M, Loeffler W, Rapp C, Kern A, Jung G and Rhizocticin A 1990 An antifungal phosphono-oligopeptide of Bacillus subtilis ATCC 6633: Biological properties *Arch. Microbiol.* **153**: 276–81

[27] Lee N K, Yeo I C, Park J W, Kang B Sand Hahm Y T 2010 Isolation and characterization of a novel analyte from Bacillus subtilis SC-8 antagonistic to Bacillus cereus *J. Biosci. Bioeng.* **110**: 298–03

[28] Yeo I C, Lee N K, Cha C J and Hahm Y T 2011 Narrow antagonistic activity of antimicrobial peptide from Bacillus subtilis SCK-2 against Bacillus cereus *J. Biosci. Bioeng.* **112**: 338–44

[29] Vijaya Kumar M L, Thippeswamy B and Vasanth Raj P 2014 Cytotoxicity and Anticancer Studies of Bacillus cereus and Bacillus pumilus Metabolites Targeting Human Cancer Cells *Appl. Biochem. Micro.* **50**: 619–23.

[30] Hoult B and Tuxford A F 1991 Toxin production by Bacillus pumilus *J. Clin. Pathol.* **44**: 455–58

[31] Lanser A C, Plattner R D and Bagby M O 1992 Production of 15-, 16- and 17-Hydroxy-9-Octadecenoic Acids by Bioconversion of Oieic Acid with Bacillus pumilus *J. Am. Oil. Chem. S.* **69**: 363–66

[32] Liu B, Luo Y, Tao X, Xie X, Ma X and Zhang L 2014 Screening and identification of Bacillus pumilus producing double active protein of anticoagulation and thrombolysis *Acta Microbiologica Sinica* **54**: 345–51

[33] Ouoba L I I, Diawara B, Christensen T, Dalgaard Mikkelsen J and Jakobsen M 2006 Degradation of polysaccharides and non-digestible oligosaccharides by Bacillus subtilis and Bacillus pumilus isolated from Soumbala, a fermented African locust bean (Parkia biglobosa) food Condiment *Eur. Food Res. Technol.* **224**: 689–94

[34] Susilowatia R, Sabdonob A and Widowatib Ita 2015 The isolat of *Bacillus pumilus* associated with the marine sponge Ircinia sp. produces surfactin-like substances, cyclic acyldepsipeptides, called bacircines 2, 3, 4, 5 and 5A *Procedia Enviro. Sci.* **23**: 240–46

[35] Felis G E, Dellaglio F and Torriani S 2009 Taxonomy of Probiotic Microorganisms *Springer Science and Business Media, LLC* **15**: 591–636

[36] Katz E and Demain A L 1977 The peptide antibiotics of Bacillus: Chemistry, biogenesis, and possible functions *Bacteriol. Rev.* **41**: 449–74

[37] Hassan S W M, Abdul-Raouf U M and Abdel-Rahiem Ali M 2015 Antagonistic interactions and phylogenetic diversity of antimicrobial agents producing marine bacteria in Suez Bay *Egy. J. of Aquat. Res.* **41**: 57–67

[38] El Amraoui B, El Amraoui M, Cohenb N and Fassouanea A 2014 Antifungal and antibacterial activity of marine microorganisms *Ann. Pharm. Fr.* **72**: 107–11

[39] Long R A and Azam F 2001 Antagonistic interactions among marine pelagic bacteria *Appl. Environ. Microbiol.* **67**: 4975–83

Biosorption of Cu^{2+} and Pb^{2+} using sophora alopecuroides residue

N Feng[1, 2], W Fan[1], M Zhu[1] and Y Zhang[1]

[1]School of Pharmacy, Ningxia Medical University, Yinchuan, Ningxia, China

Email: ncfeng@sina.com

Abstract. Sophora alopecuroides residue (SAP), a kind of traditional Chinese herbal medicine residue, was developed in an alternative biosorbent for the removal Cu^{2+} and Pb^{2+} in simulated wastewater. The morphology and surface texture of SAP were characterized by scanning electron microscopy, which showed a loose and porous structure. The biosorption experiments of Cu^{2+} and Pb^{2+} onto SAP were investigated by using batch techniques. High biosorption percentage appeared at pH values of 4.5–6.0. The experimental data followed the second-order kinetic model well. Equilibrium fit with the Langmuir isotherm model well. The maximum biosorption capacity of an adsorbent at 25℃ was respectively 60.6 mg/g Cu^{2+} and 128.1 mg/g Pb^{2+}. The findings of the present study show that SAP is an attractive and effective biosorbent for Cu^{2+} and Pb^{2+}.

1. Introduction

With the expansion of industrial activities, the increased emission of wastewater containing heavy metal ions has turned into a severe problem, and a threat to human health. Cu^{2+} is widely used in dyeing, papermaking, petroleum, plating plants, and in the production of some alloys [1, 2]. Wastewater containing Pb^{2+} mainly comes from battery factories, mining, mine tailings, the combustion of fossil fuels, sulphide ore smelting, and the drainage water of non-ferrous metal plants[3]. The permission limit of copper and lead in drinking water is 2mg/L and 0.05 mg/L [4]. Various diseases can be caused as a result of excess copper and lead in drinking water. Cu^{2+} can react with protein ligands to form coordination compounds in the human body that cause the loss of normal physiological function of protein, and even lead to death [5]. Pb^{2+} has a strong neurotoxicity that may cause a variety of neurological disorders in humans, such as memory loss, learning disabilities, and behavioral anomalies [6]. The commonly used technologies for the removal of Cu^{2+} and Pb^{2+} include chemical precipitation [7], ion exchange [8] and membrane separation [9]. These conventional technologies are neither economical nor effective, and different methods have different disadvantages

[10]. In view of this, it is significant to develop materials, or new technologies, to decrease heavy metal ions in wastewater to environmentally acceptable levels.

In numerous methods, biosorption is widely known for its efficiency, safety, low cost, and environmental protection [11]. Biosorption can utilize various different materials including microbial cells, such as bacteria, and biological materials, which contain some natural materials waste and natural material [12, 13]. Most adsorbents' adsorption mechanism, which takes plants or waste as raw materials, can be explained in the following way: plants contain a large number of rude fibre, indicating the practicability to dislodge toxic heavy metal ions from aqueous solution, as a result of presence of various functional groups on the cellulose, such as –OH, –COOH to be involved in metal ions binding [14]. *Sophora alopecuroides* L, a kind of traditional Chinese herb, contains many chemical compounds, such as: flavonoids, alkaloids, organic acid, volatile oil, protein, saccharide, amino acid, etc [15]. With *sophora alopecuroides* residue, active ingredients are extracted containing rude cellulose (31.42%), crude protein (17.66%) and a mass of amino acid, which provide potential for adsorption of metallic cations.

According to a report [16], more than 30 million tons of Chinese herb residues are produced yearly in China. Currently, they are buried or combust [17], which will cause not only a great waste of resources, but will have serious damage on the environment as well. Chinese herbal medicine residue can be made into biosorbent with appropriate treatment. The adsorbents have many advantages, such as being low cost, recyclable and biodegradable. The preparation of *sophora alopecuroides* residue adsorbent offers a new way for using the waste residue of traditional Chinese herbal medicine. In this study, the biosorption of Cu^{2+} and Pb^{2+} from simulated wastewater solution onto *sophora alopecuroides* residue at 25℃ was studied.

2. Experimental

2.1. Chemicals and equipment

Stock solution (1.000 g/L) was prepared by dissolving 3.9720 g of $CuSO_4 \cdot 5H_2O$ and 1.6228 g of $PbNO_3$ in 1 L of redistilled water. Diluting the stock solution provided a work solution for practical use. All reagents used in the experiment were of A.R grade. Surface morphological structure of *sophora alopecuroides* residue was studied using Scanning Electron Microscopy (Hitachi S-3400H). The concentrations of copper and lead were determined using atomic absorption spectrophotometer (Persee A3 series Beijing).

2.2. Preparation of biosorbent

Sophora alopecuroides residue was collected from a pharmaceutical factory in Ningxia China. *Sophora alopecuroides* waste was washed a few times with distilled water, and then dried inside a convection oven for 24 hours at 70℃. The *sophora alopecuroides* was then crushed to powder in a pulverizer, and sieved to a constant size range 0.3-0.5 mm. This treated *sophora alopecuroides* waste is called SAP.

2.3. Biosorption experiments

Batch adsorption tests were completed by shaking a series of conical flasks containing the desired dosage of biosorbent in a scheduled concentration of heavy metal solution with known pH. The pH of the Cu^{2+} and Pb^{2+} solution was adjusted to the required value by 0.1mol/L HCl and 0.1mol/L NaOH before the addition of biosorbents. 0.100 g of SAP sample was equilibrated with 25 ml of the Cu^{2+} and Pb^{2+} solution of known concentration in a sealed conical flask at 25℃ in a thermostatic shaker bath. The effect of pH value on the biosorption of Cu^{2+} and Pb^{2+} was studied by mixing 0.100g of SPA with 25mL of 100mg/L Cu^{2+} and 300mg/L Pb^{2+} solutions of pH from 2.5 to 6.0. In kinetic experiments, the adsorption time varied from 0 to 120min using 100mg/L Cu^{2+} and 300mg/L Pb^{2+} solutions. 0.100g of SAP was put in 25mL of Cu^{2+} (50-250mg/L) and Pb^{2+} (100-500mg/L) solutions at various concentrations in order to carry out isotherm adsorption experiments. In all experiments, the glass flasks were shaken at 200 rpm on a water-bathing constant temperature vibrator for a known time. After the oscillation, supernatant was filtered and the concentration of the Cu^{2+} and Pb^{2+} ion remaining was determined by flame atomic absorption spectrometry. All experiments were run in triplicate.

The biosorption capacity q(mg/g) was determined by equation (1):

$$q = \frac{(c_o - c_e)}{W}V \tag{1}$$

The adsorption percentage of heavy metal ion was determined by equation (2):

$$\text{adsorption percentage}(\%) = \frac{c_0 - c_e}{c_0} \tag{2}$$

where c_e and c_o (mg/L) are the equilibrium and initial concentrations of copper and lead, W (g) is the mass of the dry adsorbent and V (L) is the volume of copper and lead solution.

3. Results and discussion

3.1. Characterization of the biosorbent

The SEM revealed the morphology and surface structure of *sophora alopecuroides* residue (SAP) at different magnifications clearly. The SEM micrographs of SAP shown in Figure 1 indicated a loose and porous structure, which can supply a high specific surface area of SAP. It was clearly observed from the photographs that the biosorbent showed a rough structure (1000×). At 2000× magnification, the presence of asymmetric pores and irregular spiral structure in large quantity were more easily observed, which is favorable for biosorption of Cu^{2+} and Pb^{2+} ions.

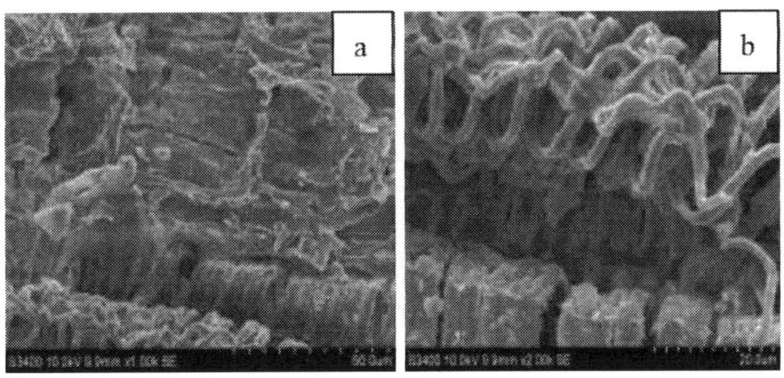

Figure 1. SEM of the SAP at different magnifications: (a) 1000×, (b) 2000×.

3.2. Biosorption of copper and lead ion

3.2.1. Effect of pH on biosorption. Figure 2 shows the impact of pH on biosorption percentage of Cu^{2+} and Pb^{2+} by SAP. Adsorption percentage of both metal ions showed a rapid increase in pH from 2.5 to 4.5 and then stabilized at higher pH values of 4.5 to 6.0. The maximum removal percentage was 95% for Cu^{2+} at pH 5.5 and 91% for Pb^{2+} at pH 5.0.

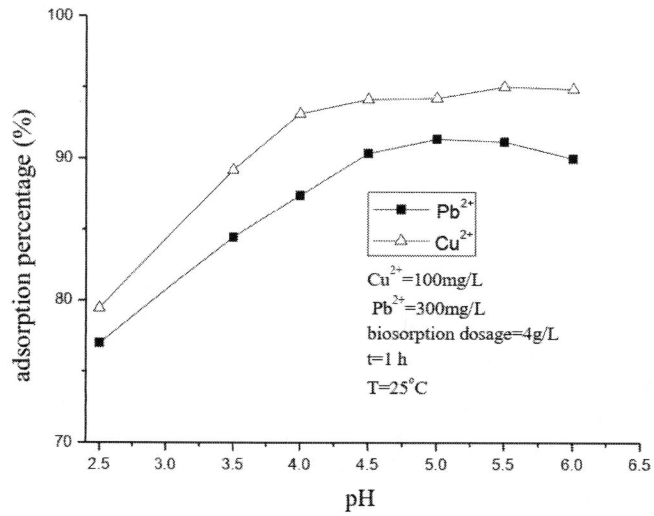

Figure 2. Effect of pH on biosorption.

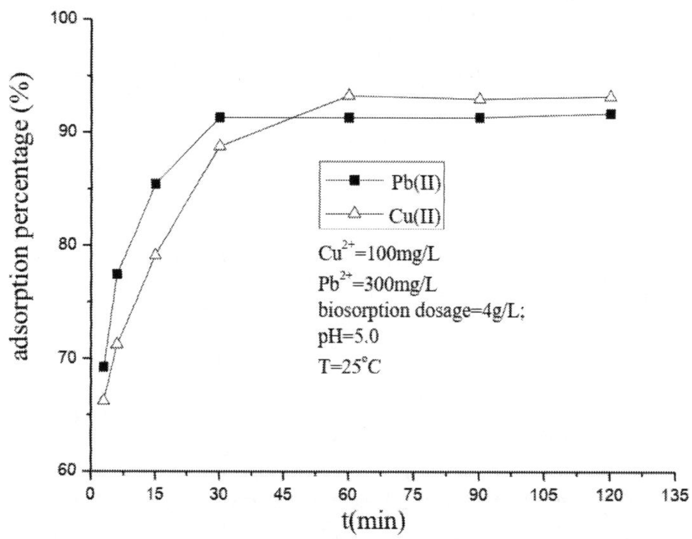

Figure 3. Effect of time on biosorption.

The solution pH was selected to avoid the possible formation of metallic precipitation that restricting the true biosorption studies showed; pH value is an independent factor in the biosorption process [18]. The metal binding sites on SAP for biosorption includes –OH and –COOH functional groups [14] in which H^+ ions can be exchanged for metal ions in solution. As the pH is reduced, H^+ will compete with Cu^{2+} and Pb^{2+} on the active sites of SAP biosorption which results in the decrease of adsorption percentage. In contrast, with pH increasing, the surface of biosorbent is negatively charged; therefore, it is favorable for the biosorption of positively charged copper and lead ions.

3.2.2. Biosorption kinetics. Figure 3 shows the adsorption kinetics of Cu^{2+} and Pb^{2+} on SAP. The biosorption percentages of Cu^{2+} and Pb^{2+} ions on SAP increase in the initial 30 min rapidly, and then attain equilibrium at about 1 h. The ultimate adsorption percentages of two metals exceed 90%. After adsorption equilibrium, the removal rate did not obviously change with time. Compared to usual adsorbents such as adsorption resin and activated carbon, the time of equilibrium of SAP was shorter [19].

Kinetic of heavy metal ions biosorption could be modeled by the first-order equation and second-order equation shown as follows:

$$\lg(q_e - q_t) = \lg q_e - \frac{k_1}{2.303}t \tag{3}$$

$$\frac{t}{q_t} = \frac{1}{k_2(q_e)^2} + \frac{t}{q_e} \tag{4}$$

where k_1 and k_2 are respectively the rate constant of first-order and second-order models, q_t (mg/g) and the amount of metal ion biosorbed at any given time t; q_e (mg/g) the amount of metal ions adsorbed at equilibrium. The kinetic constants were given in Table 1. It was found that the coefficient of correlation (R^2) for second-order model is much closer to 1 than that of the first-order kinetic model and the calculated values (q_e) agree well with experimental values ($q_{e\text{-}exp}$). As a result of the second-order model reaction mechanism, the biosorption of Cu^{2+} and Pb^{2+} processes seem to be controlled by the chemical processes involving covalent forces, sharing or exchanging of electrons between the biosorbent and copper and lead ions [20].

Table 1. Kinetic model constants for biosorption of Cu^{2+} and Pb^{2+} on SAP.

Adsorbent	$q_{e,exp}$ (mg/g)	First-order equation			Second-order equation		
		q_e (mg/g)	k_1(1/min)	R^2	q_e (mg/g)	k_2(g/mg·min)	R^2
Cu(II)	23.31	5.18	0.057	0.7300	23.72	0.022	0.9998
Pb(II)	68.48	5.30	1.754	0.1158	69.40	0.016	0.9998

3.2.3. Adsorption isotherms. Figure 4 presents the experimental biosorption isotherms of Cu^{2+} and Pb^{2+} on SAP. It showed that q_e increased as c_e increased until the equilibrium was attained, and then q_e was almost unchanged with continuing increase in c_e.

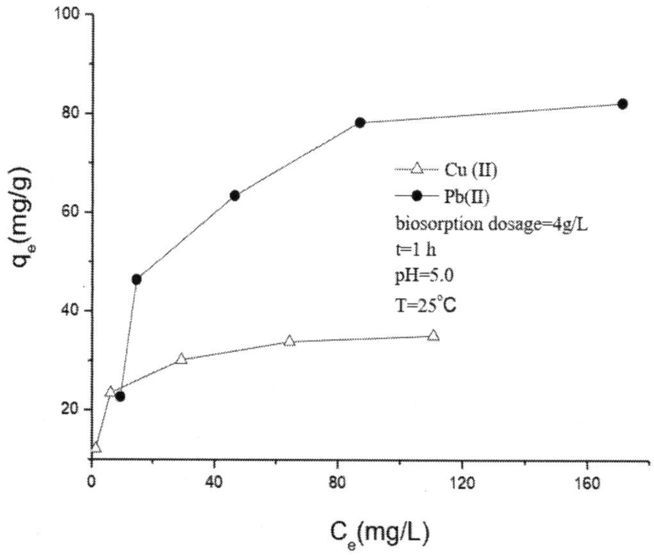

Figure 4. Biosorption isotherm of Cu^{2+} and Pb^{2+} on SAP.

Freundlich and Langmuir isotherm models are used to establish the relationship between the biosorption capacity of Cu^{2+} and Pb^{2+} on SAP and equilibrium concentration. Langmuir isotherm model is given as equation (5):

$$\frac{c_e}{q_e} = \frac{c_e}{q_m} + \frac{1}{bq_m} \tag{5}$$

where q_e (mg/g) is the biosorption capacity of Cu^{2+} and Pb^{2+} at equilibrium, q_m (mg/g) is the maximum biosorption capacity of the biosorbent, b (L/mg) is the Langmuir constant.

Freundlich isotherm model is given as below:

$$\log q_e = \log K_f + \frac{1}{n}\log c_e \tag{6}$$

where K_f and n are the Freundlich constant.

As shown in Table 2, the biosorption patterns of Cu^{2+} and Pb^{2+} on SAP are well fitted to the Langmuir model with R^2 of 0.985-0.999. In contrast, the Freundlich model has lower values of R^2, which suggests that the data follows the Langmuir model better. According to the Langmuir model, the maximum biosorption capacities (q_m) of SAP for Cu^{2+} and Pb^{2+} were respectively 60.6 mg/g and 128.1 mg/g. The constant b reflects a strong affinity between SAP and Cu^{2+} and Pb^{2+}. The K_f value of the Freundlich model shows that SAP has high adsorption density in Cu^{2+} and Pb^{2+} solutions. Compared to some other biomass used, SAP shows better adsorption capacity (table 3).

Table 2. Isotherm models parameters for the biosorption of Cu^{2+} and Pb^{2+} on SAP.

	Langmuir equation			Freundlich equation		
	q_m (mg/g)	b(L/mg)	R^2	K_f	n	R^2
Cu^{2+}	60.6	0.147	0.9874	18.98	2.343	0.9213
Pb^{2+}	128.1	0.040	0.9971	21.14	2.833	0.9047

Table 3. Maximum adsorption capacities for different biomass.

Adsorbents	q_m (mg/g) Cu^{2+}	q_m (mg/g) Pb^{2+}	References
Tobacco dust	36.0	39.6	[21]
Waste beer yeast	1.44	5.71	[22]
Cyclotella sp	26.03	37.3	[23]
S. alopecuroides residue (SAP)	60.1	96.1	This study

4. Conclusion

SAP, which is a traditional Chinese herbal medicine residue, was developed to remove Cu^{2+} and Pb^{2+} from simulated wastewater solutions. The biosorption rate was fast, and equilibrium was obtained within 60 min, The coefficient of correlation (R^2) of the second-order kinetic equation was greater than 0.999. The Langmuir and Freundlich adsorption models were used to express the sorption phenomenon, with a better result from the Langmuir model than the Freundlich model. The maximum adsorption capacities of SAP for Cu^{2+} and Pb^{2+} were 60.6 mg/g and 128.1 mg/g at pH 5.0, respectively. The results of this research indicate the feasibility and possibility that SAP can be made into an inexpensive and effective biosorbent for treatment of wastewaters containing Cu^{2+} and Pb^{2+}.

Acknowledgements

This work was supported by the National Natural Science Foundation of China (No. 21266026).

References

[1] Florido A, Valderrama C, Arévalo J A, Casas I, Martínez M and Miralles N 2010 Application of two sites non-equilibrium sorption model for the removal of Cu(II) onto grape stalk wastes in a fixed-bed column *Chemical Engineering Journal* **156** pp 298-304

[2] Zhong X L, Zhou S L, Zhu Q and Zhao Q G 2011 Fraction distribution and bioavailability ofsoil heavy metals in the Yangtze River Delta—A case study of Kunshan City in Jiangsu Province, China *Journal of Hazardous Materials* **198** pp 13-21

[3] King P, Rakesh N, Beenalahari S, Kumar Y P and Prasad V S R K 2007 Removal of lead fromaqueous solution using Syzygium cumini L.: Equilibrium and kinetic studies *Journal of Hazardous Materials* **142** pp 340-47

[4] World Health Organization 1993 *Guidelines for drinking-water quality World Health Organization* **1** pp 46 and 173

[5] Kazemipour M, Ansari M, Tajrobehkar S, Majdzadeh M and Kermani H R 2008 Removal of lead, cadmium, zinc, and copper from industrial wastewater by carbon developed from walnut, hazelnut, almond, pistachio shell, and apricot stone *Journal of Hazardous Materials* **150** pp 322-27

[6] Gupta V K and Rastogi A 2008 Biosorption of lead from aqueous solutions by green algae Spirogyra species: Kinetics and equilibrium studies *Journal of Hazardous Materials* **152** pp 407–14

[7] Djedidi Z, Bouda M, Souissi M A, Cheikh R B, Mercier G, Tyagi R D and Blais J F 2009 Metalsremoval from soil, fly ash and sewage sludge leachates by precipitation and dewatering properties of the generated sludge *Journal of Hazardous Materials* **172** pp 1372–82

[8] AL-Othman Z A, Naushad M and Inamuddin 2011 Organic–inorganic type composite cationexchanger poly-o-toluidine Zr(IV) tungstate: Preparation, physicochemical characterization and its analytical application in separation of heavy metals *Chemical Engineering Journal* **172** pp 369–75

[9] O'Connell D W, Birkinshaw C and O'Dwyer T F 2008 Heavy metal adsorbents prepared from the modification of cellulose: A review *Bioresource Technology* **99** pp 6709–24

[10] Kuyucak N and Volesky B 1998 Biosorbent for recovery of metals from industrial solutions *Biotechnology Letters* **10** pp 137–142

[11] Volesky B and Holan Z R 1995 *Biosorption of heavy metals Biotechnology Progress* **11** pp 235–50

[12] Andrès Y, Texier A C and Cloirec P L 2003 Rare earth elements removal by microbialbiosorption: a review *Environmental Technology* **24** pp 1367–75

[13] Gérente C, Lee V K C, Cloirec P L and Mckay G 2007 Application of chitosan for the removal of metals from wastewaters by adsorption–mechanisms and models review *Critical Reviews in Environment Science and Technology* **37** pp 41-127

[14] Matheickal J T, Yu U Q M and Woodburn G M 1999 Biosorption of cadmium(II) from aqueoussolutions by pre-treated biomass of marine alga DurvillAea potatorum *Water Research* **33** 335-42

[15] Gao H Y, Li G Y and Wang J H 2011 Studies on the dynamic accumulations of Sophoraalopecuroides L. Alkaloids in different harvest times and the appropriate harvest time *Journal of Chromatography* B **879** pp 1121–25

[16] Yang X Q, Yuan B and Jiang J H 2015 Utilization of residues from Chinese medicinal herbs asa kind of bioresources *Journal of Jiangsu Normal University* **33** pp 40-44

[17] Ma C F, Fan W, Zhang J B, Shi Y and Feng N C 2015 Study of Pb^{2+}, Cd^{2+} and Ni^{2+} adsorptiononto activated carbons prepared from glycyrrhiza residue by KOH or H3PO4

activation *Water Science and Technology* 72 pp 451-62

[18] Yu J X, Tong M, Sun X M and Li B H 2008 Enhanced and selective adsorption of Pb^{2+} and Cu^{2+} by EDTAD-modified biomass of baker's yeast *Bioresource Technology* **99** pp 2588–93

[19] Veli S and Pekey B 2004 Removal of copper from aqueous solution by ion exchange resins *Fresenius Environment Bulletin* **13** pp 244–50

[20] Özacar M and Şengil İ A 2003 Adsorption of reactive dyes on calcined alunite from aqueoussolutions *Journal of Hazardous Materials* **98** pp 211-24

[21] Qi B C and Aldrich C 2008 Biosorption of heavy metals from aqueous solutions with tobaccodust *Bioresource Technology* **99** pp 5595–601

[22] Han R P, Li H K, Li Y H, Zhang J H, Xiao H J and Shi J 2006 Biosorption of copper and lead ions by waste beer yeast *Journal of Hazardous Materials* B **137** pp 1569–76

[23] Schmitt D, Müller A, Csögör Z, Frimmel F H and Posten C 2001 The adsorption kinetics ofmetal ions onto different microalgae and siliceous earth *Water Research* **35** pp 779–85

Research on the water-inrush risk of coal floors based on Fisher-evaluation and AHP

D J Xu[1, 2], W X Wei[1] and S Y Xiang[1]

[1]State key laboratory of Coal Resource and safe mining, College of Geoscience and Surveying Engineering, China University of Mining & Technology, Beijing 100083, China

E-mail: xudongjinggg@126.com

Abstract. There are many factors that influence floor water-inrush. Based on the widely collected data of floor water-inrush in China, the evaluation factors in this paper consist of water pressure, aquifer type, aquiclude thickness, floor failure depth and fault throw. These are used to build a single Fisher evaluation model and a Fisher evaluation with weighting model of an analytic hierarchy process (AHP). By comparision, through AHP weighting value, the inter-class distribution of the data in the Fisher model is relatively more concentrated than the single Fisher evaluation method. It would produce higher reliability and more extensive application value.

1. Introduction

Most of the coal resources in China are underground where hydro-geological conditions in coal seams are complicated. Drilling in coal seam floors has received serious threats from limestone confined water, especially in the north China coal field. With the deepening of coal mining depth, it is dangerous for mining when thick, high water pressure and large amounts of Ordovician limestone aquifer occur. This problem has been a concern for many scholars and a lot of research has been carried out [1-3]. The results play an important role in the prediction of water inrush from floors, such as the water inrush probability index method [4], the vulnerability index method [5], and the water inrush factor analysis method [6]. In addition, the analysis method, namely fuzzy mathematics, the artificial neural network, stochastic theory, information fusion technology, the GIS system, and the expert system also have important theoretical significance to guide mine safety production. Because of the influence of geology and technology conditions, it is difficult to get the full theory and method about what is needed. Thus, the application process is largely affected [7-9].

This article uses water pressure, aquifer type, thickness of water-resisting layer, floor damage depth and fault throw information, which are commonly used in the actual production process. By introducing data weighting to set up a Fisher evaluation model of the working face of mining safety evaluation, we can provide a convenient method to evaluate inrush risk at coal mine production sites [10-13].

2. Determination of Fisher evaluation model

2.1. AHP and Fisher evaluation theory

Analytic hierarchy process (AHP) is an application of the theory of network system and multi-objective comprehensive evaluation method from Professor T L Saaty at the University of Pittsburgh in the early 1970s. The method proposes a way to solve qualitative and multi-objective complex issues, and employs a combination of quantitative level weighted decision analysis approaches. The problem is decomposed into different factors, then the factors correlated with membership can be comined into different levels to build a multi-level analysis structure model. Finally, we can obtain an important schedule in the weights among the lowest layer relative to the utmost layer [14].

The analytic hierarchy process can be roughly divided into the following steps: establishing a hierarchy model, creating a Tectonic evaluation matrix, calculating the charge quantity and doing a consistency check.

The Fisher evaluation function is one of the most influential linear evaluation functions from R. A. Fisher in 1936. In order to maximize variance between classes and minimize variance within classes, the basic theory is to determine the original vector projection direction, make the training sample projection separation and classify all the kinds of sample. Variance between classes measures the amount of difference between the average category, and variance in the class is a measure of the amount around the mean variance. The Fisher evaluation method can be divided into several steps: establishing criterion and the evaluation function, back to the generation of the sample and estimation back to the generation of error rate, and identifying new samples [15-16].

Extracting n1, n2 samples of m dimension from overall samples of two kinds is known as the training sample to build a evaluation function. By determining the value of the coefficient by analyzing the basic theory, the value of y can be acquired by generating pending approval samples into the evaluation. Therefore, we can distinguish the samples that belong after comparison with critical value y0.

2.2. Determination of two Fisher model parameters

2.2.1. Determination of evaluation index. The choice of evaluation index directly affects the evaluation result. Therefore, there are many factors that can influence the inrush in the process of coal mining. This article uses water pressure, aquifer type, thickness of water-resisting layer, floor damage depth and fault throw to build and judge the model. In addition, numbers 1 and 2 represent good and poor water abundance in aquifer type, respectively.

Table 1. Training sample data of water inrush of coal mine floor.

No.	Working Face	Water Pressure	Aquifer type	Thickness of Water-Resisting Layer (m)	Floor Damage Depth (m)	Fault Throw (m)	Actual Condition
1	Face 9901 of Taoyang Mine in Feicheng	0.6	1	17	8.6	8	II
2	Face 9204 of Dafeng Mine in Feicheng	1.08	1	16.5	16.5	3.2	II
3	Face 9906 of Taoyang Mine in Feicheng	1.42	1	25.7	15.2	0	I
4	Face 9903 of Taoyang Mine in Feicheng	0.85	1	23.1	13.9	0.4	I

5	Face 7505 of Chazhuang Mine in Feicheng	1.01	1	18	11.7	0	I
6	Face 106 of Pandong Well of Panxi Mine in Xinwen	1.7	0	10	10.7	5	II
7	Face 31104 East of Xiezhuang Mine in Xinwen	1.3	1	30	18.3	4.9	II
8	Face 1532 of Fengfeng First Mine	2.3	0	7.3	7.3	0	II
9	Face 2682 of Fengfeng Second Mine	2.9	1	40	20.9	0	II
10	Face 2671 of Fengfeng Second Mine	2.8	1	40	15	6	II
11	Face 149 of Longquan Mine in Zibo	4.06	0	65.9	16	10	II
12	Face 12031 of Jiulishan in Jiaozuo	1.8	1	23	12.3	0	II
13	Face 1441 of Wangfeng Mine in Jiaozuo	1.1	1	20	8.5	15	II
14	Floor of 33 mining area of Xieyi Mine in Huainan	2	1	30	12.9	1.5	II
15	Face 1301 of Fengying Mine in Jiaozuo	1.9	1	15	13	65	II
16	Face 2131 of Hanwang in Jiaozuo	1.1	1	23.5	8.5	0	II

2.2.2. *Preparation of the assessment grade*

Combining the actual mine drainage ability, the sixteen biggest variation range of water inrush training sample data can be collected. For the convenience of the Fisher criterion of classification, according to the largest water inrush, safe mining evaluation of working face is divided into two levels: safety of first-class (0 m3/h≤ Q <600 m3/h), Ⅰ, and safety of second-class (Q ≥600 m3/h), Ⅱ, as seen in table 1.

Table 2. Prediction samples of water inrush of coal mine floor and comparison of the results of two determining methods.

No.	Working Face	Water Pressure	Aquifer Type	Thickness of Water-Resisting Layer (m)	Floor Damage Depth (m)	Fault Throw (m)	Actual Condition	A Single Fisher Evaluation Result	AHP Weighting-Fisher Evaluation Result
1	Face 1007 of second well of	5.19	0	55.9	17	7	II	II	II

2	Xiazhuang Mine In Zibo Face 9206 of Dafeng	1.26	1	23.5	8.5	0	I	I	I
3	Face II617 of Yangzhuang Mine in Huaibei	3.11	1	44.3	14.4	3.5	II	I	II

2.3. Analysis of two Fisher evaluations

2.3.1. Analysis of one single evaluation model. Combined with the collected data sample, MATLAB software programming can be used to construct the single Fisher evaluation model. Using the single Fisher evaluation model to inspect three test samples, we can find only one sample (Face II617 of Yangzhuang Mine in Huaibei) that did not tally with the actual test results, and the other evaluation results are consistent. The accuracy can reach 66.67%, as shown in table 2.

2.3.2. Analysis of AHP weighting-Fisher evaluation model. (1) Establishing set of evaluation index weight by AHP

Using AHP, we can determine the weights of the evaluation factors. After determining a target and evaluation index, a structure can be built, followed by the judgments matrix and consistency check calculations. Following the hierarchical order sorting method, we can construct judgment matrix [17]:

$$S=\begin{bmatrix} 1 & 4 & 1/2 & 1 & 2 \\ 1/4 & 1 & 1/3 & 1/4 & 1/5 \\ 2 & 3 & 1 & 1/2 & 1 \\ 1 & 4 & 2 & 1 & 2 \\ 1/2 & 5 & 1 & 1/2 & 1 \end{bmatrix}$$

The biggest characteristic root $\lambda_{\max}=5.2859$ of the matrix can be taken from function eig in the Matlab software, and the consistency inspection and indicators can be shown as follows:

$$CI=\frac{\lambda_{\max}-n}{n-1}=\frac{5.2859-5}{5-1}=0.0715$$

According to the different order judgment matrix table, 5 order of judgment matrix and the mean random consistency index $RI=1.12$, the random consistency of the judgment matrix may be revealed by $CR=\frac{CI}{RI}=\frac{0.0715}{1.12}=0.0638<0.10$.

Thus, AHP satisfactory consistency can be acquired and normalized to:

$$A=(0.2347, 0.0573, 0.2270, 0.2991, 0.1820)$$

2.4. Results analysis

By means of the way that the weights of AHP multiply every evaluation index values, the distribution between two classes can be separated as much as possible, together with the inter-class distribution exhibited as dense as possible. When using the same method to predict test samples by Fisher, we can find out that three test samples are in accord with the actual inspection sample in complete (table 2) and that the established working face safe mining by AHP empowerment-Fisher evaluation model has relatively higher accuracy. Fisher evaluation accuracy after AHP weighting is higher than the single

Fisher evaluation, so this method has relatively broad application prospects in the mining safety inrush risk assessment.

Conclusion

- Based on the information provided, AHP analysis and Fisher evaluation theory can be used to study working face safe mining evaluation. The water pressure, aquifer type, thickness of water-resisting layer, base damage zone depth and fault throw are considered as evaluation indexes.
- Single Fisher evaluation and AHP empowerment Fisher evaluation were both applied to the mining safety mining inrush risk assessment. After comparing the evaluation results, clearly, the AHP empowerment - Fisher evaluation inrush model provided more accurate evaluation results.
- The method provides a more effective prediction method in coal mine working face safe mining, and safe mining can be effectively guaranteed by the proposed research.

References

[1] Qin, C J 2012 Research on the construction of safety system for the prevention and control of coal mine water disaster *Safety & Environmental Engineering*

[2] Jia-Lin X U, Zhu W B and Wang X Z 2011 Study on water-inrush mechanism and prevention during coal mining under unconsolidated confined aquifer *Journal of Mining & Safety Engineering*

[3] Xu D, Peng S, Xiang S, Liang M and Liu W 2015 The effects of caving of a coal mine's immediate roof on floor strata failure and water inrush *Mine Water & the Environment* 1-13

[4] Shi L, Zhang B and Zhang X 1999 Forecast of water inrush from mining floor with probability indexes *Journal of China University of Mining & Technology*

[5] Wu Q, Liu Y, Liu D and Zhou W 2011 Prediction of floor water inrush: the application of gis-based ahp vulnerable index method to donghuantuo coal mine, china *Rock Mechanics & Rock Engineering* **44**(5) 591-600

[6] Kong H L and Chen Z Q 2006 Water-inrush-factor and its application in the analysis on harmfulness of water-inrush in the longwall mining in longgu coal mine *Journal of Wuhan University of Technology* **28**(9) 80-81

[7] Wang X and Yang Y 1997 The analysis of water inrush characteristics by vertical leakage-conduction in north china-type coal mine *Coal Geology & Exploration*

[8] Zhang H Q, Yang T H, Zhao D S, Lianc-Hong L and Tang C A 2004 Seepage field analysis for water burst of roof on working face *Coal Geology & Exploration* **32**(5) 17-20

[9] Li L I and Cheng J L 2006 Floor water irruption prediction based on information fusion *Journal of China Coal Society* **31**(5) 623-62

[10] Kong H L, Miao X X, Wang L Z, Yu Z and Chen Z Q 2007 Analysis of the harmfulness of water-inrush from coal seam floor based on seepage instability theory *Journal of China University of Mining & Technology* **17**(4) 453-58

[11] Shi L, Qiu M, Wei W, Xu D and Han J 2014 Water inrush evaluation of coal seam floor by integrating the water inrush coefficient and the information of water abundance *International Journal of Mining Science and Technology* **24**(5) 677-81

[12] Shi L Q, Xu D J, Qiu M, Jing X and Sun H H 2013 Improved on the formula about the depth of damaged floor in working area *Journal of China Coal Society*

[13] Zhen-Hao X U., Shu-Cai L I, Li-Ping L I, Hou J G, Sui B and Shi S S 2011 Risk assessment of water or mud inrush of karst tunnels based on analytic hierarchy process *Yantu Lixue/rock & Soil Mechanics* **32**(6) 1757-66

[14] Chen K, Dong Q H, Zhan-Ji D U and Chu H K 2013 Prediction model for coal mine floor water inrush and its application based on fisher discrimination method *Safety in Coal Mines*

[15] Zhang W Q, Zhang G P, Li W and Hua X 2013 A model of Fisher's discriminant analysis for

evaluating water inrush risk from coal seam floor *Journal of the China Coal Society* **38**(10) 1831-36
[16] Ma Z 2008 Mine inrush water and water disaster prevention and control *China Coal*
[17] Dong-Jing X U, Shi L Q, Qiu M, Sun Q and Sun H H 2013 Fuzzy comprehensive judgment on influence degree of small fault to seam for fully mechanized coal mining *Coal Engineering*

IOP Conf. Series: Earth and Environmental Science **39** (2016) 012021 doi:10.1088/1755-1315/39/1/012021

Effect of cutter type on sediment pollutants release in channel dredging

Y R Yu[1, 2], Y Chen[1, 2, 3, 4, 5], M M Dong[1, 2] and B L Yang[1, 2]

[1] School of River and Ocean Engineering, Chongqing Jiaotong University, Chongqing 400074, China.
[2] National Inland Waterway Regulation Engineering Research Center, Chongqing Jiaotong University, Chongqing 400074, China.
[3] CRC for Water Sensitive Cities, Melbourne VIC3800, Austrialia.
[4] Faculty of Engineering, Monash University, Melbourne VIC 3800, Australia.

E-mail: cycquc@163.com

Abstract. Dredging activities are often used to maintain existing navigation channels. However, traditional dredging equipment inevitably leads to sediment resuspension and nutrient loading in water. In this work, the existing cutter used for dredging was transformed environmentally to reduce the release amount of sediment pollutants, and to avoid the formation of secondary pollution to water bodies. Simulated tests with a general cutter, a spiral cutter, along with a general and spiral cutter equipped with the anti-diffusion device were conducted respectively in this study. The change of pollutants concentration in overlying water was examined. The environmental performance of each different structure cutter was comparatively analysed as well. The result revealed that in channel dredging with a spiral cutter, the release amount of sediment pollutants was less than with a general cutter, and that a general/spiral cutter equipped with the anti-diffusion device could effectively reduce the release amount of sediment contaminants, particularly the release of the nitrogen nutrient during the 1h after the dredging treatment. The best transformation scheme for a cutter suction dredger (CSD) in its environmental-protection function may be: a spiral cutter equipped with the anti-diffusion device.

1. Introduction

Scientific research has characterized the effects of dredging as an underwater excavation process for navigational purposes or material extraction [1]. Dredging activities are necessary for the maintenance of existing navigation channels [2,3], which are important for supporting trade and economic sustainability [4].The Yangtze River, with first-ranking freight volume among all inland waterways in the globe, plays a pivotal role in the global river economic belt. In accordance with the Chinese issued document (NO.[2014]39), the guidance of the State Council on promoting the development of the Yangtze River economic belt, the Yangtze River will be transformed into a smooth golden waterway to meet the navigation requirements of million-ton class cargo ships. Accordingly, treatments of the Yangtze River channel system must be promoted to achieve this objective, especially enforcing shoal dredging, reef dredging and backwater section dredging. In general, dredging starts with the excavation of sediments at a site with a hydraulic and/or mechanical cutter [5-8]. However, traditional dredging equipment inevitably incurs sediments retaining nutrients, heavy metals, and permanent organic pollutants [1, 9-11], to suspend again in dredging. Sediments can therefore also release contaminants into the environment, as contaminants bound on sediment particle surfaces and interior matrices can be released when sediments are disturbed [12-15], causing secondary pollution to water environment. Therefore, existent dredging equipment in channel dredging will result in adverse effects on the aquatic ecosystems of Yangtze River. So far, there are a significant amount of technologies about sediment dredging applied in the treatment of polluted lakes [9,16,17], while there is still a lack of essential study about whether sediment in channel dredging will cause secondary pollution to water bodies. Most studies nowadays mainly focus on the diffusion of suspended matters [18,19], instead of pollutants release resulting from sediment resuspension.

Consequently, based on the mostly-employed CSD in channel dredging, this work will investigate the effect of a different structured cutter on the release of sediment pollutants, as well as comparatively analyze the environmental performance of each structure-different cutter, so as to provide theoretical and technical support for inventing environmental dredging equipment.

2. Materials and methods

2.1. Materials

The cutter-suction dredging device we constructed in this work consisted of a suction/drainage system and power systems which were comprised of the cutter power system and the mud pump power system as shown in figure 1. In addition to fully considering the working conditions of CSD in real practice, we set up a metering device for measuring the weight/volume of dredged sediments, a rack for the cutter-suction dredging device motion, and mud sumps for forming aquatic environment and placing dredged sediments (a main sump and a vice one, whose sizes are 60×30×50cm and 30×10×5cm respectively) in order to be as close as possible to operating in real engineering. Based on common sense and specific calculation, the main geometrical and moving parameters of the initial environmental cutter were as following: 2.5m of length, 2.0m of diameter, 7.5° of helical angle, 16° of conical angle, 40 r/min of rotation rate and 10 m/min of transverse movement speed. According to the geometric and dynamic similarity theories, those main parameters of the environmental protection cutter in this study were determined as following: 25mm of length, 20mm of diameter, 7.5° of helical angle, 16° of conical angle, 40 r/min of rotation rate and 0.1 m/min of transverse movement speed. Moreover, the main parameters of the general lab-scale cutter are 25mm of length, 20mm of diameter, 0° of helical angle, 16° of conical angle, 40 r/min of rotation rate and 0.1 m/min of transverse movement speed. Detailed parameters of the initial and the lab-scale environmental cutter are listed in table 1. Lab-scale cutters in this study are shown in figure 2.

According to the Surface Water Environment Quality Standard of the People's Republic of China (GB3838-2002), which sets the maximum limits for basic contaminants in water bodies, in order to protect aquatic life and ensure agricultural irrigation use, the water quality of the Yangtze River should meet the standard as described in table 2. Sediment samples in the study were collected from the Zhongxien County region (Chongqing, China) of the Yangtze River, the place with the most serious deposition in the perennial backwater area of the Three Gorges. To measure the maximum release amount of sediment pollutants, 25g sediments were placed in the conical flask, which was added to 500mL potable water, and shaken in a constant temperature (18.2℃, the annual average water temperature in the Three Gorges reservoir area) for 24h. Samples were taken at 1, 2, 4, 6, 8, 12 and 24h for pollutant concentration measurement after centrifugation (4000 r/min of rotation rate, 20min of durational time). The maximum release amount of sediment pollutants attained is shown in Table 3, where total nitrogen (TN), ammonia nitrogen (NH_3-N) and chemical oxygen demand (COD) were 77.8, 45.2 and 2484.8 mg/kg respectively. Potable water was used as the overlying water in this work, so as to make the release performance of sediment pollutants more remarkable.

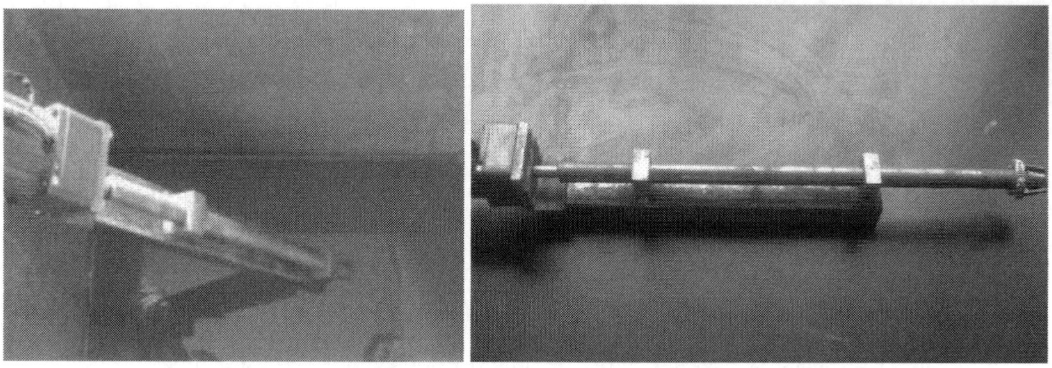

Figure 1. Cutter-suction dredging device in the study.

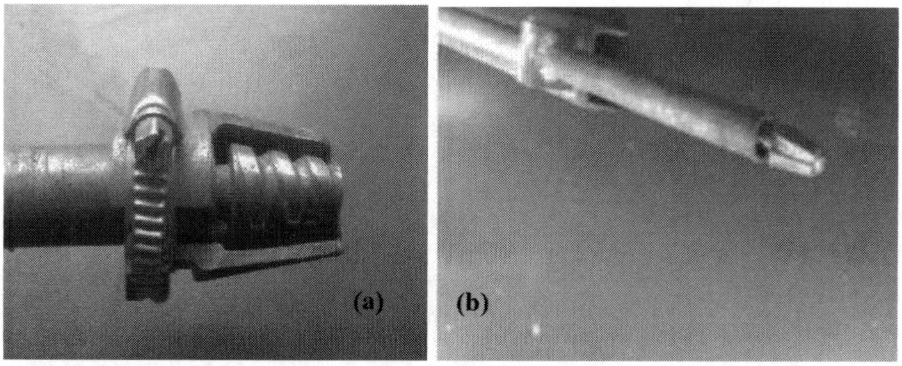

Figure 2. Lab-scale cutters in the study: (a) spiral cutter equipped with the anti-diffusion device; (b) general cutter.

Table 1. Detailed parameters of the initial and lab-scale environmental cutter.

Item	Length (m)	Diameter (m)	Helical angle (°)	Conical angle (°)	Rotation rate (r/min)	Transverse movement speed (m/min)
Initial cutter	2.5	2	7.5	16	40	10
Lab-scale cutter	0.025	0.02	7.5	16	40	0.1

Table 2. The maximum limits for basic contaminants in water bodies.

Pollutants index	TN	NH_3-N	COD
Maximum limit (mg/L)	2.0	2.0	40.0

Table 3. The maximum release amount of sediment pollutants.

Sediment pollutants index	TN	NH_3-N	COD
Maximum release amount (mg/kg)	77.8	45.2	2484.8

2.2. Methods

Simulated dredging experiments with a general cutter, spiral cutter and general/spiral cutter equipped with the anti-diffusion device, were carried out respectively in this study. The change of pollutant concentration in overlying water, including NH_3-N, TN and COD, was monitored at 0.25, 0.5, 1, 3, 5, 7, 12 and 24h, after the finishing of dredging. Corresponding analysis methods used for determining each of these items are described in more detail in table 4. The environmental performance of each different structure cutter was analysed comparatively. Based on the cumulative release amount of nutrient nitrogen, the release rate, M ($g/m^3 \cdot h$), was calculated by the following equation to analyse the release of sediment pollutants: $M = (R_i - R_{i-1}) / n$, where R_{i-1} and R_i refer to the cumulative release amount, taken at the i-1 and i times sample respectively.

Table 4. The schedule of analysis items and methods

Item	Analysis method
NH_3-N	Salicylic acid spectrophotometric method
TN	Alkaline potassium persulfate digestion-uv spectrophotometer
COD	Fast digestion spectrophotometric method

3. Results and discussion

3.1. Effect of cutter type on the release of sediment pollutants

Cumulative release performance of sediment pollutants surrounding dredging sites, when dredging with a general cutter and with a spiral cutter, was monitored as displayed in figure 3. More significantly, figure 3 clearly shows that the release of sediment pollutants can be substantially reduced by dredging with a spiral cutter. The maximum cumulative release amount of NH_3-N, TN and COD, when dredging with a spiral cutter, were 0.64, 1.28 and 40 g/m^3 respectively, whereas the maximum cumulative release amount of NH_3-N, TN and COD, when dredging with a general cutter, reached to 0.69, 1.48 and 47 g/m^3 respectively.

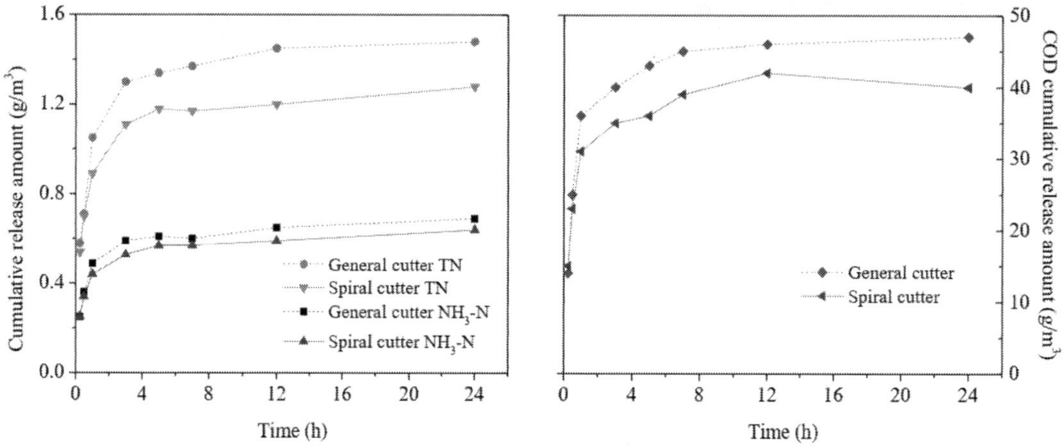

Figure 3. Release performance of pollutants contained in sediment under dredging with a spiral/general cutter: nutrient nitrogen (left); COD (right).

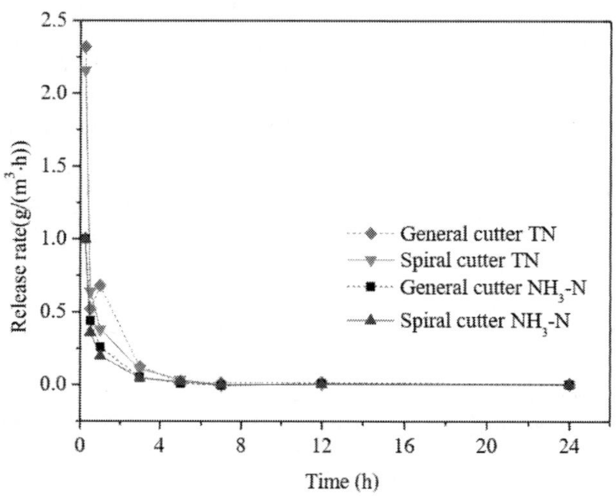

Figure 4. Nitrogen nutrient release rate under dredging with a different-structure cutter.

The nutrient nitrogen release rate under a different cutter is illustrated in figure 4. It shows that the release rate of nutrient nitrogen, after dredging, was fading over time. The difference of the nutrient nitrogen release rate between the two different-structure cutters was distinctive during the first 1h of dredging treatment, and the nutrient nitrogen release rate under dredging with spiral cutter was less than with the general one. However, there was no difference between different structure cutters after 1h of the dredging treatment. Subsequently, the declining trend of the nutrient nitrogen release rate became progressively smooth over time, approaching 0 g/(m^3·h) after 7h. Results indicated that a spiral cutter, compared with a general cutter, has many advantages when it comes to governing nutrient nitrogen release amount. For different contaminants in sediments, after overlying water became smooth, dredging with a spiral cutter could effectively abate the release resulting from the dredging operation, wherein COD, TN and NH_3-N release amount diminished approximately 14.9%, 13.5% and 7.2% respectively, compared with a general cutter.

Engineering dredging is mainly applied in channel dredging and harbor construction in order to increase the capability of channel, while environmental protection dredging aims at removing polluted sediments and providing a good condition for the advancement of the aquatic ecosystem [20]. In effect, polluted sediments cleared by environmental dredging are softer than by engineering dredging. Not only can this characteristic make sediments easy to be cut, it may also facilitate sediment suspension and diffusion. The edge of a spiral cutter is serial, which makes it stable to cut sediment and makes only small disruptions on the surface of sediment. While a general cutter, driven by the pursuit of sediment-broken and adapting abilities, abandons the serial performance. Consequently, a general cutter used in dredging would create a great shock on sediments, and disruption would rapidly foment the suspension of sediments. The shock can also release pollutants into water bodies, as pollutants bound on sediment particle surfaces and interior matrices can be released when sediments are disturbed [1,12,15].

3.2. Effect of a general cutter equipped with the anti-diffusion device on the release of sediment pollutants

The cumulative release feature of pollutants around disposal sites in the dredging experiment with a general cutter, equipped with the anti-diffusion device, is illustrated in figure 5. As shown, dredging with a housing general cutter has a great impact on diminishing the release of sediment contaminants, with the maximum cumulative release of 0.57g/m^3 for NH_3-N, 1.16 g/m^3 for TN and 34 g/m^3 for COD. The variations of the nutrient nitrogen release rate with time under dredging with a cutter that is equipped with the anti-diffusion device, is described in figure 6. As expected, the release rate of nutrient nitrogen — NH_3-N and TN — gradually declined after dredging. Meanwhile, dredging with a housing general cutter has a significant influence on the nutrient nitrogen release rate during the first hour of dredging, in which the release rate of nutrient nitrogen was less than a general cutter not equipped with the anti-diffusion device. The difference disappeared after 1h, and the decreasing trend of nutrient nitrogen release rate became increasingly smooth over time, approaching 0 g/(m^3·h) after 7h.

Results indicated that a housing general cutter used in dredging can substantially constrain the release of pollutants. Dredging by a general cutter equipped with the anti-diffusion device could reduce the release of each pollutant, which had the most significant effects on the release of COD among all, whose release rate was lighter than with a general cutter without the anti-diffusion device, decreasing around 27.7%. Moreover, it also had impacts on regulating the release of TN, of which the decreasing amount was about 21.6%. Its restriction on NH_3-N release amount was weaker than on COD and TN. Specifically, the release amount of NH_3-N decreased around 17.4%, compared to a cutter without the anti-diffusion device. Overall, this result revealed that dredging with a housing general cutter has a distinctive effect on reducing the release of sediment pollutants, and this regulating function was stronger than a spiral cutter.

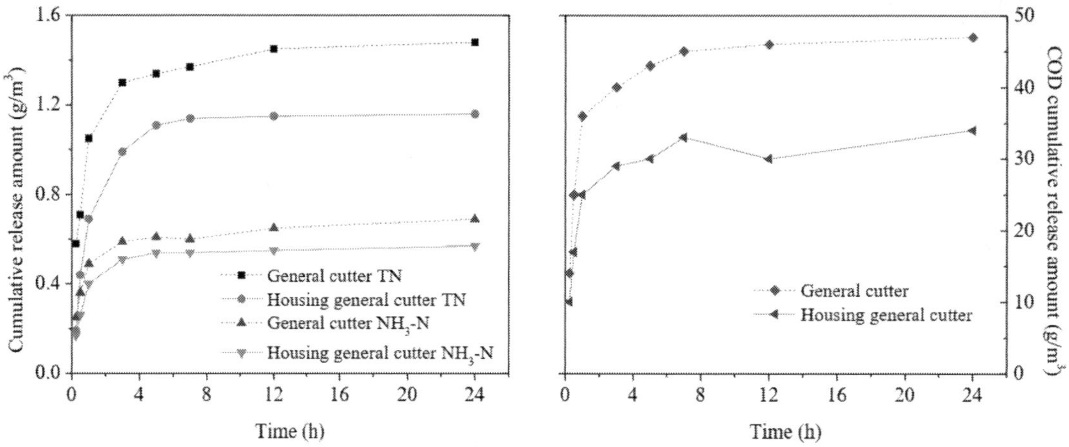

Figure 5. Release performance of pollutants contained in sediment under dredging with a housing-general/general cutter: nutrient nitrogen (left); COD (right).

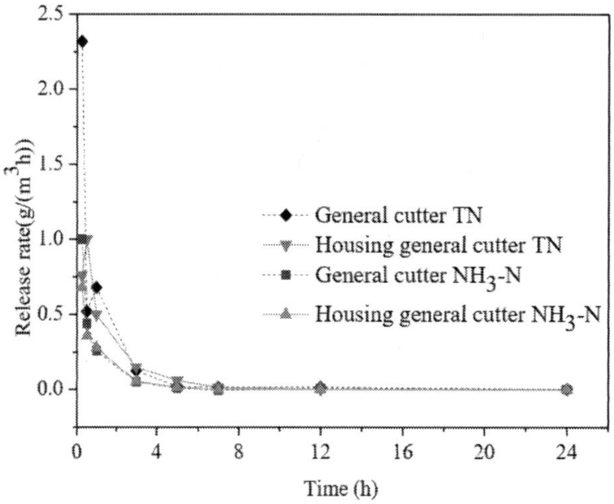

Figure 6. Nitrogen nutrient release rate under dredging with different-type cutter.

It could thus be speculated that when the cutter cuts sediments in channel dredging, water around the cutter would move as the movement of cutters, hence causing a unique flow distraction that mixes water and granulated sediment during cutting. As a result, slurry mixture will break away from the suction field of the mud pump and will diffuse in all directions. The diffusing slurry mixture therefore becomes a new source of pollution and releases contaminants contained in sediments to aquatic ecosystems, thus forming a secondary pollution in water bodies. Accordingly, when the cutter is equipped with the anti-diffusion device, the diffusion of sediments could be effectively controlled, avoiding slurry mixture to break away from the suction field of the mud pump and the formation of secondary pollution in water bodies.

3.3. Effect of a spiral cutter equipped with the anti-diffusion device on the release of sediment pollutants

When dredging with a spiral cutter equipped with the anti-diffusion device, the cumulative release characteristics of contaminants surrounding dredging points is portrayed in figure 7. It should be pointed out that dredging with a housing spiral cutter could abate the release amount of pollutants in sediment as described in figure 7. When pollutants concentration in overlying water became stable, the maximum release amount of NH_3-N, TN and COD, under dredging with a housing spiral cutter, were 0.54, 1.07 and 30 g/m³ respectively, reducing about 11.8%, 7.8% and 5.3% compared with a housing general cutter.

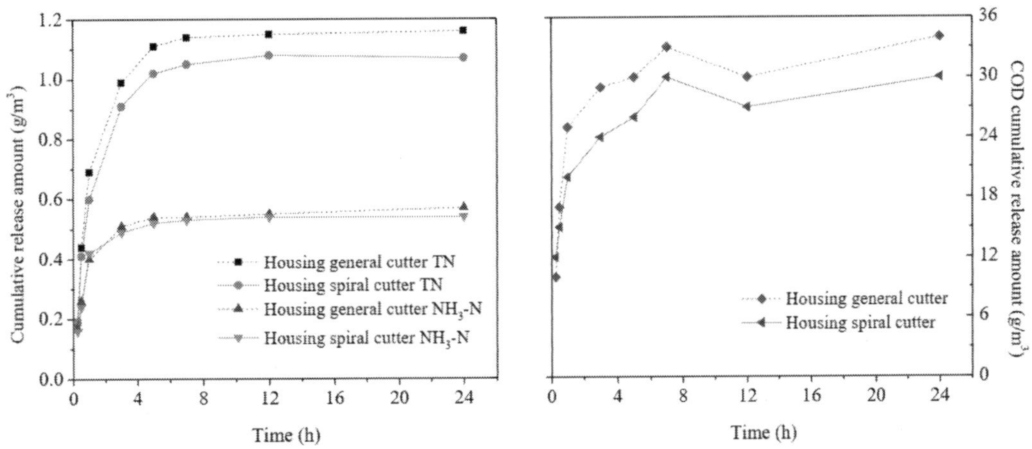

Figure 7. Release performance of pollutants contained in sediment under dredging with a housing-general/ housing-spiral cutter: nutrient nitrogen (left); COD (right).

The changes of the nutrient nitrogen release rate with time, when dredging with a housing spiral cutter, and the housing general one are shown in figure 8, from which it could be observed that the release rate of sediment nutrient nitrogen became increasingly slow over time. While dredging with a housing spiral cutter could significantly affect nutrient nitrogen release rate during the first hour of dredging. The nutrient nitrogen release rate under dredging with a spiral cutter equipped with the anti-diffusion device was relatively low compared with the housing general one. After 1h, the difference became small, and the nutrient nitrogen release rate became slow over time, approaching 0 g/(m^3·h) after 7h (figure 8).

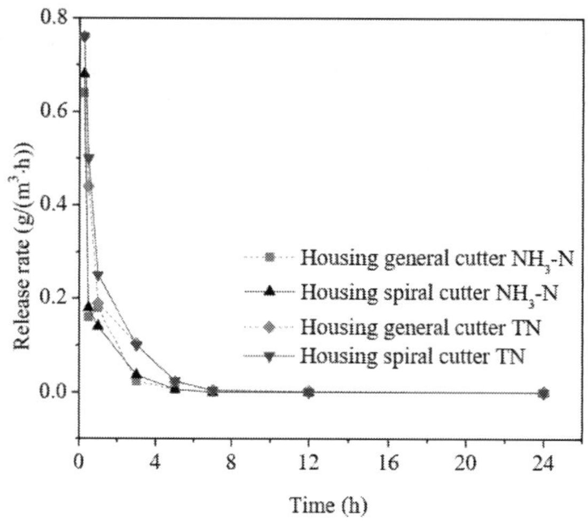

Figure 8. Nitrogen nutrient release rate under dredging with different-type cutter.

Results suggested that dredging with a housing spiral cutter could substantially restrain the release of sediment pollutants to some degree, with cumulative release of COD, TN and NH_3-N in sediments reducing around 36.2%, 27.7% and 21.7% respectively, compared with a general cutter. Therefore, given the restriction of sediment pollutant release, the most ideal transformation scheme for CSD in its environmental-protection function could be dredging with a spiral cutter that is equipped with the anti-diffusion device. In this research, for dredging with each environment-friendly cutter (the spiral and general/spiral equipped with the anti-diffusion device), the restriction on sediment NH_3-N release was less than on COD and TN. We hypothesise that each environment-friendly cutter can effectively abate disturbance to sediments and particle re-suspension, more or less in dredging practice. It can hence be speculated that COD and TN release would be diminished dramatically due to their contents in overlying water being in close relation to suspended particles derived from channel sediments. However, NH3-N exists as ionic-form easily dissolved in water, and thus has a weaker association with suspended particles compared to other contaminants, which could explain why its release restraint from environment-friendly cutters is relatively light.

4. Conclusions

This paper describes the outcomes of a research study undertaken to investigate the release performance of sediment contaminants in dredging with a different structure cutter, and the results obtained can be summarized as following: (1) when dredging with a spiral cutter, the cumulative release of NH_3-N, TN and COD in sediments, compared with the general one, reduced about 7.2%, 13.5% and 14.9% respectively; (2) when dredging with a general cutter equipped with the anti-diffusion device, the sediment NH_3-N, TN and COD cumulative release amount, compared with the general one, decreased around 17.4%, 21.6% and 27.7% respectively, and (3) when dredging with a spiral cutter that is equipped with the anti-diffusion device, the cumulative release of NH_3-N, TN and COD from sediments, compared with the housing general one, diminished approximately 5.3%, 7.8% and 11.8% respectively.

The study revealed that both dredging with each of the two cutters — general and spiral — equipped with the anti-diffusion device, and dredging with the spiral cutter, can effectively restrain the release of sediment pollutants, but the former is superior to the latter. Dredging by a spiral cutter equipped with the anti-diffusion device, compared with other different type cutters, has the best environmental performance, which may be the optimal advancing method for CSD in its environmental performance. Nevertheless, further study, especially the consideration of stream turbulence in real engineering which is relevant to sediment pollutant diffusion, is highly desirable.

Acknowledgments
The authors are grateful to the critical comments raised by the anonymous reviewers, which significantly improve this manuscript, and gratefully acknowledge the financial support provided by the Scientific & Technological Research Project of China (No. 2011BAB09B01) and the Scientific & Technological Innovation Project of Chongqing Education Commission (No. CYS15189).

References

[1] Manap N and Voulvoulis N 2015 Environmental management for dredging sediments - the requirement of developing nations *Journal of Environmental Management* **147** 338-48
[2] Pazos M, Iglesias O, Gomez J Rosales E and Sanroman M A 2013 Remediation of contaminated marine sediment using electrokinetic–fenton technology *Journal of Industrial & Engineering Chemistry* **19**(3) 932–37
[3] Wang L, Yuan X, Zhong H, Wang H, Wu Z, Chen X and Zeng G 2014 Release behavior of heavy metals during treatment of dredged sediment by microwave-assisted hydrogen peroxide oxidation *Chemical Engineering Journal* **258** 334-40
[4] Bates M E, Fox-Lent C, Seymour L, Wender B A and Linkov I 2015 Life cycle assessment for dredged sediment placement strategies *Science of the Total Environment* **511** 309-18
[5] Honmagumi K K and Chiyoda Kenki K K 1995 *Mud processing e by dredging using a closed hydraulic grab bucket, removing large solids, agitating with a high pressure pump and solidifying* JP192176:0-9
[6] Klein J 1997 *Sediment dredging and macrophyte harvest as lake restoration techniques*
[7] Antipov V V, Antipov Y V, Brakker I I, Brenner V A, Naumov Y N, Pushkarev A E, Pushkarev V A, Podkolzin A A and Zhabin A B 2006 *Channel dredging method, involves cutting ground with working blades of mechanical cutting tools installed on rotary cutting head, which performs horizontal intermittent movement in both directions transversely to axis of rotation* RU122595
[8] Du Y H and Li H W 2010 *Mechanical technology for dyke reinforcement by sediment discharge in the lower reaches of the Yellow River* Yellow River Conservancy Press
[9] Murphy T P, Lawson A, Kumagai M and Babin J 1999 Review of emerging issues in sediment treatment *Aquatic Ecosystem Health & Management* **2**(4) 419-34
[10] Ghosh U, Zimmerman J R and Luthy R G 2003 PCB and PAH speciation among particle types in contaminated harbor sediments and effects on PAH bioavailability *Environmental Science & Technology* **37**(10) 2209-17
[11] Palm A, Cousins I, Gustafsson Ö, Axelman J, Grunder K, Broman D and Brorström-Lundén E 2004 Evaluation of sequentially-coupled POP fluxes estimated from simultaneous measurements in multiple compartments of an air–water–sediment system *Environmental Pollution* **128**(1) 85-97
[12] Garrett R G 2000 Natural sources of metals to the environment *Human and Ecological Risk Assessment* **6**(6) 945-63
[13] Burton Jr G A 2002 Sediment quality criteria in use around the world *Limnology* **3**(2) 65-76
[14] Salomons W and Brils J 2004 Contaminated sediments in European River Basins *European Sediment Research Network SedNet* **71**
[15] Fluck R, Chevre N and Campiche S 2010 *Sediment Quality Assessment*
[16] Moss B, Balls H, Irvine K and Stansfield J 1986 Restoration of two lowland lakes by isolation from nutrient-rich water sources with and without removal of sediment *Journal of Applied Ecology* 391-414
[17] Ruley J E and Rusch K A 2002 An assessment of long-term post-restoration water quality trends in a shallow, subtropical, urban hypereutrophic lake *Ecological Engineering* **19**(4)

265-80

[18] Hossain S, Eyre B D and McKee L J 2004 Impacts of dredging on dry season suspended sediment concentration in the Brisbane River estuary, Queensland, Australia *Estuarine, Coastal and Shelf Science* **61**(3) 539-45

[19] Capello M, Cutroneo L, Ferranti M P, Castellano M, Povero P, Budillon G and Tucci S 2013 Mathematical simulation of the suspended solids diffusion during dredging operations on the continental shelf off the coast of Lazio (Central Tyrrhenian Sea, Italy) *Ocean Engineering* **72** 140-48

[20] Zhang S H, Jiang C B, Chen J and Hu B A 2014 Field observation and study of turbidity plume induced by dredger *Journal of Sediment Research* (4) 44-48

IOP Conf. Series: Earth and Environmental Science **39** (2016) 012022 doi:10.1088/1755-1315/39/1/012022

Effects of reservoir impoundment on fish assemblages in the Wujiang River of China

D Zhu[1, 2] and Z Yang[1]

[1]Key Laboratory of Ecological Impacts of Hydraulic Projects and Restoration of Aquatic Ecosystem of Ministry of Water Resources, Institute of Hydroecology, Ministry of Water Resource and Chinese Academic of Sciences, 578 Xiongchu Avenue, Wuhan, China

E-mail: zhudi@mail.ihe.ac.cn

Abstract. Fishery catches were investigated during the spring and autumn in the period 2006-2010. The fish assemblages' composition and dynamics were analysed and compared under the pressure of the Pengshui Reservoir impoundment. A total of 101 species were sampled in the lower Wujiang River. 78 species and 83 species were found in the upper and the lower reaches, respectively. After impoundment there were five species in the Yanhe branch and nine species in the Pengshui branch. The abundance of species was higher in the Pengshui than in the Yanhe. Fish abundance did not change in the first year of impoundment (2008) in the Lujiao-Gongtan reach, but it dramatically declined by 45.95% in 2009. The fish were subdivided into three groups and two ecotypes according to their time-series appearance in fish catches and their habitat preference. Species composition changed significantly with decreases in numbers of lotic species. Fish assemblages responded more rapidly and dramatically to impoundment in the reservoir section than the section below the Pengshui Dam.

1. Introduction

Damming can cause dramatic changes in river ecosystems and is a major factor in the decline of biodiversity and variation in aquatic animal community structures [1][2]. There has been an increase in research exploring ecological response patterns that result from damming, with particular focus on the adverse and negative impacts of damming on aquatic organisms and hydro-ecosystems [3][4][5][6]. Fish live in distinct habitats [7], and the spatial distribution of fish is affected by depth, distance from the dam, and habitat choice [8]. Further, fish are sensitive to flow regimes and hydrological conditions, which can be dramatically changed by a dam [9]. This study was conducted to identify the impact of hydropower development on fish assemblages and to understand the impact pattern in sensitive ecological areas.

2. Materials and methods

2.1. Study area

The Wujiang River is a 1037 km long canyon-shaped tributary of the Yangtze River. It contains eleven cascades that produce hydropower, and these were created as a response to the increasing electric power demand in China. The Pengshui (PS) Dam is the 10th cascade, is located in the lower Wujiang River and is 147 km from an estuary. PS Reservoir is regulated seasonally, controlling a

69000 km^2 catchment area and covering around 78.5% of the total drainage area of the Wujiang River. Its normal and dead storage levels are 293m and 278m, respectively, and its regulated storage capacity is 0.518 billion m^3. This project began in September 2005, and impoundment began in January 2008.

The Yanhe (YH), Lujiao-Gongtan (LG) and Pengshui (PS) reaches of the Wujian were selected for investigation (Figure 1). In each reach, investigations were conducted from upstream to downstream on the backwater, reservoir, and downstream sections.

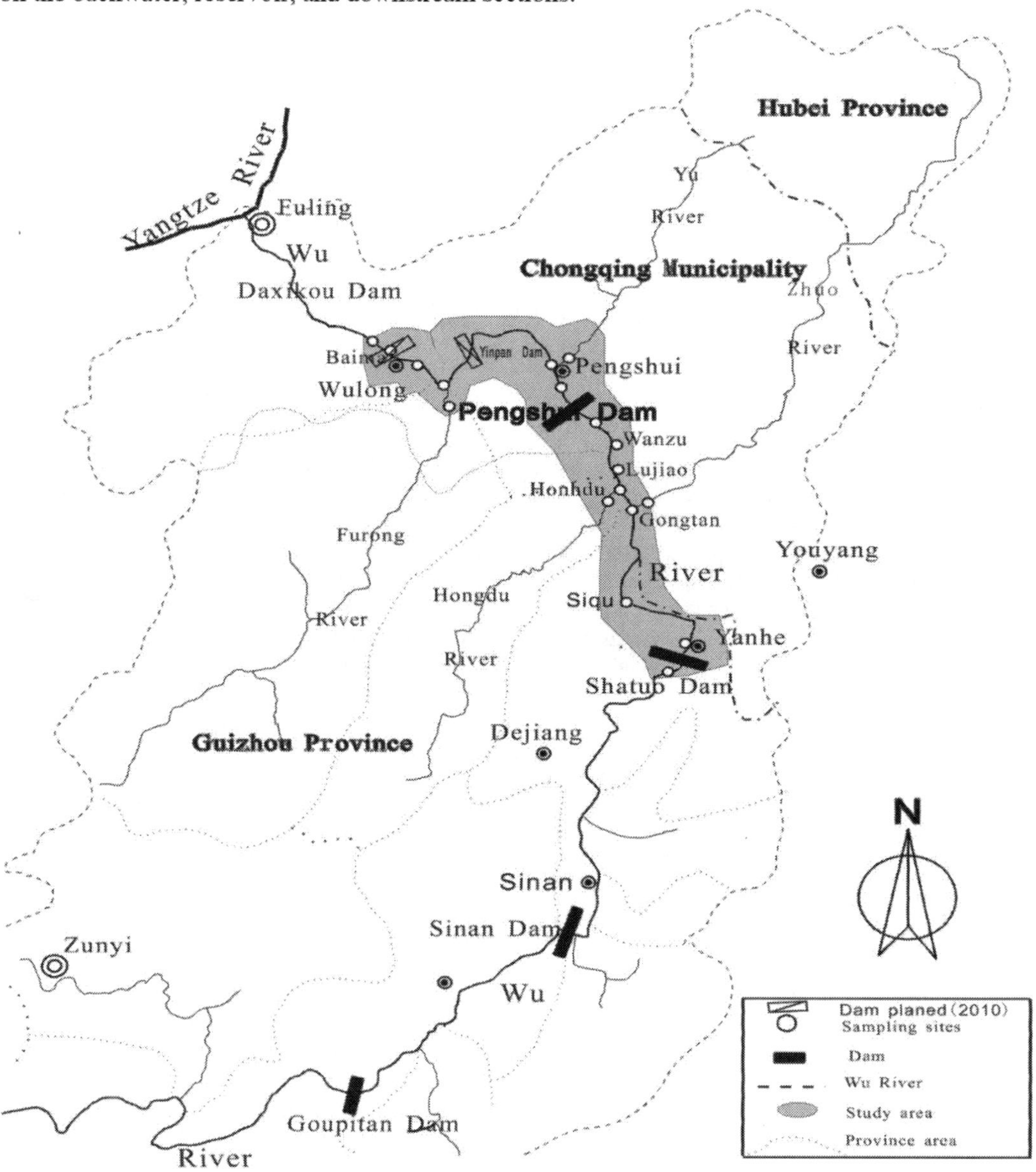

Figure 1. Distribution of three sampling reaches in the lower Wujiang River.

2.2. Data collection and transformation
Commercial fishery catches of settled gillnets and hooks were investigated in the spring and autumn seasons from 2006 to 2010. The investigation was conducted each day until no new species appeared for seven consecutive days. All the specimens collected were identified, counted, and measured.

To reduce error, transformations were made to the primary data. If a species was investigated both before and after a specific year (x), but in the year (x) the species was not collected, the mean values of the number and weight of individuals in the two years it was present are used for the species data in year (x).

2.3. Statistical methods and software
The statistical methods used to present the data included:

(1) Community dominant species

Index of relative importance IRI = $(Pi + Pm)\ F \times 10^4$

Above, Pi and Pm are the proportion of number and weight, and F is the species' frequency of occurrence. If the IRI>1000, this species is considered to be a dominant species [10].

(2) Fish diversity

Four indices have been used in the literature for estimating fish diversity, and are described below [11][12][13][14]:

- Margalef Index of community diversity $P = (S-1)(\ln N)^{-1}$
- Shannon Wiener Index $H' = -\sum P_i \ln P_i$
- Pielou evenness index $E = H'(\ln S)^{-1}$
- Simpson dominance index $D = \sum (P_i)^2$

Above, Ni is the number of individuals of i species, N is the total number of individuals, and Pi accounted for the proportion of Ni in N, that is, Pi=Ni/N, i=1, 2, 3…N

(3) Community structure similarity

Fish abundance was taken as a variation of the original matrix, and the investigational years were taken as samples of the original matrix. Similarity, in the community, structure was analyzed using the Bray-Curtis similarity coefficient[15] and Cluster analysis (the unweighted pair group method with arithmetic means (UPGMA)). First, similarity matrices between the samples were developed, and then the characteristics of fish assemblages' structure were analyzed by establishing a cluster analysis diagram.

All of the data were analyzed using PRIMER 5.0 software packages (Plymouth Marine Laboratory, UK), EXCEL 2010, and SPSS 13.0.

3. Materials and methods

3.1. Study area
From 2006 to 2010, a total of 101 species were sampled; there were 78 species in the upper reaches (LG, YH) and 83 species in the lower reach (PS). 24 endemic species of the Upper Yangtze River and three vulnerable species, *Leptobotia elongate*, *Procypris rabaudi* (which appear on the "China red data book of endangered animals species") and *Percocypris pingi* (which appear on the "China species red list"), were found.

After impoundment, fish abundance varied between five species in the YH reach and nine species in the PS reach. Fish abundance in the PS reach was higher than in the YH and LG reaches; fish abundance in the LG reach showed no difference in the first impoundment year (2008), but declined 45.95% in 2009 (Figure 2).

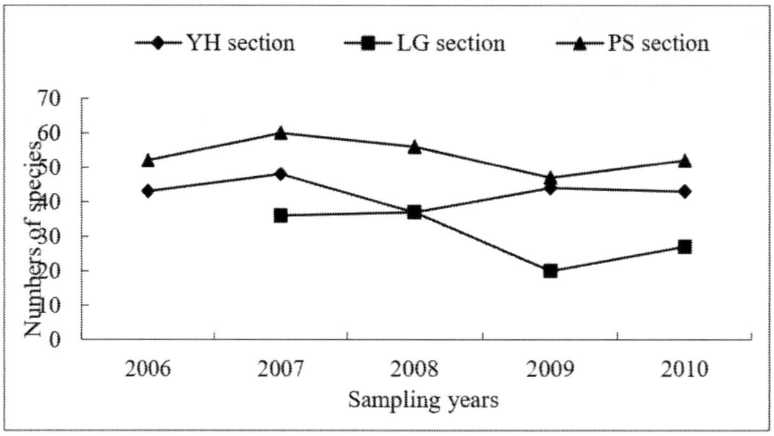

Figure 2. Dynamics in species abundance of the YH, LG and PS reaches.

In this study, fish representing more than 1% of all the species sampled in both weight and number were considered as main species, while the others were considered to be rare species. There were 20 main species in the YH reach with proportions of weight and number of 88.72% and 79.85%, respectively; there were 16 main species in the LG reach with proportions of weight and number of 89.92% and 88.99%, respectively; there were 16 main species in the PS reach with proportions of weight and number of 82.20% and 71.19%, respectively.

The fish species were classified into three groups: A) species that appeared both before and after the impoundment, B) species that appeared only before the impoundment, and C) species that appeared only after the impoundment.

The results showed that the main fish species of the YH and PS reaches did not have obvious changes in the "before and after impoundment" group; differences in species composition resulted from rare species. Fish species that appeared only after impoundment in the LG reach accounted for 34.55% of the total number of species, but in the PS and YH reaches, fish species that only appeared after impoundment accounted for smaller proportions: 19.18% and 18.07% of the total number of species, respectively (Table 1).

Table 1. Three groups of fish species based on their appearance in catches by time-series.

Reaches	YH				LG				PS			
	S[d]	PS[e]	I[f]	PI[g]	S[d]	PS[e]	I[f]	PI[g]	S[d]	PS[e]	I[f]	PI[g]
Group A[a]	50	68.49	10881	98.53	29	52.73	2576	87.65	61	73.49	8644	98.81
Group B[b]	9	12.33	40	0.36	7	12.73	56	1.91	7	8.43	32	0.37
Group C[c]	14	19.18	122	1.10	19	34.55	307	10.45	15	18.07	72	0.82

[a] Species appeared only before the impoundment
[b] Species appeared before and after the impoundment
[c] Species appeared only after the impoundment;
[d] Species
[e] Individuals
[f] Proportion of the number of species
[g] Proportion of the number of individuals

3.2. Ecological types and dominant species

3.2.1. Ecological types. The main species were classified by their habitat preference into two ecotypes: lotic and lentic species. Eleven lotic and nine lentic species were distributed in the YH reach, nine lotic and eight lentic species were distributed in the LG reach, and eight lotic and eight lentic species were distributed in the PS reach (Table 2).

Table 2. Ecotypes of main species distributed in each reach.

Lotic species	Lentic species	YH		LG		PS	
		Lotic	Lentic	Lotic	Lentic	Lotic	Lentic
Leiocassis crassilabris Günther	*Cyprinus (Cyprinus) carpio* Linnaeus	*	*	*	*	*	
Mystus macropterus (Bleeker)	*Carassius auratus* (Linnaeus)	*	*	*	*	*	*
Semilabeo prochilus (Sauvage et Dabry)	*Hemibarbus maculatus* Bleeker	*	*	*		*	
Tor (Folifer) brevifilis brevifilis (Peters)	*Zacco platypus* (Temminck et Schlegel)	*	*	*	*		*
Spinibarbus sinensis (Bleeker)	*Saurogobio dabryi* Bleeker	*	*	*			
Opsariichthys bidens Günther	*Hemiculter leucisculus* (Basilewsky)	*	*		*		*
Garra pingi pingi (Tchang)	*Siniperca kneri* Garman	*	*	*			
Onychostoma sima (Sauvage et Dabry)	*Silurus meridionalis* Chen	*	*	*			*
Leptobotia rubrilabris (Dabry)	*Pelteobagrus vachelli* (Richardson)	*	*		*		*
Pseudobagrus truncatus (Regan)	*Procypris rabaudi* (Tchang)	*			*		
Rhinogobio typus Bleeker	*Pseudobagrus ussuriensis* (Dybowski)	*			*	*	
Coreius guichenoti (Sauvage et Dabry)	*Silurus asotus* Linnaeus			*	*		
Coreius heterodon (Bleeker)	*Pseudolaubuca sinensis* Bleeker			*		*	*
Rhinogobio cylindricus Günther	*Hemiculter tchangi* Fang					*	*
Rhinogobio ventralis (Sauvage et Dabry)	*Hypophthalmichthys molitrix* (Cuvier et Valenciennes)					*	*
Leiocassis longirostris Günther						*	
Total mumber of species		11	9	9	8	8	8

There was fluctuation in fish abundance of both ecotypes in each reach during the period of 2006 to 2010 (Figure 3). In the YH reach, lotic species declined annually in terms of both number and weight, but the lentic species increased in terms of both number and weight. In the LG reach, lotic species declined annually during the first three years of impoundment and increased in the next two years, but the lentic species displayed a reverse trend. This indicates that the fish assemblages responded quickly to dam construction and were in an unsteady state in the first five years after dam construction. Below the dam, lotic species decreased annually in number, and in spite of an increase in weight after 2008, the lotic species weight was still lower than it had been in 2006.

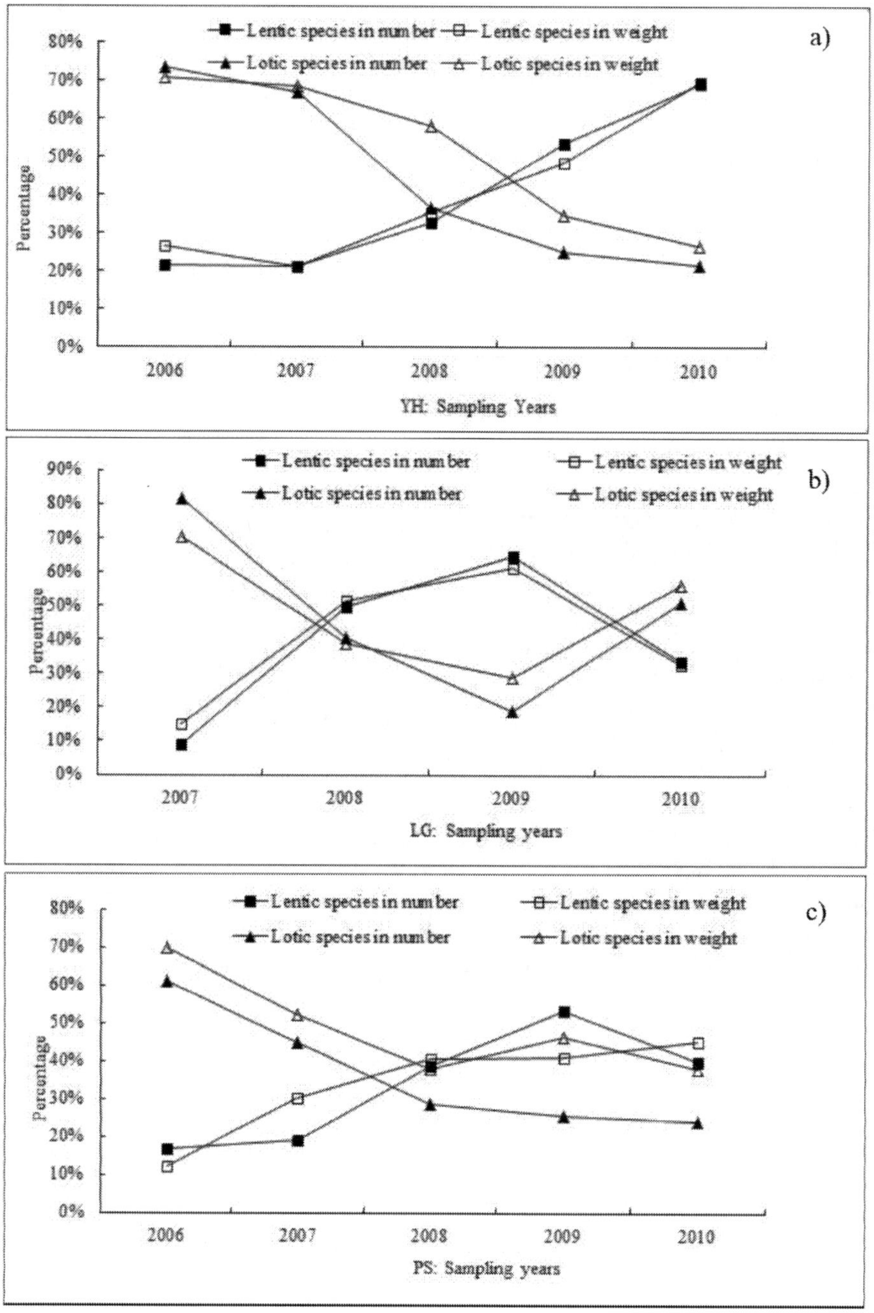

Figure 3. Lotic and lentic species in the YH, LG, and PS reaches.

3.2.2. Dominant species. Dominant species were those species for which IRI>1000 in at least one year. There were 9 dominant species in the YH reach, 12 dominant species in the LG reach and 8 dominant species in the PS reach (Figure 4).

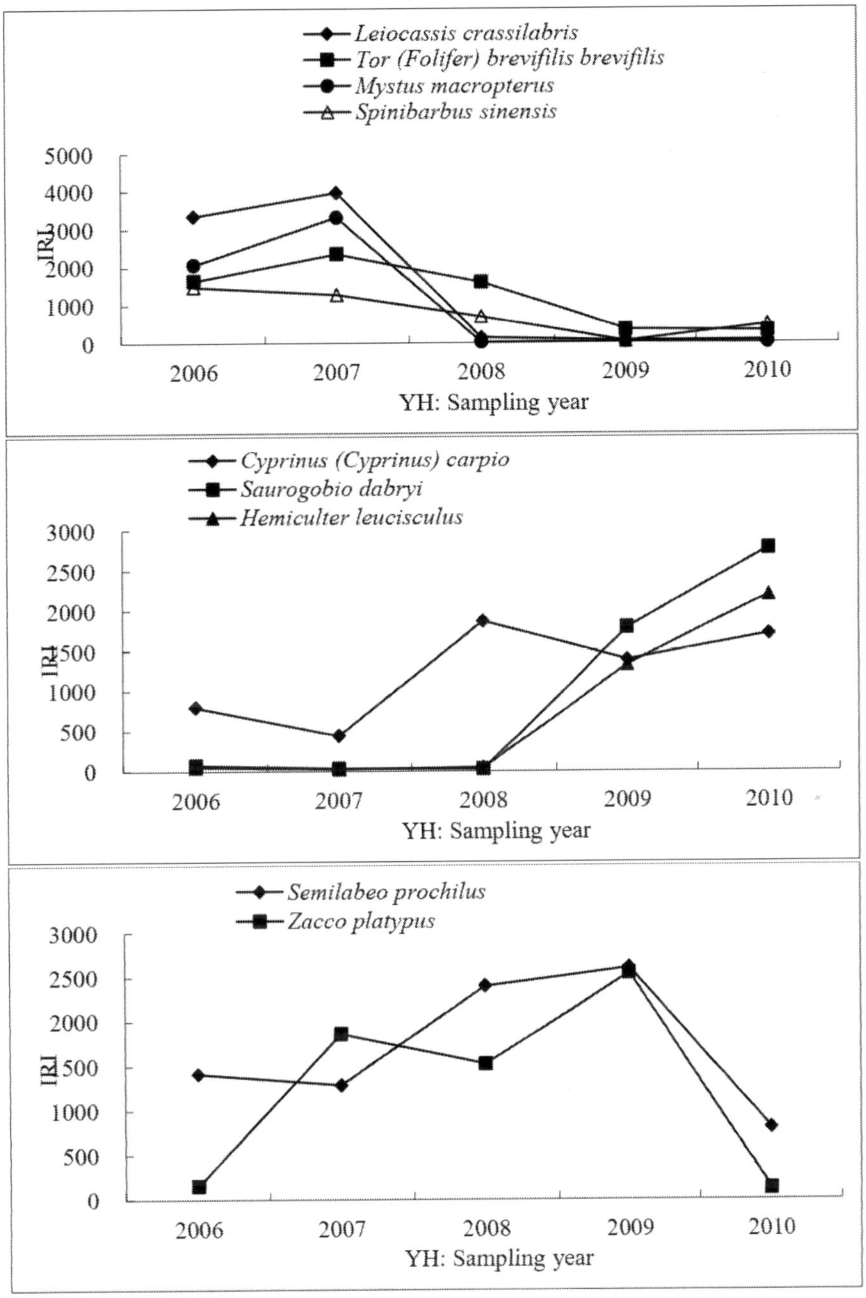

(a)YH reach

IOP Conf. Series: Earth and Environmental Science **39** (2016) 012022 doi:10.1088/1755-1315/39/1/012022

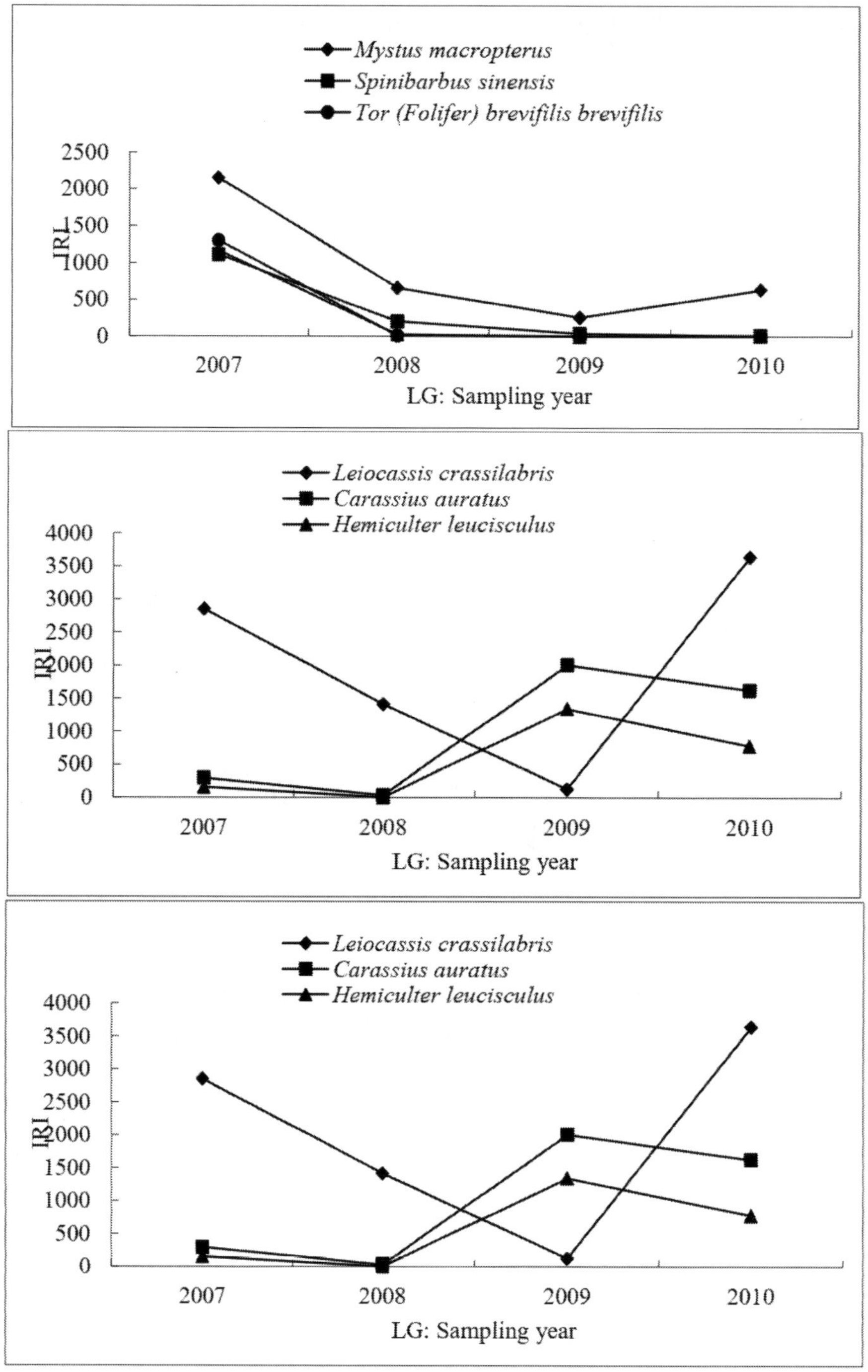

(b) LG reach

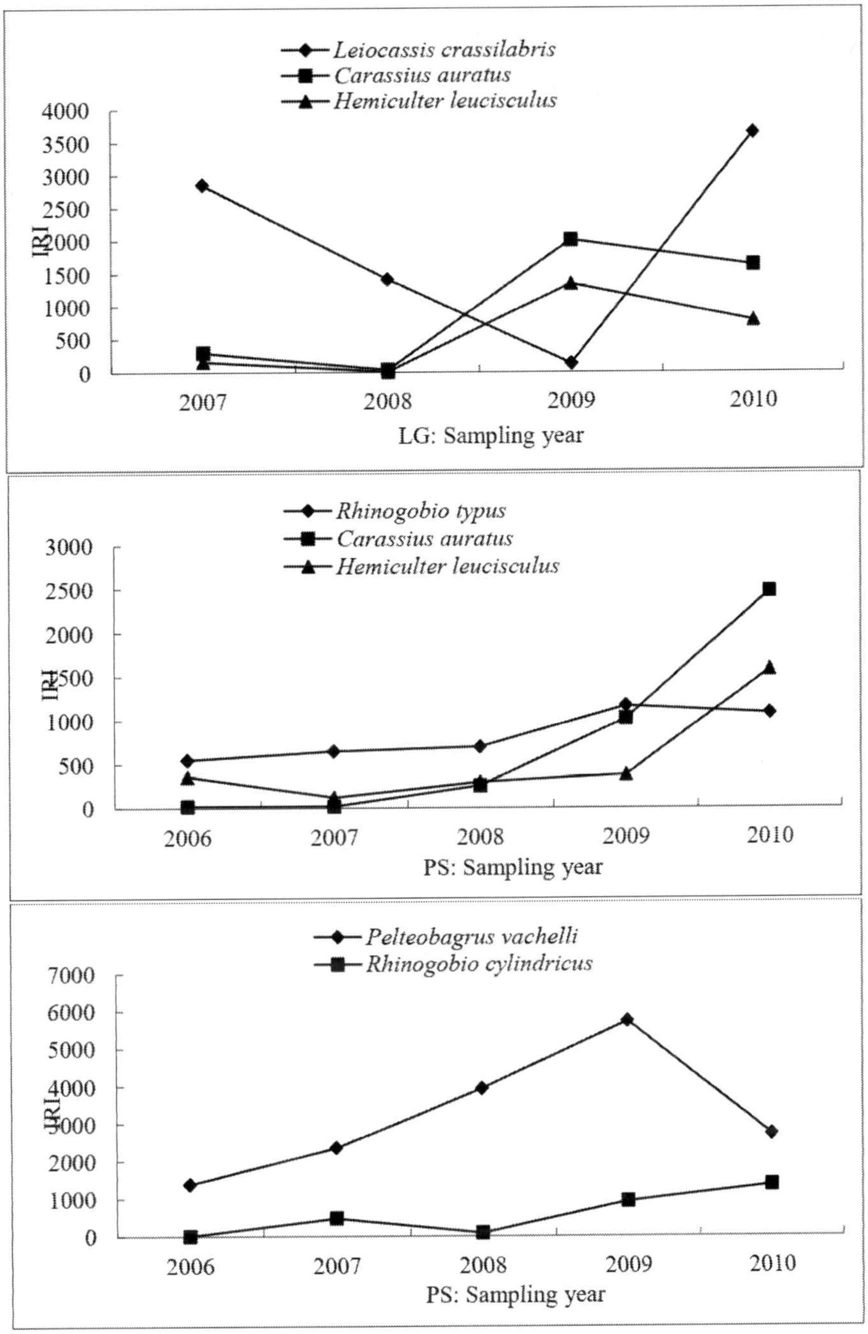

(c) PS reach

Figure 4. IRI dynamics of dominant species in the YH, LG, and PS reaches.

In the YH reach, four dominant lotic species' IRI declined obviously after the 2008 impoundment, but three lentic species' IRI increased rapidly, and they became dominant species. The IRI of the lotic species Semilabeo prochilus (Sauvage et Dabry) increased after impoundment, but decreased in 2010; lentic species Zacco platypus (Temminck et Schlegel) was dominant during 2007 and 2009.

In the LG reach, four dominant species were reduced to rare species after impoundment. The lotic species Leiocassis crassilabris Günther noticeably decreased, but it became a dominant species again in 2010, and Hemiculter leucisculus (Basilewsky), Semilabeo prochilus (Sauvage et Dabry), and Carassius auratus (Linnaeus) became dominant species after impoundment. The other species fluctuated, and most became dominant for only one year.

In the PS reach, Mystus macropterus declined after impoundment. Spinibarbus sinensis Bleeker and Leiocassis crassilabris decreased after 2007, but increased and became dominant again in 2010. Rhinogobio typus Bleeker, Carassius auratus (Linnaeus), and Hemiculter leucisculus (Basilewsky) became newly dominant species. The lentic species Pelteobagrus vachelli Richardson continuously remained a dominant species, and increased during the period of 2006-2009, but decreased in 2010.

3.3. Fish diversity

The following indices were calculated: Margalef Index of community diversity (R), Shannon Wiener diversity Index (*H′*), Pielou evenness index (E) and Simpson dominance index (D) (Figure 5).

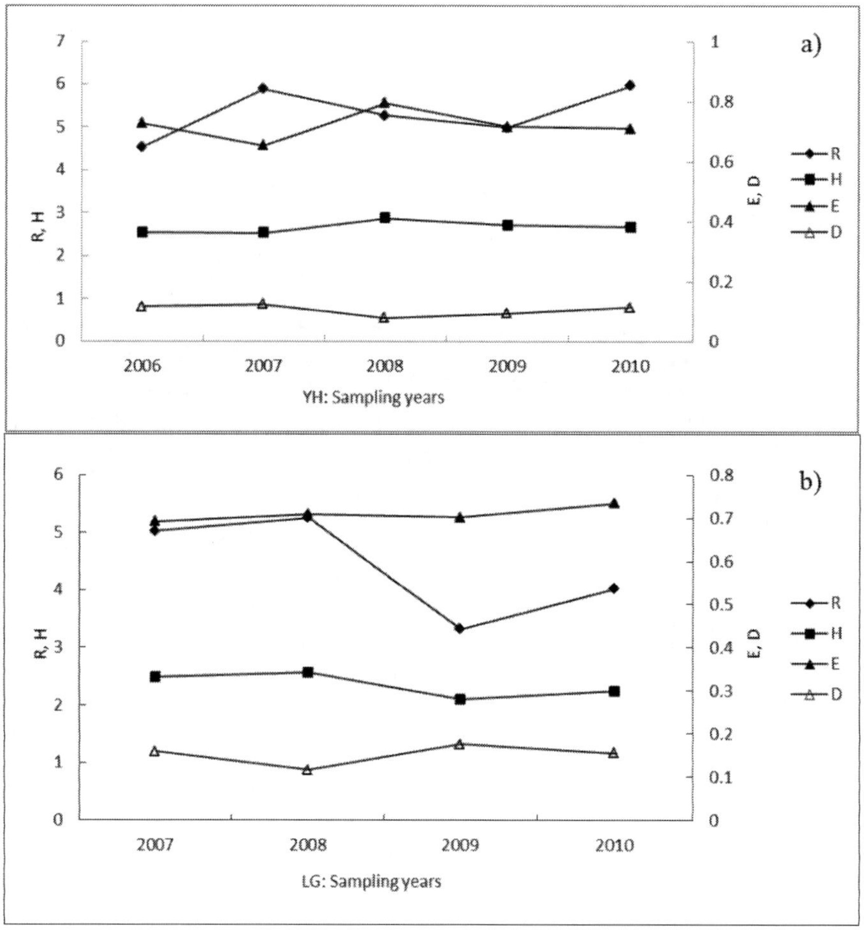

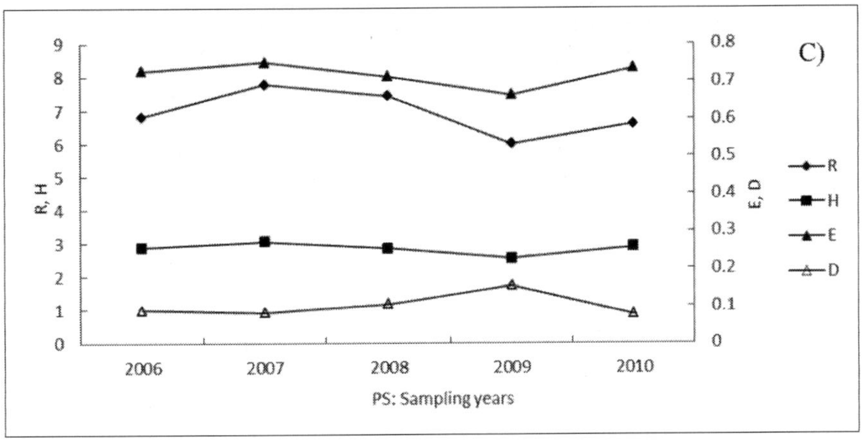

Figure 5. Biodiversity attributes of the YH, LG and PS reaches.

In the YH reach, the averages of the four indices were: Margalef Index (R)=5.3331, Shannon Wiener Index (H)=2.6636, Pielou evenness index (E) =0.7201, and Simpson Index (D)=0.1052. It appeared as though there were no obvious differences before and after impoundment. In the LG reach, the averages of the four indices were: R=4.4045, H=2.3465, D=0.1515, and E=0.7096. During the period of 2006-2010, the R and H indices declined, but the E and D indices increased. In the PS reach, the averages of the four indices were: R=6.9165, H=2.8416, D=0.8992, and E=0.7172. There was no obvious difference before and after the PS Reservoir impoundment.

3.4. Community structure similarity

The Principal Component Analysis (PCA) results of the different years of the study period were analyzed using the Bray Curtis similarity coefficient (Figure 6).

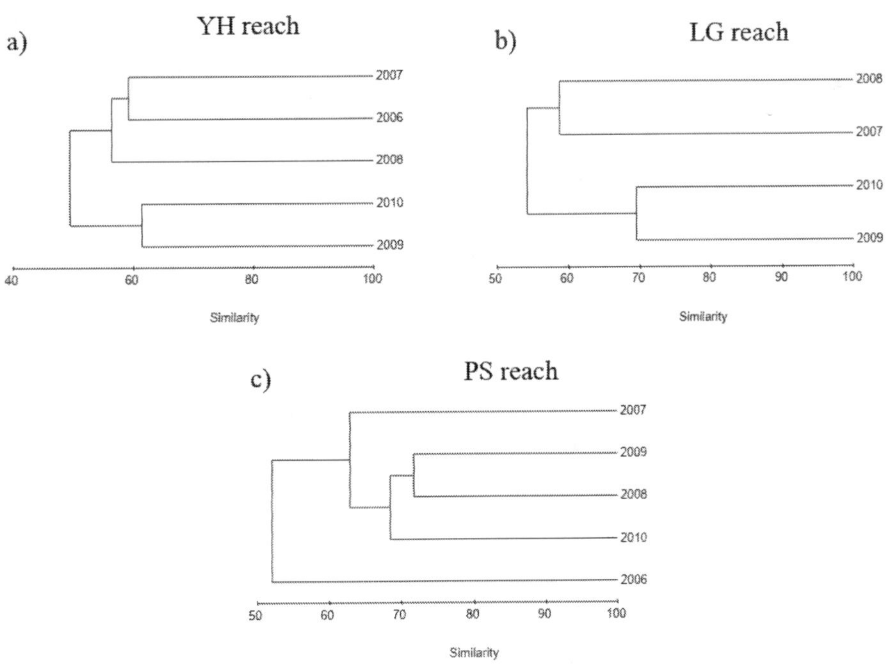

Figure 6. CA plots of fish assemblages in the YH, LG, and PS reaches.

In the YH reach, the five investigated years were divided into two groups with a coefficient value of 49.42%: Group 1 included the years of 2006, 2007, and 2008; Group 2 included the years of 2009 and 2010. In the LG reach, the four investigated years were divided into two groups using the coefficient value of 54.23%: Group 1 included the years of 2007 and 2008; Group 2 included the years of 2009 and 2010. In the PS reach, the five investigated years were divided into two groups using the coefficient value of 52.08%: Group 1 included the year 2006; Group 2 included the years of 2007, 2008, 2009 and 2010. The results demonstrated that in the reservoir and upstream fish assemblages the structure changed obviously after 2009, but in the downstream fish assemblages, the structure changed in 2007 and 2008 and then became stable after 2009.

SMPER was used to rank each group of fish assemblages according to their contributions (>4%) to the communities' structural differences. This indicated that nine species had a combined contribution of 69.88% in the YH reach, while twelve species had a combined contribution of 72.31% in the LG reach. Eight species had a combined contribution of 48.43% in Groups 1 and 2 and ten species had a combined contribution of 56.50% in Groups 2 and 3 in the PS reach. Both lotic and lentic species contributed to changes in fish assemblages in the upper stream, but only lotic species contributed significantly to changes in fish assemblages below the dam.

4. Discussion

4.1. Habitat and fish assemblages

Habitat is very important to river ecosystems. Habitat diversity and high heterogeneity, which are linked through the species habitat, affect the structure and function of fish communities [16]. Animal habitat selection is influenced by genetics, acquired experiences, and individual development stages. A fish species finds suitable living conditions only in a selected stretch of an entire river. Kondolf et al.[17] believe that habitat is one of the three most important factors in assessing ecological restoration. As early as the 19th century, ichthyologists used this observation as the basis of a zonation system for river courses[18].

In the Yangtze River Basin, the anadromous Chinese sturgeon (*Acipenser sinensis*) and several river migratory fishes (*Coreius guichenoti, Leptobotia elongate, Rhinogobio ventralis* and four major Chinese carps) return during the breeding season in search of suitable breeding habitats for spawning. The Gezhouba Dam and the Three Gorges Dam have blocked spawning migration, and now the Chinese sturgeon' spawning ground is located in only about a 3km section below the Gezhouba dam. These species, which produce pelagic eggs, lost their spawning grounds because of submersion of the reservoir, and then disappeared in the reservoir. Further, many studies have found that fish assemblage structure and elevation were significantly correlated with drainage area, flow, runoff, slope of cultivated land, grassland, and fish abundance[19][20] .

Prior to dam construction, the Wujiang River was a typical canyon-type river. Before the river was dammed, most fish species in the Wujiang River were lotic inhabitants. Those lotic species varied greatly between species for their swimming capacity and duration[21]. Fish species have formed unique ecological habits through a long-term adaptation and interaction process [22][23][24].

The characteristics of the fish assemblages are: 1) most dominant species are carnivores and omnivores feeding on invertebrates, and 2) most populations are composed of juniors and small fish (weight<50g). The reasons for this could be: 1) overfishing with small mesh nets, hooks, and settled gillnets causing catches to be concentrated with junior and small fish; 2) fish species composition may have changed as a result of changes in food organisms and physical habitats caused by the dams; 3) although their habitat needs are similar, those species with broader diets have a greater chance of survival in a degraded environment.

4.2. Fish assemblages' dynamics

Fish community structure varied both through space (longitudinal) and time (annual). In the PS reach, in 2006 and 2007 (the years before the PS reservoir impoundment), the fish community structure similarity index was 46.60%, but the similarity index between 2007 and 2008 was 61.57%. The number of dominant species decreased by 2 species from 2006 to 2007. The PS reach was under pressure from human activity starting in 2007, and the stress increased after impoundment in 2008.

In the LG reservoir area, fish assemblages' structure similarity indices differed significantly between 2007 and 2008 (18.79%), while the lotic population size among dominant species decreased, and some lentic species were introduced as new dominant species. This means that fish assemblages in the reservoir were influenced more seriously than the fish population in the lower reach, which responded quickly (even in as short a time frame as nine months) to changes in the flow regime caused by dam construction.

Therefore, fish assemblages changed as the lotic species decreased in population and the lentic species became dominant. Fish assemblages may need a longer time to attain a steady state, and a long-term ecological effect to identify this will require further monitoring of fish assemblages.

4.3. Impacts of dam construction and impoundment

Human activities such as hydro-engineering, overfishing, and water pollution have brought serious consequences for the ecological integrity of fish assemblages[25][26]. River damming can completely destroy aquatic ecosystem function[27][28][29]. Dams alter water flow, temperature, and sediment regimes of lotic systems[4][30]. Dams can enhance the clarity and reduce the variability of the river flow, which can also cause an increase in periphyton abundance. Invertebrates below the dam experience reductions in species richness due to the reduction of habitat heterogeneity[31]. Additionally, dams can fragment river systems and isolate fish populations[24]. Finally, river cascade development can amplify these impacts.

It has been proven that dams can cause decreases in habitat diversity and degeneration of fish populations[3][32][33][34]. The effects of the Gezhouba and Three Gorges Reservoirs constructed in the Yangtze River have also been studied. Hu et al.[4] preliminarily reported that aquatic ecosystem dynamics were influenced by the impoundment of Three Gorge Reservoir. However, Wu et al.[5] compared fish assemblages before and after the Gezhouba Dam construction and argue that fish assemblages have not changed radically because ecological conditions below the Gezhouba Dam did not change much. However, the authors suggest that the Gezhouba Dam is different due to its low-head and runoff Hydropower Station.

Usually, it is very difficult to separate the effects of hydro-engineering from other disturbances. Fish catch data showed that most fish populations were composed of junior individuals with body lengths between 100-200mm. This could be because small mesh-size nets are commonly used in this Yangtze River basin. People attempt to gain the greatest economic output and neglect natural fishery resource protection. Long-time and accumulative malpractice has caused damage to native fish populations [25][35]. Therefore, it is important that fishery standards and management strategies be established for the demand of future socioeconomic development and the need for natural resource conservation in the Yangtze River basin.

Acknowledgements

We are grateful to our colleagues Li Wan, Haitao Zheng, Juxiang Hu, Meihua Xiong and Fang Shi for their work and assistance in the field investigation. The authors also appreciate the comments and suggestions of Dr. Elias Dimitriou, who contributed to the clarity and readability of the paper. The use of fish in this study was approved by Hubei Province Association of Laboratory Animal Sciences.

References

[1] Francis J M and Keith H N 2005 *Changes in hydrologic regime by dams Geomorphology* **71** 61–78

[2] Tadeusz P, et al. 1998 A long term study (1985–1995) of fish populations in the impounded *Warta River Hydrobiol.* **368** 157–73

[3] Aníbal T A, et al. 2005 Multi-impact evaluation of new medium and large hydropower plants in Portugal centre region *Renew Sust Energy Rev.* **9** 149–67

[4] Hu Z and Cai Q 2006 Preliminary report on aquatic ecosystem dynamics of the Three Gorge Reservoir before and after impoundment *Acta Hydrobiol. Sinica* **30**(1) 1-6
[5] Wu G, et al. 1988 The fish fauna and species composition in Yichang Reach below the Gezhouba hydro-project *Freshw. Fishery* **3** 8–13
[6] Xiong M, et al. 2010 Study of the influences of Three Gorges project's initial operation on river regime of the middle and lower Yangtze River *J. Hydro. Eng.* **29**(1) 120–25
[7] Šmejkal M, et al. 2013 Associations of fish with various types of littoral habitats in reservoirs *Ecol. Freshw. Fish* **23** 405–23
[8] Prchalová M, et al. 2009 The effect of depth, distance from dam and habitat choice on the spatial distribution of fish in a canyon-shaped reservoir *Ecol. Freshw. Fish* **18** 247–60
[9] Perkin J and Bonner T H 2011 Long-term changes in flow regime and fish assemblage composition in the Guadalupe and San Marcos Rivers of Texas River *Res. Applic.* **27** 566–79
[10] Du J, et al. 2010 Changes of fish diversity and trophic levels in Quanzhou Bay *Biodiv. Sci.* **18** (4) 420–27
[11] Clarke K R 1993 Non–parametric multivariate analyses of changes in community structure *Aust. J. Ecol.* **18** 117–43
[12] Clarke K R and Gorley R N 2001 *PRIMER 5 User Manual/Tutorial* Plymouth UK Primer
[13] Pielou E C 1975 *Ecological Diversity* (New York: John Wiley) p 165
[14] Qian Y and Ma K 1994 *Theory and Method of Study on Biodiversity* (Beijing: Chinese Science and Technology Press) p 237
[15] Bray T R and Curtis J T 1957 An ordination of the upland forest communities of southern Wisconsin *Ecol. Monog.* **27** 325–49
[16] Franco A, et al. 2009 A habitat–specific fish–based approach to assess the ecological status of Mediterranean coastal lagoons *Mar. Pollut. Bull.* **581** 704–17
[17] Kondolf G M, et al. 2003 *Geomorphic Classification of Rivers and Streams* (Tools in Fluvial Geomorphology) ed Kondolf G M & Piégay H (UK Chichester: John Wiley & Sons) pp 171–204
[18] Holčík J 1989 *The Freshwater Fishes of Europe* AULA-Verlag Wiesbaden
[19] Allan J D 1995 *Stream Ecology: Structure and Function of Running Waters* (London: Chapman and Hall) p 388
[20] Bram G W A and Piet H N 2003 Fish zonations and guilds as the basis for assessment of ecological integrity of large rivers *Hydrobiol* **500** 157–78
[21] Angelier E 2003 *Ecology of Streams and Rivers Science* Publishers Inc. Enfield. p 215
[22] Chang J and Cao W 1999 Fishery significance of the river-communicating lake and strategies for the management of fish resources *Resour. Envir. Yangtze Basin* **8**(2) 153–57
[23] Margalef D R 1957 Information theory in ecology *Gen. Syst.* **3** 36–71
[24] Ming H, Yao S and Lin C 1999 Effect of Dams on Fish Assemblages of the Tachia River, Taiwan *Acta Zool.***10**(2) 77–90
[25] Chen D, et al. 2002 On the dynamics of fishery resources of the Yangtze River and its management *Acta Hydrobiol. Sinica.* **26**(2) 685–90
[26] Zhu D and Chang J 2008 Annual variations of biotic integrity in the upper Yangtze River using an adapted index of biotic integrity *Ecol. Indic.* **8** 564–72
[27] Baxter R M 1997 Environmental effects of dams and impoundments *Annu Rev. Ecology Syst.* **8** 255–83
[28] Xie P 2003 *Three-Gorges Dam: Risk to Ancient Fish Science* **302** 1149–50
[29] Young S P, et al. 2003 Conservation strategies for endemic fish species threatened by the Three Gorges Dam *Conserv. Boil.* **17**(6) 1748–58
[30] Giller S and Malmqvist B 1998 *The Biology of Streams and Rivers* (Oxford:Oxford University Press) p 296
[31] Cushing C E and Allan J D 2001 *Streams: Their Ecology and Life* (San Diego:Academic Press) p 366

[32] Conceição K S and de Oliveira V M 2010 Habitat fragmentation effects on biodiversity patterns *Physica. A* **389** 3496–502
[33] Guo W, et al. 2009 Effects of Three Gorges Cascade Reservoirs on River Hydrological Regimes of the middle and lower reaches of the Yangtze River *China Rural Water Hydropower* **12** 7–12
[34] Wootton R J 1990 *Ecology of Teleost Fishes* (London: Chapman and Hall) p 404
[35] Liu S, et al. 2005 Study on the fishery resources in the middle reach of the Yangtze River *Acta Hydrobiol. Sinica* **26**(9) 708–11

Aquifer water abundance evaluation using a fuzzy-comprehensive weighting method

Z Wei[1, 2]

[1]Faculty of Resources and Safety Engineering, China University of Mining & Technology, Beijing, No. Ding 11, Xueyuan Road, Haidian District, Beijing 100083, P.R.China

E-mail: 25328482@163.com

Abstract. Aquifer water abundance evaluation is a highly relevant issue that has been researched for many years. Despite prior research, problems with the conventional evaluation method remain. This paper establishes an aquifer water abundance evaluation method that combines fuzzy evaluation with a comprehensive weighting method to overcome both the subjectivity and lack of conformity in determining weight by pure data analysis alone. First, this paper introduces the principle of a fuzzy-comprehensive weighting method. Second, the example of well field no. 3 (of a coalfield) is used to illustrate the method's process. The evaluation results show that this method is can more suitably meet the real requirements of aquifer water abundance assessment, leading to more precise and accurate evaluations. Ultimately, this paper provides a new method for aquifer water abundance evaluation.

1. Introduction

Coal mining and tunnel construction inevitably expose, damage, and/or disturb roof aquifers. Although it will not induce water inrush when the process exposes the low water abundance zone, it is capable of causing severe water bursts in the high water abundance zone during mining [1]. Therefore, scientific and accurate evaluation of aquifer water abundance distribution features is essential for safe coal mine production and management [2-3]. There are many methods to evaluate questions that contain uncertain factors such as analytic hierarchy process (AHP), fuzzy mathematics, gray theory, entropic weight method, and a neural network method [4-9]. Many scholars have applied these methods in water abundance evaluation, but the assessment methods are still flawed [10-12]. This paper establishes an aquifer water abundance evaluation model that combines fuzzy evaluation with a comprehensive weighting method based on collecting the main controlling factors of aquifer water abundance. The study's results serve as a theoretical guide for coal mine water inrush prevention and treatment.

2. Fuzzy evaluation method

2.1. Principle of fuzzy evaluation method

The fuzzy evaluation method, based on fuzzy mathematics, is one of the most widely used decision methodologies. It analyzes the relevant factors and their detailed attributions and then grades or classifies them by the maximum membership degree principle. The fuzzy evaluation method can consider many factors and the relationship between them. It is, in fact, a fuzzy mathematics of multi-target decision-making method. The principle of fuzzy evaluation method is A•R=B (1), where A is the fuzzy vector of factors, representing the weight of each factor set X={x1,x2,•••,xn}; and R is the fuzzy matrix of factor sets and target set. The target set is defined as U={u1,u2,•••,un}, R=(r*ij*) n*m, where r*ij* is the membership degree of x*i* to u*i*, *i*=1,2, •••,n, *j*=1,2, •••,m. B is the judging standard that represents the membership degree of the factor sets to target sets when taking a comprehensive consideration of all the factors.'•' represents matrix multiplication.

2.2. Membership function defining

For a specific fuzzy set, membership functions can operate and process the factor sets quantitatively. Based on the specific conditions of the study area, this paper opted for the Ridge membership function [13].

$$\mu_{\text{IV}}(u)=\begin{cases}1 & u_1<u\\ 0.5+0.5\sin\dfrac{2\pi}{u_1-u_2}(u-\dfrac{3u_1+u_2}{4}) & \dfrac{u_1+u_2}{2}<u\le u_1\\ 0 & u\le\dfrac{u_1+u_2}{2}\end{cases} \tag{2}$$

$$\mu_{\text{III}}(u)=\begin{cases}0 & u_1<u\\ 0.5-0.5\sin\dfrac{2\pi}{u_1-u_2}(u-\dfrac{3u_1+u_2}{4}) & \dfrac{u_1+u_2}{2}<u\le u_1\\ 0.5+0.5\sin\dfrac{2\pi}{u_1-u_3}(u-\dfrac{u_1+2u_2+u_3}{4}) & \dfrac{u_2+u_3}{2}<u\le\dfrac{u_1+u_2}{2}\\ 0 & u\le\dfrac{u_2+u_3}{2}\end{cases} \tag{3}$$

$$\mu_{\text{II}}(u)=\begin{cases}0 & \dfrac{u_1+u_2}{2}<u\\ 0.5-0.5\sin\dfrac{2\pi}{u_1-u_3}(u-\dfrac{u_1+2u_2+u_3}{4}) & \dfrac{u_2+u_3}{2}<u\le\dfrac{u_1+u_2}{2}\\ 0.5+0.5\sin\dfrac{2\pi}{u_2-u_3}(u-\dfrac{u_2+3u_3}{4}) & u_3<u\le\dfrac{u_2+u_3}{2}\\ 0 & u\le u_3\end{cases} \tag{4}$$

$$\mu_{\text{I}}(u)=\begin{cases}1 & 0<u\le u_3\\ 0.5-0.5\sin\dfrac{2\pi}{u_2-u_3}(u-\dfrac{u_2+3u_3}{4}) & u_3<u\le\dfrac{u_2+u_3}{2}\\ 0 & \dfrac{u_2+u_3}{2}<u\end{cases} \tag{5}$$

Where μi(u) is the membership function of each factor; u is the actual value of each factor; u1, u2, u3 are the standard threshold value of each grade (respectively). The value is determined according to related specification and criterion.

2.3. Factor weight definition

There are two methods to get the weight of each factor: the subjective weighting method and the objective weighting method. The subjective weighting method is easily affected by user experience (or lack thereof). When there are too many factors, it is too difficult for users to give an accurate judgment. With the objective weighting method, the weight of each index is obtained mainly by searching for potential links between initial data sets and then analysing any variation trends among them. The objective weighting method can avoid the influence of subjective factors, but it still has a few shortcomings. For instance, the weight of each index may be inconsistent with the actual conditions – in other words, the primary index may not get the largest weight by simple data analysis. A new comprehensive weighting method is proposed in this study; a method that remains as objective as possible and therefore overcomes the subjectivity of previous weighting methods.

2.3.1. Subjective weight obtaining. Subjective weight is obtained by AHP. First, collecting the main controlling factors of the problem and classifying them into several hierarchies, where the primary factors belong to the lowest hierarchy, and the target object belongs to the highest hierarchy. An evaluation system and mathematical model are constructed, based on which a quantitative indicator or a scale indicator is to be established indicating the influence that the various factors have on their superior hierarchy. The user is then asked to grade for each factor; the weight of each factor can be obtained from the judgment matrix of each hierarchy. Finally, the highest hierarchy evaluation target is assessed and we can get the weight of each factor to meet the overall goal [14].

2.3.2. Objective weight obtaining. Taking the most important factor as the reference factor, denoted as X0=(x10,x20,•••,xn0)T, the comparison factor denoted as X*j*=(x1*j*,x2*j*,•••,xn*j*)T (*j*=1,2, •••,m), initializing X0 and Xj to obtain the factor matrix x*i*0'=x*i*0/x10,x*ij*'=x*ij*/x1*j* (6), X0'=(x10',x20', •••,xn0')T, X*j*'=(x1*j*',x2*j*',•••,xn*j*')T (*j*=1,2, •••,m)[15]. The relational coefficient of X0 and X*j* is represented by

$$x_{ij}=\frac{\min\limits_{1\le i\le n}\min\limits_{1\le j\le m}\left|x_{i0}'-x_{ij}'\right|+\rho\max\limits_{1\le i\le n}\max\limits_{1\le j\le m}\left|x_{i0}'-x_{ij}'\right|}{\left|x_{i0}'-x_{ij}'\right|+\rho\max\limits_{1\le i\le n}\max\limits_{1\le j\le m}\left|x_{i0}'-x_{ij}'\right|} \tag{7}$$

The relational degree of reference factor X0 and comparison factor Xj are represented [16] by

$$x_j=\frac{1}{n}\sum_{i=1}^{n}x_{ij}\text{ , } j=1,2,\ ...m \tag{8}$$

The larger the value of x*j*, the closer the relationship between the comparison factor and the reference factor, and the greater the weight of this factor compared to others [17].

The weight of each factor *aj* will be obtained by

$$a_j=x_j\Big/\sum_{j=1}^{m}x_j,\ j=1,2,\ ...m \tag{9}$$

2.3.3. Synthetic weight obtaining. Using multiple synthetic normalization methods to calculate the synthetic weight of each factor, the synthetic weight can be represented by:

$$\beta_j=\left(\alpha_j+\omega_j\right)/2 \tag{10}$$

Where αj is objective weight, ωj is objective weight, and βj is synthetic weight, (j=1, 2, •••, m).

2.3.4. Fuzzy-comprehensive weighting method model. The roof aquifer water abundance evaluation model combines the fuzzy comprehensive evaluation method with the synthetic weighting method [18]. This can be represented by:

$$B = [(\alpha_1 \bullet \omega_1)/2(\alpha_2 \bullet \omega_2)/2 \bullet\bullet\bullet, (\alpha_j \bullet \omega_j)/2] \bullet [r_{i1}, r_{i2} \cdots, r_{im}]^T \tag{11}$$

Where αj is objective weight, ωj is objective weight, and rij is membership degree, i=1,2,•••,n; j=1,2,•••,m. According to the maximum membership degree principle, Bj= max (B1, B2, •••, Bm) and the evaluation result is the grade corresponding to j.

3. Applications

To illustrate the process of the fuzzy-comprehensive weighting method, the no. 3 well field of a particular coal field was used as a case study. The study area is located in the Dongsheng District of Erdos City. It shows an irregularity rectangle, covers an area of 125.65 km2, and spreads 12.2 km from north to south and 10.3 km from east to west (on average). According to the hydrogeological prospecting data, the water-bearing medium can be divided into five groups: (I) the loose unconfined Quaternary aquifer, (II) the lower Cretaceous Zhidan group porous confined aquifer, (III) the middle Jurassic Zhiluo Formation fragmental rock porous-fissure confined aquifer, (IV) the mid-lower Jurassic Yanan Formation fragmental rock porous-fissure confined aquifer, and (V) the upper Triassic Yanchang Formation fragmental rock porous-fissure confined aquifer. Mudstone and sandy mudstone in Anding Formation function as a stable water-resistant layer. Folds and faults structures are poorly developed, and the stratum trend to the west in general with an inclination of 1~3°. The no. 3 well field mainly mines 2-2# coal seam. The direct water filled aquifer, which threatens the mine's safety, is the aquifer of Zhiluo formation above 2-2# coal seam. The aquifer is thick, rich in water abundance, and it covers the entire well field. The aquifer mainly consists of medium-grained sandstone; some sections are coarse-grained sandstone. The aquifer water abundance of the Zhiluo formation is to be evaluated using fuzzy-comprehensive weighting method.

3.1. Classification of roof aquifer water abundance

After analyzing the geological and hydrogeological conditions of the study area, the main controlling factors of aquifer water abundance were obtained. These include specific capacity, hydraulic conductivity, aquifer thickness, flushing fluid consumption, the thickness ratio of friable rocks to plastic rocks, and the rock quality designation (RQD). Then the roof aquifer water abundance was divided into four grades: (I) low water abundance, (II) medium water abundance, (III) high water abundance, and (IV) higher water abundance. The National Industry Standard, Industry Specifications, and previous studies were also considered to establish the corresponding relationship between different factors and water abundance grade (shown in table 1).

Table 1. Main controlling factors of water abundance and its classification.

Factors		Water Abundance Grade			
		I	II	III	IV
Aquifer	Specific capacity	0.001	0.01	0.1	1
Hydrogeological	Hydraulic conductivity	0.01	0.1	1	5
Condition	Aquifer thickness	1	50	100	150
Aquifer	Flushing fluid consumption	0.1	0.3	0.7	1
Geological	Thickness ratio of friable rocks to plastic rocks	0	1	10	20
Condition	RQD	0.6	0.7	0.8	0.9

3.2. Membership degree of each factor

The author selected 15 typical borehole datum in the study area (seen in table 2), each of them including six main controlling factors of aquifer water abundance. The membership degree of each factor was calculated using the fuzzy evaluation method of equation (2) to equation (5). For example, the membership degree of specific capacity in the area located by the borehole XJ-10 is (0, 0.013521953, 0.986478047, 0). Similarly, the membership degree of other factors can be obtained. Then the fuzzy matrix R can be established.

Table 2. Main controlling factors and parameter values of the Zhiluo group aquifer.

Drill name	specific capacity	Hydraulic conductivity	Aquifer thickness	flushing fluid consumption	thickness ratio of friable rocks to plastic rocks	rock quality designation
XJ-10	0.05045	0.07451	142.66	0.15	9.634	0.7
XJ-11	0.07145	0.10445	113.37	0.12	10.476	0.829
XJ-22	0.08771	0.13375	175.71	0.4	7.02	0.518
XJ-23	0.0495	0.06033	119.67	0.38	2.268	0.739
XJ-24	0.05995	0.09238	167.75	0.14	0.841	0.711
XJ-25	0.06158	0.15521	99.65	0.66	6.794	0.612
XJ-26	0.08158	0.12384	98.13	0.26	20.861	0.743
XJ-27	0.07416	0.11203	54.19	0.67	2.7	0.688
XJ-28	0.06516	0.07497	167.75	0.35	6.826	0.778
XJ-29	0.06011	0.08313	195.08	0.23	2.829	0.829
XJ-30	0.05622	0.06697	90.56	0.47	1.619	0.836
XJ-31	0.0836	0.09183	135.52	0.31	5.172	0.827
XJ-32	0.06933	0.13187	79.21	0.26	1.904	0.865
XJ-40	0.0273	0.03054	86.6	0.11	2.259	0.808
XJG-8	0.12402	0.25916	201.89	0.66	4.078	0.603

3.3. Weight of each factor

The subjective weight of each factor can be obtained via the AHP method. The objective weight can be obtained by equation (6) to equation (9). The synthetic weight uses the arithmetic mean of subjective weight and objective weight according to equation (10). Table 3 shows the results.

Table 3. Weights of aquifer water abundance controlling factors.

	Factors	Subjective Weight	Objective Weight	Synthetic Weight
Aquifer water abundance	Specific capacity	0.3919	0.2057	0.2988
	Hydraulic conductivity	0.1801	0.1815	0.1808
	Sandstone thickness	0.1949	0.1629	0.1789
	Flushing fluid consumption	0.0738	0.1392	0.1065
	Thickness ratio of friable rocks to plastic rocks	0.0872	0.1704	0.1288
	Rock quality designation	0.0721	0.1401	0.1061

3.4. Water abundance evaluation result

Adopting a fuzzy-comprehensive weighting method model and equation (11), the standard that represents the membership degree of the factor sets to target sets is obtained. Then, according to the maximum membership degree principle, fourteen of the fifteen boreholes' water abundance belong to high water abundance grade III, while XJG-8 borehole belongs to the middle water abundance grade II.

4. Discussion

This paper proposes a new revision of the fuzzy-comprehensive weighting method to evaluate aquifer water abundance. This method is an improvement over previous methods in at least two ways. First, it employs a fuzzy method that can evaluate the aquifer water abundance in a quantity that can also include some qualitative factors. Second, the weight of each main controlling factor considered both the subjective and the objective weights. It is a more scientific and accurate method than prior methods for aquifer water abundance evaluation. However, there are some deficiencies in the evaluation method; to circumvent these, the evaluation standard must be created in conditions where there is no standard threshold value of each grade. This point will be further examined in future research.

5. Conclusion

This paper establishes an improved aquifer water abundance evaluation model by combining fuzzy evaluation with a comprehensive weighting method. The method's main objective is to collect the main controlling factors of aquifer water abundance. It provides a more effective model of aquifer water abundance evaluation for other areas.

This paper also proposes the application of a synthetic weighting method to factor weight calculation in aquifer water abundance evaluation. This synthetic weighting method overcomes not only the subjectivity but also the lack of conformity of subjective weighting methods, and therefore becomes more suitable to meet the actual requirements of aquifer water abundance evaluation.

References

[1] Wu Q and Wang M 2006 *Hydrogeol J.* **14** pp 882–93
[2] Wu Q, Wang M and Wu X 2004 *International Journal of Rock Mechanics and Mining Sciences* **41** pp 557–71
[3] Wu Q 2014 *Journal of China Coal Society* **39** pp 795-805
[4] Zhang W Q and Yu H L 2006 *Mining Safety & Environmental Protection* **33** pp 50-52
[5] Zhao X E, Qi Y M and Luo B 2012 *China Coal* **38** pp 35-9
[6] Ip W C, Hu B Q, Wong H and Xia J 2009 *Journal of Hydrology* **379** pp 284–90
[7] Zheng D Z and Yang B B 2012 *Safety in Coal Mines* **43** pp 132-35
[8] Gao W D 2012 *Mining Safety & Environmental Protection* **39** pp 22-24
[9] Song B 2010 *Journal of North China Institute of Science and Technology* **7** pp 1-3
[10] Luo B 2014 Comprehensive Analysis and evaluation of water hazards conditions on coal seams roof based on GIS techniques and multisource information—a case study in 39110 working face of 3rd underground coal mine, Shuozhou city, Shanxi province *China University of Mining & Technology*
[11] Wu Q, Fan Z L, Liu S Q, Zhang Y W and Sun W J 2011 *Journal of China Coal Society* **36** pp 1124-28
[12] Xu L L 2010 Research method and its application on water damage of roof at Pingshuo No.1 Shaft *China University of mining & Technology Beijing*
[13] Li B, Guo J, Li H K,Ding J P,Qiu H T and Wang X Q 2012 *China Coal* **8** pp 51-54
[14] Wu Q, Fan S K, Zhou W F and Liu S Q 2013 *Mine Water Environ.* **32** pp 229-38
[15] Deng J L 1989 *Journal of Grey System* **1** pp1-24
[16] Huang Y P and Huang C C 1996 *Fuzzy Sets and Systems* **78** pp 107-19
[17] Li X H 2007 *Journal of Shandong University of Science and Technology (Natural Science)* **42** pp 33-41
[18] Li B, Guo X M, Xu S, Liu W S, Wang Z H and Li H K 2014 *Journal of Henan Polytechnic University (Natural Science)* **33** pp 6-11

Degradation of phenol wastewater by a new electromagnetic induction photo-catalytic reactor

X C Yuan[1, 2], Q H Meng[1], J Y Sun[1], Y Yan[1], L Li[1], G C Li [1] and D Li [1]

[1]School of Chemistry and Chemical Engineering, Jiangsu Key Laboratory of Green Synthetic Chemistry for Functional Materials, Jiangsu Normal University, Xuzhou 221116, P.R. China

E-mail: yxingcheng@163.com

Abstract. A new inductive photo-catalytic reactor was obtained by the alternative magnetic field and optical coupling, which was driven by AC supply. In the cylinder reactor, UV-LED lights with the wavelength of 375-380nm were evenly distributed, and the phenol solution was used as simulated wastewater. The effects of initial phenol concentration, pH, TiO_2, H_2O_2, alternative magnetic frequency, current, and reaction time on the phenol degradation were investigated under an imposed alternative magnetic field. The optimized conditions and results were as follows: phenol concentrations of 15mg/L, pH of 7, H_2O_2 of 15μL, TiO_2 of 0.18g and alternative magnetic frequency of 12 KHz and current of 2A. With these conditions, the phenol degradation ratio reached 47.1% in 1 h reaction time. The new reactor is very promising for the effective treatment of refractory organic pollutants.

1. Introduction

Contamination of surface water due to refractory organic pollutants is posing a direct threat to human health and environment, which has become an important research topic about how to deal with these pollutants in the global. Many studies have shown that semiconductor photo-catalytic technology has many advantages to remove refractory organic pollutants, such as high removal efficiency, no secondary pollution, quick decomposition rate, and easy operation [1-3]. This technology is a kind of extremely promising purification technology and has received widespread attentions.

The core technology of semiconductor photo-catalytic oxidation is the light source, and developing light sources with long lifetime and uniform illumination has attracted much interest [4-6]. There are two types of light sources for photo-catalytic reactors: solar and artificial light; the latter source includes a high-pressure mercury lamp, fiber, and LED. Traditional photo-catalytic reactors usually use high-pressure mercury lamps as the light source, whose illumination intensity is high, but decays very fast, resulting in short lifetime. Moreover, the light from mercury lamps irradiates only from a side or surface, resulting in low light efficiency. Ollis and Marinangel proposed that fiber should be used as the transmission medium and photo-catalyst substrate [7]. The photo-catalytic process can be more environment-friendly, safer due to the strong and flexible fiber, lower energy radiation, and long distance lightning [8]. However, fiber reactors are expensive and cannot provide uniform illumination, i.e., the closer the light source, the stronger the illumination by the fiber. Recently, LED lights have become one of the most potential light sources. Compared to traditional mercury lamps, LED lights can convert electric power into optical energy. Moreover, when DC is used as the power source, LED lights have simple modulation, long lifetime, high photon efficiency, no toxicity, high safety, and low volume; they have been widely used in the degradation of pollutants by photo-catalytic.

In this study, UV-LED lights with wavelength of 375-380nm were used to manufacture inductive photo-catalytic reactors. Based on electromagnetic induction, the micro-inductive LED light sources were evenly distributed in a cylinder reactor wrapped up with coils. Thus, an inductive photo-catalytic reactor was obtained by the alternative magnetic field and optical coupling, which was driven by AC supply. Phenol solution was used as the simulated wastewater. The effects of initial phenol concentration, pH, TiO_2, H_2O_2, alternative magnetic frequency, current, and reaction time on the phenol degradation were investigated under an imposed alternative magnetic field.

2. Materials and methods

2.1. Chemicals and reagents

AEROXIDE TiO_2 P_{25} was purchased from Beijing Entrepreneur Science and Trading Co., Ltd. CH_3OOH were purchased from Sigma-Aldrich and HPLC grade. HCl, NaOH, H_2O_2, C_6H_5OH were analytical-grade and obtained from Shanghai Chemicals Factory, China. The laboratory water was obtained from Wahaha drinking pure water.

2.2. Experimental set-up

The new electromagnetic induction photo-catalytic reactor was shown in Figure 1.

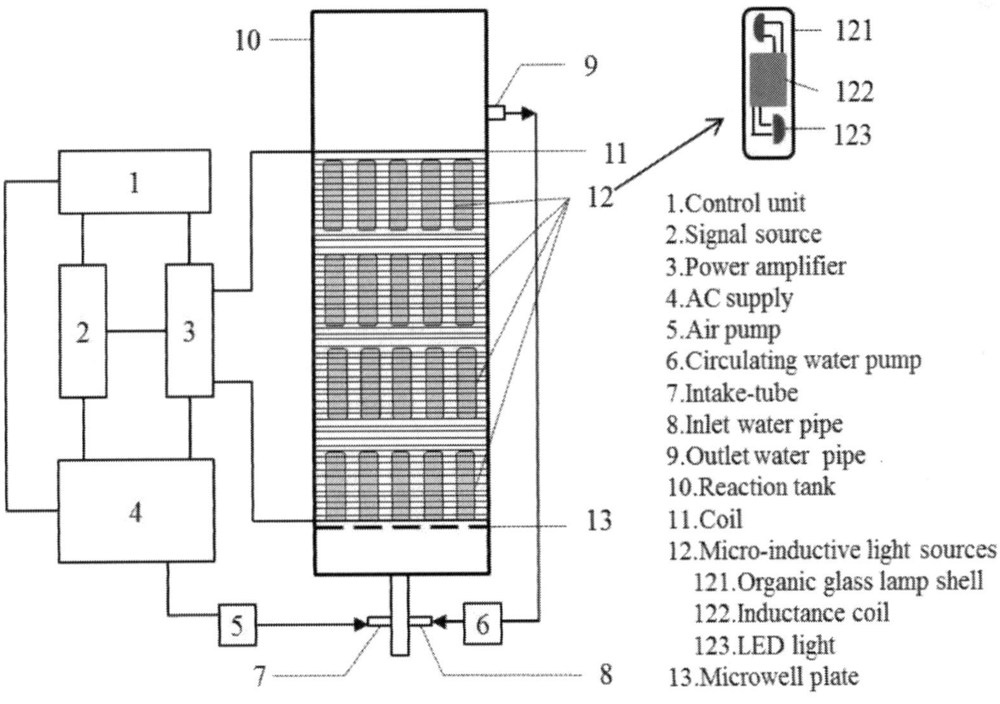

Figure 1. Schematic diagram of experimental set-up.

This photo-catalytic system was composed mainly of two parts: high-frequency AC (alternating current) supply and photo-catalytic reactor. The core was the inductive photo-catalytic reactor. The signal source of the high-frequency AC supply provided sinusoidal alternative signal; after amplification using the power amplifier, the signal worked on the coils that wrapped on the external wall of the reactor, thus generating an alternative magnetic field. Now, the inductive coils inducted energy from the alternative magnetic field on evenly filling micro-light sources in the reactor, which generated current to power the UV-LED lights and emitted UV light in a certain wavelength. The organic wastewater could be catalytically degraded by the alternative magnetic and optical fields. The reaction liquid was removed through the top outlet using a circulation pump and added through the bottom inlet. The photo-catalytic reaction was repeated continuously. An air pump was used to introduce air from the bottom of the reactor to provide sufficient air and prevent the concentration of catalysts. Because of the even distribution of inductive light sources in the reactor, the efficiency of light sources was enhanced.

The structure of micro-inductive light sources was shown in Figure 1 (label 12). This was constituted by two UV-LED pins with parallel anti-phase connection, then connected to two output pins of inductive coils, and sealed by transparent synthetic glass. Cost-effective LED lights with the wavelength of 375-380 nm were selected for the device as the light sources. The micro-inductive light sources were densely placed in the reactors at multiple layers. When electromagnetic induction was applied in photo-catalytic reactors, it was not necessary to connect to power supply. The light sources could be evenly distributed in the reactors. Thus, the disadvantage of the breakdown of one LED light leading to the malfunction of all of them could be avoided.

2.3. Experimental procedures

The phenol solution was used as the simulated wastewater. The effects of the initial phenol concentration, initial pH, TiO_2, H_2O_2, alternative magnetic frequency, current, and reaction time on the phenol degradation were investigated under imposed alternative magnetic field. The experiment was designed as follows:

TiO_2 and H_2O_2 were added into 250 mL phenol solution. The temperature was controlled by circulating water at 30 ℃. The air flow rate was controlled using a compressor. The light source intensity was controlled using a regulated power supply. A cycling pump was used to circulate the reaction liquid. The current intensity and frequency of alternative magnetic field were adjusted. A stirrer was used to mix the reaction liquid. After 20 minutes, a sample was collected from the reactor and centrifuged using the high-speed centrifuge at 18,000 rpm for 5 min. The phenol was analyzed in the supernatant to calculate the degradation ratio.

2.4. Analysis

Analysis of phenol was performed using a HPLC system (Agilent, USA, 1100 Series high-performance liquid chromatography). The column temperature was 30℃ and a sample volume of 20 μL was injected from the sampler. Chromatographic separation was achieved on a Kromasil 100-5C_{18} (250 ×4.6mm, E56403, Alltech, USA), the mobile phase consisted of 55% water and 45% methanol, the flow rate was 1.000 mL/min. The determination wavelength was set at 270 nm.

The phenol degradation efficiency was calculated using the following equation:

$$\eta = \frac{A_0 - A}{A_0} \times 100\%$$

η —The degradation efficiency of phenol (%);
A0 — The peak area of phenol before degradation;
A —The peak area of phenol after degradation.

3. Results and discussion

3.1. Effect of initial concentration and pH on the phenol degradation

The phenol solutions with initial concentrations of 10 mg/L, 15 mg/L, 20 mg/L, 25 mg/L respectively and the initial pH of 7 were used about 250ml. And other operating parameters were TiO2 of 0.20 g, H2O2 of 50μL, air flow rate of 60 L/h, current of 1 A, alternative magnetic field of 8 KHz, and the reaction time of 1 h respectively. The results were shown in Figure 2. The initial concentration was shown an effect on the phenol degradation with a certain extent. When the phenol concentration was 15 mg/L, the degradation percentage reached as high as 25.4%. Although the concentration of organic pollutants in industrial wastewater is usually unknown, a proper pollutant concentration can enhance the degradation.

The phenol solutions with initial pH of 4,5,6,7,8,9,10,11 respectively and initial concentration of 15mg/L were used about 250ml. And other operating parameters were TiO2 of 0.20 g, H2O2 of 50μL, air flow rate of 60 L/h, current of 1 A, alternative magnetic field of 8 KHz, and the reaction time of 1 h respectively. The results were shown in Figure 3. The initial pH of the solution was shown significantly effect on the phenol degradation. In an acidic solution, the photo-catalytic degradation became more efficient at higher pH. The highest photo-catalytic degradation was achieved around neutral pH. However, the higher pH would cause lower photo-catalytic degradation in alkaline solutions. As the pH>9, OH— and ArO— would be adsorption on the TiO2 surface adsorption, which resulted in competing with each other and reduced the amount of •OH. Therefore, when the pH value is relative taller, phenol degradation rate can be reduced.

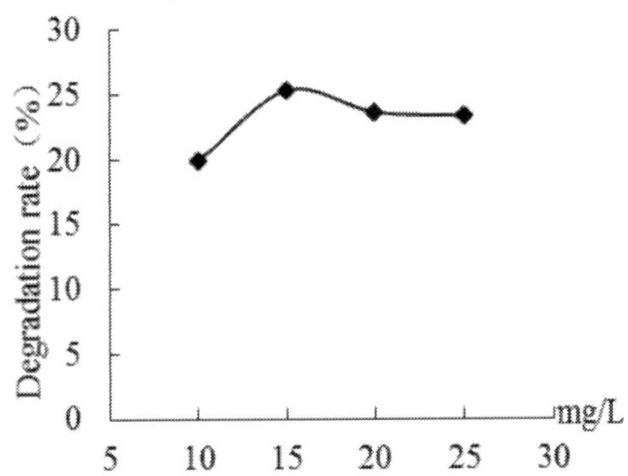

Figure 2. Effect of initial concentration on the phenol degradation.

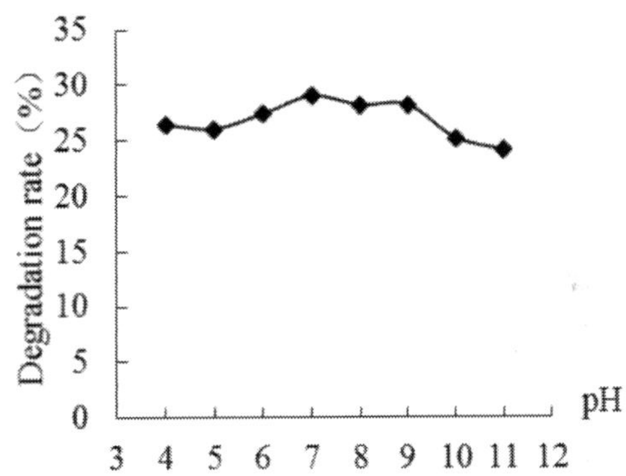

Figure 3. Effect of pH on the phenol degradation.

3.2. Effect of TiO_2 and H_2O_2 additives on the phenol degradation

The catalyst TiO_2 was added to the phenol solutions with 0.14g, 0.16g, 0.18g, 0.20g, 0.22g respectively. And other operating parameters were initial concentrations of 15mg/L, pH of 7, H_2O_2 of 50μL, air flow rate of 60 L/h, current of 1 A, alternative magnetic field of 8 KHz, and the reaction time of 1 h respectively. As shown in Figure 4, the photo-catalytic degradation ratio first increased with the amount of catalyst and then decreased after reaching its peak value. Under the UV irradiation, the higher quality of TiO_2 concentration in the solution produced more electron/hole and more •OH, which would speed up the reaction [9-10]. However, excessive TiO_2 concentration reduced the light energy utilization and made the degradation effect weaker due to the incident ultraviolet shielding and scattering effect. The optimum amount was determined to be 0.18 g in this study.

Similarly, the catalyst H2O2 was added to the phenol solutions with 5μL, 10μL, 15μL, 25μL, 50μL, 100μL, 150μL respectively. And other operating parameters were initial concentration of 15mg/L, pH of 7, TiO2 of 0.18g, air flow rate of 60 L/h, current of 1 A, alternative magnetic field of 8 KHz, and the reaction time of 1 h respectively. The amount of H2O2 was shown to affect the photo-catalytic degradation of phenol by TiO2 to a certain extent. The photo-catalytic degradation ratio first increased with the amount of H2O2 and then decreased after reaching the peak value at a dosage of 15μL, as shown in Figure 5. With the increase of H2O2 dosage, the amount of •OH increased in the solution, which promoted the organic matter degradation under the condition of light. Excessive H2O2, however, not only consumed •OH, but the decomposition of H2O2 was also invalid [11].

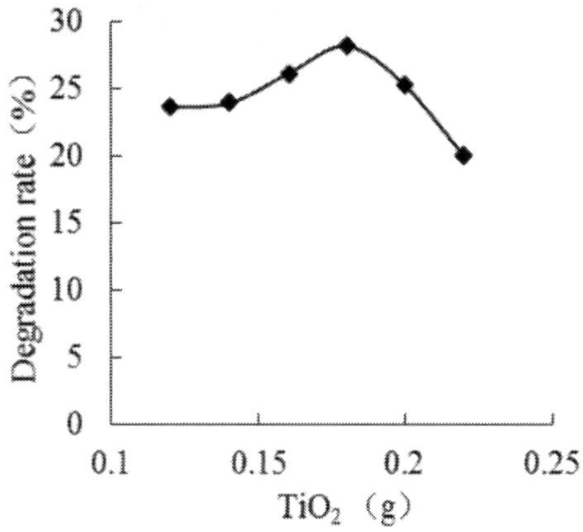

Figure 4. Effect of TiO_2 on the phenol degradation.

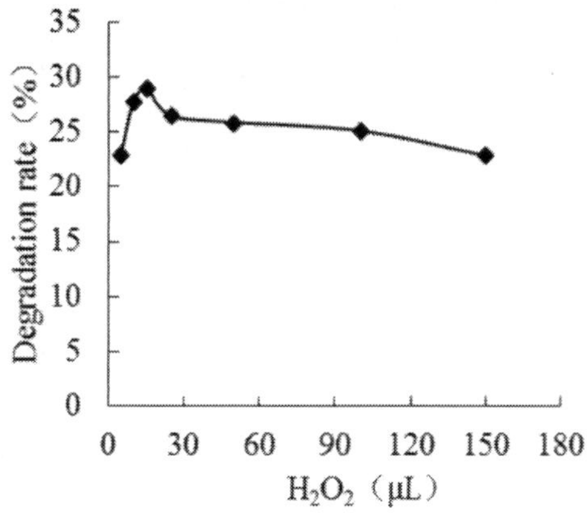

Figure 5. Effect of H_2O_2 on the phenol degradation.

3.3. Effect of magnetic field and coil current on the phenol degradation

The frequency of alternative magnetic field was changed from 1 to 20 KHz. And other operating parameters were initial concentration of 15mg/L, pH of 7, H_2O_2 of 15μL, TiO_2 of 0.18g, air flow rate of 60 L/h, current of 1A and the reaction time of 1 h respectively. As shown in Figure 6, the applied alternative magnetic field was shown to affect the photo-catalytic degradation of phenol. At low frequency, the effect of applied alternative magnetic field was not obvious. At the higher frequency, the magnetic field intensity became higher, resulting in the stronger UV light, a stronger coupling between the magnetic field and UV light, and the higher photo-catalytic degradation ratio of phenol.

In the same way, the high-frequency current of coil was changed from 0.4A to 2A by increasing 0.2A. And other operating parameters were initial concentration of 15mg/L, pH of 7, H2O2 of 15μL, TiO2 of 0.18g, air flow rate of 60 L/h, alternative magnetic field of 12 KHz and the reaction time of 1 h respectively. As shown in Figure 7, the coil current was shown to affect the photo-catalytic degradation of phenol. With the stronger coil current, the intensity of the alternative magnetic field became stronger, and the phenol degradation became more efficient. When the frequency of magnetic field was 12 KHz and the coil current was 2.0 A, the phenol degradation ratio reached as high as 47.1%.

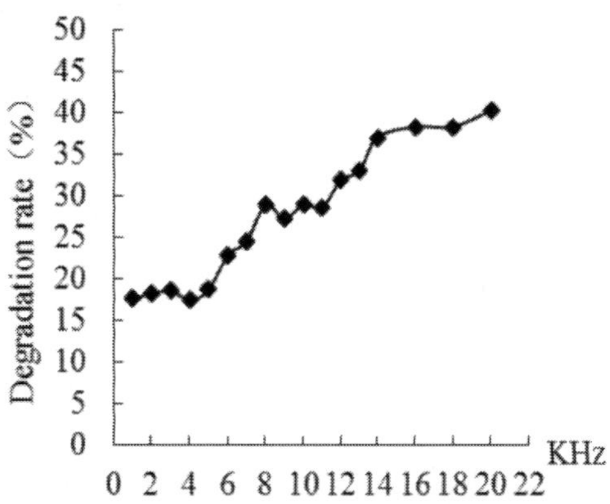

Figure 6. Effect of magnetic frequency on the phenol degradation.

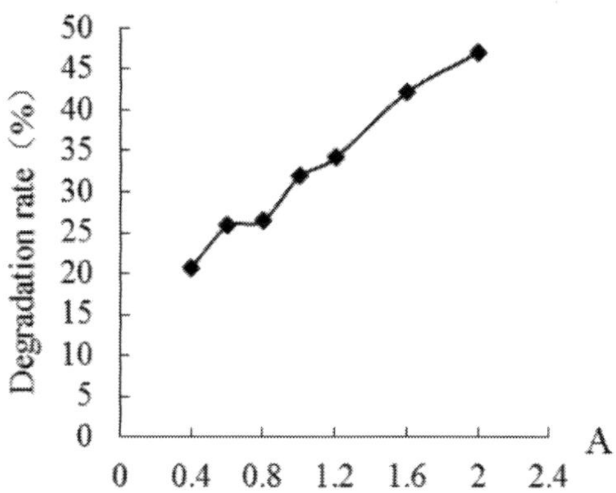

Figure 7. Effect of coil current on the phenol degradation.

3.4. Effect of reaction time on the phenol degradation

According to the above optimization of conditions, the operating parameters were initial concentration of 15mg/L, pH of 7, H_2O_2 of 15μL, TiO_2 of 0.18g, air flow rate of 60 L/h, alternative magnetic field of 12 KHz respectively. And then the reaction time was changed with 0.5h, 1h, 2h, 3h, 5h, 7h, 13h and 23h. As shown in Figure 8, the reaction time was shown to affect the photo-catalytic degradation of phenol. With the longer reaction time, the degradation of phenol became much faster. Then, the photo-catalytic degradation decreased after a certain point. The phenol degradation reached as high as 97.9% after 23 h.

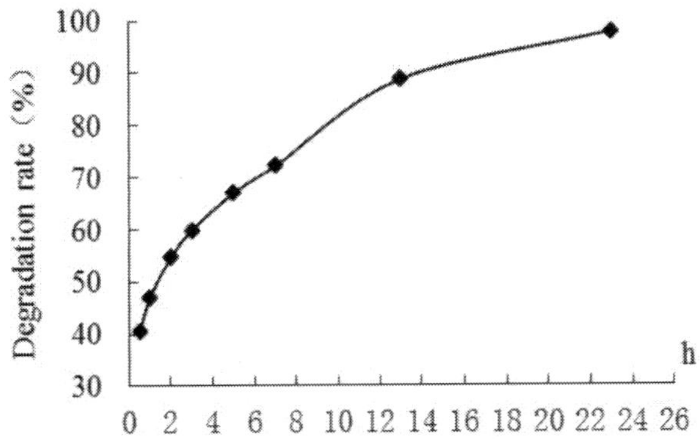

Figure 8. Effect of reaction time on the phenol degradation.

4. Conclusion

The new electromagnetic induction photo-catalytic reactor worked very well for the phenol degradation using the simulated wastewater. The optimized conditions and results were as follows: phenol concentration of 15mg/L, pH of 7, H_2O_2 of 15μL, TiO_2 of 0.18g, air flow rate of 60 L/h and alternative magnetic field of 12 KHz and current of 2A. With these conditions, the phenol degradation ratio reached 47.1% in 1 h reaction time.

The new reactor generated electricity through electromagnetic induction without any external power. The photo-catalytic efficiency could be enhanced by coupling an alternative magnetic field and UV light and evenly distributing micro-light sources in the reactor. Thus, the lifetime of micro UV-LED light sources was extended, and the wavelength could be adjusted. So, the new reactor is very promising for the effective treatment of organic pollutants.

References

[1] Gao T and Qian C 2000 *Industrial Water Treatment* **20** 10
[2] Salmin D and Shamali S A 2013 *Australian Journal of Basic and Applied Sciences* **7** 172
[3] Harraz F A, Mohamed R M, Rashad M M, Wang Y C and Sigmund W 2014 *Ceramics International* **40** 375
[4] Li K, Xu Y L and Jia J P 2011 *Environmental Pollution and Control* **33** 78
[5] Wang Z P, Cai W M, Hong X T, Zhao X L, Xu F and Cai C G 2005 *Applied Catalysis B: Environmental* **57** 223
[6] Shahrezaei F, Mansouri Y, Zinatizadeh L A and Akhbari A 2012 *Powder Technology* **221** 203
[7] Natarajan T S, Thomas M, Natarajan K, Bajaj H C and Rajesh J T 2011 *Chemical Engineering Journal* **169** 126
[8] Marinangeli R E and Ollis D F 1977 *AIChE J.* **23** 415
[9] Qamar M, Muneer M and Bahnemann D 2006 *Journal of Environmental Management* **80** 99
[10] Chen S F and Liu Y Z 2007 *Chemosphere* **67** 1010
[11] Zhang F, Li M amd Li W 2011 *Chemical Engineering Journal* **175** 349

Acknowledgements
The authors would like to specially thank the support from the project on the natural science foundation of the colleges and universities in Jiangsu province (Foundation item No. 15KJD610003) and the National Natural Science Foundation of China (Foundation item No. 21173180).

Enhanced formation of aerobic granular sludge with yellow earth as nucleating agent in a sequencing batch reactor

Q L He[1], S L Zhang[1], Z C Zou[1] and H Y Wang[1, 2]

[1]School of Civil Engineering, Wuhan University, Wuhan 430072, China

Email: hywang96@126.com

Abstract: Enhanced formation of aerobic granulation was investigated by adding yellow earth as a nucleating agent in a sequencing batch reactor with a constant setting time of 10 min. As a result, granules with an average diameter over 1 mm were obtained on the 4th day. The mature granules behaved better than the seed sludge in the water content, specific gravity, sludge volume index, settling velocity, and specific oxygen uptake rate. The yellow earth stimulated the secretion of extracellular polymeric substances, especially proteins. Both chemical oxygen demand and ammonia nitrogen had a removal rate over 90%, and more than 80% of the total inorganic nitrogen was removed even under aeration conditions due to simultaneous denitrification. The enhancement effects of the yellow earth might be based on the unique physicochemical characteristics and short settling time. A settling time of 10 min or more turned out not to be a prerequisite for a rapid granulation process.

1. Introduction

Aerobic granular sludge technology has been proposed as a novel and efficient approach in both domestic and industrial wastewater treatment. With the excellent settling velocity, separate settling tanks are not required, and an 80% area reduction is possible [1]. Aerobic granules are characterized by their compact structure, dense microbial structure, high biomass, and tolerance to high organic loading rate and toxicity. Additionally, the layered structure of aerobic granular sludge provides oxic, anoxic, and anaerobic zones from the outer layer to the deep core, thus creating shelter for co-existence of nitrifying microbes, denitrifying phosphorus removal organisms, as well as the anaerobic organisms. Therefore, simultaneous nitrogen, phosphorus, and chemical oxygen demand (COD) removal can take place from the liquid during an aeration period [2].

However, the operational conditions of aerobic granulation are strictly limited by multiple parameters, including reactor configuration, media and seed sludge, settling time, organic loading rate, volume exchange ratio, hydrodynamic shear force, feast-famine regime, cycle time, and other environmental conditions (i.e. pH and temperature) [3]. Significant efforts have been made to enhance the formation of aerobic granular sludge as well as to reveal the mechanism of aerobic granulation, which impedes the optimization and application of this technology for real use. Distinctly, aerobic granulation in both pilot-scale and full-scale reactors took much longer than in lab-scale reactors [4-7]. Thus, many studies have been conducted in the last two decades to find a way to enhance the formation of aerobic granular sludge [8].

It has been widely acknowledged that a short settling time is essential for fast granulation. To date, almost all granular sludge has been obtained at a settling time lower than 8 min, regardless of an initial low settling time of about 2-5 min or an alternation between higher and lower values [9]. Li et al. [6]

in particular realized aerobic granulation in lab-scale, pilot-scale, and full-scale tests with settling time durations as high as 10, 20 and 40 min. The seed sludge used in their study was obtained from Yancang Wastewater Treatment Plant (WWTP) in China. In this particular sludge, small particles with a low sludge volume index (SVI) of 60-79 mL·g-1 were observed. Short settling time helps discharge sludge flocs with poor settling velocity and enriches the granules. Meanwhile, the short settling time always leads to low concentrations of sludge due to sludge loss by discharge, especially those with an initial low time duration [10]. Therefore, an appropriate settling time is essential for aerobic granulation.

Multiple substrates have been added to accelerate the start-up of aerobic granular sludge by providing nucleating agents first for bacteria to attach. Metal ions (particularly calcium, magnesium, and aluminum) [11-12], metal (like zero-valent iron) [13], and mature granules were believed to be able to enhance the formation of granular sludge. Liu et al. [14-15] studied the enhancement effect of poly-aluminum chloride (PAC) on aerobic granulation and found PAC could accelerate the granulation process. Li et al. [16] reported that dried sewage sludge micropowder as a nucleating agent could enhance formation of aerobic granules. Although these strategies could shorten the start-up period to different extents, the additional cost of adding these external carrier media should be considered. Thus, applying a more economical approach to enhance the aerobic granulation process is desirable and promising for further applications.

Yellow earth, as a common and inexpensive material [17], contains mainly silicon dioxide and aluminum oxide. Contrastingly with the agents mentioned above, yellow earth provides a novel and cost-effective alternative in the enhancement of aerobic granular sludge. For the first time, an initial settling time of a constant 10 min was selected for the granulation process. Ultimately, then, this paper proposes a new approach involving yellow earth in the acceleration of granulation. The present study's main objectives were to (1) confirm whether a short settling time was a prerequisite for granulation, and (2) whether the granulation process could be accelerated with yellow earth as a nucleating agent.

2. Materials and methods

2.1. Reactor and operation

A column reactor made of Plexiglas was used in the present study (shown in figure 1). In brief, the reactor was 120 mm in diameter and 800 mm in total height, giving an effective volume of 7.0 L with a working height of 700 mm. Thus, the reactor's effective ratio of height to diameter (H/D) was about 5.2. 3.5 L of synthetic wastewater was fed into the reactor at the beginning of each cycle at a volume exchange ratio of 50%. Air was introduced by a fine-bubble pump at the bottom of the reactor at a constant rate of 2.5 L min^{-1}. A mechanical stirrer was configured for the reactor at a constant speed of 250 rpm. The experiment was conducted at a water temperature of 18 ± 0.5 ℃. The reactor was operated sequentially in 3 h per cycle, consisting of 2 min of feeding, 166 min of aeration (stirring meanwhile), 10 min of settling, and 2 min of effluent discharge.

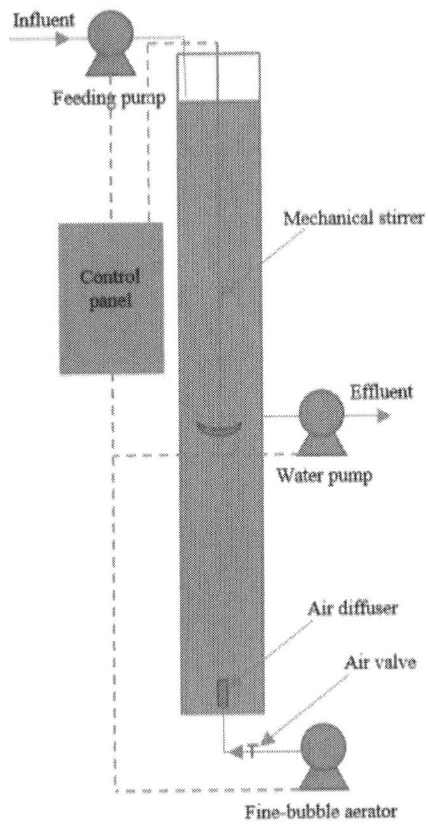

Figure 1. Schematic diagram of the sequencing batch reactor.

2.2. *Seed sludge, agent and media*

The seed sludge obtained from Shahu Wastewater Treatment Plant (China) operated with an A/A/O process configuration. The seed sludge was inoculated into the reactor with an initial mixed liquor suspended solid (MLSS) of 7440 mg L^{-1} after appropriate pretreatment. After a 2-cycle operation (6 h equally), 1 g L^{-1} of the yellow earth passed through a 50-mesh (corresponding to a diameter of 0.27 mm) sieve was added to the reactor.

Synthetic wastewater used in the present study was made up of the following compositions (per liter): 500 mg COD, 25 mg ammonia nitrogen (NH_4^+-N), 5 mg total phosphorus (TP), and 1 mL trace solution per 15 L influent as our previous compositions [18]. Influent pH was adjusted around 7.5 without control during the operation.

2.3. *Analytical methods*

COD, NH_4^+-N, nitrite (NO_2^--N), nitrate (NO_3^--N), TP, settling velocity (SV), SVI, MLSS, mixed liquor volatile suspended solids (MLVSS), water content and specific gravity were measured according to standard methods [19]. A pHS-25 meter and YSI5000 meter were used to measure the pH and DO. Extant respirometry was used to measure the specific oxygen uptake rate (SOUR), following the procedures by Abdullah et al. [20]. Total inorganic nitrogen (TIN) was the sum of NH_4^+-N, NO_2^--N and NO_3^--N. The morphology was observed, photos were taken, and granule size analysis was conducted via an image analysis system (Image-Pro Plus, V6.0, Media Cybernetics) with an Olympus Microscope (SZX9, Japan). EPS was extracted following a modified heat extraction method by Yang et al. [21]. Protein (PN) content was determined by a modified Lowry method, and the content of polysaccharides (PS) was analyzed using a phenol-sulphuric acid method [9]. EPS was regarded as the sum of PN and PS.

3. Results

3.1. Formation and characterization of granules

Yellow earth was added to the reactor to investigate its effects on the enhanced formation of aerobic granular sludge. The yellow earth distributed itself well in the water solution, and particles of different sizes could be seen clearly by the Telegraph (shown in figure 2a). Small granules (figure 2b and figure c) were observed on the 2nd and 3rd day. On the 4th day, no obvious sludge flocs could be seen in the reactor. Regular and bright yellow granules with an average of 1.06 mm were visible from the outside of the reactor, indicating the completion of the start-up of the granulation process. During the whole operation period, biofilm adhesion phenomenon was observed on the surface of the reactor, implying a moderate biological selector [9]. The granule size analysis was conducted on the aerobic granular sludge on the 4th day and the results are shown in figure 3. As seen in figure 3, the average diameter of the granules was 1.06 mm, and a majority of particles (83%) were in the range from 0.7 to 1.3 mm.

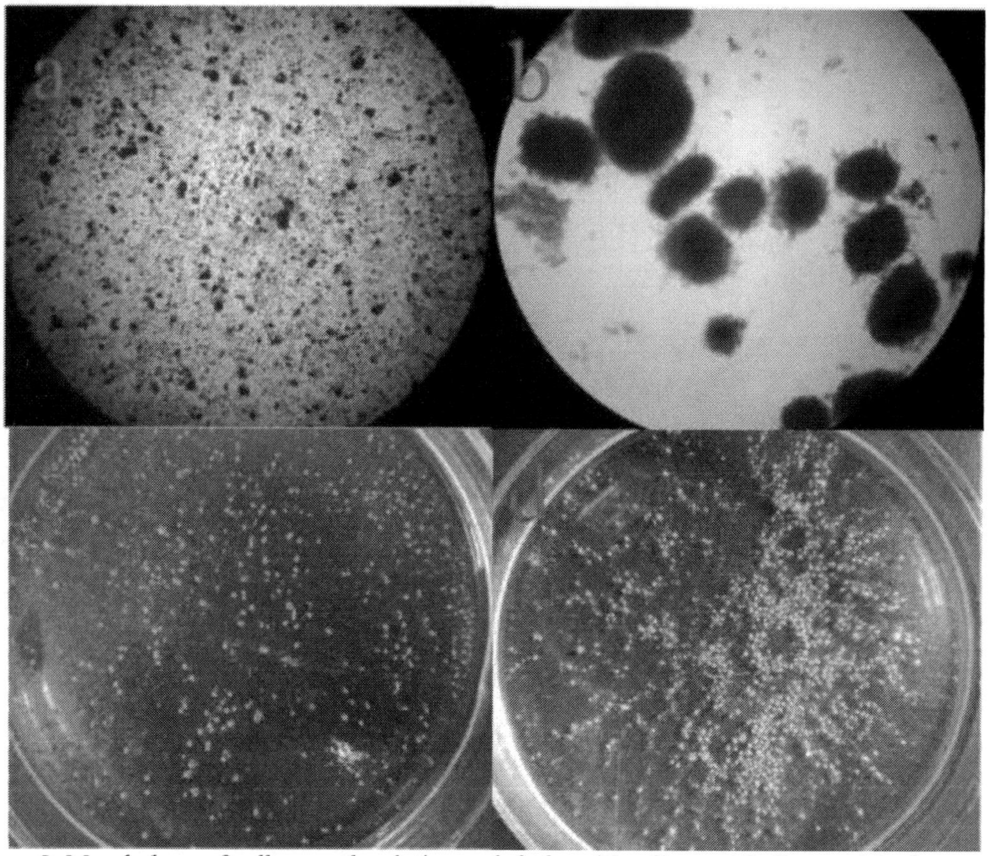

Figure 2. Morphology of yellow earth solution and sludge: (a) yellow earth dispersed in the water phase; (b) sludge flocs on day 2; (c) granules on day 3; and (d) mature granular sludge on day 4.

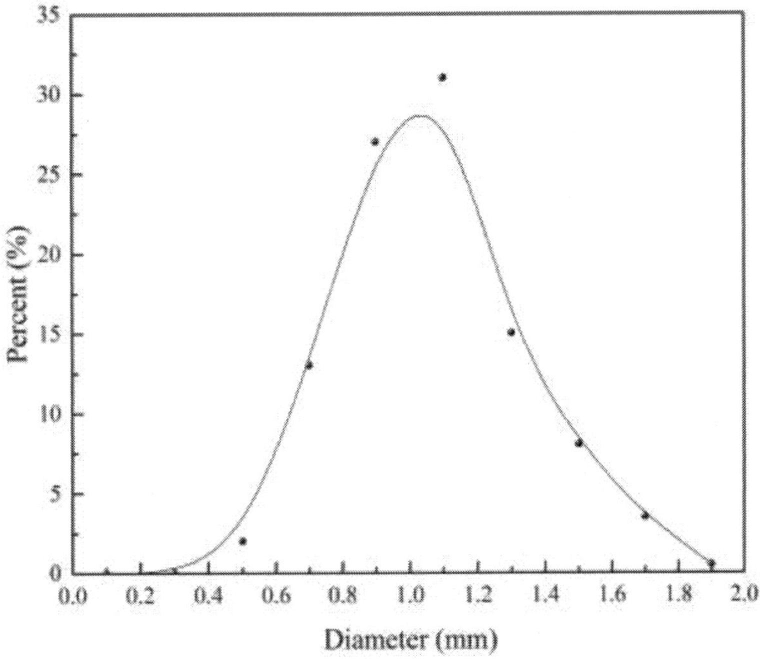

Figure 3. Diameter analysis of the granular sludge on day 4.

Compared with the seed sludge, the granular sludge showed improved physiochemical and biological characteristics (Table 1). As an important index for settling performance of activated sludge, the sharp decrease of SVI from 75 to 23 mL g^{-1} indicated the tight structure of granules and successful granulation, as does the rising of values of SV_{30}/SV_5 from 0.65 to 0.96 [22]. Water content revealed the profile of hydrophilic and hydrophobic cells within granules. In the present study, the water content fell from 99.67% to 98.26%, indicating the enrichment of hydrophobic cells [23]. The increase of specific gravity also confirmed the results [22]. SOUR suggested microbial activity in granular sludge and the significant enhancement of SOUR from 13.58 to 42.8 mg O_2 g $MLVSS^{-1}$ h^{-1}, indicating that granules were stronger in the organism's activity than in inoculated activated sludge [20].

Table 1. Attributes of activated sludge in present study (granules on day 4).

Sludge	SVI (mL/g)	SV_{30}/SV_5	Water content (%)	Specific gravity	SOUR (mg O_2 g $MLVSS^{-1}$ h^{-1})
Seed	75	0.65	99.67	1.008	13.58
R_e	23	0.96	98.26	1.022	42.8

The profile of MLSS and MLVSS/MLVSS is shown in figure 4. By the visible insight from outside the reactors, the reactors suffered a loss of sludge during the first 7 days. Then it rose slightly afterward. However, the ratio of MLVSS to MLSS increased gradually from 0.47 for the seed sludge to 0.733 for granular sludge on the 11th day, which indicated the higher bioactivity and lower inorganic composition of the granular sludge than the flocs [24].

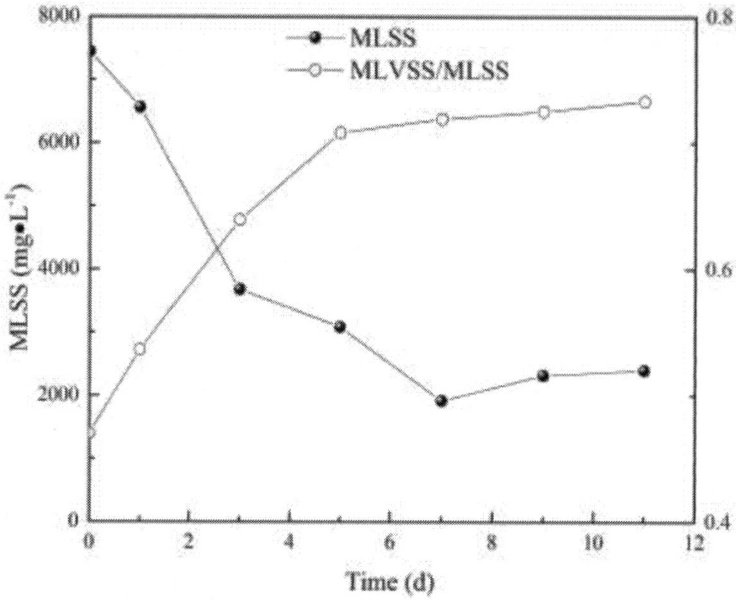

Figure 4. Profile of MLSS and MLVSS/MLSS.

3.2. EPS and PN/PS

EPS is a vital substance that has long been believed to aid the granulation process [25]. The variations of EPS and PN/PS over operation were shown in figure 5. Compared with the EPS content of seed sludge (92.16 mg g MLSS^{-1}), it rose sharply during the formation of granular sludge, which reached a peak (345.92 mg g MLSS^{-1}) on the 7th day. In general, adding earth on the first day significantly enhanced the amount of EPS and maintained this trend throughout the operation. The ratio of PN to PS (PN/PS) greatly increased during the granulation process from 1.21 to about 2.45.

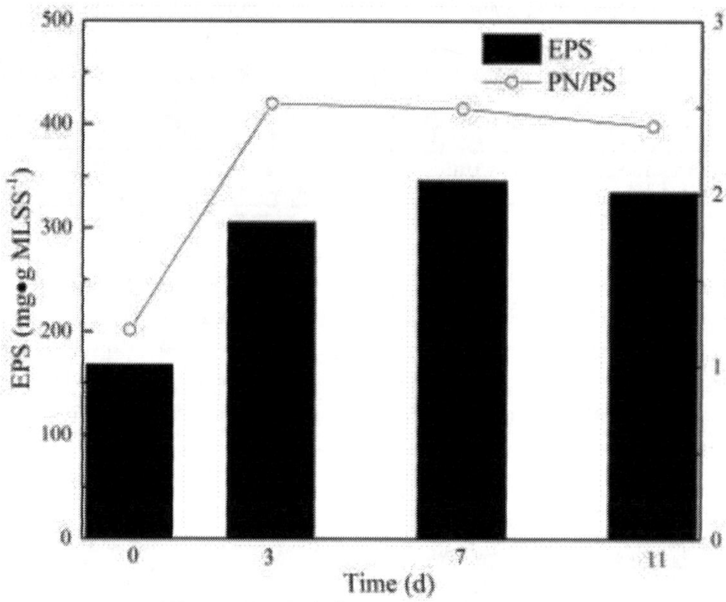

Figure 5. Variations of EPS and PN/PS.

3.3. Reactor performance

COD and nitrogen removal performance are presented in Figure 6. As shown in Figure 6(a), over 90% of influent COD was removed throughout the operation with the stable effluent concentration of about 20 mg L^{-1}. As for nitrogen removal (Figure 6(b)), 95% of the influent NH_4^+-N was reduced while about 80% of TIN was eliminated as the effluent NO_3^--N, and NO_2^--N were about 1.6~5.3 and 0~0.4 mg L^{-1} respectively. At constant aeration conditions, though, obvious nitrogen loss was observed and it could be concluded that denitrification took place during this oxic phase.

Interestingly, the TP removal rate varied from 17 to 46% (data not shown), demonstrating the simultaneous nitrogen and phosphorus removal (SNPR) capacity of both reactors [2].

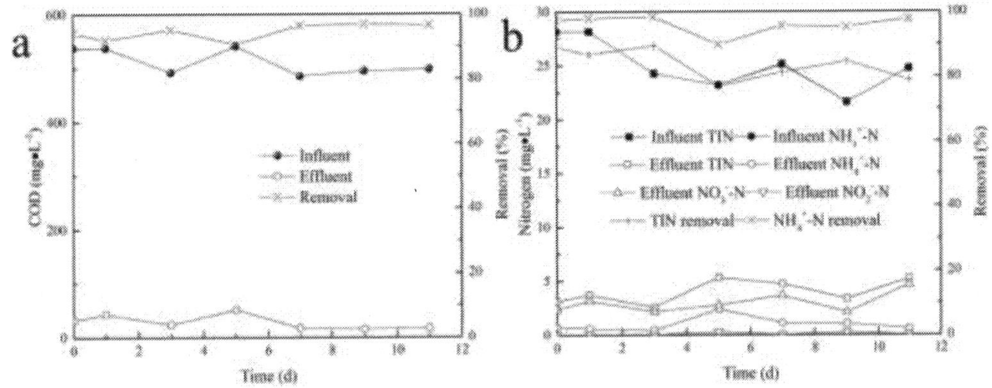

Figure 6. Reactor performance: (a) COD removal and, (b) nitrogen removal.

4. Discussion

A short settling time under 8 min has long been assumed as a prerequisite for aerobic granulation in SBR. The present study proposes a different view. Successful aerobic granulation was realized with a constant settling time of 10 min. Settling time, as a hydraulic selection pressure, was once believed to drive the hydraulic retaining of the granular type biomass [26] by discharging the sludge flocs with poor settling ability. The settling time in combination with the height of the reactor allowed only particles with a settling velocity larger than 2.1 $m \cdot h^{-1}$ to be maintained in the reactor. However, mature aerobic granules cultivated in lab-scale (51 $m \cdot h^{-1}$) or pilot-scale (18~40 $m \cdot h^{-1}$) reactors had a much higher settling velocity. The results of the present work suggest that a short settling time is not a prerequisite for an aerobic granulation process under such conditions.

The profile of MLSS was not synchronous with the granulation process. Complete granulation was obtained on the 4th day, while the MLSS kept dropping until the 7th day. This phenomenon is aligned with the findings of Liu et al. [10], who also detected a time lag between granulation and biomass growth. According to Li et al. [27] and Liu et al. [14], granules with external carrier media like PAC or dried sludge contained less organic compounds, explaining the relatively low MLVSS/MLSS value for mature granules in the present study.

This study's granulation speed is comparable to previous studies [9, 12, 27-29]. First, the readily available and inexpensive sources of the yellow earth used as a nucleating agent in the present study provide an economically-viable alternative for wider application. The unique characteristics of the yellow earth used were key for rapid aerobic granulation. According to the formation process of this study, a yellow-earth-based process was proposed for its enhancement effects, as shown in figure 7. Yellow earth's chemical components (mainly made up of silicon dioxide and aluminum oxide) and physical size (above 0.27 mm) make it ideal as a nucleating agent to induce bacteria attachment and aggregation of granular-type microbes, as well as the shaping of particles in combination with the constant aeration and stirring conditions. Namely, the small flocs act as the nuclei for granular-type biomass to attach, and stimulates the secretion of EPS (especially the PN). Ones with media sizes are embedded, and the larger ones help shape the small and irregular granules into regular ones [6, 15]. In addition, the yellow earth particles may have a bridging effect among particles, microbes and EPS [27]. EPS also plays a vital role in aggregation and cohesion of different types of biomass as well as yellow earth particles [30].

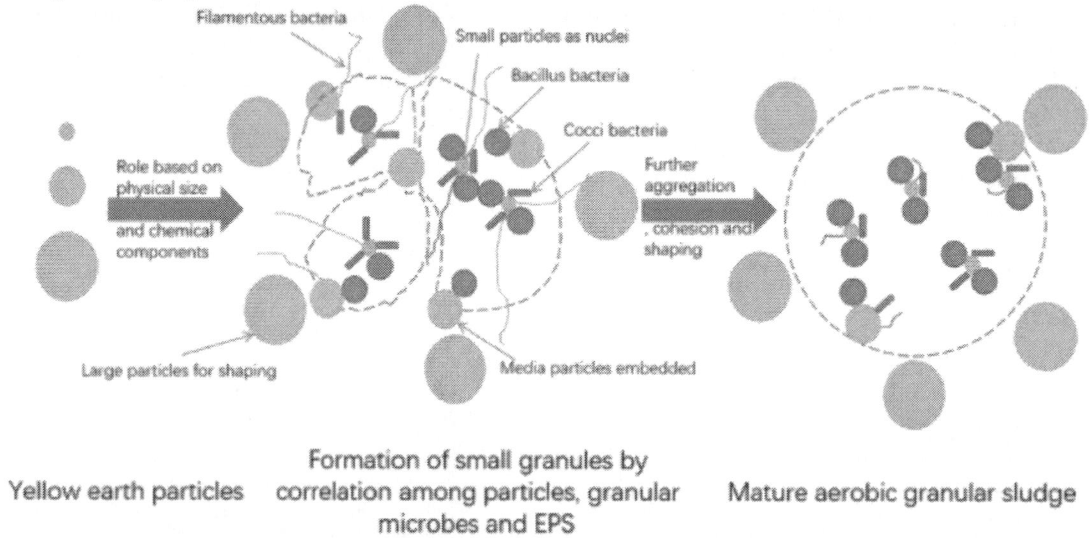

Figure 7. Formation of aerobic granular sludge enhanced by yellow earth.

With a settling time of 10 min, aerobic granulation was realized within 4 days in a sequencing batch reactor with yellow earth as the nucleating agent. The granular sludge showed distinct attributes including well-settling performance, low water content and high specific gravity, high microbial activity, and abundant EPS content (especially PN content). Excellent COD and nitrogen removal were achieved during the start-up period, and the denitrification process was detected due to the layered structure of the sludge. This study proves that settling is not a prerequisite for aerobic granulation in the aforementioned conditions. Rapid granulation was enhanced by the physicochemical characteristics of the yellow earth. The research presented in this paper offers an improved and cost-effective alternative for the enhanced formation of aerobic granular sludge.

References

[1] Bruin L M, Kreuk M K, Roest H F, Uijterlinde C and Loosdrecht M C 2004 Aerobic granular sludge technology: an alternative to activated sludge? *Water Sci. Technol.* **49**: 1-7

[2] Kreuk M K, Heijnen J J and Loosdrecht M C 2005 Simultaneous COD, nitrogen, and phosphate removal by aerobic granular sludge *Biotechnol Bioeng* **90**: 761-69

[3] Jang A, Yoon Y H, Kim I S, Kim K S and Bishop P L 2003 Characterization and evaluation of aerobic granules in sequencing batch reactor *J. Biotechnol* **105**: 71-82

[4] Ni B J, Xie W M, Liu S G, Yu H Q, Wang Y Z, Wang G and Dai X L 2009 Granulation of activated sludge in a pilot-scale sequencing batch reactor for the treatment of low-strength municipal wastewater *Water Res.* **43**: 751-61

[5] Isanta E, Suárez-Ojeda M E, Val Del Río Á, Morales N, Pérez J and Carrera J 2012 Long term operation of a granular sequencing batch reactor at pilot scale treating a low-strength wastewater *Chem. Eng. J.* **198-199** 163-70

[6] Li J, Ding L B, Cai A, Huang G X and Horn H 2014 Aerobic sludge granulation in a full-scale sequencing batch reactor *Biomed Res. Int.* **2014** :2687-89

[7] Pronk M, Kreuk M K, Bruin B, Kamminga P, Kleerebezem R and Loosdrecht M.C 2015 Full scale performance of the aerobic granular sludge process for sewage treatment *Water Res.* **84**: 207-17

[8] Liu Z, Liu Y, Kuschk P, Wang J, Chen Y and Wang X 2016 Poly aluminum chloride (PAC)

enhanced formation of aerobic granules: Coupling process between physicochemical–biochemical effects *Chem. Eng. J.* **284** :1127-35

[9] Long B, Yang C Z, Pu W H, Yang J K, Jiang G S, Dan J F, Li C Y and Liu F B 2014 Rapid cultivation of aerobic granular sludge in a pilot scale sequencing batch reactor *Bioresource Technol* **166**: 57-63

[10] Liu Y Q and Tay J H 2015 Fast formation of aerobic granules by combining strong hydraulic selection pressure with overstressed organic loading rate *Water Res.* **80**: 256-66

[11] Wan C, Lee D J, Yang X, Wang Y, Wang X and Liu X 2015 Calcium precipitate induced aerobic granulation *Bioresource Technol* **176**: 32-37

[12] Wang S, Shi W, Yu S, Yi X and Yang X 2012 Formation of aerobic granules by Mg^{2+} and Al^{3+} augmentation in sequencing batch airlift reactor at low temperature *Bioproc Biosyst Eng.* **35**: 1049-55

[13] Kong Q, Ngo H H, Shu L, Fu R S, Jiang C H and Miao M S 2014 Enhancement of aerobic granulation by zero-valent iron in sequencing batch airlift reactor *J. Hazard Mater* **279**: 511-17

[14] Liu Z, Liu Y, Kuschk P, Wang J, Chen Y and Wang X 2016 Poly aluminum chloride (PAC) enhanced formation of aerobic granules: Coupling process between physicochemical biochemical effects *Chem. Eng. J.* **284**: 1127-35

[15] Liu Z, Liu Y, Zhang A, Zhang C and Wang X 2014 Study on the process of aerobic granule sludge rapid formation by using the poly-aluminum chloride (PAC) *Chem. Eng. J.* **250**: 319-25

[16] Li Y, Zhang Y, Zhao Z, Sun S, Quan X and Zhao H 2016 Enhancement of sludge granulation in hydrolytic angiogenesis by denitrification *Appl. Microbiol Biotechnol* **100** :3313-20

[17] Bindhu B and Madhu G 2015 Influence of three selection pressures on aerobic granulation in sequencing batch reactor *India J. Chem.. Techn..* **22**: 241-47

[18] He Q, Wang H, Yang X, Zhou J, Ye Y, Chen D and Yang K 2015 Culture of denitrifying phosphorus removal granules with different influent wastewater *Desalin Water Treat* 1-8.

[19] APHA 2005 Standard methods for the examination of water and wastewater *American Public Health Association (APHA)*: Washington, D.C, USA

[20] Abdullah N, Yuzir A, Curtis T P, Yahya A and Ujang Z 2013 Characterization of aerobic granular sludge treating high strength agro-based wastewater at different volumetric loadings *Bioresource Technol.* **127**: 181-87

[21] Yang Y C, Liu X, Wan C, Sun S and Lee D J 2014 Accelerated aerobic granulation using alternating feed loadings: alginate-like exopolysaccharides *Bioresource Technol.* **171**: 360-66

[22] Kreuk M K, Pronk M and Loosdrecht M C 2005 Formation of aerobic granules and conversion processes in an aerobic granular sludge reactor at moderate and low temperatures *Water Res.* **39**: 4476-84

[23] Lochmatter S, Gonzalez G G and Holliger C 2013 Optimized aeration strategies for nitrogen and phosphorus removal with aerobic granular sludge *Water Res.* **47**: 6187-97

[24] Yu S, Sun P, Zheng W, Chen L, Zheng X, Han J and Yan T 2014 The effect of COD loading on the granule-based enhanced biological phosphorus removal system and the recoverability *Bioresource Technol.* **171**: 80-87

[25] Verawaty M, Pijuan M, Yuan Z and Bond P L 2012 Determining the mechanisms for aerobic granulation from mixed seed of floccular and crushed granules in activated sludge wastewater treatment *Water Res.* **46**: 761-71

[26] Zhou D, Niu S, Xiong Y, Yang Y and Dong S 2014 Microbial selection pressure is not a prerequisite for granulation: dynamic granulation and microbial community study in a complete mixing bioreactor *Bioresource Technol.* **161**: 102-08

[27] Li J, Liu J, Wang D, Chen T, Ma T, Wang Z and Zhuo W 2015 Accelerating Aerobic Sludge Granulation by Adding Dry Sewage Sludge Micro powder in Sequencing Batch Reactors *Int .*

J. Environ. Res. Public Health **12**: 10056-065
[28] Moy B Y, Tay J H, Toh S K, Liu Y and Tay S T 2002 High organic loading influences the physical characteristics of aerobic sludge granules *Lett. Appl. Microbiol.* **34**: 407-12
[29] McSwain B S, Irvine R L and Wilderer P A 2004 The influence of-settling time on the formation of aerobic granules *Water Sci. Technol.* **50**: 195-202
[30] Gao D, Liu L, Liang H and Wu W M 2011 Aerobic granular sludge: characterization, mechanism of granulation and application to wastewater treatment *Crit. Rev. Biotechnol* **31**: 137-52

The model test of restoration project of the gravel beach of Chen Village fishing port

D X Wang [1, 4], J S Gui [2] and J W Sun [3]

[1] Department of ocean and civil engineering, DaLian Ocean University, DaLian 116000, China

[2] Department of ocean and civil engineering, DaLian Ocean University, DaLian 116000, China

[3] Key Laboratory of Sea-Area Management Technology.SOA, National Bureau of Oceanography, DaLian 116000, China

E-mail: 21140216@qq.com

Abstract. Gravel beach is a case in coastal landform by wave action. It is more and more crucial for the environment of coastal engineering in recent years. However, it is poorly studied for it in China. And this paper which is based on the model test of Restoration Project of the Gravel Beach of Chen Village Fishing Port, uses two dimensional normal physical models, aiming at exploring the movement of gravel beach under wave action and verifying the stability of the gravel beach section. The test depends on different water levels (designed high water level, designed low water level, and extreme high water level) and return periods (2, 5, 10, 25, 50 years once). Finally, two distinct experimental sections are got under the changed conditions and the movement law of gravels is obtained.

1. Introduction

Gravel beach is generally formed near the bedrock coast, and it is also a typical coastal landform effected by wave action. Because of the function of coastal protection and the great tourism resources [1], a large number of people are taking it accounts into the construction of artificial beach [2]. There have been some examples of the application of artificial gravel beach in shoreline protection [3], but in China things are different: only some scholars have made some surveys on the natural gravel beach [4, 5]. And even worse, the lack of information and attention leads to fewer physical model tests on gravel beach. The good news is the case that the coastal environment has gradually become a new hot spot in recent years along with the national attention to the environment. LaoTieShan, a famous tourist attraction in Dalian, is located at the border of the Yellow Sea and the Bohai Sea, and Chen Village Fishing Port is located at the foot of LaoTieShan. In recent years, due to wave erosion and man-made influence, the south side of the port has been a certain degree of damaged. Withal, relevant departments take active measures, intending to fix the gravel beach. In order to achieve better engineering effect, a physical model test is needed in order to forecast the effect of the project and the stability of the gravels. Experiment was conducted in the wave flume of the Key Laboratory of Coastal Engineering of Dalian Ocean University, Liaoning Province. The representative section was extracted from the given terrain, and the gravity similarity criterion was adopted to this text. The goal is to reach a static equilibrium in the gravel beach of Chen Village Fishing Port such that erosion or sediment problems could be avoided or diminish. The initial wave conditions are got from the mathematical wave model test results, including different return periods (2, 5, 10, 25, 50 years once) and different water level (designed high water level, designed low water level, and extreme high water level).

2. Initial conditions

2.1. Water level

Extreme high water level: 2.91m

Designed low water level: 0.10m

Designed high water level: 2.27m

2.2. Wave conditions

Wave conditions could be seen in Table 1.

Table 1. Wave conditions.

Water level(m)	Return periods(year)	Significant wave height(m)	Wave period(s)
2.91	2	0.97	6.4
	5	1.50	8.3
	10	1.72	8.7
	25	1.95	9.1
2.27	2	0.94	6.4
	5	1.42	8.3
	10	1.65	8.7
	25	1.89	9.1
0.1	2	0.80	6.4
	5	1.24	8.3
	10	1.40	8.7
	25	1.62	9.1

2.3. The text terrain

In accordance with the relevant specifications [6, 7, 8], some research experience from scientific institutions [9, 10, 11] and measured topographic map, three typical sections in the project area were selected. On the basis of the representative of the section, the slope and the feasibility of putting in place in the actual wave tank, a typical section is chosen from those three (Figure 1). After the terrain construction is completed, the cement is used for getting a hard surface, which is convenient for the initial wave. After that, the screened gravel is putting on the original terrain as required, which forming the sand fill surface of construction plan (Figure 2). Since then_terrain construction for the physical model test is completed.

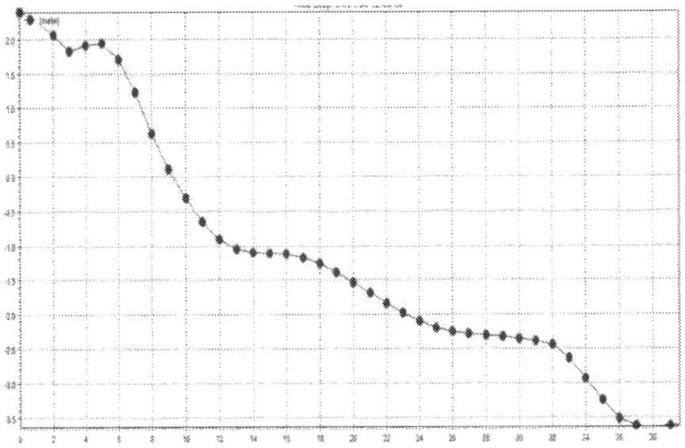

Figure 1. Typical section.

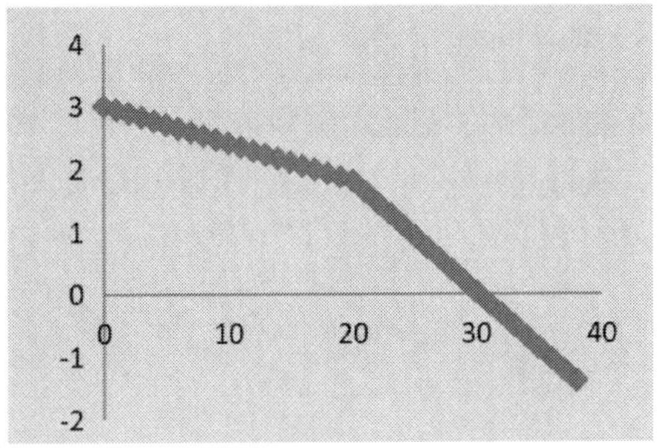

Figure 2. The designed sand fill surface(points 0-38,1m per point).

3. Similarity criterion

In coastal engineering, the main external loads acting on the building are taken from the waves. Because the propagation of the wave is mainly dependent on gravity, it is usually called the sea wave as the gravity wave, so the wave simulation should follow the gravity similarity criterion. As a result, in coastal engineering, the gravity similarity criterion is the most commonly used criterion, especially in the experimental study of wave and its effect on engineering structures. When the system is under the action of gravity, gravity is:

$$G = mg = \rho V g$$

The above 'm' is mass, 'g' is the acceleration due to gravity, ρ is density, V is the volume. The subscript P indicates the value of each physical quantity in the prototype. The subscript m indicates the values of each physical quantity in the model. Such as:

$$F_{gp} = m_p g_p = \rho_p V_p g_p \qquad F_{gm} = m_m g_m = \rho_m V_m g_m$$

If λ refers to the ratio of prototype and model value:

$$\lambda_{F_g} = \lambda_m \lambda_g = \lambda_\rho \lambda_l^3 \lambda_g = \lambda_\rho \lambda_v^2 \lambda_l^2 \qquad \text{or:} \frac{\lambda_v^2}{\lambda_g \lambda_l} = 1$$

Also:

$$\left(\frac{v^2}{gl}\right)_p = \left(\frac{v^2}{gl}\right)_m = F_r$$

Here F_r is a dimensionless number called the gravity similarity criterion, or the Froude number.

Under the condition of gravity similarity criterion, as the result of $\lambda_g = 1$:

IOP Conf. Series: Earth and Environmental Science **39** (2016) 012026 doi:10.1088/1755-1315/39/1/012026

The ratio of velocity: $\lambda_v = \lambda_l^{1/2}$

The ratio of pressure: $\lambda_p = \lambda_l^2$

The ratio of time: $\lambda_t = \lambda_l^{1/2}$

The λ_l is the ratio of length.

The ratio of force and mass: $\lambda_F = \lambda_m = \lambda_l^3$

Taking into account the test equipment and relevant literature [6, 7, 8], as well as other factors such as technical requirements [12], determining the ratio of the length is between model and prototype is 1:10.

4. The text

4.1. Text materials

Because the test is based on the actual project, prototype stones used in the text should be selected from the quarry so that we could get the gradation of these prototype stones. After that, the stone is divided into three parts which are 'small', 'medium' and 'large' respectively. From those above working, in the light of the ratio of length, the parameter of the model gravel could be gotten: the ratio of that three parts is 3:4:3 and the average diameters of the model gravel are 2.2mm, 3.7mm and 4.27mm respectively. According to average diameters, three sieves diameter at 1.25mm, 2.5mm and 5mm are selected to get the model gravel. The parameters of model gravel are given in table 2.

Table 2. The parameters of model gravel.

	Major axis (mm)	Middle axle (mm)	Short axis (mm)	Quality (g)	Average diameter (mm)
Small	3.5	2.33	1.35	0.02	2.2
Medium	5.28	3.94	2.62	0.07	3.7
Large	6.1	4.51	2.92	0.11	4.27

4.2. The process

Because of different water levels and wave conditions, the text is done by this process:

First of all, water levels are used from low to high, it means that the order is designed low water level, designed high water level and extreme high water level. Secondly, waves are, under different water levels, from 2 years once to 50 years once which is to say that it is also following the small-to-large. At last, each wave simulates three times (two times under extreme high water level). After each simulation is completed, the terrain is not restored. And should not start next simulation before the water gets still.

5. Results

Owing to the limit of the experimental flume, the waves two years once and five years once under designed low water level cannot be reached. In fact, the influence under designed low water level, no matter which wave, is very little. Moreover, the terrain is not restored until the cycle over so the missing waves have little effect on the final results. The conclusive results under extreme high water level are shown in Figure 3-6. (The red lines are the designed sand fill surfaces and blue lines the new profile)

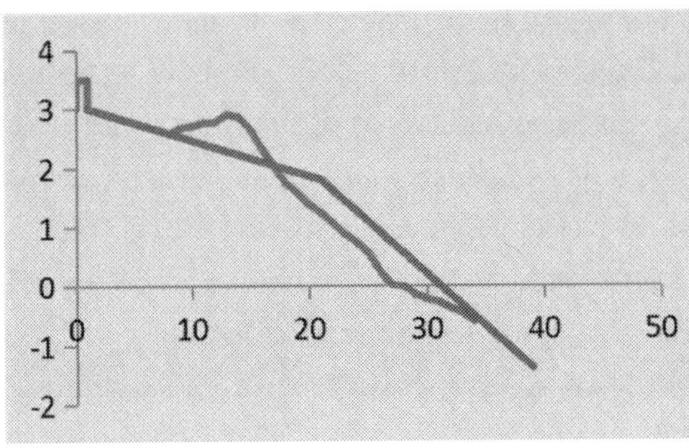

Figure 3. The result of 2-year-once wave.

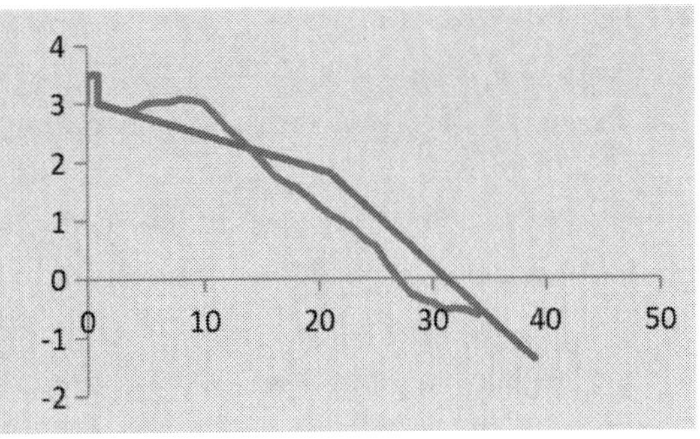

Figure 4. The result of 5-year-once wave.

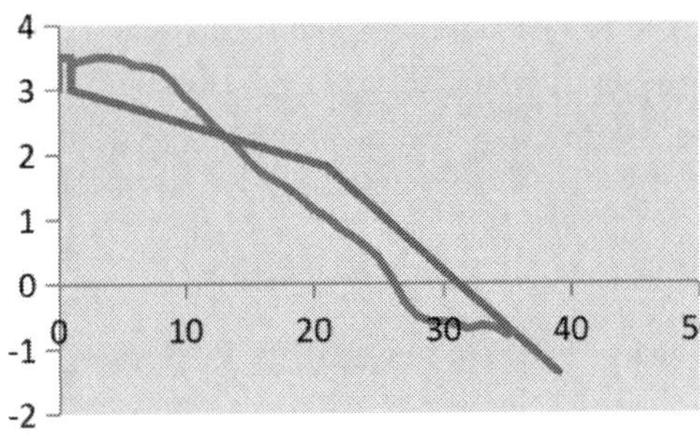

Figure 5. The result of 10-year-once wave.

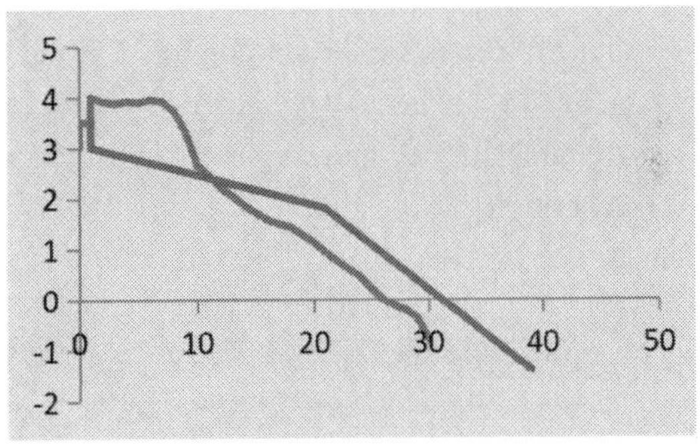

Figure 6. The result of 25-year-once wave.

6. Discussion

- Under the designed low water level, wave has little effect and cannot affect the beach, so this is not the point we follow closely.
- Under the designed high water level, the effects of wave increases slowly as time went by. With the action of wave, gravels at the wave breaking part are emptied. Some go downward with water. The other part goes upward and accumulates, forming a small beach berm. Due to the increasing of the wave and the passage of time, a 1:6 slope is established which roughly parallel to the designed filling surface.
- Under the extreme high water level, because of higher water level, the influence of the wave is not only similar as before but also more obvious than before. It is worth stressing that, at the measurement point 26-30 (that is, the distance from the wave wall 26-30m), a pit is formed. And with the increase of the wave, the bunker is inclined to increase as well.

7. Conclusion

- This text carries out a considerable number of simulations under various working conditions, but it actually do not get a real stable section. It is difficult to form a stable section on the beach, which can only form a relatively stable section. However, due to the consideration of this section test is to simulate the most unfavorable situation, and the actual situation in the gravel beach cannot always be the most unfavorable situation, so this result can meet the requirements.

- Under the action of wave, gravels located at the measurement point 15-30 (i.e. Distance gear wave wall 15-30m) will be displaced away, and finally a slope which is parallel to the original design of filling sand surface could be formed. Even so, the height of the surface declines by about 1m. To avoid this, in actual construction, larger gravels should be utilized to ensure the stability of this wave breaking part. It is no doubt beneficial to the project and also numerous other adverse phenomena, such as erosion and the sediment load, will decrease.
- The joint point of the two slopes needs to move towards the seaward direction. It is to say that the length of the upper slope should be increased. Because we could see it from the results that gravel beach, under the waves washed, has the same law of motion in reference literature [13,14] : With the action of wave, the bunker is formed and this bunker continues 'climbing' and finally a platform emerges. The advantage of the movement towards seaward is that it can make the starting point of the bunker away from the beach so that the 'climbing' could be limited. At the same time, the extensions of the upper slope also could ease the accumulation of gravels so that the bunker behind the platform could be less steep and towering.

Reference

[1] Wang A J and Gao S 2005 The role of gravel beach in coastal protection *Proceedings of the twelfth China Coastal Engineering Symposium* pp 385-290

[2] Evan H 1974 Equilibrium profiles of coarse material under wave attack *Proceedings of 4th International Conference of Coastal Engineering* pp 939-57

[3] Zenkovich V P and Schwartz M L 1987 Protecting the Black Sea-Georgian S.S.R. gravel coast *Journal of Coastal Research* **3** pp 201-9

[4] Wang A J and Gao S 2004 Sediment Distribution and Shape Characteristics of Gravel Beaches, Zhujiajian Island, Zhejiang Province *Journal of Nanjing University(Natural Sciences)* **40**(6) pp 747-58

[5] Zhou L Y and Wang F F 2011 The characteristics and causes of Hainan Wanning Hua Kok boulder stone beach *Marine Geology Frontiers* **28** 11

[6] Ministry of Communications of the PRC 2001 *Wave Model Test Regulation*

[7] Ministry of Communications of the PRC 2010 *Technical Regulation of Modelling for Tidal Current and Sediment on Coast and Estuary*

[8] Ministry of Communications of the PRC 2015 *Code of Hydrology for Harbour and Water Way*

[9] Dou X P and Zhang X M 1999 Summary of sediment research in recent ten years by Nanjing Hydraulic Research Institute *Journal of Sediment Research* pp 17-20

[10] Xiong Z L and Zeng J 2010 Various problems of sediment physical models of tidal estuaries *Advances in Science and Technology of Water Resources* **30** pp 19-23

[11] Wang X L 2012 Design method of physical model of sediment deposition *Friends of Science Amateurs* pp 43-44

[12] Wang Y H and Jiang W G 2012 Study of scale effect in scour physical-model experiments *Journal of Sediment Research* pp 31-34

[13] Liu F 2014 *The Preliminary Experimental Study on the Processes of Water Flow and Sediment Bed-forming*

[14] Dong L H 2012 *The Numerical and Experimental Application Research of Equilibrium Profile Design In Beach Nourishment*

Analysis and predication of urban water security: a case study of Chengdu City, China

L Chen[1, 3] and J Shi[2]

[1] China Institute of Water Resources and Hydropower Research, Beijing, China
[2] State Key Laboratory of Hydroscience and Engineering, Tsinghua University, Beijing, China

E-mail: linscu@163.com

Abstract. To investigate the sustainable state of the water resources in Chengdu, the pressure-state-response model was adopted to establish the index system of the water security, in which the index weights were calculated by the analytic hierarchy process. The comprehensive values of the urban water security in recent years have been calculated. The results showed that the water security level of Chengdu was grade IV from 2005 to 2013, which indicated a safe state. To ensure the coordinated development among the society, economy and environment, several suggestions were proposed. The water-saving consciousness of the public and the water-saving city could be improved through increasing the water-saving facilities and techniques. The environmental investment could be increased for improving the treatment rate of the municipal and industrial wastewater. The management and warning system could be improved to strengthen the ability of coping with the accidental problems relating to the water environment and water resources.

1. Introduction

The urban water security has become increasingly prominent with the urban development. The urban water problems, such as the shortage of water resources, water pollution, water incidents and sustainable security, have aroused the wide concern. The water resources are significant factors in the planning and layout of the city. Thus, it is important to realize the sustainable utilization of water resources and guarantee the water security [1]. The urban water security is not only an important part of the regional water security, but also an essential component of the urban ecological security. The assessment of urban water security is the foundation for establishing the strategy of water security and making decisions in the management [2].

The definition of urban water security has different expressions. Jia [3] explained that the essence of water resources security was whether the water supply could meet the reasonable demand. Han [4] stated that in terms of negative points of the water security, it could be defined as the deterioration and harmful effects of regional water conditions under natural and human factors. Cheng [5] regarded the water security as a social state where everyone had access to the safe and clean water, and the water could meet the demand of life, production and ecological environment. Tong [6] suggested that the urban water security require the adoption of a long view, in order to be able to account for the slow unfolding of some hydrological, ecological and social processes, and the necessary time for waterworks investments to yield their fruits.

The urban water security is a comprehensive system which usually consists of various factors. Many evaluation methods have been proposed for such system with multiple factors, such as the multi-level fuzzy method, system dynamics method, set pair analysis method and pressure-state-response (PSR) structure model [7]. The PSR model has advantages in evaluating the resource utilization and sustainable development. Thus, it is able to reflect the comprehensive relationship among the nature, society, economy, environment and resources [8]. The pressure indexes are used to reflect unsustainable development caused by human activities, consumption patterns and economic system. The state indexes are used to represent the current system status. The response indexes are used to characterize the countermeasures to promote the sustainable development. The PSR model has been widely used for the establishment of evaluation systems, such as the ecological security [9], wetland ecosystem health [10], water environment security [11] and urban water security [2].

In this paper, the concept of urban water security is regarded as the capacity for sustainable development of the city under the influence of water resources, which could ensure the development of society, economy, ecological environment, citizens and humanistic environment. The water security state of Chengdu is evaluated with the quantitative system, which could provide a reference for decision makers. It could also be beneficial for the rational utilization of water resources and promotion of the sustainable urban development.

2. Overview of the study area

Chengdu is the capital of Sichuan province, China. The Chengdu plain is located at the transition zone from the northwest mountains to the basin. The topography around the Chengdu plain is high in the northwest area which is close to the Longmen and Qionglai Mountains, and low in the southeast area which is near to the Longquan Mountain, as shown in figure 1. The city is located at the area with monsoon-influenced humid subtropical climate, which has distinct seasons and abundant rainfall. There are many rivers in Chengdu with a total area of 700 km^2, which mainly belong to the Minjiang and Tuojiang River. According to the water resources bulletin of Chengdu from 2005 to 2013, the annual amount of surface water resources is 8.12 billion m^3 and the annual runoff depth is 668.81 mm.

The demand of water resources, especially the industrial water, has largely increased with the urban development. Thus, the contradiction between the water supply and demand could become serious. In recent years, the deterioration of ecological environment in the upstream of the Minjiang River and the decrease of precipitation have influenced the inflow and the water amount of Chengdu. With the

IOP Conf. Series: Earth and Environmental Science 39 (2016) 012027 doi:10.1088/1755-1315/39/1/012027

urbanization of Chengdu moving forward, the water security issues of this city could become more prominent.

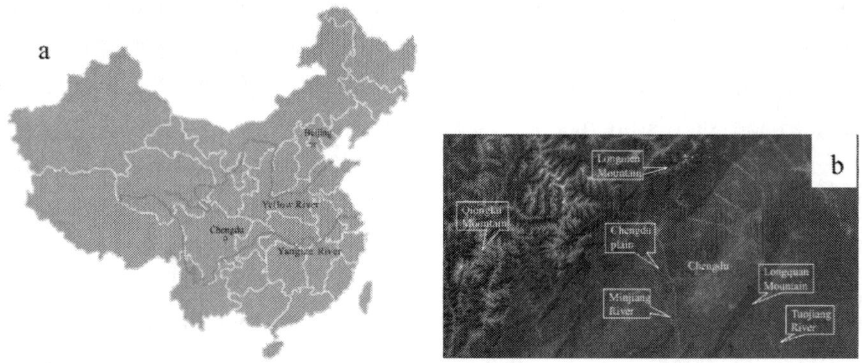

Figure 1. The location and topography of Chengdu: (a) Schematic map of the city location. (b) A snapshot of the topography at the Chengdu plain.

3. Evaluation index system

3.1. Establishment of index system

With the consideration of regional conditions in Chengdu, the index system of urban water security is established on the basis of improved PSR model, which could be divided into 3 subsystems and 3 levels, as shown in table 1.

3.2. Weight calculation

The analytic hierarchy process (AHP) is used to calculate the index weight. A complex system could be decomposed into sequential and hierarchical structure with the AHP method, in which qualitative judgments are quantified through the rank of decision schemes. Thus, it is a decision assessment method combined of qualitative and quantitative analyses, which is suitable for the system evaluation with multiple criteria [16]. The analysis process of the AHP method generally consists of the following 4 steps:

- Analyze the relationship between factors of the system and establish a hierarchical structure.
- Compare the importance of items on the current level based on the same criterion on the previous level and construct the judgment matrix of pairwise comparison.
- Calculate the relative weights of the items to the criterion through the judgment matrix.
- Calculate the synthetic weight of the items to the overall system and sort the weights.

Thus, the index weights could be obtained by the process of the expert scoring, calculation and consistency judgment, as illustrated in table 1.

4. Calculation of standardization

The principle of diminishing marginal benefit [13] is used to evaluate the effect of index variation. Assuming the short term production and other conditions unchanged, such as the technology level, the benefit could start to decrease when the input amount of the production factor has reached a certain degree.

Under the principle of diminishing marginal benefit, the index system of urban water security could

Table 1. Index weights in the PSR model of the urban water security.

Criteria level	Item level	Index level	Weight (‰)
Pressure index	Water resources	Daily water consumption per capita (L)	22.64
		Water resource amount per capita (m^3)	68.54
		Utilization ratio of water resources (%)	22.64
	Water environment	Discharge amount of specific pollutants in industrial wastewater (mg/L)	107.09
		Discharge amount of domestic wastewater (104 m3)	43.43
	Social economy	Industrial water demand per 10,000 RMB output (m^3)	68.54
		Population density (people / square kilometers)	12.17
	Flood and drought	Disaster probability (time/year)	9.31
		Ratio of economic loss to gross domestic product (GDP) (%)	22.64
State index	Environment	Water quality of drinking water source (%)	24.41
		Public green space per capita (m2)	43.01
	Social economy	GDP per capita (10,000 RMB)	60.87
		Engel coefficient (%)	24.41
		Urbanization rate (%)	34.12
	Resources	Surface water resources (10^8 m^3)	60.87
		Green coverage rate (%)	84.31
Response index	Environmental protection and control	City sewage treatment rate (%)	31.53
		Ratio of investment in treatment of industrial wastewater to GDP (%)	95.22
		Discharge attainment rate of industrial waste water (%)	66.30
		Reuse ratio of industrial water (%)	17.50
	Social economy	Per capita net income of rural residents (RMB)	17.50
	Management	Disaster prevention and emergency response capability	10.09
		Water saving consciousness of citizens	7.04
		Laws, regulations and management in water environment	45.82

be divided into three categories, namely the bigger with the better index, the smaller with the better index and linear variation index [14]. Then, the calculation models are chosen for three categories. The model of the bigger with the better index is the power function: $y = a + bx^{1/2}$ $(y \in [0,1])$ .Similarly, the model of the smaller with the better index adopts the power function: $y = a + bx^2$ $(y \in [0,1])$. The model of linear variation index uses the linear function: $y = a + bx$ $(y \in [0,1])$.

Taking into account the relevance of society, economy and water resources, the calculation range of indexes are assigned and the index calculating functions are determined after obtaining the coefficient a and b, as shown in table 2.

5. Water security assessment of Chengdu

5.1. Classification of water security level

Considering the current social and economic development in China, the range of different levels of urban water security could be determined on the basis of the national standards, international standards, urban planning and critical values of specific indexes. Thus, the water security standard could be divided into 5 levels, namely I, II, III, IV and V, as shown in table 3.

Table 2. Calculating standards and models of the index system of urban water security.[a]

Index	Worst	Middle	Best	Function
Daily water consumption per capita (L)	60		200	$Y=(-1.21103)+(0.15634)X^{0.5}$
Water resource amount per capita (m^3)	50		1000	$Y=(-0.288)+(0.041)X^{0.5}$
Utilization ratio of water resources (%)	60		10	$Y=1.02857-(0.0002857)X^2$
Discharge of COD in industrial wastewater (mg/L)	1000		100	$Y=1.01010-(0.000001010)X^2$
Discharge of A-N in industrial wastewater (mg/L)		50	15	$Y=1.04945-(0.00022)X^2$
Discharge amount of domestic wastewater (10^4 m^3)	50000		15000	$Y=1.0989-(0.00000000043956)X^2$
Industrial water demand per 10,000 RMB output (m^3)	300		6	$Y=1.0004-(0.000011116)X^2$
Population density (people / square kilometers)	1300		500	$Y=1.17361-(0.0000006944)X^2$
Disaster probability (time/year)	5		0	$Y=1-(0.2)X$
Ratio of economic loss of disaster to GDP (%)	2	1	0	$Y=1-(0.5)X$
Water quality of drinking water source (%)	0		100	$Y=(0.01)X$
Public green space per capita (m^2)	0		16	$Y=(0.25)X^{0.5}$
GDP per capita (10,000 RMB)	1.3		5.3	$Y=(-0.98122)+(0.86059)X^{0.5}$
Engel coefficient (%)	60		30	$Y=1.33333-(0.00037)X^2$
Urbanization rate (%)	0		85	$Y=(0.10847)X^{0.5}$
Surface water resources (10^8 m^3)	5		100	$Y=(-0.28801)+(0.128801)X^{0.5}$
Green coverage rate (%)		30	60	$Y=(-0.786)+(0.231)X^{0.5}$
City sewage treatment rate (%)	0		100	$Y=(0.01)X$
Ratio of investment in treatment of industrial wastewater to GDP (%)	0		0.2	$Y=(2.2361)X^{0.5}$
Discharge attainment rate of industrial waste water (%)	0		100	$Y=(0.01)X$
Reuse ratio of industrial water (%)	0		100	$Y=(0.01)X$
Per capita net income of rural residents (RMB)	0		8000	$Y=(0.01118)X^{0.5}$
Disaster prevention and emergency response capability	0		1	Y=X
Water saving consciousness of citizens	0		1	Y=X
Laws, regulations and management in water environment	0		1	Y=X

[a] The index range refer to relevant industrial standards and index values of developed cities.

Table 3. Levels of urban water security.

Water security level	Very dangerous	Dangerous	Generally safe	Safe	Very safe
Range of comprehensive index	[0, 0.2)	[0.2, 0.5)	[0.5, 0.7)	[0.7, 0.9)	[0.9, 1]

- Level I: the water security state is very dangerous. With the full deterioration of water resources, as well as the water shortage, drought, flood and severe water pollution, the water environment function is disabled. The social and economic sustainability of the city is seriously hindered. Thus, the first level and most important alert would be suggested for the government.
- Level II: the water security state is dangerous. With the influence of water shortage, serious water pollution and degradation of water environment function, the water resources and water environment system could threaten the social and economic sustainability of the city. Thus, the second level alert would be suggested for the government.
- Level III: the water security state is generally safe. The water resources and water environment system are able to coordinate with the city development, and the satisfaction degree of urban social and economic sustainability is ordinary. The water supply is maintained at the critical state and the water environment function still has the certain recoverability under the water pollution. Thus, strengthening the management of water security would be suggested for the government.
- Level IV: the water security state is safe. The water resources and water environment system are able to coordinate with the city development, and the satisfaction degree of urban social and economic sustainability is moderate. The water amount could meet the demand and the water is slightly polluted. The water environment function has the sufficient recoverability. Thus, maintaining the current management of water security would be suggested for the government.
- Level V: the water security state is very safe. The water resources and water environment system are able to coordinate efficiently with the city development, and the satisfaction degree of urban social and economic sustainability is high. The water amount is abundant and the water quality is good. The water environment function has the strong recoverability and is not susceptible to external pollutions. Thus, maintaining the current management of water security would be suggested for the government.

5.2. Water security assessment

The score of each index y_{ij} can be obtained by substituting the index value into the corresponding calculation function, as illustrated in table 4.

With the index weights (table 1) and scores (table 4), the comprehensive value of water security could be obtained as follows:

$$S_j = \sum_{i=1}^{n} w_i y_{ij} \tag{1}$$

where S_j is the comprehensive value of water security, w_i is the index weight and y_{ij} is the index score after the standardization of the index datum. Thus, the comprehensive value of water security of Chengdu could be obtained from 2005 to 2013, as listed in table 5. The results show that the water security states of Chengdu from 2005 to 2013 reach the safe level IV, in which the water amount could meet the demand and the water environment function has the sufficient recoverability with slight pollution.

6. Prediction of urban water security

To predict the future water security state of Chengdu, the index data of GDP per capita from 2008 to 2013 are analyzed and the fitting function is obtained with the correlation coefficient of 0.984. Thus, the calculated vale of GDP per capita in 2014 is 82,900 RMB, which is close to the real value of 83,064 RMB. Meanwhile, the predicted value of GDP per capita in 2015, 2016 and 2017 are 88,300, 93,200 and 97,700 RMB, respectively.

Table 4. Scores of each index of urban water security from 2005 to 2013 in Chengdu.[b]

Index	2005	2006	2007	2008	2009	2010	2011	2012	2013
Daily water consumption per capita	1.000	1.000	1.000	0.678	0.697	0.926	0.821	0.875	1
Water resource amount per capita	0.801	0.676	0.718	0.875	0.679	0.911	0.942	0.796	0.911
Utilization ratio of water resources	0.000	0.000	0.000	0.010	0.000	0.000	0.084	0.000	0.220
Discharge amount of specific pollutants in industrial wastewater	0.868	0.829	0.825	0.788	0.955	0.429	0.673	0.871	0.774
Discharge amount of domestic wastewater	0.996	1.000	0.991	0.509	0.519	0.000	0.000	0.000	0.000
Industrial water demand per 10,000 RMB output	0.786	0.851	0.903	0.912	0.957	0.963	0.975	0.983	0.991
Population density	0.644	0.623	0.614	0.582	0.568	0.550	0.534	0.523	0.507
Disaster probability	0.000	0.800	0.800	0.400	0.600	0.800	0.400	0.800	0.400
Ratio of economic loss of disaster to GDP	0.966	0.924	0.999	0.982	0.990	0.996	0.985	0.992	0.989
Water quality of drinking water source	0.997	0.999	0.977	0.977	0.992	1.000	1.000	1.000	1.000
Public green space per capita	0.740	0.759	0.818	0.815	0.824	0.909	0.917	0.924	0.917
GDP per capita	0.298	0.382	0.420	0.532	0.729	0.910	1.000	1.000	1.000
Engel coefficient	0.872	0.908	0.771	0.816	0.827	0.827	0.827	0.870	0.877
Urbanization rate	0.840	0.858	0.858	0.861	0.874	0.878	0.888	0.897	0.904
Surface water resources	0.811	0.698	0.758	0.929	0.729	0.972	1.000	0.873	1.000
Green coverage rate	0.547	0.609	0.629	0.638	0.657	0.665	0.660	0.664	0.678
City sewage treatment rate	0.798	0.793	0.822	0.795	0.885	0.907	0.967	0.951	0.917
Ratio of investment in treatment of industrial wastewater to GDP	0.548	0.625	0.632	0.707	0.592	0.866	0.583	0.592	0.600
Discharge attainment rate of industrial waste water	0.962	0.842	0.964	0.987	0.997	0.998	1.000	1.000	1.000
Reuse ratio of industrial water	0.813	0.822	0.837	0.874	0.880	0.880	0.876	0.823	0.897
Per capita net income of rural residents	0.749	0.783	0.824	0.900	0.944	1.000	1.000	1.000	1.000
Disaster prevention and emergency response capability	0.600	0.630	0.680	0.700	0.750	0.800	0.820	0.850	0.880

Index	2005	2006	2007	2008	2009	2010	2011	2012	2013
Water saving consciousness of citizens	0.600	0.630	0.650	0.680	0.700	0.720	0.750	0.760	0.780
Laws, regulations and management in water environment	0.700	0.800	0.850	0.870	0.900	0.910	0.920	0.940	0.950

[b] The data refer to the water resources bulletin and statistical yearbook of Chengdu [15, 16].

Table 5. Comprehensive value of water security from 2005 to 2013 in Chengdu.

Time	2005	2006	2007	2008	2009	2010	2011	2012	2013
Comprehensive value	0.7257	0.7348	0.7622	0.7672	0.7755	0.7794	0.7871	0.7969	0.7995

Through the correlation analysis between the index of GDP per capita and comprehensive value of water security from 2008 to 2013, the corresponding exponential function could be fitted as $y = 0.74e^{0.01x} - 10.21e^{-2.53x}$ with the correlation coefficient of 0.953. Thus, the comprehensive value of water security from 2014 to 2017 could be predicted, as shown in table 6.

Table 6. Predicted values of water security in Chengdu.

Time	2014	2015	2016	2017
Comprehensive value	0.8040	0.8083	0.8122	0.8159

The results show that the comprehensive value of water security is 0.8040 in 2014 and the predicted values are 0.8083, 0.8122 and 0.8159 in 2015, 2016 and 2017, respectively. The water security state is safe, and the comprehensive values show a steadily rising trend, as illustrated in figure 2. Thus, the water resources and water environment system are able to coordinate with the city development, and the satisfaction degree of urban social and economic sustainability is moderate.

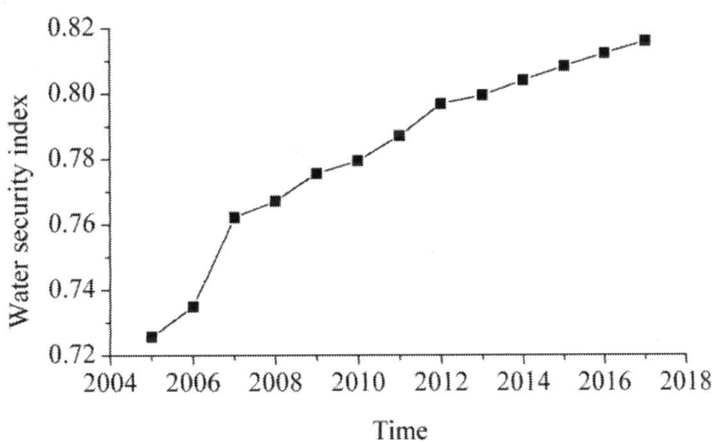

Figure 2. Comprehensive values of water security in Chengdu.

7. Conclusion

Based on the improved PSR model and AHP method, the urban water security of Chengdu is evaluated. The water security states are safe from 2005 to 2013, which is consistent with the actual situation. However, the index values of public green space per capita and ratio of investment in treatment of industrial wastewater to GDP are slightly low, which need more improvements. With the urban development, the index value of discharge amount of domestic wastewater is bad and the sewage treatments need to be improved. Through the correlation analysis, the water security states are predicted to be safe from 2014 to 2017 with a steadily rising trend. Thus, the overall water security state of Chengdu is safe and optimistic.

With the development of the city, the scale of society, economy and urbanization need to be planned in the existing bearing capacity of the water environment, and the water resources should be regarded as an important strategic resource. To ensure the coordinated development among society, economy, and environment, several suggestions are proposed. The water-saving consciousness of the public and the water-saving city could be improved through increasing the water-saving facilities and techniques. The environmental investment could be increased to improve the treatment rate of municipal and industrial wastewater. The management and warning system could be improved to strengthen the ability of coping with the accidental problems relating to the water environment and water resources.

Acknowledgments

The authors wish to thank Professor Hu Chunhong and Li Kefeng for their helpful discussions and suggestions to improve the presentation of this paper. The authors are also grateful to the editor and the reviewer for their helpful comments.

References

[1] Shi Z T and Liu X Y 2008 Concept and connotation of urban water security *China Hydrol.* **28** 24-27

[2] Jin C L and Gong L 2009 On the urban water security assessment based on the Pressure-State -Response model *Saf. Env.* **9** 104-8

[3] Jia S F, Zhang J Y and Zhang S F 2002 Regional water resources stress and water resources security appraisement indicators *Prog. Geogr.* **21** 538-45

[4] Han Y P and Ruan B Q 2003 Research on evaluation index system of water safety *Acta Scien. Circum.* **23** 267-72

[5] Cheng J G, Yang X L, Wei C J and Zhao W 2004 Discussion on water security *China Water Resour* 21-23

[6] Tong F and Dong Z C 2009 Water security scenario analysis: methodology and a case study *Proc. of 16th IAHR-APD Congress and 3rd Symp. of IAHR-ISHS (Nanjing)* vol **1** (Berlin: German Springer) pp 307-13

[7] Zhang G L, Wang L B and Ding S G 2007 A study of water security assessment of Jinan city based on analytic hierarchy process *J. of Shandong Normal Univ.*(Nat. Sci. Ed.*)* **22** 94-99

[8] Chen S J 2006 *Evaluation Forecasting and Regulation Research in the Water Safety System* (Beijing: China Water Power Press)

[9] Bai X and Tang J 2010 Ecological security assessment of Tianjin by PSR model *Proc. Env. Sci.* **2** 881-7

[10] Mai S Z, Xu S J and Pan Y J 2005 Application of the PSR model to the evaluation of wetland ecosystem health *Trop. Geogr.* **25** 317-21

[11] Peng J Q 2010 Evaluation for water environment security of urban lakes *Env. Prot. Sci.***36** 62-4

[12] Wei S X and Zhou X Z 1998 *Multiple-attribute decision making-theory, methods and applications in C3I Systems* (Beijing: National Defense Industry Press) pp 155-70

[13] Zhou H Z 1999 *Microeconomics* (Shanghai: Shanghai People's Publishing House) pp 110-20

[14] Huang Y, Liu X Y, Shi Z T, Fang S D and Bai S G 2009 Study on assessment methods of evaluate indexes for complex system: taking urban water safety as case *J. China Hydrol.* **29** 45-48

[15] *Chengdu Water Resources Bulletin* 2005-2013 (Chengdu: Chengdu Water Conservancy Bureau)

[16] *Statistical Yearbook of Chengdu* 2005-2013 (Beijing: China Statistics Press)

The protection and development of historic Lougang water resource features along Lake Tai

J Deng[1, 2, 3], X M Tan[1] , Y P Li[1, 2], J H Wan[1, 2], J G Liu[1, 2] and B Zhou[1, 2]

[1]Department of Water History Research, China Institute of Water Resources and Hydropower Research, Beijing 100048, China
[2] Research Center on Flood and Drought Disaster Reduction of the Ministry of Water Resources, Beijing 100048, China

E-mail: dengj@iwhr.com

Abstract. Overcoming the conflicts between protecting water resource heritage features and the sustainable development of modern society is a major undertaking. Lougangs (rivulets) and polders are Lake-Tai-Basin-specific water resource engineering projects that hold a significant historic position. This paper presents background on water resource heritage features, based on the origins and evolution of Lougangs along Lake Tai. Through analysis of their value and current standing, we propose countermeasures for protection and development of the Lougang water resource heritage features along Lake Tai.

1. Introduction

The water resource sector has provided increasing attention to the cultural and scientific values of ancient water resource engineering projects since the designation of the Dujiangyan Irrigation System as China's first water conservancy world cultural heritage feature in 2000. Over recent decades, hundreds of water resource engineering projects have been the focus of national, provincial, municipal and county-level cultural relic protection organizations. During the last ten years, the West Lake, Hani Terrace, Grand Canal, and other water resource projects have been placed on the UNESCO World Heritage List or the UNESCO World Heritage Tentative List.

In order to learn more about ancient Chinese water resource engineering and heritage features, the Ministry of Water Resources determined plans for a general survey of such projects during China's first general survey of water resource heritage features which took place from 2010 to 2012. According to the survey results, more than 600 ancient water resource engineering projects of high cultural value maintained function, benefiting communities. Included among first-line water resource systems are many historic relics that showcase ancient hydrological measuring, water rights identification, and memorials. These water resource heritage features are evidence of the advanced nature of the ancient Chinese civilization and are an important part of China's cultural heritage [1]. As a natural cultural strategy and a water resource management plan are undertaken, issues related to cultural heritage protection and water resource infrastructure construction will increasingly converge and become more closely associated with urban and rural land management, transportation concerns, and historic preservation. Conflicts may occur between the demands of protection and those of development.

"Lougang" is a local term that refers to the densely-distributed and continuous rivulets that discharge flood waters into the West Tiaoxi River and the East Tiaoxi River. Lougangs and polders are water resource engineering projects found in the Lake-Tai-Basin, and which are primarily distributed in the eastern, northern, and southern regions of Lake Tai and more sparsely distributed in the north and northeast parts of the lake (figure 1).

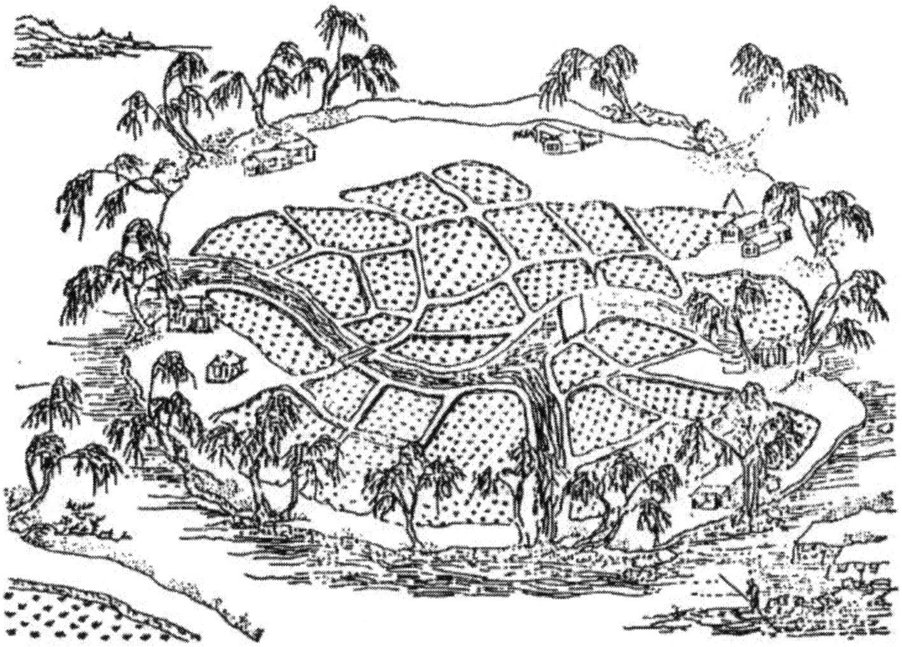

Figure 1. Schematic diagram of Lougang and Dike Paddy Field [2].

The water system of the Lake Tai Basin is centered around Lake Tai, and is divided into the upper and lower reaches. The primary sources of the upper reach are the southern Tiaoxi River water system and the western Jingxi River water system. The lower reach is comprised of three rivers: the Wusong River, the Dong River, and the Lou River, which sluice water out to the sea from the northeast, east and southeast, respectively. Both the Tiaoxi River water system and the Jingxi River water system feature hydrologic characteristics of mountain streams, namely torrents with water sources located short distances away. The Lake Tai Basin, in which water levels are generally higher than in fields, is quite vulnerable to flood disasters as a result of its special terrain, which creates large catchments and high levels of ground water. These specific conditions make it imperative to construct and regularly maintain water conservancy, irrigation, and drainage facilities in order to facilitate agricultural harvests. In order to accomplish this, ancient people adopted the unique measures of "latitudinal Tangs" and "longitudinal Lous" in the lower reaches of the Tiaoxi River and Jingxi River to create riptide cushions. These tactics were able to eliminate drought and flood disasters in order to make full use of water resources. The transverse rivulets located on the eastern, southern, and western sides of Lake Tai are known as "Latitudinal Tangs". The small longitudinal ditches which drain into Lake Tai are referred to as "Lou," "Gang," "Pu," or "Du," (described in further detail below) and are collectively referred to as "Lougangs."

Prior to the formation of the Lake Tai embankments, no distinct boundaries existed between the eastern and southern parts of Lake Tai. The embankments in the South Lake Tai began to take shape during the spring and autumn periods (771-476 BC), and ultimately stretched to the Wujiangtang Road during the tenth century AD. In order to meet the many demands of regional flood control which include drainage, irrigation, and shipping, the lakeshore bottomlands and plains have undergone continuous expansion. Over time, the Lougangs along Lake Tai have extended and moved closer to one another. Presently, Lougangs are distributed at intervals of 500-840 meters. Excavation and maintenance of the Lougangs, as well as land reclamation, farming, and sericulture, stimulate one another for common development, thus forming relatively independent mulberry embankments which exhibit independent irrigation systems and agricultural production systems. The Lougang system which came into existence near the mid-Tang Dynasty (approximately the 8th century AD), was solidified during the Wu and Yue States from the Five Dynasties (10th century AD), and have relatively declined since the Ming and Qing Dynasties (12th-20th century AD)[3].

Huzhou City in Zhejiang Province is home to the oldest, most complete, and ideal Lougangs, which criss-cross the pond and polder system in the Lake Tai Basin. This paper uses the Lougangs along Lake Tai in Huzhou City as an example system, and examines the water resource heritage features of the Lougangs along Lake Tai. The status quo and existing issues in preservation are presented, as well as their scientific value. By taking into consideration the sustainable development of water conservancy and the need to improve both the ecology and the quality of human life, this paper seeks optimized solutions to the balance of utilization, protection of cultural heritage, and reasonable scientific protection of water resource heritage features.

2. Primary content of Lougang water resource heritage features along Lake Tai

In the Lake Tai Basin, beneficial interactions exist among the Lougangs, Latitudinal Tangs (transverse rivulets) and embankments, mulberry fields, fish ponds, and criss-cross ponds, which create a unique ecological system of connected rivers and lakes. The multiple interactions between dredging, manure storage, irrigation, and farming form a unique cultural and natural environment. Lougangs, which are a cultural heritage feature specific to the Lake Tai Basin, have existed over two thousand years of regional political, economic, and cultural history.

The development of Lougangs was synchronous with the construction of the Lake Tai embankments, which have since served as a foundation of regional environmental improvement and agricultural development. Lougangs have been relevant to two thousand years of water control and agricultural development in the Lake Tai Basin, the existence of which has spanned the evolutionary processes of human life and the natural history of the Lake Tai Basin circle. The Lougangs are an important part of China's water resource heritage. Lougangs and polders are specific to the Tai Lake water resource engineering projects, and have played a crucial role in regional agricultural economy.

The Lougangs and polders were developed in the wake of the construction of Latitudinal Tangs on the shores of Lake Tai. Latitudinal Tang construction resulted in the emergence of relatively independent polder areas and irrigation ditch systems. Under the Wu-Yue Hegemony in the late Warring States period (475-221 BC), military settlements facilitated the formation of the canal system, in addition to the development of polders. By the tenth century AD, an engineering system for the Lougangs along Lake Tai was completely established, and which laid the foundation for regional agricultural development. Throughout the Yuan, Ming, and Qing Dynasties, the Lake Tai Basin became China's primary area of grain and textile production, and was a major place of origin for tribute grain sent to Beijing and armies through water transport. From the 13th century AD onward, it also served as a major economic center of South China.

The Lougang water resource heritage features along Lake Tai consist of four aspects: the engineering of the Lake Tai embankment, the Lougang and Latitudinal Tang system, the polder and canal system, and other related heritage features (ancient bridges, embankments, and polders surrounded by mulberry embankments). These features are described in table 1 and figure 2.

Table 1. Composition of Lougang water resources along Lake Tai.

Category		Heritage feature names
Lake Tai embankment engineering projects		
	Latitudinal Tangs	Di Tang (Changhushen Line Waterway), North Latitudinal Tang（Ancient Grain Transport River、South Latitudinal Tang（Litang River、Medium Latitudinal Tang（Zhongtang River, Daishan Gang.
Lougang and Latitudinal Tang system	Lougangs	73 Lougangs, including Daqian Gang, Xiaomei Gang, Xintang Gang–Changdou Gang（former Zhangpo Gang, Hanghuxi Waterway)、Luo Lou（Yijiayang Gang to the South), Huan Lou（Huanshen Waterway), Tang Lou（Lake Taijia Project, Pu Lou（Yacun Gang and Xingyao Tang to the South、Baimi Tang（Dongzong Line Waterway. A total of 30 Lougangs including Hu Lou are priority protection targets.
	Criss-cross pond	Dadang Criss-cross Pond, Songxi Criss-cross Pond, Qingdun Criss-cross Pond, Lujia Criss-cross Pond, Changtian Criss-cross Pond, Xishan Criss-cross Pond, etc.
	Control engineering projects	Openings, culvert gates, sluice gates, revetments, river ports.
	Polders	Yigao Polder, Dalou Polder, Dongqiao Polder, Xulou Polder, etc.
Lougang and polder system	Irrigation and drainage canal system	Lateral canals, sublateral canals, ditches.
	Control engineering projects	Sluice gates, culverts.
	Ancient bridges	Shuangjia Bridge（Qiao Lou、Shangyi Bridge（Yigao Lou、Xiangwang Bridge（Tang Lou）, etc.
Other related heritage features		
	Temples	Tianhou Niangniang Temple in Daqian Village, General Xu

		Temple in Yangduqiao Village, etc.
	Worship activities	Sanguan Temple Fair, the release of water lanterns, a rite to the rain and sun gods, dragon boat races, 36 Lou songs (performed by farmers and boatmen to combat drought and to drain waterlogged fields), Jin Lou Barn Lantern, Yang Lou Dragon Lantern, etc.
	Inscriptions and historical classical works	

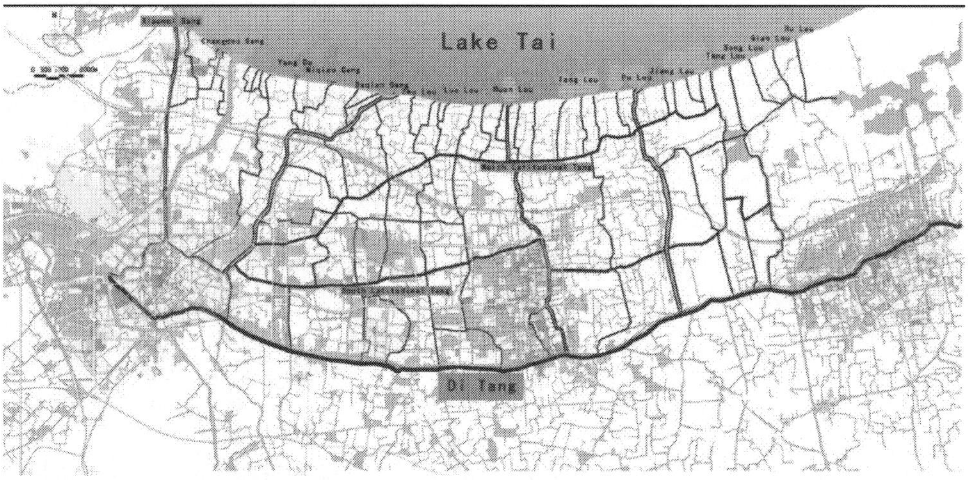

Figure 2. Distribution of water resource heritage features of Lougangs along Lake Tai.

2.1. Lake Tai embankment engineering

Construction of and improvements to the Lake Tai embankment system were basic conditions for formation of the Lougang water network and polder construction (figure 3). The Lake Tai embankment in Huzhou City is approximately 65 km long, and has contributed to fundamental change in the regional water environment of the Lougangs and polders. Seasonal flooding of the tidal flats dries out the nearby land, which has created the basic conditions required for the development of irrigation agriculture, the regional economy, and local culture.

Figure 3. Lake Tai embankment constructed during the 14th century.

2.2. Lougang and Latitudinal Tang system

The Lougangs and Latitudinal Tangs of historic importance include three Latitudinal Tangs and 73 Lougangs in the region (figure 4). The Lougang and Latitudinal Tang system, which uses the Latitudinal Tangs as latitude markers and the Lougangs as longitude markers, is the principal component of this heritage feature. There are also 16 ponds which criss-cross the longitudinal Lougangs, Latitudinal Tangs, and sluices.

Figure 4. The Zhu Lou Lougang.

Criss-cross ponds are large areas of water located between Lougangs and Latitudinal Tangs with approximate areas of 200-600 thousand square meters. Examples of large criss-cross ponds include Shengjia Criss-cross Pond, Dadang Criss-cross Pond, Songxi Criss-cross Pond, Qingdun Criss-cross Pond, Lujia Criss-cross Pond, Changtian Criss-cross Pond, and Xishan Criss-cross Pond. These ponds represent critical water tanks and wetlands along Lake Tai.

2.3. Lougang and polder system

The size of the Lougangs and polders generally ranges from dozens to thousands of acres. A large-scale polder merge occurred during the years 1950-1990, and there is evidence of recent construction of small, medium, and modern polders. As a result of these events, on 75 polders (with total areas exceeding ten thousand acres) remain, which accounts for 4.48% of the total number of polders. There are a total of 1181 polders with areas of less than one thousand acres; these account for 70.46% of the total number of polders. All polders encompass complete canals within the fields, inner and outer harbors, and pumping stations.

2.4. Other related heritage features

Other related water resource heritage features include ancient bridges on the Lougangs, water deity temples in the vicinity of the Lougang openings, and water-related worship activities.

3. Analysis of the value of the Lougang water resource heritage features along Lake Tai

Huzhou City is the only area in the Lake Tai Basin with complete and ideal extant Lougangs which are crisscrossed by pond and polder systems. This is the area in which the system originated, and contains the most distinctive characteristics of any system within the Lake Tai Basin. Lougangs and polders combine to form a water resource engineering system designed to respond to the characteristics of the low-lying topography, dense river network, and the conditions of the water and soil resources along the southern shore of Lake Tai. This system was created by ancient indigenous peoples in pursuit of social and economic development who sought to combat natural disasters. This system serves as both a model of China's traditional water conservancy and as an outstanding representative of harmony between humans and water during periods of ecological, ecomonic, transportation, cultural and social development. This system has played an extremely important role in Chinese civilization and water conservation history.

The Lougangs along Lake Tai are extremely valuable to science and technology. Their features include scientific sites, rational layouts, complete engineering systems, water storage and discharge through gate dams and sluice gates, and comprehensive utilization of water resources. Furthermore, the water heritage system of Lake Tai reflects more than a millennium of development and historical-cultural characteristics including profound cultural values; systems, mechanisms, laws and regulations related to water management; ancient bridges; records; inscriptions; poetry; irrigation festivals; and folk customs. In addition, as the system evolved, it created a variety of colorful ecological environment landscapes, which highlight the complex relationship between water and land and the harmony between humans and water.

3.1. Lougangs (Tangs and Pus) and polders created the farming civilization of Lake Tai

Within the South Lake Tai Basin, the most integrated extant farming and water resource system comprised of Lougangs (Tangs and Pus) and polders dates back to the spring and autumn periods. Based on three millennia of history, this system is an important foundation for Tangs, Pus and polders, in addition to the fish ponds surrounded bymulberry embankments in the Lake Tai Basin. The system not only represents the highest level of water resource engineering technology in the agricultural civilization era, but also created the farming civilization of the Lake Tai Basin known as the "world's granary." The system was a crucial means of transport which promoted the area's reputation as "Wu-Yue Culture," the "Land of Fish and Rice," "Silk City," and "Financial Revenue Stream," and provided a solid framework for the sustainable development of an economy and culture in a feudal society which positioned agriculture as the backbone of the country.

The Lougang and polder system in Huzhou City is endowed with an appropriate size, scientific management, and rational layout, and was incorporated into the natural water network while maintaining the integrity of the traditional pattern. The system is comparable to the Dujiangyan Irrigation System of the Zhengguo Canal in terms of uniqueness and peculiarity.

"Densely-distributed and continuous rivulets running to a lake" are referred to as "Lous." Specifically, these uninterrupted rivulets flow to Lake Tai from south to north to enable flood discharge from the West Tiaoxi River and East Tiaoxi River, and are densely distributed to the east of the Daqian Gang. "Gang" refers to a rivulet located in or to the west of Daqian Gang that is connected to lakes and rivers. "Tang" refers to an "embankment or dike, such as a pond or seawall. Tangs, which are referred to in historical records, are linked to lands with water on the left and right side." In the Lake Tai Basin, "tang" typically refers to an artificial river water system that runs in an east-west direction and joins with longitudinal rivers. Specific examples include the North Latitudinal Tang, South Latitudinal Tang, Di Tang, Shuanglin Tang, and Lianshi Tang. Pu refers to a "canal that leads to a large river." In "Water Control," Wujun Tu Jing Xu Xi states: "a longitudinal Pu is either five or seven lis (one li is a Chinese unit of length equal to 500 metres) in length, and a Latitudinal Tang is either seven or ten lis long; because embankments are built using soil from Tangs and Pus, Pus and Tangs get deeper and wider, while embankments get higher and thicker, protecting them from water hazards and guiding water to flow into rivers." In ancient times, Gangs, Pus, and Dus were rivulets and ditches that linked to rivers and lakes. Polders are "also called wei fields (wei meaning embankment); polders are fields surrounded by embankments," specifically farmlands with "water running outside of embankments and fields taking shape inside of the embankments." Centuries of manual labor created a chessboard-like water network and polder system that has gradually formed the Lake Tai Basin, featuring "longitudinal Pus (Lous) and Latitudinal Tangs" interlinked with one another [3].

3.2. *World cultural heritage value*

Lougangs and polders play an important role in water conservancy. According to Kai (2007), "There used to be polders south of the Yangtze River; in each polder, there was a big city that occupied an area as large as dozens of square lis, with waterways inside and sluice gates outside. In the event of droughts, the gates opened up to let in river water, while in the event of deluges, the gates were closed to shut out floods"[4].

The Lougang and polder system has created a unique urban and rural settlement system in the southern region of Lake Tai City which features fortification by ditches, the creation of mulberry silk fabrics, farming and fisheries, as well as a network of intertwined canals [5]. These elements have given the area south of the Yangtze River a reputation as a "land of fish and rice" and "the world's granary" and have influenced the rich cultures of the Wu and Yue feudal states.

"Polders" and "enclosed fields" have several important differences. Polders are intended to manage farmland by flood control. Water and farmland are inseparable and complement one another, as seen in traditional Chinese philosophical concepts such as "harmony between man and nature" and "adroit utilization according to circumstances." Enclosed fields can destroy the natural ecological environment by taking land away from rivers and lakes, thus resulting in damage to lakeshores, decreasedwater area, and the blockage of watercourses. The Lougang and polder system continues to function into the present day as an important part of the renowned history and culture of Huzhou City, and holds important historical and cultural heritage value. Its significance and value are comparable to that of the Dujiangyan Irrigation System.

3.3. The Lougang and polder system has witnessed the development of and changes to the northern region of Zhejiang Province

Human civilization on the south bank of Lake Tai developed along endorheic rivers and inlands. The peoples of this region were reliant on the natural topography, and included the Majiabang culture and Songze culture in the swampy areas, the Liangzhu culture along the rivers and mountains, and Gu City in the Warring States. Prior to the construction of hydraulic facilities including the Beijing-Hangzhou Grand Canal and the Lougangs and polders, the early settlers of Huzhou City attempted to construct criss-cross ponds throughout Lake Tai. These efforts artificially transformed Tingzhou Township, and gradually integrated water resource development with city management, farming, fishing, and hunting. This formed symbiotic, interdependent relationships among the cities, townships, forests and fields, that shaped the urban and rural development with the unique economic and social characteristics of Huzhou City.

In the South Lake Tai region today, the Lougang and polder system combined with water environmental governance continues to play a role in storage, drainage, farming, and irrigation, and forms an important component of lakefront eco-conservation efforts, holding considerable historical significance.

3.4. An important part feature of the water network system in the Lake Tai Basin

Ling Jiexi, a famous Qing Dynasty hydrologist and author of the Southeast Water Resources Strategy, reported that the Latitudinal Tang system on the south bank of Lake Tai covered an extensive area with Jinshuanglin Tang as the southern boundary, the lakeside as the northern boundary, Huzhou City as the western boundary, and Nande as the eastern boundary. Huzhou City was located at the center of this water system and Lake Bilang, situated to the south of the city, was a water storage hub that gathered inflows from several counties. Tang lines and Lougangs divided the flows in order to discharge the water. Water resource patterns of the entire area functioned in concert with one another. The water system of the Lougang and polder area was downstream of the East Xitiaoxi River and West Xitiaoxi River, at the end of the river which flowed into the lake. Dai Biaoyuan. a Yuan Dynasty poet, described this in his poem entitled "Huzhou:" "Tianmu Mountains are clustering, and water flows into different Gangs around Lake Tai." This is a vivid portrayal of the water network system [6].

Huzhou City is located at the junction of the East Tiaoxi River and the West Tiaoxi River. Its inflows come from the west and south of the city, and gather at the Longxi Gang. These inflows run through the eastern and northern gates of the city, down to the Lougang and polder system at the lakeside, and lead to Lake Tai. Huzhou City gathers inflows from the south of the city, and divides the inflows into discharges. These features make the ancient city of Huzhou a gateway from the Tiaoxi Rivers to the lake and acts as a water allocation hub.

3.5. Unique urban and rural water pattern: mountains integrated with water featuring urban and rural symbiosis

Water diversion in the ancient city of Huzhou formed a water city pattern found only in the southern regions of the Yangtze River. This system consisted of inner, middle, and outer rings of moats and multiple land and water gates, namely "three interlinked rings" and "six gates and six joints." It had a total of six water gates and a city pattern. The Yuliang River (known today as the People Road) forms the axis of the city. This system is not found in any of the other water cities south of the Yangtze River.

The countryside exhibits many more water features than urban areas. This is exemplified by the old saying, "you will come across a bridge every ten steps." According to the "Huzhou City Annals" compiled in 1993, Huzhou City had over 8,200 bridges in 1990, with an average density of 1.43 bridges per square kilometer; this is more than twice the number of bridges found in Venice, Italy, which is known as the "City of Water".

Throughout the urban and rural areas of Huzhou are nodes of an agricultural water conservancy ecosystem that comprises the entire Tiaoxi Watershed, separated only by a strip of water. These form part of a complete water system that has been an integral part of the socio-cultural fabric of the region from ancient times to the present. This integration of mountains, water, cities and the countryside still remains intact, as a model of traditional urban and rural integration in the southern regions of the Yangtze River.

3.6. A compound ecosystem carrying rich resource elements

Invented by the early settlers of Lake Tai, Tangs, Pus, and polders are unique creations that turn sludge into fertile soil. They also represent a major feat of systematic engineering that integrate water conservancy, economy, ecology, and culture, and played an important role in the creation of the cultures of the Wu and Yue Feudal States. They are forbearers of the unique Lougangs and polders located on the western and southern banks of Lake Tai, and are based on the original water network which included natural embankments and piers. These features are located on the southern bank of Lake Tai and feature unique designs and layouts, and demonstrate the ideals of water engineering construction and plains-area water network management. Lougangs and polders are comprehensive historical and cultural resources of the compound ecological system.

3.6.1. Spatial structure of Lougangs and polders. Lougangs and polders allowed for the creation of a farming civilization on the lakeshores of South Lake Tai. The primary elements of the cultural landscape in this region are Lous, Gangs, embankments, fields, bazaars, sluices, stone bridges, temples, pavilions, water tankers, quays, and revetments. The countryside distinctively integrates natural and artificial facilities [7].

3.6.2. Distribution of cultural relic resources. Cultural resources on the lakeshores consist mostly of rural construction and include villages, bridges, dwellings, memorial archways, and ancient ruins. There are few intact ancient villages. The geographical dispersion of the villages make them difficult to protect, which increases the importance of protecting the historical environment of the cultural relics. After the third national survey on cultural relics, a number of traditional constructions were added to the protected list. This adds to the significance of exploration and enhancement of the historical and cultural values of lakeshores. There are examples of renowned cultural heritage in the Lougangs along Lake Tai, such as the Cliff Inscriptions of Huanglong Hole, Bachelor Fei's Cemetery at Taishi Bay, Bailian Bridge, Daqian Chastity Memorial Archway, and ancient lake bridge groups.

3.6.3. Distribution of tourism resources. Rural tourism resources are divided into two categories: natural human and rural human landscapes. These resources can be divided into eight districts, according to their distributions and agglomerations along the lakeshores: Jia Pu, Sheng Jiadang, Xintang, Bianshan Mountain, Daqian, Huan Lou, Yigao, and Yangxi. The largest number of tourism resources are intensively grouped to the east of Bianshan Mountain [8].

4. Evaluation of current situations

4.1. Evaluation of current preservation efforts

Lougangs are ancient water resource engineering projects that continue to function. Those located in Huzhou City remain the regional backbone of irrigation, flood control, and flood drainage irrigation projects. In recent years, issues have resulted from urbanization including blockages of the Lougangs, excessive gardens, and intensive real estate development. These issues affect the functional performance of the Lougangs and destroy the historic landscape.

4.1.1. Lougangs. The overall Lougang system pattern still exists. The watercourse is an average of approxiamtely 2-3.5 meters wide. The northern ends of the Lougangs primarily lead to Lake Tai, or are indirectly connected to Daishan Gang or Di Tang.

The siltation in the Lougongs is gradually worsening as a result of nearly 30 years of the present market economy and the dredging up of river sludge with nets and other traditional dredging measures.

A number of the Lougangs connected to the lake exhibit signs of decline. In the 1990s, there was at total of 73 Lougongs connected to Lake Tai; by 2000, that number dropped to 47, and in 2015, only 39 Lougangs existed, of which 34 did not lead to the lake. Portions of the embankments, ancient bridges, and water deity temples located on the ancient Lougangs have been ravaged, and lake inlets for Lougangs have been blocked. This is the result of lacking water resources, limited protection for cultural relics, and insufficient mutual cooperation between urban construction and other departments that impact the Lougangs. Ancient lake floodgates associated with the Lougangs have disappeared and the ancient stone bridges of the Lougangs have been destroyed. Occasionally, the former silting sections of the lake inlets occupied by the Lougangs have also been used for residential construction.

4.1.2. Latitudinal Tangs. A majority of the Latitudinal Tangs have been blocked and subsequently lost their integrity. Three primary Latitudinal Tangs remain largely intact: Di Tang, Northern Latitudinal Tang, and Southern Latitudinal Tang.

4.1.3. Crisscross ponds. The large-scale expansion of urban construction and high–profile engineering projects of recent years have resulted in rivers and ponds filled with mud and the removal of land from waterways in order to satisfy land consolidation and transportation construction needs. This phenomenon is becoming more and more impactful, and has resulted in the weakening of pond functions such as adjustment, storage, and riptide cushions. Industrial development has also introduced water pollution problems.

4.1.4. Water resource engineering facilities. A large number of sluice gates and culverts in Huzhou City have been closed, including Pu Gang, Chen Lou, and Shen Lou; only ten functional sluice gates remain. A total of 18 gates have been disabled, but relics of the watercourses still exist. Only 21 gates retain flood control, irrigation, and transportation functions.

4.1.5. Polders. Urbanization has led to substantial reduction of polders. The traditional farming mode of ponds surrounded by mulberry embankments has disappeared. Under the traditional mode, mulberries were planted on the embankments of ponds and fish were kept in the ponds; silkworms fed on the mulberry leaves, fish fed on silkworm fece, and mulberries were cultivated in the pond mud. The embankment agricultural system is increasingly shrinking. Watercourses and other primary water resource systems have long been in a state of disrepair and subject to severe silting. Environmental pollution also exists, resulting in important historic sites now on the verge of disappearing.

4.2. Evaluation of management status

Water resource heritage features are located in two administrative districts of Huzhou City: Wuxing District and Changxing County. Water systems are under the management of departments at the provincial, municipal, and county levels. Farmland systems are supervised by competent land, resource, and agricultural departments. Heritage feature protection involves the activity of multiple departments including planning, cultural relic protection, water conservancy, and environmental protection, among others. Multi-departmental management involving administrative jurisdictions and administrative subjects can lack professional coordination at the municipal level. Without over-arching planning efforts, unified management and control, and guidance for the protection and utilization of the Lougang water resource heritage features along Lake Tai, harmonious development is difficult to promote.

4.2.1. Foundation of systems. In recent years, the value of the Lougang water resource culture heritage features has been widely recognized by both the government and the public, both of which have begun to pay increasing attention to heritage protection. Massive amounts of manpower, materials and financial resources have been contributed to protection of the heritage features, and considerable success has been achieved. The social and economic benefits of the Lougang water resource heritage feature protection are remarkably increasing.

4.2.2. Major issues.

Lougangs are the purview of the departments of water conservancy, transportation, municipal administration, and cultural relics. The administrative system must be better organized, as it currently lacks targeted management measures. The protected areas are over-developed without approval of the governmental departments, and the first line of heritage management personnel lack necessary professional qualifications. The socio-cultural heritage of the Lougangs will be destroyed by new construction if proper steps for their protection are not enacted immediately.

4.3. Evaluation of current utilizations

Lougangs are ancient water resource engineering projects that continue to function into the present day. Latitudinal Tangs and longitudinal Gangs still play a functional role in flood control, flood drainage, water supply, and water conservancy. The polder irrigation and drainage system still functions. Continued work must be done in order to preserve the status and raise awareness of the historic value of the Lougangs. Care must be taken to balance preservation efforts with local development efforts. The general history and culture of Huzhou has a stronger local and international reutation than the Lougangs. Local first-line government organizations and rural communities are greatly aware of the need to protect the Lougang culture. This can be capitalized on while working toward the historic preservation of these important heritage sites.

5. Countermeasures of protection and utilization

The overall goals of protection and utilization are to:

- Scientifically protect the remarkable historical and cultural water conservancy heritage features of the region,
- Protect the unique regional agricultural irrigation system,
- Thoroughly understand the cultural influences of Lougang water resources,
- Arrange engineering and non-engineering measures for Lougang water resource heritage feature protection,
- Restore natural and landscape environments in the Lougangs and embankment area,
- Achieve coordinated development of protection and utilization, and
- Realize permanent protection for the Lougang water resource heritage features.

The features discussed below are important to the realization of these goals.

5.1. Weighing the spatial levels of point, line, and surface in Lougang heritage

It is important to protect and develop according to a point-line-surface system. In this context, point refers to the use of an opening or a control engineering project of the Lous, or an ancient bridge, as a key node; line refers to a linear engineering system, such as a watercourse or an embankment; and, surface refers to a regional polder system comprised of typical plots interspersed with points and lines. Combining point-line-surface featrues allows for the protection and utilization of the Lougang water resource heritage features at all spatial levels.

When balancing the technical, economic, social, and landscape aspects of heritage preservation, it is important to identify and distinguish between the different value characteristics of important sections. Systematic protection should be applied to relics and their background environments. Core protection, utilization, and exhibition efforts should focus on specific items, such as the openings of the Lous, criss-cross ponds, key projects, typical polders, historical bridges, cultural heritage, and ancient villages and blocks. The spatial network of the Lougang water resource engineering system should be built using Latitudinal Tangs and longitudinal Gangs as the backbone of watercourses and embankments. When these efforts are combined with the polder system located between Lougangs, the surface of the Lougang water resource heritage features along Lake Tai will begin to take shape. These steps can be undertaken to develop a systematic, complete, and explicitly-structured heritage plan for protection and utilization.

5.2. Functions come first with reasonable protection

Heritage protection and management measures should be based on sustainable development of water conservancy and water transport functions, such as flood control and irrigation. The integrity of the Lougang irrigation system should be maintained, and continued protection of ancient features should be enacted. Preservation of historical information should be maximized, with minimal intervention to the heritage features. Advanced global concepts and technical means of protection should be introduced to promote multi-disciplinary cooperation. The disciplines of water conservancy history, heritage protection, and historical geography should collaborate on in-depth research into the scientific and cultural value of the heritage features. Both engineering and non-engineering protection measures should be employed in water resource feature studies.

The public should be encouraged to participate in the protection of heritage features, and multi-objective development should be promoted and prioritized during the coordination and development of efficient agriculture, ecological agriculture, and tourism agriculture in addition to heritage protection. Furthermore, in the water environment and ecology of the Lougangs and criss-cross ponds, a comprehensive restoration should be performed, with reasonable utilization of lakeshores and planning of greenway or landscape belts that function both for recreation and for the display of the Lougangs.

Traditional construction technology and traditional materials should be used in the maintenance and repair of the Lougang water resources heritage features. Furthermore, maintenance and repair should be undertaken by knowledgeable and qualified organizations for the preservation of water resource engineering cultural relics protection. These organizations should attempt to maintain the status quo.

No restoration efforts should be undertaken at this time to restore the missing components of Lougang heritage features. These steps should only be enacted after careful assessment of the scientific, historic, and cultural value of the missing components. Rehabilitation must be implemented in consultation with experts and with the approval of the municipal government.

5.3. Adaptation to local conditions and cultural experiences

A museum devoted to the Lougang water resource heritage features should be established. At this museum, indoor exhibitions will focus on the Lougang heritage features as well as the regional culture and folk customs. There will also be two outdoor display areas for Lougang water resource heritage features, one display area for the Yigao Lou, Chen Lou, Pu Lou and Daqian Gong water management features, and another for polders or fish ponds surrounded by mulberry embankments.

Key Lougang engineering projects must undergo emergency maintenance and repair. Conscientious conservation plans should be made for the water resource heritage features of Lougangs and polders, and further exploration should be conducted into the profound culture. Emergency repairs should be conducted on a number of Lougang hydraulic buildings.

A cultural tourism brand for the Lougangs should be created. Tourism efforts should focus on the construction of greenways, bicycle races around the lakes, and other activities to draw attention to the unique features of Lake Tai [9].

6. Conclusion

Lougangs along Lake Tai are part of China's unique water conservancy heritage system. This system includes the Lake Tai embankment engineering projects, the Lougang and Tang system, the polder and canal system, and other related heritage features. Lougangs along Lake Tai are of important scientific and technological significance because of their rational layouts, complete engineering systems, use of gate dams and sluice gates for water storage and discharge, and their comprehensive utilization of water resources. Furthermore, Lougangs possess profound cultural value and systems which represent more than a millennium of historical and cultural development. This includes their mechanisms, laws, and water management regulations; ancient bridges; records; inscriptions; poetry; irrigation; festivals; and folk customs. In addition, their evolution created a variety of colorful ecological environments and landscapes which highlight the complexities of the harmony between humans, land, and water.

Lougangs are ancient water conservancy projects which still function and serve as major regional irrigation, flood control and drainage projects in Huzhou City. However, the level of awareness of Lougangs has yet to reach a level that will effectively protect these unique cultural heritage features. This paper offers countermeasures using principles, restrictions, and specific measures regarding water resource protection and utilization that aim to promote the coordinated protection and utilization of the water heritage sites and to realize the permanent protection of the Lougang water conservancy heritage features.

Acknowledgments
This work was supported by IWHR Research& Development Support Program (JZ0145B572016).

References

[1] Tan X M 2012 Elaborations on definitions, characteristics, types and values of water cultural heritages *Water Resources in China* **21**: pp 1-2
[2] Zhou K Y 2002 *Science and civilization in China* Water Engineering Science Press Beijing: 224
[3] Miao Q Y 1982 Formation and development of enclosing farmlands with embarkments in Taihu District *Agricultural History of China* **1**: p 14
[4] Guo K 2007 A study of history of enclosing farmlands with embarkments in Taihu District *Proceedings of the 7th International Conference of the East—Asian Agriculture History* p 411
[5] Zhou Q 2009 Specialized mulberry fields and their economic and ecological environment of Lake Jia Plain in in late Ming and early Qing *Journal Of Chinese Historical Geography* **4**: 27
[6] An J S 2006 Construction and changes of water landscape system in historical regions south of the Yangtze River -- Based on historical data of Jiaxing area *Journal of Chinese Historical Geography* **21**:18
[7] Wei Y P, Guo S Y and Lu W 2002 Report of urban water culture and water landscape and its policies -- a case study of Huzhou *Planners* **1**: 39
[8] Cui D S and Sun Y 2011 Research on spatial structure optimization of Huzhou tourism scenic spot system based on fractal dimension *Scientia Geographica Sinica* **3**: 341
[9] Yin Z H and Liu Q Y 2014 Study on protection of cultural heritage and tourism development of rice in Lake Tai Basin *Agricultural History of China* **5**: 125

IOP Conf. Series: Earth and Environmental Science **39** (2016) 012029 doi:10.1088/1755-1315/39/1/012029

Water level fluctuations in Guangxi near coast caused by typhoons in South China Sea

B Chen[1,4], M C Shi[2], X Y Chen[1], Y Ding[2], B X Zheng[3], D X Dong[1] and S F Qiu[1]

[1]Guangxi Key Laboratory of Costal Marine Environmental Science, Guangxi Academy of Sciences, 530007 Nanning, Guangxi, China

[2]Ocean University of China, 266003 Qingdao, Shandong, China

[3]Third Institute of Oceanography, State Oceanic Administration, 361005 Xiamen, Fujian, China

E-mail: cbgxkxy@163.com

Abstract. The increasing and decreasing water level in Guangxi near the coast is mainly caused by typhoon-induced storm surges. The maximum values of increase and decrease recorded at the Beihai Station over the last 42 years are 1.45 m and –1.87 m. The water level first decreases and then increases in Guangxi near the coast. Apart from the direct influence of typhoons, the westward coastal-trapped wave along the Guangdong continental shelf is also greatly affected. A portion of the westward coastal-trapped wave passed directly through the Qiongzhou Strait into Beibu Gulf, and the other portion passed through the southern part of Hainan Island into the Beibu Gulf. On July 2, 2001, the westward coastal-trapped wave caused by Typhoon Durian induced a strong westward flow along the Guangxi coast, and the geostrophic flow velocity on the surface reached 92cm/s. On July 6, 2001, though Typhoon Utor landed at the northern part of the Pearl River Estuary, it widely induced a 20-cm increase in the water level in Guangxi near the coast on July 8. Apart from the direct and indirect influences of typhoons on the fluctuation of the water level along the Guangxi coast, the particular topography and atmospheric gravity wave also have affect the fluctuation of the water level.

1. Introduction

Typhoons cause abnormal changes in water levels along the coast, which is usually regarded as increased or decreased water level or the storm surge. Storm surge is a typical maritime disaster. When a storm surge caused by a typhoon exceeds the normal tidal level, i.e., the astronomical tidal level, it is referred to as a storm surge increase; the alternative decrease is called a storm surge decrease. Since the 1950s, storm surge theory and its numerical simulation have made much progress due to researchers both home and abraod [1-12, 22-23].

The Guangxi shore is located in the northern part of South China Sea, and its geographical environment is unique. Typhoons landing or passing by the coastal areas of Guangxi greatly affect the coastal areas each year. Storm surges caused by typhoons demonstrate a particular pattern, which has been rarely reported. In view of this pattern, this paper studies the mechanisms which cause fluctuation of the water level in Guangxi near the coast, as well as the typical current and residual current of coastal-trapped waves.

2. Increased water level in harbors and disasters caused by typhoons landing near the Guangxi Coast from 1965 to 2012

Eachyear, storm surges caused by landing typhoons lead to severe disasters in Guangxi coastal areas. According to the previous study[13-15], about 90 typhoons have affected Guangxi coastal areas from 1965 to 2012, 18 of which led to tremendous disasters caused by direct landing (figure 1). The most notable change in water level is due to the direct landing of typhoons with an average increase of 111.2 cm, which is 2.6 times greater than the increase induced by typhoons which did not land. For example, the water level increased by 2.33 m in Tieshan Port and 1.03 m in Weizhou Island induced by Typhoon No. 7109, 2.00 m in Fangchen Port and 1.86 m in Pearl Port induced by Typhoon No. 8303, and 1.61 m in Beihai Port and 1.53 m Longmen Port induced by Typhoon No. 6509. During the landing period, the measured water level in major ports along the Guangxi coast is 5.08 m - 8.33 m, which represents 2 m higher than the normal water level[16-17], and storm surges caused by landing typhoons led to extremely heavy losses[18-20]. For example, on September 9, 1996, the Guangxi coastal areas were affected by Typhoon No. 15, which inflicted damage to 1.1148 million people including 61 deaths and 88 missing persons, the collapse of 34,700 houses, damage to 1,099 vessels with 173 vessels sank, and a direct economic loss of 2.555 billion yuan in Beihai City. Meanwhile, 20,000 houses collapsed and 300 m of seawall were destroyed in Qinzhou City. For another example, on June 26, 2001, Typhoon No. 3 Durian directly affected Beihai City, leading to a heavy rain in the city and a daily rainfall at the level of 425 mm. This heavy rain caused an increase in the water level along the coast, most of the roads in the city were destroyed, and streets and houses were drowned. Such a heavy rain over such a short time hasn't occured in a century. More than half a million people were affected and 150,000 houses were flooded, 200 m of seawall were broken in 30 places, river banks collapsed in 150 places, and 16 weirs and water gates were destroyed. The water overflowed 16,900 mu fish and shrimp ponds including the destruction of more than 950 mu of shrimp ponds, 21,920 mu of crop fields and 548 km of roads.

In recent years, many typhoons have landed in Guangxi coastal areas. For instance, Typhoon No. 1330 Haiyan, which occurred on November 11, 2013, landed in the Philippines on November 8, passed by the northern part of Beibu Gulf and entered the Guangxi Region. The wind speed reached 12-13 scale, and rainfall reached 100 - 230 mm in Guangxi coastal areas, while the rainfall reached 451 mm in Fangcheng

Port. Heavy rainfall made the water level rise by 2.23 m in major river estuaries along Guangxi coastal areas. A total of 12,000 people were affected by Typhoon Haiyan, including one death in the cities of Beihai, Qinzhou and Fangcheng Port, and direct economic losses of about ten million yuan.

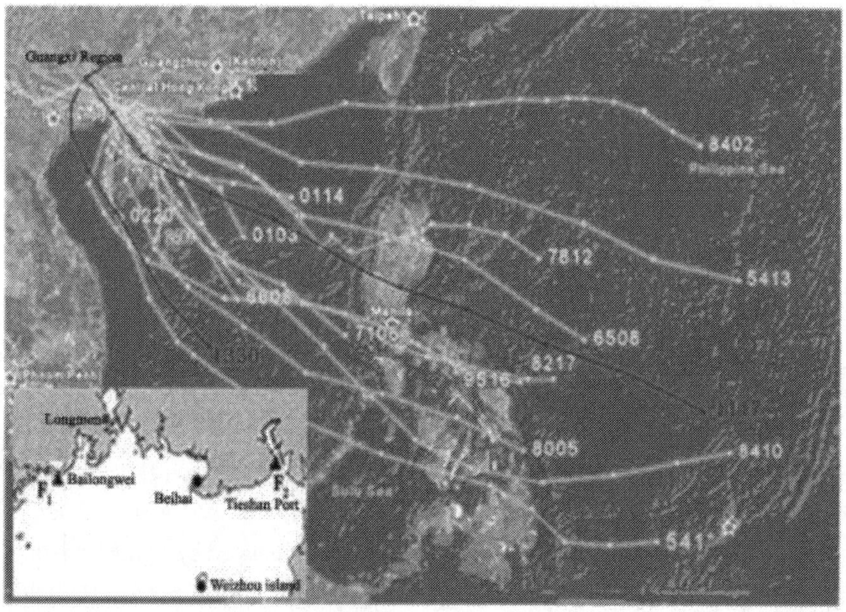

Figure 1. Roadmap of typhoons that landed in Guangxi coastal areas from 1965 to 2013.

3. Case Study of current, residual current and water level fluctuations at the Guangxi near shore

3.1. The Influence of Typhoon Nesat

On the morning of September 24, 2011, Typhoon Nesat generated on the surface of the northwestern Pacific Ocean. At 7:00 on September 27, it landed on the eastern coast of Luzon Island of the Philippines. Around 14:30 on September 29, it extended into the coastal area of Wengtian Town, Wenchang City, Hainan Province. At approximately 21:15, it landed again on Jiaowei Town, Xuwen, Guangdong Province. At 11:30 on September 30, it landed on the coast of Quang Ninh in northern Vietnam (figure 2). Influenced by Typhoon Nesat, heavy storms affected Hainan, Guangdong and Guangxi. According to the reported data, the rainfall reached 332 mm in Fangcheng Port from 20:00 of September 28 to 14:00 of September 30. The wind speed reached 11-14 scale on the southern coast of Guangxi, representing it was the strongest typhoon which ever landed on the Guangxi coast in the past six years. Affected by Nesat, the current, residual current and water level along Guangxi coast changed greatly.

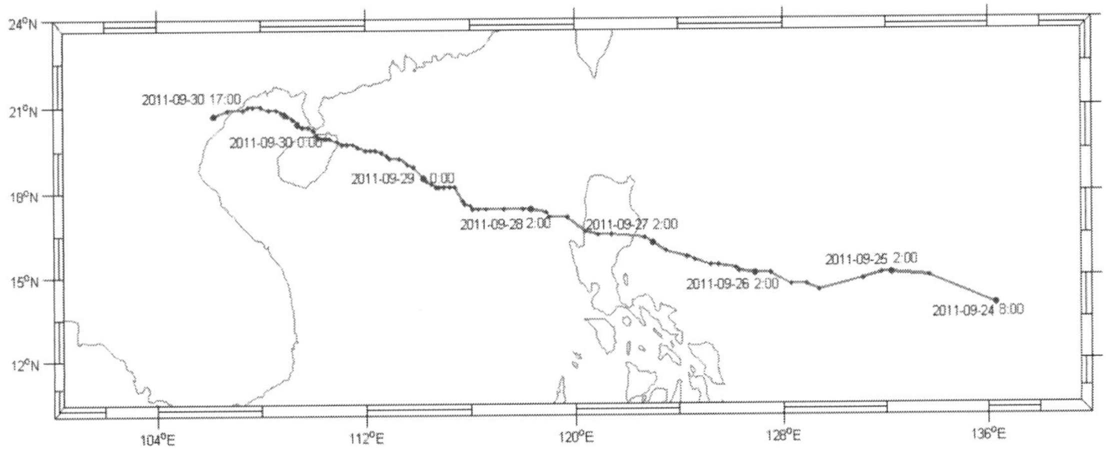

Figure 2. Path of Typhoon Nesat.

3.1.1. Changing Characteristics of current and residual current along Guangxi near shore under the influence of typhoons. During the Typhoon Nesat landing in the Guangxi Region, an AWAC acoustic Doppler current profiler implanted at a depth of 8 m near the southern Bailong Peninsula to monitor the current, wave and tide (F1 station in figure 1). The observation layer spacing of the current velocity and direction was set to 0.5 m. The profile data were stratified as surface (underwater 1 m), middle and bottom (approximately 1.5 m above the sea floor) for analysis. In the absence of typhoons, the actual velocity at the observation station is relatively low, and the maximum velocity at each layer was below 50 cm/s. The average velocity in the surface, middle and bottom layers were 10.8 cm/s, 8.6 cm/s and 8.4 cm/s, respectively. However, during Typhoon Nesat, the annual maximum velocity was 103.7 cm/s, 94.1 cm/s and 71.0 cm/s in the surface, middle and bottom layer, respectively. Additionally, outside of typhoon season, the velocity of the residual current is also relatively slow, and the maximum velocity generally falls below 20 cm/s with the average velocities in the surface, middle and bottom layer equal to 5.7 cm/s, 3.5 cm/s and 3.1 cm/s, respectively. However, the maximum velocities reached 39.7 cm/s, 32.4 cm/s and 20.7 cm/s, respectively, during the landing period of Typhoon Nesat.

3.1.2. Changing characteristics of the water level in Guangxi near shore under the influence of typhoons. On September 29, 2011, under the influence of Typhoon Nesat, the wind direction was primarily northwest along the coast of the Bailong Peninsula; the average wind speed reached 8.0 m/s in the four hours from 17:00 to 20:00 and the direction shifted to the north. At 8:00 on September 30, the typhoon center moved west and entered Vietnam, and the wind direction turned south around Bailongwei (figure 3). However, when the wind shifted to the north around Bailongwei, a decrease in the water level occurred near the observation station as a result of the wind moving away from the coast. At 2:00 on September 20, the decreased water level reached its maximum of –92 cm.

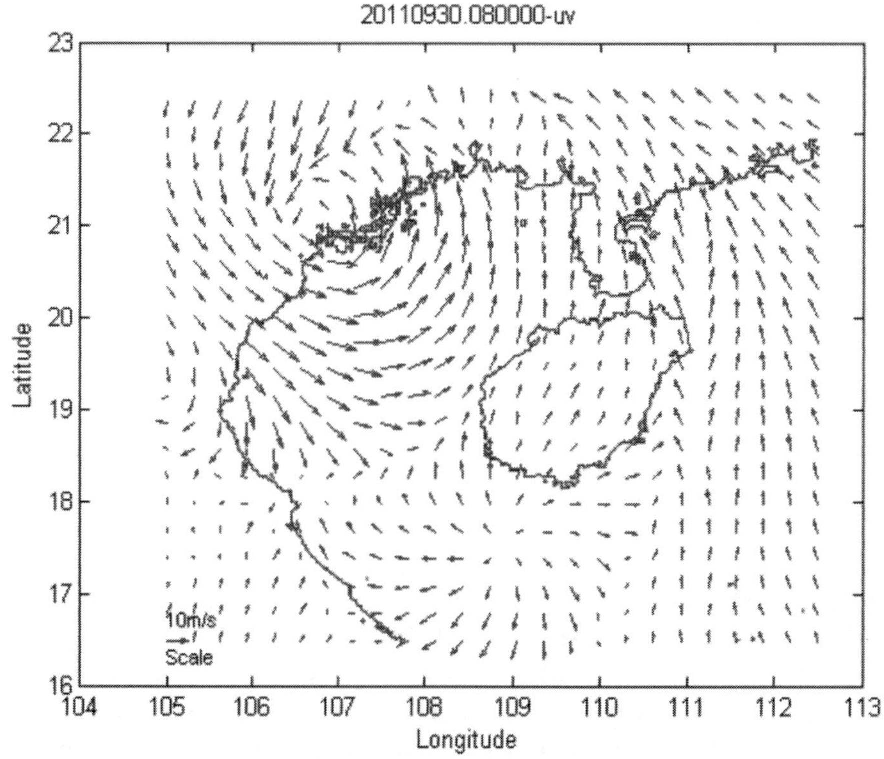

Figure 3. Wind fields at the time of the maximum increase in water level in 2011.

As the typhoon moved westward, the wind gradually turned to a southern direction, while the surface water simultaneously flowed toward the shore but was blocked by the coast. The water level therefore began to increase near the observation station. At 10:00 on September 30, water level increase reached its maximum of 65 cm (Figure 4), and then quickly decreased. Several fluctuations followed; the first water level increase was 45 cm, and the second increase was 20 cm.

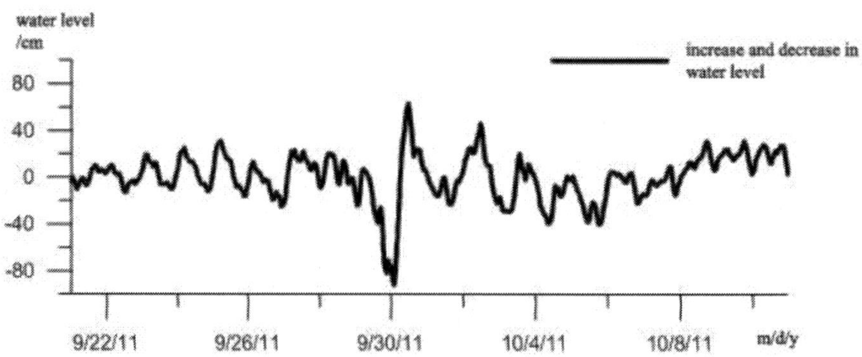

Figure 4. Increased and decreased water level near Bailongwei Observation Station (F1) during Typhoon Nesat.

3.2. Influence of Typhoon Durian

In order to further study the increased and decreased water level during the typhoon landing along the coastal areas in the Guangxi Region, the water level fluctuation data at Shitoubu (F2 station in figure 1), Tieshan Port was selected and which recorded the details of the effects of Typhoon Durian from July 1 to July 3, 2001.Tieshan Port, located on the eastern coast of the Guangxi Region, is surrounded by land in the east, west and north, while it faces the sea in the south Typhoon Durian moved from east to west and crossed over Tieshan Portat at 6:00 on July 2 (figure 5); the southern wind gradually strengthened to a maximum windspeed between 25 and 30 m/s under the influence of the strong onshore wind around the Shitoubu area .

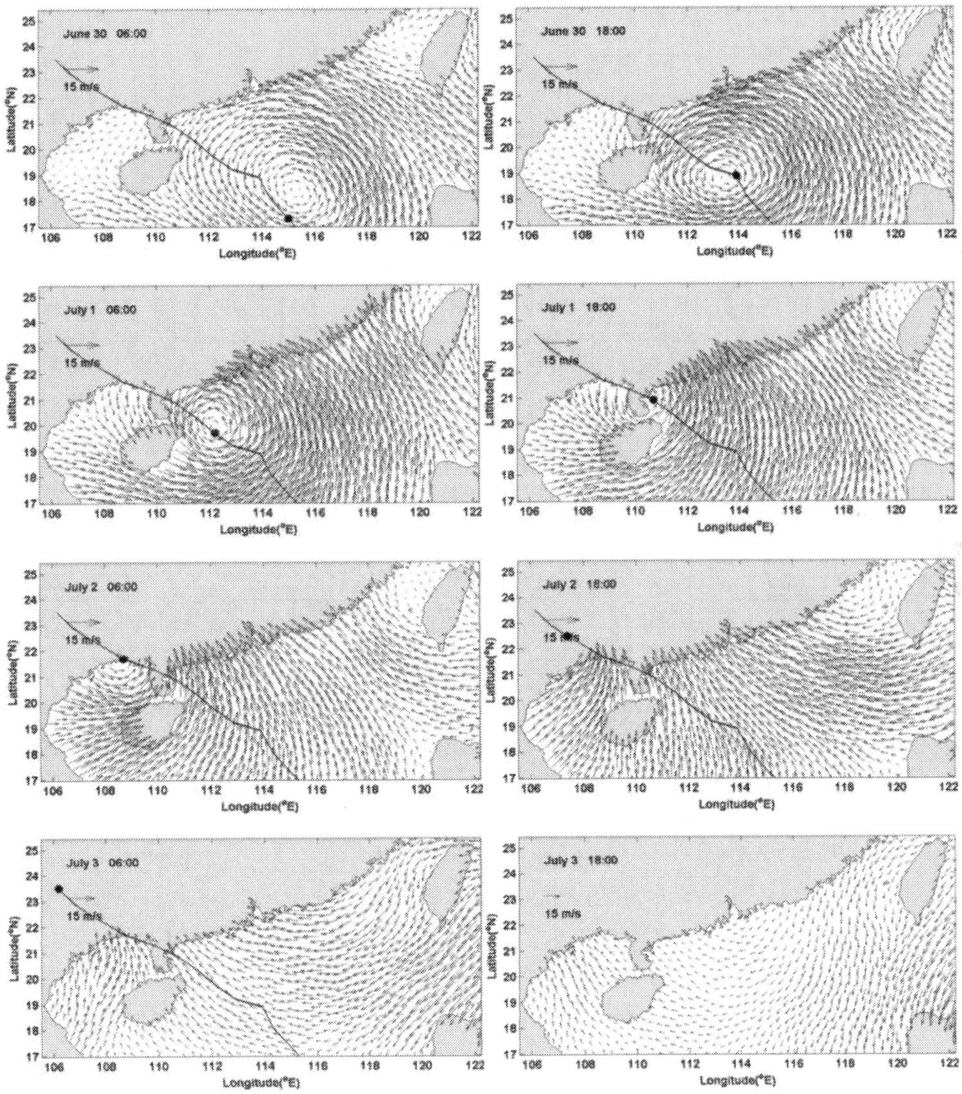

Figure 5. Path of Typhoon Durian and its induced wind fields.

While Typhoon Durian passed this area, the water level around the observation station of Shitoubu experienced a water level decrease followed by an increase. As the low pressure approached to Shitoubu from the east, the water level began to decrease around Shitoubu to a maximum decrease of–48 cm at 8:00 on July 2. At approximately 10:00, the water level returned to the origin. The south wind then increased, and the water level around Shitoubu increased (figure 6) to a maximum value of 140 cm between 14:00 and 16:00 on July 2. The water level then gradually returned to normal as the low pressure moved westward and the wind weakened. The increase/decrease in water level accompanied the typhoon, and was subject to the distribution and variations of the typhoon field and the air pressure field.

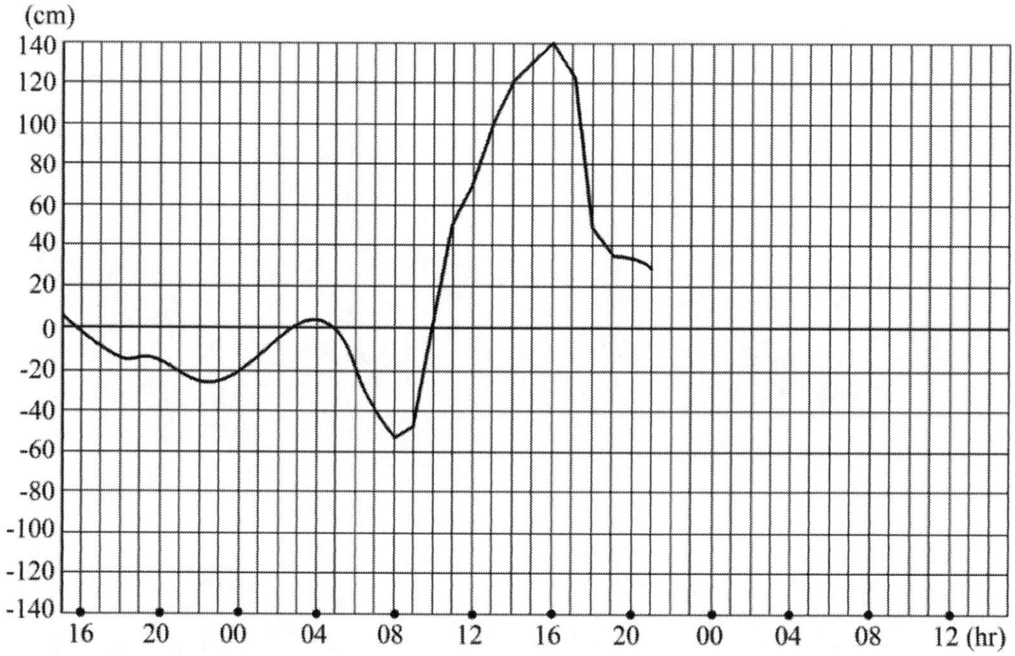

Figure 6. Increase/decrease in water level around observation stations Shitoubu and Tieshan Port during Typhoon Durian as it passed the Guangxi Region.

4. Fluctuations in water level caused by coastal-trapped waves

Great attention has recently been paid to the study of coastal-trapped waves in the northern continental shelf of the South China Sea (SCS) caused by typhoons. Ding[21] performed a quantitative analysis of the fluctuations of water level in the coastal areas of the northern part of SCS of Typhoon Durian on June 30, 2001, Typhoon Ute on July 5, 2001, and the landing of Typhoon Yutu on on Guangdong Province on July 23, 2001. They reported that the wave tends to spread from northeast to southwest along the coast and its signal was very strong under the influence of typhoon when they came. The wave magnitude gradually decreased as the typhoon vanished. The spreading wave speeds caused by Typhoon Ute along the coast of the northern part of the SCS was approximately 4.2-11.4 m/s. When this wave propagated to the east coast of Hainan Island, the spreading speed significantly reduced to approximately 4.2 m/s.

A portion of the coastal-trapped wave caused by Typhoon Ute passed across the Qiongzhou Strait and entered the Beibu Gulf (figure 7); the other portion crossed the east mouth of Qiongzhou Strait and was blocked by Hainan Island, and then continued to spread in a clockwise direction around the southern part of Hainan Island and entered the Beibu Gulf. On July 6, Typhoon Ute landed in the north part of the Pearl River estuary, but caused an increase in water level of more than 20 cm in the Guangxi near shore on July 8. This increase in water level involved the entire coast.

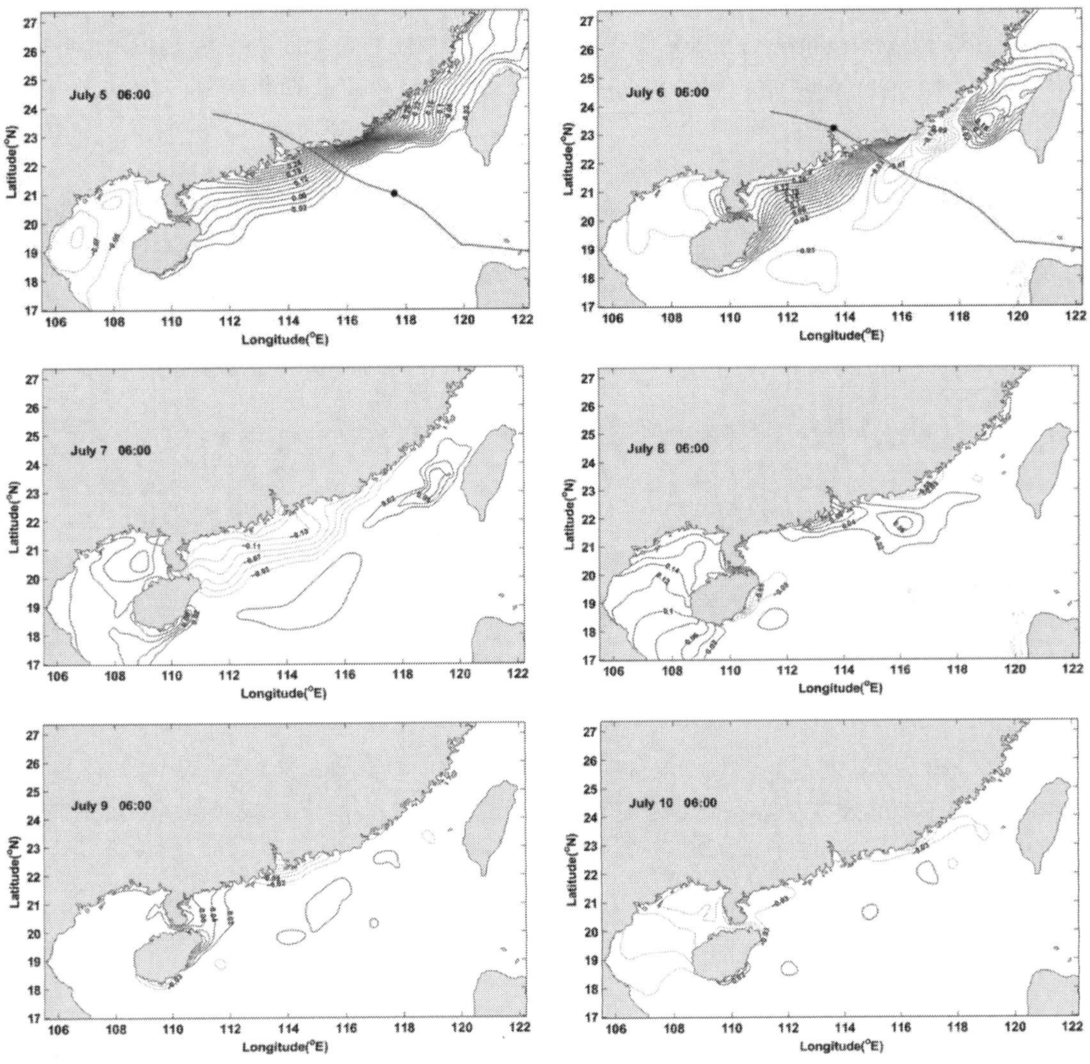

Figure 7. Contours of abnormal distributions of water level along the coastal areas in the northern part of the South China Sea during Typhoon Ute (Ding, 2012).

Typhoon Ute landed in the east of Pearl River estuary, and caused a 20-plus cm increase in the water level. Therefore, the typhoons that landed in the western part of Guangdong Province should cause coastal-trapped waves, which have more significant impact on the Guangxi near shore.

Typhoon Durian generated within the SCS, formed a tropical depression approximately 650 km south of Hong Kong in the morning of June 30, and then moved to the northwest and developed into a typhoon. At 05:00 on July 2, it landed near Zhanjiang with a maximum wind speed reaching 35 m/s. A part of the coastal-trapped wave caused by Typhoon Durian crossed the Qiongzhou Strait and entered the Beibu Gulf (figure 8), while the other portion crossed the eastern mouth of the Qiongzhou Strait and was blocked by Hainan Island, then continued spreading in a clockwise direction around Hainan Island and entered the Beibu Gulf.

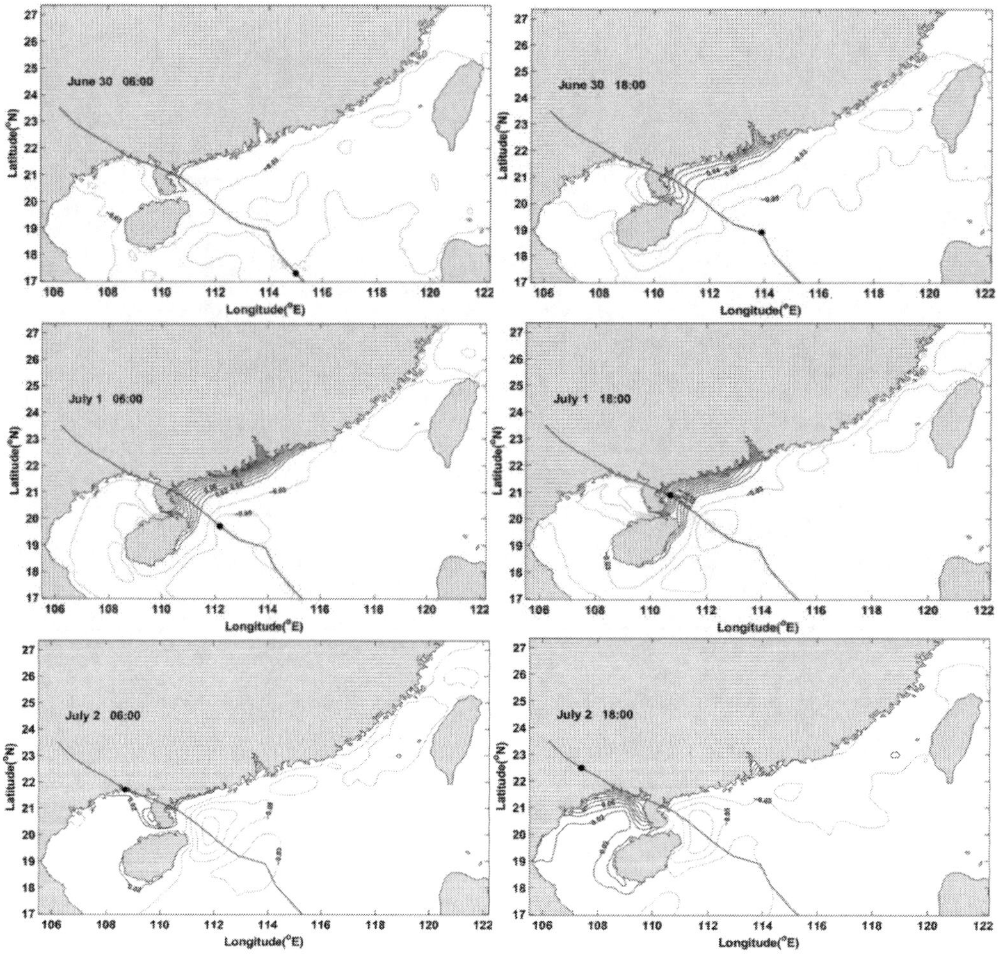

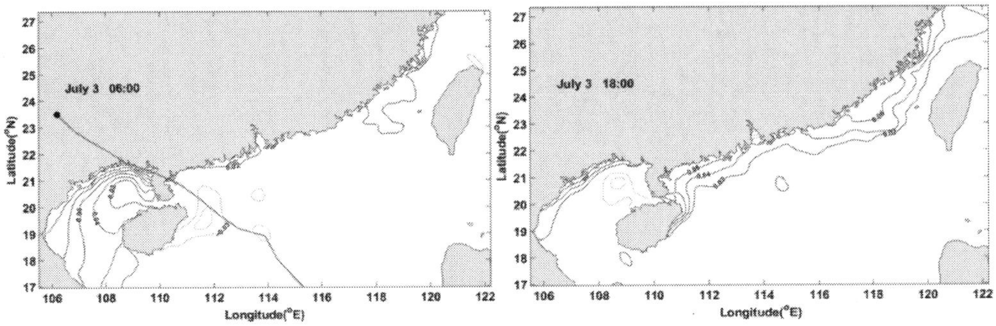

Figure 8. Contours of abnormal distributions of water level along the coastal areas in the northern part of the South China Sea during Typhoon Durian.

As shown in figure 8, a very important phenomenon can be observed at 18:00 on July 2. Not only does an increase in water level appear in the Guangxi near shore, but a strong westward flow also occurred along the Guangxi coast. Computational results indicate that its surface geostrophic flow velocity approached 92 cm/s. This remarkable westward flow along the coast would transport a huge body of water and would have an invaluable role in the diffusion of pollutants, water refreshment, and environmental protection in Guangxi coastal areas. The impacts of coastal-trapped waves on the water level and flow in the Guangxi near shore have not yet been comprehensively investigated, and represent a direction for future research..

5. Calculation of water level fluctuation in the Guangxi near shore

5.1. Increase of Water Level

The Beihai Observation Station was the most representative among the observation stations in Guangxi coastal areas because it reported long-term tidal data. Thus, the data observed at the Beihai Observation Station over 42 years from 1965 to 2006 was analyzed, and demonstrated the extreme value of water level increases each year, as shown in figure 9. Results indicate that the largest water level increase occurred in 1996 with an increase of 1.45 m, but there were several years without water level increases, i.e., 1969, 1970, 1972, 1975 and 2004. The average of water level increase was 0.71 m over 42 years.

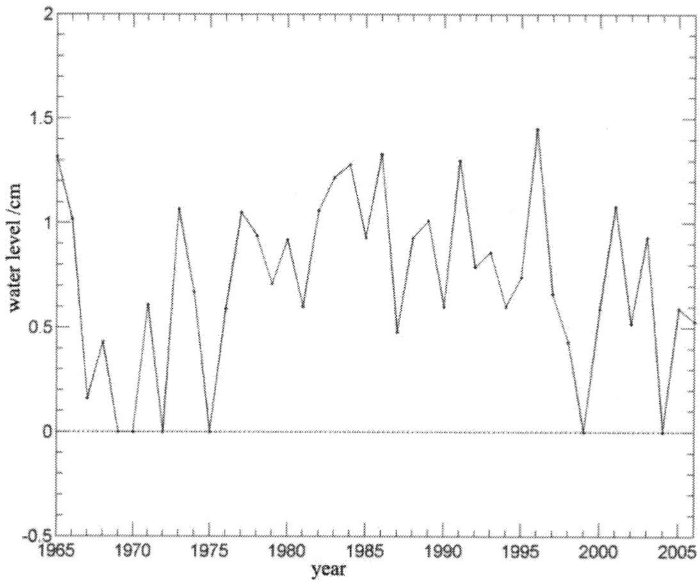

Figure 9. Maximum annual increase in water level at Beihai Observation Station.

Figure 10 depicts the cases in which the return levels of water level increases in different return periods could be acquired, with the use of Gumbel form (or an extreme type I distribution). Because the results obtained from Weibull and Pearson-III are basically identical, the pictures are not presented. Results indicate that a 100-year return from water level increase was 1.78 m in Beihai. According to the synchronized observation statistics, an equation was built to calculate a 100-year return level in Bailongwei, which was equal to 1.69 m of water level increase.

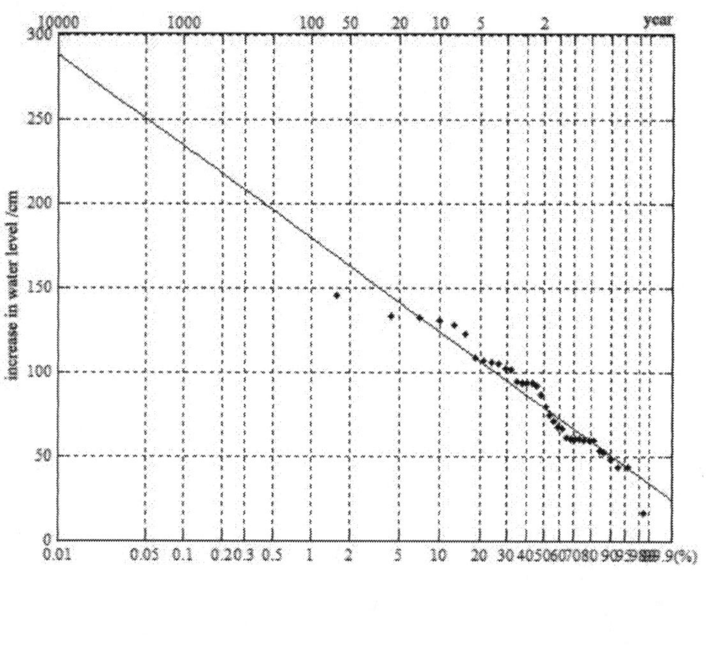

Figure 10. Distribution of Gumbel extreme values for water level increases based on the data recorded at Beihai Observation Station.

5.2. *Water level decrease*

Figure 11 depicts the extreme value of water level decrease each year as observed at the Beihai Observation Station over 42 years from 1965 to 2006. The maximum extreme decrease occurred in 1985, when the water level decreased by -1.87 m. The minimum water level decrease occurred in 1965 when the water level decreased by -0.53 m. The average water level decrease was -1.11 m over 42 years.

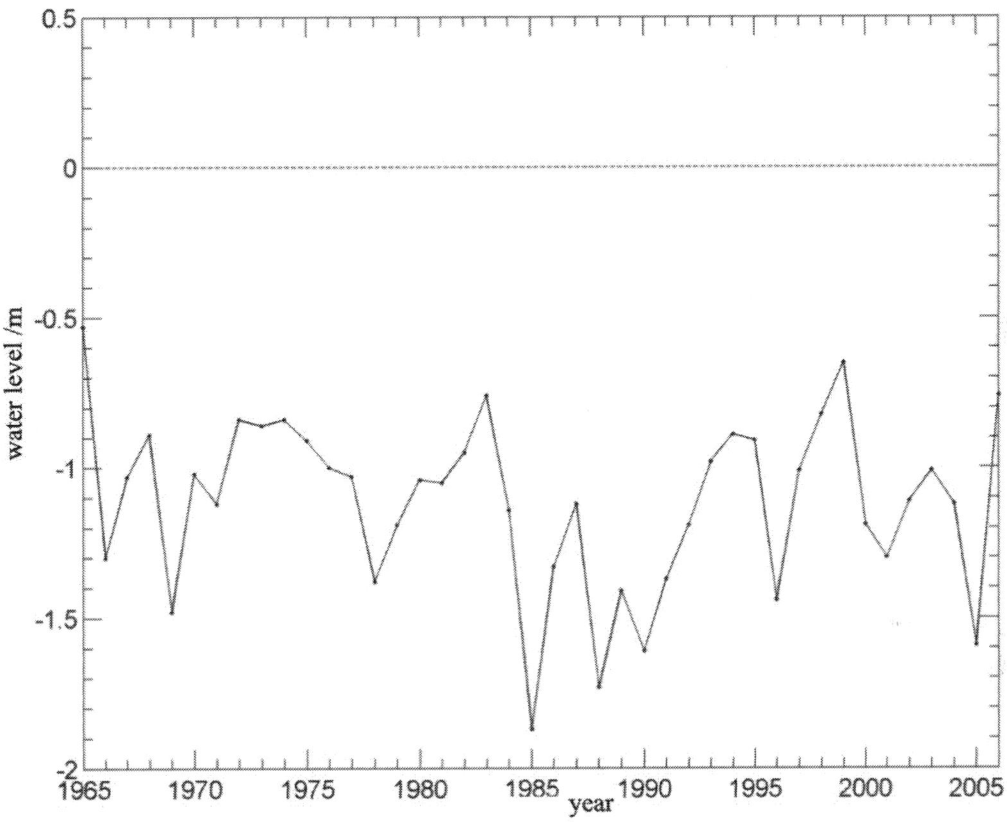

Figure 11. Maximum annual decrease in water level observed at Beihai Observation Station.

Figure 12 depicts the cases in which the return levels of water level decrease in different return periods could be acquired, with the use of the Gumbel form (or the extreme type I distribution). Because the results obtained from Weibull and Pearson-III are basically identical, the pictures are not presented. As shown in Figure 11, results indicate that a 100-year return level of water level decrease was –2.15 m in Beihai. According to the synchronized observation statistics, an equation was built to calculate a 100-year return level in Bailongwei, equal to a –1.75 m water level decrease.

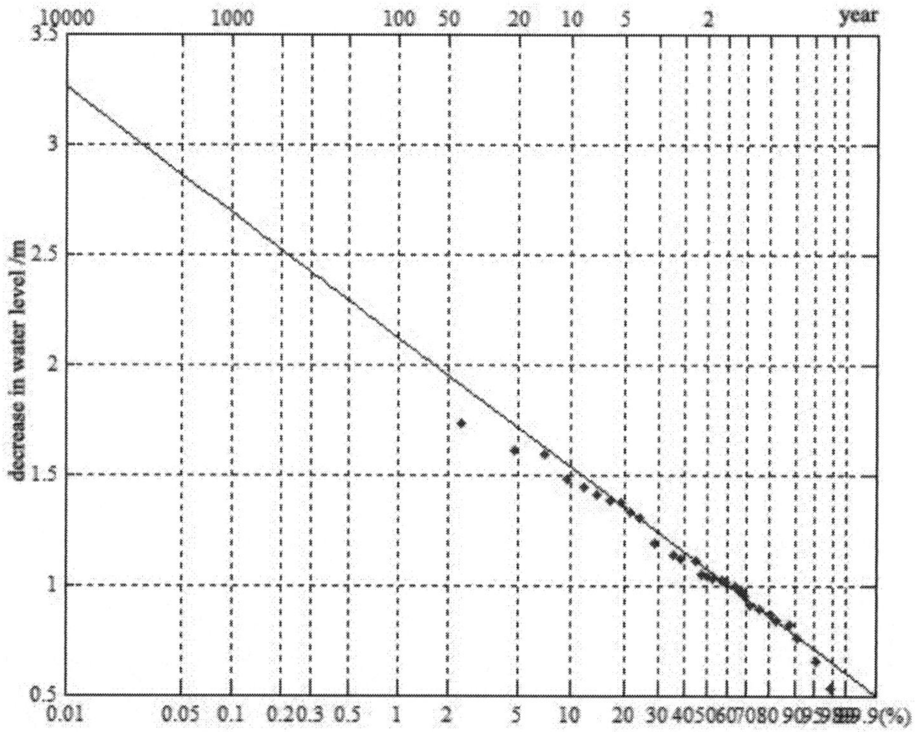

Figure 12. The Gumbel distribution based on the maximum water level decrease observed at Beihai Observation Station.

6. Analysis of major factors affecting water level fluctuation in the Guangxi near shore

6.1. Wind as a major affecting factor

According to the data recorded by the Dongxing Weather Station (table 1), the wind in Bailongwei Peninsula began to turn northward under the influence of Typhoon Nesat on September 29, 2011. The average wind speed was 3.3 m/s from 9:00 to 12:00, 4.9 m/s from 13:00 to 16:00, 8.0 m/s from 17:00 to 20:00, 5.8 m/s from 21:00 to 24:00, and the wind then gradually turned westward. Due to the effect of the north wind, the water level began to decrease along the coast of the Bailongwei Peninsula at 6:00, the water level reduced sharply at 20:00, and the water level decrease reached a maximum at 2:00 on September 30. Obviously, this decrease was directly related to the role of the northerly wind. However, the wind then turned southerly from 1:00 to 4:00 on September 30, with an average speed of 7.3 m/s, and the maximum wind speed exceeded 14 m/s. From 5:00 to 8:00 the average wind speed was 7.7 m/s and the maximum wind speed exceeded 16 m/s. The average wind speed decreased to 5.5 m/s from 9:00 to 12:00, and the wind turned to southwest after 12:00 when the wind speed dropped to approximately 5 m/s. Because the wind turned southerly, the wind blew towards the shore, and the water accumulated along the coast resulting in the water level increase, which reached a maximum of 65 cm at 10:00.

Table 1. The wind speed and major daily wind direction in Dongxing from September 28, 2011 to October 1, 2011.

Time (Month-Day)	AverageDaily Wind Speed (ms^{-1})	Major Daily Wind Direction	Average 4-Hour Wind Speed (ms^{-1})					
			21-24	1-4	5-8	9-12	13-16	17-20
9-28	1.5	NW	1.0	1.2	1.1	1.3	2.1	1.6
9-29	4.2	N	1.0	1.5	2.4	3.3	4.9	8.0
9-30	5.9	N-S-SW	5.8	7.3	7.7	5.5	6.5	2.7
10-1	1.2	W	1.9	1.0	0.9	1.8	1.5	1.0

6.2. Terrain as a constraining factor

Numerous semi-closed bays exist along the Guangxi coast, which deeply penetrate into the land with complicated geographical and environmental conditions. Moreover, the Beibu Gulf area is small, so the reflection of waves along the coast of Vietnam greatly impact the Guangxi coastal areas. Therefore, the fluctuation of water level in the bays of Guangxi demonstrate unique characteristics in which a water level decrease appeared prior to the sharp increase in the water level. The magnitude of the increase is large, and the water level rapidly inceased. Every typhoon-induced water level increase generally reaches over 1 m, while the time necessary to decrease the water level is long, and which can continue for more than 10 to 20 hours. For example, during strong Typhoon No. 3 in 1983, the water level increased by approximately 2.0 m in less than one hour in Fangchen Port, but it took more than 10 hours for the water level to return to normal. Sometimes this process can last for more than 20 hours [16], and this phenomenon is rarely seen in other coastal areas.

Additionally, the speed and direction of flow recorded at observation stations is also subject to terrain conditions. Under the influence of Typhoon Nesat from September 29 to September 30, 2011, for example, the direction of the main flow was NE ~ E and SSW～WSW due to the terrain constraints of the Bailong Peninsula (F1 station), and the frequency of the northward flow increased gradually from the sea surface to the bottom. During this period, the maximum flow speed reached 103.7 cm/s on the sea surface, while the flow speed was small in other directions, usually less than 50 cm/s.

6.3. Atmospheric gravity wave as a strengthening factor

It was observed that the strengthening, distribution and extreme values of water level fluctuation were closely related to atmospheric gravity waves in coastal bays[19]. This study demonstrated that, under the same weather conditions and even on the same side of a bay, the water level fluctuations were also very different. This paper chose the total statistics of certain bays in which the water level fluctuation lasted for several days during the typhoons, and the water level energy spectra were analyzed. Results demonstrate that there many extreme values of the energy spectra correspond to different frequencies of oscillation of waves in bays, but the maximum extreme value was always more closely associated with the intrinsic oscillation frequency of bays. For example, Typhoon No. 9 in 1986 was not very strong but the water level increased by 2 m, while the periodicity of the corresponding maximum extreme value of energy spectrum was 102 min, which was approximate to the bay's intrinsic oscillation periodicity of 99 min. Similarly, this paper analyzed the energy spectra of simultaneous changes in air pressure, and also found that when the maximum extreme value of the energy spectrum was approximately equal to the bay's intrinsic oscillation periodicity, the maximum extreme value of the increase or decrease in water level also appeared. In conclusion, the resonance between the atmospheric gravity wave and the bay's intrinsic oscillation periodicity plays an important role in the occurrence of the maximum extreme value of the increase or decrease in water level, and also affects the strength and distribution of water level fluctuation in bays.

7. Conclusions

- The fluctuation of the water level in the Guangxi near shore is primarily caused by typhoon-induced storm surges. Using Beihai Observation Station as an example, the maximum extreme value of the water level increase was 1.45m, and occurred in 1996. Over the last 42 years, the average water level increase was 0.71m. The 100-year return level was equal to 1.78 m in Beihai, and 1.69 m in Bailongwei.
- The maximum extreme value of the water level decrease was -1.87m and occurred in 1985, while the minimum was -0.53m and occurred in 1965. The average of water level decrease was -1.11 m over the last 42 years. The 100-year return level of water level decrease was equal to -2.15 m in Beihai and -1.75 m in Bailongwei. The trend analysis showed that typhoon-induced storm surges led first to water level decreases followed by increase in the Guangxi near shore. Apart from the direct effects of the typhoon, the water level fluctuation in the Guangxi near shore was also indirectly affected by the westward transmission of coastal-trapped waves from the Guangdong coast. This westward transmission had two paths: one directly passed into the Beibu Gulf through the Qiongzhou Strait, and the other bypassed the eastern and southern coasts of Hainan Island and then entered the Beibu Gulf from the north. On July 6, 2001, Typhoon Utor landed at the north shore of the Pearl River estuary, causing a 20-cm increase in the water level in the Guangxi near shore on July 8. At 18:00 on July 2, Typhoon Durian landed at Guangxi, the westward transmission of the coastal-trapped wave induced a strong westward flow in the coastal areas of Guangxi, and the velocity of the surface geostrophic flow reached 92 cm/s.
- Besides the direct and indirect effects of typhoons, the terrain and the atmospheric gravity waves as a strengthening factor also played important roles in the water level fluctuation in the Guangxi near shore.

Acknowledgements

This study was supported by National Natural Science Foundation of China (41266002), Key Project of Guangxi Natural Science Foundation (2012GXNSFEA0533001) and Guangxi Academy of Sciences Foundation (13YJ22HY07)

Reference

[1] Jelesnianski C P 1965 A numerical calculation of storm tides induced by a tropical storm impinging on a continental shelf *Month. Wea. Rev.* **93**: 343 -58

[2] Jelesnianski C P and Shaffer J W A 1992 SLOSH (Sea, Lake, and Overland Surges from Hurricanes.) *NOAA Technical Report* NWS 48

[3] Mastenbroek C, Burgers G and Janssen P 1993 The dynamical coupling of a wave model and a storm surge model through the atmospheric boundary layer *J. Phys. Oceanogr.* **23**:1856-66

[4] Weisberg R H and Zheng L 2006 Hurricane storm surge simulations for Tampa Bay *Estuaries Coasts* **29** (6A): 899-913

[5] Jones J E and Davies A M 2009 Storm surge computations in estuarine and near-coastal regions: the Mersey estuary and Irish Sea area *Ocean Dynamics* **59** (6):1061-76

[6] Feng S Z 1982 *Introduction to Storm Surge* Science Press, Beijing

[7] Wang X N 1985 Study on a simple method for prediction of storm surge caused by typhoon *Acta Oceanologica Sinica* **7**(2):233-39

[8] Sun W X 1987 A further study of ultra-shallow water storm surge model *Shangdong Institute of Ocean* **17**(1):34-45

[9] Yin B S, Wang T, Hou Y J, et al. 2001 Numerical study of the influence of waves and tide surge interaction on waves in Bohai Sea *Oceanologia et Limnologia Sinica* **32**(1):109-16

[10] Tang L, Zhan J M and Chen Y Z 2011 Typhoon process and its impact on the surface circulation in the northern south China sea *Journal of Hydrodynamic* **23**(1):95-104

[11] Zhao P and Jiang W S 2011 A numerical study of the effects of coastal geometry in the Bohai sea on storm surge induced by cold-air outbreaks. J. Ocean Univ. China **10**(1): 9-15.

[12] Han S Z and Pan S 2013 Numerical simulation and analysis of storm surge in the Hangzhou Bay *Periodical of Ocean University of China* **43**(7):1-6

[13] Nong Z L, Li W Q and Wang D 2009 *Ocean and sea disaster division in Guangxi Zhuang Autonomous Region* Guangxi Surveillance and Forecasting Center

[14] Chen X Y, Liu H, Dong D X, Qiu S F and Chen B 2013 Analysis of Guangxi marine disaster risk *Guangxi Sciences* **20**(3): 248-53

[15] Chen X Y, Chen B, Liu H, Dong D X and Chen M 2013 Guangxi coastal storm surge disasters and countermeasures *Transactions of Oceanology and Limnology* **4**: 17-23

[16] Qiu S F and Chen B 2000 Impacts of typhoon course and landform on water level fluctuation in the coastal bays in Guangxi *Guangxi Sciences* **7**(4): 282-5

[17] Chen B, Dong D X, Chen X Y, Liu H and Qiu S F 2014 Analysis on causes of tropical cyclones and their disasters affecting on Guangxi coast based on historical data *Marine Science Bulletin* **16**(1): 2-14.

[18] Chen B and Qiu S F 2000 Reckoning of extremal water level fluctuation caused by storm surge in every year in Beihai Port *Journal of Guangxi Academy of Sciences* **6**(3):12-15

[19] Chen B and Shi M C 2001 Numerical simulation of storm surge in Lianzhou Bay *Marine Science Bulleten* **20**(3):88-92

[20] Chen B and Wei G S 2002 Numerical calculation of storm surges near Guangxi coast *Transaction of Oceanology and Limnology* **2**:1-8

[21] Ding Y, Bao X W and Shi M C 2012 Characteristics of coastal trapped waves along the northern coast of the South China Sea during year 1990 *Ocean Dynamics* **62**:1259-85

[22] Yang W K , Yang Q Y, Yin B S, et al. 2016 Study on the storm surge characteristics of NO.1409 typhoon Rammasun along Tieshan Bay *Marine Forecasts* **33**(1): 80-85

[23] Li M J, Wu S H, Liu S X, et al. 2015 Impacts of storm surge and spring tide on the beach erosion of the southwestern Weizhou Island, Guangxi Province *Acta Oceanologica Sinica* **37**(9): 126-37

IOP Conf. Series: Earth and Environmental Science **39** (2016) 012030 doi:10.1088/1755-1315/39/1/012030

Impacts of feeding strategy on microbial community structure diversity in vertical flow constructed wetlands

W L Jia[1, 3], J Zhang[2] and Q Wang[1]

[1] School of Chemistry and Chemical Engineering, Jiangsu Normal University, Xuzhou, China

[2] School of Environmental Science and Engineering, Shandong University, Jinan, China

E-mail: jiawenlin@jsnu.edu.cn

Abstract. The impacts of feeding strategy (intermittently or continuously) on contaminant removal performance and microbial community structure in vertical flow constructed wetlands (VFCWs) were evaluated. The results showed that intermittent feeding strategy improved the removal of COD, TP and ammonium in VFCWs, although TN removal was weakened correspondingly The bacterial diversity decreased with the increase of substratum depth in all CWs. The intermittent feeding favored the growth of microorganisms due to the enhancement of oxygen content in the substratum. The feeding strategy had little impact on the microbial community in the surface substratum. However, in the bottom substratum, the impacts were of great significance. The microbial community structure similarity between the CWs with different feeding strategies was low.

1. Introduction

In the present day, vertical flow constructed wetlands (VFCWs) have been widely developed in many countries to treat municipal wastewater, due to their high removal efficiency and low operation cost compared to other wetlands [1-3]. In the CWs, nutrients can be removed in alternate anaerobic and aerobic environments. However, the VFCWs are faced with problems during nitrogen removal, which are primarily caused by the low oxygen availability and incomplete nitrification in the substratum [4].

In order to improve the nutrient removal of VFCWs, many strategies have been studied in recent years, such as artificial aeration and step feeding [5]. Moreover, an intermittent feeding strategy has been proved as an effective method to enhance the nutrient removal in VFCWs [6-8]. The redox conditions of the wetland substratum could be improved by intermittent feeding, according to the operation of intermittent flooding and drainage. During the cyclic flood and drain operation, the substratum is periodically exposed to the atmosphere, and the oxidation condition in the substratum is improved due to the fast oxygen diffusion [9]. It has been reported in previous literature that the removal efficiency of ammonium in intermittently fed horizontal subsurface flow CWs was 14% greater than that of traditional, continuously fed CWs [10]. Compared to artificial aeration, the intermittent feeding strategy is of low cost and simpler operation.

Microorganisms are considered to be responsible for the transformation and degradation of pollutants in CWs, and the living environment is important to the diversity of microbial community structure. Truu, et al., analyzed the distributions of the microbial community in a horizontal flow CW and found that the microbial community varied with bed depth and chemical properties [11]. In addition, it was found that bacteria were more abundant on root surfaces due to the better oxygen diffusion [12]. Since the intermittent operation can alter the redox condition of substratum, the microbial community structure would change correspondingly. However, few studies to date have focused on the impacts of feeding strategy on microbial community structure diversity in VFCWs.

In the present study, VFCWs were conducted and fed continuously or intermittently. The contaminant removal performance was determined and the microbial community structures were analyzed using the denaturing gradient gel electrophoresis (DGGE) technique. The aims of the present study are to evaluate the effects of feeding strategy on microbial community structure and to reveal the spatial distribution of bacterial communities in the VFCWs.

2. Materials and methods

2.1. Characterization of experimental system

The VFCWs used in this study were constructed in circular barrel polyethylene containers with surface areas of 0.24 m^2 and depths of 65 cm. The substratum consisted of three successive layers: at the bottom was a 5-cm layer of river gravel with particle diameter of 4-5 cm, and then a 40-cm layer was paved using gravel with particle diameter of 1-3 cm. At the top, washed sand was used with a depth of 15 cm. The CWs were fed with synthetic wastewater, and the complete influent contained 350 mg/L COD, 40 mg/L NH_4^+-N and 5 mg/L TP. The structure of the experimental system and the components of the synthetic wastewater can be found in previous literature [6].

Sequencing fill and the draw batch mode was employed in the CWs as the influent mode. The wastewater was pumped into the CWs within 10 minutes during the feeding period. The wastewater was then kept in the CWs for a certain number of days to be purified, and the effluent was then discharged from the bottom. In order to ensure the purification performance of the CWs, the hydraulic retention time was seven days. After the CWs were well-established, three different feeding strategies were applied to three systems. Two CWs (W1 and W2) were operated with the intermittent feeding mode and one (W3) was operated with the continuous feeding mode. For W1, the substratum was flooded for four days, the water was then discharged completely and the CWs were dried for three days to obtain an F/D ratio of 4:3 during one running cycle. For W2, the F/D ratio was 6:1. For W3, the substratum was flooded for seven days during one cycle. After running for approximately three months, the purification performance of all CWs was stable.

2.2. *Sampling*

The influent and effluent of all CWs was determined and then COD, ammonium, total nitrogen (TN) and total phosphorus (TP) concentrations were analyzed using the standard methods described in APAH-AWWA-WPCF [13].

The substratum in the upper and lower layers of all CWs were sampled using sampling apparatus, to investigate the microbial community structure at different layers. The sampling depths were 10-15 cm and 50-55 cm, respectively. In addition, in order to evaluate the impacts of fill and draw operation on microbial community structure, the sampling was conducted on the 4th and 7th days, respectively. The samples are listed in table 1.

Table 1. Sample list of VFCWs.

Samples	S1	S2	S3	S4	S5	S6	S7	S8	S9	S10	S11	S12
CWs	W1	W2	W3	W1	W2	W3	W1	W2	W3	W1	W2	W3
Layer	Upper			Lower			Upper			Lower		
Time	4th day						7th day					

2.3. *Microbial analysis*

The total genomic DNA from the substratum microbes of all samples was extracted using the DNA Isolation Kit. F357-GC clamp and R518 primers were used to amplify the bacterial 16S rDNA by touchdown polymerase chain reaction (PCR) [14], using the protocol described by Wu, et al., [15] which is described briefly as: 94°C for 5 minutes, 30 cycles of 94°C for 1 minute, 65°C to 55°C for 50 seconds, and 72°C for 40 seconds; the first 20 cycles with 0.5°C decrease of the annealing temperature every cycle, and a final extension cycle at 72 °C for 10 minutes. DGGE analysis was then performed using the Bio-Rad Dcode system (Bio-Rad, USA). Electrophoresis was performed at 120 V for eight hours in 1×TAE buffer at a constant temperature of 60 °C, and the liner gradient was 29-50%. In order to visualize the DGGE bands, the gels were stained by ethidium bromide and photographed under UV-light after electrophoresis.

2.4. *Data analysis*

The DGGE profiles were analyzed using Quantity One software (Bio-Rad, USA) to reveal the banding patterns. The diversity index and dominance index were then calculated, respectively [16, 17], according to the number of bands in each lane of the DGGE profiles. The richness (R) values of the microbial communities were represented by the total numbers of DGGE bands, and the evenness (E) was calculated according to the formula $E = H'/\ln R$. The Dice coefficient was calculated to analyze the community similarities, and cluster analysis was conducted and expressed by dendrograms, which was calculated by an unweighted pair group method with the arithmetic average (UPGMA).

3. Results and discussion

3.1. *Impacts of feeding strategy on contaminant removal performance of VFCWs*

The contaminant removal efficiencies of each wetland were determined and the results are shown in Figure. 1. It can be observed that the COD removal efficiency in all CWs exceeded 85%. However, the removal efficiencies of W1 and W2 were higher than that of W3, indicating that intermittent feeding can improve the removal of COD. The intermittent feeding increases the oxygen content in the substratum, and the activities of aerobes are enhanced due to the alternant flooding and drying [6]. Moreover, the organic matter absorbed by the substratum can also be degraded during the drying period. Therefore, the organic matter is removed quickly in the intermittently-fed VFCWs. The removal of TP was also improved by intermittent feeding. It is possible that the physical-chemical accumulation processes, such as adsorption, precipitation or accumulation in the substratum were the dominant phosphorus removal processes [18]. The intermittent feeding supplied an alternate anaerobic-aerobic condition which was conducive to the metabolism of polyphosphate accumulating organisms (PAOs), and the phosphorus adsorbed by the substratum could be utilized by PAOs under aerobic conditions during the drying period of W1 and W2, and the adsorption capacity of the substratum would be correspondingly increased.

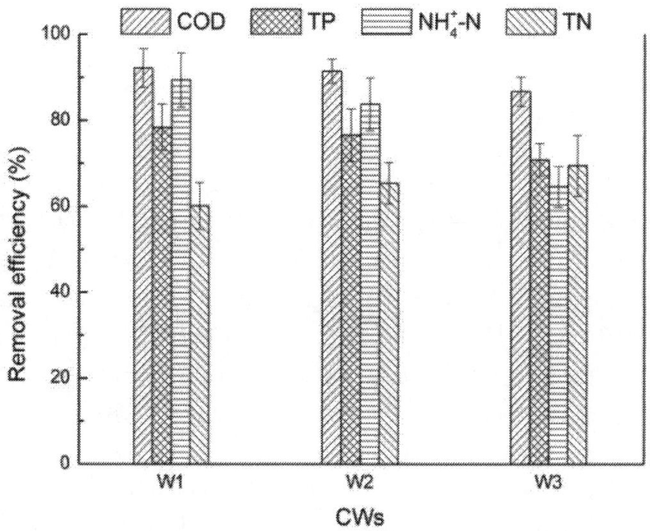

Figure 1. Contaminant removal efficiencies of the VFCWs.

It is noteworthy that the removal of ammonium was greatly improved by intermittent feeding. The NH_4^+-N removal efficiency of W1 and W2 was 30%-38% higher than that of W3. This is because the intermittent feeding increased the oxidization conditions in the substratum, thus facilitating the nitrification process. However, TN removal differed from that of ammonium, and the TN removal efficiency decreased as a result of intermittent feeding. In W1, the TN removal efficiency was only 60%, while it reached 69.5% in W3. Although some previous studied have reported that intermittent loading can increase TN removal [19, 20], this was not observed in the present study. It is thought that the denitrification process may be inhibited by the intermittent operation during the drying period.

3.2. Impacts of feeding strategy on bacterial communities diversity

The DGGE banding patterns of 16 S rDNA fragments of all samples are shown in figure 2. Results indicate that the DGGE patterns from CWs with different feeding strategies were distinct. The quantity of DGGE bands in the upper layers of all CWs (S1-S3 and S7-S9) was greater than that observed in the lower layers (S4-S6 and S10-S12), indicating that the bacteria species in the upper layers were more abundant than those in the lower layers for all CWs. Some bands, such as band 1 and

2, were commonly detected in all CWs, suggesting that microbes with these bands were stable, and were not affected by the feeding strategy. However, some bands also emerged or vanished with individual CWs. Band 3 emerged in the upper layer of W2 and W3 with longstanding flooding. However, band 5 appeared in both the upper and lower layers of W1, and vanished in W2 and W3. These results indicate that the different feeding strategies changed the living environment of microbes, and that some bacteria were eliminated due to their inability to adapt to changes.

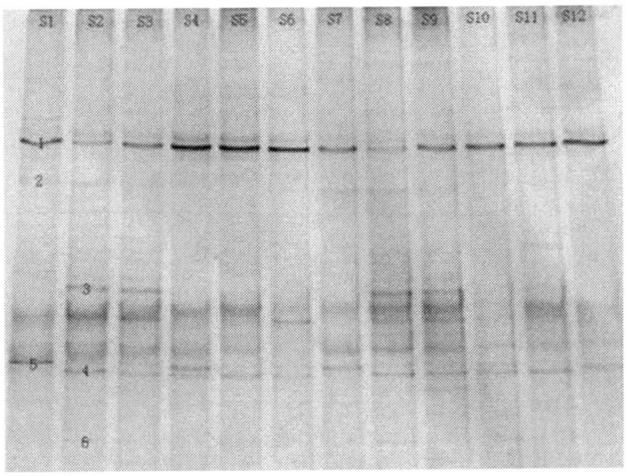

Figure 2. DGGE profiles of 16S rDNA from CWs with different feeding strategies.

Table 2. The bacterial community richness, diversity index, community dominance and evenness from different CWs.

Samples	*R*	*H'*	*D*	*E*
S1	26	3.11	0.05	0.95
S2	24	2.66	0.08	0.84
S3	28	2.67	0.09	0.80
S4	14	1.09	0.46	0.41
S5	17	1.07	0.49	0.38
S6	12	0.69	0.66	0.28
S7	15	2.47	0.09	0.91
S8	25	2.64	0.10	0.82
S9	24	2.21	0.16	0.70
S10	12	1.67	0.23	0.67
S11	19	1.69	0.26	0.57
S12	11	1.06	0.40	0.48

The richness, diversity, dominance and evenness of bacterial communities in CWs with different feeding strategies were calculated based on the DGGE patterns, and the results are shown in table 2. It was presumed that each band in the DGGE profiles represented one phylogenetically distinct population. Therefore, *R* was equal to the total numbers of the bands, and ranged from 11 to 28 bands per lane during the operation. Results indicate that the indices of *H'* and *E* in the upper layers were significantly higher than those observed in the lower layers, indicating that the species of microorganisms in the upper layers of all CWs were more abundant. Alternatively, the dominance indices (*D*) of the lower layers were greater. In the subsurface flow CWs, the pollutant would be held and adsorbed first by the surface substratum, which is more conducive to the growth of microorganisms [12]. As the depth increased, the organic matter and nutrients in the substratum decreased gradually, and some microorganisms would be eliminated resulting in correspondingly lower diversity.

As shown in table 2, the diversity in the upper layers of W1 on day four was much higher than that observed in W2 and W3 (S1 vs. S2 and S3). Moreover, indices of H' and E in the lower layers of W1 and W2 were significantly greater than that of W3. These results indicate that intermittent feeding favored the growth of microorganisms in both the lower and upper layers due to the enhancement of oxygen content in the substratum.

Table 3. Dice coefficient within and between samples of CWs with different feeding strategies.

	S1	S2	S3	S4	S5	S6	S7	S8	S9	S10	S11	S12
S1	1											
S2	0.754	1										
S3	0.722	0.710	1									
S4	0.499	0.472	0.484	1								
S5	0.477	0.524	0.528	0.629	1							
S6	0.476	0.536	0.505	0.594	0.655	1						
S7	0.496	0.528	0.517	0.508	0.418	0.451	1					
S8	0.507	0.498	0.518	0.432	0.578	0.427	0.559	1				
S9	0.547	0.524	0.502	0.390	0.564	0.412	0.574	0.741	1			
S10	0.405	0.366	0.382	0.460	0.507	0.447	0.224	0.326	0.318	1		
S11	0.578	0.639	0.602	0.539	0.493	0.525	0.553	0.458	0.412	0.262	1	
S12	0.373	0.460	0.428	0.481	0.382	0.525	0.524	0.305	0.259	0.277	0.683	1

Dice coefficients were calculated based on the DGGE profiles to reveal the similarity among all samples, and the results are shown in table 3. Moreover, cluster analysis based on the Dice coefficients was conducted and expressed byvisualized dendrograms to investigate the community similarity among different wetlands (figure 3). Results indicate that the Dice coefficient of all samples varied greatly. The similarity coefficient of the upper-layer samples was greatest, indicating that the feeding strategy had little impact on the microbial community in the surface substratum. The similarity between the upper samples and lower samples was low, mainly caused by the differing oxygen content. In the surface substratum, the oxidizing condition favored the existence of aerobes. However, the bottom substratum it was mainly reductive, where amphimicrobes and anaerobes are the dominant microorganisms. The S5 and S6 samples demonstrated high community similarity (65%), and the similarity between S4 and S6 was low (figure 3). This is because W1 had a longer drying period than W2, and the bottom substratum of W1 could maintain oxidizing conditions. Although intermittent feeding was also used in W2, the F/D ratio was 6:1, and the drying time was only one day. Therefore, the similarity between the bottom substratums of W2 and W3 was higher. It is noteworthy that the samples taken on days four and seven had different coefficients. This may be because on day seven, W1 has been dried for three days, and some biofilm existing in the substratum had decayed when sampling. Therefore, the coefficient varied.

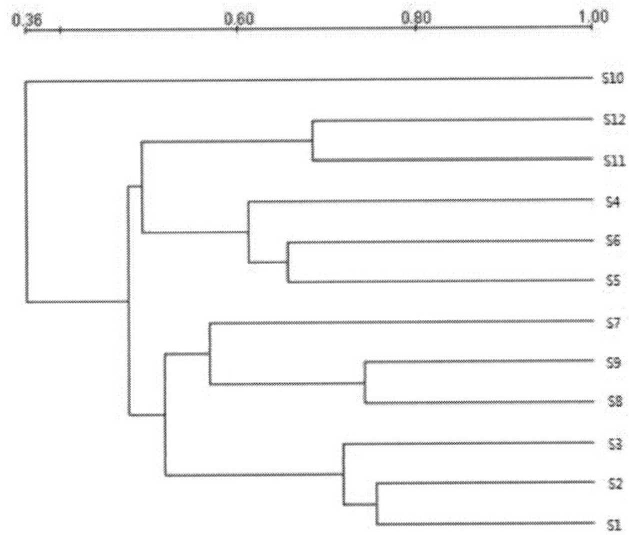

Figure 3. Cluster analysis of the 16S rDNA DGGE patterns.

4. Conclusions

Intermittent feeding canenhance the removal of COD, TP and ammonium in VFCWs. However, TN removal was weakened by intermittent feeding due to the inhibition of denitrification. A decrease in bacterial diversity with increasing substratum depth was found in all CWs, independent of which feeding strategy was used. In the substratum, the growth of microorganisms was enhanced by intermittent feeding due to the long drying time. The diversity in the upper layers of W1 was much greater than that observed in W2 and W3. In the bottom substratum, the feeding strategy had significant impacts on microbial community. The microbial community structure similarity between the CWs with different F/D time ratios was low.

References

[1] Brix H and Arias C A 2005 *Ecol. Eng.* **25** 491-500
[2] Wu H, Zhang J, Li P, Zhang J, Xie H and Zhang B 2011 *Ecol. Eng.* **37** 560-8
[3] Zheng Y, Wang X, Dzakpasu M, Zhao Y, Ngo H H, Guo W, Ge Y and Xiong J 2016 *Bioresource Technol.* **207** 134-41
[4] Brix H, Arias C A and del Bubba M 2001 *Water Sci. Technol.* **44** 47-54
[5] Kirui W K, Wu S B, Kizito S, Carvalho P N and Dong R J 2016 *J. Environ. Manage.* **166** 38-44
[6] Jia W, Zhang J, Wu J, Xie H and Zhang B 2010 *Desalination* **262** 202-8
[7] Jia W, Zhang J, Li P, Xie H, Wu J and Wang J 2011 *Ecol. Eng.* **37** 1815-21
[8] Wu H, Fan J, Zhang J, Ngo H H, Guo W, Hu Z and Liang S 2015 *Bioresource Technol.* **176** 163-8
[9] Behrends L, Houke L, Bailey E, Jansen P and Brown D 2001 *Water Sci. Technol.* **44** 399-405
[10] Caselles-Osorio A and Garcia J 2007 *Sci. Total. Environ.* **378** 253-62

[11] Truu J, Nurk K, Juhanson J and Mander U 2005 *J. Sci. Environ. Heal. A* **40** 1191-200

[12] Wang Q, Xie H, Ngo H H, Guo W, Zhang J, Liu C, Liang S, Hu Z, Yang Z and Zhao C 2016 *Environ. Sci. Pollut. R.* **23** 4036-45

[13] APAH-AWWA-WPCF 2001 *Standard methods for the examination of water and wastewater* 20th ed. (Washington: American Public Health Association)

[14] Muyzer G, Dewaal E C and Uitterlinden A G 1993 *Appl. Environ. Microb.* **59** 695-700

[15] Wu J, Zhang J, Jia W, Xie H and Wu H 2009 *Int. Conf. on Environmental Science and Information Application Technology* (Wuhan: China) pp 160-163

[16] Shannon C E and Weaver W 1963 *The mathematical theory of communication* (Illinois: The University of Illinois Press)

[17] Simpson E H 1949 *Nature* **163** 688

[18] Moortel A M K, Rousseau D P L, Tack F M G and de Pauw N 2009 *Ecol. Eng.* **35** 175-83

[19] Merlin G, Pajean J L and Lissolo T 2002 *Hydrobiologia* **469** 87-98

[20] Li Y, Li H, Sun T and Wang X 2011 *J. Hazard. Mater.* **189** 336-41

IOP Conf. Series: Earth and Environmental Science **39** (2016) 012031 doi:10.1088/1755-1315/39/1/012031

Pore pressure propagation in a permeable thin-layer coal seam based on a dual porosity model: A case of risk prediction of water inrush in coalmines

B Zhu[1,3], F Gao[1], J W Yang[2] and G Q Zhou[1]

[1]College of Earth Sciences, Guilin University of Technology, Guilin, Guangxi 541004, China

[2]Department of Earth and Environmental Sciences, University of Windsor, Windsor, ON N9B 3P4, Canada

E-mail: zhubin1969@qq.com

Abstract. Thin-layer coal seams, a type of filling coal rock body, are considered aquifer systems made up of dual porosity medium with immediate floor. A numerical simulation for the pore pressure propagation along a thin-layer coal seam was carried out for the case of the Zhaogezhuang coalmine in China. By valuing the permeability (K_f) of the thin-layer coal seam, pore pressure variation with time was simulated and compared to the analytical solutions of a dual porosity model (DPM). The main conclusions were drawn as follow: (1) Seepage in the thin-layer coal seam was predominant in the whole process, and the distance of seepage was lengthened and the pore pressure decreased with increased K_f; (2) A series of simulated hydraulic graphs demonstrated that the pore pressure characteristics of peak-occurring and time-lag effects agreed with the analytical solutions of DPM; (3) By adjusting the parameters of DPM, two results of analytical solutions and numerical solutions fit well, particularly in the thin-layer coal seam; (4) The power law relationship between the peak-values and lag time of pore pressure were derived statistically under consideration of the K_f parameter in the range of 10^{-8} to 10^{-10} m^2/pa·s orders, and it was reasonable that the K_f of the thin-layer coal seam was in the range of 10^{-8} m^2/pa·s orders. The results were significantly helpful in decision-making for mining water prevention and prediction in practice.

1. Introduction

Recently, water inrush from a coal seam is commonly occurring water disaster that poses a serious threat to the safety of coal mining. The risk of water inrush from a coal floor layer causes geologists and the public great concern.

Usually, a series of initial or induced fractured spaces, such as fissures, sandwiched between thin layers and fault zones, are the main pathway for groundwater flow. Water inrush incidents occurring through these shortcuts generally have a certain "time-lag effect" [10, 11]. However, systematic research on pore pressure propagation preceding water inrush incidents, particularly in those exposed to subsurface open spaces, has not been conducted. According to recent studies, a coal seam intersecting with a fault zone might become a secondary pathway that conducts groundwater to subsurface open spaces [15].

In recent years, many research achievements regarding groundwater flow in fracture families have been accomplished. The studies generally concentrated on the seepage theory model of rock mass, such as the single fracture model (SFM) by Louis [5] and dual porosity model (DPM) by Barenblatt, et al. [1]. Onder [7] obtained analytical solutions of transient flow in a DPM by the Fourier transformation. Cornaton and Perrochet [2] derived a semi-analytical solution on the same issue usingthe Laplace transforms. Based on the in-depth hydraulic analysis of seepage through fracture networks, equivalent continuum, and dual porosity models, a comparative research project on seepage in different media models was summarized by Samardzioska and Popov [8].

In this paper, the study area is described by means of hydrogeology, boundary and initial conditions, and solution scheme in Section 2. A model that properly simplified and conceptualized the groundwater seepage process was established by embedding an aquifer system with a sandwiched thin-layer coal seam and fractured porous confining rock (buffer layer), as shown in figure 1. In Section 3, a series of numerical and analytical solutions are provided. The results of seepage evolution based on DPM analytical solutions are illustrated in Section 4. A comparison between simulated and analytical pore pressures was assessed in Section 5. Furthermore, a suite of statistic patterns of hydraulic characteristics acquired from simulation results and graphically presented in accordance with the correlation analysis between peak-values and lag-time of pore pressure are offered. Additionally, a novel strategy for risk prediction of water inrush in North China coalmines was proposed under consideration of the permeability of thin-layer coal seam as a parameter.

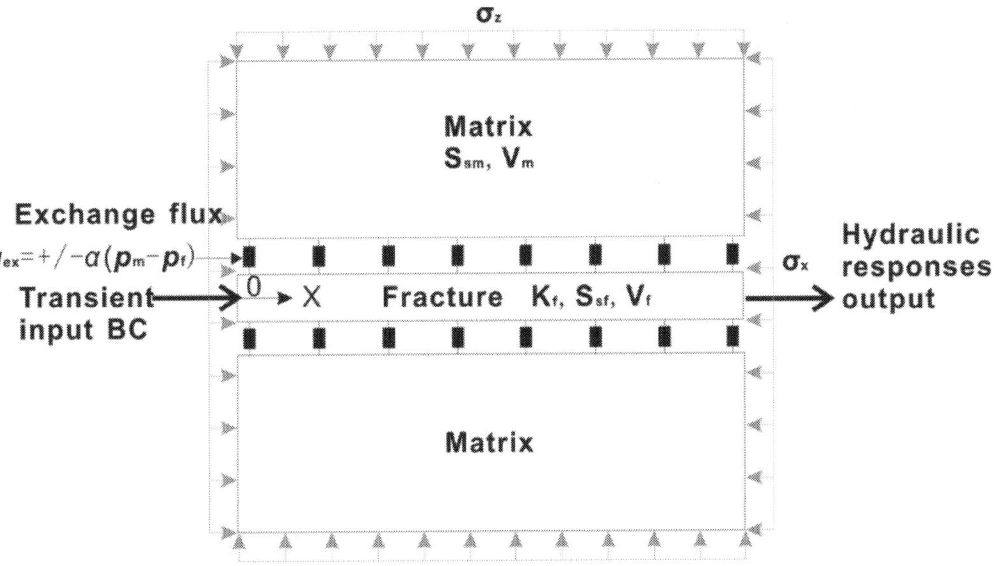

Figure 1. Conceptual scheme of the analytical model.

2. Establishment of numerical model

2.1. *Hydrogeologic framework in study area*

Zhaogezhuang Coalmine is located in the northeastern suburb of the Guye region, in the city of Tangshan, Hebei Province, China (figure 2). It covers an area of approximately 40 km^2, and feeatures a semiarid climate, warm temperates, and distinct seasons. Mean annual precipitation is about 614.7 mm in the region. The surface covering is mainly composed of alluvial-eluvial deposits with a thickness of up to 40 m. As a result of the Neocathaysian structural system, a series of compression-shear folds and faults along the NNE direction occur alternately with representation of the main tectonic features of this area. Due to the influence of the local vortex structure, a small number of extensional folds and faults are emerging in the east wing of Kaiping syncline (the coalmine-hosting area of Kaiping coalfield).

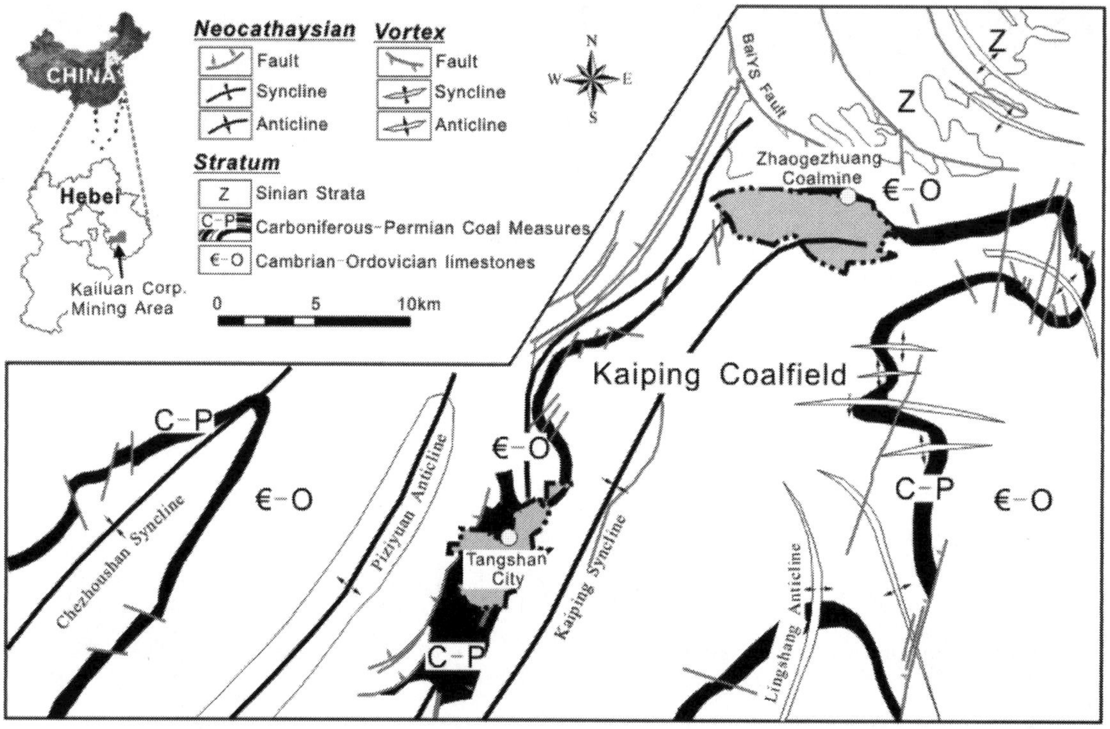

Figure 2. Geological map of the Kaiping coalfield with locations of sites, strata, and tectonics mentioned in the text.

The Zhaogezhuang coalmine is situated just at the Kaiping syncline's tip, the end of the syncline axis turning sharply from N34°E to an almost E-W direction due to a strong squeezing and twisting tectonic stress (figure 2). The Ordovician Majiagou formation predominates the potentially flooding aquifer that is composed of thickly bedded and pale grey carbonate with a thickness of about 400 m. The aquifer hosts a thick sequence of upper Ordovician carbonates known as the Ordovician limestone aquifer [14].

The hydrogeologic framework model used in this study was based on stratigraphic and permeability data from the Kaiping Coalfield Basin (figure 3). There are a group of Carboniferous-Permian (C-P) strata that contain a series of coal measures beneath the basin. The Ordovician limestone aquifer is under a depth of 1300 meters and is separated by a confining unit (Bauxitic mudstone). The coal measures lie in the range of approximate 200 m thicknesses. The Ordovician limestone aquifers are the underlying formation of coal measure strata and are the main flooding aquifer in the study area. A typical comprehensive stratigraphic column of the Zhaogezhuang coalmine is depicted in figure 3. There exists a thin layer, named 14_{3S} thin-layer coal seam, in the carboniferous formation. It is the ending segment of a seepage system consisting of a series of rock masses and fractures. Within this seepage system, the groundwater mainly comes from the Ordovician aquifer through the F8 fault zone and finally to the thin-layer coal seam [16].

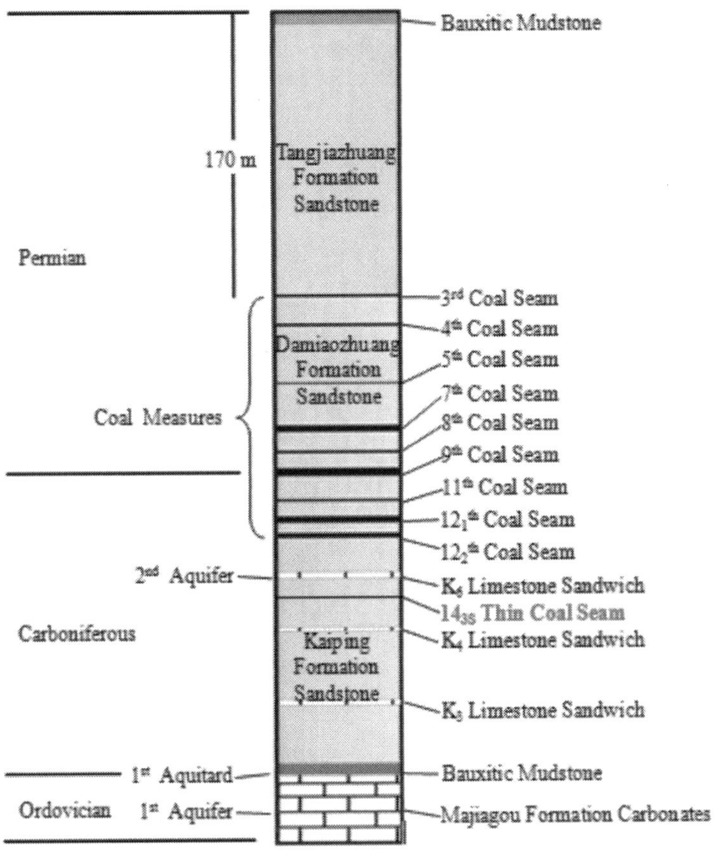

Figure 3. Stratigraphic column representative of hydrogeologic conditions within the Kaiping coalfield (In order to simplify our graphical presentation of model results, many of the thin middle and upper lithologic units are lumped together into the coal measures.).

2.2. Solution schemes, boundaries, and initial conditions

The study site consisted of a segment of laneway located in the 14th level (-1225 m depth) in the Zhaogezhuang coalmine. The 14_{3S} thin-layer coal seam with 0.2-0.5 m thickness is hosted in a carboniferous formation and exposed out of the laneway on the lateral sides due to excavation, as shown in figure 4. It is not a water-bearing layer; however, when there is injection from a conductive fault zone nearby, it becomes a part of the seepage system by motivation of pore pressure propagation. Then, as the pressure accumulates, the water inrush incidents occur in time, as a case of small-scale water inrush ensued in the 13th level No.1 crosscut (-1100 m depth) in 2005 [16].

Figure 4 illustrates schematic maps of the model in which the main stratigraghic units and boundary conditions are discretized by grid elements with a scope of 200×100×50 m in a Cartesian coordinate system with a depth ranging from -1200 m to -1250 m. Two layers of aquitard strata, the immediate roof and immediate floor, called buffer layers in coalmining sciences, had 3.0 m thicknesses for convenience in this paper. The 14_{3S} thin-layer coal seam was sandwiched between two buffer layers. The buffer layers together with the 14_{3S} thin-layer coal seam formed the main body of the dual porosity (permeability) model (DPM). Furthermore, the body of the DPM was sealed by overlying and underlying impervious carboniferous fine sandstone, referred to as main roof and main floor in coalmining sciences, and each was 21 m thick. The top mechanic boundary of the model domain was loaded by 27.6 Mpa, which approximated the lithostatic pressure from the overlying strata.

In order to capture the pressure during the process of seepage, a series of observation points were set up at 10 m intervals along the DPM body.

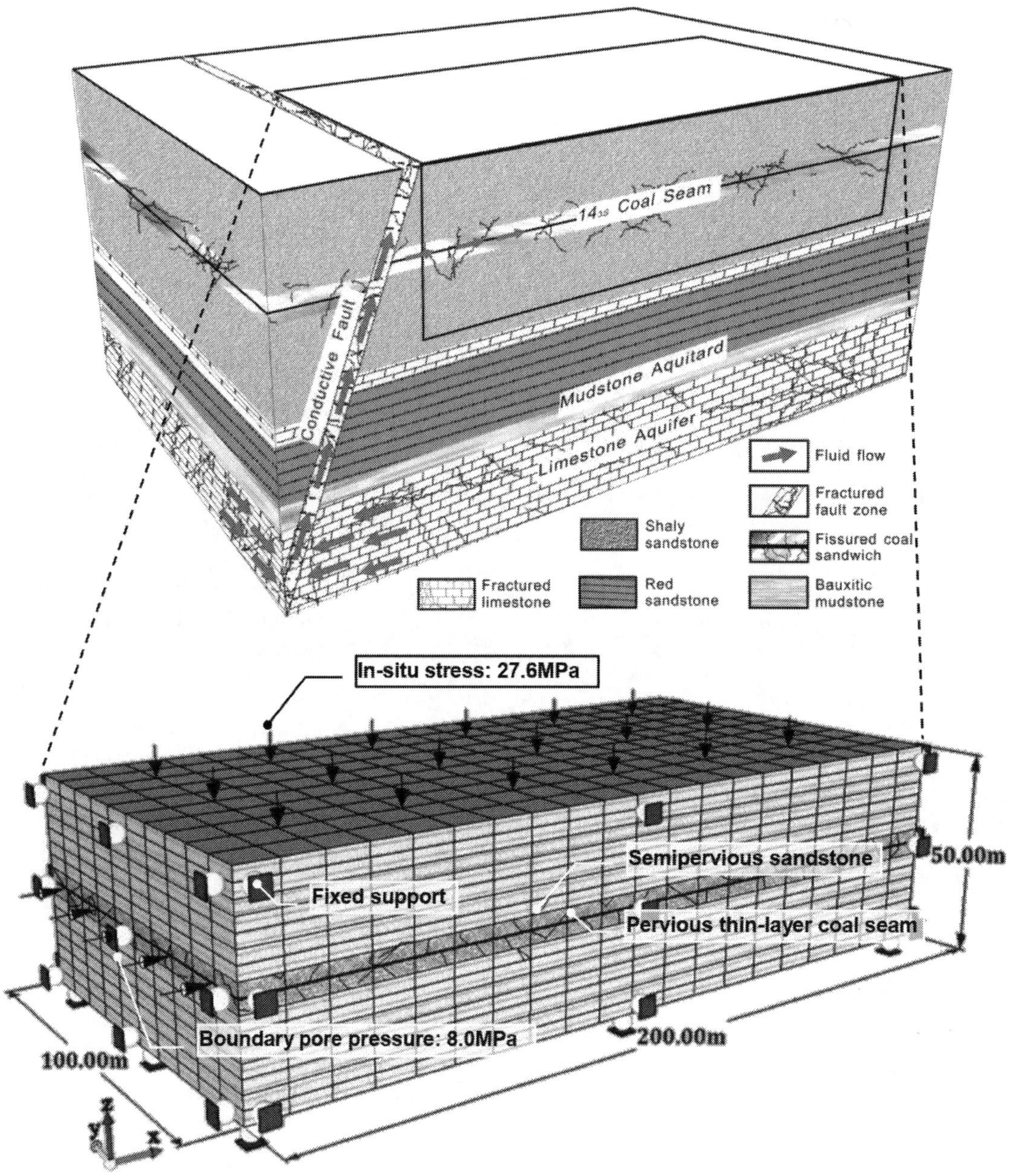

Figure 4. Top diagram: Conceptual model; Bottom diagram: Numerical model of study area.

Based on long-term observation data, the maximum amplitude of groundwater level fluctuations was about 250 m, in the range of -260 to -110 m under the ground surface, approximately corresponding to 9.3 to 10.8 Mpa pressure in the study site at a depth of -1225 m. Considering the head loss in the process of seepage propagation along the fault zone, a specified-head boundary of the lateral side in the scope of the DPM body was set to be 8.0 Mpa from beginning to end. The other sides of the model domain were set to be no-head and no-flux boundaries.

The numerical model was mainly composed of the following three types of rock material: (1) Impervious shaly sandstone, which were overlying and underlying the fissured coal seam sandwich, i.e. the DPM body; (2) Semipervious immediate roof and floor, which were sandstone, named buffer layers herein; and (3) Pervious 14_{3S} thin-layer coal seam. The latter two comprised the DPM body in which the 14_{3S} thin-layer coal seam was considered the fracture continuum and the buffer layer was the porosity continuum. The mechanical parameters of the medium materials are shown in Table 1. The hydraulic behaviors of seepage propagation along the DPM body was summarized in consideration of the K_f changes from 10^{-8} through 10^{-10} m^2/s, which was equal to 8.64×10^{0} to 8.64×10^{-2} m/d.

Table 1. Physical and mechanical parameters of rock mass.

Property	Shaly sandstone	Buffer layers	14_{3S} thin-layer coal seam
Density /$kg{\cdot}m^{-3}$	2600	2000	1500
Elastic modulus/GPa	19.0	0.2	0.2
Poisson ratio	0.24	0.25	0.4
Cohesion /MPa	2.00	0.013	0.07
Internal friction angle /°	42	18	27
Tensile strength /MPa	3.0	1.0	0.8

3. Theoretical basis

3.1. *Differential equations of dual porosity (permeability) model*

As previously mentioned in Section 1, many valuable studies have been conducted on analytical solutions of unsteady flow in dual porosity models of rock mass. Two differential equations controlling two flows in the fracture and porosity continuum are described by the two diffusion equations below [1]:

$$K_f \frac{\partial^2 p_f}{\partial x^2} = S_{sf} \frac{\partial p_f}{\partial t} - \alpha (p_m - p_f) \tag{1a}$$

$$K_m \frac{\partial^2 p_m}{\partial x^2} = S_{sm} \frac{\partial p_m}{\partial t} + \alpha (p_m - p_f) \tag{1b}$$

where K_f and S_{sf} are the permeability and specific storage of the high-permeability medium (fracture), respectively; K_m *and* S_{sm} are the permeability and specific storage of the low-permeability medium (matrix), respectively; p_m and p_f are the hydraulic pressures (converted by hydraulic heads) of the matrix and the fracture, respectively; and α is the water-exchanging coefficient.

Considering the equal water-exchanging term, the weakening of permeability in a matrix due to its small value and the weighting of fracture-porosity, can be expressed as follows:

$$K_f \frac{\partial^2 p_f}{\partial x^2} = S_{sf} \frac{\partial p_f}{\partial t} + \beta S_{sm} \frac{\partial p_m}{\partial t} \tag{2a}$$

$$S_{sm} \frac{\partial p_m}{\partial t} = -\alpha (p_m - p_f) \tag{2b}$$

If ϕ is the fracture total porosity, then the matrix total porosity is $1-\phi$; thus, the following is obtained:

$$\phi = \frac{V_f}{V_t}, \quad 1-\phi = \frac{V_m}{V_t}$$

$$\beta = \frac{1-\phi}{\phi}$$

where V_f is the fracture volume, V_m is the matrix volume, V_t is the apparent total volume, and β is the porosity ratio.

Subject to the boundary conditions, the following is obtained:

$$p_f(0,t) = BC_1(t) \ \text{ or } \ -K_f \frac{\partial p_f}{\partial x}(0,t) = BC_2(t), \text{ and } \ p_f(+\infty, t) = 0$$

and the initial conditions are as follow:

$$p_f(x,0) = p_m(x,0) = 0$$

3.2. Analytical solutions of dual porosity (permeability) model

The analytical solution of the problem was achieved by applying integral transformations successively with regard to space and time variables. After using the Laplace transformations and required inversion procedures, the solutions for hydraulic pressure of the fracture continuum were obtained as follow:

$$p_f(x,t) = \frac{2P_{f0}}{\sqrt{\pi}} \int_{\frac{x}{2}\left(\frac{S_{sf}}{tK_f}\right)^{1/2}}^{\infty} \left[\exp\left(-\xi^2 - Y\beta\alpha\right)\right] \cdot \exp\left(-\frac{\alpha}{S_{sm}} t\right) \left\{\delta(t) + \left[\sqrt{A} J_0\left(2\sqrt{At}\right) / \sqrt{t}\right]\right\} \mathrm{d}\xi \tag{3a}$$

Finally, the solution for hydraulic pressure of the matrix continuum was as follows:

$$p_m(x,t) = \frac{2P_{f0}}{\sqrt{\pi}} \int_{\frac{x}{2}\left(\frac{S_{sf}}{tK_f}\right)^{1/2}}^{\infty} \left[\exp\left(-\xi^2 - Y\beta\alpha\right)\right] \cdot \exp\left(-\frac{\alpha}{S_{sm}} t\right) \left[J_0\left(2\sqrt{At}\right)\right] \mathrm{d}\xi \tag{3b}$$

where $A = Y\beta\alpha^2 / S_{sm}$ and $Y = x^2 / 4\xi^2 K_f$, $J_0(\cdot)$ is the zero-order Bessel function of the first kind.

The derivation of equations 3a and 3b were omitted here. There was a series of results from them, which were similar to previous analytical solutions of DPMs proposed by other literatures [6,7,9]. By adjusting the coefficients, such as K_f, S_{sf}, S_{sm}, α, and β, the seepage processes along the matrix continuum were calculated with equation 3b. The seepage processes along the fracture continuum obtained with equation 3a were characterized by the occurrence of obvious pressure peaks. The following numerical simulation demonstrated the phenomenon.

4. Results of simulation studies

4.1. Features of pore pressure propagation during seepage evolution

As shown in figure 5, a group of seepage evolution nephograms along the fissured coal sandwich are illustrated at three phases of pore pressure propagation from three simulated scenarios. Permeability was determined by assigning K_f, i.e. 10^{-8} m^2/ pa·s, 10^{-9} m^2/ pa·s, and 10^{-10} m^2/ pa·s. The simulated pore pressure durations were recorded automatically by the history function embedded in the FLAC^{3D} software. Some typical phenomena were as follow:

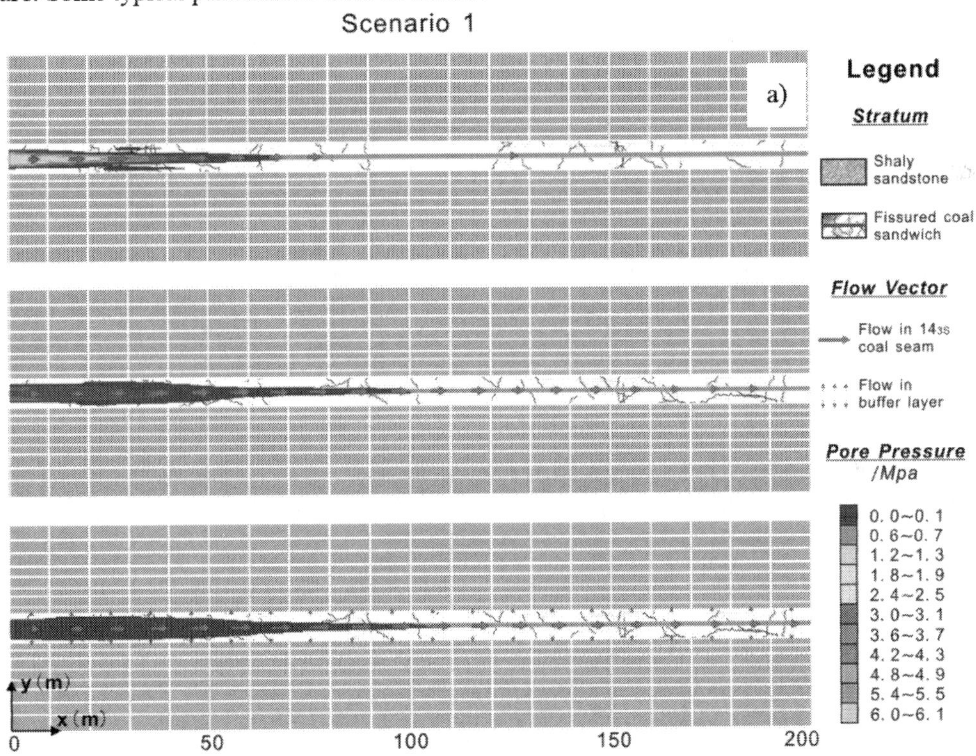

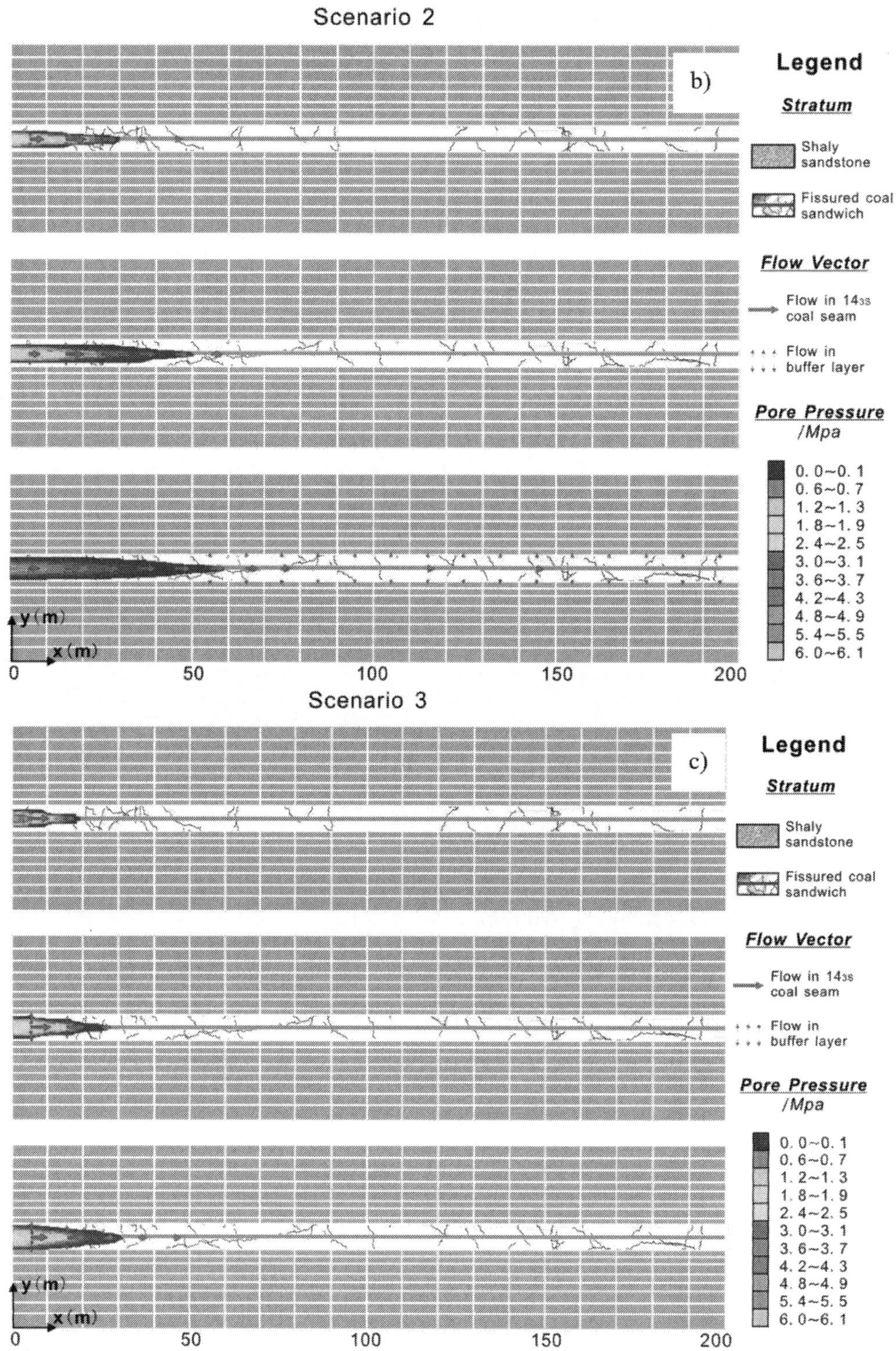

Figure 5. Nephograms of simulation of pore pressure propagation and fluid flow of seepage evolution corresponding to various scenarios in terms of K_f: (a) 10^{-8} m^2/ pa·s, (b) 10^{-9} m^2/ pa·s, (c) 10^{-10} m^2/ pa·s.

The following observations were made according to the nephograms in figure 5 above:

- From the flow vectors, seepage evolved mainly along the 14_{3S} thin-layer coal seam. The nephogram of pore pressure was the shape of a cluster of 'flames.' The pore pressures within the 'flame core' were higher than the surroundings. The pore pressures were descent steeply over the transition between two media, i.e. thin-layer coal seam and the buffer layer. It was also observed that there were water exchanges between two media, especially in the later time period (more than 500 days), and the exchanges increased in scope and intensity under the circumstance of each simulated scenario.
- In the thin-layer coal seam, at early time periods (less than 100 days), the pore pressures in the flow inlet nearby were higher, e.g. maximum pressure up to about 6.0 Mpa (figure 5c), and decreased gradually over time. As a result, the curve's slopes of pore pressure duration varied from steep towering to gentle sloping with the progression of time. Eventually, it was seen that the curves became a flat slope along the seepage pathway in the intermediate phase (about 500 days) and final phase (around 1, 000 days) (figures 5a-5c). Permeability determined the pore pressure magnitude and the dissipation speed.
- With decreased permeability of the 14_{3S} thin-layer coal seam, it was observed that the pore pressure levels rose and conversely, the distance of seepage decreased. For example, the maxima of water pressure at the 10 m location were 1.91 Mpa, 1.84 Mpa, and 1.31 Mpa, corresponding to the three scenarios of K_f, 10^{-8} m^2/ pa·s, 10^{-9} m^2/ pa·s, and 10^{-10} m^2/ pa·s, respectively. The phenomenon relative to the permeability could potentially prevent water inrush disasters if the correlation between them could be discovered.

In short, seepage mainly occurred in the 14_{3S} thin-layer coal seam. The pore pressure levels and the seepage distance depended upon the value of the permeability coefficients in the fissured rock sandwich. These phenomena were simulated effectively by the numerical model in which the fluid-mechanical interaction embedded function in FLAC^{3D} was applied by tuning a few mechanical (table 1) and hydraulic parameters.

4.2. Hydraulic characteristics of pore pressure duration

According to K_f values, a series of pore pressure duration curves along the 14_{3S} thin-layer coal seam and buffer layers were illustrated successively (figures 6a-6c) by exporting their historical records of pore pressure determined by hypothetical history points in the numerical model of FLAC^{3D}. Some hydraulic characteristics were concluded as follow.

IOP Conf. Series: Earth and Environmental Science **39** (2016) 012031 doi:10.1088/1755-1315/39/1/012031

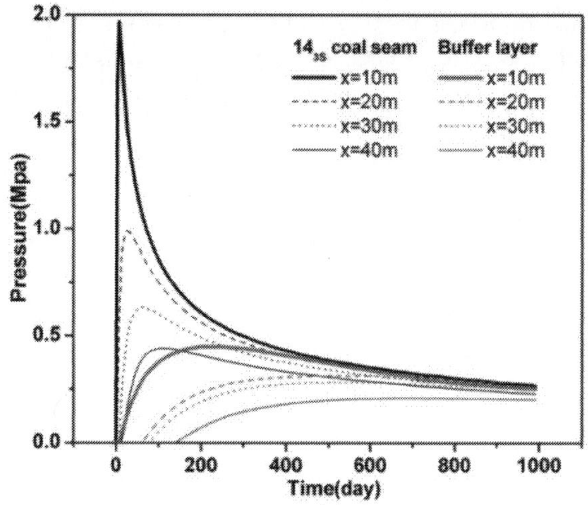

(a)

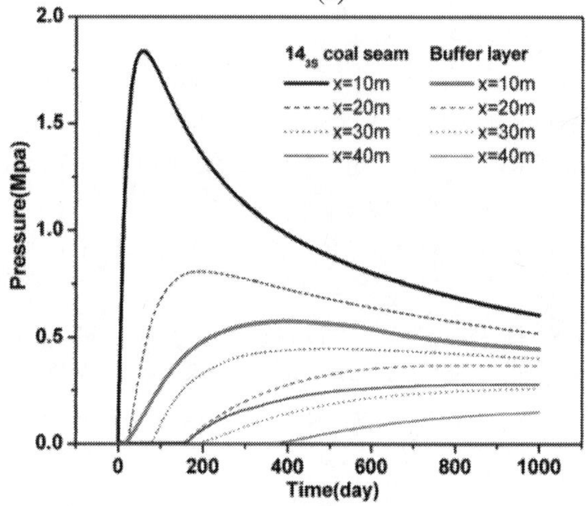

(b)

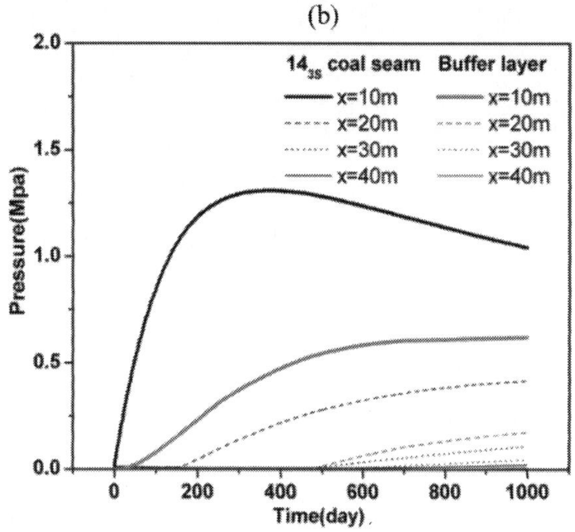

(c)

Figure 6. Curves of pore pressure duration of seepage evolution in different locations along the 14_{3S} thin-layer coal thin and buffer layer corresponding to the various scenarios in terms of K_f: (a) 10^{-8} m^2/ pa·s; (b) 10^{-9} m^2/ pa·s; (c) 10^{-10} m^2/ pa·s.

Referring to the figures above, the following observations were made:

- In the 14_{3S} thin-layer coal seam, there were obvious peak values in every pore pressure duration curves; however, they were less obvious in the buffer layers. Accordingly, the pore pressure duration curves presented a series of leptokurtic-shape lines with heavy tails in the nearby inlet (closer than 10 m) or platykurtic shapes at a distance (longer than 30 m) for the whole simulation period.
- The closer to the inlet of 14_{3S} thin-layer coal seam, the higher the pore pressure peak values and, as a result, the shorter the peak-occurring time lag period. Every group of pore pressure peak value in a scenario showed a descending trend along the thin-layer coal seam. In the segments of depletion, i.e. the intermediate or final phase, the pore pressure duration curves gradually slowed down in either the thin-layer coal seam or the buffer layer.
- The results of simulation indicated that the time-lag effects of the pore pressure peaks appeared successively in the 14_{3S} thin-layer coal seam. Whether the time lag period of pore pressure peaks in each of historical points was postponed mostly depended upon the variations of K_f value. Generally, with a lower K_f value, the peak-occurring lag times were delayed.

5. Analysis and discussion

5.1. Comparison of numerical and analytical solutions

Using the results obtained by numerical simulations, a series of pore pressure graphs of the 14_{3S} thin-layer coal seam (filling fracture medium) showed obvious peak-occurring and time-lag effects, which did not appear or were less obvious in the buffer layers (porosity medium).

In comparison to many existing analytical [6] or semi-analytical solutions [2] using DPMs, the novel solutions performed better for a variety of challenging parameter fields, including K_f, K_m, S_{sf}, S_{sm}, β, ϕ and structural and hydraulic parameters of rock mass. By tuning these parameters, the analytical results were very close to the simulated results, which were only obtained by valuing the mechanical and hydraulic parameters in the numerical model. Estimations of the fitting results in the various simulated scenarios were assessed by means of the root-mean-square error (RMSE), computed as follow:

$$RMSE = \sqrt{\frac{\sum_{i=1}^{n}\left(P_{sim,i} - P_{anal,i}\right)^2}{n}} \tag{4}$$

where n is the total number of moment of the discretized time domain, $P_{sim,i}$ is the simulated pore pressure at the i-th moment, and $P_{anal,i}$ is the analytical pore pressure at the i-th moment.

Figure 7 illustrates the fitting results in which the RMSE was regarded as the evaluation index. It can be seen that the fitting results of 14_{3S} thin-layer coal seam, RMSE 0.025 to 0.084 MPa, were better than that of the buffer layer, RMSE 0.065 to 0.121 MPa. Hydraulic parameters for various scenarios used in the analytical solutions of DPM are shown in Table 2.

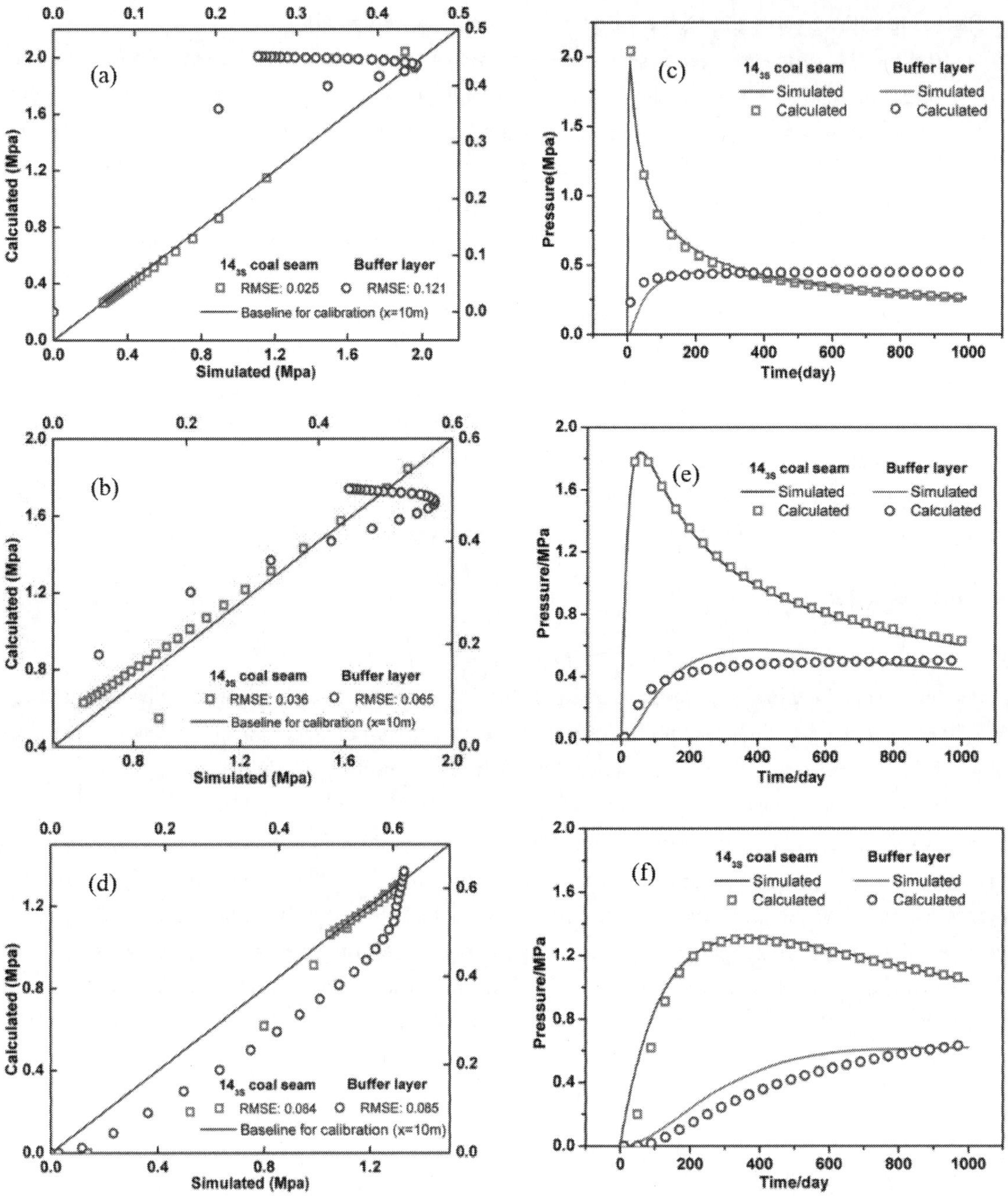

Figure 7. *Figures a, b, d*: Comparison of simulated (lines) and calculated (scatters) water pressure in x=10 m by numerical models and analytical solutions, respectively; *Figures c, e, f*: Scatter plot of simulated and calculated water pressures for calibration. (Each point represents a value of simulated/calculated water pressure for a time point. The performance evaluation index values (RMSE) are also shown in the figures.)

The results of the analytical solutions of the DPM provided accurate and analytical representations of peak-occurring and time-lag effects. This may have been caused by the following: (1) By use of Bessel function, the analytical solutions were more reasonable from a technical point of view; (2)

More impacting factors were parameterized in the form of coefficients, such as K_f and K_m, S_{sf} and S_{sm}, α, β, ϕ and so on. These parameters provided powerful capability and flexible options for fine tuning the shape of calculated curves. Therefore, the results obtained by the analytical solutions were more flexible and likely to achieve the desired fitting effects. Of course, they should correspond to reality.

Table 2. Hydraulic parameters used in the analytical solutions of DPM.

Parameters	Values
Scenario 1	
Permeability in 14_{3S} thin-layer coal seam (K_f)	1.0×10^{-8} m^2/ pa·s
Storage in 14_{3S} thin-layer coal seam (S_{sf})	$1.0\times10^{-0.15}$ MPa^{-1}
Storage in buffer layer (S_{sm})	$1.0\times10^{3.82}$ MPa^{-1}
Exchange coefficient (α)	$1.0\times10^{-6.06}$ m^2
Porosity-weighting coefficient (β)	$1.0\times10^{2.6}$
Scenario 2	
Permeability in 14_{3S} thin-layer coal seam (K_f)	1.0×10^{-9} m^2/ pa·s
Storage in 14_{3S} thin-layer coal seam (S_{sf})	$1.0\times10^{-0.15}$ MPa^{-1}
Storage in buffer layer (S_{sm})	$1.0\times10^{3.0}$ MPa^{-1}
Exchange coefficient (α)	$1.0\times10^{-6.44}$ m^2
Porosity-weighting coefficient (β)	$1.0\times10^{3.87}$
Scenario 3	
Permeability in 14_{3S} thin-layer coal seam (K_f)	1.0×10^{-10} m^2/ pa·s
Storage in 14_{3S} thin-layer coal seam (S_{sf})	$1.0\times10^{-0.32}$ MPa^{-1}
Storage in buffer layer (S_{sm})	$1.0\times10^{3.88}$ MPa^{-1}
Exchange coefficient (α)	$1.0\times10^{-4.85}$ m^2
Porosity-weighting coefficient (β)	$1.0\times10^{3.74}$

Through the analysis of several occurrences of typical lagging water inrush incidents at different depths in the past decade, i.e., No. 1 crosscut in the 9th level (approximately -730 m in 1972) [12] and in the 13th level (almost -1100 m in 2005) [13], it was found that the groundwater dynamics in both hydraulic discharge rate and water pressure presented a declining trend in the five-year observation period [17]. Examples are as follow: (1) The hydraulic discharge from 0.15 m^3/min to 0.10 m^3/min; and (2) The in-situ monitored water pressure from 4.0 Mpa to 2.0 Mpa. However, some problems arose while analyzing the in-situ monitoring hydraulic graph. First, the in-situ monitored water pressures contributed by both porosity and fracture seepages were mixed and undistinguishable from each other. Second, because of the lack of knowing when and where a water inrush incident will occur, it could be speculated that a peak value along the hydraulic graph should appear in the initial phase over which the in-situ hydraulic graphs are impossible to be recorded thereby. Notably, the fact that the occurrence of peak values exist is an important clue for water-inrush predictions, which is worth investigating for issues of mining security. The numerical model approach provides a promising and feasible solution to solving these problems. Meanwhile, novel analytical solutions are worth considering as a testing approach to verify the precision and reasonableness of numerical simulation solutions.

5.2. *Correlation analysis of hydraulic behaviors for risk prediction of water inrush*

A reasonable estimation of the permeability coefficients should be determined before risk prediction of water inrush is carried out. In reference to previous studies [3, 4], the permeability K_f range of 0.06 to 3.72 m/d was proposed. After all, it is reasonable that the permeability coefficient of the 14_{3S} thin-layer coal seam is at the range of 10^{-8} m^2/pa·s order of magnitude.

There exists a relationship between the pore pressure peak-value and peak-occurring lag time by considering the hydraulic coefficients as parameters such as K_f, K_m, and so on. As a case study, the correlation patterns of hydraulic characteristics under the influence of K_f were analyzed. By varying the value of K_f from 10^{-7} to 10^{-9} m^2/pa·s, the correlation between the pore pressure peak value and peak-occurring lag time was obtained based on K_f as an impacting parameter. Ultimately, the power law relationship between them can be found statistically. The results are shown in figure 8.

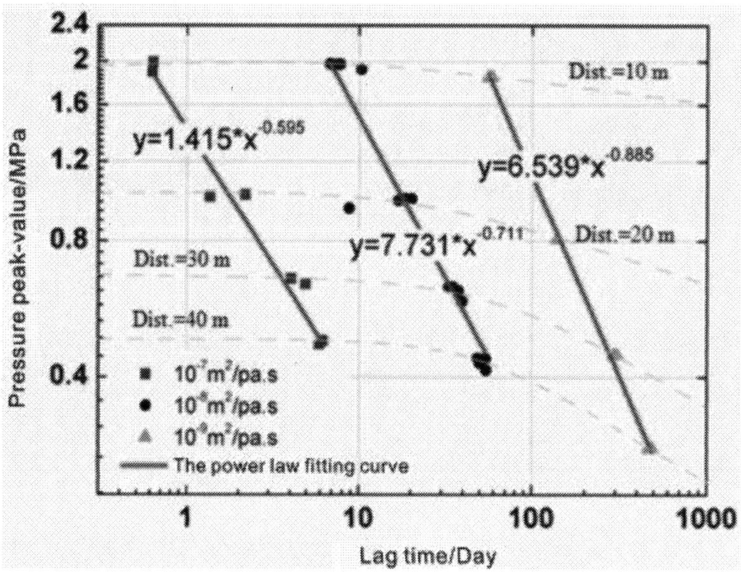

Figure 8. Correlation between the pore pressure peakvalue and the peak-occurring lag time within the range of 10^{-7}~10^{-9} m^2/pa·s orders of magnitude of K_f.

Table 3. Risk prediction chart of critical time for water inrush along the fissured coal sandwich.

Distance /m	Permeability K_f /$m^2 \cdot (pa \cdot s)^{-1}$	Pore pressure peak value /Mpa	Peak-occurring lag time /day	Critical time for water inrush prevention /day
10	10^{-8}	1.96	8.1	8.1
20		0.98	16.1	24.2
30		0.62	36.2	60.4
40		0.43	51.6	87.8

The data of figure 8 can be applied to risk prediction of water inrush to demonstrate a schematic example for practical application. A risk prediction chart for critical time for water inrush prevention along a fissured coal sandwich corresponding to the K_f as 10^{-8} m^2/pa·s order of magnitude is provided in Table 3. This approach can be applied in any coalmine with potential risks of water inrush occurring.

6. Conclusions

This paper conducted simulation analyses of the seepage evolution along the 14_{3S} thin-layer coal seam with pore pressure propagation in an assumed DPM. The pattern of pore pressure propagation was related to the permeability of the rock mass and agreed well with the analytical solution proposed by this paper. By considering the K_f as a parameter, the correlations between pore pressure peak-value and peak-occurring lag time were derived statistically and applicable to risk prediction of water inrush events. With the efforts made in this paper, the following conclusions were drawn.

- Seepage mainly took place in full-filled fractures, i.e. 14_{3S} thin-layer coal seam. The phenomenon of longer seepage distance and lower pore pressure occurred in the process with increased values of K_f.
- The pore pressure graph simulated in the 14_{3S} thin-layer coal seam had obvious hydraulic characteristics of peak-occurring and time-lag effects instead of the porous medium, which was representative of the buffer layers. With increased K_f, the time-lag effect became more remarkable. By assessing with RMSE value, the simulated results matched well with the theoretical calculated results using analytical solutions proposed by the DPM.
- There was a significant correlation between pore pressure peak-value and peak-occurring lag time. The power law relationship between them was found statistically within the range of 10^{-8} to 10^{-10} m^2/pa·s orders of K_f.
- In accordance with the previous study, the K_f of the 14_{3S} thin coal-seam was within the range of 10^{-8} m^2/pa•s, and it corresponded in reality to the Zhaogezhuang coalmine. A scheme for risk prediction of water inrush was tabulated and proved applicable in decision-making for risk prevention and prediction in practice.

Acknowledgments
This work was supported totally by the Natural Science Foundation of China (41462014) and the Kailuan Group Corporation Grant. The research was also partially supported by the Natural Science Foundation of China (41162011).

References

[1] Barenblatt G, Zheltov I P and Kochina I N 1960 Basic concepts in the theory of seepage of homogeneous liquids in fissured rocks *Appl. Math. Mech.* **24**(5): 1286-303
[2] Cornaton F and Perrochet P 2002 Analytical 1D dual-porosity equivalent solutions to 3D discrete single-continuum models. Application to karstic spring hydrograph modeling *Journal of Hydrology* **262**: 165-76
[3] Feng Z C, Zhao Y S and Wen Z M 2005 Percolation mechanism of fractured coal rocks as dual-continua *Chinese Journal of Rock Mechanics and Engineering* **24**(2): 236-40
[4] Li X G and Gao Y F 2003 Influence of mining on floor infiltration coefficients *Chinese Journal of Rock Mechanics and Engineering* **22**(7): 1078-82
[5] Louis C and Rock H 1974 *Rock Mechanics* ed. by L. muller, Udine
[6] Moutsopoulos K N and Tsihrintzis V A 2009 Analytical solutions and simulation approaches for double permeability aquifers *Water Resources Management* **23**: 395-415
[7] Onder H 1982 One-dimensional transient flow in a finite fractured aquifer system *Hydrological Sciences Journal* **43**(2): 243-65
[8] Samardzioska T and Popov V 2005 Numerical comparison of the equivalent continuum, non-homogeneous and dual porosity models for flow and transport in fractured porous media *Advances in Water Resources* **28**(3): 235-55
[9] Wilson R K and Aifantis E C 1982 On the theory of consolidation with double porosity *International Journal of Engineering Science* **20**(9): 1009-35
[10] Wu Q, Liu J T, Dong D L, et al. 2002 Trial analyses and numeric simulations of water-bursting time-effect under coal bed *Acta Geologica Sinica* **75**(4): 554-61
[11] Wu Q, Zhou Y J, Liu J T, Zhong Y P, Yin, Z R, Li J M, Hong Y Q and Zhou R G 2003 The mechanical experiment study on lag mechanism of water-bursting of fault under coal seam *China Coal Society* **28**(6): 561-5
[12] Wu Q, Liu J T, Zhong Y P, et al. 2002 The numeric simulationsof water-bursting time-effect for faults in Zhaogezhuang coal-mine, Kailuan, China *China Coal Soc.* **27**(5):511–6
[13] Wu Q, Zhu B and Liu S Q 2011 Flow-solid coupling simulation method analysis and time identification of lagging water-inrush near mine fault belt *Chinese Journal of Rock Mechanics and Engineering* **30**(1): 93-104
[14] Xiao Z R 1977 Expanding geological report of mining in both Linxi and Zhaogezhuang coalmines (Tangshan: Kailuan Group Co. Ltd.) ed Zhu S Z, Fu X M and Liu T W chapter 2 pp 10-18
[15] Zhang Y X, Tu S H, Bai Q S and et al. 2013 Overburden fracture evolution laws and water-controlling technologies in mining very thick coal seam under water-rich roof *International Journal of Mining Science and Technology* **23**(5): 693-700

[16] Zhu B 2009 Research on fluid-mechanical interaction simulationand application on lagging water inrush near fault zone. Workingreport of post-doctoral fellow *China University of Mining and Technology* Beijing: pp1–10

[17] Zhu B, Wu Q, Yang J W and Cui T 2014 Study of pore pressure change during mining and its application on water inrush prevention: a numerical simulation case in Zhaogezhuang coalmine, China *Environmental Earth Science*

IOP Conf. Series: Earth and Environmental Science **39** (2016) 012032 doi:10.1088/1755-1315/39/1/012032

The hydrogeochemical characteristic of different sulfide-containing gangue during leaching

Y Q Sun [1, 2, 3], L Duan[1, 2], M Yao [1] and W P Li [1]

[1]School of Environmental Science and Engineering, Chang'an University, Xi'an 710054, China

[2]Key Laboratory of Subsurface Hydrology and Ecological Effect in Arid Region of Ministry of Education, Xi'an 710054, China

E-mail: sunyaqiao@126.com

Abstract. This investigation examined the hydrogeochemical characteristics of different sulfide-containing gangue samples from the Hancheng mining area in Shaanxi, China. The changes in the hydrogeochemical properties during the leaching experiments on gangue samples containing different levels of sulfur (LSG-Low sulfur gangue, MSG-Moderate sulfur gangue and HSG-High sulfur gangue) were investigated by analyzing the major physical and chemical constituents (pH, EC, TDS, Ca^{2+}, Mg^{2+}, Na^{+}, K^{+}, Cl^{-}, SO_4^{2-}, HCO_3^{-} , F^{-}, and Fe). The results indicated that the release of SO_4^{2-} and H^{+} from the dissolution of sulfide minerals resulted in low Ca^{2+} and HCO_3^{-} concentrations at pH levels below 4.7 for the HSG leachate. Also for the HSG samples, the molar ratio of SO_4^{2-}/Fe exceeded twice that which was expected, meaning the higher SO_4^{2-} concentrations in the HSG samples result from more than just the oxidation of pyrite. Additionally, the leachate of three sulfide-containing gangue samples exhibited changes in the hydrogeochemical facies with the majority of changes occurring in the cation-type and anion-type facies. The HSG samples were of a [Cl-SO_4] type for sulfide dissolution, the unaffected LSG samples were of a [Ca-HCO_3] type, and the MSG samples were of a [Ca-HCO_3-SO_4] type.

1. Introduction

Coal gangue is the solid waste that is generated during coal mining [1-2]. There are approximately 1500 coal gangue dumps in China that together contain a combined weight of four billion tons. This is highly problematic as this waste has the potential to severely impact the surrounding environment as well as human health due to the harmful elements contained within the gangue [3-5]. However, due to the high sulfide content of this gangue, there are limited applications for this waste. The hydrogeochemistry of this sulfide-rich substance plays a key role in its environmental impact [6-8]. More specifically, the aqueous oxidation processes associated with the sulfide minerals in the gangue produce sulfuric acid, which causes water pollution and soil acidification in addition to harming human health [9-10].

There is a need for investigations to focus on the hydrogeochemical behavior of gangue-water, such as its low pH, resulting drainage with sulfuric acid, and the dissolution of heavy and toxic metals. Sulfur-rich gangue has been classified into the following three types by the National Coal Board (NCB) depending on its sulfur content [11]: low content, with total sulfur <1%; moderate content, with total sulfur between 1-2%; and high content, with total sulfur>2%. In contrast, the Chinese determined the standard sulfur content ranges in coal grade divisions according to its different utilizations. Additionally, not only has sulfur coal gangue been identified in China [12-13], but it has also been identified in several other countries, such as Upper Silesia[14], Northern Iran[15], and India[9]. Furthermore, it is important to continue investigating this area because a large amount of sulfide-containing coal is currently being used as fuel, and consequently, the amount of oxidized sulfur is high [16-18]. Furthermore, sulfur coal gangue is associated with the atmospheric dispersion of fine particles, the spontaneous combustion of gangue dumps, the leaching of elements, and the formation of acid drainage as caused by natural weathering and/or oxidation processes.

The chemistry of gangue leachate depends upon, amongst other factors, the hydrogeological environment, the accompanying gangue material, and the interaction time in gangue-water. Geochemical processes are responsible for releasing sulfur and carbonate buffering into the gangue-water. High sulfide-containing coal gangue may lead to the leaching of harmful elements as well as the release of organic pollutants. Moreover, cations and silica are released for the gangue-water interaction of primary and secondary minerals [19-20]. Contaminants/pollutants from gangue typically flow through the common pathways in gangue dumps during dispersion. Investigations of gangue pollutants are often limited to identifying the release extent of heavy metals as well as their source, cause, and prevention [21-23]. Thus, investigations into the geochemical processes that affect sulfide-containing gangue during gangue-water interactions are often overlooked.

Over the last forty years, due to an increase in coal mining, there has been a significant increase in the volume of gangue extracted from the Hancheng coal mine and stored on-site. The lack of stringent environmental protection measures, coupled with increased stacking, has resulted in adverse environmental impacts on the ground water and soil in the study area. This study aims to investigate the hydrogeochemical characteristics controlling the leaching of harmful substances from the gangue and to evaluate the evolution of sulfur oxidation in gangue-water. To undertake this research, the leaching of quality parameters were investigated, including parameters such as pH, electrical conductivity (EC), total dissolved solids (TDS), total alkalinity, and total hardness (TH). Various cations/anions (Ca^{2+}, Mg^{2+}, Na^{+}, K^{+}, Cl^{-}, SO_4^{2-}, HCO_3^{-}, F^{-}, and Fe) were determined. A hydrogeochemical facies evaluation was also undertaken by plotting a Piper diagram in order to understand the gangue-water geochemistry in the leaching experiment. Overall, a systematic effort was made to investigate the hydrogeochemical characteristics of sulfur and the other chemical elements released into gangue-water.

2. Materials and Methods

2.1. Sample Collection

Coal gangue was collected from three sites in the Hancheng coal mining area, Shaanxi province, China (figure 1) in May 2014. The sample locations were chosen to provide gangue that contained low, moderate, and high sulfur contents. The low sulfur gangue (LSG, 0.04% sulfur content) was collected from the Permian Shanxi formation #3 coal bed; the moderate sulfur gangue (MSG, 1.29% sulfur content) was collected from the Permian Shanxi formation #2 and #3 coal beds; and the high sulfur gangue (HSG, 3.40% sulfur content) was collected from the Carboniferous System Taiyuan formation #11 coal bed. The gangue samples were air-dried in the laboratory and sieved through 0.147 mm nylon sieve meshes.

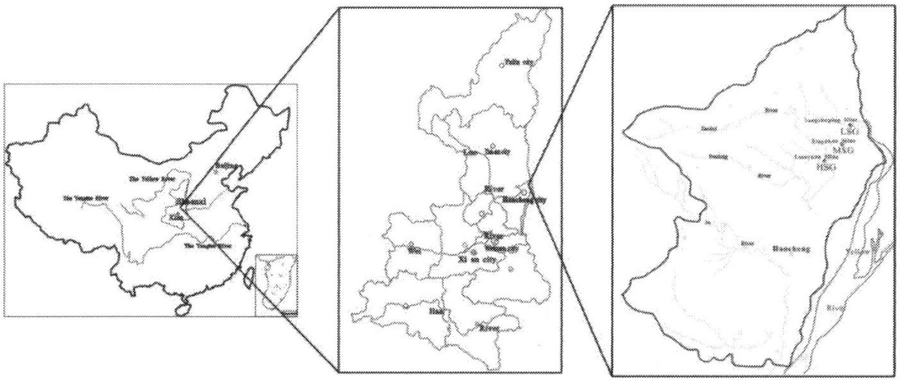

Figure 1. Sampling site locations within the Hancheng, Shaanxi province.

2.2. Leaching tests

To investigate the effect of sulfide on the hydrogeochemical processes in gangue, leaching experiments using deionized water were conducted at liquid-solid ratios (L:S) of 10:1, 8:1, 6:1, 4:1, and 2:1. The samples were rotated in a tumbler at 40 rpm for 48 h before being filtered for chemical composition analysis. Leached samples for cation determination were filtered through a 0.45 um Millipore nitrocellulose filter. Then, 10mL of the filtered sample was acidified with ultrapure HNO_3 (10%) until pH<2 was attained. All samples were stored at 4℃ in HDPE bottles prior to analysis.

2.3. Analytical Methods

The samples prepared using the leaching method were analyzed for pH, EC, TH, and cations/anions (including Ca^{2+}, Mg^{2+}, Na^+, K^+, Cl^-, SO_4^{2-}, Cl^-, HCO_3^- , F^- and Fe). Major cations (K, Na, Ca, and Mg) and other trace elements were determined using inductively coupled plasma atomic emission spectrometry (ICP-OES) (American thermoelectric icap6300). The pH was measured using METTLER TOLEDO SG2-ELK, and the EC was measured using METTLER TOLEDO FG3. The total S was determined using an infrared carbon sulfur analyzer (CS-8800), and the mineral composition of the gangue was analyzed by XRD (D/max-2500).

The analytical precision for the cations and anions, as indicated by the ionic balance error, was calculated to be within a ±5% limit for all of the samples. The quality of the analysis was ensured by standardization using blank, spike, and duplicate samples.

3. Results and Discussion

3.1. Chemical characteristics

The mean pH values in the samples were 9.18, 7.46, and 2.96 for LSG, MSG, and HSG, respectively (Table 1). The recorded pH of the LSG and MSG samples were alkaline. The LSG samples had a pH ranging from 8.88 to 9.40, and the MSG samples ranged from 7.37 to 7.58. In contrast, the pH results of the HSG samples were very acidic, ranging from 2.38 to 3.62 (Table 2). The HSG leachates had the highest average EC values (3976.00us/cm), thereby indicating that the acidic condition of the samples resulted in high concentrations of dissolved material during the leaching process. The leachates from the LSG and MSG samples had EC ranges of 172.20-435.00us/cm and 1371.00-2420.00us/cm, respectively. This indicated that in alkaline conditions, the leachate had lower solute concentrations as compared with the leaching process [8, 24]. The average hardness values were 10.75, 23.97, and 8.21mg/L for LSG, MSG, and HSG, respectively.

Table 1. Chemical data of leaching samples.

Sulfur content	Low (0.04%)			Middle (1.29%)			High (3.40%)		
	mean	max	min	mean	max	min	mean	max	min
PH	9.18	9.40	8.88	7.46	7.58	7.37	2.96	3.62	2.38
EC (μs/cm)	274.84	435.00	172.20	1,831.40	2,420.00	1,371.00	3,976.00	7,330.00	1,830.00
TH (mg/l)	10.95	16.15	5.44	23.97	32.01	6.01	8.21	14.21	4.23
K^+ (mg/L)	2.332	2.916	1.720	2.416	3.916	1.900	1.781	2.365	1.466
Na^+ (mg/L)	4.788	11.630	1.365	1.992	3.441	0.413	0.614	1.077	0.241
Ca^{2+} (mg/L)	8.978	13.220	4.666	20.949	27.110	2.795	6.675	12.460	2.758
Mg^{2+} (mg/L)	1.977	3.146	0.770	3.023	4.896	1.187	1.537	1.745	1.449
HCO_3^- (mg/l)	105.287	157.713	55.465	67.812	89.949	49.677	0.000	0.000	0.000
Cl^- (mg/L)	3.412	4.555	1.914	16.491	29.000	9.233	212.466	398.500	115.230
SO_4^{2-} (mg/L)	5.126	5.871	4.107	36.021	59.550	24.141	100.236	127.770	85.980
F^- (mg/L)	2.182	3.963	1.190	1.503	1.584	1.446	0.008	0.010	0.006
Fe (mg/L)	9.533	11.290	7.851	36.028	46.700	17.190	24.890	27.390	23.400

The results from the cation analysis indicated that Ca^{2+} and Mg^{2+} were the dominant elements; together they averaged 61%, 84%, and 77% of the total cations in the LSG, MSG, and HSG samples, respectively (Figure 2). Ca^{2+} was the major constituent with average concentrations of 8.978, 20.949, and 6.675mg/L, respectively (table 1). For comparison, the dominance of Ca2+ in the samples followed the following sequence: MSG (74%) > HSG (63%) > LSG (50%). Together, K^+ and Na^+ averaged 39%, 15%, and 23% of the total cations in the LSG, MSG, and HSG samples, respectively (figure 2).

Bicarbonate (HCO_3^-) is the dominant anion in the LSG and MSG samples with average values of 105.287mg/L (ranging from 55.465-157.713mg/L) and 67.812mg/L (ranging from 49.677-89.949mg/L), respectively. For the dissolution of pyrite during the HSG leaching process, the pH value is lower than 4.7 (ranging from 2.38-3.62 mg/L; mean of 2.96 mg/L) (table 1). The contribution of HCO_3^- is 0% (figure 2). The dissolution of gypsum and pyrite minerals (Table 4) in the samples contributes to the SO_4^{2-} concentration; thus, the SO_4^{2-} results increased to 4.107-5.871mg/L (mean of 5.126mg/L), 24.141-59.550mg/L (mean of

36.021mg/L), and 85.980-127.770mg/L (mean of 100.236mg/L) in the LSG, MSG, and HSG samples, respectively. Therefore, the SO_4^{2-} percentage in the samples followed the following sequence: HSG (32%) > MSG (30%) > LSG (4%) (figure 2). Additionally, the Cl^- concentration in the samples followed the following sequence: HSG (68%) > MSG (13%) > LSG (3%). In contrast, the low F- contribution was the following: LSG (2%) > MSG (1%) > HSG (0%).

Table 2. Mineral composition of the gangue used in the leaching experiments.

Mineral phase	LSG (wt%)	MSG (wt%)	HSG (wt%)
Quartz	40	35	16
Plagioclase	2	1	2
Potassium feldspar	1	/	1
Calcite	4	/	/
Chlorite	/	/	7
Montmorillonite	/	1	2
Illite	11	12	13
Ammonium Illite	/	/	
Kaolinite	26	34	32
Dolomite	1	/	/
Gypsum	/	8	8
Pyrite	/	/	8
Anatase	/	/	1
Jarosite	/	4	5
The amorphous phase	15	5	5

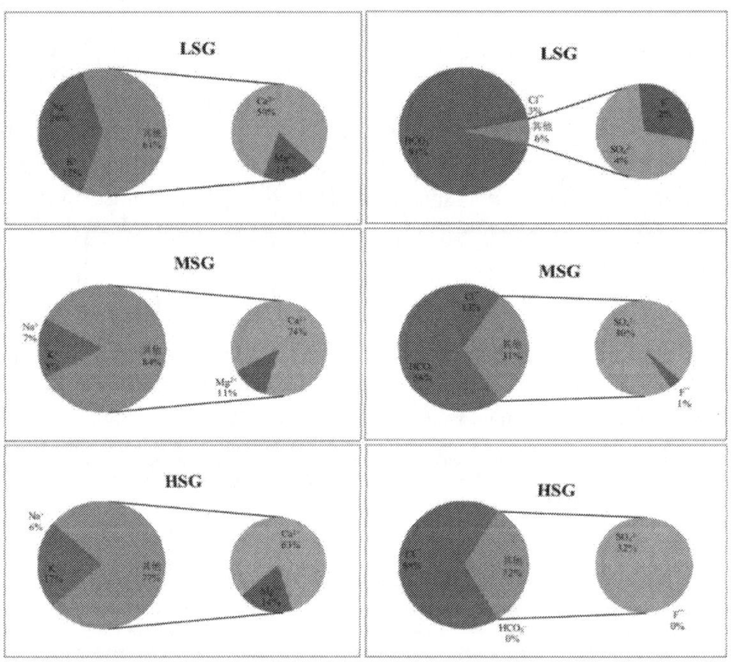

Figure 2. Average percent contribution of individual ions.

3.2. *Hydrochemical facies of leaching*

The Piper diagram (figure 3) reveals the similarities and differences in the sample compositions from the three gangue sites. The results are consistent with the other results in this study, and they confirm the hydrochemistry characteristics from the different sulfur-containing gangue samples in the experiment. As indicated in the left ternary diagram of Figure 3, the LSG and MSG samples were [Ca] type. he right ternary diagram indicates that the [HCO_3], [HCO_3-SO_4], and [Cl-SO_4] types are present in the LSG, MSG and HSG samples, respectively. Therefore, the LSG and MSG samples are predominantly [Ca-HCO_3] and [Ca-HCO_3-SO_4] types, respectively. The HSG samples are dominated by the sulfur mineral dissolution (pyrite, gypsum, and jarosite; table 2), which, as noted previously, results in these samples having lower pH values and a lack of carbonate formation. The chemical type of the HSG samples is chloride-sulfate water. Furthermore, it should be noted that the differences in the hydrogeochemical facies in the LSG, MSG, and HSG samples of [Ca-HCO_3], [Ca-HCO_3·SO_4], and [Cl-SO_4], respectively, are observed in the ternary diagram (Figure. 3). The differences are possibly influenced by the mineral composition and the sulfur contained in the gangue as well as by the interaction with water during the leaching process. This will affect the chemical characteristics and the water-gangue interactions.

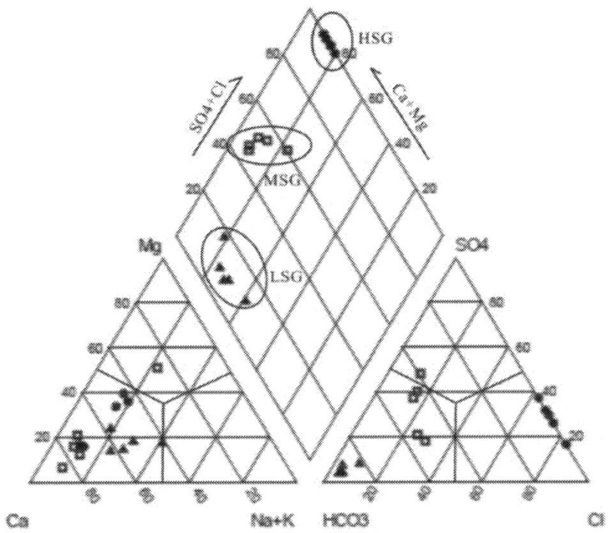

Figure 3. The Piper Diagram for gangue leaching samples.

3.3. *pH changes with pyrite oxidation*

During the leaching experiments, pyrite was identified as one of the main sulfide minerals in the HSG samples (table 2). Pyrite oxidation releases free H+ into the solution, which results in low pH values and a high SO_4^{2-} concentration leached from the gangue. The chemical reaction to produce sulfuric acid in the solution is:

$$FeS_2(s)+7/2O_2+H_2O \rightarrow Fe^{2+}+2SO_4^{2-}+2H^+ \quad (1)$$

$$3FeS_2(s)+11.25O_2+10.5H_2O \rightarrow Fe(OH)_3\downarrow+6H_2SO_4 \quad (2)$$

The gangue leaching increased the sulfate content and lowered the pH in the leachate as well as elevated the iron and SO_4^{2-} concentrations (Chemical quation.1).

In the HSG leaching samples, a high sulfide-containing mineral concentration (table 2) and a low pH indicated that leaching of these samples was affected by acid mine drainage and the pyrite oxidation. The maximum acidification produced by the complete oxidation of the sulfur minerals in the gangue leaching was investigated. In all of the HSG samples, the pH of the samples ranged between 2.38 and 3.62 for about 8% pyrite oxidation and dissolution. In these samples, a lower concentration of dissolved total Fe was detected (table 1); this being an indication of the rapid precipitation of the mineral phases of Fe (Chemical quation.2) for the FeS_2 oxidation [22].

High Fe and SO_4^{2-} concentrations (figure 2) in the HSG samples may be due to the oxidation of pyrite, the dissolution of iron-bearing minerals and gypsum, and the subsequent transformation of oxidation products under acidic conditions. The SO_4^{2-}/Fe molar ratio recorded in these samples was more than twice the expected levels from pyrite dissolution alone, which is a result that suggests the occurrence of the dissolution of gypsum and jarosite (Table 2). The saturation indices of gypsum were less than zero, thus indicating that gypsum was unsaturated in the HSG samples. As the HSG samples were strongly corrosive to sulfur-containing minerals during leaching, minerals, such as gypsum and pyrite, dissolved during leaching further reduce the pH of the solution. During the HSG samples leaching, the neutralization process resulting from carbonate mineral dissolution was much less than that resulting from the pyrite dissolution. Therefore, the pH values of the HSG samples were lower, even though more silicate minerals were available for dissolution buffering during leaching.

4. Conclusions

The hydrogeochemistry characteristics of the LSG, MSG, and HSG leached samples were investigated in order to analyze the reaction between sulfur-containing minerals in the gangue and water. Ca^{2+} was found to be the major ion released from the gangue during the experiments. Furthermore, it was found that the pH values of the HSG samples were very acidic (2.38-3.62) due to the dissolution of pyrite. Additionally, the molar ratio of SO_4^{2-}/Fe in the HSG samples was more than twice that which was expected, which suggested that the higher SO_4^{2-} concentrations in the HSG samples affected the oxidation of pyrite as well as the dissolution of other sulfur minerals in the gangue. Furthermore, the [Ca-HCO_3], [Ca-HCO_3·SO_4], and [Cl-SO_4] facies were the main hydrogeochemical facies in the LSG, MSG, and HSG samples, respectively. Sulfur mineral dissolution and lower pH levels were responsible for a lack of carbonate formation and chloride-sulfate type water formation in the HSG samples. The present results have the potential to aid in providing a scientific basis during coal gangue landfill treatments and comprehensive applications.

Acknowledgment

This work was supported by the National Natural Science Foundation of China (41102150, 41002086, and 41372258), the natural science foundation of Shaanxi Province (2013JM5003 and 2016JM4002), and the Opening Foundation of Shaanxi Key Laboratory of Exploration And Comprehensive Utilization of Mineral resources (2014HB006). We thank Key Laboratory of Subsurface Hydrology and Ecological Effect in Arid Region of Ministry of Education, Chang'anUnversity, China.

References

[1] Shealy M and Dorian J P 2010 Growing Chinese coal use: Dramatic resource and

environmental implications *Energy Policy* **38**(5): 2116–22

[2] Shujie Y 2005 Some Aspects of the Coal Industry Sustainable Development in China In ESobczyk & J KickiEds, International mining forum 2005 new technologies in underground mining, safety and sustainable development Proceedings of the Sixth International Mining Forum 2005 CRC Press *Cracow-Szczyrk-Wieliczka* Poland: 5–12

[3] Fan J S, Sun Y Z, Li X Y, Zhao C L, Tian D X, Shao L Y and Wang J X 2013 Pollution of organic compounds and heavy metals in a coal gangue dump of the Gequan Coal Mine China *Chinese Journal of Geochemistry* **32**: 241-47

[4] Meng L, Feng Q, Zhou L, Ping L and Meng Q 2009Environmental cumulative effects of coal underground mining Procedia *Earth and Planetary Science* **11**: 1280–84

[5] Boularbah A, Schwartz C, Bitton G and Morel J L 2006 Heavy metal contamination frommining sites in South Morocco: 1 Use of a biotest to assess metal toxicity of tailings and soils; 2 Assessment of metal accumulation and toxicity in plants *Chemosphere* **635**: 802–17

[6] Chandra A P and Gerson A R 2011 Redox potential Eh and anion effects of pyrite FeS2 leaching at pH1 *Geochimica et Cosmochimica Acta* **75**: 6893-911

[7] Romero A, Gonzalez I and Galan E 2011 Stream water geochemistry from mine wastes in Pena de Hierro, Ritinto area, SW Spain: a case of extreme acid mine drainage *Environmental Earth Sciences* **62**: 645-56

[8] Tao X Z, Wu P, Tang C Y, Liu H and Sun J 2012 Effect of acid mine drainage on a karstbasin: a case study on the high-As coal mining area in Guizhou province, China *Environmental Earth Sciences* **65**: 631-38

[9] Murkute Y A 2014 Hydrogeochemical characterization and quality assessment of groundwater around Umrer coal mine area Nagpur District, Maharashtra, India*Environmental Earth Sciences* **72**: 4059-73

[10] Kuenzer C, Zhang J, Tetzlaff A, Van Dijk P, Voigt, S and Mehl H 2007 Uncontrolled coal fires and their environmental impacts: Investigating two arid mining regions in north-central China *Applied Geography* **271**: 42–62

[11] Spears D A, Rippon J H and Cavender P F 1999 Geological controls on the sulphur distribution in British Carbonigerous coals: a review and reappraisal *International Journal of Coal Geology* **401**: 59-81

[12] Wen X Q, Zeng F G and Zhang H 2013 Geochemical characteristics of the coal gangues from the Du'erping coal mine,Xishan coalfield, North China *Chinese Journal of Geochemistry* **32**: 227-34

[13] Xue Q, Lu H J, Zhao Y and Liu L 2014 The metal ions release and microstructure of coal gangue corroded by acid-based chemical solution *Environmental Earth Sciences* **71**: 3235-44

[14] Szczepanska J and Twardowska I 1999 Distribution and environmental impact of coal-mining wastes in Upper Silesia, Poland*Environmental Geology* **383**: 249-58

[15] Shahhoeiny M, DoulatiArdejani F, Shafaei S Z, Noaparast M and Hamidi D 2013 Geochemical and mineralogical characterization of a pyritic waste pile at the AnjirTangen Coal Washing Plant, Zirab, Northern Iran *Mine Water and the Environment* **32**: 84-96

[16] Tang Z and Yang W 2000 Efficient and environment friendly use of coal *Fuel processing Technology* **62**: 137-41

[17] Zhang J, Han C and Xu Y Q 2003 The release of the hazardous elements from coal in theinitial stage of combustion process *Fuel processing Technology* **84**: 121-33

[18] Su S, Li B, Cui S and Tao S 2011 Sulfur Dioxide emissions from combustion in china: From 1990 to 2007 Environmental Science and Technology American Chemical Society Publications Beijing *Peking University*: 1–8

[19] Medina A, Gamero P, Querol X, Moreno N, Leon B D, Almanza M and Vargas G, Izquierdo M and Font O 2010 Fly ash from a Mexican mineral coal Ⅰ : mineralogical and chemical characterization *Journal of Hazardous Material* **181**: 82-90

[20] Sun W, Wu Q, Dong D and Jiao J 2012 Avoiding coal–water conflicts during the development of China s large coal-producing regions *Mine Water and the Environment* **311**: 74–78

[21] Ribeiro J, Silva E F D, Jesus A P D and Flores D 2011 Petrographic and geochemical characterization of coal waste piles from Douro coalfield *International Journal of Coal Geology* **87**: 226-36

[22] Silva L F O, Querol X, Boit K M, Vallejuelo F O and Madariaga J M 2011 Brazilian coal mining residues and sulphide oxidation by Fenton's reaction: an accelerated weathering procedure to evaluate possible environmental impact *Journal of Hazardous Materials* **186**: 516-25

[23] Zhang Y, Feng Q Y, Meng Q J, Lu P and Meng L 2012 Distribution and Bioavailability of metals in subsidence land in a coal mine China *Bulletin of Environmental Contamination Toxicology* **89**: 1225-30

[24] Lee J E and Kim Y 2008 A quantitative estimation of the factors affecting pH changes using simple geochemical data from acid mine drainage *Environmental Geology* **55**: 65-75

Evaluation and analysis of underground brine resources in the southern coastal area of Laizhou Bay

M Tian[1], H T Zhu[2], J Feng[1,3] and Q S Zhao[1]

[1]Department of Environmental Science and Engineering, Qingdao University, Qingdao 266000, China
[2] Qingdao No.2 High School, Qingdao 266000, China

E-mail: qdufengjuan@126.com

Abstract. The southern coastal districts of Laizhou Bay are some of the most important areas for underground brine exploitation in Shandong Province. Recently, these areas have been gradually developed by the underground brine mining industry. Such economic interest has led to brine exploitation so that underground brine resources are running out. Based on this phenomenon, this study describes the supply, runoff and draining conditions of the area by collecting and organizing the background information of the studied area. Hydrogeological parameters are then calculated according to pumping tests, and the amount of sustainable resources in the coastal areas of the Southern Bank of Laizhou Bay are then calculated based on the uniform distribution of wells. Under the circumstances of underground brine mining, the exploitation potential of the underground brine is evaluated in accordance with the calculation results of exploitation quantum. Finally, suggestions are provided for the sustainable exploitation of underground brine in the area.

1. Introduction

Underground brine, one of the most accessible and exploited underground water resources, contains high-concentrations of Na and Cl ions and varying amounts of elements such as Br, I, K, Li, Mg, Ca, Sr, and Ba, with a mineralization exceeding 50 g/L. Underground brine has been widely-used in the salt, chemical, aerospace, and nuclear industries [1]. Its formation is primarily determined by natural geographical conditions and climatic characteristics.

Underground brine in the study area was formed in the late quaternary period and has become a massive metallogenic belt distributed along the contemporary coastal line [2]. The general mineralization varies from 50g/L to 150g/L, with the highest reaching 217g/L. Generally, the buried depth of the underground brine varies from 0m to 60m while the deepest ranges from 70 to 80m, including approximately four brine layer groups, which can be categorized as latent underground brine andpressurized underground brine. According to incomplete statistics, there are 228 mining enterprises which produce crude salt and bromine, more than 70 of which are located in Shouguang City and Hanting District, Weifang City. The gross area covers 189.79 km^2 with 7084 brine-producing wells, and the annual exploitation quantity is approximately equal to 2 9489.7 million m^3/a, in addition to an annual crude-salt production of approximately 8.099 million tons and bromide production of 31.5 thousand tons. The overall production quantity of Shouguang City is ranked the highest in the world, producing enormous economic profits.

However, while environment-unfriendly development has brought considerable economic profit, over-exploitation leads to great disparity in exploitation quantity and recharge rate of underground brine. The amount of mined underground brine has increased in recent years, despite a simultaneous decrease in groundwater level which has forced underground brine production into a depression, resulting in huge impact to the geological and ecological environment of the area. Therefore, evaluating and analyzing the underground brine resources in the southern coastal area of Laizhou Bay can contribute to the protection and rational utilization of brine resources [3-5].

2. Hydrological and geological conditions of underground brine resources

2.1. The division of underground brine layers

The phreatic and faint confined brine layer, as well as the unconfined and faint confined aquifer, is located in the Holocene marine sedimentary strata in the quaternary region. Due to the constant mining of underground brine, this aquifer only exists in areas with a buried groundwater depth less than 10 m.

Confined brine layers, located in the Pleistocene marine sedimentary strata in the Quaternary, are separately distributed throughout the entire area, as well as an impermeable layer of certain thickness located between the aquifer and the unconfined and faint confined aquifer above. In this region, the confined aquifer is the main brine layer, composed of two to four fairly stable brine layers in the coastal area north of Weifang and three fairly stable brine layers in Tushan, Laizhou. The buried depth of the confined brine layers in the north coastal district of Weifang is the shallowest and thinnest.

There is a compact aquifuge with a stable horizon located between the phreatic and faint confined brine layer and the unconfined and confined aquifer; the permeability coefficient of the aquifuge varies from 2.5×10^{-7} to 6.7×10^{-7}cm/s, and exhibits good water isolating properties.

2.2. Distribution of underground brine layers

The underground brine resources in the study area are of particular zonation. To the west of Yangzi Town, Hanting District, the concentration of brine is low-high-low from the coast to inland, which indicates lower brine concentrations both near shore and far off-bank, and higher brine concentration in the middle region. To the east of Yangzi Town, Hanting District, the concentration of brine is high-low going from the coast to inland [6], which indicates a higher brine concentration near shore and lower brine concentrations far off-bank.

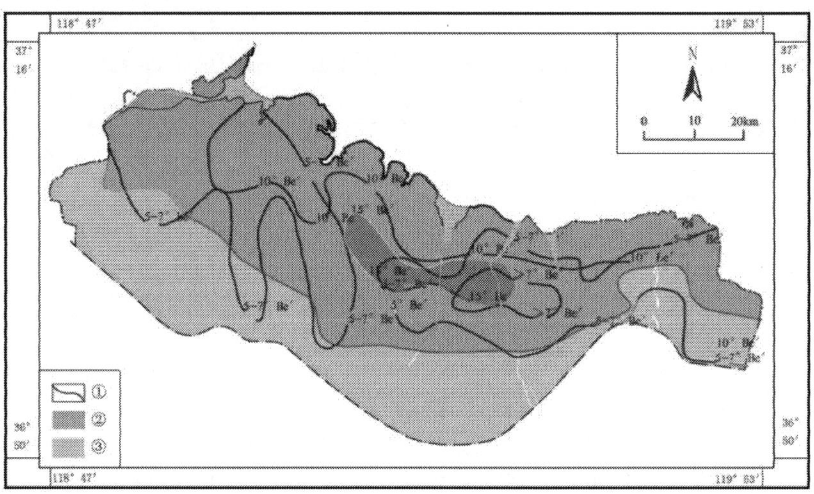

①Isoline of brine concentration ②Brine area ③Salt-water area

Figure 1. Brine concentration distribution.

Lower brine concentration near shore is frequently affected by tidal action. This area is approximately 10km wide, and the brine concentration varies from approximately 5°Be′ (°Be′ is used to denote the concentration of solution) to 10°Be′. The higher concentration area in the middle is roughly equivalent to the position of the 2.5m above sea level and varies from 10km to 20km in width. This area is generally unaffected by tidal action except for the equinoctial circle spring tide every month or year. Brine in this area exhibits a high and stable concentration, which generally ranges from 10°Be′ to 15°Be′ with a maximum value of 17.5°Be′. The area of lower brine concentration far off-bank is farther away from the coast, and ranges from 5km to 10km in width. This area is mostly unaffected by seawater but is affected fresh water coming from land and meteoric water. The brine concentration ranges from 5°Be′ to 10°Be′ and gradually changes into saltwater with a concentration less than 5°Be′. Underground brine in the studied areas is distributed along the Xiaoqing River and Sha River. This area is parallel to the coast and distributed in a northwest-to-southeast direction, and is located 75km to 90km EW-trending, and from 6km to 20km NS-trending. The total area is approximately 1348.48km^2.

2.3. Hydrological and geological characteristics

2.3.1. Recharge, runoff and drainage. The recharge of the brine in this area is accomplished by artificial recharge. Seawater flows through artificial saving pools and is poured vertically into the shallow brine layer. The runoff of natural brine is primarily caused by production activities. In the brine extraction area, when a large amount of mined brine changes the brine level and forms a cone of depression, brine around the mining area or peripheral brine will runoff into the mining area. Brine, seawater and the surrounding saltwater is basically in dynamic equilibrium with slow runoff in areas where brine is conserved, such as Wudi, the Jiaozhou Bay coast, etc. [7, 8]. The main drainage route natural brine is artificial exploitation, followed by evaporation discharge, which concentrate the brine concentrated allow infiltration of atmospheric precipitation to contribute to the dilution of natural brine.

2.3.2. Water abundance. Experimental pumping tests were conducted on mining wells in salt fields. Results indicate that when drawdown ranges from approxiamtely 3.3m to 15.0m, water outflow from one certain well varies from 11.72m^3/d.m to 255.68m^3/d.m, indicating that water abundance is medium to good. The results also indicate that the coefficient of permeability ranges from 0.85m/d to 13.94m/d, and that the radius of influence ranges from 31.85m to 172.49m. Units-inflow of the brine layer increases, and the water abundance improves from west to east.

2.3.3. Characteristics of water level dynamic. According to underground brine statistics released by the geological environment monitoring station in Weifang, the level of brine gradually decreased as the exploitation quantity of brine increased. Overall, the trend decreased annually. From 1991 to 2005, the water level of the brine decreased slowly with a rate of 0.13m/a. After 2005, due to increased prices of crude salt and bromine, a substantial increase of the exploitation quantity of brine resources occurred which, combined with the influence of brine exploitation in other regions, resulted in a water level decrease of 1.39m/a. This indicates that brine resources are limited and lack the ability to recharge. Therefore, long-term exploitation will certainly cause the brine weater level to decrease [9].

3. Evaluation of underground brine resources

3.1. Exploitable amount of underground brine resources

Assuming that the group well pumps do not impact a single well, the exploitable amount of underground brine in the study area can be estimated using the method of uniform distribution of wells [10] and the exploitable amount of underground brine in the entire area can be calculated based on the theory of unsteady flow, according to the following equation (1), (2), (3), (4), (5), we could calculated the exploitable amount of underground brine resource [11, 12].

$$Q_z = NQ_d t \tag{1}$$

$$N = \frac{F}{D^2} \tag{2}$$

$$Q_d = \frac{Ts}{0.183\lg\frac{2.25at}{r_0^2}} \tag{3}$$

$$s = H_0 - H \tag{4}$$

$$a = \frac{T}{\mu} \tag{5}$$

where Q_z is the exploitable amount of underground brine (m^3/d), N is the well spacing, Q_d is the water yield of a single well (m^3/d), t is the time elaspsed since pumping began (d), F is the division area (m^2), T is the transmissibility coefficient (m^2/d), s is the drawdown in a piezometer measured (m), a is the hydraulic conductivity coefficient (m^2/d), r_0 is the distance from the axis of the piezometer wells (m), μ is the specific yield, H_0 is the initial head, and H is the hydraulic head;

The calculated parameters are substitued into the equation to obtain different beneficial components in brine; the results ranking from highest to lowest are as follows: NaCl, $MgCl_2$, $MgSO_4$, $CaSO_4$, KCl, and Br, and their respective exploitable amounts are 1653.6×10^4t/a, 285.7×10^4t/a, 126.6×10^4t/a, 80×10^4t/a, 16.6×10^4t/a and 4.254×10^4t/a. According to the brine concentration partition data, the largest exploitable amount of brine resources is greater than 10 °Be′, which also contains the largest useful set of components. The results are shown in table 1 and table 2. In addition, the exploitation quantity and the extraction of useful components in Shouguang based on the division of administration are much greater than those indicated in the other three cities, followed by Changyi, Hanting and laizhou. Laizhou has the smallest exploitation quantity among the studied cities.

Table 1. Exploitable amount of resources in different concentrations.

Concentration	Brine $10^4m^3/a$	NaCl $10^4t/a$	Br $10^4t/a$	$MgCl_2$ $10^4t/a$	$MgSO_4$ $10^4t/a$	$CaSO_4$ $10^4t/a$	KCl $10^4t/a$
5~≤7°Be′	5243.5	239.2	0.8	49	17.2	15	3.2
7~10 °Be′	7518.4	561.2	1.2	91.3	41.6	29.6	8.2
≥10 °Be′	8357.7	853.2	2.2	145.4	67.8	35.4	16.6
Sum	21118.8	1653.6	4.2	285.7	126.6	80	28

Table 2. Exploitable amount of resources at different distances.

Exploitation amount (10^4t)	NaCl	Br	$MgCl_2$	$MgSO_4$	$CaSO_4$	KCl
Shouguang	874.0	2.3	157.3	66.4	51.2	12.7
Hanting	255.6	0.6	43.0	18.6	8.6	4.6
Changyi	444.7	1.1	71.1	35.9	16.6	8.7
Laizhou	79.3	0.2	14.3	5.7	3.6	2.0

3.2. Exploitation potential of underground brine resources

The present study indicated a number of factors in brine exploitation including the over exploitation of brine so that the underground brine level decreases at an unexpected speed. It is necessary to protect the environment and sustain resources evaluate the potential of underground brine. Brine resources are evaluated by analyzing the exploitation potential coefficient, calculated as equation (6):

$$P = \frac{Q_z}{Q_k} \tag{6}$$

where P is the exploitation potential coefficient, Q_z is the exploitable amount of underground brine ($10^4m^3/a$), and Q_k is present amount of underground brine ($10^4m^3/a$).

According to the present level of brine resource exploitation, the exploitation potential coefficient in each division is calculated and compared to the standard value. The results are shown in figure 2. The exploitable amount of brine resources is $21118.8\times10^4m^3$, and the present amount is $29489.7\times10^4m^3/a$, with an average exploitation potential coefficient of 0.91.

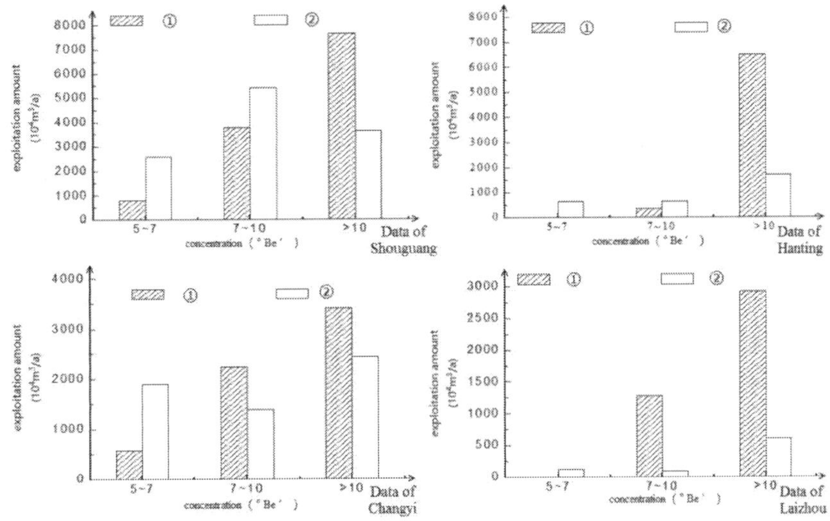

①exploitable amount ② recent amount

Figure 2. Exploitable amount and present exploited amount in each area.

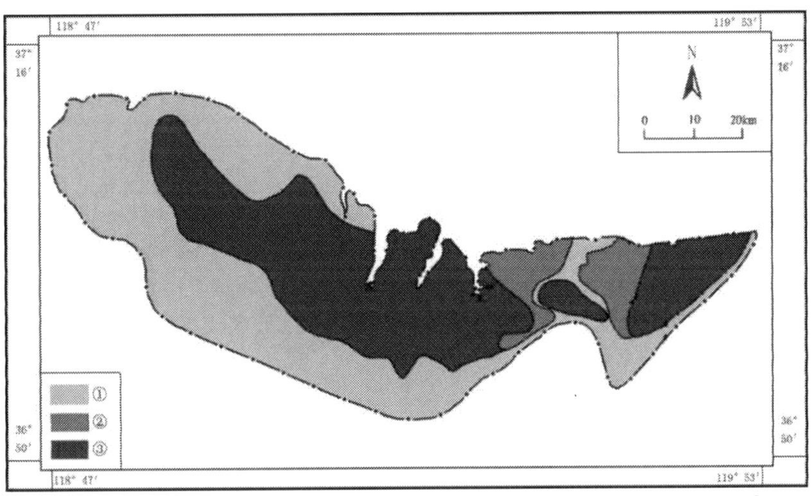

①Potential region($P\geq1.2$)②Equilibrium region of exploitation and recharge($0.8<P<1.2$)③Over-exploitation region($P<0.8$)

Figure 3. Distribution of exploitation potential.

However, the different exploitation situation result in over-exploitation region in some areas, especially the regions brine concentrations exceeding 10°Be′. The over-exploitation is $14501.9\times10^4m^3/a$. Due to over-exploitation, regional underground brine areas have formed a cone of depression. Based on the exploitation potential coefficient of underground water, the study area can be divided into a potential region, an equilibrium region of exploitation and recharge, and an over-exploited region. The results are shown in Figure 3.

4. Summary

Results indicate that underground brine resources in the study area are being exploited on a large scale with some minor areas of overexploitation, based on calculation and analysis, but the area as a whole maintains the balance of discharge and recharge, indicating that the area still possesses comparatively tremendous exploitable capacity. Results indicate that mining enterprises around the study area are implementing an irrational mining policy and potential resources in some areas that have not been fully utilized. Some suggestions are described below.

According to the evaluation of the exploitation potential of underground brine resources in the studied area, mining enterprises should adjust exploitation areas reasonably, e.g., by lowering the upper limit of brine extraction in over-mined areas and enlarging the mining capacity in the regions with abundant mining potential.

Comprehensive dynamic e-systems for the monitoring of underground brine are required to monitor the long-term water level, exploitation quantity, and water quality of underground brine, and more accurate calculations should be performed to study hydrogeology parameters, which are necessary to provide more scientific support for the exploration and development of underground brine in this region.

Appropriate funds should be invested to improve current exploitation and extraction technology in order to conduct the production mode so that mining enterprises reap greater profits by increasing exploitation quantity and utilizing the efficiency of brine resources, thus reducing the cost to utilize the limited resources fully and sustainably, especially for $MgCl_2$, $MgSO_4$, $CaSO_4$ and KCl.

References

[1] Kong Q Y, Zhang T Z and Yu X F 2004 Deposits in Shandong Province *Shandong Science and Technology Press* pp 522-36

[2] Liu F F, Zhao Q S and Xu Y 2013 Environmental problem and countermeasures from the brine in Weifang City *China Science and Technology Information* **25**(11) 56-66

[3] Liu G, Han X J, Feng S T and Tian Q A 2011 Study on relation between dynamic change and water exploitation amount of underground brine water in south coast of Laizhou Bay *Land and Resources in Shandong Province* **27**(2) 9-15

[4] Tagorti M A, Essefi E, Touir J, Guellala R and Yaich C 2013 Geochemical controls of groundwaters upwelling in salineenvironments: Case study the discharge playa of Sidi El Hani *Journal of African Earth Sciences* **86** 1-9

[5] Zou Z G, Zhang D S and Tan Z R 2008 Ground brine resource and its exploitation in Shandong Province *Geological Survey and Research* **31**(3) 214-21

[6] Zheng Y M, Gao M S, Liu S. *et al.* 2014 Distribution characteristics of subsurface brine resources on the southern coast of laizhou bay since late Pleistocene *Hydrogeology and Engineering Gelogy* **41**(5) 14-5

[7] An L S, Zhao Q S and Xu Y 2013 Dynamic characteristics of the shallow groundwater table and its genesis in the Yellow River Delta *Environmental Science & Technology* **36**(9) 53-5

[8] Demin A P 2001 Trends in water Supply an Water Conservation in Russia *Water Resources* **27**(6) 670-87

[9] Li S N, Wang G X, Deng W and Li Y S 2008 Variationsof groundwater depth in the Yellow River Delta in recent twodecades *Progress in Geography* **27**(5) 49-56

[10] Feng S T, Tan Y F and Liu G 2013 Pontentiality analysis on uderground brine resources in Shandong Province *Land and Resources in Shandong Province* **29**(9) 74-8

[11] Gao S D 2005 The estimation of the groundwater resources in Weifang City *Hohai University* pp 22-25

[12] Tagorti M A, Guellala R, Gallala W, Essefi E and Tlig S 2014 Geochemical and hydrogeological studies of a sodium sulphate deposits: the case of Sabkhet El Ghine Oum El Khialate, Southeast Tunisia *Carbonates Evaporites* **29** 301-2

Hydraulic fracturing model based on the discrete fracture model and the generalized J integral

Z Q Liu[1], Z F Liu[1, 3], X H Wang[1] and B Zeng[2]

[1]Department of Thermal Science and Energy Engineering, University of Science and Technology of China, Hefei, Anhui 230026, P. R. China
[2]Research Institute of Petroleum Exploration & Development-Langfang, Langfang, Hebei 065007, P. R. China

E-mail: zfliu13@mail.ustc.edu.cn

Abstract. The hydraulic fracturing technique is an effective stimulation for low permeability reservoirs. In fracturing models, one key point is to accurately calculate the flux across the fracture surface and the stress intensity factor. To achieve high precision, the discrete fracture model is recommended to calculate the flux. Using the generalized J integral, the present work obtains an accurate simulation of the stress intensity factor. Based on the above factors, an alternative hydraulic fracturing model is presented. Examples are included to demonstrate the reliability of the proposed model and its ability to model the fracture propagation. Subsequently, the model is used to describe the relationship between the geometry of the fracture and the fracturing equipment parameters. The numerical results indicate that the working pressure and the pump power will significantly influence the fracturing process.

1. Introduction

Unconventional resources reserves such as shale or tight gas are huge. However, their natural deliverability is insufficient to meet industrial needs due to their low permeability [1]. As hydraulic fractures can effectively increase the drainage area, hydraulic fracturing technique has become a widely-used stimulation for low-permeability reservoirs [2]. To ensure that the hydraulic fractures meet the stimulation requirement, it is necessary to simulate the hydraulic fracturing process and analyze the factors which affect the fracturing treatment.

Fracturing simulation focuses on rock deformation caused by a fracturing fluid and the fracture propagation criteria. Fluid leak-off was neglected by the early fracturing models [3-4], in which the ability of the fracturing fluid to deform the rock is overestimated. Subsequently, other models [5-8] employ the empirical "Fluid-loss coefficient" to account for fluid leak-off. Assuming that the calculation domain is infinite, the fracture geometry is calculated in the models mentioned above using the analytical solution of the displacement. In actuality, the deformation of the rock framework and the seepage interact with one another. Fluid-solid coupling theory is applied to describe this phenomenon in recent fracturing models [9-11], in which rock deformation and fluid flow are solved simultaneously. Thus, an accurate method of calculating the fluid flux across the fracture surface is necessary. To achieve high precision, the discrete fracture model (DFM) [12] is used to calculate the flux in the present work.

In linear elastic fracture mechanics, the failure criteria is met when the stress intensity factor (SIF) exceeds the fracture toughness. The commonly-used methods to calculate SIF are summarized as follows. Advani, et al., [13] calculated the SIF by substituting the stress on the fracture surface into the analytic formulas of the SIF for infinite space. Similarly, Dean, et al., [9] obtained the SIF by using the analytic solution of the infinite solid. The J integral [14] is another method used to calculate the SIF. An integral path far from the fracture tip can be chosen in order to avoid the numerical error resulting from the stress singularity. Considering fluid-solid coupling, the generalized J integral is introduced into the present work.

Based on the discussion above, a hydraulic fracturing model based on the discrete fracture model and the generalized J integral is proposed.

2. Model description

2.1. Reservoir-flow model

The mass conservation equations of the fluid flow for the matrix and the fracture are given below, respectively [15]:

$$\frac{1}{r}\frac{\partial}{\partial r}\left(-\rho\frac{K_{\mathrm{m}}}{\mu}r\frac{\partial P_{\mathrm{m}}}{\partial r}\right)+\frac{\partial}{\partial z}\left(-\rho\frac{K_{\mathrm{m}}}{\mu}\frac{\partial P_{\mathrm{m}}}{\partial z}\right)+\rho q_{\mathrm{fm}}\delta(\mathbf{r}-\boldsymbol{\tau}_{\mathrm{f}})=0 \quad (1)$$

$$\frac{1}{r}\frac{\partial}{\partial r}\left(-\rho w\frac{K_{\mathrm{f}}}{\mu}r\frac{\partial P_{\mathrm{f}}}{\partial r}\right)-\rho q_{\mathrm{fm}}+\frac{\rho Q}{2\pi r_{\mathrm{w}}}\delta(\mathbf{r}-\mathbf{r}_0)=0 \quad (2)$$

where P_{m} is the fluid pressure in the matrix, P_{f} is the fluid pressure in the fracture, μ and ρ are the viscosity and the density for the fluid, respectively, $\boldsymbol{\tau}_{\mathrm{f}}$ is the fracture strike, w is the fracture width, $\mathbf{r}_0$ and r_{w} are the location and the radius for the well, Q is the well injection, r and z are the radial coordinate and the axial coordinate, respectively, $\delta(\mathbf{r})$ is the Dirac function, K_{m} is the matrix permeability, K_{f} is the fracture permeability equal to $w^2/12$, and q_{fm} is the fluid flux across the fracture surface.

In the present work, the embedded discrete fracture model is introduced. The fluid flux across the fracture surface is calculated according to the method proposed by Zhou, et al., [12] as follows:

$$q_{\mathrm{fm}}=-\frac{K_{\mathrm{fm}}}{\mu}\frac{\partial P}{\partial n_{\mathrm{f}}^{+}}-\frac{K_{\mathrm{fm}}}{\mu}\frac{\partial P}{\partial n_{\mathrm{f}}^{-}}=-\frac{K_{\mathrm{fm}}}{\mu}\frac{P_{\mathrm{f}}-(P_{\mathrm{m}})_{\mathrm{M}^{+}}}{d^{+}}-\frac{K_{\mathrm{fm}}}{\mu}\frac{P_{\mathrm{f}}-(P_{\mathrm{m}})_{\mathrm{M}^{-}}}{d^{-}} \quad (3)$$

where $d^{\pm}$ is the distance from the fracture grid to the reference point $\mathrm{M}^{\pm}$ in the matrix, and K_{fm} is the permeability between the fracture and the matrix which is the harmonic mean of K_{m} and K_{f}.

2.2. Geo-mechanical model

The system of equations representing the solid mechanics includes the equilibrium equation, the constitutive equation and the geometric equation. Considering the effect of in-situ stress, variations, such as the displacement, are decomposed into the initial portion caused by in-situ stress and the effective portioncaused by other external forces. For instance, the decomposition of ϕ is $\phi=\phi'+\phi''$, where ϕ' is the initial portion and ϕ'' is the effective portion.

The equations for the initial portion are given as follows:

$$\frac{\partial\sigma'_{ji}}{\partial x_j}=0 \quad (4)$$

$$\sigma'_{ij}=2G\varepsilon'_{ij}+\lambda\theta'\delta_{ij} \quad (5)$$

$$\varepsilon_{ij}' = \frac{1}{2}\left(\frac{\partial u_i'}{\partial x_j} + \frac{\partial u_j'}{\partial x_i}\right) \tag{6}$$

The effective portion satisfies the following equations:

$$\frac{\partial \sigma_{ji}''}{\partial x_j} + f_i'' = 0 \tag{7}$$

$$\sigma_{ij}'' = 2G\varepsilon_{ij}'' + \lambda\theta''\delta_{ij} + \sigma_{0,ij}\delta_{ij} \tag{8}$$

$$\varepsilon_{ij}'' = \frac{1}{2}\left(\frac{\partial u_i''}{\partial x_j} + \frac{\partial u_j''}{\partial x_i}\right) \tag{9}$$

where σ_{ij}' is the initial stress, σ_{ij}'' is the effective stress, ε_{ij}' is the initial strain, ε_{ij}'' is the effective strain, u_i' is the initial displacement in the i-direction, u_i'' is the effective displacement in the i-direction, θ' is the initial volume strain, and θ'' is the effective volume strain, $\sigma_{0,ij}$ is the in-situ stress, G is the shear modulus, λ is the Lame coefficient, δ_{ij} is the Kronecker delta, and f_i'' is the body force of the rock or the impaction of the fracturing fluid on the rock. Based on the equilibrium equation of the rock and the fluid-solid coupling theory [16], the expression of f_i'' is described as follows:

$$f_i'' = -\frac{\partial\left(\alpha P_{\mathrm{m}}\delta_{ij}\right)}{\partial x_j} \tag{10}$$

where α represents Biot's coefficient.

2.3. Fracture propagation model

From a linear elastic fracture mechanics perspective, the fracture will extend when the SIF exceeds the fracture toughness [17]. In this work, the generalized J integral is used to calculate the SIF as follws [14, 18]:

$$K_{\mathrm{I}} = \left(\frac{E}{1-\mu^2}J\right)^{1/2} \tag{11}$$

$$J = \frac{1}{a}\left[\int_{\Gamma} \mathrm{w}\,r\mathrm{d}z - \int_{\Gamma+l^{+}+l^{-}} \sigma_{ij} n_j \frac{\partial u_i}{\partial r} r\mathrm{d}\Gamma - \iint_D f_i \frac{\partial u_i}{\partial r} r\mathrm{d}r\mathrm{d}z\right] \tag{12}$$

where K_{I} is the SIF; E is the elastic modulus; μ is Poisson's ratio; w is the strain energy density; a is the distance from the fracture tip to the axis of symmetry,; Γ, l^{+} and l^{-} are the integral contours; D is the integral area; n_j is the unit vector in the j-direction; u_i is the actual displacement in the i-direction, which equals $u_i' + u_i''$; f_i is the body force of the rock, which equals f_i''; and σ_{ij} is the actual stress.

3. Model validation

In this section, the reliability of the model is verified by comparing the numerical predictions with analytic results of specific examples. The calculation domain is the cylindrical body shown in figure 1.

There is a concentric horizontal well with a radius of r_w in the domain. The radial fracture extends from the middle of the inner surface. The specific parameters are listed in table 1.

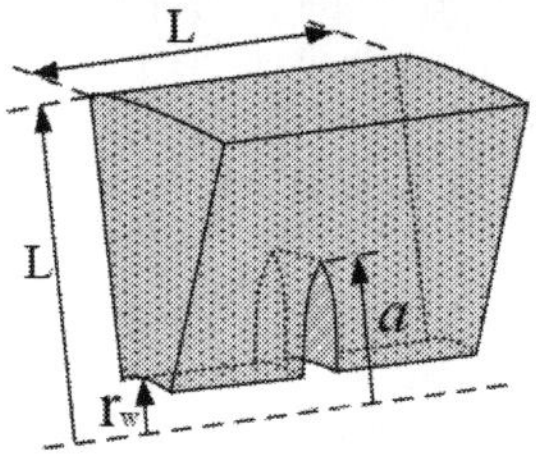

Figure 1. Schematic diagram of the calculation domain.

Table 1. Specific parameters of the model.

Fluid density (kg/m^3)	982.6
Elastic modulus (GPa)	35.0
Fluid viscosity (mPa·s)	0.456
Poisson's ratio	0.22
Well radius (m)	0.07
Injection (m^3/min)	0.5
Fracture toughness (MPa·m$^{1/2}$)	1.5

3.1. Comparison with the radial fracture model

The simulation results of the radial fracture propagation are compared with the analytic solution. It is assumed that: (a) the rock deformation is quasi-static; (b) the lag of the fluid is neglected; (c) the fluid leak-off is neglected.

Since the fracture width tends to zero near the fracture tip, the pressure drop primarily occurs in this area. Therefore, the assumption of a uniform pressure P along the fracture is reasonable. When the fracture is in a critical state, the SIF is equal to the fracture toughness. Using the fracture volume and the SIF analytic formals in infinite space [19], the relationship between the fracture radius and time is obtained as follows:

$$\frac{8\sqrt{\pi}K_{IC}\left(1-\mu^2\right)}{3EQ}a^{1/2}\left(a^2-r_w^2\right)=t \tag{13}$$

where K_{IC} is the fracture toughness, a is the fracture radius, r_w is the well radius, and t is time.

The simulation results are shown in figure 2. Results indicate that the numerical results will tend to the analytic solution as the calculation domain increases. However, when the size of the calculating domain is insufficiently large, there is significant difference between the numerical result and the analytic solution. This indicates that the influence of the outer boundary condition must be taken into consideration.

3.2. Influence of the solid boundary conditions

The influence of the boundary conditions on the SIF are studied in this example. The fracture radius is 20m. The stress applied on the fracture surface is 15MPa. Other parameters are listed in table 1. The outer boundary conditions are set as zero stress and zero displacement, respectively.

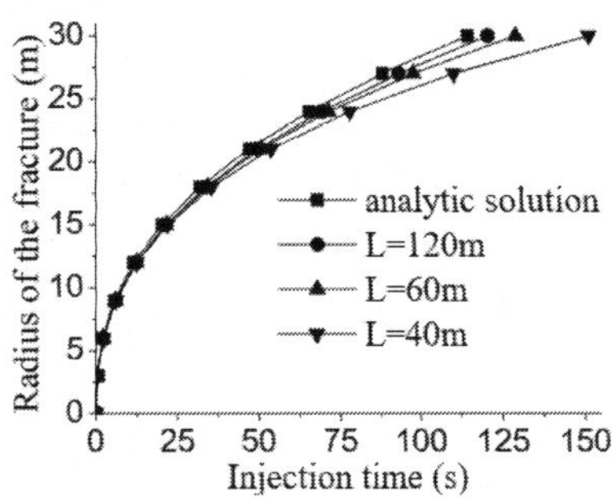

Figure 2. Comparison of the proposed model with the radial fracture model.

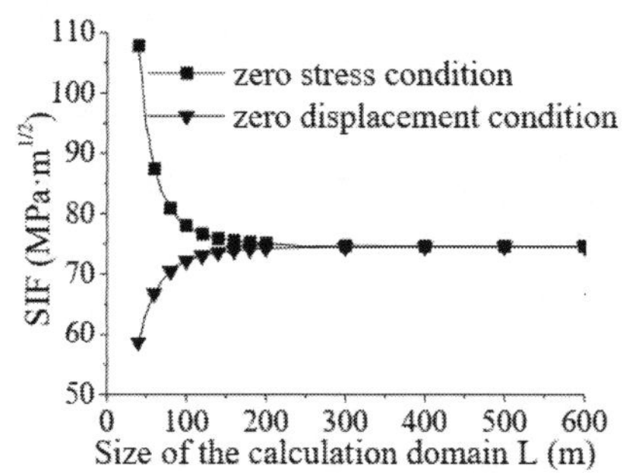

Figure 3. Influence of the solid boundary conditions on the SIF.

The results are presented in figure 3, which indicates that the SIF under the two different conditions tends to the same value as the calculation domain increases. However, when the size of the calculating domain is insufficiently large, the SIF under the zero stress condition is always greater than that under the zero displacement condition. Moreover, the difference will increase significantly when the size of the calculation domain decreases.

4. Application

The influence of the fracturing equipment parameters on the fracturing process is analyzed in this section. The calculation domain is shown in figure 1. The radius of the outer boundary is 300m and the axial thickness is 200m. The matrix permeability is 2 mD, the Biot's constant is 0.89, and the other property parameters are listed in table 1. The fluid pressure exerted on the axial outer boundary is 14.06MPa. The radial outer boundary is assumed to be impermeable. The in-situ stresses in the axial and the radial directions are assumed to be -12 and -22 MPa, respectively. The simulation results are shown in figure 4.

Figure 4(a) and figure 4(b) show the variations of the SIF and the pump power with the fracture radius, respectively, when the well pressure is fixed. This indicates that the SIF increases monotonically as the fracture radius increases. However, to maintain a fixed well pressure, the necessary pump power will increase rapidly as the fracture radius increases. Under the limitation of the maximum output power of the fracturing equipment, the fracture will cease to extend in some radius.

Figure 4(c) and figure 4(d) show the variations in the fracture radius and the fracture width at the wellbore with the well pressure, respectively, when the pump power is fixed. Results indicate that low well pressure corresponds to long and narrow fractures, while high well pressure corresponds to short and wide fractures.

This numerical example indicates that the working pressure and the pump power will influence the fracturing process significantly.

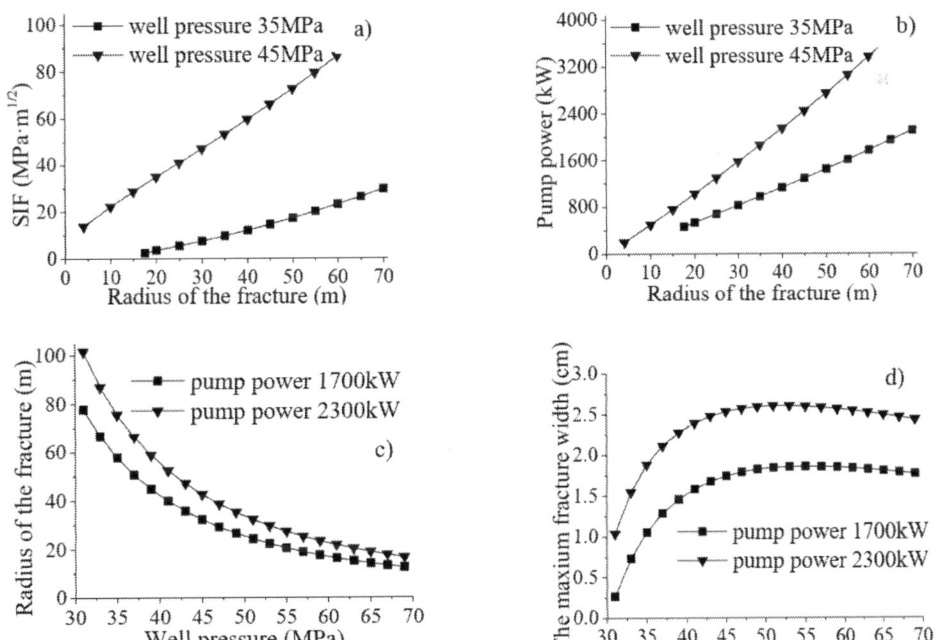

Figure 4. Influence of fracturing equipment parameters on the fracturing process.

5. Conclusions

A hydraulic fracturing model which combines the discrete fracture model and the generalized J integral has been presented. Compared to traditional models, the proposed model calculates the fluid flux across the fracture surface by avoiding the use of the empirical "fluid leak-off coefficient". In addition, by using the generalized J integral to calculate the SIF, both the assumption of infinite space and the numerical error resulting from the stress singularity are avoided. These ideas can improve the accuracy of the hydraulic fracturing model effectively, and are thus practically important for the optimization of fracturing design.

Several examples have been provided in this paper. The primary conclusions are as follows: (a) the simulation result agrees well with the analytic solution. Results indicate the reliability of the proposed model. (b) The model can be used to describe the relationship between the geometry of the fracture and the fracturing equipment parameters. The numerical results indicate that the working pressure and the pump power will significantly influence the fracturing process.

The proposed model has the ability to accurately and effectively simulate the model I fracture propagation, and has practical value in fracturing design. In further research, the model can be extended to simulate mixed model fracture propagation by introducing mixed model stress intensity factors, thus providing a much wider range of applications.

Acknowledgements

This work is supported by the CNPC-CAS Strategic cooperation Research Program (2015A-4812).

Reference

[1] Zou C N, et al. 2012 *Petroleum Exploration and Development* **39**(01) 13-26
[2] Wang Y H, et al. 2012 *Acta Petrolei Sinica.* **33**(A01) 149-58
[3] Khristianovic S A and Zheltov Y P 1955 *Proc. 4th world petroleum congress* (Rome)
[4] Perkins T K and Kern L R 1961 *Journal of Petroleum Technology* **13**(09) 937-49
[5] Geertsma J and De Klerk F 1969 *Journal of Petroleum Technology* **21**(12) 1571-81
[6] Nordgren R P 1972 *Society of Petroleum Engineers Journal* **12**(04) 306-14
[7] Clifton R J and Abou-Sayed A D 1979 *Symp. on Low Permeability Gas Reservoirs* (Denver, Colorado, USA)
[8] Settari A and Cleary M P 1986 *SPE Production Engineering* **1**(06) 449-66
[9] Dean R H and Schmidt J H 2009 *SPE Journal* **14**(04) 707-14
[10] Zhang G M, et al. 2010 *Rock and Soil Mechanics* **31**(05) 1657-62.
[11] Biao F J, et al. 2011 *Engineering Mechanics* **28**(10) 228-35
[12] Zhou F Q, Shi A F and Wang X H 2014 *Petroleum Exploration and Development* **41**(02) 262-6
[13] Advani S H and Lee J K 1982 *Society of Petroleum Engineers Journal* **22**(02) 209-18
[14] Kou X D and Zhou W 2000 *Chinese Journal of Rock Mechanics and Engineering* **19(01)** 18-23
[15] Kong X Y 2010 *Higher Seepage Mechanics* (Hefei: University of Science and Technology of China Press) 48-74
[16] Biot M A 1955 *Journal of Applied Physics* **26**(02) 182-85
[17] Adachi J I, et al. 2010 *International Journal of Rock Mechanics and Mining Sciences* **47**(04) 625-39
[18] Zhang X M, et al. 2012 *Fracture Mechanics* (Beijing: Tsinghua Press) 109-12
[19] Tada H, et al. 2000 *The stress analysis of cracks handbook* (New York: ASME Press) 342-50

IOP Conf. Series: Earth and Environmental Science **39** (2016) 012035 doi:10.1088/1755-1315/39/1/012035

Using the DEM-based Xin'anjiang hydrologic model to simulate hydrologic characteristics in SiheRiver Basin

C S Qi[1], D D Liu[1, 3], L Y Tan[1] and L Zhu[2]

[1]Shandong Survey and Design Institute of Water Conservancy, Shandong 250013, China

[2]School of Civil Engineering and Water Conservancy, Ningxia University, Ningxia 750002, China

Email: east_liu@foxmail.com

Abstract. In order to consider the spatial distribution of precipitation, the basin is generally divided by the Thiessen polygon method with the same concentration parameters in all divided units in Xin'anjiang hydrologic model, without considering the impact of terrain. The runoff concentration characteristics depend on the terrain condition to a great extent.Therefore, it is necessary to consider the impact of terrain condition in the model. The purpose of this article is to consider the impact of terrain condition to improve the Xin'anjiang hydrologic model. The basin is divided, and then the slopes of the divided sub-basins and the distances to the basin outlet are extracted based on the DEM. Twofunction relations between the terrain slope and the runoff concentration are established, and the application results of Sihe river basin show that the Xin'anjiang hydrologic model simulation precision is improved greatly based on the DEM,and it is more reasonable that the recession coefficient and the average gradient of the basin is an exponential relationship.

1. Introduction

Since the Xin'anjiang hydrologic model was developed, it has been widely used globally, especially in the humid and sub-humid lands of China[1]. Unfortunately, the modeler only considered the spatial distribution of rainfall without considering other distributions, especially the terrain factor. Also, the same runoff generation and concentration parameters are used in the model calculation units. In fact, the terrain features not only affect the runoff generation characteristics, but also bring impacts on the basin runoff concentration process [2, 3]. In addition, the computing units are divided by the Thiessen polygon method with multi-outlets of runoff in each unit; thus, the runoff generation processes are disrupted.

At present, the studies of the Xin'anjiang hydrologic model have focused on three aspects. One is the optimization method of parameter calibration[4], followed by the storage capacity curve line[5]. The third is how to add a new module so that the Xin'anjiang hydrologic model can be adapted to infiltration-excess runoff conditions [6]. The development of the computer, GIS technology, especially the DEM, provides technical support for the hydrologic model based on the topographic effects. There are many watershed hydrologic models that have been developed to adapt DEM data structure or directly are based on DEM structure, such as the SHE model [7, 8], TOPMODEL [9], and HMS model [10,11].

One of the main development methods for Xin'anjiang hydrologic model is to take full advantage of the new data sources to improve the accuracy of the simulation. The purpose of this article is to divide the sub-basin and determine the flow recession coefficient to develop the Xin'anjiang hydrologic model based on the DEM and GIS.

2. The Xin'anjianghydrologicmodel introduction

The Xin'anjiang hydrologic model is a kind of distributed model in which the basin is divided into many units and the runoff generation and concentration process are modelledfor each unit[1]. For each unit area, the total runoffis determinedwiththe saturation-excess runoff method. Then, the runoff is divided intosurface runoff RS, interflow RI and underground runoff RG according tothe storage capacity theory of the free water reservoir. Finally, the discharge hydrograph of the basin can be obtained by the channel flood routing based on the each unit export flow. To take into account the uneven distribution of rainfall rather than others, the Thiessen polygon method is used to divide the basin unit with a rainfall station as the general centre. In practical applications, there are other classification methods, such as the natural sub-basins division method[12]or the grid cell division method[13].

2.1. Model structure

The model structure can be divided into four parts, which are evapotranspiration calculation, runoff producing calculation, runoff division calculation and runoff concentrationcalculation. The model calculation process is shown in figure 1.

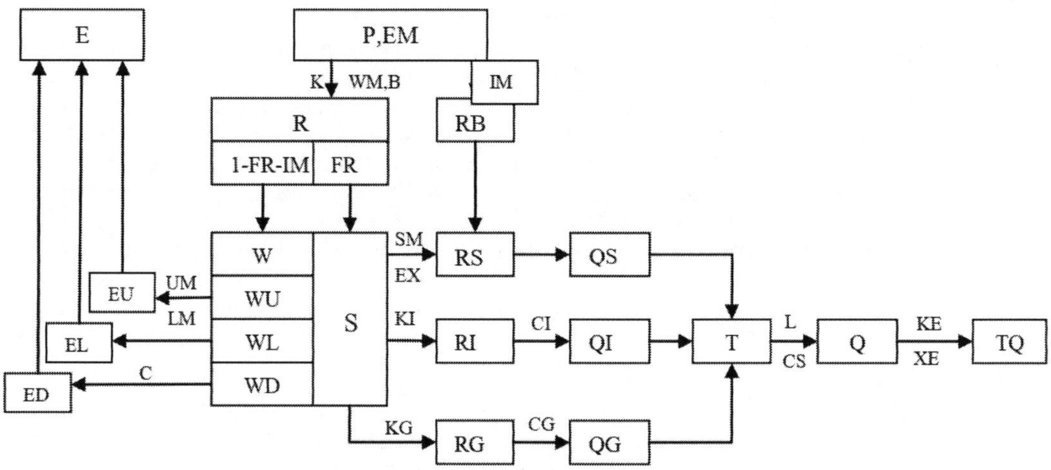

Figure 1. Xin'anjiang hydrologic model structure.

The Internal variables outside theblocksare parameters, and allsymbolsinsidethe blocksare variablesincludinginputs,outputs, and state variables. Model inputs include the measured rainfall process P(t) and pan evaporation process EM(t). Also, model outputsfor the basin are the exit section flow process Q(t) and the actual evapotranspiration process E(t).The function, method and corresponding parameters of each level structure for the model are shown in table 1.

Table 1. Function, method and the corresponding parameters forXin'anjiang hydrologic model.

Level	Function		Method	Parameters
Level 1	Evapotranspiration calculation		Three-layer model	KC、UM、LM、C
Level 2	Runoff producing calculation		Saturation-excess runoff	WM、B、IM
Level 3	Runoff division calculation		Free water reservoir	SM、EX、KG、KI
Level 4	Runoff concentration calculation	Slope flow concentration	Slope flow concentration	CS、CI、CG
		River flow concentration	River flow concentration	KE、XE

2.2. Model parameters

In the Xin'anjiang hydrologic model, each parameter has a definite physical meaning, in principle, so that most of them can be directly determined.However, due to the lack of measurement and experimentation, in reality some parameters can only be determined with system identification methods based on measured outflow. According to the classification of the model parameters, the model physical parameters are measured directly, generally without calibration. For process parameters, the initial value or range are determined according to the hydrological and hydraulic characteristics of the basin. Then, the optimal parameter values are calibrated by the trial and error method or the automatic optimization method. In actual work, model parameter calibration usually is executed with the manual trial and error method combined with theautomatic optimization method, which was used in thisresearch[1]. In the model, KCis the ratio of potential evapotranspiration to the pan evaporation; generally, KC <1.0. UM is the tension watercapacity of the upperlayer, which is about10-20mm, and LM is the tension watercapacity of the lowerlayer, generally 60-90mm. C represents the evapotranspiration coefficient of the deeperlayer, which is about 0.15-0.20 in the humid southern region and 0.09-0.12 in the north sub-humid region,depending on the deep root plant coverage area. WMis the areal mean tension watercapacity, generally 120mm. B is the exponential of the distribution of the tension water capacity, which is determined by the uneven distribution of tension water capacity. In the hilly area, it is generally 0.1 for a small watershed, 0.2-0.3 for a slightly larger watershed, and 0.3-0.4 for alarge basin. IM is the ratio of impervious area to the total area of the basin, which is determined according to the specific circumstances. SM is the free waterstorage capacity, generally about 10-30mm. SM determines the surface runoff, so it is important for the calibration parameters. EXis the exponential of distribution water capacity, which is generally about 1.0-1.5mm, as determined by the inhomogeneity conditions of the surface free water capacity. KG and KIare the outflow coefficient of the free waterstorage to the groundwater flow and the interflow respectively, withKG+KI=0.7～0.8. CS is the recession constant of the lowerinterflow storage, which is generally about 0.2 depending on the river geomorphology. CI and CG are the recession constant of the groundwater storage and the channel network storage respectively,and the corresponding values are 0.6-0.9 and 0.99-0.998. KE and XE are the Muskingum coefficients, which were calculated according to the hydraulic characteristics of the river. The significance and range of the parameters are shown in table 2.

2.3. Flow concentrationcalculation

Asthis research focused on the impact of topography on the flow concentration, the theory of flow concentration was mainly introduced. In the Xin'anjiang hydrologic model, the runoff concentration calculation consists of two stages:the slope flow concentration and the river flow concentration[1].

- Slope flow concentrationcalculation

Through the runoff generation calculation, the surface runoff (QS) is acquired and flows into the river directly. All the surface runoff becomes the total inflow of the surface runoff on the river network (RS). After the soil water reservoir subsides (regression coefficient CI), the subsurface runoff

(QI) of all units flows into the river network as the total subsurface runoff (RI). Similarly, through the underground reservoir regulation, the underground runoff (QG) of all units flows into the river network as the total groundwater runoff (RG). The calculation formulas are as equation (1),(2),(3),(4).:

$$RS(t) = QS(t) \times U \tag{1}$$

$$RI(t) = RI(t-1) \times CI + QI(t) \times (1 - CI) \times U \tag{2}$$

$$RG(t) = RG(t-1) \times CG + QG(t) \times (1 - CG) \times U \tag{3}$$

$$TR(t) = RS(t) + RI(t) + RG(t) \tag{4}$$

whereU is a unit conversion factor.

- River flow concentrationcalculation

In the model, the division-unit water entering the river to the unit export is simulated by the dimensionless unit hydrograph[1]. A basin with certain hydrological data and similar area to the studied sub-basin can be selected to analyze the unit hydrograph of the surface runoff in the studied basin or the adjacent basin.

Table 2. The significance and range of the parameters.

Type	Notat-ion	Parameter meaning	Value range
Evapotranspiration parameters	KC	Ratio of potential evapotranspiration to the pan evaporation	Generally KC <1.0
	UM	Tension watercapacity of upperlayer	About10-20mm
	LM	Tension watercapacity of lowerlayer	About 60-90mm
	C	Evapotranspiration coefficient of deeperlayer	About 0.15-0.20 in the humid southern region and 0.09-0.12 in the north sub-humid region
Runoff producing parameters	WM	Areal mean tension watercapacity	About 120mm
	B	Exponential of the distribution oftension watercapacity	Generally 0.1 for a small watershed, 0.2-0.3 for a slightly larger watershed, 0.3-0.4 for the large basin
	IM	Ratio of impervious area to the total area of the basin	Determined according to the specific circumstances
Runoff separation parameters	SM	Free waterstorage capacity	Generallyabout 10-30mm
	EX	Exponential of distribution watercapacity	Generallyabout1.0-1.5mm
	KG	Out flow coefficient of free waterstorage to the groundwater flow	KG+KI=0.7～0.8
	KI	Out flow coefficient of free waterstorage to the interflow	KG+KI=0.7～0.8
Runoff concentration parameters	CS	Recession constant of lowerinterflow storage	generally CS=0.2
	CI	Recession constant of groundwater storage	About 0.6-0.9
	CG	Recession constant of channel network storage	About 0.99-0.998
Muskingum coefficients	KE	Geometry factor	Calculated according to the hydraulic characteristics of the river
	XE	Residencetime of water	Calculated according to the hydraulic characteristics of the river

In practical applications, a larger basin should be divided into many units, and it is unrealistic to analyze the surface runoff unit hydrograph for each unit. Therefore, the routing calculation is simplified. Through the regression process, the soil water and groundwater flow away from the unit. Also, the surface runoff is regulated by the surface water reservoir (regression coefficient CS) as the Unit export flow. Formula (1) can be rewritten as equation (5).

$$RS(t) = RS(t-1) \times CS + QS(t) \times (1 - CS) \times U \tag{5}$$

In equations (1) to (5), the parameters CS, CI and CG are determined by the parameter calibration in the Xin'anjiang model, and are used in all units.

3. Methodology

3.1. Unit division

In order to keep the integrity of the runoff generation and the operability of the method, the divided unit should be kept consistent with the watershed boundary as far as possible. Additionally, the distribution of the precipitation stations should be taken into account to ensure that there is a station within each divided unit.

3.2. Analysis of runoff concentration characteristics

As the Watershed topography characteristics determine its convergence characteristics, the terrain characteristics are expressed by using the average slope, which is calculated by the grid DEM and GIS. The slope of each grid is extracted by the GIS, and the arithmetic mean slopes of all grids represent the runoff concentration characteristics.

3.3. Runoff concentration calculation improvement

In the improved model, the evapotranspiration calculation, runoff producing calculation, and runoff divisioncalculation still use the original model calculation method, only to improve the surface runoff concentration.For simplicity, the surface runoff concentration is taken as the example to demonstrate the runoff concentrationimprovement.The average slope determines the runoff concentration characteristics with the same recession coefficients. The larger the basin mean slope, the higher the runoff concentration speed,and the lower the recession coefficients. The relationship between the recession coefficient and the average gradient of the basin is the key factor to reflect the influence of different gradients on the runoff concentration. Based on the relationship between the recession coefficient and the average gradient, it is assumed that the recession coefficient and average gradient of the basin is an exponential relationship and an inverse proportion function relationship.The performance of the model under two cases was analysed.

- Case 1: Assume that the recession coefficient and average gradient of the basin is an exponential relationship.

$$CS = CS_0 \times e^{-(g-1)} \tag{6}$$

where,CS_0 is the regression coefficient when the slope is 1 and PD is the mean speed.

The soil runoff and underground runoff are expressedusing the same relationship. Then, the runoff concentration calculation formula can be improved as equation (7),(8),(9),(10).

$$RS(t) = RS(t-1) \times CS_0 \times e^{-(g-1)} + QS(t) \times (1 - CS_0 \times e^{-(g-1)}) \times U \tag{7}$$

$$RI(t) = RI(t-1) \times CI_0 \times e^{-(g-1)} + QR(t) \times (1 - CI_0 \times e^{-(g-1)}) \times U \tag{8}$$

$$RG(t) = RG(t-1) \times CG_0 \times e^{-(g-1)} + QG(t) \times (1 - CG_0 \times e^{-(g-1)}) \times U \tag{9}$$

$$TR(t) = RS(t) + RI(t) + RG(t) \tag{10}$$

where the parameters ofCS_0, CI_0andCG_0 should be calibrated.

- Case 2: Assume that the recession coefficient and the average gradient of the basin is an inverse proportion function relationship.

$$CS = CS_0 \times g^{-1} \tag{11}$$

where CS0 is the regression coefficient when the slope is 1 and g is the mean speed.

The soil runoff and underground runoff are expressedusing the same relationship. Then, the runoff concentration calculation formula can be improved as equation (12),(13),(14),(15).

$$RS(t) = RS(t-1) \times CS_0 \times g^{-1} + QS(t) \times (1 - CS_0 \times g^{-1}) \times U \tag{12}$$

$$RI(t) = RI(t-1) \times CI_0 \times g^{-1} + QR(t) \times (1 - CI_0 \times g^{-1}) \times U \tag{13}$$

$$RG(t) = RG(t-1) \times CG_0 \times g^{-1} + QG(t) \times (1 - CG_0 \times g^{-1}) \times U \tag{14}$$

$$TR(t) = RS(t) + RI(t) + RG(t) \tag{15}$$

where the parameters CS_0,$CI_0$and CG_0 should be calibrated.

The model performance wasanalyzed to determine the merits of the improved model under the above two cases.

3.4. Model performance evaluation

The performances of model areanalysed by three evaluation indexes, including the relative error of the runoff, the relative error of the peak flow and the deterministic coefficient that is recommended by the World Meteorological Organization.

$$RE = \frac{SR - MR}{SR} \times 100\% \tag{16}$$

$$PFE = \frac{SPF - MPF}{SPF} \times 100\% \tag{17}$$

$$DC = 1 - \frac{\sum_{1}^{n} (x_{0i} - x_{ci})^2}{\sum_{1}^{n} \left(x_{0i} - \bar{x}\right)^2} \tag{18}$$

where RE is the relative error of the runoff, MR is the actual runoff, SR is the simulated runoff, PFE is the relative error of the peak flow, MPF is the actual peak flow, SPF is the simulated peak flow, DC is thedeterministic coefficient ($0 \le DC \le 1.0$), x_0 is the actual daily runoff, x_c is the simulated daily runoff, $\bar{x}$ is the average data of x_0 , and n is the length of the series. The smaller the absolute values of the relative errorsare, the better the simulation resultsare. In contrast, the greater the deterministic coefficient is, the better the simulated results are.

4. Application

4.1. Unit division

The study basin is the uppercatchment of Shuyuan hydrologic stationon Siheriver in Shandong province of China with an area of 1542km^2. There are 10 precipitation stations in the basin. The extracted river network, the precipitation stations and the hydrologic stations are shown in figure 2. The DEM map and the basin division are given in figure 3. The slope of the basin is shown in figure 4.

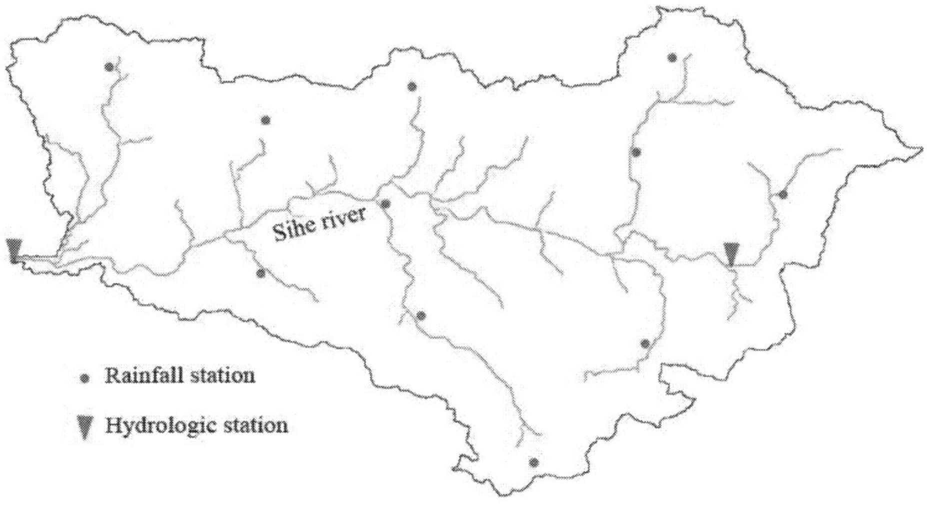

Figure 2. The river network, precipitation stations and hydrologic stations of Sihe river basin.

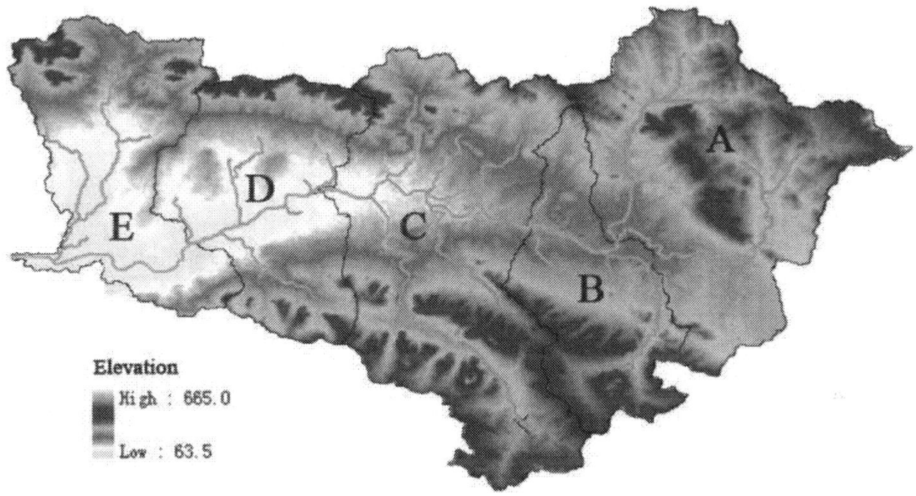

Figure 3. The DEM and division units of Sihe river basin.

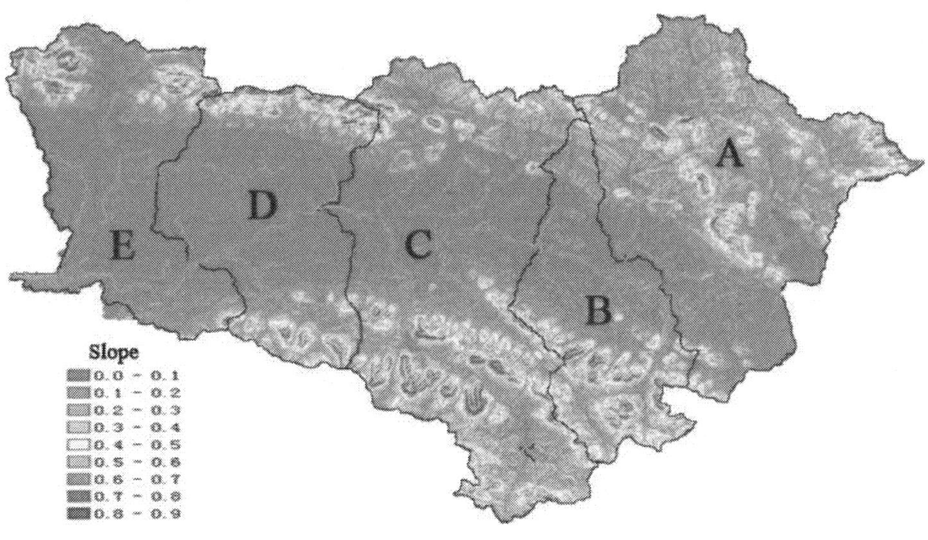

Figure 4. The slope of Sihe river basin.

4.2. *Unit characteristics*

Based on the DEM, the area, mean slope and distance to the basin outlet can be calculated by GIS (shown in table 3).

Table 3. The area, mean slope of sub-basin and along-channel distance to basin outlet of basin.

Sub-basin	A	B	C	D	E	Total
Area(km^2)	400.9	185.1	385.5	277.6	292.9	1542
Weight	0.26	0.12	0.25	0.18	0.19	1
Mean slope	0.39	0.36	0.35	0.28	0.23	(a)
Distance to basin outlet of basin(km)	58.8	49.3	29.4	15.1	0.5	(b)

4.3. *Parameter calibration*

In order to compare the model performance based on DEM with original model performance without considering the terrain factors, the parametersCS_0, CI_0 and CG_0 are calibrated using natural meteorological and hydrological data of the 8 years from 1965 to 1972.The other parameters remain unchanged. The calibrated parameters of the original model and improved model are shown intable 4. The performance of the improved modelunder two function relationships between the recession coefficient and the average gradient of the basin named case 1 and case 2 are presented in table 5.

Table 4. Model calibration parameters.

Parameter	Original model	Improved model (Case 1)	Improved model (Case 2)	Parameter	Original model	Improved model (Case 1)	Improved model (Case 2)
K	0.55	0.55	0.55	KI	0.3	0.3	0.3
WM	120	120	120	KG	0.4	0.4	0.4
WUM	20	20	20	CS_0	0.2	0.1	0.13
WLM	70	70	70	CI_0	0.8	0.4	0.2
B	0.3	0.3	0.3	CG_0	0.99	0.8	0.7
C	0.15	0.15	0.15	KE	1	1	1
SM	20	20	20	XE	0.48	0.48	0.48
EX	1.5	1.5	1.5	L	1	1	1

Table 5. Daily rainfall-runoff simulation results during the calibration period from 1965 to 1972.

Year	Relative error of runoff(%)			Relative error of peak flow(%)			Deterministic coefficient		
	Original model	Improved model (Case 1)	Improved model (Case 2)	Original model	Improved model (Case 1)	Improved model (Case 2)	Original model	Improved model (Case 1)	Improved model (Case 2)
1965	9.72	6.52	6.48	-16.56	-12.35	-12.45	0.822	0.877	0.882
1966	4.36	3.24	3.54	-12.62	-9.85	-10.04	0.854	0.823	0.814
1967	5.45	3.56	3.95	-15.87	-5.74	-7.45	0.795	0.815	0.804
1968	7.96	2.65	2.3	-10.25	-6.34	-7.32	0.756	0.864	0.813
1969	2.31	-1.26	0.96	-7.89	-5.12	-6.61	0.825	0.912	0.851
1970	3.57	2.65	2.84	-12.63	-7.86	-8.98	0.814	0.935	0.864

1971	1.21	1.02	1.07	-2.65	-3.57	-3.45	0.625	0.814	0.758
1972	0.98	1.36	1.26	5.68	1.25	1.13	0.726	0.862	0.757
Average	4.45	2.47	2.80	-9.10	-6.20	-6.90	0.777	0.863	0.818

4.4. Results and discussions

With the calibrated parameters, the daily runoff of 12 years of Shuyuan hydrological station from 1973 to 1984 is simulated to validate the performance of the DEM-based Xin'anjiang hydrologic model. The simulated results under the original model, improved Case 1 and improved Case 2, are showed in table 6. The simulated and observed runoffs from 1975 to 1977 for three models are selected to show the performances.

In the calibration and validation period, the average relative error of annual runoff and peak flow for two improved models were smaller than those of the original model, and the deterministic coefficient was greater, which suggests that the model considering topographic effects has better performance, with better simulated runoff.

In addition, the average relative error of the annual runoff and peak flow for improved model Case 1 were smaller than those of improved model case 2, and the deterministic coefficient was greater, which demonstrates that the improved model case 1had better performance, with better simulated runoff. In improved model case 1, the recession coefficient and average gradient of the basin was an exponential relationship. However, it is an inverse proportion function relationship in improved model case 2. The performance of the two improved models shows that it is more reasonable that the recession coefficient and the average gradient of the basin is an exponential relationship. Why was this phenomenon produced? The recession coefficient variations under the two relationships are plotted in figure 5 and figure 6. As can be seen from the figure, when the recession coefficient and watershed average slope is an exponential relationship, the dot distribution is smoother with smaller gradients. Conversely, when the recession coefficient and the average slope of the basin are an inverse proportion function, the gradient is larger, especially between the first and the second point. This may be the reason for the better performance of the improved model case 1.

Table 6. Daily rainfall-runoff simulation results.

Year	Relative error of runoff(%)			Relative error of peak flow(%)			Deterministic coefficient		
	Original model	Improved model (Case 1)	Improved model (Case 2)	Original model	Improved model (Case 1)	Improved model (Case 2)	Original model	Improved model (Case 1)	Improved model (Case 2)
1973	3.23	0.56	2.15	-8.96	-5.21	-6.75	0.754	0.871	0.853
1974	5.34	3.28	4.23	-6.47	-6.33	-6.54	0.698	0.795	0.784
1975	4.35	4.22	4.01	-12.24	-9.63	-10.98	0.812	0.825	0.821
1976	5.65	2.36	5.12	-8.77	-7.24	-8.57	0.864	0.811	0.799
1977	4.34	3.56	3.74	-9.89	-5.32	-7.49	0.722	0.795	0.787
1978	6.89	5.12	5.03	-9.56	-4.48	-4.01	0.698	0.802	0.714
1979	7.65	3.26	3.11	-5.94	-3.23	-3.12	0.755	0.834	0.806
1980	2.32	-0.23	0.75	-11.63	-10.98	-10.78	0.891	0.884	0.887
1981	6.48	3.58	3.79	-6.23	-5.65	-5.77	0.755	0.857	0.798
1982	4.21	4.62	4.12	-9.25	-8.34	-8.44	0.902	0.918	0.92
1983	-1.23	2.01	1.57	-6.34	-5.65	-5.93	0.685	0.854	0.855
1984	4.12	2.64	2.13	-10.89	-6.14	-8.21	0.772	0.813	0.811
Average	4.446	2.92	3.31	-8.848	-6.517	-7.216	0.776	0.838	0.820

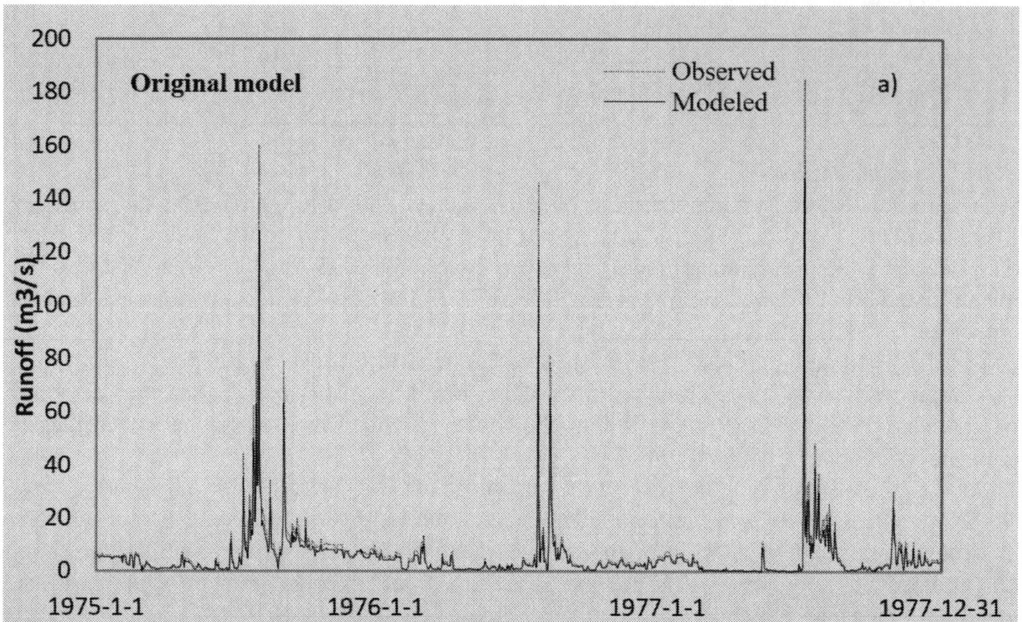
Original model
Observed
Modeled
a)
Runoff (m3/s)
200
180
160
140
120
100
80
60
40
20
0
1975-1-1
1976-1-1
1977-1-1
1977-12-31

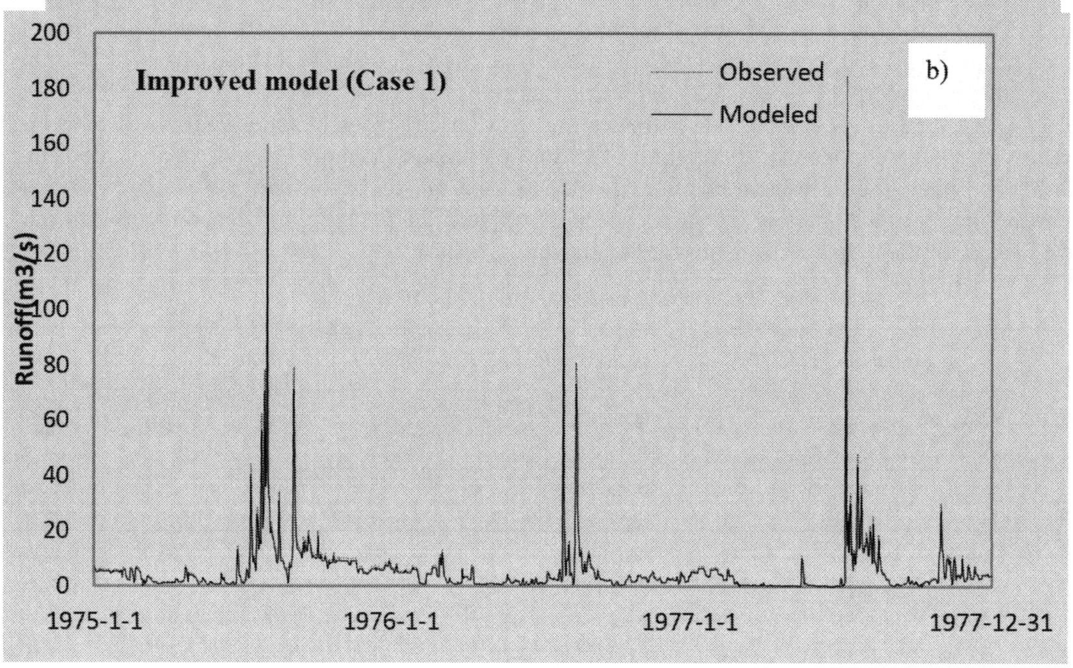
Improved model (Case 1)
Observed
Modeled
b)
Runoff(m3/s)
200
180
160
140
120
100
80
60
40
20
0
1975-1-1
1976-1-1
1977-1-1
1977-12-31

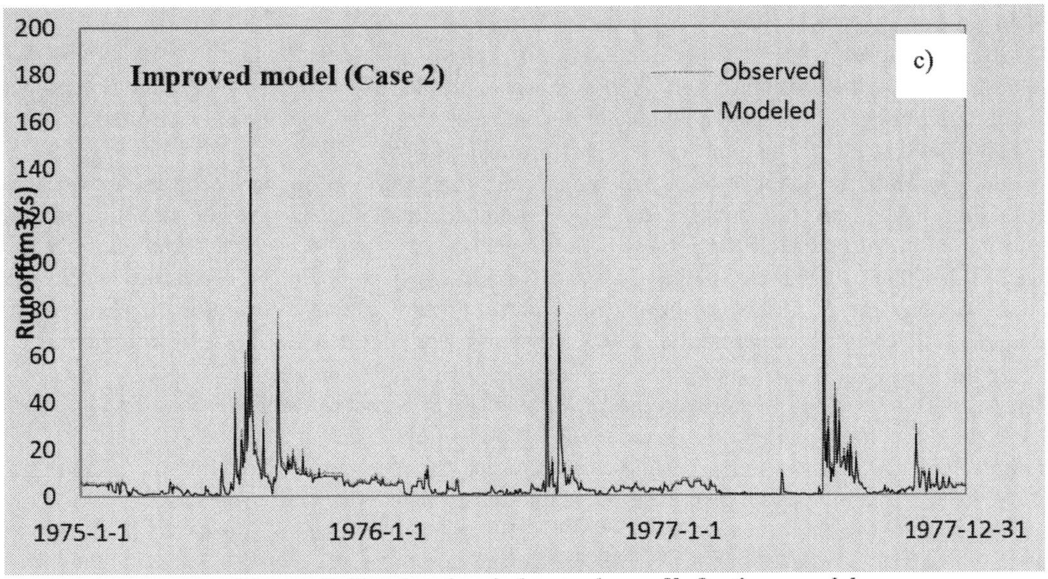

Figure 5. Simulated and observed runoffs for three models.

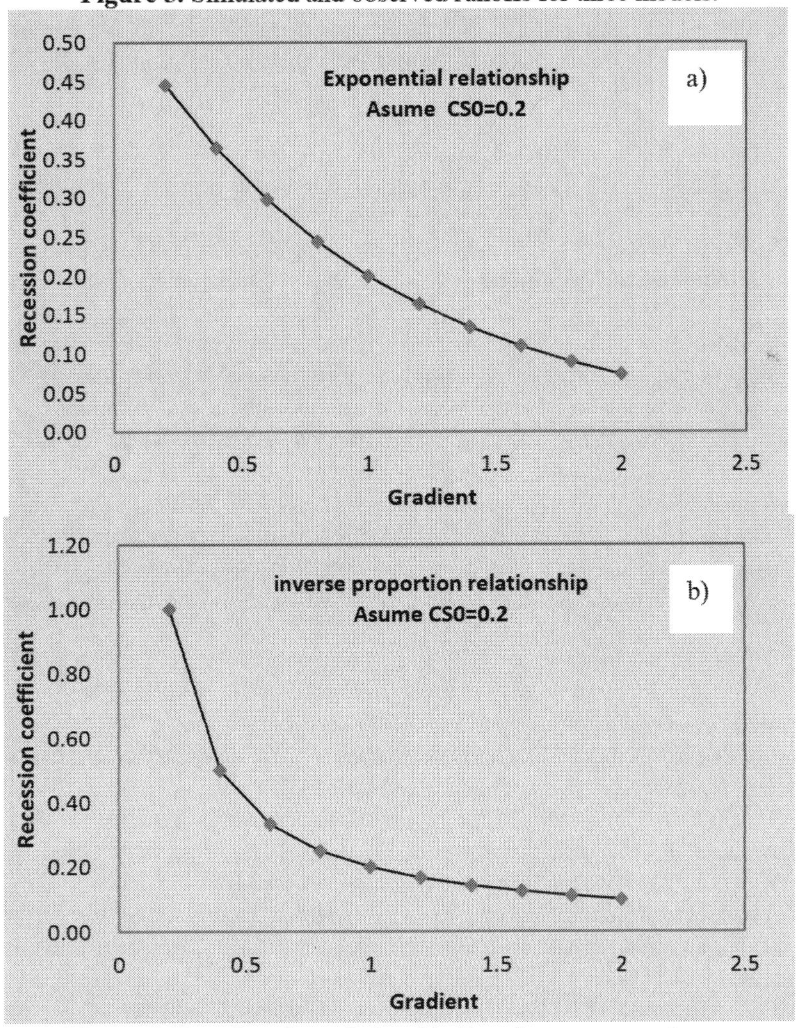

Figure 6. Two function relationship lines betweenrecession coefficient and gradient.

5. Conclusions

In the Xin'anjiang model, if the units are divided by the Thiessenpolygons method, it is likely that the unit are segmented by the basin boundary, and there are multiple runoff outlets in one unit. Based on the utilization of DEM, the unit can be divided along the basin boundary, and it can be ensured that the there is one outlet on each divided unit with a clear physical concept.

The rainfall runoff formation process, in fact, is the temporal and spatial redistribution of the rainfall of the basin, so the runoff formation characteristics depend largely on the terrain characteristics of the watershed. Using the DEM of the basin, the slope of each sub-basin unit can be calculated. Through establishment of the quantitative relationship between the recession coefficient and the average gradient of the sub-basin, the impact of the spatial distribution of the terrain characteristics on the runoff concentration can be fully considered in the Xin'anjiang hydrologic model, which gave the DEM-based Xin'anjiang hydrologic model better performance.

The performance of the two improved models shows that it is more reasonable that the recession coefficient and the average gradient of the basin is an exponential relationship.

Whether there are other functional relationships between the recession coefficient and the average gradient of the sub-basinthat could give the model better performancerequires further study.

Acknowledgments

This research was financially supported by the Non-profit Industry Financial Program of MWR(No. 201501054) and the Water conservancy scientific research project of Shandong Province(No. SDSLKY201404).The authors would like to acknowledge valuable discussions with Jeff Pompeo.

References

[1] Zhao R J 1992 The Xinanjiang model applied in China *Journal of Hydrology* **135**(1-4) 371-81
[2] Wu S, Li J and Huang G H 2008 A study on DEM-derived primary topographic attributes for hydrologic applications: Sensitivity to elevation data resolution *Applied Geography* **28**(3) 210-23
[3] Kenward T, Lettenmaier D P, Wood E F and Eric F 2000 Effects of digital elevation model accuracy on hydrologic predictions *Remote Sensing of Environment* **74**(3) 432-44
[4] Cheng C T, Zhao M Y, Chau K W, Wu X Y 2006 Using genetic algorithm and TOPSIS for Xinanjiang model calibration with a single procedure *Journal of Hydrology* **316**(1-4) 129-40
[5] Jayawardena A W and Zhou M C 2000 A modified spatial soil moisture storage capacity distribution curve for the Xinanjiang model *Journal of Hydrology* **227**(1-4) 93-113
[6] Li Z J, Kong X G and Zhang C W 1998 Improving Xin'anjiang model *Journal of China Hydrology* **4** 19-22
[7] Abbott M B, Bathurst J C, Cunge J A, O'Connell P E and Rasmussen J 1986 An introduction to the European Hydrologic SystemSysteme-HydrologiqueEuropeen, SHE, 1: History and philosophy of a physically-based, distributedmodeling system *Journal of Hydrology* **87**(1-2) 45-59
[8] Christine E, Michaela M, Hopeb A S and Loaicigac H A 2006 Distributed hydrological modelling in California semi-arid shrublands: MIKE SHE model calibration and uncertainty estimation *Journal of Hydrology* **317** (3-4) 307-24
[9] Beven K J and Kirkby M J 1979 A physically-based,variable contributing area model of basin hydrology *Hydrological Sciences Bulletin* **24**(1) 43-69
[10] Yu Z and Schwartz F W 1998 Application of integrated basin-scale hydrologic model to simulate surface water and groundwater interactions in Big Darby Creek Watershed,Ohio *Journal of the American Water Resources Association* **34**(2) 409-25
[11] Liu D D, Yu Z, Hao Z, Yang C C and Ju Q 2007 Groundwater simulation in theYangtze river basin with a coupled climate-hydrologic model *Journal of China University of Geosciences Special Issue* **18** 155-57
[12] Srinivasan R and Arnold J G 1994 Integration of a basin-scale water quality model with GIS *Journal of the American Water Resources Association* **30**(3) 453–62
[13] Yu Z B, Pollard D and Cheng L 2006 On continental-scale hydrologic simulations with a

coupled hydrologic model *Journal of Hydrology* **331**(1-2) 110-24

IOP Conf. Series: Earth and Environmental Science **39** (2016) 012036 doi:10.1088/1755-1315/39/1/012036

A study of ecological red-line area partitioning in the Chishui River Basin in Guizhou

S F Yang [1] and Y L An [2, 3, 4]

[1]Soil and Water Conservation Technical Consulting and Research Center of Guizhou Province, Guiyang, Guizhou, China
[2]School of Geographic and Environmental Sciences, Guizhou Normal University, Guiyang, Guizhou, China
[3]Guizhou Mountain Resources and Environmental Remote Sensing Application Laboratory, Guiyang, Guizhou, China

E-mail: anyulun@126.com

Abstract. Maintaining ecosystem balance and realizing the strategic goal of sustainable development are key objectives in the field of environmental sciences. Accordingly, drawing ecological red lines in sensitive and vulnerable environmental areas and important ecological function areas, determining the distribution range of ecological red-line areas, providing scientific guidance for developmental activities, and effectively managing the ecological environment are significant work tasks supported by policy guidance from the State Council and from knowledge gained in educational circles. Taking the Chishui River Basin in Guizhou as the study object, this research selected water and soil loss sensitivity, as well as assessments of karst rocky desertification sensitivity as background assessments of the eco-environment. Furthermore, the functions of soil conservation, water conservation, and biodiversity protection were integrated with exploitation-prohibited areas, and an organic combination of ecological needs and social service functions was created. Spatial comprehensive overlay analysis and processing revealed that the combination marked nine major ecological red-line areas in a total area of 5,030.58 km^2, which occupied 44.16% of the total basin area. By combining the current eco-environmental situation of the Chishui River Basin with the marked out red-line areas, this research proposed corresponding ecological red-line area management suggestions. These suggestions are expected to provide a scientific foundation for eco-environmental protection and subsequent scientific research in Chishui River Basin.

1. Introduction

The Opinion on Reinforcing Key Work of Environmental Protection No.35 (2011) issued by the Chinese State Council recommended drawing ecological red lines in areas such as major ecological function areas, sensitive and vulnerable land, and marine ecological environment areas and formulated corresponding environmental standards and policies for all main functional areas [1]. It was in this published opinion that the concept of ecological red lines was first proposed. Ecological red line is a term used to represent the ecological bottom line required by a sustainable development strategy and ecological civilization. This ecological bottom line encompasses the implications of ecological and social development scopes. It is an organic combination of the two scopes, both of which provide mutual benefits and win-win results under the requirements of determining special geographical environmental conditions and objective human activities.

Studies and partition works of ecological red lines have been carried out successively in educational circles, although only in a small number as they are still in the exploratory stage. Na Fu, *et al.* pointed out that ecological red-line areas refer to areas with vulnerable ecosystems or important ecological functions or those that require comprehensive protection [2]. Taking Guigang City in Guangxi as an example, Zhili Zuo provided an effective approach to realizing the sustainable utilization of soil resources by dividing ecological red-line areas [3]. With Hulunbuir grassland as the study object, Yu Feng marked out red-line areas and proposed corresponding red-line management measures [4]. Yan Xu, *et al.,* on the basis of their analysis of eco-environmental features of the Bohai Sea, studied and determined a technical method suitable for ecological red-line partitioning in the area.

The current study conducted an investigation and partitioning work of ecological red-line areas of the Chishui River Basin in Guizhou, China. From the basin scale, the eco-environmental restoration task for the Chishui River Basin, in terms of the upstream water source conservation area, is significant. The development of the midstream wine industry has a high degree of dependence on the natural environment, and is accompanied by the rapid development of Maotai Town and the surrounding industries, cities, and towns. Consequently, the pressure borne by the eco-environment in the basin areas is increasing. The brewing environment for brand-name and high-quality wines, such as "Guizhou Moutai Liquor," as well as the local economic and social development are also greatly influenced. The lower reaches are mostly distribution zones of national-level nature reserve areas and forest parks with significant ecological service functions as scientific research and investigation values. In recent years, as the population has continuously expanded, the influence of human activities in this area has intensified and the eco-environment has been seriously affected. Hence, conducting research on ecological red-line area partitioning and management methods in the Chishui River Basin in Guizhou is imperative and will be of significant guidance to protecting a continually deteriorating eco-environment.

2. Data and methods

2.1. Research area profile

The Chishui River Basin is located at E105°13′19″-106°58′34″ and N27°13′16-28°45′58″ in the northern part of Guizhou Province, which is located in the southwest region of China. Its landscape mainly consists of karst, and the ecological environmental is sensitive and vulnerable. The overall length of the main stream within Guizhou territory is 268.40 km. It occupies 61.5% of the total length of the Chishui River main stream and the basin area is 11,392.44 km^2 in Guizhou territory, which occupies 55.74% of the total basin area of the Chishui River. The basin belongs to a mid-subtropical monsoon climate, which is warm and humid. The variations of annual temperatures are minor; it is warm in winter and cool in summer, so the climate is pleasant. Vegetation types are mainly artificial *Pinus yunnanensis*, Masson pine, and subtropical secondary evergreen broad-leaf forests. Soil types are complicated and various, mainly yellow earth, lime earth, and purple soil. Danxia landform within the basin features a complete geomorphic development, where biological diversity is abundant and various old and endemic plants abound.

Figure 1. Location of Guizhou province in China.

2.2. *Data sources and preprocessing*

2.2.1. Data sources. Fundamental thematic data involved in this study included landform, soil type, geological background, and vegetation type from the Guizhou basic geological database. Land-cover data, a 30 m-resolution Landsat^{TM} remote-sensing image, and NPP data of 2010 were provided by the Ecology 10 Year Project group; 30 m-resolution DEM data were derived from the International Scientific Data Service Platform; and hydro-meteorological data information (1981–2010) was obtained from the China Meteorological Data Sharing Service Network platform. NDVI and bedrock bareness rate data were attained from remote-sensing image inversion. All data were uniformly placed under an equivalent secant conic (Albers) projection.

2.2.2. Data preprocessing. Supervised classification interpretation in ERDAS was utilized to determine the land-use type, which includes 10 types, namely, forest, spinney, grass, open woodland, water, paddy field, dry farm, building, rock, and bare soil. Vegetation coverage was calculated from the NDVI via dimidiate pixel model, which has been proven simple and effective in the case study region. NPP was provided by the Ecology 10 Year Project group using 10-day NPP data calculated from the 2010 annual NPP. The rate of exposed bedrock was calculated based on the rate of vegetation coverage and bare soil according to its relation. Rainfall data were used to calculate the precipitation erosion force according to the formula, and then spatial interpolation was applied to obtain the raster data of precipitation erosion force.

2.3. *Study method*

An analysis of the scientific value and connotation of ecological red lines provided in the Introduction section of this paper indicated that, at present, educational circles mainly adopt a certain degree of ecological sensitivity and vulnerability to define an unbridgeable bottom line in ecology. Based on the special ecological environment and eco-environmental background in the Chishui River Basin, this study took the Chishui River Basin as the study object and selected water and soil loss sensitivity, as well as assessments of karst rocky desertification sensitivity as background assessments of the eco-environment. By integrating the functions of soil conservation, water conservation, and biodiversity protection with the exploitation-prohibited area, an organic combination of ecological needs and social service functions was created. This combination completely marked out ecological red-line areas in the Chishui River Basin in Guizhou Province.

2.3.1. *Assessment content and standard of ecological red lines in the Chishui River Basin*

- Evaluation of water and soil loss sensitivity

Water and soil loss sensitivity in the karst region was shown to have a significant correlation with soil formation rate in addition to precipitation, soil, landform, and vegetation factors, which is closely related to acid-insoluble substance content in carbonatite [6–7]. This study selected five indexes, namely, precipitation erosion force, soil erodibility, topographic relief degree, aboveground vegetation coverage, and soil loss tolerance, to conduct an analysis and evaluation of water and soil loss sensitivity in the Chishui River Basin. A comprehensive overlapping calculation of water and soil loss was also conducted using Formula (1) [8–10].

$$SS_j = \sqrt[n]{\prod_{i=1}^{n} S_i} \tag{1}$$

In the formula, SS_j denotes the water and soil loss sensitivity index of space unit j, S_i is the sensitivity grade of factor i, and n is the number of factors. The natural parting method was used to compare the evaluation results. This method used demarcation points obtained by statistical Jenks optimization to make the sum of internal variance of grades reach the minimum value. The concrete grading standard (Table 1) refers to the Temporary Regulations on Ecological Function Areas Partitioning issued by the Ministry of Environmental Protection [11], the National Ecological Protection Red Lines—Technical Guide of Ecological Function Red-Line Partitioning (Trial) [12], and other relevant studies [13–14].

Table 1. Evaluation index grading standard of water and soil loss sensitivity.

Index	Precipitation Erosion Force	Soil Erodibility	Topographic Relief Degree	Vegetation coverage	Soil Loss Tolerance	Grading Valuation	Grading Standard
No	<25	<0.20	0–20	≥0.8		1	1.0–2.0

IOP Conf. Series: Earth and Environmental Science **39** (2016) 012036 doi:10.1088/1755-1315/39/1/012036

Sensitivity							
Mild Sensitivity	25–100	0.20–0.25	20–50	0.6–0.8		3	2.1–4.0
Medium Sensitivity	100–400	0.25–0.30	51–100	0.4–0.6	Inter-bedding Area of Carbonatite	5	4.1–6.0
High Sensitivity	400–600	0.30–0.40	101–300	0.2–0.4	Interlayer Area of Carbonatite	7	6.1–8.0
Extreme Sensitivity	>600	>0.40	>300	≤0.2	Continuity Carbonatite Area	9	>8.0

- Evaluation of karst rocky desertification sensitivity

Karst rocky desertification, which results from the irresponsible social and economic activities of humans acting on a vulnerable eco-geological environmental background, is one of the most prominent eco-environmental problems in Guizhou Province [15]. This study chose four factors, namely, bedrock bareness rate, topographic slope, vegetation coverage, and lithology, as evaluation factors of karst rocky desertification sensitivity. Furthermore, an evaluation of the single factors of karst rocky desertification sensitivity in the Chishui River Basin was conducted with grading standards determined in table 2 [16–17]. Next, according to formula (1), a comprehensive overlapping calculation was conducted and a grade map of karst rocky desertification sensitivity was developed.

Table 2. Grading standard table of karst rocky desertification sensitivity evaluation indexes.

Index	Bedrock Bareness Rate	Topographic Slope	Vegetation Coverage	Lithology	Grading Valuation	Grading Standard
No Sensitivity	≤0.3	≤5°	≥0.8	Non Carbonatite	1	1.0–2.0
Mild Sensitivity	0.3–0.5	5°–8°	0.6–0.8	Secondary Impure Limestone	3	2.1–4.0
Medium Sensitivity	0.5–0.6	8°–15°	0.4–0.6	Limestone and Clasolite	5	4.1–6.0
High Sensitivity	0.6–0.7	15°–25°	0.2–0.4	Limestone and Dolomite	7	6.1–8.0
Extreme Sensitivity	≥0.7	≥25°	≤0.2	Continuity Limestone	9	>8.0

- Evaluation of soil conservation function significance

According to suggestions put forward in Temporary Regulations on Ecological Functional Areas Partitioning, the evaluation of soil conservation significance should be based on soil erosion sensitivity, as well as on the analysis of the endangered degree and scope it may cause to the downstream bed and water source [11]. The universal soil loss equation contains the following five factors that can be used to conduct a quantitative calculation of soil conservation function: precipitation erosion force (R), soil erodibility factor (K), slope factors (S), slope length factor (L), and vegetation coverage factor (C). This study used a rectified universal soil loss equation to conduct a quantitative calculation of soil conservation while only taking the soil conservation amount in its natural state into consideration. The model structure was arranged as follows [20–22]:

(2)

$$A_p = R \cdot K \cdot LS \tag{3}$$

In the above equation, *Ac* denotes the soil conservation amount, *Ap* is the potential soil erosion amount, and *Ar* is the actual soil erosion amount. A spatial overlapping calculation of grid maps of the above factors was conducted, and then fractile quantile sorting method was adopted to divide overlapping calculation results into five grades, namely, generally significant, relatively significant, moderately significant, highly significant, and extremely significant. Then, a grading distribution graph of soil conservation significance of the study area was obtained.

- Evaluation of water source conservation function significance

Water source conservation is an important service function of an eco-system. Different eco-systems have different water source conservation capacities. Specifically, the water storage effects of eco-systems are used to measure their functions of intercepting precipitation and conserving moisture. This study adopted a precipitation storage method to conduct a quantitative evaluation of water source conservation [23, 24]. The formula is shown below:

$$Q = A \times J \times R \tag{4}$$

$$J = J_0 \times K \tag{5}$$

$$R = R_0 - R_g \tag{6}$$

In the above formula, *Q* denotes an increment of the precipitation and moisture intercepted and conserved by eco-systems compared to bare land (unit is m^3); *A* is the area of eco-system types in the calculation region; *J* is the average runoff-producing precipitation over the years in the calculation region (P>20 mm); J_0 is the average total precipitation over the years in the calculation region (mm); and *K* is the proportion of runoff-producing precipitation in total precipitation of the calculation region. According to relevant study results, 0.6 was selected for the *K* value, and *R* was the efficiency coefficient of precipitation interception and runoff reduction of eco-systems when compared to bare land (or clear-cutting slash) [25, 26]. According to measured data and relevant study results, as well as a combination of the distribution, vegetation coverage, soil, topographic features, and relevant features of the corresponding bare land of eco-system types, this study determined the *R* values of vegetation types of eco-systems [27-31]. R_0 is the precipitation/runoff ratio of bare land under runoff-producing precipitation conditions, and R_g is the precipitation/runoff ratio of eco-systems under runoff-producing precipitation conditions.

- Biodiversity protection function evaluation

Biodiversity refers to a great variety of living organisms within a specific spatial scale; it is an embodiment of the complicated relationship between organisms and between them and the environment, as well as a sign of rich and colorful biological resources [32]. According to the National Ecological Protection Red Lines—Technical Guidance of Ecological Function Red-Line Partitioning (Trial), the present study adopted the method based on biodiversity to conduct a biodiversity protection significance evaluation. The model is shown below:

$$S_{bio} = NPP_{\mathrm{mean}} \times F_{pre} \times F_{tem} \times (1 - F_{alt}) \tag{7}$$

In the equation, S_{bio} is the biodiversity protection significance, NPP_{mean} is the net primary productivity, NPP_{mean} is the precipitation, F_{tem} is the temperature parameter, and F_{alt} is the altitude parameter.

- Exploitation-prohibited area

Exploitation-prohibited areas in the Chishui River Basin include nature reserve area/scenic spots, exploitation-prohibited areas of wine production, and special water source protection areas for national wine. Nature reserve area and scenic spot distribution diagrams are from the directories of national-level and province-level nature reserve areas and scenic spots, as well as their distribution diagrams in Guizhou Province. After researching relevant data [33], the exploitation-prohibited areas for wine production were determined to be located mainly on the upper reaches of the wine production base and surrounding regions of the Chishui River in Renhuai City. According to the Upstream Ecological Function Protection Area Partitioning in the Chishui River (within Guizhou territory) [34], special water source protection areas for national wines refer to basin areas within the Guizhou territory, from the Jiucang River at the right bank of the Chishui River to the original water intake of Yuwo Moutai brewery.

2.3.2. Ecological red-line partitioning method in chishui river of guizhou province. Using GIS spatial analysis technology, and on the basis of the evaluation of eco-environmental sensitivity and ecological service function significance, extremely sensitive eco-environmental areas and extremely significant ecological service function areas were extracted to conduct broken patch processing. Low-grade patches were contained inside patches with extremely sensitive grade and extremely significant grade. All patch scopes with a lower grade were incorporated into patches with a higher grade. Based on this and comparing the results with land coverage data in 2010, the main construction lands were separated from the red-line areas. The determination of red-line borders should take full consideration of information, such as mountain ranges, rivers, geomorphic units, and vegetation to maintain the integrity of the eco-system and the connectivity of landscapes as best as possible. Furthermore, it should give priority to marking out areas with little artificial interference and easy mastery and form a spatial distribution diagram containing all ecological red lines.

3. Analysis of ecological red-line area partitioning results in the Chishui River Basin in Guizhou Province

3.1. Ecological red-line area types in the Chishui River Basin in Guizhou Province

According to the above partitioning method, ecological red-line areas in the Chishui River Basin were divided into nine major types as expounded in the basic definition of national ecological red lines including water and soil loss, karst rocky desertification, soil conservation, water source conservation, biodiversity, and nature reserve area/scenic locations. For the local typical ecological industry, namely, the wine making industry, this study also marked out exploitation-prohibited red-line areas of wine production and special water source protection red-line areas for national wines. Given the particularity of the geo-ecological environment in the Chishui River Basin, the same unit of some regions within the basin involved multiple red-line areas. Overlapped areas in red-line areas were uniformly called comprehensive ecological red-line protection areas. The situation of ecological red-line areas partitioning in the Chishui River Basin is shown below (table 3, figure 2).

Table 3. Scheme of ecological red-line area partitioning in Chishui River Basin.

Name of Red-Line Area	Area (km^2)
Karst Rocky Desertification Sensitivity Red-Line Area	1.42
Water and Soil Loss Sensitivity Red-Line Area	295.70
Soil Conservation Red-Line Area	156.09
Water Source Conservation Red-Line Area	61.06
Biodiversity Red-Line Area	787.86
Nature Reserve Area/Scenic Spot Red-Line Area	273.54
Exploitation-prohibited Red-Line Area for Wine Production	254.24
Special Water Source Protection Red-Line Area for National Wines	489.79
Comprehensive Ecological Protection Red-Line Area	2710.88
Total	5030.58

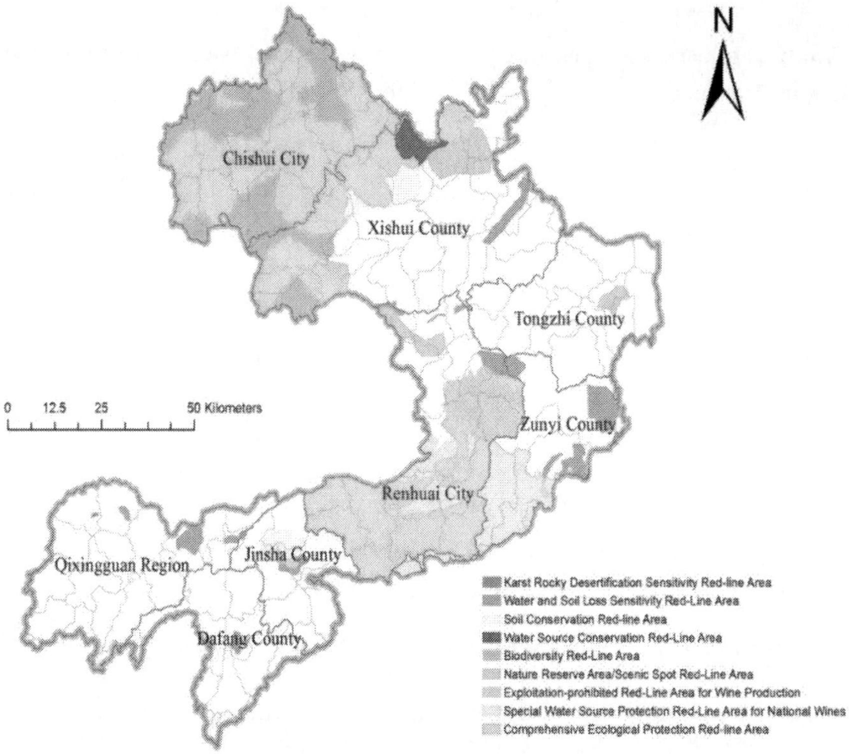

Figure 2. Distribution diagram of ecological red-line areas in the Chishui River Basin.

3.2. Distribution of ecological red-line areas in the Chishui River Basin in Guizhou Province

- Karst rocky desertification sensitivity red-line areas

One karst rocky desertification sensitivity red-line area (1.42 km2) is intensively distributed in Bijie. The lithological characters of these areas are mainly continuity limestone with geological conditions for karst rocky desertification, and the land-cover types are mainly barren grassland and dry cropland, upon which karst rocky desertification can easily happen under artificial interference. Although this red-line area is relatively small and presents random and crushing-state distribution, the influence of karst rocky desertification on the agriculture and the eco-environment cannot be underestimated. To protect this area, man-made land reclamation and soil vegetation damaging behaviors should be strictly restricted. In addition, the government should actively give guidance for the restoration and construction of vegetation and restrain further expansion of karst rocky desertification.

- Water and soil loss sensitivity red-line areas

There are 13 of this type of red-line area (295.70 km2) with a broad distribution range, including the Qixingguan region, Dafang County, Jinsha County, Zunyi County, and Renhuai City. Areas sensitive to water and soil loss usually have many sloping croplands with low vegetation coverage. These are major areas of water and soil conservation, as well as eco-environmental construction in the Chishui River Basin. Governance and protection projects centering on slope remediation should be carried out, a complete basic farmland water conservancy infrastructure constructed, and basic agricultural conditions improved. Furthermore, farmland should be actively returned to forests or grasslands, a planned development of economic forests conducted, forest and grass coverage increased, the base of soil conservation stabilized, and full advantage should be taken of the self-repairing capacity of nature to effectively control water and soil loss.

- Soil conservation red-line areas

Three soil conservation red-line areas (156.09 km2) are mainly distributed in the middle of Jinsha County and a small part of northwestern areas of Xishui County. In terms of geological lithology, these are areas where limestone mixed with clasolite and non-carbonatite are distributed. Their soil layers are relatively thick, and the vegetation cover types are mainly forests and grasslands, with good conditions for water and soil conservation. The protection of soil conservation work in these areas should be the focus, reasonable allocation of water and soil resources, and comprehensive forbiddance implemented for a large area for land reclamation and vegetation damage.

- Water source conservation red-line areas

There is one water source conservation red-line area (61.06 km2) mainly distributed in Chengzhai Village of Xishui County. This area is one part of the national-level mid-subtropical evergreen broad-leaved forest nature reserve, which has good basic conditions and natural conditions for water source conservation. Reinforcing the forbiddance, management, and protection of the nature reserve area are important measures for protecting water source conservation red-line areas.

- Biodiversity red-line areas

One biodiversity protection red-line area is mainly distributed in most parts of Chishui City and the Xishui nature reserve area located in southwestern Xishui County. The area is 787.86 km2 in size and occupies 6.92% of the total basin area. This area has superior heat-quantity conditions, with clear characteristics of plant diversity and various old and endemic plants. It is of great research value and ecological protection significance for biodiversity protection, especially for rare animals and plants. The government should prohibit capturing and killing of wild animals, as well as ban destruction of the environment of rare animals and plants by explicit order.

- Nature reserve/scenic spot red-line areas

There are six types of red-line areas mainly distributed in nature reserve areas and scenic spots in Chishui City, Xishui County, Renhuai City, and Tongzi County totaling 273.54 km2. The protection of nature reserve areas/scenic spots needs to rationally deal with the relationship between tourism development and nature reserve, develop suitable tourism destinations, develop ecological tourism, and protect the integrity of the eco-system within the reserve areas.

- Exploitation-prohibited area red-line area for wine production

There is one exploitation-prohibited red-line area for wine production that is mainly located in the wine production base of Renhuai City and the surrounding areas, which totals 254.24 km2 and occupies 2.23% of the total basin area. This red-line area is the wine production base and forbids development of new enterprises with serious pollution based on original construction, discharge of industrial sewage and domestic sewage into the Chishui River, and reinforces the protection of the water source environment in the Chishui River.

- Special water source conservation red-line area for national wines

One special water source conservation red-line area for national wines is located in the upper reaches of the Chishui River in Renhuai City and the surrounding areas of production base, which totals 254.24 km2 and occupies 4.30% of the total basin area. The protection of this red-line area is of great significance to the production of "Guizhou Moutai" national wine. This area has the water sources necessary for a national wine brewery and provides qualified water environment for national wine production. Environmental and eco-environmental water protection should be intensified on the upper reaches. Furthermore, the development of chemical industries and high-pollution industries, such as mining, metallurgy, and papermaking, all of which affect water quality, should be forbidden, water pollution discharge standards should be tightened, and enterprises that pollute water should be shut down.

- Comprehensive ecological protection red-line areas

As some areas have various red-line areas that mutually overlap in geographical spaces, these areas are extremely significant comprehensive protection areas because of their multiple protection objectives including comprehensive protection red-line area consisting of eight major red-line areas, water and soil loss sensitivity and exploitation-prohibited red-line areas for wine production, water source conservation and biodiversity protection red-line areas, and soil and water source conservation red-line areas. These overlapping areas are mainly distributed in the national-level Cyathea spinulosa nature reserve area and the mid-subtropical evergreen broad-leafed forest nature reserve area, which are at the border between Chishui City and Xishui County, and RenHuai City and its surrounding areas. This type of red-line area covers 2,719.88 km2, which occupies 23.80% of the total basin area. Protection of the eco-environment in this area should aim at biodiversity and the special habitat of C. spinulosa nature reserve area to completely eradicate interference and damage from human activities.

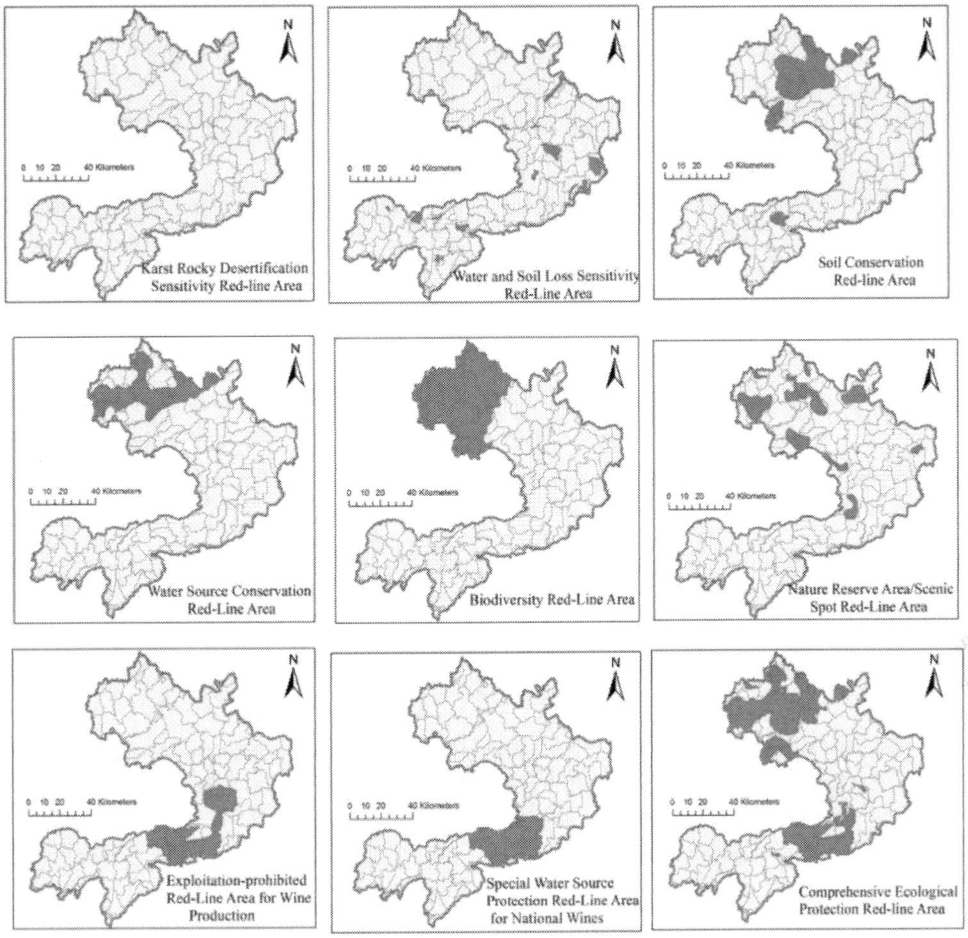

Figure 3. Distribution diagrams of all types of ecological red-line areas.

4. Conclusion and suggestions

Taking into consideration the comprehensive features of natural factors and human activities in the Chishui River Basin, as well as the objective requirement of industry and life, this study specified nine major ecological red-line areas with a total area of 5,030.58 km^2. This area makes up 44.16% of the total basin area. Owing to the particularity of the physico-geographical environment in the Chishui River basin, red-line areas are mainly ecological service red-line areas and exploitation-prohibited red-line areas, which are liberally distributed in Chishui City, the national-level nature reserve area in Xishui County, Renhuai City, and its surrounding areas.

According to features of the distribution situation for all types of ecological red-line areas, the principle of “priority given to special water source protection, equal attention paid to ecological protection and ecological construction” must be prioritized. This principle must be combined with a sorting management and comprehensive administration of ecological red lines in the Chishui River. Access to industrial development must be restricted and a mechanism for ecological benefit compensation must be established to give priority to organizing land use in red-line areas, implementing agricultural production on the condition that eco-environmental quality is not affected, and constructing environmentally-friendly land use modes. Importantly, a sustainable development path must be achieved to realize healthy and harmonious existence between the regional economy and the environment.

This study discussed the technical method of ecological red-line partitioning, which basically reflected the ecological features of the study region in a theoretical sense. For practicability of ecological red-line areas, it is necessary to conduct further study according to the population and social and economic development situation of the study region. By combining suggestions of the local government, harmony between red-line areas and regional ecological protection planning, as well as land use planning can be guaranteed. This study was an objective embodiment and scientific presentation of the local ecological environment, but the far-reaching significance of partitioning ecological red-line areas is not limited to this. More detailed protection regulations and measures should be formulated during subsequent real life management of red-line areas. Tangible ecological red-line protection assessment indexes must also be established to truly embody the science and authority of ecological red lines, as well as their core values, which are essential for social development and environmental protection.

Acknowledgments

Foundation item: The National Natural Science Fund of China (41161002) and the Natural Characteristics and Ecology in the fragile district of the "Semi-Karst" in Guizhou; Guizhou Special No.46 (2011) (Natural features and utilization of an ecologically fragile district in Guizhou "Semi-Karst"); Guizhou Special subject (2012-2013) on remote sensing investigation and assessment of the National 10-years (2000 - 2010) Ecological Environment Change.

References

[1] The state council general office 2011 *The comment on strengthening environmental protection work of the state council*

[2] Fu N and Li X B 2007 Study on the method of ecological red line area of land use planning geographical society of China *The Academic Conference Paper Sets*

[3] Zuo Z L 2010 *The study on land-use layout based on ecological red-line areas—A case study of Guigang City in Guangxi Zhuang Autonomous Region* Guangxi: Guangxi Teachers Education University

[4] Feng Y 2013 *Research on Method of Ecological Red Line Area Delineation in Hulunbeir Grassland* Beijing: China environmental science research institute

[5] Xu Y, Liang B, Bao C G, et al. 2013 Research on the index system and the technical methods of ecological red line division for the Bohai Sea *Marine Science Bulletin* **32** 361-7

[6] Wang S J, Ji H B, Ouyang Z Y, et al. 1999 The preliminary study on weathering pedogenesis of carbonate rock *Science in China (Series D)* **29** 442-9

[7] Li R L, Wang S J, Zhou D Q, et al. 2003 The Correlation between Rock Desertification and Lithology in Karst Area of Guizhou *Acta Geographica Sinica* **58** 314-320

[8] Fan F D, Wang K L, Xiong Y, et al. 2011 Assessment and spatial distribution of water and soil loss in karst regions, southwest China *Acta Ecologica Sinica* **31** 6353-62

[9] Li D M, Gao Z W , Fu X, et al. 2010 Characteristic of ecological sensitivity in Yunnan ecological zones *Acta Ecologica Sinica* **30** 138-45

[10] Liu C X, Li Y C, Yang H, et al. 2011 RS and GIS-based assessment for eco-environmental sensitivity of the Three Gorges Reservoir Area of Chongqing *Acta Geographica Sinica* **66** 631-42

[11] The State Environmental Protection Administration 2002 *The interim regulation of Ecological function zoning* Beijing: Chinese academy of sciences

[12] Department of environmental protection of China 2014 *National ecological protection line -ecological function red line drawn technology guidelines (trial)* Beijing: China environmental science research institute

[13] Yang G B, Li Y Q and An Y L 2006 Pixel-Based Assessment and Spatial Distribution of Sensitivity of Soil Erosion in Guizhou *Carsologica Sinica* **25** 73- 8

[14] Luo J, Wang K L, Chen H S, et al. 2009 Sensitivity of Water and Soil Loss of Northwest Karst Area in Guangxi Province *Resources and Environment in the Yangtze Basin* **18** 579- 84

[15] Xiao R B, Ouyang Z Y, Wang X K, et al. 2005 Sensitivity of rocky desertification and its spatial distribution in southwestern China *Chinese Journal of Ecology* **24** 551-4

[16] Zhou X, Tong X H, Hua C, et al. 2006 The ecological environment sensitivity comprehensive evaluation and spatial distribution of Guangxi *Journal of Guangxi Teachers Education University (Natural Science Edition)* **23** 1-8

[17] Li Y G, Jiang Y J and Shan N 2009 Analysis of Karst Rocky Desertification Actuality and Sensitivity in Nanchuan of Chongqing Based on RS and GIS *Journal of Anhui Agricultural Sciences* **37** 2193-95

[18] Zhou H Y, Wang X Z and Li H X 2011 The land desertification sensitivity evaluation in north of Guangdong mountainous *Soil and Water Conservation in China* **10** 56-57+64

[19] Liu C X, Li Y C and Yang H 2011Sensitivity assessment of karst rocky desertification in three gorges reservoir area of Chongqing and its spatial differentiation characteristics *Resources and Environment in the Yangtze Basin* **20** 291-7

[20] Xiao Y, Xie G D and An K 2003 The function and economic value of soil conservation of ecosystems in Qinghai-Tibet Plateau *Acta Ecologica Sinica* **23** 2367-78

[21] Liu M C, Li D Q, Wen Y M, et al 2005 The spatial analysis of soil retention function in Sanjiangyuan region and its value evaluation *China Environmental Science* **25** 627-31

[22] Han Y W, Gao J X, Wang B L, et al. 2012 Evaluation of soil conservation function and its values in major eco-function areas of Loess Plateau in eastern Gansu province *Transactions of the Chinese Society of Agricultural Engineering* **28** 78-86

[23] Li J C 1999 *The theory of ecological value* Chongqing University Press

[24] Zhang S H, Zhao G Z, Tian Y Z, et al 2001 Study on value the ecological environment valuation of forestry resources—For case by Hunchun forestry in Changbai Mountain *Journal of Yanbian University (Natural Science Edition)* **27** 126-34

[25] Huang Q R, He X Y, Zhou M Q, et al. 1998 Preliminary observation on climatic effect of ecosystems of fruit trees and crops in red soil of Hilly Region *Acta Agriculture Jiangxi* **10** 76-83

[26] Duan S S and Lin Q Q 2000 Effect of the soil and water conservation of fruit tree and grass Intercropping patterns on gentleslopeland in Guangdong *Grassland of China* **5**

35-40

[27] Zhou X F 1994 *Chinese forest ecosystem positioning research* Haerbing: Northeast Forestry University press

[28] Yu X X, Qin Y S, Dong L H, et al. 2002 The Forest Ecosystem Services and Their Valuation of Beijing Mountain Areas *Acta Ecologica Sinica* **22** 783-6

[29] Zhang J, Li D L and Wang W 2008 Influence of terrain on precipitation in Qinghai-Tibet plateau during summer monsoon *Scientia Geographica Sinica* **28** 235-40

[30] Yan S G, Zhou C X, Shen W S, et al. 2010 Evaluation of ecological compensation based on leading ecosystem service function value-A case study in Ruoergai *Ecological Science* **06** 543-6

[31] Zhao X Y and Dong X 2010 Application minimum-data approach in ecological compensation—A case of Huanghe River water supply area of Gannan *Scientia Geographica Sinica* **05** 748-54

[32] Wu J G, Lv J J and Ai L 2009 The impacts of climate change on the biodiversity *Ecology and Environmental Sciences* **18** 693-703

[33] Zhao Z 2013 *Guizhou environmental protection action-The ecological protection "red line" of Chishui river Basin in Renhuai*

[34] The environmental protection agency of Guizhou 2006 *The instruction of ecological function protection division of Chishui river upstream (Guizhou)* Guizhou: Environmental Protection Agency

Rural sewage treatment processing in Yongjia County, Zhejiang Province

W H Wang[1] and T H Kuan[2, 3]

[1] Professor, Zhejiang University of Water Resources and Electric Power, Hangzhou, China
[2] Ph.D., P.E. Retired Engineer, Public Works, City of Los Angeles, Los Angeles, CA, USA

Email: tsunkuan@yahoo.com

Abstract. Issues regarding water pollution in rural areas of China have garnered increased attention over the years. Further discussion on the circumstances and results of existing domestic sewage treatment methods may serve as an appropriate reference in solving these important issues. This article explored the current conditions of water contamination in rural areas of China, introduced the characteristics and effects of applicable sewage treatment technology, and summarized the results of the planning, installation, and operation of rural sewage treatment facilities in Yongjia County in Zhejiang Province. However, relying on a single technical design rule is not adequate for solving the practical problems that these villages face. Instead, methods of planning rural sewage treatment should be adapted to better suit local conditions and different residential forms. It is crucial, ultimately, for any domestic sewage treatment system in a rural area to be commissioned, engineered, and maintained by a market-oriented professional company.

1. General water environment situation in rural areas of China

China's recent rapid economic development has greatly increased quality of life for its population, but it has come with a serious tradeoff. Chinese people bear a double burden of air and water pollution. In studies from the summer of 2011, the Chinese government reported that 43 percent of state-monitored rivers were so polluted that they are unsuitable for human contact, and over half of China's population is threatened by seriously polluted water. Due to the long-standing urban-rural dualistic structure and overemphasis on economic development policies negative results, such as ignoring water scarcity, overextending water resources, misusing unsafe pesticides, polluted drinking water sources, and arbitrary disposal of wastewater from local enterprises and villagers, water crises have arisen throughout rural areas of China. According to the Annual Report of 2011 published by the Chinese Ministry of Environmental Protection, the total discharge of pollutants in rural areas makes up approximately half of the national total. These statistics are ominous. This Chinese saying summarizes well the water environment and sanitary situation in rural China: "Sweeping trash depends on the wind blowing, drying sewage relies on the sun shining even as drinking fresh water is determined by the rain" [1].

IOP Conf. Series: Earth and Environmental Science **39** (2016) 012037 doi:10.1088/1755-1315/39/1/012037

However, in 2010, the Chinese People's Congress posted "Three Red Guide Lines" as tough thresholds for water resource management principles in China; all levels of Chinese Government are rigorously carrying out steps to protect water resources nationwide. After struggling for several years, the lessening rural waste is improving ecological and human health conditions in the Chinese countryside. The idyllic scene of clean creeks and green hills in many local areas is gradually returning because the Chinese government has done much to address the high levels of pollution, including raising environmental priorities through the cadre evaluation system, and budgeting RMB 5.1 trillion for key pollution control in 2011-2015 [2]. Many measures for the mitigation of pollution and treatment of domestic sewage have been undertaken in rural regions. Consequently, Yongjia County, Zhejiang is a typical example to demonstrate the green awakening of the water environment in rural China.

2. Introduction of water resource conditions in Yongjia County

Yongjia County is located in Wenzhou District in the southeast of Zhejiang province, which is one of the most developed regions in China and famous for its scenic splendor. Yongjia is 15 kilometers north of Wenzhou city proper, which administers the county. The Nanxi River Scenic Area is located within this county, and the river has been tentatively nominated for the list of UNESCO World Heritage Sites.

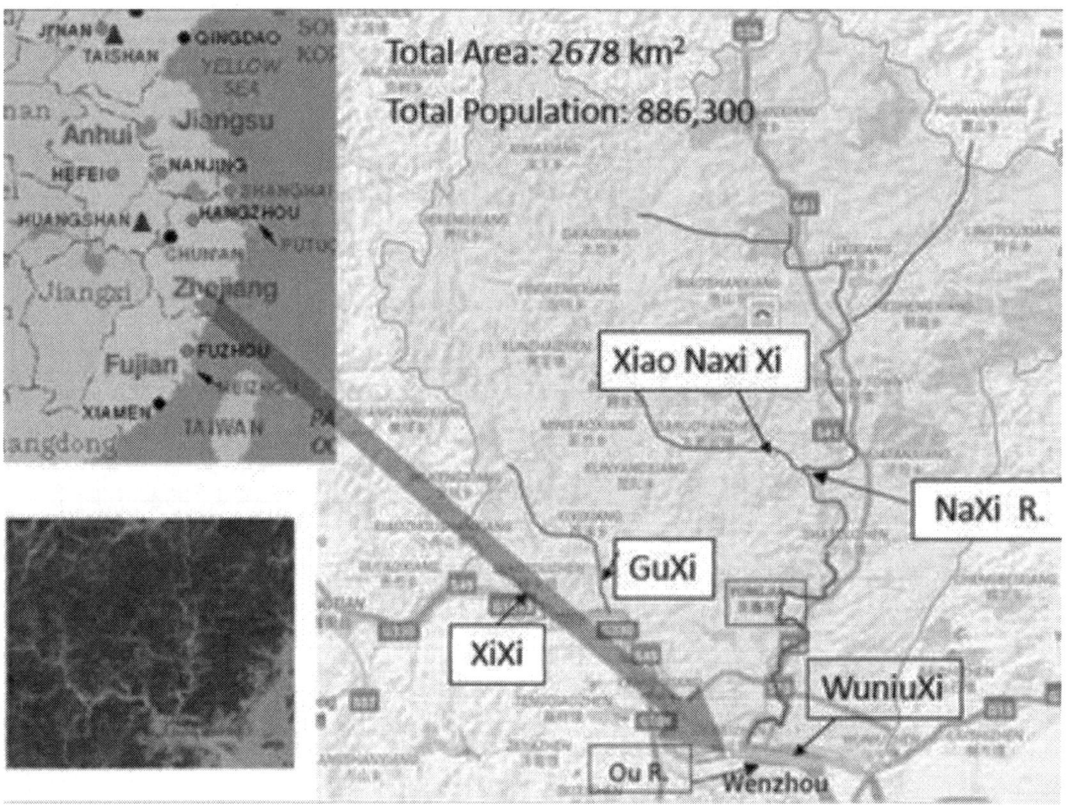

Figure 1. Maps of Yongjia County.

Yongjia County had a total population of 886, 300 in 2007. Among this population, 63% (555, 640) were farmers living in 166, 360 separate households. The whole county is divided into 8 sub-districts, including 10 towns, 650 administrated villages, and 1881 native villages. Yongjia County is situated downstream of the southern Ou River, crossing the Ou River from north to south ending in Wenzhou City, and east of Yueqing City (See Figure 1). The total area of Yongjia County is 2698 square kilometres, of which 2308.5 square kilometres are mountainous, 277 square kilometres are plains, and 112.7 square kilometres consist of rivers and lakes. Yongjia also has 36.16 hectares of cultivated land. The area is known as "Eight Parts Mountain, One Part Water, and One Part Farmland."

Yongjia generally belongs to a subtropical monsoon climate, with mild seasons of full rainfall, an average annual temperature of 18.2° C, and average annual rainfall of about 1702.2 mm, which creates an average annual total of more than 3.6 billion cubic meters of water resources. All rivers in this region belong to the Ou River system, of which Nanxi, Xixi, Guxi, and Wuniuxi are the four major tributaries. The county has a total of 3377 rivers with a total length of 4302 km, including 3294 km of mountain rivers (including Nanxi River) with 1.57 km/km2 density; there are 83 strings of river network totaling about 114 km with a drainage density of 1.76 km/km2.

In recent years, with the developments of society and economy, the county's waste water emissions have increased year after year. For example, in 1998, 1999, 2000, the county's waste water emissions were 0.57, 0.59, 0.62 million tons, respectively. Thus, pollution in the inhabited areas along the river plains and inland towns has become more serious, making river water quality below Grade IV or V due to industrial and domestic wastewater discharge. Since 2012, water quality of the county's surface water has improved slightly, but the outlook for overall water quality is not optimistic in Yongjia.

3. Challenges of water resource management for Zhejiang Province and Yongjia County

In 2010, China's Communist Party Central Committee and State Council promulgated a "Three Red Lines" policy intended to establish clear and binding limits on water quantity usage, efficiency, and quality. The development of socialist ecological civilization is necessary in implementing the sustainable development of the diversification of propulsion strategy. Under the progress of promoting the construction of water ecological civilizations, the Government of Zhejiang Province launched a program named "The Government of Five Water" (Five Water Project), means that the Government will try their best to govern polluted water, treat sewage, prevent flooding, supply clean water for citizens, and encourage people to save water.

Zhejiang is famous for water scenery, which makes it beautiful and prosperous. Due to the dramatic growth of urbanization and rapid development of industrialization in past decades, many beautiful rivers and lakes are covered with a layer of black shadow, and many sightseeing destinations have been destroyed. This somber reality has made people understand that water is vital to agriculture and is the lifeblood of the city. Moreover, the polluted water resources have become a hurdle for growing the economy and improving livelihood.

From 2013 to 2014 in Zhejiang Province, the data show that the "Five Water Project," as a breakthrough point, forced economic transformation and upgrades, and remedied severe environmental pollution. Under the program, 5058 kilometers of black rivers were mitigated; rural sewage treatment facilities were built for more than 3.3 million residents, and a rainwater pipe network of over 2845 km was constructed. Domestic sewage and agricultural contamination are two major sources of rural pollution. Zhejiang comprehensively managed the construction of rural sewage treatment facilities; the Province completed 6.66 million sewage treatment stations and laid 2036 million meters of village pipe, which benefited over 330 million farmers. The program also enhanced source control of agricultural contamination and remediation of farm pollution, which lead to the rehabilitation of infrastructures and standardization of new villages [3, 4].

Meanwhile, the Yongjia county government placed "Five Water Project" at a high priority and started a project for the purpose of sustainability of water resources and mitigation of rural wastewater pollution.

First, they invested RBM 7.137 billion to complement local subjects of the "Five Water Project," including to resolve black, smelly rivers, build sewage treatment plants, and complete village sewage treatment facilities. Under the project urban sewage pipe networks were constructed and arranged, which raised the sewage treatment rate from 34% to 90% and benefited an additional 60,000 farming households, and resolved and improved 284,800 rural drinking water safety issues. The hard work paid off; Yongjia County was twice named "The Most Attractive County" in Zhejiang Province.

4. Strategy and plan for achievement of domestic wastewater treatment in Yongjia County

From 2014 to 2016, the household sewage treatment coverage rate of Yongjia County nearly tripled from 34.1% to 90%, and construction on 1661 village sewage treatment facilities (covering 88% of the entire county's villages) was completed. This great achievement did not come easily; it was based on the diligent efforts of those working on an intelligent propulsion strategy and comprehensive development plan.

As in humid areas of southern China, rural areas of Yongjia County historically had a very low sewage treatment rate for domestic wastewater. Most villages lacked a sewage system, and households relied on natural conditions to discharge untreated wastewater. Existing limited rural sewage systems often malfunctioned and were expensive to operate due to poor construction, significant leaks, and lack of management personnel. As such, existing rural sewage treatment facilities do not provide the best example in sanitary control. Most villages in vast rural areas of Yongjia are scattered because the terrain is complex, and the economic conditions of many villages are limited, which makes financing improvements and construction of treatment facilities extremely difficult. Therefore, a suitable and strategic management project is necessary to overcome these difficult conditions.

After actively pursuing "Five Water Project" and broadly congregating opinions and demands from the citizens of Yongjia, a rational project goal was assembled, which is to integrate environmental management and protection of water resources, reform rural villages, and consider eco-tourism development as the primary means to finance the construction of the Domestic Wastewater Facility Project ("Project"). Under this common goal, the Government of Yongjia County can finance RMB 0.74 B from the State subsidy, private investments, and personal donations; and the Zhejiang Design Institute of Water Conservancy & Hydroelectric Power is invited to act as their consultant to prepare a technical development plan in a combination of centralized processing and distributed processing to accomplish the Project. The centralized process will use pipe lines to collect discharged wastewater into nearby big treatment plants suitable for large villages where the households number in the hundreds; the distributed process will build a small facility and treat domestic wastewater in place, as appropriate for dispersed households. The estimated total cost to complete the Project is about RMB 0.92 B for construction and remediation of treatment facilities in 650 administrated villages and 1700 native villages, based on the average cost of 5, 200 RMB/per household. The planning stage is divided into a three-year period. The first year, 2014, was spent consolidating and amending existing treatment facilities and launching a pilot program to accumulate experience. The second year, 2015, was the key period to extensively develop the project, and 2016 was designated to solve complicated areas and finish the project [5].

5. Analysis of process model and summary of results

Most villages in Yongjia County are in areas of rolling hills, and the layout of the villages is dispersed. So, the chosen principle of rural sewage treatment processing adhered to the idea of a "mainly decentralized, but supplemented by centralized system" based on an anaerobic + wetland decentralized processing model. Remote and dispersed rural villages apply the decentralized processing mode, and villages where geographical conditions allow use centralized processing, which is mainly determined by local conditions.

In 2015, thirty-five percent of the villages in Yongjia County adopted the decentralized governance model. This is defined by two or more households' toilet water, wash water, and kitchen water collected into a nearby anaerobic tank (septic tank), which is then treated to discharge into the nearest wetland, farmland, or woodland. The decentralized processing mode has advantages of low investment costs, savings of 20% to 30% compared to the centralized processing model, low operation and maintenance costs, and no need to occupy farmland. Practical experience has proved that determining the approach according to local conditions is not only widely accepted by farmers, but also conforms to the Level One Class B of National Urban Sewage Treatment Standard for use in farm irrigation.

The decentralized processing model of "Anaerobic + Wetland/Ecological Pond" can be demonstrated as follows: (See figure 2)

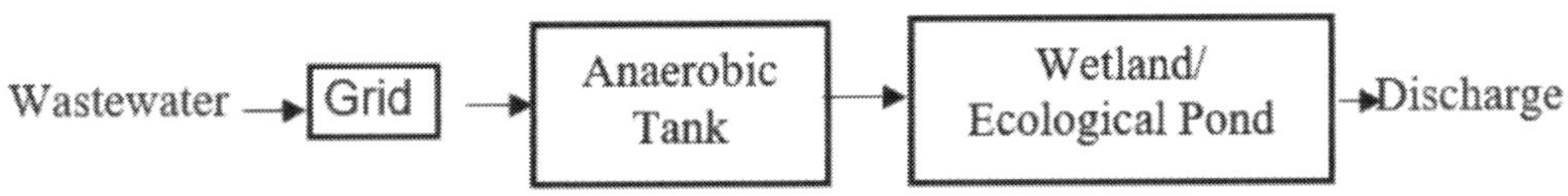

Figure 2. Simple chart of decentralized processing.

Wastewater flows through the grid and solid waste is removed into the unpowered anaerobic tank, which is a masonry or concrete underground tank, for processing. The anaerobic tank is equipped with a multi-grid, correspondingly playing adjustment, sediment, and anaerobic oxidation effects. Anaerobic treatment technology is an effective removal process of organic contaminants, which causes a mineralization effect that transforms organic compounds into methane and carbon dioxide. The effluent runs into wetlands or an eco-system pond for reprocessing. Wetlands and ecological ponds are chosen mainly based on existing idle conditions such as fish ponds, reservoirs, and other low-lying land. When using artificial wetlands or ponds, the type of drift processing is selected as either surface flow or sub-flow, depending on the water quality, site topography, and other geological conditions; subsurface flow is usually constructed in wetlands to avoid clogged pipes. In wetland ponds, various aquatic vegetation is planted to act in aerobic and anaerobic states through absorption and adsorption functions, which combine physical, chemical, and biological functions for further sewage treatment.

According to the official posted data, the recent implementation results from the last three years are summarized in table 1 below.

As the achievements demonstrate in the above table, the coverage rate of the household wastewater treatment eventually reached 90%, which increased rural sewage treatment capacity to 31, 480 t/d (almost double the existing capacity), benefitted 80% of all people living in Yongjia County, effectively improved the source control of surface water and the health of the rural environment, reduced the spread of disease, and made the rivers and streams become beautiful again. Hence, the Yongjia Government not only firmly adheres to the principle "To complete one, to check one, and to operate one," making sure every treatment facility, whether in a centralized or decentralized process, is utilized under the regulation of Zhejiang Rural Sewage Treatment Facilities Management Implementation Plan, but also issues the policy for local administration offices to contract out the maintenance of facilities to dependable private companies to warrant that each facility can be monitored, examined, and kept in regular service.

Certainly, the social and economic consequences of improvements to Yongjia rural sewage treatment are significant. Through specific implementation of the project and the improvement of quality of life, people's environmental awareness awakened in what produced a qualitative leap. Thus, the protection of the environment and conservation of resources became a conscious act of the villagers.

Incidentally, it is hard to analyse and evaluate the exact revenue obtained from rural sewage treatment by reducing the loss or damage to the county's economic development caused by the ecological impacts of environmental pollution, but indirect profits solely gained through tourism was RMB 4.923 B in 2015, an increase of 25.74% [6, 7].

Table 1. Results of Domestic Wastewater Treatment Project in Yongjia County (Summary of Official Data).

	Decentralized Processing		Centralized Processing			Invest	Benefitted People
	A Villages	N Villages	Treat Plant	Pipe Ext	Village	¥x10^8	Household
Planning Goal	650	1661	12	142 Km	39	9.2	148, 585
2013 Existing	188	435	3^a	26 Km			52, 290
2014	178	(505)^b	8	58 Km	(17)^b	2.17	31, 300
2015	215	(588)^b	2	72.7 Km	(6)^b	2.45	34, 000
2016 (Cont.)	191	(568)^b	1	100 Km	(16)^b	(2.0)^b	20, 000
Subtotal In 3 Years	584	(1661)^b	11	230.7 Km	(39)^b	6.62	85, 300

[a] Three total treatment plants existed, two already operating and one not fully operating.
[b] Figures in ( )brackets should be further verified.
*A Villages= Administration Villages; N Villages= Native Villages

Incidentally, it is hard to analyze and evaluate the exact revenue obtained from rural sewage treatment by reducing the loss or damage to the county's economic development caused by the ecological impacts of environmental pollution, but indirect profits solely gained through tourism was RMB 4.923 B in 2015, an increase of 25.74% [6, 7].

6. Conclusion

Successfully completing a well-managed rural sewage treatment project in Yongjia County in a short three-year period, while properly addressing the daily demands of around 120, 000 village households, is an illustrious example for China.

Therefore, the assessment of accomplishing a rural sewage treatment in Yongjia County can be concluded as follows:

First, the described Project is one of the sustainable water resources management events successfully completed by the Government of Yongjia County. This Project illustrated that reform against severe rural sewage contamination should be well-planned, strategically managed, and faithfully executed. In addition, the reform system must be supervised by a unified department, and the importance of promoting the market-oriented operation to contract out the project should be appropriately stressed.

Sufficient financial resources are the second most important factor in this project. As described previously, the Yongjia Government amassed about 80% of the estimated cost from Government subsidies, and 10% from private investments for the project budget due to a serious shortage of capital investment. Without ample financial support, it is impossible to complete the project.

Strengthened construction management, long-term maintenance, and reinforced promotion of market-oriented operations are the last factors needed to ensure long-term operation. In order to further develop the project, the county government promulgated five regulations, one after another, and ultimately opted to contract out the continuing operation and maintenance to private sectors for completion.

In conclusion, the method or strategy successfully applied for treatment of domestic sewage in rural areas of Yongjia County has certain representativeness because not only are the physical and geographical conditions in Yongjia County representative of most rural areas in China, but the economic and social circumstances are comparable to other countryside populations of this nation. However, it is only the dawn of a green awakening, and there is a long way to go to overcome the bitterly opposed change to the environmental pollution in China.

References

[1] Yu X M, Geng Y, Heck P, and Xue B 2015 *A Review of China's rural water management Sustainability 2015* ISSN 2071-1050
[2] Zhu J L, Yan Y, He C and Wang C 2015 China's environment: Big issues, accelerating effort, ample opportunities *Goldman Sachs-Asia*
[3] United States-China Economic and Security Review Commission 2015 *China Ahead of the 13th Five-Year Plan: Competitiveness and Market*
[4] Chun G 2015 "Five Water Project" Zhejiang Has Drawn a Beautiful Picture *People's Daily,* D797
[5] Zhejiang environmental protection science design institute and Yongjia County Rural Affairs Office 2014 *Yongjia County Rural Sewage Management Planning (2011-2016)*
[6] Jiang J F 2016 *The Yongjia County Government Annual Working Report- 2015*
[7] Jiang J F 2015 *The Yongjia County Government Annual Working Report- 2014*

Mechanism of toppling instability of the human body in floodwaters

C W Shu[1], S S Han[1, 3, 4], W N Kong[2] and B L Dong[3]

[1]State Key Laboratory of Water Resources and Hydropower Engineering Science, Wuhan University, Wuhan 430072, China
[2]Hubei Provincial Water Resources and Hydropower Planning Survey and Design Institute, Wuhan 430072, China
[3]School of Water Resources and Hydropower Engineering, Wuhan University, Wuhan 430072, China

E-mail: sshan9202@whu.edu.cn

Abstract. Extreme urban flood events occur frequently in China, often leading to heavy casualties. Thus, it is of great importance to study the mechanism of the instability of the human body in floodwaters. The results of such research can provide scientific reference for city flood control standards. In this paper, a formula for the incipient velocity of the human body, during toppling instability in floodwaters, was derived based on mechanical characteristics, instability mechanism, and critical conditions during instability. A series of flume experiments were conducted to investigate the incipient velocity of two 3D printed human body models of different sizes; the resultant experimental data was used to determine parameters in the derived formula. Additionally, grip strength was taken as a standard of a person's ability to withstand floodwaters. Finally, crowd factors were introduced, and based on this study, a criterion for the toppling instability of different subjects in floodwaters was proposed. Compared to the results of previous studies, the proposed formula can better predict the instability of the human body in floodwaters.

1. Introduction

Statistics from the *Emergency Events Database* indicate that the loss of life and property caused by floods, ranks highest among all natural disasters [1]. According to an in-depth investigation conducted by the *Ministry of Housing and Urban-Rural Development*, of the 351 cities investigated in China, 62% of them were hit by urban water disasters between 2008 and 2010. Even worse, 137 of these cities were struck more than three times [2, 3]. Unfortunately, the situation became more serious in 2011 when many cities, including Beijing, Chengdu, Wuhan, and Hangzhou, suffered rainstorms, which lead to severe urban water-logging disasters [4]. On July 21, 2012, Beijing suffered its biggest rainfall since the establishment of PRC [5, 6]. In recent years, urban floods have become more frequent, leading to traffic paralysis, serious adverse effects on the normal operation of cities, and an increasing threat to human lives. The safety of the inhabitants on floodways is a big problem of urban storm water design and flood plain management. According to the statistics provided in 2011 by *China's State Flood Control and Drought Relief Headquarters and Ministry of Water Resources*, the number of dead and/or missing, as a result of flash flooding, is on the rise; in 2010 alone, the death toll reached 3,222. Populations living in flood zones are expected to increase in the coming decades due to the effects of climate change, the rapid growth in population, and the expansion of urbanization [8]. Hence, the instability mechanism of the human body during a flood disaster is an important issue to be studied.

Existing research on the body's instability in floodwaters is insufficient, and the topic is only vaguely mentioned in flood risk management protocols. Most references to this serious issue are expressed by the relationship between incipient velocities and water depths. The main existing research ideas are as follow.

- *Experimental testing.* Foster, *et al.* [9] performed experiments on human stability in a flume 6 meters long, 0.6 meters wide, and 0.75 meters deep, with the subjects consisting of six male children aged 9 to 13 years old. However, there was no quantitative evaluation method, because the criteria developed for stability and instability flow conditions were based on the personal tendency of the experimenter. Based on the earlier work of Foster and Cox, Yee [10] expanded on the research by experimenting the stability of four children. Abt, *et al.* [11] undertook laboratory experiments on human instability in a flume 61 meters long, 2.24 meters wide, and 1.22 meters deep with different surfaces; the subjects consisted of twenty adults. A formula defining the critical value of instability of a human body in floodwaters was obtained, which included the height and mass of the experimenter, and the flow depth and velocity of the water. The equation was as follows:

$$hv_c = 0.0929(e^{0.001906Lm+1.09})^2$$

According to experiments on human stability, Suetsugi [12] believed that conditions became unsafe when the product of flow depth and velocity reached 0.5 m^2/s. A movable platform, 1.17 meters long, inside a test water basin was used to test the stability of rescue workers by Karvonen, *et al.* [13]. Rescue workers were represented by seven participating adults. The research method of Karvonen, *et al.* was unique because the platform was moving through stagnant water, as opposed to the subjects being exposed to flowing water. Based on their experimental data, the equation derived by Abt, *et al.* was improved upon, as follows:

$$hv_c = 0.004Lm + 0.2$$

Jonkman, *et al.* [14] reported on a study conducted by FHRC in the UK, in which a professional stuntman was subjected to different flowing water within the channel. In all of the tests, instability reportedly occurred after the subject slipped backwards in the relatively high velocities and low water depths. The experimental results indicated that the condition of flow, describing instability or maneuverability of a human body, is closely related to a subject's height and weight. However, there is no known theory for the parameters of friction and drag coefficients.

- *Empirical and theoretical analysis.* Based on the experimental results of Abt, *et al.*, Takahashi, *et al.* [15] developed a computational model for human stability which included friction, drag force, and weight. For any given person's weight and height, computational resolution of frictional force, drag force, and weight facilitated the calculation of critical velocity for either sliding or toppling modes of instability in a given water depth. In water depths less than 0.48 of a subject's height, only the feet and legs were exposed to drag forces, which led to discrepancies between computational results and the experimental results obtained by Foster and Cox. Keller, *et al.* [16] undertook a theoretical analysis of human body stability, and one formula was derived. However, this purely theoretical method was highly dependent on the friction and drag coefficient values. The friction coefficient and drag coefficient were 0.3 and 1.2 respectively in the study, which did not match the general values of friction and drag coefficient measured by Takahashi, *et al.* [15]. Lind, *et al.* [17] considered the toppling instability of a circular cylindrical body, a square parallel pipe, and composite cylinders for the torso and two legs, assembled to represent a person immersed in floodwaters and subjected to buoyancy and drag forces; four formulae were established. The experimental data of Abt, *et al.* [11] and Karvonen, *et al.* [13] was used to calibrate and compare these formulae. It should be noted that the geometrical similarities of the study objects were not accurate. Consequently, Ramsbottom, *et al.* [18] and Penning-Rowsell [19] proposed an equation to judge the risk level of a human body in floodwaters, found by semi-quantitative assessments including flow depth and velocity.

According to the equation, several risk levels were ranked, including low risk, moderate risk, significant risk, and extreme risk, and a division chart of risk level for Britishmen was developed. Compared to other equations, the equation written by Penning-Rowsell was superior because it was clear that the stability of a human body in floodwaters was not only about the product of their weight and height, but was also related to the ground features. Cao, *et al.* [20] identified two types of instability mechanisms of a person standing still in floodwaters, which included sliding and toppling instability. A formula for the incipient velocity of a human body for toppling and sliding instability in floodwaters was derived. At the same time, a dynamic equation for a person walking in floodwaters, considering only one generalized coordinate, was derived based on the Lagrange method. The numerical solution was presented through the higher-order Runge-Kutta method. Cao, *et al.* asserted that sliding instability usually occurred when the flow velocity was <0.25 m/s while toppling instability generally occurred when the flow velocity was >2.5 m/s. They suggested values of flow depths of 0.8-0.9 m for human walking instability research in floodwaters.

- *Combination of theoretical analysis with flume experiments.* Recently, Xia [21,22] achieved results by conducting research on a human model, with a scale of 1:5.54, and by considering the threshold of motion of coarse sediment. The derived method was mentioned by Jianheng Xie [23] in Mechanics of Sediment Transport. However, there was no theory for the values of height and weight of the model in the experiments.

Unfortunately, the decrease in flood diversion areas, such as rivers and farmland, caused by increasing urbanization and economic growth, and the increase in impermeable surfaces, further exacerbates urban flood risks and places people's lives and property in growing danger. All of the above-mentioned research, based on experimental tests using actual people, was significantly dependent on the psychological factors and physical attributes of the test subjects. In addition, the research based on laboratory experiments, using models of human bodies, was less precise because the geometrical attributes of the study subjects were dissimilar. Research projects based on theoretical analysis, combined with flume model experiments, are more scientific than other research methods. However, there is still no known theory for the selection of values for height and weight of a model in experimentations, and the existing research does not take varied subjects into account. Thus, it is necessary to further study the mechanism of instability of the human body in floodwaters, in order to provide scientific reference for the implementation of city flood control standards.

2. Force analysis and mechanism of instability

2.1. Force analysis

If a person stands facing upstream in floodwater, the water depth doesn't surpass their head. Therefore, the stability of a human body subjected to flooding is determined by the drag force, frictional force, its own gravitational force and buoyancy force. Figure 1 shows the forces acting on a submerged human body during a moment of instability.

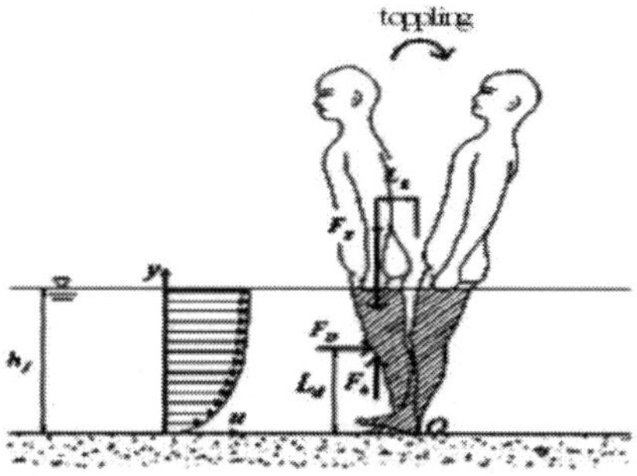

Figure 1. Forces acting on a submerged human body for toppling instability.

- Drag force. In the horizontal direction, the drag force (F_D) acting on a flooded human body can be expressed as follows [21, 22]:

$$F_D = A_d C_d \gamma_f u_b^2 / 2g \tag{1}$$

where u_b is the near-bed velocity; C_d is the drag force coefficient; γ_f is the unit weight of the flow; and A_d is the stressed area, $A_d = a_d(b_p \times h_f)$. It is well known that the value of C_d is closely related to the shape of the subject and the object Reynolds number. According to the test of drag force, the value of C_d is inversely proportional to the Reynolds number when the Reynolds number is <20,000, and the value of C_d is regarded as a constant number, 1.08, when the Reynolds number is >20,000.

- Buoyancy force. It is assumed that the distribution of the volume of the human body in the height direction is a smooth curve. As such, F_b can be expressed by the following:

$$F_b = \rho g V_b \tag{2}$$

According the study conducted by Xia [21, 22], V_b can be expressed by the following:

$$V_b / V_p = 0.633 \left(\frac{h_f}{h_p} \right)^2 + 0.367 \frac{h_f}{h_p} \tag{3}$$

where $V_p = 0.0008 m_p + 0.0092$ (according to the research conducted by Zhao, 2005; ρ is the density of water, g is the gravitational acceleration, V_b is the submerged volume of the human body, V_p is the volume of the human body, and h_p is the height of the person. Equation (3) proposes that the action of gravity will reduce, because the buoyancy force increases as the depth of flow rises; therefore, buoyancy exerts a significant effect on a submerged human body, which cannot be ignored.

- Gravity force. The gravity force can be expressed as follows:

$$F_g = m_p g \tag{4}$$

Where m_p is the mass of human body.

- Frictional force. According to the research of Xia [21,22], the frictional force expression can be written as follows:

$$F_R = \mu g (m_p - \rho V_b) \tag{5}$$

2.2. Mechanism of instability of a submerged human body

Existing research shows that sliding and toppling instability are the two instability mechanisms. Sliding instability generally occurs in cases of low water depths and high flow velocities, meaning $F_D > F_R$. Toppling instability usually occurs in cases of high water depths and low flow velocities, meaning $F_D > F_g$. Hydrologic features of urban floodwaters are usually high water depths at low flow velocities; as such, this study focused on toppling instability.

When a human stands facing upstream in the urban flood, as shown in figure 1, the toppling instability critical condition of the human body would be the heel (O) and would overturn the total torque backward around the pivot point O ($\sum M_O = 0$), namely $F_D \cdot a_h h_f - F_g \cdot a_g h_p = 0$, where a_h and a_g are the correction coefficients [21,22]. Using equations (1) and (5), a deformation equation can be derived for u_b, as follows:

$$u_b = \sqrt{\frac{2ga_g}{a_d a_p a_h}} \sqrt{\frac{1}{h_f^2}(\frac{m_p}{\rho} - V_b)} \sqrt{\frac{1}{C_d}} \tag{6}$$

Analogously, using the average velocity instead of the representative near-bed velocity, which can be expressed as $u_b = (1+\beta)(a_b h_p / h_f)^\beta U$ and by substituting $u_b = (1+\beta)(a_b h_p / h_f)^\beta U$ into the equation (6), the incipient velocity of the toppling instability for a submerged human body can then be expressed as follows:

$$U = \alpha(\frac{h_f}{h_p})^\beta \sqrt{\frac{1}{h_f^2}\left(\frac{m_p}{\rho} - V_b\right)} \sqrt{\frac{1}{C_d}} \tag{7}$$

where $\alpha = \frac{1}{1+\beta}\frac{1}{a_b^\beta}\sqrt{\frac{2ga_g}{a_p a_h a_d}}$ and the unit is $m^{0.5}s^{-1}$. Parameters α and β are closely related to the human body friction coefficient and drag coefficient; the values of α and β can be determined according to the experimental data of the flume.

3. Flume experiments

3.1. Model design

According to *The China National Physique Monitoring Report* in 2010 and 2014, the height range of adult males in China was 1.65 m-1.72 m, with an average of about 1.69 m (figure 2). The weight range of adult males in China was 61.5 kg-71.5 kg, with an average of about 68.9 kg (figure 3). The physical model for the instability of a submerged human body was designed into two different models, with geometric scales of λ_L=4 and λ_L=8. The outstretched arms of the model were designed with 70 degrees of angle on the left arm and 10 degrees of angle on the right arm, with both arms forming an angle of 100 degrees, according to the action of inertia of a human body in floodwaters (figure 4). The size and scale of the model was in line with the average size of a Chinese adult, taking national standard *GB1000-88*, *The China National Physique Monitoring Report* in 2014, and the monograph written by Qingshan Guo [25] for reference. Figures 5 and 6 illustrate the human body models used in this study, which were created by 3D printing using photosensitive resin. The hollow insides of the models were filled with steel balls with a diameter of 2 mm, and sealed with waterproof material in order to mimic the distribution of gravity on a real human. The basic parameters are shown in table 1.

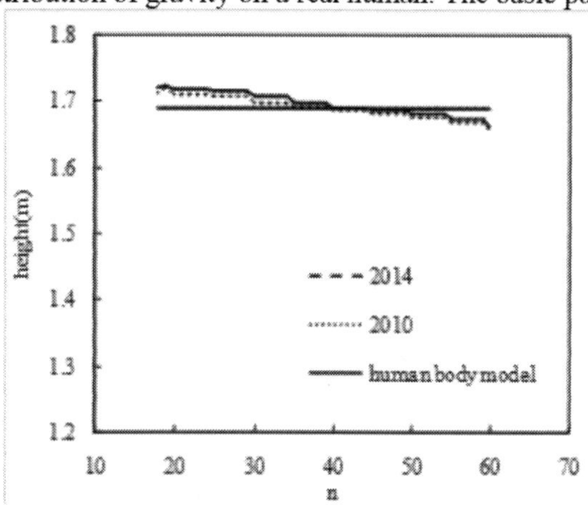

Figure 2. Average height distribution of adult males in China.

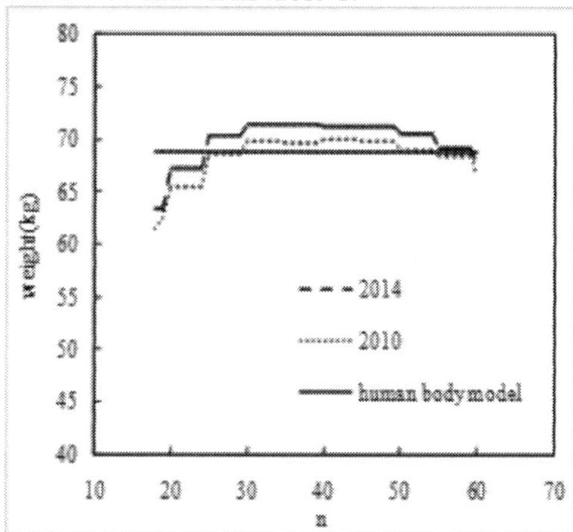

Figure 3. Average weight distribution of adult males in China.

Figure 4. Action of inertia on a human body in floodwaters.

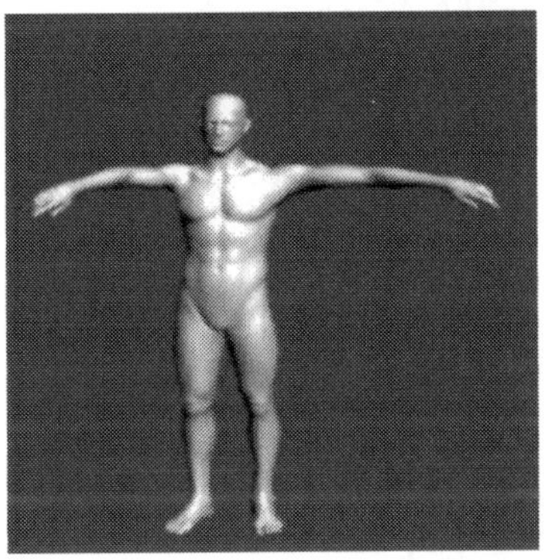

Figure 5. Model of the human body.

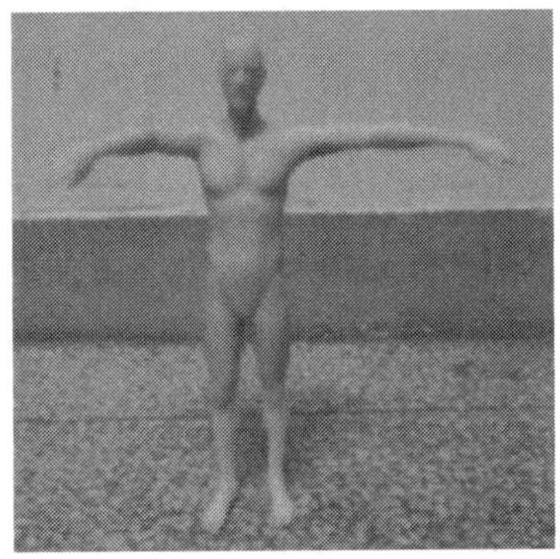

Figure 6. 3D printed model of the human body.

According to the study conducted by Xia [21, 22], the prototype and the physical model must conform to the principles of geometric, kinematic, and dynamic similarities. The model used in this study strictly followed geometric similarities; therefore, the scales of length, width, and height were equal. For example, the model with the scale of λ_L=4 can be expressed as follows:

$$\lambda_L = \lambda_W = \lambda_H = 4 \tag{8}$$

The model needed to satisfy the principles of kinematic similarities, so the scale ratio for the velocity λ_U can be written as follows:

$$\lambda_U = \lambda_L^{1/2} = 2 \tag{9}$$

According to the principles of dynamic similarities, the scale ratio of the force λ_F was written as follows:

$$\lambda_F = \lambda_L^3 = 64 \tag{10}$$

The density of the model and the prototype almost unanimously, hence the following was true:

$$\lambda_{F_g} = \lambda_{F_b} = \lambda_F = 64 \tag{11}$$

According to Xia's [21, 22] analysis:

$$\lambda_{F_D} = \lambda_F \tag{12}$$

According to the principles of the friction roughness similarities, the following was true:

$$\lambda_{F_R} = \lambda_F = 64 \tag{13}$$

Table 1. Basic parameters of the model.

Scale	Prototype	Model	
	1:1	1:4	1:8
Height (cm)	169.02	42.26	21.13
Weight (g)	68900	1076.56	134.57
Palm volume	$V_1 = 0.00566V_p$		
Forearm volume	$V_2 = 0.01702V_p$		
Upper volume	$V_3 = 0.03495V_p$		
Thigh volume	$V_4 = 0.0924V_p$		
Shank volume	$V_5 = 0.04083V_p$		
Trunk volume	$V_6 = 0.6132V_p$		
Palm length	$L_1 = 0.109h_p$		
Forearm length	$L_2 = 0.157h_p$		
Upper length	$L_3 = 0.1172h_p$		
Thigh length	$L_4 = 0.232h_p$		
Shank length	$L_5 = 0.247h_p$		
Trunk length	$L_6 = 0.300h_p$		

3.2. Experiments

Experiments were conducted in a flume in the State Key Laboratory of Water Resources and Hydropower Engineering Science at Wuhan University to investigate the critical condition of toppling instability for the model of the human body. The horizontal flume was 6.0 meters long, 1.2 meters wide, 0.8 meters deep with a slope of 0.5%, and was equipped with a glass bed and glass sides, and an adjustable tail gate; the water temperature was 21.5° C to 21.6° C. The model was placed at the base of the flume, in the center, in order to reduce impacts on the flume walls. The adjustable quantity of flow in this experiment was 0-100 L/s, which was gradually increased every 90 s from the initial depth of 0.01 m. At the same time, the tail gate was adjusted and the state of the model was observed. Each test was conducted three times; the flow and corresponding water depth and velocity were recorded as the submerged model went from stable to unstable, and the arithmetic average was calculated.

3.3. Analysis of experimental results

Thirty groups of effective data were collected from the models for toppling instability in this study, including fifteen groups of data using the model with λ_L=4 and fifteen groups using the model with λ_L=8. The relationship of the incipient velocities for different water depths at toppling instability is shown in figure 7, with the horizontal axis representing critical depth of the water h_f(m) and the vertical axis representing the corresponding velocity U_c(m). The following information can be drawn from figure 7:

- The incipient motion of the two models was basically the same, and the incipient velocity for the models decreased with increases in the water depth. This was because when the water depth increased, the wetted area increased, resulting in an increasing drag force. When the effective gravity decreased, friction for resisting sliding decreased.
- The smaller the mass of the model, the easier it was to move when the models faced the same direction, because the effective gravity becomes smaller when the mass of model decreases with the proximity of the buoyancy force of the models in the same water depth.

The law of incipient motion, concluded from the experimental results, was consistent with the changing law of the curve of incipient velocity formula for a submerged human body. Therefore, the structure of equation (7) was reasonable.

4. Criteria for toppling instability of a human body in floodwaters

4.1. Parameter calibration

A formula for toppling instability was derived, which was relatively complex because of more variables. Therefore, equation (7) was transformed to the following:

$$U/(\sqrt{\frac{1}{h_f^2}\left(\frac{m}{\rho}-V_b\right)}\sqrt{\frac{1}{C_d}})=\alpha(\frac{h_f}{h_p})^\beta \tag{14}$$

where the relation between $U/(\sqrt{\frac{1}{h_f^2}\left(\frac{m}{\rho}-V_b\right)}\sqrt{\frac{1}{C_d}})$ and $\frac{h_f}{h_p}$ represents a power function.

According to Xia's [21, 22] research, the prototype, and the physical model, must conform to the following scale ratios of water depth and velocity:

$$\begin{cases} h_{fp}=h_{fm}\times\lambda_L \\ U_{cp}=U_{cm}\times\sqrt{\lambda_L} \end{cases} \tag{15}$$

where the subscripts p and m refer to the prototype parameters and model parameters. The scaled-up experimental data, obtained from 30 tests using equation (15) for the prototype, is shown in the points of figure 7. The experimental parameters $h_p=1.7m$ and $m_p=68kg$ were used for real human bodies in the study by Jonkman and Penning-Rowsell, while $h_p=1.69m$ and $m_p=68.9kg$ were used in this study. Hence, the experimental data obtained from Jonkman and Penning-Rowsell was applicable in this study. A statistical analysis software package, SPSS, was used to determine α and β, as follows: $\alpha=1.915$ and $\beta=-0.391$.

Figure 7 shows that the critical instability conditions, obtained using the scale ratios, matched well with the calculations from equation (7), confirming the accuracy of the critical conditions for the prototype.

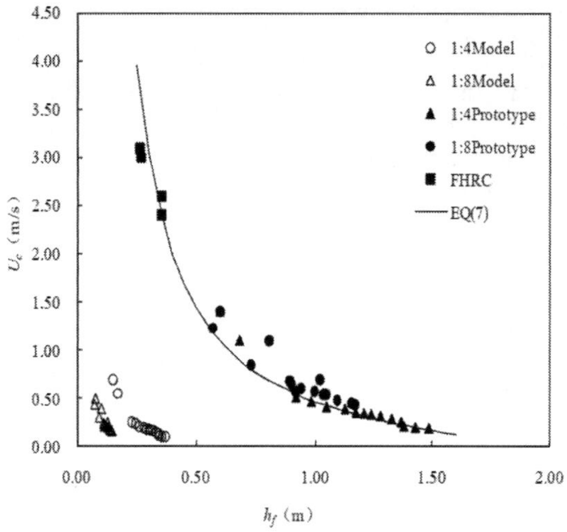

Figure 7. Comparisons the critical conditions between the experimental data and the calculations using the derived formula.

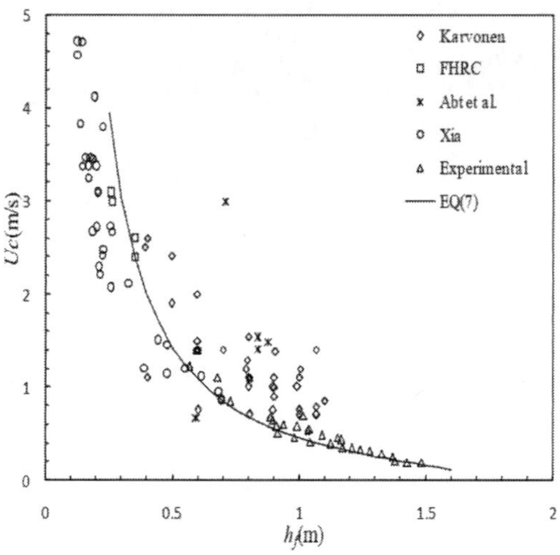

Figure 8. Comparison to previous results.

4.2. Comparison with previous achievements

Most existing research on the instability of the human body in floodwaters neglected to investigate the effects of body buoyancy. However, buoyancy exerts a significant effect on a submerged human body, as confirmed by Xia. Buoyancy can be ignored only when the depth of floodwater is shallow enough. Thus, research that does not consider buoyancy is not applicable when the depth of the floodwater is higher, such as in the results obtained by Abt, *et al.*, Takahashi, *et al.* and so on.

Existing experimental data for the instability of human bodies in floodwaters using prototypes are mainly represented by the results of Abt, *et al.* [11], Karvonen, *et al.* [13], Jonkman and Penning-Rowsell[14], and Xia[21,22]. Twenty real humans were used to conduct experimental research on the instability of the human body in floodwaters in the study of Abt, *et al.*[11]. Flume studies were undertaken on four different bottom surfaces in over 70 tests. In the experimental program of Karvonen, *et al.* [13], seven real humans wore survival suits and safety helmets. Four controlled field tests of human body stability in a natural channel were conducted by FHRC in the UK using a stuntman as the test subject. The subject completed maneuvers, including standing and walking at right angles and walking into the flow of the water (Jonkman and Penning-Rowsell), Xia conducted 54 tests on the instability of humans in a flume using a human body model with $\lambda_L = 5.54$. Before the model became unstable, it was placed in two positions in the flowing water including facing upstream in floodwater and with its back to the oncoming flow direction. Comparisons to the results of the above-mentioned studies are shown in figure 8. As shown in figure 8, in conditions of the same depth, the scattered points of the previous experimental data for real humans in flumes were higher than the points of the experimental data in this study, which meant that the incipient velocities in this study were generally lower than those of the experimental data regarding real humans. The reason may be that the model lacked the ability to resist floodwaters subjectively and actively, making it less stable than a real human in floodwaters. The scattered points of the experimental data given by Xia were below the points in this study, which may be because the weight of the prototype model used by Xia was 60 kg, which was less than the weight of prototype used in this study (68.9 kg). Therefore, compared to the stability of the model in this study, the stability of the model used by Xia was worse. Hence, the formula derived in this study was more reliable for predicting the toppling instability of the human body in floodwaters.

4.3. Extensive application

Because the human body model used in this study was based on the average height and weight of adult males, equation (7) derived herein based on the experimental data cannot be used to reflect the instability of the entire population in floodwaters because the values of children, adults, elderly, males, and females are different. As such, crowd factors K_r were introduced into the formula derivation of the toppling instability for varied population, in order to make the formula closer to reality.

The strength of the upper limb muscles can be represented by grip strength. Therefore, grip strength, which is generally related to the weight of the tested person, can be evaluated in a physical fitness test, through which a scientific physical assessment can be made [26, 27]. Thus, the grip strengths of different people were used as criteria for the resistant ability of a human in floodwaters. The measurement of grip strength was conducted with a hand grip dynamometer, which tested 218 volunteers of different ages, consisting of 106 males and 112 females. The grip strength values of the males' and females' left and right hands are plotted in figure 9. As shown, the grip strengths of the left and right hands were evenly distributed close to the line y=x, which showed that the grip strengths of the left and right hands were approximate to each other. Thus, the average grip strengths of the left and right hands were used in this study.

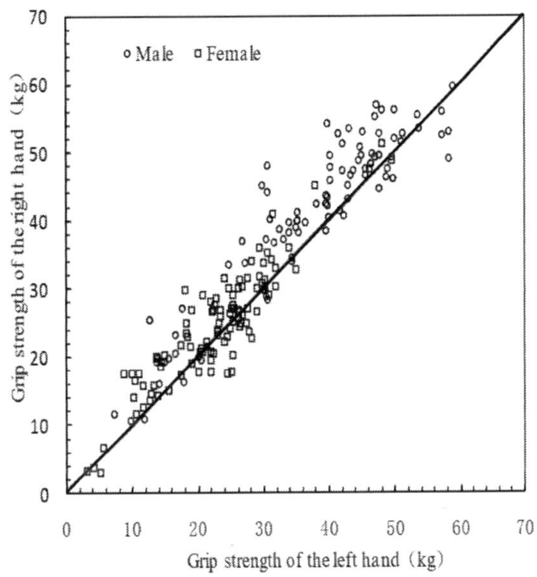

Figure 9. Grip strength of left and right hands.

According to the *United Nations Convention on the Rights of the Child*, which was passed in November 20th, 1989, individuals under the age of 18 are deemed as children, and above the age of 60 are deemed as elderly [28]. Therefore, people aged 18 to 60 were deemed as adults in this study.

Figure 10 shows the distribution of average grip strength of males and females, among which the grip strength of children basically met the linear relationship. The average grip strength for adult males and females was 44.1 kg and 25.38 kg, respectively, all fitting within the standard of grip strength set in The China National Physique Monitoring Report in 2014 released by the General Administration of Sport of China. Therefore, the crowd factors K_r of varied population were calculated according to the distribution of the grip strength, which is shown in table 2.

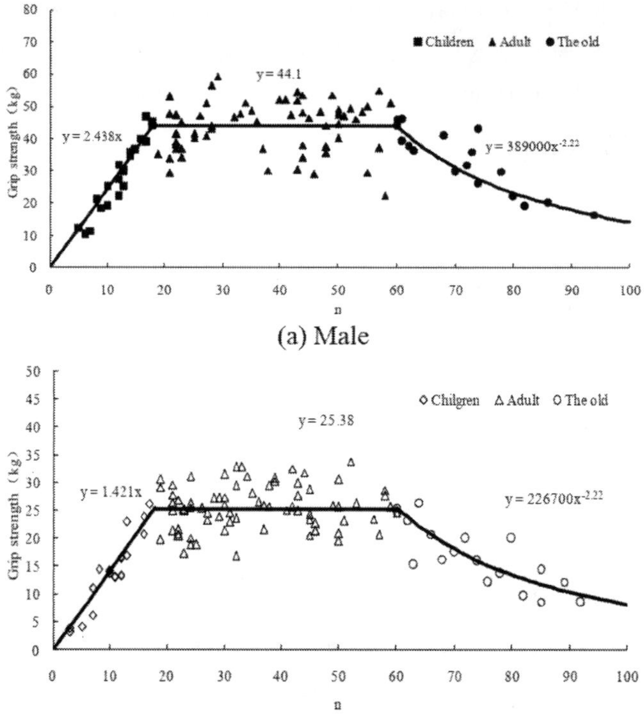

(a) Male

(b) Female

Figure 10. Distribution of grip strength.

Table 2. Crowd factors K_r

Population \ Crowd factors		K_r
Male	Children	0.0553 n
	Adult	1
	Elderly	8821 $n^{-2.22}$
Female	Children	0.0323 n
	Adult	0.576
	Elderly	5140 $n^{-2.22}$

Note: n-age

Substituting crowd factors K_r into equation (7), the incipient velocity for a submerged human at toppling instability can be expressed as follows:

$$U = 1.915K_r(\frac{h_f}{h_p})^{-0.391}\sqrt{\frac{1}{h_f^{\,2}}\left(\frac{m}{\rho}-V_b\right)}\sqrt{\frac{1}{C_d}} \tag{16}$$

In 2010, there were 85.02 million disabled persons in China, accounting for 6.34% of the total population [29]. Therefore, equation (16) has wide application.

5. Conclusions

In this study, the mechanism of toppling instability of the human body in floodwaters was researched using theoretical analysis and flume experiments combined with a force analysis. A formula was derived based on data acquired from the flume experiments undertaken to establish the incipient velocity on two human body models of different scales. The formula can be used to predict the incipient conditions (the flow depth and velocity of the water) of the toppling instability of the human body in floodwaters. The following statements refer to the analytical and experimental tests conducted in this study:

- In this paper, the experiment assumed people were uninjured during the flood. The outstretched arms of the model were designed with 70 degrees of angle in the left arm and 10 degrees of angle in the right arm with both arms together forming an angle of 100 degrees. The slope of the flume was 0.5%, the water temperature was 21.5° C to 21.6° C, and human body model was without clothes.
- The strength of the upper limb muscles can be represented by the grip strength; therefore, grip strength, which is generally related to the weight of the tested person, was evaluated in physical fitness testing through which scientific physical assessments were made. With this data, the grip strengths of different people were used as criteria for the resistant ability of the human body in floodwaters.
- Considering the human body's action of inertia in floodwaters, it was more practical to make a human body model with 3D printing, which was closer to reality, thus ensuring the experimental data was more accurate.
- Various forces acting on a submerged human body were analyzed and the crowd factors were further improved and became more reasonable, forming a wide formation from children to the elderly. A formula of incipient velocity for a submerged human body at toppling instability was derived consisting of crowd factors K_r and the drag coefficient C_d, and parameters of the formula were derived using the experimental data. Compared to previous research results, it was confirmed that the derived formula was reliable in predicting the toppling instability of the human body in floodwaters.
- In addition to the factors of flow conditions, the gender and the physical characteristics of the subject, such as height and weight, there are many other factors that influence the stability of a human body in floodwaters, including the clothing worn by the subject, the slope of the flume, turbulence characteristics, water temperature, fatigue, and so on. Hence, further research with additional experiments on large-scale human body models, or real human bodies, in real floodwaters is needed to further improve the formulae of incipient velocity of a submerged human body in different environmental conditions.

Acknowledgments

The study reported herein was partly supported by the National Natural Science Foundation of China [grant numbers 51309180], and the State Key Program of National Natural Science Foundation of China [grant numbers 51339004]. Special Funds for Public Industry Research Projects of the National Ministry of Water Resources [grant numbers 201401038].

References

[1] Chen Y 2009 The Flood disaster and its defense *Overview of Disaster Prevention* **04** 44-51
[2] Li L F 2013 Refused to urban waterlogging *Informatization of China Construction* **9** 4-9
[3] Jin C L 2015 The analysis and research on urban flood control *Taiyuan University of Technology Master's degree thesis*
[4] Ju N S and Gong K 2011 Discussion on reason and countermeasure of urban waterlogging *Jiangsu Construction* **Z1** 90-93
[5] Wang G Z, Li X, Chen J B and Wu X H 2015 The rainstorm indirect economic loss assessment of multi departments based on IO Model—a case study on the rainstorm on July 21st in Beijing *Journal of Catastrophology* **2** 94-99
[6] Wang N Y and Wang M L 2012 Public opinion analysis of Beijing "7.21" storm-related disasters and reflections about it *Water Resources Development Research* **12**(11) 13-15
[7] China's State Flood Control and Drought Relief Headquarters and Ministry of Water Resources 2011 *Bulletion of Flood and Drought Disasters in China*

[8] Zhang J Q and Lou C W 2014 Study on the metropolis waterlogging prevention in the view of urban governance *Journal of Shanghai administration institute* **15**(4) 31-39

[9] Foster D N and Cox R J 1973 Stability of children on roads used as floodways *Technical Report* (Water Research Laboratory, The University of New South Wales, Manly Vale, NSW, Australia) No. 73/13

[10] Yee M 2003 Human stability in floodways *Undergraduate Honours Thesis* (School of Civil and Environmental Engineering, University of New South Wales, Sydney, Australia)

[11] Abt S R, Wittier R J, Taylor A and Love D J 1989 Human stability in a high flood hazard zone *Water Resources Bulletin* **25**(4) 881-90

[12] Suetsugi K 1998 Control of floodwater and improvements of evacuation system for floodplain management *Floodplain Risk Management* (Proceedings of an International Workshop) ed S Fukuoka and A A Balkema (Hiroshima) pp 191-207

[13] Karvonen R A, Hepojoki A, Huhta H K and Louhio A 2000 The use of physical models in dam-break analysis *Rescdam Final Report* (Helsinki University of Technology, Helsinki, Finland)

[14] Jonkman S N and Penning-Rowsell E 2008 Human instability in flood flows *Journal of the American Water Resources Association* **44**(4) 1-11

[15] Takahashi S, Endoh K and Muro Z-I 1992 Experimental study on people's safety against overtopping waves on breakwaters *Report on the Port and Harbour Institute* **34** (4) pp 4-31

[16] Keller R J and Mitsch B 1993 Safety aspects of design roadways as floodways *Research Report* **69** (Urban Water Research Association of Australia)

[17] Lind N, Hartford D and Assaf H 2004 Hydrodynamic models of human instability in a flood *Journal of the American Water Resources Association* **40**(1) 89-96

[18] Ramsbottom D, Wade S, Bain V, Hassan M, Penning-Rowsell E, Wilson T, Fernandez A, House M and Floyd P 2004 *R&D outputs: flood risks to people. Phase 2. FD2321/IR2* (Department for the Environment, Food and Rural Affairs/Environment Agency, London, United Kingdom)

[19] Penning-Rowsell E C, Floyd P, Ramsbottom D and Surendran S 2005 Estimating injury and loss of life in floods: a deterministic framework *Natural Hazards* **36** 43-64

[20] Cao L W, Zhong G H, Liu S G and Wu X G 2013 Analysis of human instability in flood flow *Journal of Tongji University(Natural Science)* **41**(11) 1675-81

[21] Xia J Q, Gu A C, Shu C W, and Guo P 2014 Criterion of human stability in floodwaters based on theoretical and experimental studies *Journal of Catastrophology* **29**(2) 4-11

[22] Xia J Q, Falconer R A, Wang Y J and Xiao X W 2014 New criterion for the stability of a human body in floodwaters *Journal of Hydraulic Research.* **52**(1) 93–104

[23] Xie J H 1993 *Sedimentation research in China* (Beijing: Water and Power Press)

[24] State Administration of Sports 2015 *The China National Physique Monitoring Report in 2014*

[25] Guo Q S and Wang Y H 1995 *Ergonomics design* (Tianjin: Tianjin University Press)

[26] Fan H B, Sun Y P and Ji L 2015 Enlightenment and international comparison of evaluation indexes and test methods of strength physical fitness tests *China Sport Science* **35**(1) 80-87

[27] Ye C L and Zhang C H 2014 Status and progress of research on hand grip dynamometer *Contemporary Sports Technology* **4**(10) 176-78

[28] Cohen C P and Gunawardana A D Z 1993 The United Nations Convention on the rights of the child *American Journal of International Law* **87**(3) 101-04

[29] Li Y S, Sun P and Zhang C X 2008 The status-quo of social security of China's handicapped persons and strategies of perfection *Hebei Academic Journal* **28**(5) 7-13

Study of distribution and characteristics of the time average of pressure of a water cushion pool

Y H Guo[1] and J F Fu[1, 2]
[1]Faculty of Electric Power Engineering, Kunming University of Science and Technology, Kunming 650000, China

E-mail: 2631524@qq.com

Abstract. When a dam discharges flood water, the plunging flow with greater kinetic energy, will scour the riverbed, resulting in erosion damage. In order to improve the anti-erosion capacity of a riverbed, the cushion pool created. This paper is based on turbulent jet theoryto deduce the semi-empirical formula of the time average of pressure in the impinging portion of the cushion pool. Additionally, MATLAB numerical is used to conduct a simulation analysis according to turbulent jet energy and watercushion depth when water floods into the water cushion pool, to determine the regularities of distribution and related characteristics.

1. Introduction

A stilling basin is a traditional method of underflow energy dissipation. When jet freefalls into a stilling basin where hydraulic jump takes place, turbulent fluctuation action transforms, kinetic energy into heat energy and potential energy. The effect of a stilling basin is advantageous for its dissipation effect, because it is adaptable to the geological conditions and the amplitude of tailwater variatio. However, it is also widely used for moderate-lower head projects. However, the water cushion pool is usually formed by downstream barrages or check dams, which belongs to a new type of energy dissipater [1]. The mainstream of kinetic energypassesdown a turbulent fluctuation(for turbulence generation and turbulent dissipation) and a large-scale water vortex in the pool is continually exposed to the turbulent shearing and diffusion effect. Accompanied by these processes, energy is consumed by the viscous dissipation. This is a better method of energy dissipation, mainly used in high arch dam spillways and energy dissipation. For example, this method is already used in the Ertan and Xiluodu projects, etc., as shown in figure 1.

In cushion poolenergydissipation,energy is dissipated byturbulentfluctuation,collision andhydraulic jumps in the water cushion pool [2].Therefore, compared to traditional methods of energy dissipation, the reduction of the huge impactand severe fluctuating pressure is necessary, as well as the reduction of erosion damage on the bottom of the water cushion pool. At present, water cushion pool impact resistance design is only measured by the maximum allowable impact pressure difference at the bottom of the water cushion pool; in asense, this method is not very rigorous [3]. Therefore, the analysis of time-continuouspressure spatial distribution in the cushion pool is necessary. Meanwhile, differenthydraulic factorsmight affect the time average of pressure, requiring appropriate adjustment for corresponding hydraulic factors in order to reduce the pressure threshold and reduce the probability of erosion damage to the cushion pool to ensure the safety of the downstream riverbed.

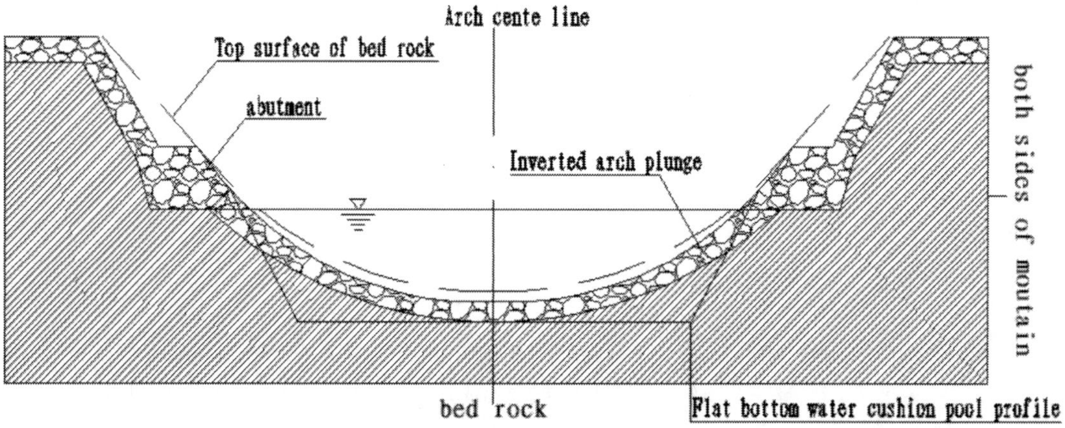

Figure 1. The flat and counter-arch water cushion poolprofile.

1.1. The significance of study on time-average pressure of the cushionpool impingingportion

When the main stream of a submerged jet reaches the bottom of the water cushion pool, part of its kinetic energy is converted to pressure energy, and a huge impinging pressureis generated. Meanwhile, sand carried in the water jet impacts the bottom directly. Therefore, the impinging portion is the greatest area of cavitationerosion and scour damage. Similarly, the impinging is the main energy dissipation area [4], which plays an important role in the improving of the energy dissipation effect. Thispaper analysesthe time average of pressure distribution of the impinging portion of a water cushion pool, and estimates corresponding hydraulic factors that might affect the time average of pressure. This is significant to the design of a water cushion pool, by analyzing the compressive strength of bottom concrete which must be more stronger than the maximum time average of hydrodynamic pressure[5].On one hand, the water cushion pool must be subjected to erosion damage andmaintain the safety of the downstream slope; alternatively, this can effectively improveenergy dissipation of the water cushion pool. The present research conducts simulation analysis to study the time average of pressuredistribution of the impinging portion of a water cushion pool.

A large portion of energy dissipation is related to the aeration of the plunging jet so that the pressure on bottom decreases;however, under such conditions, the mainstream is refracted by the riverbed in an impinging area, the flow rate decreases rapidly, and pressure increasessharply.These water flow characteristics are similar to those around the flow stagnation point [6-7].

A numerical calculation model is as shown in figure 2, in which the flow direction is designated by the X axis, the cross initiative transversal section and the water inlet is set to the origin ofcoordinates, and the vertical direction is designated by the Y axis. The water cushion pool length is set to 150cm, and the width is 100cm.

Generally, the mean pressure is described by $\Delta\overline{p_m} = \overline{p_s} - \overline{p_{\min}}$, $\overline{p_s}$ is the maximumtime average of pressure at the stagnation point, $\overline{p_{\min}}$ is the minimum bottom time average of pressure, and, $\Delta\overline{p_m}$, is typically described by equation(1).

$$\Delta\overline{p_m} = f(u_0, h_0, H, \rho) \tag{1}$$

Among these, ρ is the water density, obtained by dimensional as equation(2):

$$\Delta\overline{p_m}\left(\frac{1}{2}\rho u_0^2\right)^{-1} = f(H, p_0^{-1}) \tag{2}$$

For a submerged impinging jet, the experimental result reported by Dr. Duoming Xu [8] is β = 40° —50°, as equation(3):

$$\Delta\overline{p_m}\left(\frac{1}{2}\rho u_0^2\right)^{-1} = 0.475\exp\left[-0.088\left(Hh_0^{-1}\right)^2\right] \tag{3}$$

In order to deduce the formula of time average of pressure distribution under experimental conditions, and to investigate whether the impinging portion is corresponds to a normal distribution [9-11] of the time average of pressure,the maximum flow velocity at the centeral axis of the cushion pool is set as follows:

$$u_0 u_{\max}^{-1} = kxh_0 + A \tag{4}$$

In equation (4), and as shown in figure 2 below, the jet direction is designated by the X axis, the vertical jet direction is designated by the Y axis, $u_{\max}$ is the maximum flow velocity of the impinging jet center, h_0 is the thickness of the incident flow, andu_0 is the flow rat into the pool. The measured data and simulation data are shown in figure 3.

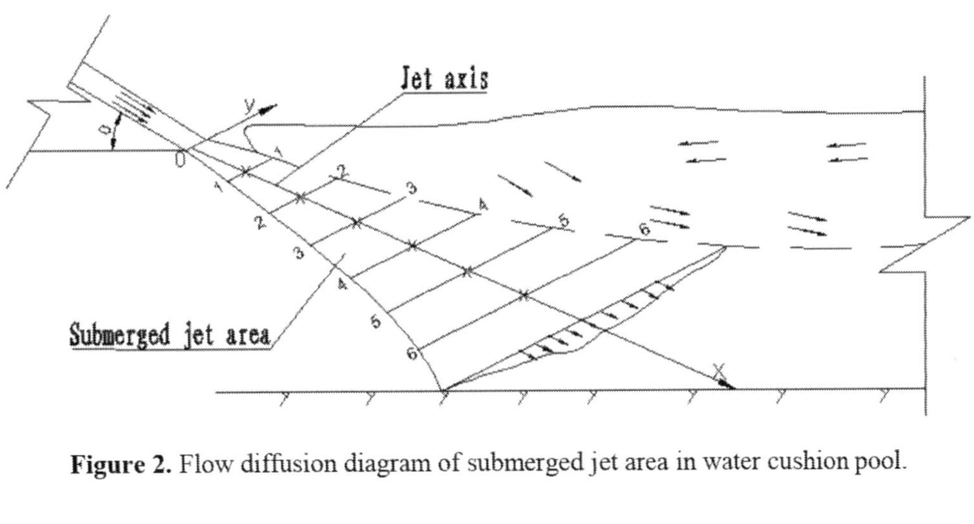

Figure 2. Flow diffusion diagram of submerged jet area in water cushion pool.

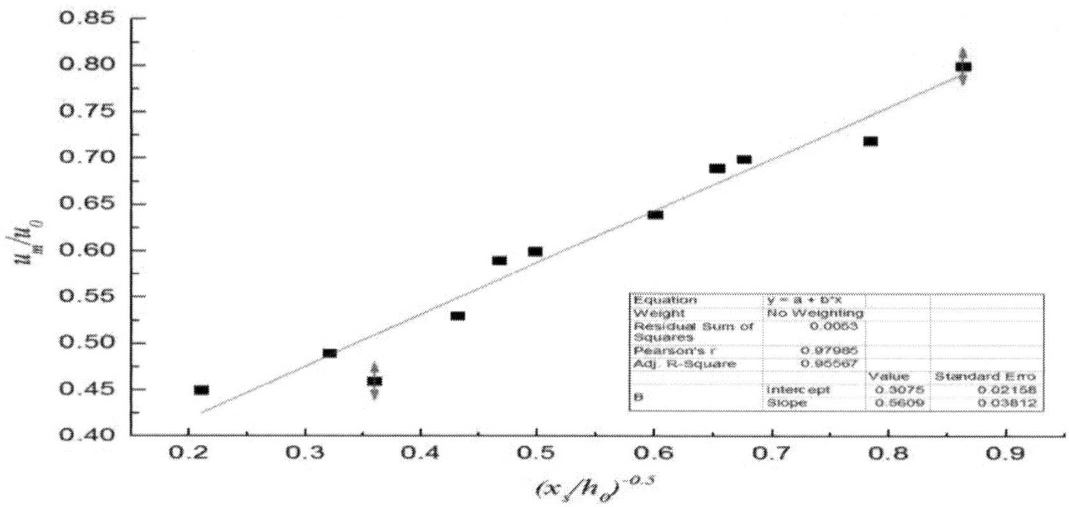

Figure 3. Maximum velocitydistribution of submerged jetaxial line.

Using the planar diffusion equation [13], the formula of maximum flow velocity is obtained as follows:

$$u_0 u_{\max}^{-1} = 0.56 x h_0^{-1} + 0.31 \tag{5}$$

According to the law of conservation of energy:

$$\sin\theta u_{\max} = (2 p_{\max} \rho^{-1})^{1/2} \tag{6}$$

As deduce from equation (5) and equation (6) :

$$p_{\max} = \frac{\rho \sin^2\theta u_0^2}{2(0.56\frac{x}{h_0} + 0.31)^2} \tag{7}$$

The time average of pressure when a two-dimensionalsubmerged turbulent jet impacts the bottom of a water cushion pool, making theprocess dimensionless, there is a right similarity on test points near the normal distribution curve, expressed as follows:

$$p_w p_{\max}^{-1} = \exp[-0.625 (x b_p^{-1})^2] \tag{8}$$

In equation (8), p_w is the time average of dynamic pressure of each measuring point, p_{max} is the distance from the maximum measuring time average of the pressure point, and b_p is the jet flow half width at the$p_w = p_s/2$ position.

Figure 4 and figure 5 respectively represent the dynamic pressure distribution at different turbulent energies and water cushion depths. According to the previous research results[14], the impinging area ranges from approximately $\frac{x}{H} > 0.70$(the x axis is the mainstream distance measurement, and H presents water depth), to the downstream wall jet area $\frac{x_l}{H} \approx 0.35$ ($x_l$is the wall jet area). The diagrams indicates that theexperimental data not only corresponds to the pressure of the normal distribution at the impinging portion, but also indicates similar impinging portion ranges, verifying the rationality of equation (8). According to the article, the impinging portion range is selected along the center axis from 50cm-80cm.

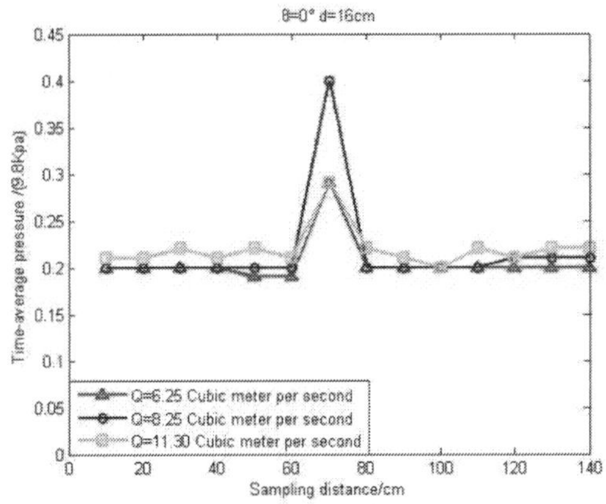

Figure 4. The time average of dynamic pressure energies distribution diagramfor different turbulent.

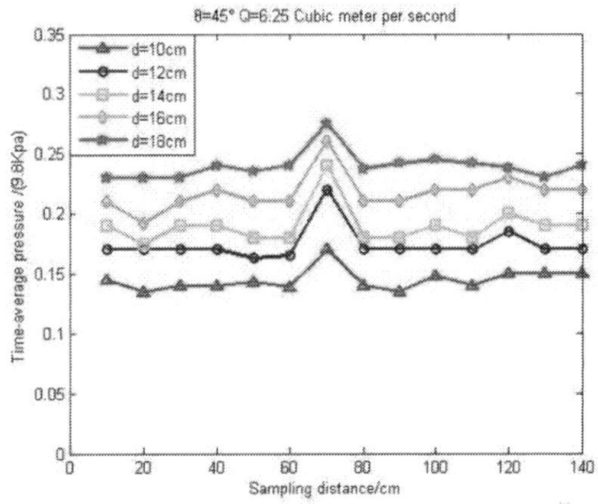

Figure 5. The time average of dynamic pressure distribution diagram for different water cushion.

2. Experimental model

2.1. Measuring point setup.

On the basin floor of the water cushion pool, there are 6 rows of 84 pressure taps along the path of the layout; the specific setup is shown in figure 6. The floor length is 150cm, floor length is 100cm, the impact area measuring points are selected along the axis direction D5-D8, and the origin of coordinates are selected from the product of the length and width.

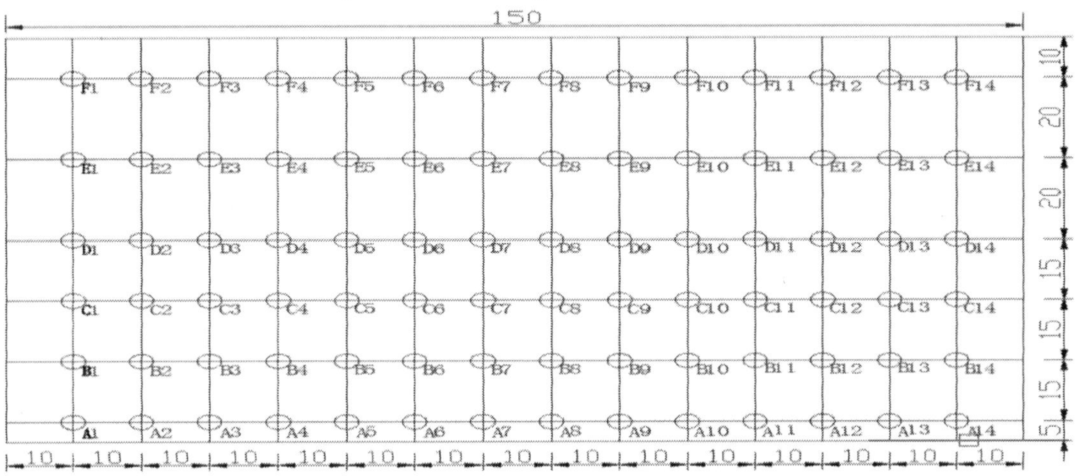

Figure 6. Arrangement diagram of pressure measuring points atthewater cushion pool floor.

2.2. Experimental working conditions.

The working conditions in our experiments are as follows: the turbulent energy was set to $6.25\times10^{-3}m^3/s$, $8.25\times10^{-3}m^3/s$, $11.30\times10^{-3}m^3/s$, water cushion depth was 12cm, 14cm, and 16cm. By changing the 16 working conditions of turbulent energy and water cushion depth, the influence of the conditionson the time average of pressure were investigatied, while the different turbulent jet energy and water cushiondepth impacted the cushion pool impinging portion.The specific working conditions are shown in table 1.

Table 1. Different working conditions.

Woking condition	Incident flow angle(°)	Depthof water cushion (cm)	Turbulent jet energy $(10^{-3}m^3/s)$	Time average of pressure(9.8*kpa*)			
				D_5	D_6	D_7	D_8
1			6.25	0.13	0.13	0.12	0.13
2	0	10	8.25	0.15	0.13	0.13	0.13
3			11.30	0.33	0.14	0.15	0.14
4		12		0.17	0.14	0.15	0.16
5	0	14	6.25	0.17	0.17	0.17	0.17
6		16		0.20	0.19	0.19	0.20

2.3. The influence of different turbulent jet energy on time average of pressure at impinging portion.

Using MATLAB, a simulated analysis of the time average of pressure of the impinging portion in a water cushion pool was conducted, in order to demonstrate that under identical conditions, the time average of pressure at the impinging portion of the basin floor was much greater than that at other portions of the flow structure. Therefore, the impinging portion is more prone to suffer impact damage [15]. The experimental data consists of 16 numerical test values along the center axis of the basin floor, respectively drawing the nephogram and 3D space diagram distribution of time average of pressure under different working conditions with regard to the influence of different turbulent jet energies on the time-averaged pressure, set as from condition 1 to condition 9. By observing the pressure distribution, the pressure variation trend was analyzed as well as its influence on hydraulic factors under corresponding working conditions.

2.3.1. The influence of different turbulent jet energies on the time average of pressure at the impinging portion. See figure 7 to figure 12.

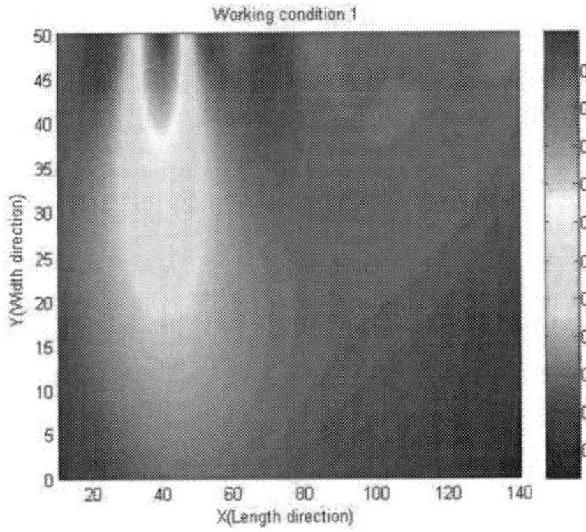

Figure 7. Working condition1: θ=0° d=10cm Q=6.25m³/s.

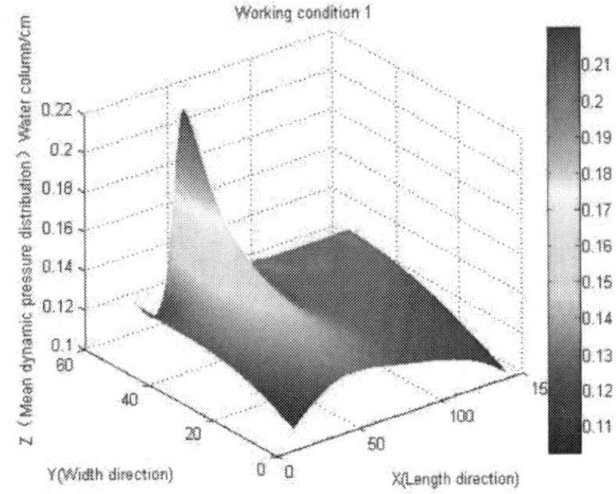

Figure 8. Working condition 1: θ=0° d=10cm Q=6.25m³/s.

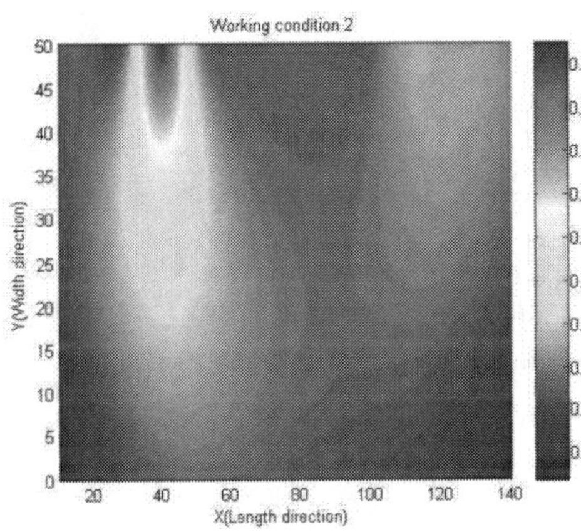

Figure 9. Working condition2: θ=0° d=10cm Q=8.25m³/s.

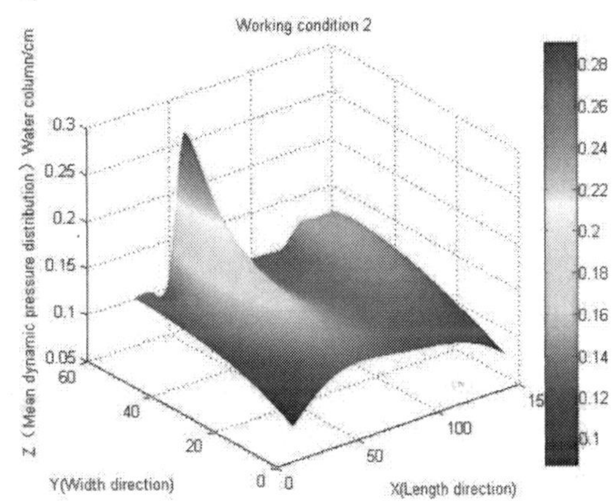

Figure 10. Working condition 2: θ=0° d=10cm Q=8.25m³/s.

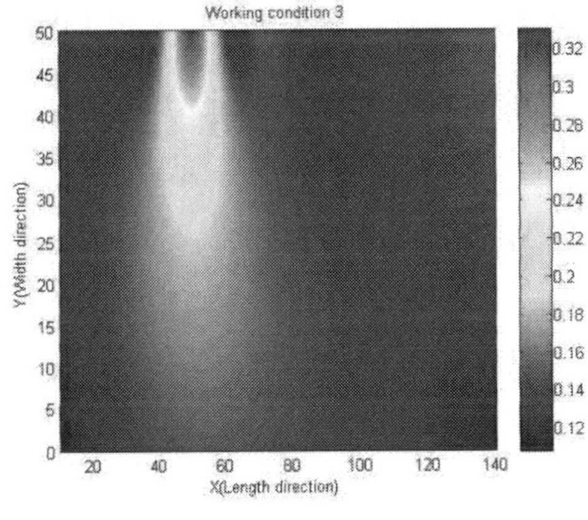

Figure 11. Working condition3: θ=0° d=10cm Q=11.30m³/s.

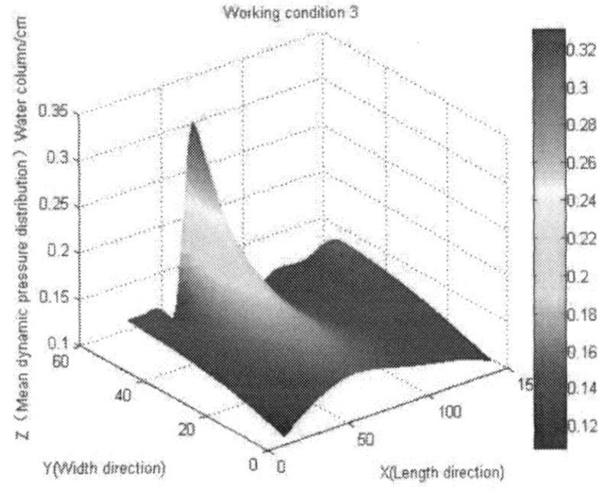

Figure 12. Working condition 3: θ=0° d=10cm Q=11.30m³/s.

2.3.2. The time average of pressure analysis.

- For any condition, under identical conditions, the time average of pressure at the impinging portion of the basin floor was much greater than that at other flow structure portions, and the maximum pressure appeared in the impinging portion, indicating that the impinging portion is more prone to suffer impact damage. In the design of a water cushion pool, the compressive strength of the bottom concrete slabs should be much greater than the time average of the dynamic pressure maximum.
- From condition 1 to condition 3, the incident angle was 0 degrees, the water cushion depth was 10cm, the turbulent jet energy increased in turn. From condition 4 to condition 6, the incident angle was 0 degree, the water cushion depth was 12cm, the turbulent jet energy increased in turn. From condition 7 to condition 9, the incident angle was 0 degree, the water cushion depth was 16cm, the turbulent jet energy increased in turn. The three conditions were compared and analysis was conducted. The average dynamic pressure at the impinging portion increased with turbulent jet energy, up to a peak value at the impingement point. The peak difference were large, but the corresponding impact on the upstream and downstream area and the time average of dynamic pressure tends to be approximately stable and equal.
- Once the incident angle and the depth of the water cushion are defined, the pressure is more concentrated and the peak value increases with turbulent jet energy. A great visible difference was observed on maximum impact pressure from the turbulent jet energy; for practical project applications, great attention must be paid to the control of turbulent jet energy.

2.4. The influence of different water cushion depths on time average of pressure at the bottom floor.
The influence of different depths of water cushion on the time average of pressure took into account nine working condictions for analyze. Two coniditions were identical as conditions 1 and 4 mentioned above, and are there removed from the following analysis (See figure 13 to figure 18).

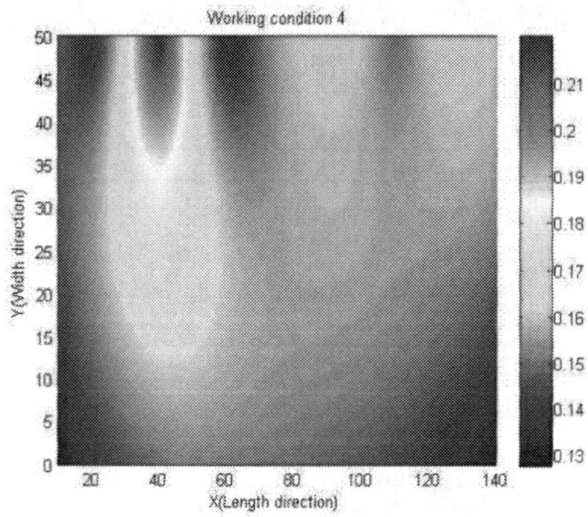

Figure 13. Working condition4: θ=0° d=12cm $Q=6.25m^3/s$.

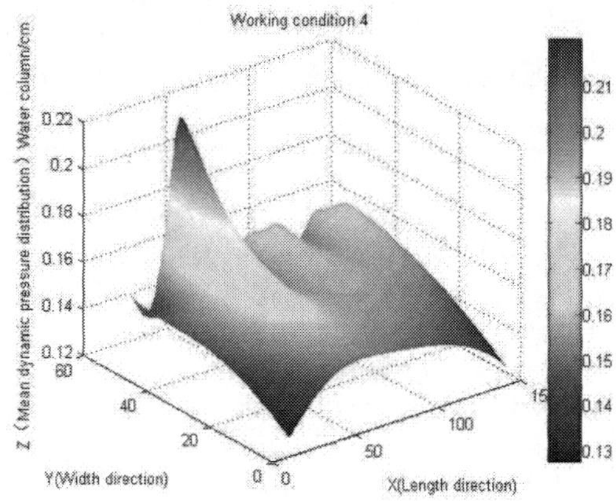

Figure 14. Working condition 4: θ=0° d=12cm $Q=6.25m^3/s$.

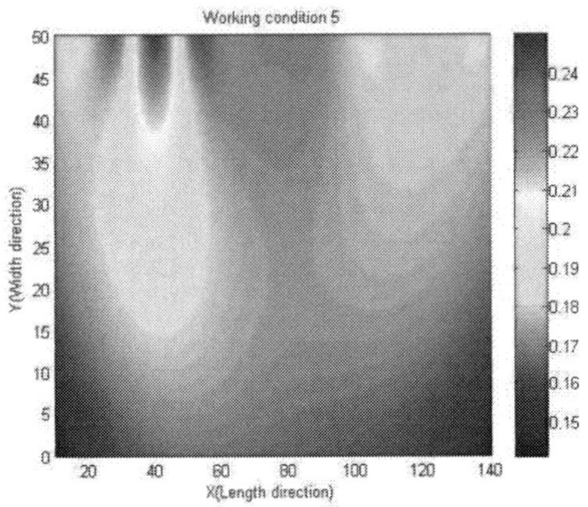

Figure 15. Working condition5: θ=0° d=14cm $Q=6.25m^3/s$.

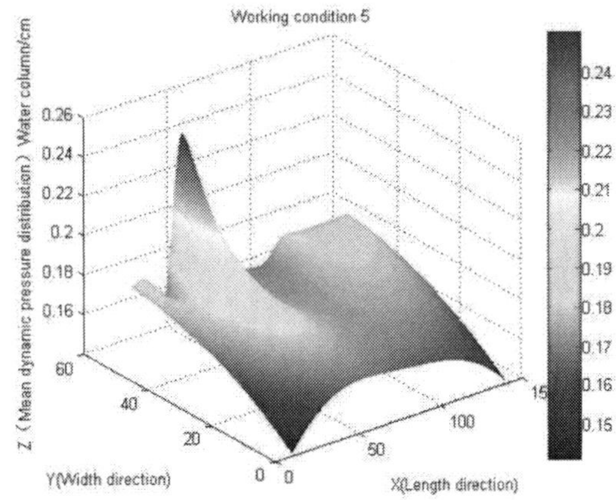

Figure 16. Working condition 5: θ=0° d=14cm $Q=6.25m^3/s$.

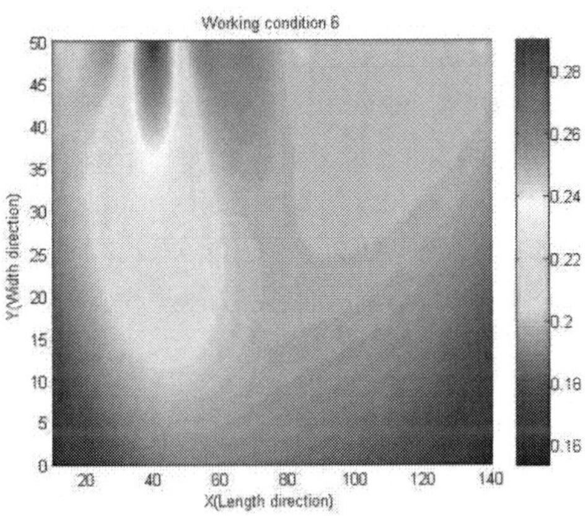

Figure 17. Working condition6: θ=0° d=16cm $Q=6.25m^3/s$.

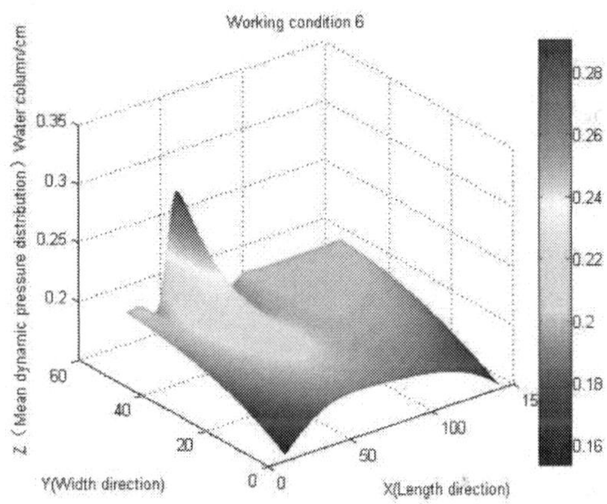

Figure 18. Working condition 6: θ=0° d=16cm $Q=6.25m^3/s$.

2.4.1. The time average of pressure analysis.

- Under changing depths of water cushion, the peak value of the timeaverage of pressure still appears in the impinging portion, indicating that the impinging portion is most prone to suffer scour damage.
- Conditions 1, 4, 10 and 11, all included incident angles of 0 degrees, turbulent jet energy of $6.25\times10^{-3}m^3/s$, and increasing water cushion depth. From condiction 12 to condiction 16, the incident angle was 45 degrees, the turbulent jet energy was $6.25\times10^{-3}m^3/s$, depth of water cushion increased in turn. The two conditions were compared for analysis. Results cleary indicate that with different water cushion depths, when the time average of dynamic pressure reached the maximum value, the corresponding impact on the upstream and downstream area, and the time average of dynamic pressure tends to be approximately stable and equal. Additionally, the average dynamic pressure of the impinging portion increases with the depth of water cushion; however, under different working conditions, the pressure difference value near the maximum

pressure portion is more obvious than that of turbulent jet energy, and tends to increase with water cushion depth, though the peak value difference is not large.

- Once the incident angle and the turbulent jet energy are defined, as the depth of the water cushion increases, the pressure becomes more concentrated. There is little influence on the impingement portion time average of dynamic pressure exerted by the depth of the water cushion.

3. Conclusions

- Using MATLAB combined with numerical analysis and simulation, respectively for turbulent jet energy and the depth of water cushion, the distribution of time average of pressure on water cushion pool impinging portion was analyzed. Base on the corresponding pressure nephograms and 3D space diagrams, the water cushion impinging portion was verified as the most prone to suffer erosion damage. Therefore, subsequent recommendations can be made for water cushion pool anti-scouring design. The specific conclusions of this paper are as follows. Based on turbulent jet theory, dimensional analysis and the corresponding experimental data, semi-empirical formula for the impinging portion at any point pressure under the condition of this experiment can be determined and compared to the corresponding time average of pressure distribution curve, proving that the time average of pressure on the impingement area is in accordance with normal distribution characteristics. The present study demonstrated that the formula correctly, deduces the approximate range under experimental working conditions.
- In the experiment, incident angle and the depth of water cushion were defined, by adjusting the turbulent jet energy, and obvious wave change peaks were observed. Stronger turbulent jet energy resulted in greater peak values, and the time average of dynamic pressure tended to be stable and approximately equal.
- The incident angle and turbulent jet energy were defined by adjusting the depth of the water cushion. Results indicate that the time average of pressure on the impinging portion increases with the depth of the water cushion, but the peak of the wave change does not vary greatly. The time average of dynamicpressure distribution on the upstream and downstream areas indicate clear variation as compared to that observed under the turbulent jet energy working conditions. As the depth of water cushion increased, the time average of dynamic pressure tended to increase generally under each working condition in the upstream and downstream areas.

The calculated results of the semi-empirical formula are in good agreement with the experimental results, and confirm the relevance of parameters based on analysis of the time average of pressure on the impinging portion. Compared to other areas, the impinging portion is most prone to suffer erosion damage. According to the described experimental research, aeration decreases the dynamic pressure at the bottom of the water cushion pool, while the fluctuating pressure increases. Therefore, optimal design of water cushion pools should conduct self-aeration and artificial aeration appropriately, to reduce scour damage to the impinging area of the water cushion pool.

Acknowledgments
This work was financially supported by Applied basic research project of Yunnan province (Research and application of the numerical simulation of two - phase flow, 2010ZC049).

Reference

[1] Lv Y Q, Li X Y and Dong J R 2001 Study on flow characteristics in water cushion pool and mechanism of energy dissipation *Journal of Yangtze River Scientific Research Institute* pp 10-12

[2] Liu P Q 2003 Probe into design criteria on energy dissipation and scour-prevention for plunge-pools *Journal of Yangtze River Scientific Research Institute* pp 3-6

[3] Rae P J 1994 Hydraulic design of spillway plunge pool linings *Proceeding National Conference on Hydraulic Engineering* pp 396–400

[4] Abramovich G and Schindel L 1963 *The Theory of Turbulent Jets* MIT Press

[5] Yang Y Q and Xu W L 1991 Numerical simulation of submerged jet within a water cushion pool *Journal of Hydrodynamics* pp 36–44

[6] Liu P Q, Dong R J and Yu C Z 1993 Investigation on rock bed scour by ski-jump water jets *Journal of Yangtze River Scientific Research Institute* pp 1–9

[7] Chen Y C, Dong J R and Wang Y M 1998 Study of the flow characteristics in the scour of downstream of Three Gorge spillway *Journal of Tsinghua University(Sci& Tech)* pp 115-18

[8] Xu D M and Yu C Z 1983 Surface water jet and its impact on the bottom of the channel pressure pulsation characteristics *Shuili Xuebao* pp 52–58

[9] Rajaratnam N 1967 Hydraulic jumps *Advances in Hydroscience* pp 197–280

[10] Rajaratnam N 1976 *Turbulent jets* Amsterdam: Elservier Sc PublCo

[11] Frost W and Moulden T H 1977 *Handbook of turbulence* New York: Plenum Press

[12] Liu P Q 2008 *The free turbulent jet theory* Beihang University press

[13] Wang H Q and Zhang G B 2008 Hydraulic structures rushed abrasion and cavitation erosion failure mechanism and countermeasures *Yunnan Hydroelectric Power Generation* pp 89–92

[14] Dong Z Y, Yang Y Q and Wu C G 1994 Effect of aeration on dynamic water pressure at the bottom of the water cushion pool *Science in China (Series A)* pp 431-9

[15] Liu P Q, Li F T and Wang Y 2003 Experimental study on the change of flow pattern and pressure in the flow structure of the water cushion pool *Shuili Xuebao* pp 25-8

IOP Conf. Series: Earth and Environmental Science **39** (2016) 012040 doi:10.1088/1755-1315/39/1/012040

Research and application of ecological river courses restoration technology

Z C Cui[1, 3], Y Z Yang[1], X L Gao[1], H Xiao[1] and H C Liu[2]

[1]Shandong Water Polytechnic, Rizhao, Shandong Province, China
[2]Dongying Water Conservancy Survey Design Institute, Dongying, Shandong Province, China

E-mail: cuiizc@sina.com

Abstract. In this paper, a practical method of ordered binary comparison determined by weight vector is proposed, as based on correlative concepts of the dualistic relative comparative method in fuzzy mathematics. By taking advantage of the proposed method, subordinated degree of evaluation indicators can be defined, such as weightiness and the degree of importance of ecological restoration of river courses, and a mathematical model can be established. The proposed mathematical model is clear in its physical conception and offers convenient calculations, and provides a theoretical foundation for the ecological restoration of river courses. This paper employs "standard values" of the evaluation index system (EIS) of ecological river networks as derived by previous literature [1] as the theoretical basis for the ecological restoration river courses.

1. Introduction

In Dongying City, there are a total of thirty key river channels in addition to the Yellow River, and the basin area of each river exceeds 100 square kilometers. According to the size of basin area, the quantity of flood drainage flow and waterlogging drainage flow combine with the management and development of the river network in Dongying City; some rivers located in Dongying as part of the primary river course are listed. Every key river must satisfy the following requirements: its basin area must exceed 100 square kilometers, and its intra-regional length must be greater than 20 kilometers. Numerous EIS studies of the key ecological river network in Dongying City have been conducted based on this simulation [1].

Based on survey data and analysis of the current natural hydrographic network and the character of the ditch network, it was found that the hydrological integrity, chemical integrity, structural integrity, biological integrity and service integrity were unable to meet the requirements of an ecological river network. For this reason, previous literature [1] proposed the fundamental principle of constructing the ecological river network indicator system for Dongying City, and provided an EIS of ecological river courses and "standard values". Whereas the ecological river courses are described by numerous indicators, each indicator has unique impact on the ecological river courses, and varies in importance in regard to the repair of ecological river courses.

Kernal technologies involved in the restoration of ecological river courses includes description of impact, the development of repar strategies according to the importance of indicators, and methods for the application of standard indicator values as derived from previous literature [2].

In addition, many previous studies [1-5] demonstrate the application of the index system.

2. Weight vector of the ecological river course index system

In order to identify key points in the restoration of ecological river courses, the importance (weight) of each indicator reported in previous literature [1] must be determined. This paper refers to the method of ordered binary comparison in fuzzy mathematics, and proposes the practical method of ordered binary comparison driven by the weight vector.

A certain river course has two evaluation indicators: flood prevention ability and flood draining ability. To determine their relative importance, consider that flood prevention ability (identified as x_1) is more important than flood draining ability (identified as x_2), for fuzzy concept importance of B, the membership degree of this river course's flood prevention ability which belongs to the "importance" is larger than that of the flood draining ability, from the fuzzy set point of view, as follows:

$$\mu_{\underset{\sim}{B}}(x_1) > \mu_{\underset{\sim}{B}}(x_2)$$

That is the concept of binary comparison.

Previous studies have detailed the theoretical demonstration of ordered binary comparison determined by weight vectors [6-10]. This paper details the concrete steps to determine the practical weight vectors of the evaluation index.

- Step 1: The weight (i.e., importance) of m evaluation indicators requires serious consideration. The number of the most important indicator is set to 1, and so on.
- Step 2: According to the features of the metric matrix (1) which reflect the importance of an ordered binary comparison, the values of the first row elements in the matrix (1) have been carefully constructed.

$$\mu = \begin{bmatrix} \mu_{11} & \mu_{12} & \cdots & \mu_{1m} \\ \mu_{21} & \mu_{22} & \cdots & \mu_{2m} \\ & \cdots\cdots & & \\ \mu_{m1} & \mu_{m2} & \cdots & \mu_{mm} \end{bmatrix} = (\mu_{ij}) \tag{1}$$

The values of the elements should satisfy the following conditions:

$$\begin{cases} 0 \le \mu_{ij} \le 1 \quad i \ne j \\ \mu_{ij} = 0.5 \quad i = j \\ \mu_{ij} + \mu_{ji} = 1 \\ \text{if } \mu_{hi} \le \mu_{hg},\ \mu_{ig} \ge \mu_{gi} \end{cases} \tag{2}$$

That is, the "importance" of binary comparisons between evaluation indicator 1 and evaluation indicator 2 must be specified, as well as the importance of comparisons between evaluation indicators 1 and evaluation indicator 3, ⋯, between evaluation indicators 1 and evaluation indicator m, as follows: $\mu_{12}, \mu_{13}, \cdots, \mu_{1m}$, and there is a such condition: $0.5 < \mu_{12} < \mu_{13} < \cdots < \mu_{1m}$. Then, according to the condition $\mu_{ij} + \mu_{ji} = 1$ in Equation (2), the values of the elements in the first column of matrix (1) can be determined.

- Step 3: Based on the values of the indicators in the first row and the first column in matrix (1), the values of the elements in the first column of matrix (3) can be determined.

$$\alpha = \begin{bmatrix} 1 & \mu_{12}/\mu_{21} & \cdots & \mu_{1m}/\mu_{m1} \\ \mu_{21}/\mu_{12} & 1 & \cdots & \mu_{2m}/\mu_{m2} \\ \cdots & \cdots & \cdots & \cdots \\ \mu_{m1}/\mu_{1m} & \mu_{m2}/\mu_{2m} & \cdots & 1 \end{bmatrix} \tag{3}$$

Formula (4) is deduced from the values of the elements in the first column of matrix (3).

$$\begin{aligned} \vec{W}' &= (w_1', w_2' \cdots, w_m') \\ &= (1, \frac{1-\mu_{12}}{\mu_{12}}, \cdots, \frac{1-\mu_{1m}}{\mu_{1m}}) \end{aligned} \tag{4}$$

The values of the elements should satisfy the following conditions:

$$0.5 < \mu_{12} < \mu_{13} < \cdots < \mu_{1m} \tag{5}$$

After normalizing formula (4), the weight vectors of m evaluation indicators is obtained.

In order to obtain the weight vectors of m evaluation indicators that conform to actual situations, the following strategies have been adopted in the current paper.

Relevant experts and leaders are invited to take an active part in certain organized activities; a well-designed questionnaire is employed. After careful consideration, the experts and leaders determine the "importance" of m evaluation factors according to a unified format, and determine the "importance" measurement value of binary comparisons based on the sorting result: $\mu_{12,}\mu_{13},\cdots,\mu_{1m}$, and a condition exists so that: $0.5 < \mu_{12} < \mu_{13} < \cdots < \mu_{1m}$.

After transferring the related data into Formula (4) and normalizing the data, each expert provides their determined weight vectors of m evaluation indicators. Then, the average value of each evaluation indicator is employed as the weight vector for further calculation.

Since each expert may consider a problem from individual aspect, this method is not the same as determining the "importance" of m evaluation indicators, so it cannot be based on the same number of evaluation indicators to determine the average value.

The construction of ecological river courses will depend on the weight values of the evaluation indicators. Greater values for evaluation indicators indicate more important functions, and greater impact on the ecological performance of a river course.

3. Degree of importance of river courses ecological restoration index system

If the value of a certain indicator is identical to or near its standard value under current conditions, though this indicator may be of great importance to the ecology of the river course, and the indicator weight may be very high, the priority harness of this river course is not encouraged.

In order to prioritize the implementation of river course ecological restoration and to highlight the significant values of the important indicators, the proposed method for determining the degree of importance of the river ecological management index system is provided under current conditions.

Suppose that the present quantitative value of a river ecological evaluation indicator is x_i, and its standard value is s_i. Then the degree of importance of ecological management is defined as follows:

$$w_i \frac{|x_i - s_i|}{s_i} \tag{6}$$

The significance of each element is described as follows:

w_i: The weight of the i-th indicator;

x_i : The current actual value of the i-th indicator;

s_i : The standard value of the i-th indicator

i : 1, 2, $\cdots$, n.

According to the importance of ecological management, the sequence of ecological and river course restoration is determined.The feature function is used to describe the descriptive indicators (such as slope shape, bank structure, embankment type, etc). The characteristic function of subset A of the domain U is expressed as follows:

$$\chi_A(u) = \begin{cases} 1 & u \in A \\ 0 & u \notin A \end{cases} \quad (7)$$

For example, suppose that subset A represents the slope shape. When the slope shape of a river course belongs to the standard value of subset A, the value is equal to 1 and the rest of the values are equal to zero. Therefore, in the ecological river network EIS, for the current observed values $\vec{X} = (x_1, x_2, \ldots, x_n)$ and the standard values $\vec{S} = (s_1, s_2, \ldots, s_n)$, the degree of importance of each current monitoring indicator of ecological management can be calculated according to Formula 2.1 and Formula 2.2. The degree of importance will be an important basis for ecological restoration and restoration.

4. Case study

4.1. The ecological status of Yihong River

Yihong River derives its origin from Sheng Tuo Zhen Cui Jia, west of Kenli County, and runs south-east before, turning into the Guangli River and finally flowing into the Zhimai River. The length of the Yihong River is 48km, including a valley area of 312 square kilometers.

Under current conditions, its river slope is l/7000-1/10000, water depth is between 1 and 2.5 meters, and the bottom elevation ranges from 5.4m to -1.0m. Its slope is approximately l:3, and the maximum flow drainage is 110m3/s. According to the two measured results from 2014, its water quality is shown in table 1.

Table 1. The water quality monitoring of a section of Yihong River (Nnit: mg/L).

Base information	Field number of the sample points	
	Spillway Q-1	Spillway Q-2
Sample point site	S231bridge(XiSi Road, HuangHe Road)	DongBa Road Yihong River bridge
Longitude	NL: 37° 34′ 20"	NL: 118° 32′ 55"
Latitude	EL: 37° 28′ 26"	EL: 118° 45′ 20"

In the above table, two sample points are employed. The field number of one sample point is Spillway Q-1, and the other sample point is assigned the number Spillway Q-2. The basic information recorded at the two sample points is shown in table 2.

Table 2. The basic information of two sample points.

Time (Year.Month)	Field number	Inspection result		Surface water class
		NH4+-N	CODcr	
2014.08	Spillway Q-1	<0.05	107.30	Poor V
	Spillway Q-2	13.94	379.60	Poor V
2014.10	Spillway Q-1	0.28	100.39	Poor V

Spillway Q-2	16.27	162.98	Poor V

Note: NL is the abbreviation of Northern latitude. EL is the abbreviation of east longitude.

The Delphi method is used to conduct research using "Back-to-Back" communication to consult with five local experts and seven community residents. After two rounds of consultation, the expert opinions and suggestions tend to conclude that the "Yihong River landscape coordination degree is very poor" In regard to the ecological environment of this river course, there is a great gap between the demands of the country people and local residents. Simultaneously, the environmental protection consciousness of local residents is not strong, and the emission dissatisfaction phenomenon of enterprises is a common occurreance. Based on the analysis of the overall ecological functions of the Yihong River, the primary task of the river's ecological restoration is exceptionally arduous, and the goal of becoming an ecological river course has not been completely achieved.

4.2. The ecological EIS of Yihong River

According to actual investigation and the consultation of local experts and residents, there are 16 primary indicators in the ecological EIS of the Yihong River. Based on the methods for determining the weight vector of the indicators used in this paper, the weight of the evaluation indicators are shown in Table 3.

Table 3. The weight table of the evaluation indicators.

Criterion layer	Primary evaluation indicators	Secondary evaluation indicators
Nice natural functions (0.5)	1. Storage of surface wate (0.2)	1. Flood control capacity (0.07) 2. Drainage capacity (0.07) 3. Ecological water requirement (0.06)
	2. Water quality (0.2)	4. COD (0.1) 5. NH3-N (0.1)
	3. Soil and water conservation (0.15)	6. Vegetation coverage (0.05) 7. Slope shape (0.02) 8. Bank structure (0.05) 9. Embankment type (0.03)
	4. Corridor value (0.15)	10. Biodiversity corridors (0.08) 11. Riparian vegetation zone (0.07)
Sustainable social and economic functions (0.5)	1. Sustainable use of surface water resources (0.16)	12. Rate of surface water resources development and utilization (0.16)
	2. Cultural leisure (0.7)	13. Landscape coordination degree (0.04) 14. Life satisfaction degree (0.03)
	3. Environmental health (0.7)	15. Environmental protection consciousness (0.04) 16. Health condition (0.03)

IOP Conf. Series: Earth and Environmental Science **39** (2016) 012040 doi:10.1088/1755-1315/39/1/012040

4.3. The ecological restoration of Yihong River

According to the standard value of ecological river network evaluation indicators in Hekou District [1] as combined with current ecological conditions of the Yihong River, the degree of importance of each current monitoring indicator of ecological management can be calculated according to Formula 2.1 and Formula 2.2. The degree of importance will be an important basis for ecological restoration and restoration.

Due to data limitations and based on the weight table of the evaluation indicators presented in Table 3, the present research provides the optimal order of current ecological restoration for the Yihong River, and proposes a plan for its ecological construction.

4.3.1. Flood control capacity and flood drainage ability. The flood control capacity and the flood drainage capacity of each section must not be less than a given value, according to previous literature [1].

4.3.2. Ecological water requirement. Because Yihong River passes through the county and in consideration of expert advice, the ecological water requirement of the Yihong River should increase by 10% based on a given value in previous literature [1].

4.3.3. Water quality. Considering the actual social economic development, the primariy task is to achieve the following goals: COD≤35, NH3-N≤1.8, so as to achieve a water quality standard level Ⅳ or Ⅴ. The long-term task is to achieve the following goals: COD≤30, NH3-N≤1.5, so as to achieve a water quality standard level Ⅲ or Ⅳ.

4.3.4. Vegetation coverage ratio. According to the present conditions of Yihong River, the vegetation coverage rate is greater than 50%, and the long-term vegetation coverage rate is greater than 70%.

4.3.5. The slope shape, the bank structure and the embankment type. In the county town, the river slope shape should be a convex structure, and should use ecological slope protection measures so as to eventually become the natural revetment. For other river sections, step-by-step methods are adopted to reduce and eliminate the straight wall. During implementation, the following provisions should be observed.

- The distribution of the shoreline in the ecological protection area should ensure that the ecological shoreline ratio is not less than 80% and the production shoreline ratio is not greater than 5%.
- The distribution of the shoreline in scenic areas should ensure that the ecological shoreline ratio (according to the level of the scenic area) is not less than 30-50%, and that the production shoreline is not greater than 5-10%.
- As much as possible, the shorelines in the county should be living shorelines.
- The river slope is less than 1:3.

4.3.6. Biological diversity. A short-term goal is the realization of biological diversity greater than 0.5. A long-term goal is the realization of biological diversity greater than 0.7.

4.3.7. Bank vegetation belt. The bank vegetation belts within 20 meters on both sides of the Yihong River must be realized. Long-term goals include implementing bank vegetation belts within 30 meters on both sides of the Yihong River.

4.3.8. Development and utilization ratio of surface water resources. The rate of surface water resource development and utilization should be less than 60%.

4.3.9. Landscape coordination degree and life satisfaction degree. Further expansion of the biological treatment measures in the river course must be achieved so that the river becomes a micro -sewage treatment site which will degrade some of the water pollutants and ensure water quality. The water body will achieve twists, turns and variety, in accordance with natural ecological changes, so that the value of ornamental scenery is greatly improved.

Local experts and residents are invited to participate in survey activities and sensory evaluation activities to achieve the following long-term goals: improved harmony between the Yihong River and the surrounding environment, and life satisfaction of local people that exceeds 90%.

4.3.10. Environmental protection consciousness and health condition. Higher percentages of environmental consciousness indicate stronger environmental protection consciousness of local residents. Higher percentages of health condition indicates greater health improvement to local residents. The ecosystem of the regional river course is an important symbol of this region, and is likely to indicate whether the residents are happy and comfortable.

With the help of ecological construction to the Yihong River, the local residents will have good health and strong environmental awareness. The ultimate goal is that people with environmental protection consciousness exceed 90%, and those with good health exceed 95% of the population.

5. Conclusion

This paper provides a comparison to previously reported literature [1]. The "standard values" of the EIS of the ecological river network presented in previous literature [1] are used as the theoretical basis for ecological river network restoration.

The ecological features of a river course are described by numerous indicators, and are restricted to the economy, industry and downstream conditions. It is difficult to meet the standard values of an ecological river course perfectly. The ecological restoration process of rivers is gradual in practice, and greater importance placed on any particular indicator should be assigned priority in processing.

A practical method of ordered binary comparison determined by weight vectors is proposed, based on correlative concepts of the dualistic relative comparative method in fuzzy mathematics. By taking advantage of the practical method, the subordinated degree of evaluation indicators (i.e., weights) can be established, and the degree of importance of the river course in ecological management can be determined, as well as its mathematical model. The proposed mathematical model provides clear physical implications and convenient calculations, and provides a theoretical foundation for the determination of the ecological restoration order of river courses.

Acknowledgments

Funded projects: Ministry of Water Resources 2012 Public welfare industry scientific research and special funded projects (NO.201201114).

References

[1] Cui Z C, Yang Y Z, Gao X L, *et al.* 2015 Research on evaluation index system of the key ecological river network in Dongying city *Journal of Shandong Water Polytechnic* (3) 9–11

[2] Yang L R, Chen L D, *et al.* 2009 Research and development of ecological system features and self-purification capacity *Journal of Ecology* **29**(9) 5067–72

[3] Wang H, Zhang M K, *et al.* 2008 Summary of river ecosystem restoration techniques *Jiangxi Agricultural Sciences* **20**(6) 105–08

[4] Li G Y and Sheng L X 2011 Analysis of water demand of The Yellow River estuary ecological system *Journal of Northeast Normal University (Natural Science)* **43**(3) 138–44

[5] Dong Z R, Sun Y D, *et al.* 2007 *Ecological hydraulic engineering principles and techniques* Beijing: China Water Power Press pp 22–114

[6] Chen S Y 1998 *Engineering fuzzy theory and application* Beijing: National defence of Industry Press pp 45–139

[7] Cui Z C and Tian W L 2002 Analysis of the problem of fuzzy mathematics in the application of water and electricity energy *Hydropower and Energy Science* (2) 15–19

[8] Cui Z C, Guo L H and Tian W L 2000 A fuzzy optimizing multi-dimensional dynamic programming model and its application for water resources systems *Advances in Water Science* **11**(2) 186–93

[9] Cui Z C, Bai H, Zhao H L, et al. 2009 Evaluation model for water resources carrying capacity and its application based on fuzzy programming *Engineering Journal of Wuhan University* **42**(4) 139–43

[10] Cui Z C, Wang Q T and Yu J Y 2012 Application of fuzzy constraint linear programming to assessment of carrying capacity of Rizhao regional water resources system *Journal of Hydroelectric Engineering* **32**(2) 98–101

Simulated Asian–Australian monsoon with a spectral element atmospheric general circulation model

X Y Liu[1, 2]

[1]College of Meteorology and Oceanography, PLA University of Science and Technology, Nanjing 211101, China

E-mail: lxy@escience.cn

Abstract. A low-top version of SEMANS (Spectral Element Model with Atmospheric Near Space resolved) has been used to carry out numerical simulation on characteristics of Asian-Australian Monsoon (A-AM) in the work. The simulation results are validated with ERA-Interim reanalysis dataset and precipitation data from satellite remote sensing. It's shown that the model can reproduce the major climatic features of A-AM with stronger easterly in the tropical Eastern Pacific, and a weaker northerly component in the Northern Hemisphere.The simulated precipitation rate is larger and the double ITCZ (Inter-Tropical Convergence Zone) in the tropical Eastern Pacific in the northern spring is not reproduced. A due to the absence of variation longer than a year in the bottom boundary conditions, the model cannot reproduce the relationships between the monsoon indexes and the surface air temperature in the broad area near the equator.

1. Introduction

The area dominated by monsoon is the key region where the atmosphere gets momentum and water vapor, and the activities of monsoon have direct connections with variations of global general circulation,weather, and climate [1]. Asian-Australian Monsoon (A-AM) is the largest monsoon circulation system on the earth, and it influences the weather and climate of China to a great extent. With reanalysis datasets, Chinese researchers have performed many studies, such as the relationship between the westerly jet stream (and the front associated with it) over the northern China and the low frequency variation of summer Australian Monsoon [2]. Alongside the feature of water vapor transportation in the A-AMregion[3], the climatic characteristics of flow filed structure and the evolution of the summer A-AM subsystems [4]. The quasi-biennial oscillation of water vapor sink in the A-AMregion[5], and the relationships between A-AM and the summer precipitation of Yangtze River [6]. Numerical modeling is a necessary means to carry out scientific research. With the numerical model, the influence of surface land situation on A-AM has been investigated [7].Forced with observed sea surface temperature, atmospheric general circulation model (AGCM) can reproduce some characteristics of interannual variabilities associated with A-AM and ENSO [8,9], whereas the ability of prediction of Indian summer precipitation is weak. This may be due to that the contribution of ENSO to interannual variability is comparable to the strength of regional perturbation caused by the inner perturbation of numerical model, which is independent of sea surface temperature anomaly [10]. Some researchers discussed the connections of the ability of reproducing climatic precipitation with that of reproducing interannual variation in numerical simulation [11,12]. Wang et al.(2004) [13] and Zhou et al. (2009) [14] evaluated the capacity of the atmospheric model in reproducing the interannual variation of A-AM, showing that AGCMs share similar weakness in some aspects such as modeling the pattern of abnormal precipitation. To deduce differences in the monsoon simulations run with observed sea surface temperatures and with ocean–atmosphere coupling, Meehl et al.(2012) [15] studied the Asian–Australian Monsoon regimes and processes in CCSM4 (Community Climate System Model, version 4). The ability of CCSM4 as well as its atmospheric component CAM (Community Atmosphere Model) to simulate the Asian summer monsoon, focused particularly on the inter-model comparison and the role of air–sea interaction, were also examined by Islam et al. (2013) [16]. Xue et al. (2015) [17] reviewed the major progress that has been made in monsoon studies in China.

In numerical modeling,both the discretization method and the parameterization of physical processes influence the results. In current AGCMs, finite difference and spectral approximations are mostly used to discretize the equations describing the movement of the atmosphere, and finite element and finite volume discretizations are also used in a few ones.The finite difference has the merit of being intuitive and easy to implement for parallel computing. It has the disadvantage of low accuracy and slow convergence to the real solution.The spectral expansion used in AGCM usually adopts spherical harmonic function as the basis. Since the approximated function is calculated in the whole domain of computation, the local error will be propagated to a broad area. Furthermore, the implementation of parallel computation algorithm based on the idea of area division is difficult. However, the spectral method is still popular in atmospheric modelling since it gives higher precision compared to other traditional methods. Compared to the spectral method, the finite element method divides the computation domain into many elements, in each of them, different local approximation function is used. When the finite element method is used, the division of computation domain can be very flexible. So it is suitable for the problem with complex geometry. Although the finite element method has the merit of versatility, it is less precise than the spectral method. A finite volume method constructs discrete scheme based on the idea of regional integration and the finite difference method can thus be seen as one particular case of it.Compared to the finite difference method, the finite volume method can be used to construct discrete schemes much more flexibly and the constructed schemes can meet the constraint of integration relations better. The finite volume method has similar drawbacks as that of the finite difference method.The spectral element method combines the accuracy of the spectral method and the flexibility of finite element method [18]. It has more advantages than each method. The basic idea of spectral element method is to decompose the computation domain into finite elements and then get the solution within each element with the spectral method. So the spectral element method possesses the merits of the finite element method and the spectral method. It is desirable to use spectral element AGCM to perform a numerical study of monsoon.

There are connections between the variations of different atmospheric levels. To explore the influence of middle atmosphere on the simulation of monsoon, twin experiments with low and high top models should be setup. As a first step, a low top version of a Spectral Element Model with Atmospheric Near Space resolved (SEMANS) [19] is used to perform numerical study on the A-AM in the work. The model and design of a numerical experiment will be given in section 2. Simulation results and analysis will be presented in section 3. In the last part, the conclusion will be given (section 4).

2. Numerical model,datasets and design of experiment

The low top version of SEMANS [19] is used in the work. The spectral element approximation and finite difference method are used in the horizontal and vertical respectively in SEMANS. The model uses quadrilateral element [20, 21], although triangular element may also be used in many applications [22]. To solve the polar problem, which is unavoidable when the longitude-latitude grids are used, a cubed sphere projection is used in SEMANS. To improve the calculation accuracy of pressure gradient at high levels of the atmosphere, a mixed p- σ coordinate in the vertical, is adopted in the model.

The data for low boundary conditions, initial values, and comparison in the analysis are all from the ERA-Interim [23], which is a widely used reanalysis dataset from ECMWF (European Centre for

Medium-Range Weather Forecasts). The global precipitation at one-degree daily resolution from multi-satellite observations [24] for 1997-2007 is also used.

In the experiment, each projected cube face is decomposed into 81 elements, and an interpolation polynomial of 8 degrees is used for spectral approximation in each local element [19]. The atmospheric state on 00:00:00 GMT, January 1, 1997, is used for the initial values. The low boundary conditions (land surface temperature, sea surface temperature, sea ice concentration, albedo, snow thickness and so on) are means of years from 1997 to 2007 with seasonal variations considered whereas interannual variations not. In the vertical, the model atmosphere is divided into 28 layers, with about 5 layers within the planetary boundary layer. The model top situates at the 0.4hPa.

Schemes of physical processes used are identical to that of SEMANS. The model has been integrated for 20 years, and monthly mean model results saved for analysis.During the integration, global mean total energy (sum of sensible heat, potential energy, latent heat for phase change between cloud water and cloud ice, latent heat for phase change between vapor and cloud ice)and air mass(represented by surface air pressure) are monitored every step. There is no strange apparent value found in theses variables.

3. Simulation results and analysis

3.1. Winds and precipitations in summer and winter

The simulated and observed precipitation in summer (mean of June, July, and August) is given in Figure 1. It's shown that the model reproduced the precipitation extrema associated with the south Asian monsoon region, ITCZ (Inter-Tropical Convergence Zone) and SPCZ (South Pacific Convergence Zone) with the amount of precipitation bigger than that of observation. The simulated precipitation in the East China is a bit smaller than the observed. The model simulated the low-level southerlies from the Arabian Sea and Bengal Bay to the South Asia and the two extrema of precipitation associated with them well. The model also reproduced the intense precipitation center in the South China Sea adjacent to the Indonesia. Compared to the result from the reanalysis dataset, the simulated easterly over the tropical eastern Pacific Ocean is stronger. Besides, the simulated subtropical high circulation in the Pacific Ocean is weak, and its location is northward, giving a contribution to the smaller northerly component in the tropical Pacific Ocean.

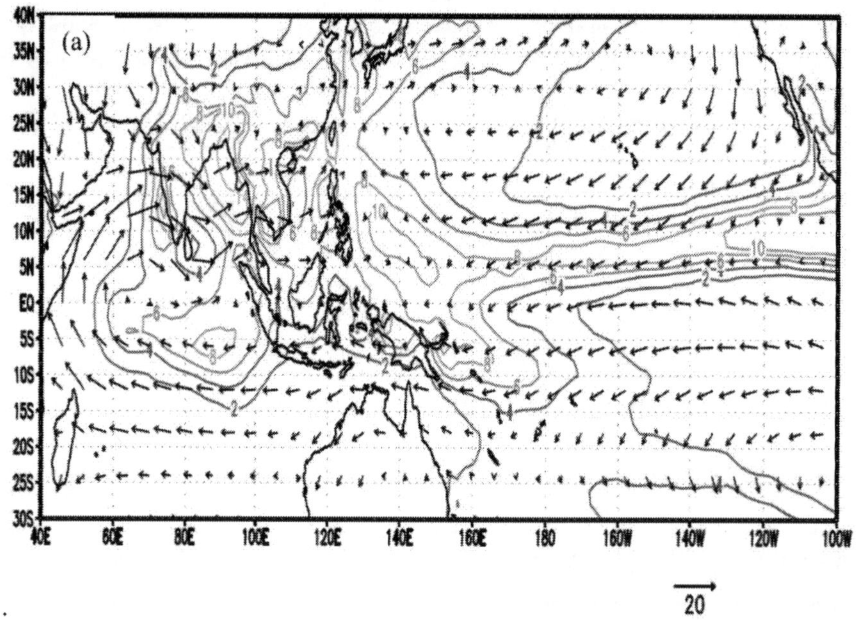

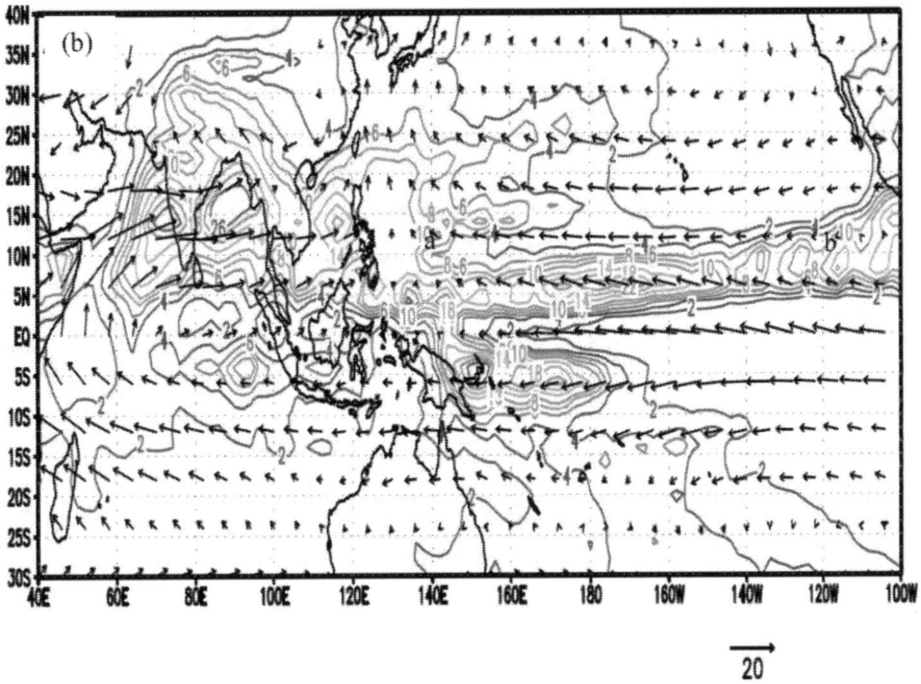

Figure 1. Averaged precipitation rate and the wind at 850 hPa in summer from (a)reanalysis and (b) simulation. Contour interval for precipitation rate is 2 mm/day.

In winter (mean of December and subsequent January and February), the model reproduced the principal feature of precipitation in tropical Indian-Pacific Ocean. The intense precipitations in ITCZ, SPCZ and over the Indonesian and Australian were all reproduced well (see Figure. 2). But the model produced a false precipitation center in the South China Sea and the simulated northerly component in the tropical easterly zone is weak. Also, the location of the simulated transition region from easterly to westerly in the subtropical Pacific Ocean is much northward, and the simulated northerly component in the Arabian Sea and Bengal Bay is weak (compare Figure 2b. to Figure 2a.).

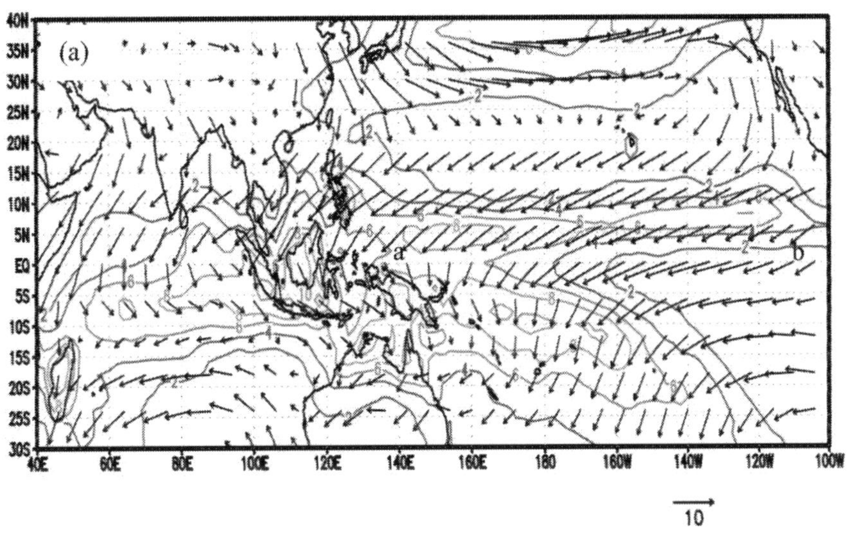

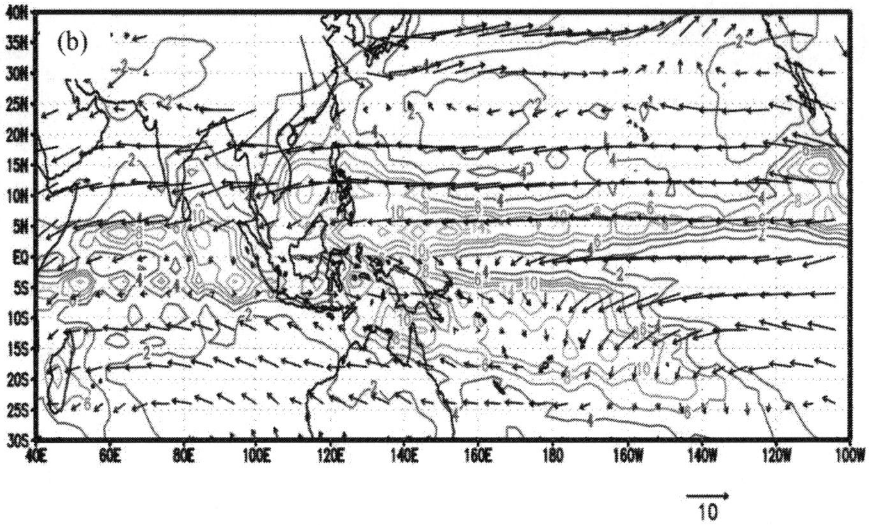

Figure 2. Same as Figure 1 except for winter.

3.2. The annual variation of precipitation

To analyze the precipitation feature in regions of the Indian monsoon and the Northwest Pacific, two zones, one for the Indian monsoon (60°E-105°E, 7°N-27°N) and another for the Northwest Pacific (105°E-150°E, 7°N-22°N), are selected. The variations of areally averaged precipitation rate in the two regions are shown in Figure 3. It can be seen that the simulated precipitation rate in the Indian monsoon, is bigger (this is consistent with the results in Figure 1 and Figure 2), and the bias is the biggest in summer with a value of difference over 5 mm/day. From February to May, the simulated areal averaged precipitation rate is close to observation for the Northwest Pacific region. But in other months, the model produced bigger precipitation, which is also consistent with the results in Figure 1 and Figure 2.

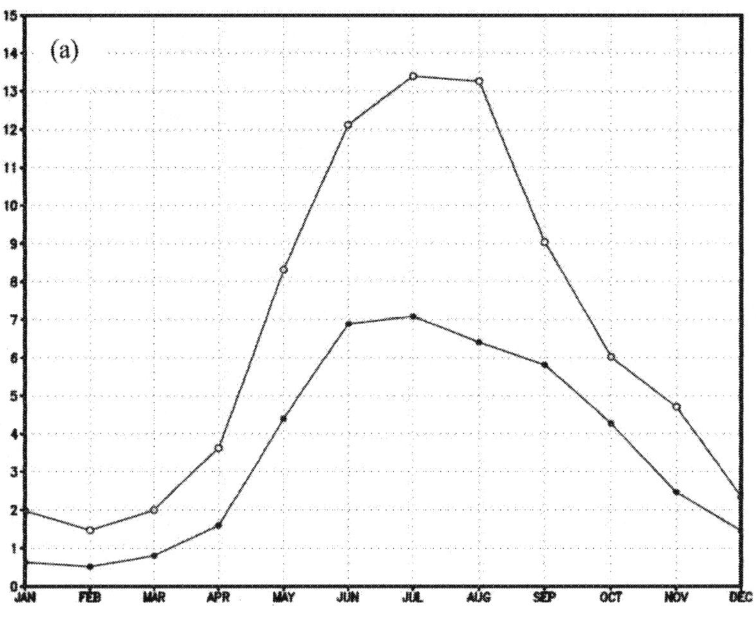

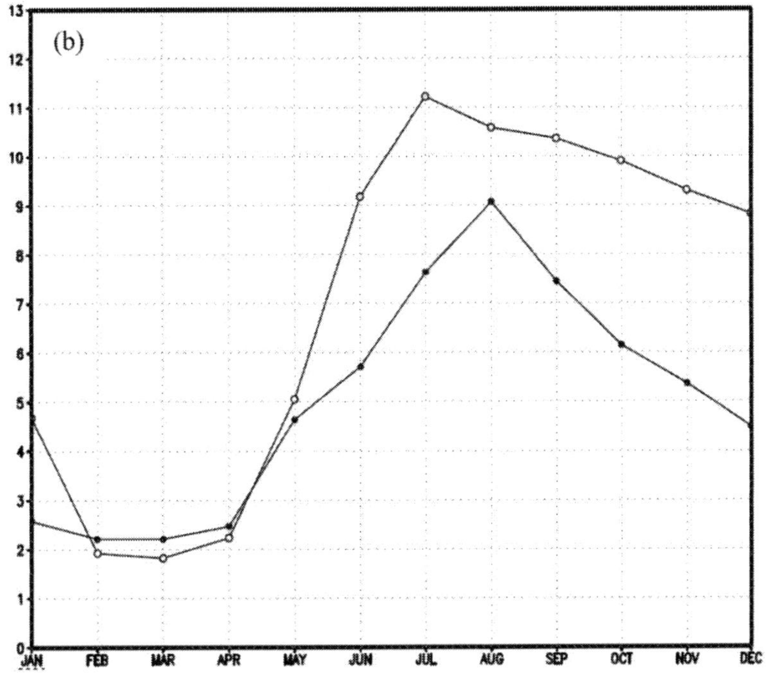

Figure 3. Annual variation of areally averaged precipitation rate for (a)Indian Monsoon region and (b) Northwest Pacific Ocean region. The solid circle and empty circle denote result from observation and simulation respectively. The unit of the vertical axis is mm/day.

From the distribution of zonal mean precipitation rate from 150°W to 100°W (Figure 4), it can be seen that there are extreme ITCZ precipitation north of the equator in the second half year with the most significant value in August and the second biggest value in November. There is double ITCZ precipitation bestriding the equator in March and April. This is consistent with the result from the Xie and Arkin (1996) dataset [25]. The model only reproduced the ITCZ precipitation north of the equator and the simulated occurring time of intense precipitation does not match that of observation.

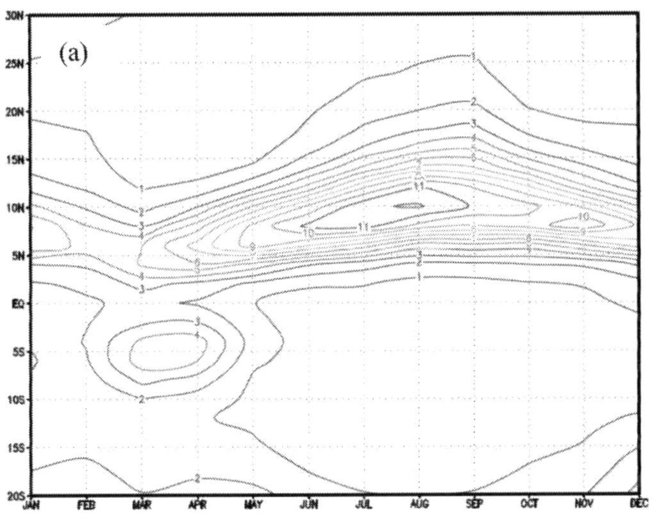

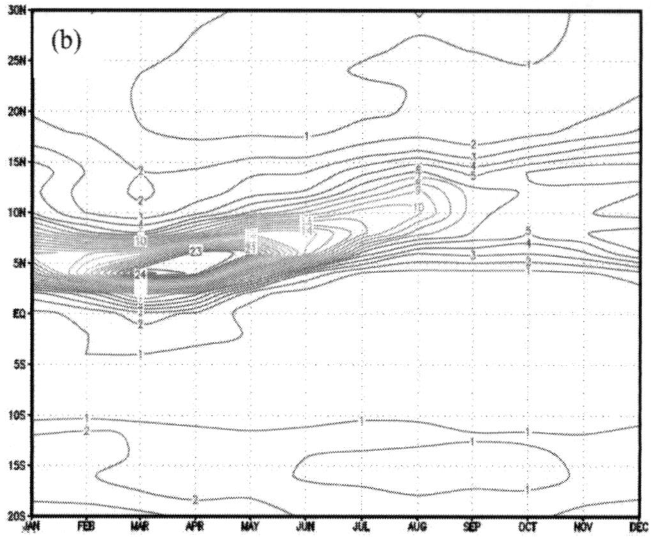

Figure 4. Annual variation of zonal mean precipitation rate over the eastern Pacific Ocean from (a) observation and (b) simulation. The contour interval is 1 mm/day.

3.3. Influences of A-AM on the global atmosphere

To analyze the influences of A-AM on the global atmosphere, correlations between the index of circulation of South Asian monsoon,the global surface air temperature in the northern summer (see Figure 5), the index of precipitation of Australian monsoon, and global surface air temperature in the northern winter (see Figure 6);are calculated respectively. Here, the index of circulation of South Asian monsoon, is defined as the averaged zonal wind difference over the area (0°-20°N, 40°-110°E), between the 850hPa and the 200hP [26]. The index of precipitation of Australian monsoon is defined as the precipitation rate averaged over the area (15°S-0°, 110°-150°E) [27]. In most areas near the equator and south of India the two correlations are negative (see figures 5a and 6a). The model could not reproduce these relationships. This is due to that the low boundary conditions such as sea surface temperature and land surface temperature of the model do not include the variation longer than a year. The A-AM is the effect of complicated interactions between the ocean, land, and atmosphere. Among them, ENSO is the primary element determining the interannual variability of A-AM. So, the correlations in other areas will not be analyzed.

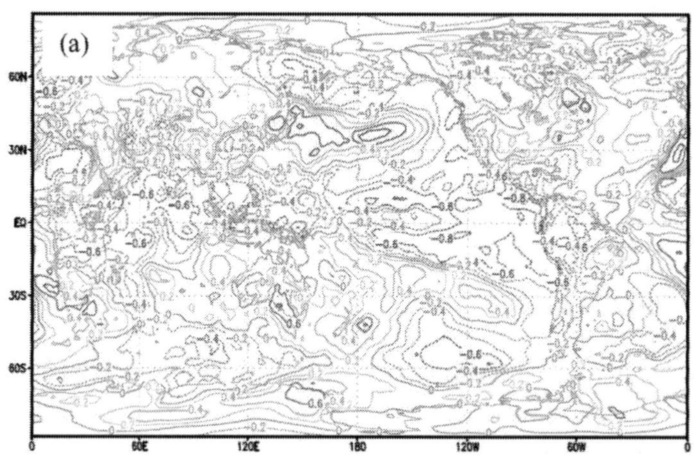

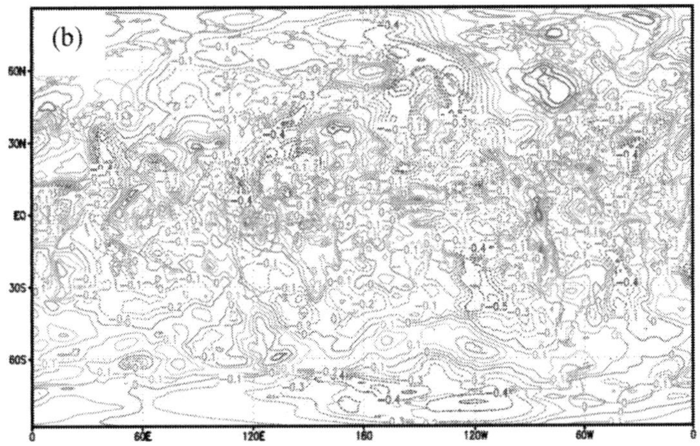

Figure 5. Correlation between circulation index of South Asian monsoon and the global surface air temperature in the northern summer from (a) reanalysis dataset and (b) simulation.

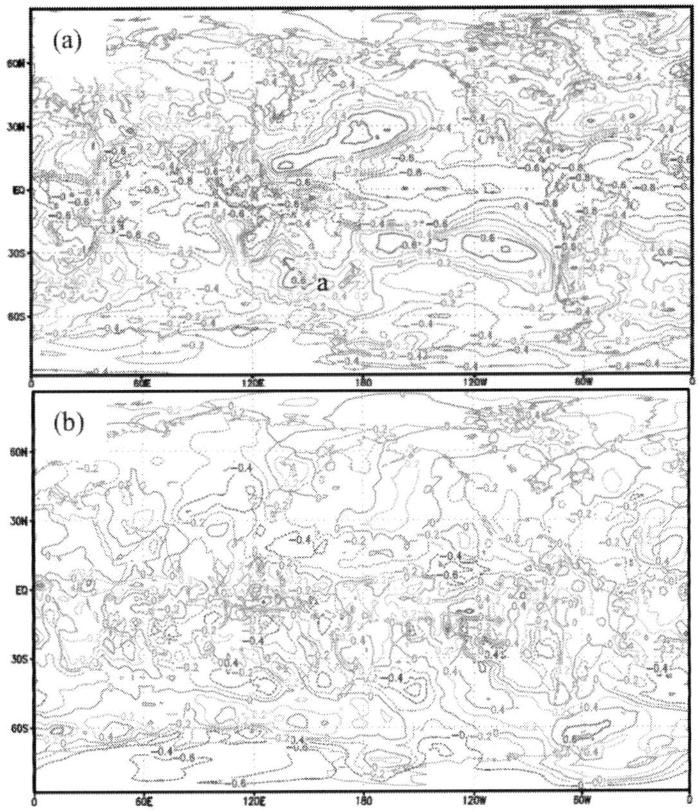

Figure 6. Same as Figure 5.but for the precipitation index of Australian monsoon in the northern winter.

4. Conclusion

A numerical experiment on Asian-Australian monsoon has been carried out with a global spectral element atmospheric general circulation model. Reanalysis dataset ERA-Interim and precipitation dataset from satellite observation has been used to validate the simulation results. It's shown that the model can reproduce the principal climatic features of Asian-Australian monsoon with stronger easterly over the eastern tropical Pacific Ocean and weaker northerly over the northern tropical ocean. The reproduced precipitation rate is bigger than the observed in general, and the feature varies in different regions. Due to the exclusion of variation longer than a year, the model cannot reproduce the negative correlation between the monsoon index and the surface air temperature in the most area near the equator.

Acknowledgments
The work is sponsored by the Natural Science Foundation of China (Grant No. 41276190) and the China Special Fund for Research in the Public Interest (Grant No. GYHY201506011). The ERA-Interim reanalysis datasets are downloaded from the ECMWF data server website.

References
[1] Ding Y 1996 The South China Sea monsoon experiment and the East Asian monsoon study *Frontier and Prospect of Modern Atmospheric Science* (Beijing： Meteorology Press) pp 43–46
[2] He J, Wang J and Miao C 1993 Low-frequency coupling of winter Asian-Australian Monsoon circulation *Tropical Meteorology* **9** 193–201
[3] Qiao Y and Lin M 2006 The transportation of water vapor in the Asian-Australian monsoon region(in Chinese) *Journal of Sun Yat-Sen University* (Natural Science Edition) **45**102–5
[4] Chen J and Huang R 2006 Comparative study of climatic feature of components of Asian-Australian monsoon-I flow field structure *Chinese Journal of Atmospheric Science* **30** 1091–102
[5] Qiao Y, Jian M and Luo H 2005 The quasi-biennial osscillation of water vapor sink in the Asian-Australian monsoon region and its connection with the atmospheric general circulation(in Chinese)*Journal of Sun Yat-Sen University* (Natural Science Edition) **44** 98–101
[6] Zhang L, Zhang L and Zeng P 2011 The connections of the Asian-Australian monsoon and the summer precipitation over the middle and lower reaches of Yangtze River *Tropical Meteorology* **27** 189–201
[7] Bao Y, Lv S and Liu H 2006 Numerical study of the influences of Australian land vegetation on the Asian-Australian monsoon (I): Impacts on the climate and winter monsoon of Australia *Plateau Meteorology* **25** 763–9
[8] Palmer T N, Brankovic C, Viterbo P and Miller M J 1992 Modeling interannual variations of summer monsoons *Climate* **5** 399–417
[9] Sperber K R and Palmer T N 1996 Interannual tropical rainfall variability in general circulation model simulations associated with the Atmospheric Model Intercomparison Project *Climate* **9** 2727–50
[10] Goswami B N 1998 Interannual variations of Indian summer monsoon in a GCM: External conditions versus internal feedbacks *Climate* **11** 501–22
[11] Gadgil S and SajaniS 1998 Monsoon precipitation in the AMIP runs *Climate Dyn.***46** 59–89
[12] Soman M K and Slingo J M 1997 Sensitivity of Asian summer monsoon to aspects of sea surface temperature anomalies in the tropical Pacific Ocean *Quart. Roy. Meteor. Soc.***123** 309–36
[13] Wang B, Kang I and Lee J 2004 Ensemble simulations of Asian–Australian monsoon variability by 11 AGCMs. *Climate* **17** 803–18
[14] Zhou T,Wu B and Wang B 2009 How well do Atmospheric General Circulation Models capture the leading modes of the interannual variability of Asian-Australian Monsoon *Climate* **22** 1159–73
[15] Patera A T 1984 A spectral element method for fluid dynamics: laminar flow in a channel expansion *Comput. Phys.* **54** 468–88
[16] Meehl G A, Arblaster J M, Caron J, Annamalai H, Jochum M, Chakraborty A and Murtugudde

R 2012 Monsoon regimes and processes in CCSM4 Part 1: The Asian-Australian Monsoon *Climate* **25** 2583–608

[17] Islam S, Tang Y and Jackson P L 2013 Asian monsoon simulations by Community Climate Models CAM4 and CCSM4 *Climate Dynamics* **41** 2617–42

[18] Xue F, Zeng Q, Huang R, Li C, Lu R and Zhou T 2015 Recent advances in monsoon studies in China*Adv. Atmos. Sci.* **32** 206–29

[19] Liu X Y, Liu C Y and Yao S S 2015 A spectral element model with atmospheric near space resolved (SEMANS) *Chinese Journal of Computational Physics* **32** 40–50

[20] Giraldo F X and Rosmond T E 2004 A scalable spectral element Eulerian atmospheric model (SEE-AM) for NWP: Dynamical core tests *Mon. Wea. Rev.* **132** 133–53

[21] Thomas S and Loft R 2005 The NCAR spectral element climate dynamical core: semi-implicit eulerian formulation *Journal of Scientific Computing* **25** 307–22

[22] Liu X Y 2011 Numerical Simulation of Typhoon Movement with a Regional Spectral Element Barotropic Atmospheric Model *Chinese Journal of Computational Physics* **28** 35–40

[23] Simmons A, Uppala S, Dee D, et al. 2007 ERA-Interim: New ECMWF reanalysis products from 1989 onwards *ECMWF Newsletter* **110** 25–35

[24] HuffmanG J, Adler R F, Morrissey M, et al. 2001 Global precipitation at one-degree daily resolution from multi-satellite observations *Hydrometeor* **2** 36–50

[25] Xie P and Arkin P A 1996 Global precipitation: a 17-year monthly analysis based on gauge observations, satellite estimates, and numerical model outputs *Bull. Amer. Meteor. Soc.* **78** 2539–58

[26] Webster P J and Yang S 1992 Monsoon and ENSO *Quart. J. Roy. Meteor.Soc.* **118** 877–926

[27] Meehl G A and Arblaster J 1998 The Asian-Australian monsoon and El Nino-Southern Oscillation in the NCAR Climate System Model *Climate* **11** 1357–87

IOP Conf. Series: Earth and Environmental Science **39** (2016) 012042 doi:10.1088/1755-1315/39/1/012042

Application of the dynamic ant colony algorithm on the optimal operation of cascade reservoirs

X X Tong[1,2], W S Xu[1], Y F Wang[1], Y W Zhang[1] and P C Zhang[1]

[1] Changjiang River Scientific Research Institute of Changjiang Water Resources Commission, Wuhan, China

E-mail: ckytxx@163.com

Abstract. Due to the lack of dynamic adjustments between global searches and local optimization, it is difficult to maintain high diversity and overcome local optimum problems for Ant Colony Algorithms (ACA). Therefore, this paper proposes an improved ACA, Dynamic Ant Colony Algorithm (DACA). DACA applies dynamic adjustments on heuristic factor changes to balance global searches and local optimization in ACA, which decreases cosines. At the same time, by utilizing the randomness and ergodicity of the chaotic search, DACA implements the chaos disturbance on the path found in each ACA iteration to improve the algorithm's ability to jump out of the local optimum and avoid premature convergence. We conducted a case study with DACA for optimal joint operation of the Dadu River cascade reservoirs. The simulation results were compared with the results of the gradual optimization method and the standard ACA, which demonstrated the advantages of DACA in speed and precision.

1. Introduction

The reservoir optimal operation is a complex combinatorial optimization problem that is non-convex, nonlinear, and has high dimensions. Traditional algorithms, such as dynamic programming, successive approximation of dynamic programming, step-by-step optimization algorithm have obvious defects such as "dimension disaster", slow convergence speed and initial solution dependency, which have modest effect on joint dispatching of cascade reservoirs. Continued research development in biological intelligence, artificial intelligence, genetic algorithms, particle swarm optimization, and the differential evolution algorithm of intelligent optimization algorithms have been widely used in optimizing cascade reservoir operations.

Chen et al [1] proposed hypercube floating-point encoding and an adaptive adjustment control strategy of parameters to improve the genetic algorithm, and to enhance the local optimization ability of the genetic algorithm. Zheng et al [2] proposed a method of chaos optimization to generate an initial solution; this approach controls parameters' adaptive adjustment according to individual fitness and dispersion degree population, which overcomes premature GA and improves convergence performance. Ji et al. [3] introduced the catfish effect mechanism to the standard particle swarm algorithm; this particle swarm optimization algorithm (CE-PSO) improves the algorithm's ability to jump out of the local optimal solution. Qin et al. [4] brought the external file update strategy to the differential evolution algorithm and to modify operator, the improved algorithm can improve the uniformity of the distribution of population. Zheng et al. [5] embedded the hybrid leapfrog algorithm in the differential evolution algorithm, which can improve the global optimization ability of the algorithm through the interaction and evolution of the upper and lower double algorithm. Based on the optimal operation schedule made by the deterministic hydropower station optimal operation, Li et al. [6] used the idea of aggregation-decomposition and applied genetic programming to simulate the total output and the allocation of total output in cascade hydropower stations. This improved the cascade hydropower stations' total output reliability. Although these algorithms have improved the optimal capacity of the original algorithm to some extent, they all failed to create dynamic balance between the global and local search.

The Ant Colony Algorithm (ACA) [7, 8] is an optimization method based on the swarm intelligence algorithm. This algorithm uses the walking path of artificial ants to represent the feasible solution of a specific problem, which makes use of the ant colony's self-organization. It switches path information through pheromones to form collectively autocatalytic behavior and ultimately finds the optimal path. The algorithm has few control parameters and high robustness. However, this algorithm suffers from an imbalance between global search and local optimization, and the local optimal problems caused by quick reduce of population diversity in the process of evolution.

Therefore, this paper advances a new algorithm named Dynamic Ant Colony Algorithm (DACA) with improved parameter control and optimization strategies over ACA. DACA uses a cosine decreasing strategy and dynamically adjusts the heuristic factor to balance the ant global search and local optimization. At the same time, disturbing path found by each iteration of ant colony algorithm and using the randomness and ergodicity of chaotic search to improve the ability of local optimization algorithm and avoid local optimum.

2. Mathematical model

As a combined operation, cascade hydropower stations exhibit power, compensation, capacity compensation, and hydrological compensation. This paper establishes the optimal the long-term operation model of a cascade hydropower station to maximize energy output.

2.1. *The objective function*

$$E = \mathrm{Max} \sum_{t=1}^{T} \sum_{i=1}^{N} N_{i,t} \bullet \Delta t \tag{1}$$

$$N_{i,t} = A_i \bullet Q_{i,t} \bullet H_{i,t} \tag{2}$$

where t is short for time; E is the total power's generating capacity of the scheduling cascade; $N_{i,t}$ is the output in period t of power station i; A_i is the comprehensive efficiency coefficient of the power station in period t; $Q_{i,t}$ is the average output current of power station i in period t; $H_{i,t}$ is the average power output head of power station i in period t; T is the hours of the scheduling period; N is the number of cascade hydropower stations; Δt is the length of a single time period.

2.2. *The constraint*

- Water balance constraints

$$V_{i,t} = V_{i,t-1} + (I_{i,t} - Q_{i,t} - q_{i,t}) \bullet \Delta t \tag{3}$$

- Water level constraints

$$ZL_{i,t} \leq Z_{i,t} \leq ZU_{i,t} \tag{4}$$

- Runoff constraints

$$QL_{i,t} \leq Q_{i,t} \leq QU_{i,t} \tag{5}$$

- Output constraints

$$NL_{i,t} \leq N_{i,t} \leq NU_{i,t} \tag{6}$$

where t is time; $ZL_{i,t}$, $ZU_{i,t}$ are the lowest and highest water level of power station i in period t, respectively; $QL_{i,t}$, $QU_{i,t}$ are the lower and upper limit of power runoff of power station i in period t, respectively; $NL_{i,t}$, $NU_{i,t}$ are the lower and upper limit of power output of power station i in period t, respectively; $V_{i,t-1}$, $V_{i,t}$ are the storage capacity of reservoir i in the beginning and the end of period t, respectively; $I_{i,t}$ is the average inflow of reservoir i in period t; $q_{i,t}$ is the surplus water flow.

3. Ant colony algorithm

3.1. *The basic principle*

ACA is a heuristic random search method based on group collaboration. The artificial ants' path attempts to solve the problem; the artificial ants choose the walk path according to the pheromones left behind by other hands. In doing so, they leave behind pheromones for subsequent ants, forming a positive feedback mechanism (Dorigo M et al. 1996). As the algorithm advances, pheromones in the optimal path increase while pheromones in the poor path decrease due to volatilization. Eventually, the whole ant colony would focus on the optimal solution under the positive feedback mechanism.

3.2. *Ant colony algorithm in the application of the optimal reservoir operation*

- Assume that reservoir water level at the end of each time period as decision variables, discrete the feasible domains of water level from normal storage level to dead storage level, that is to say, to solve the reservoir optimization scheduling model is to seek an optimal decision $\left[x_1, x_2, \cdots, x_T\right]$ which can get maximum generating volume of cascade reservoirs. Therefore, the initial solution $X = \left[X_1, X_2, \cdots, X_{NP}\right]$ is generated randomly within the scope of the feasible solution, in which NP is the ant colony size. $X = [X_1, X_2, \cdots, X_{NP}]$ is the route for ant k; it represents the solution of the problem. D is the space dimension of the solutions.

- Generally, in the marching process of ant k, the path selection is based on the principle of high pheromone concentrations in accordance with high transition probability. The probability calculation formula is as follows:

$$p_{m,n}^{k}(t) = \begin{cases} \frac{[\tau_{m,n}(t)]^{\alpha} \cdot [\eta_{m,n}(t)]^{\beta}}{\sum_{s \subset allowed_k} [\tau_{m,n}(t)]^{\alpha} \cdot [\eta_{m,n}(t)]^{\beta}}, & if\ j \in allowed_k \\ 0 & else \end{cases} \tag{7}$$

where $p_{m,n}^{k}(t)$ is the state transition probability of ant k from water level m to n; $\tau_{m,n}(t)$ is the residual amount of pheromones on path(m,n) in stage t. Each path during initialization pheromone amount are equal, is set as a constant; $\eta_{m,n}(t)$ is the expected value of the transition amount from m to n in stage t, referred to as prior knowledge; it is identified as the reciprocal of generating capacity $d_{m,n}$ on the water level from m to n, namely $\eta_{m,n}(t) = 1/d_{m,n}$; $allowed_k$ is the water level that ant k can choose in stage t; α is the inspired information factor, which reflects the relative importance of residual pheromones as the ant colony moves; β is the expected heuristic factor, which reflects the relative importance of expectation.

- To avoid expectation failure caused by excess residual pheromone, the pheromone left by each ant volatilizes and gradually decreases. The pheromone updating method is as follows:

$$\tau_{m,n}(t+n) = (1-\rho) \bullet \tau_{m,n}(t) + \Delta\tau_{m,n}(t) \tag{8}$$

$$\Delta\tau_{m,n}(t) = \sum_{k=1}^{NP} \Delta\tau_{m,n}^{k}(t) \tag{9}$$

where ρ is the volatilization factor; $\Delta\tau_{m,n}(t)$ is the pheromone increment in this cycle on the path (m,n) at initial moment $\Delta\tau_{m,n}(t) = 0$; $\Delta\tau_{m,n}(t)$ is the amount of pheromone left by ant k on path (m,n) in this cycle. The updating method of Ant Cycle Model in this article is:

$$\Delta\tau_{m,n}(t) = \begin{cases} Q/L_k, & \text{ant } k \text{ passes through path}(m,n)\ in\ this\ circle \\ 0, & \text{other} \end{cases} \tag{10}$$

where Q is the total amount of released pheromones by the ant on the path of one cycle or one process; L_k is the path length of ant k in this cycle.

4. Improved strategy and implementation steps of the ant colony algorithm

4.1. Dynamic adjustment strategy of heuristic factor

In general, the algorithm always has strong developmental ability in the early stage, and can search on the global scope; it increases its exploration abilities around the best solutions and improves local optimization abilities. The balance between global exploration and local optimization is an important aspect that affects the algorithm's performance.

Formula 7 shows the ants' routing probability. The heuristic factor α reflects the relative importance degree of pheromones accumulated by ants in guiding ants to choose the best path. The greater α is, the greater the influence of accumulated pheromones in the past is, and the greater the possibility that the ant chooses the past path.The weaker randomness would be of the search algorithm. The smaller α is, the lower the influence that the past path would have for the ant, and the more inclined it would be to perform a local search. Due to the same initial pheromone value in various paths, a much larger α would not significantly decrease the randomicity of the algorithm search. Accordingly, the value of α has dynamic adjustments in this paper; it uses a larger value early in the process to make the algorithm explore on a global scale, and it uses a smaller value later in the process to improve the local optimization ability. This study used a cosine decreasing strategy, to keep a smaller value early and a larger value later for α, as follows:

$$\alpha = \alpha_{min} + (\alpha_{max} - \alpha_{min}) \bullet \frac{\cos(\frac{iter}{iter\,max}\pi)+1}{2} \tag{11}$$

where $\alpha_{\max}$, $\alpha_{\min}$ are the upper and lower limits of the pheromone heuristic factor, respectively; $iter$, $iter_{\max}$ are the iterations times and the maximum iteration times, respectively.

Figuer 1. shows the comparison of the cosine decreasing with linear decrease of α.

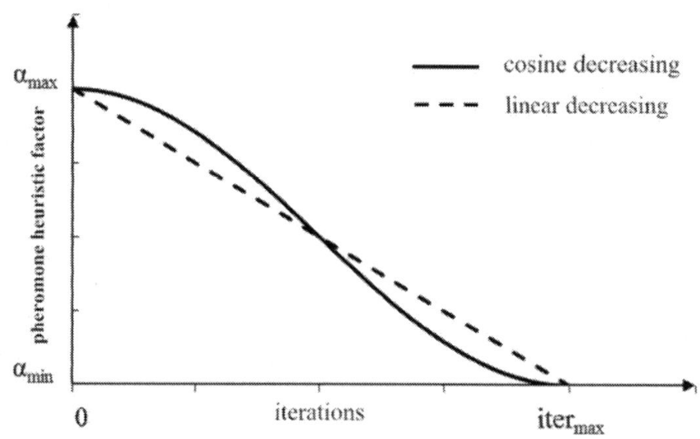

Figure 1. the comparison of the cosine decreasing with a linear decrease.

4.2. Local chaos disturbance strategy

On the basis of the dynamic adjustment on heuristic factor, to avoid the local optimal solution of rapid aggregation in ACA, this paper utilized the ergodicity and randomness of a chaotic search and conducted local chaos optimization of ant colony in poor individual in the form of chaos disturbance, which improves the ability of ACA to jump out of local optimal solution and continue to find the optimal solution.

This study uses the one-dimensional logistic model to design chaotic mapping. The iterative equation is:

$$R_i^{g+1} = c \bullet R_i^g \bullet \left(1 - R_i^g\right),\ R_i \in [0,1] \tag{12}$$

where R_i^g is the chaotic variables, which represents a certain number generated by the chaotic equation between $[0,1]$ and $R_i^1 \in [0,1]$. g is the number of iterations; c is the control parameters and $0 \le c \le 4$. When $c > 3.57$, $R_i^1 \in [0,1]$ and $R_i^1 \ne \{0.25, 0.5, 0.75\}$, sequence will never repeated "wandering" in $(0,1)$, not to any point of convergence, and sensitively depends on the initial point.

Assuming that the new ant colony X^g is obtained after iteration i, disturb the ant colony according to formula (13) and (14), colony Y^g is got after disturbance. Where R^{g+1} is the matrix of D chaotic sequences and γ is the step disturbance.

$$X^g = \begin{bmatrix} x_{1,1}^g, x_{1,2}^g, \cdots, x_{1,D}^g \\ x_{2,1}^g, x_{2,2}^g, \cdots, x_{2,D}^g \\ \cdots \\ x_{NP,1}^g, x_{NP,2}^g, \cdots, x_{NP,D}^g \end{bmatrix} \qquad R^{g+1} = \begin{bmatrix} R_{1,1}^{g+1}, R_{1,2}^{g+1}, \cdots, R_{1,D}^{g+1} \\ R_{2,1}^{g+1}, R_{2,2}^{g+1}, \cdots, R_{2,D}^{g+1} \\ \cdots \\ R_{NP,1}^{g+1}, R_{NP,2}^{g+1}, \cdots, R_{NP,D}^{g+1} \end{bmatrix} \tag{13}$$

$$Y^g = X^g + \gamma \cdot R^{g+1} \tag{14}$$

After obtaining the disturbance group Y^g, based on the greedy selection strategy, compare the adaptation value of the corresponding individual X_i^g and Y_i^g of the original ant colony X^g and the disturbed ant colony Y^g; take the individual with the better adaption value for the next generation of ants, as follows:

$$X_i^{g+1} = \begin{cases} X_i^g, & if \quad f(X_i^g) \geq f(Y_i^g) \\ Y_i^g, & if \quad f(X_i^g) < f(Y_i^g) \end{cases} \tag{15}$$

4.3. Steps for improved ant colony algorithm for solving joint optimization dispatching of cascade reservoirs

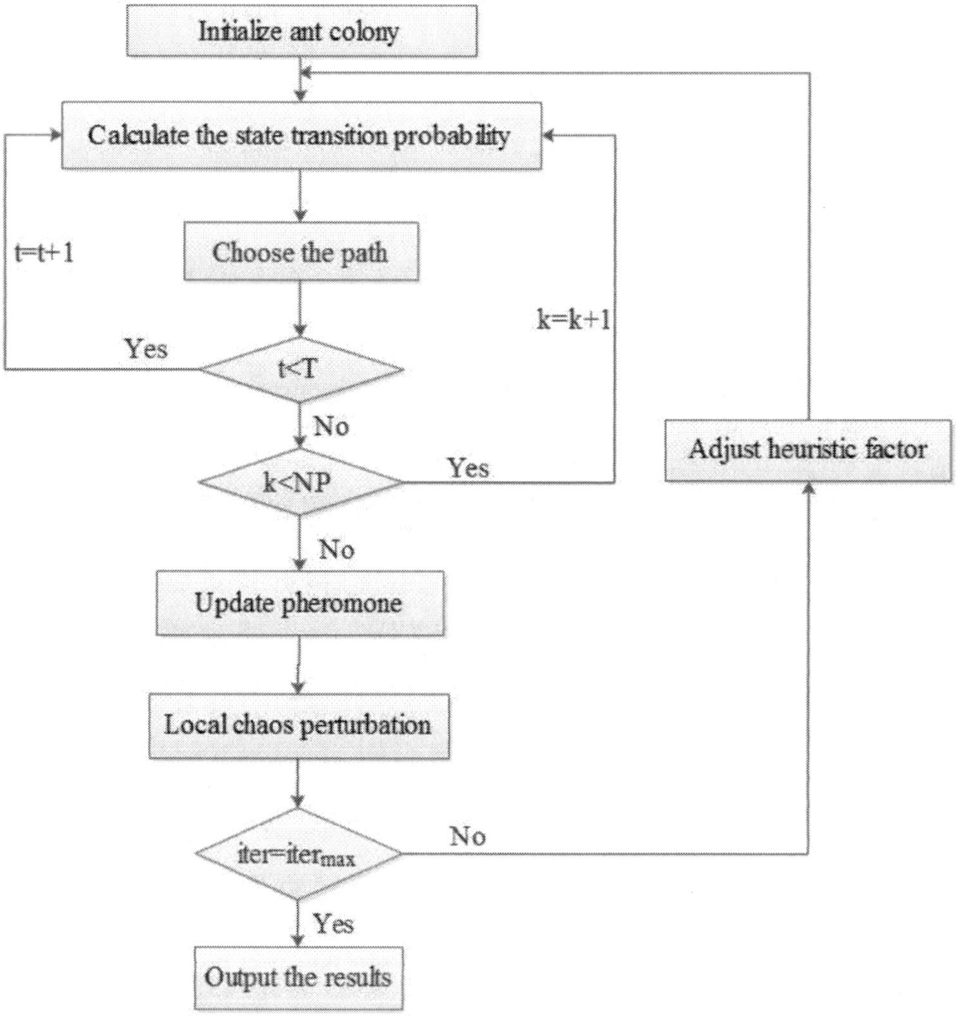

Figure 2. The flowchart of steps.

5. Case study

5.1. Introduction of cascade reservoirs and preferences of the algorithms

To validate the feasibility and effectiveness of the proposed DACA, and to apply the improved algorithm to solve the cascade reservoirs optimization scheduling problem, we conducted a case study on the cascade reservoirs of Dadu River. Figure 3 shows its cascade topology structure; table 1 shows the basic data of each plant. Reservoirs A and B have annual adjustment capabilities, while the other reservoirs only have daily adjustment capabilities; therefore, this multi-reservoir can be treated as a cascade reservoirs system with 2 reservoirs and 9 power stations.

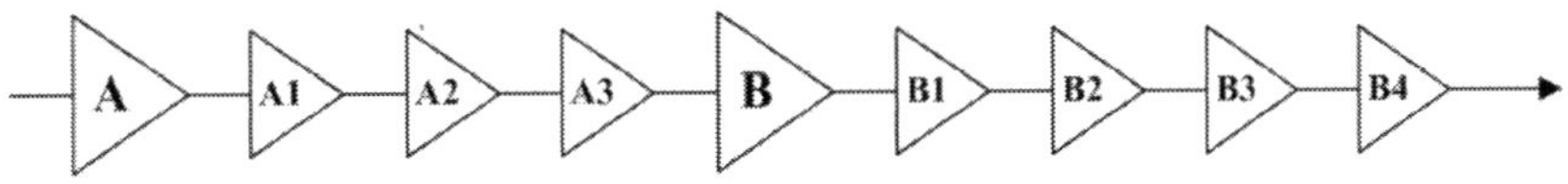

Figure 3. cascade topology.

Table 1. basic data of each power station.

Item	Unit	A	A1	A2	A3	B	B1	B2	B3	B4
Normal water level	m	2930	2315	1692	955	850	660	528	474	432
Dead water level	m	2886	2307	1687	952	790	655	520	469	430
Installed capacity	MW	240	240	240	700	3600	660	730	600	480
Warranted output	MW	77.4	87.8	82.7	136.1	926	162	197	138	151
Efficiency coefficient		8.5	8.2	8.3	7.9	8	8.5	8.2	8.8	8.5
Regulation performance		Year	Day	Day	Day	Year	Day	Day	Day	Day

Take the measured runoff data as the input sequence of each month in a certain year. Use the three algorithm types, POA, ACA, and DACA, to optimally simulate the cascade reservoirs system. Refer to the parameter setting range of the Ant colony algorithm in the literature Intelligent Optimization Methods [9], and continuously adjust the parameter set to the test. Eventually, set the ACA algorithm parameter to the following: NP = 200, $\alpha=5$, $\beta=5$, $\rho=0.3$, $Q=5000$, $iter_{max}=1000$. For DACA, $\alpha_{max}=5$, $\alpha_{min}=1$, $\gamma=0.1$, and the other parameters are set to be the same as in the ACA.

For the POA, ACA, DACA three kinds of algorithms of the initial solution are randomly generated, the latter two algorithms in ant colony routing is according to certain probability to choose, also has a certain randomness. Therefore, compare that many times' simulation results of the calculation of average results.

5.2. Calculation and discussion

Table 2 shows the calculation of the three algorithms, including the average total generating capacity and the average computing time.

Table 2. Different algorithms income average cascade total generating capacity and computing time.

Algorithm	average total generating capacity (10^8kW·h)	average computing time (s)
POA	372.8	310
ACA	364.5	64
DACA	374.9	75

The calculation found that ACA achieved (364.5•10^8kW·h) total generating capacity, which was lower than that of POA at (372.8•10^8kW·h). In the calculation process, ACA uses local optimization with an inadequate global search because of the increase of iterations times to quickly reduce ant colony diversity. Due to the strong guidance of the individual optimal path, the ant colony will be trapped in the local optimal value and cannot jump out. The cascade power generation solved by DACA (374.9•10^8kW·h) is greater than those of POA (372.8•10^8kW·h) and ACA (364.5•10^8kW·h). This is because the DACA cosine decreasing strategy for introducing the heuristic factor can be as the iteration and the size of the dynamic adjustment heuristic factor, which makes the algorithm better balance the global search and local optimization. At the same time, chaotic disturbance strategies can strengthen the local mining algorithm ability, increase ant colony diversity, and improve the algorithm's ability to jump out of local optimal value.

As shown in table 2, ACA had a 64s computing time, while DACA took 75s and POA took 310s. DACA's computing time was slightly longer than that of ACA because DACA added the descending strategy and the ant colony stimulating factor of the cosine chaotic disturbance strategies. Considering the increased cascade power generation, DACA's increased computing time is acceptable.

As for the algorithm structure, even though the POA algorithm only optimized the two stages each time, a single calculation amount will still exponentially increase with the increasing of number of reservoir (dimension), so the computing time must rise sharply. The calculation amount of ACA and DACA increased linearly with the efforts of the reservoir (dimension), so the influence of the power station number on the time-consuming was relatively small. To improve the quality of the ACA optimization result, DACA uses the cosine decreasing strategy of the heuristic factor chaotic disturbance strategies. As a result, increasing the reservoir quantity or the computing time has a slight effect on DACA performance.

The cascade reservoir water level was calculated by three algorithms, as shown in figure 4 and figure 5

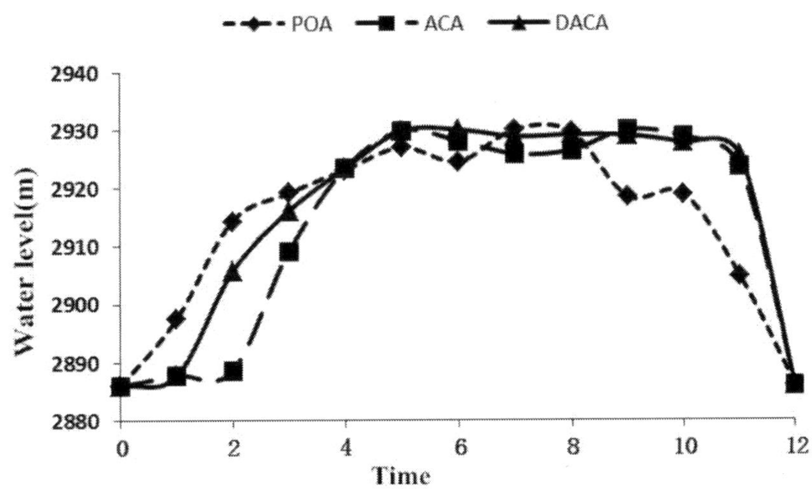

Figure 4. Reservoir water level process of A.

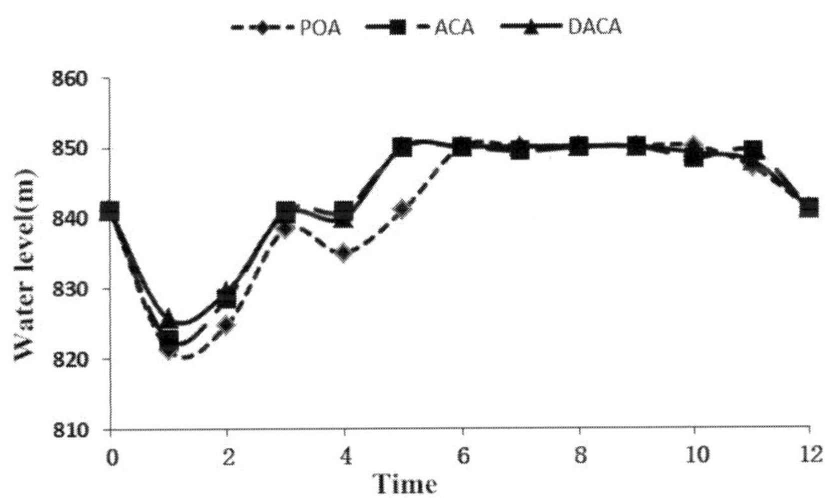

Figure 5. Reservoir water level process of B.

Observing the reservoir water level changes shown in figure 4, it is obvious that the POA and DACA water storage processes have significant differences. POA rapidly stores water early, resulting in more abandoned water. The early released water loses the high water head advantage. Meanwhile, DACA begins a water storage later, which takes full advantage of the large capacity, maintains high water for a longer period of time, and uses the high head to increase generating capacity. The design head of power station unit A is 560 meters, is the high water head, and flow rate of the change on the influence of the generating capacity is relatively more obvious, DACA's calculation results is more suit for the power station feature than POA's results.

The adjusted water level of reservoir B is relatively high, so it must increase the discharge to generate electricity in the early stages. At the same time, the unit design head is 148 m, the average flow rate for many years is 1230 m3/s, so the effect of the water head change of the cascade total generating capacity is more apparent than the flow rate change. DACA's generating capacity solution is higher than that of POA. DACA's calculation result is more suitable for maintaining high water power generation.

6. Conclusions

In the proposed DACA, the cosine decreasing strategy of the heuristic factor is used to adjust the balance between global searches and local optimization of ACA. The randomness and ergodicity of the chaotic search are used to make chaos disturbancen ant colony search results, and improve the ability to jump out of local optimal algorithm. The case study showed that the dynamic ant colony algorithm has stronger global searching ability and higher convergence stability than that of conventional algorithms.

Acknowledgements

The research work was supported by Open Foundation of Basic Scientific Research Operating Expenses of Central-Level Public Academies and Institutes (CKSF2016040/TB) and National Natural Science Foundation of China under Grant No.51209014.

References

[1] Chen L, Mei Y, Dong Y and Yang N 2008 *Journal of Hydraulic Engineering* **39**(5) 550-56
[2] Zheng J, Yang K, Ni F and Liu G 2013 *Journal of Hydraulic Engineering* **44**(2) 205-11
[3] Ji C, Liu F, Peng Y and Xiang T 2013 *Journal of Hydroelectric Engineering* **32**(1) 70-6
[4] Qin H, Zhou J, Wang G and Zhang Y 2009 *Journal of Hydraulic Engineering* **40**(5) 513-19
[5] Zheng H, Mei Y, Hu T, Huang C and Tang H 2013 *Journal of Hydroelectric Engineering* **32**(1) 54-62
[6] Li L, Liu P, Zhang Z, Zhao J and Shen Y 2013 *China Rural Water and Hydropower* **2** 134-7
[7] Dorigo M, Maniezzo V and Colorni A 1996 *IEEE Transactions on Systems, Man, and Cybernetics – Part B* **26(1)** 29-41
[8] Dorigo M and Gambardella L M 1997 *IEEE Transactions on Evolutionary Computation* **1**(1) 53-56
[9] Wang D, Wang J, Wang H, Zhang R and Guo Z 2007 *Intelligent Optimization Methods* (Beijing: Higher Education Press)

Flood evolution assessment and monitoring using hydrological modelling techniques: analysis of the inundation areas at a regional scale

M Podhoranyi[1, 2], **S Kuchar**[1] **and A Portero**[1]

[1] IT4Innovations, VSB - Technical University of Ostrava, 17. listopadu 15/2172, 708 33 Ostrava, Czech Republic

E-mail: michal.podhoranyi@vsb.cz

Abstract. The primary objective of this study is to present techniques that cover usage of a hydrodynamic model as the main tool for monitoring and assessment of flood events while focusing on modelling of inundation areas. We analyzed the 2010 flood event (14th May – 20th May) that occurred in the Moravian-Silesian region (Czech Republic). Under investigation were four main catchments: Opava, Odra, Olše and Ostravice. Four hydrodynamic models were created and implemented into the Floreon+ platform in order to map inundation areas that arose during the flood event. In order to study the dynamics of the water, we applied an unsteady flow simulation for the entire area (HEC-RAS 4.1). The inundation areas were monitored, evaluated and recorded semi-automatically by means of the Floreon+ platform. We focused on information about the extent and presence of the flood areas. The modeled flooded areas were verified by comparing them with real data from different sources (official reports, aerial photos and hydrological networks). The study confirmed that hydrodynamic modeling is a very useful tool for mapping and monitoring of inundation areas. Overall, our models detected 48 inundation areas during the 2010 flood event.

1. Introduction

The floodplains of rivers play a significant role in hydrological cycles, and an understanding of floodplain hydrodynamic processes is therefore important for assessing and monitoring of water movements within river catchments [18]. Inundation areas may have different patterns and different histories of formation within the floodplains. They are very dynamic and their evolution is often very fast during flood events. Characterization of inundation areas and their dynamics over local geographical regions has been investigated using different hydrological techniques with different degrees of accuracy. In general, these techniques are based on the detection and estimation of the extent of inundation areas. The timing, extent, duration, and inter-annual variability of floodplain inundation are major drivers of floodplain ecosystem function [10].

There are a few major approaches for mapping inundations: (a) application of remote sensing techniques (e.g. microwave and optical satellite sensors – these two types of remote sensors are suitable for inundation detection) [16, 9], or (b) usage of hydrological models (e.g. hydrodynamic models) [13, 1]. Both approaches may provide information on the inundation dynamics of the region [16]. The combination of these approaches has been investigated by using assimilation methods in a few studies [8, 6]. This study is focused on the second approach, and thus, we used hydrodynamic model HEC – RAS ver.4.1 for our experiments. Specifically, we used a combination of hydrodynamic and rainfall-runoff models (HEC-HMS). The common outputs from hydrodynamic models are the spatial and temporal changes of inundated areas [12].

There are many studies that use hydrodynamic models for mapping inundation areas on the global or local scale. Rudorff et al. (2014) [13] used the hydrodynamic model LISFLOOD-FP for studying of the inundation dynamics for the large Amazon river floodplain (area 2440 km^2). Yamazaki et al. (2011) [17] proposed a new global river routing model, CaMa-Flood, that explicitly parameterizes the subgrid-scale topography of a floodplain, thus describing floodplain inundation dynamics. Several studies have used hydrodynamic models to estimate the dynamics of water for small areas based on gauge data [5, 4].

The primary objective of this study is to present techniques that cover usage of hydrodynamic modelling as the main tool for monitoring and assessment of flood event with focus on inundation areas. Data from local water networks and remote sensing data were used for analysis.

2. Study area

The Moravian-Silesian region (MSK) is located in the north-eastern part of the Czech Republic. The geography of the region is varied and contains many different landforms (e.g. the Moravian-Silesian Beskids and the Moravian Gate). Hydrologically, the area is divided into four main catchments: Odra, Ostravice, Olša and Opava catchments (figure 1).

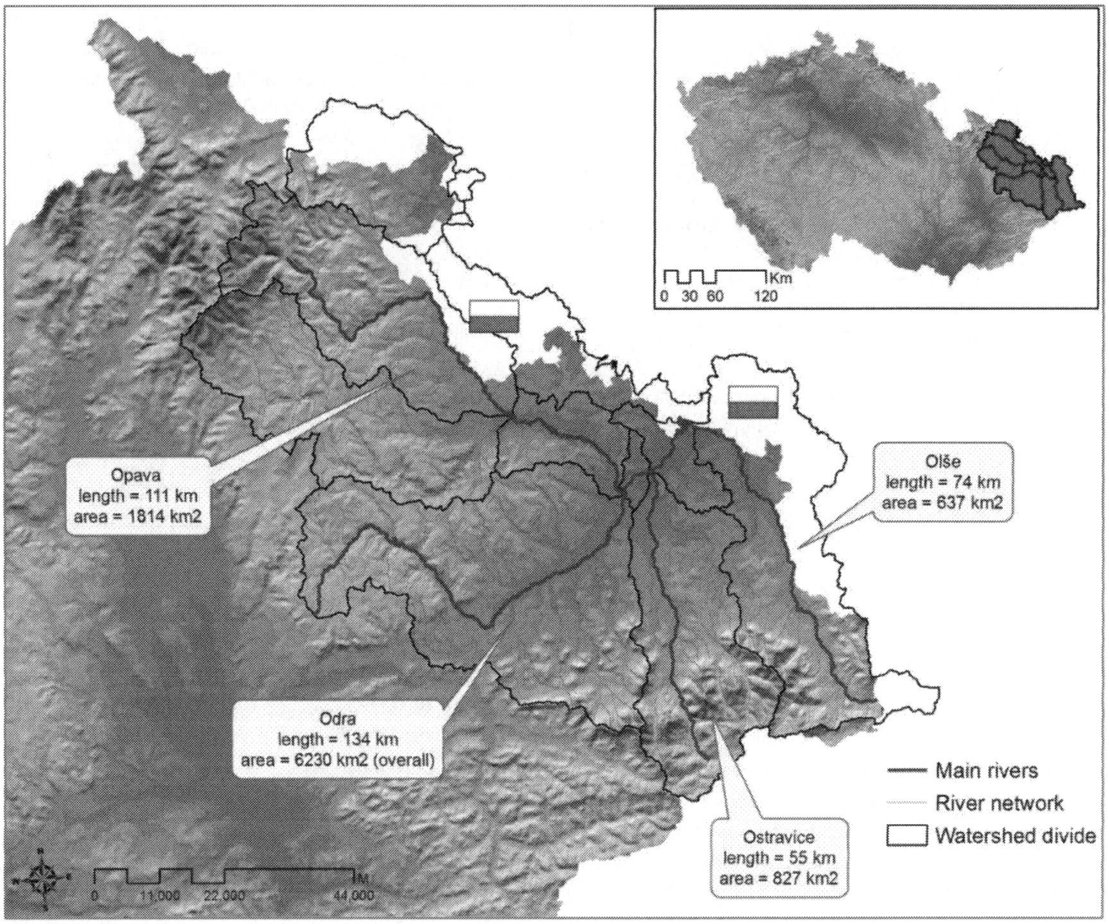

Figure 1. Study area with highlighted major catchments.

2.1. Precipitation situation during flood event

For our experiment, we used data of a flood event that occurred from 14th May to 20th May 2010 in the Moravian-Silesian region. The precipitation situation during the flood event was described by [2]: "More than 100mm fell in the 24 hours to the morning of 17 May in the border region of Poland and the Czech Republic in the Western Beskids, not far from the border with Slovakia. The highest rates were approx. 180 mm (figure 2). The following day brought precipitation averaging about 80mm with a maximum of 115 mm, followed by a further 50 mm next day. At the stations, Lysa Hora (catchment of the river Odra) and Bielsko-Biala (river Vistula) about 320mm was measured during this period. The area of precipitation was small and stationary. Only 22mm fell south of the Western Beskids."

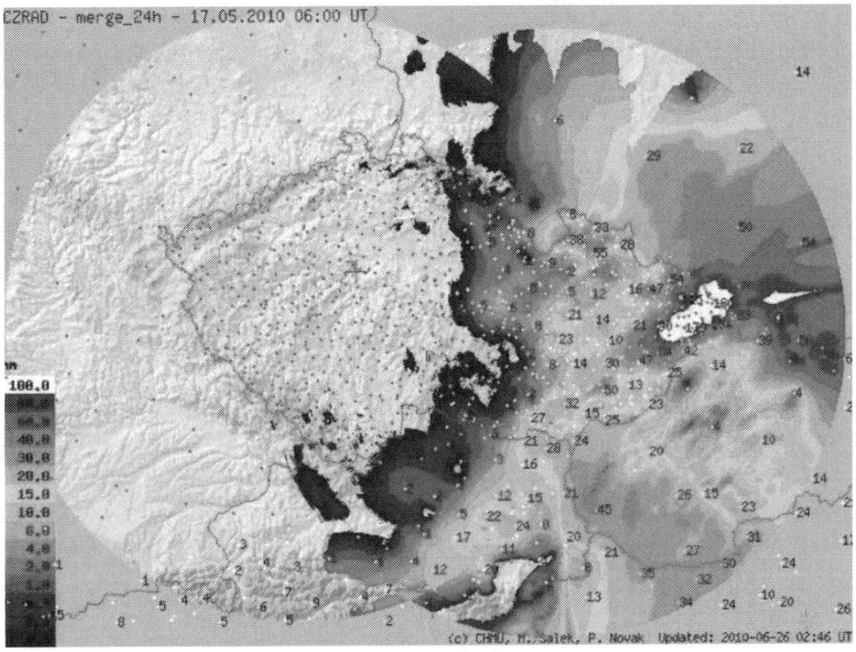

Figure 2. Daily rainfall intensity (from 16th May 8:00 CET to 17th May 2010 8:00 CET) source: CHMI 2010.

3. Methods

3.1. Topographical and hydrological data

The topographic component plays a key role in hydrodynamic flood modelling accuracy. A typical approach used to display topography is the LIDAR [3] or photogrammetry method. The input DEM in this study was created using photogrammetry data from 2010 (resolution 10x10m). The topographical dataset was evaluated by means of RMS (Root Mean Square) error in two parts: (a) floodplain area, and (b) land out of the floodplain. Overall, nearly 250 GCPs (Ground Control Points) were used for evaluating (150 points were placed within the floodplain area). The vertical accuracy of the DEM reached 0.49 m in the floodplain area and 0.75 m out of the floodplain.

Evaluated DEM was used as a background for deriving further data, which are necessary for numerical modelling. Cross-sections were derived from input DEM, and then they were incorporated into the 1D hydrodynamic models. The created cross-sections were compared and completed with real field measurements (approx. 50 cross-sections were geodetically measured). Values of the Manning roughness coefficient were attached to the river channel (range 0.035 – 0.04) and overbank areas (taken from aerial photographs).

Tributaries to the main rivers were recorded via 85 gauging stations. In this research, hourly observations of rainfall discharges water levels and temperatures at the above gauging stations for the period from 14th May to 20th May 2010 were used. The runoff from areas without any water stations was computed by simple rainfall-runoff simulations.

3.2. Hydrodynamic modelling

The calculation of river overflow and inundation areas was achieved by four hydrodynamic models (one model = one catchment). In order to study the dynamics of the water, we applied unsteady flow simulation to the entire area (HEC-RAS 4.1). For unsteady flow, a mechanism was based on Saint-Venant equations. The relation (1) is applied to the continuity equation, whereas relation (2) is applied to dynamics [14].

$$\frac{\partial A}{\partial t}+\frac{\partial Q}{\partial x}=q \tag{1}$$

$$\frac{\partial \mathrm{y}}{\partial \mathrm{x}}+\frac{\partial \mathrm{y}}{\partial \mathrm{x}}\left(\frac{\mathrm{aQ}^2}{\mathrm{A}}\right)+\mathrm{gA}\frac{\partial}{\partial \mathrm{x}}\mathrm{h}+\frac{\mathrm{gQ|Q|}}{\mathrm{C}^2\mathrm{AR}}=0 \tag{2}$$

where: h = lateral inflow, Q = discharge, a = momentum distribution coefficient, C = Chezy resistance coefficient, A = flow area, g = gravitational acceleration, and R = hydraulic or resistance radius.

Boundary conditions were entered to all open ends of the model schematization. The upstream boundary conditions were chosen as flow hydrographs (relationship discharge and time) whilst downstream conditions were entered as a rating curve (relationship discharge and stage). Flow hydrographs and rating curves were constructed from real measurements provided by water stations.

3.3. Rainfall-Runoff modelling

Rainfall-Runoff modelling results were used for ungauged parts of the catchment. At this point, semi-distributed model HEC-HMS was used as an appropriate approach for such a task. The method SCS-CN (Soil Conservation Service) was selected for determining the amount of direct runoff from a rainfall [15]. Basically, the curve number is an empirical parameter that is based on the hydrologic soil group, land use and hydrologic conditions (range of CN was from 30 to 98).

3.4. Platform for monitoring

The hydrological models made in this study were incorporated into the Floreon+ platform [11]. In this case, Floreon+ was mainly used as a tool for semi-automatic monitoring of inundation areas within the study area. The Floreon+ system is developed for monitoring, modelling, prediction and support in disasters, and it is primarily intended for the Moravian - Silesian Region. As it is a web-based platform, its modularity and responsiveness allows easy integration of various thematic sections, regions and data. Its main contribution is at the area of disaster management support, not only the processes but also in the way of simulations. The system is currently comprised of: (1) prediction and monitoring of flooding, (2) modelling of air pollution caused by leakage of dangerous substances and (3) monitoring of real-time traffic situations [11].

4. Results

We analyzed the 2010 flood event (14th May – 20th May) that occurred in the Moravian-Silesian region. Under investigation were four main catchments: Opava, Odra, Olše and Ostravice. Four hydrodynamic models were created and implemented into the Floreon+ platform in order to map the inundation areas that arose during the flood event. However, the hydrodynamic models did not cover the entire domain because the mainly upstream parts of the catchments are not covered by important data (lack of information about channel geometry, no water stations, streams too small for modelling-insufficient DEM resolution).

Inundation areas were monitored, evaluated and recorded semi-automatically by means of the Floreon+ platform. We focused on information about the extent and presence of the flood areas. The modelled data obtained from the semi-automatic process are depicted in table 1 (For detailed information about methods and algorithms for flood areas detection, please visit Floreon+ internet pages – floreon.it4i.eu).

Table 1. Results of semi-automatic process for inundation areas.

Catchment	Modelled area (%)	No. of detected flood areas	First occasion (time/date)	Flood area < 1 km^2	Flood area 1 km^2<x<5 km^2	Flood area > 5 km^2
Ostravice	70	20	01:00/17th May	14	2	4
Odra	70	14	07:00/17th May	9	5	0
Olše	40	8	20:00/16th May	6	1	1
Opava	20	6	08:00/17th May	6	0	0

First, inundation areas were detected after two days of rain in the north-east part of the study area (specifically – Olše and Ostravice catchment). This was expected because the storm was stationary over this part of the region. As an example, we introduced the situation within the inhabited area in the Ostravice catchment (figure 3). In this example is depicted the inundation dynamics every 6 hours during May 17th. This inundation area was confirmed (presence as well as an area of inundation), and thus, in this case, the hydrodynamic model worked well. Fortunately, this area was densely covered by all kinds of hydrological data, and therefore the validation process was fully completed (mainly, aerial images were helpful).

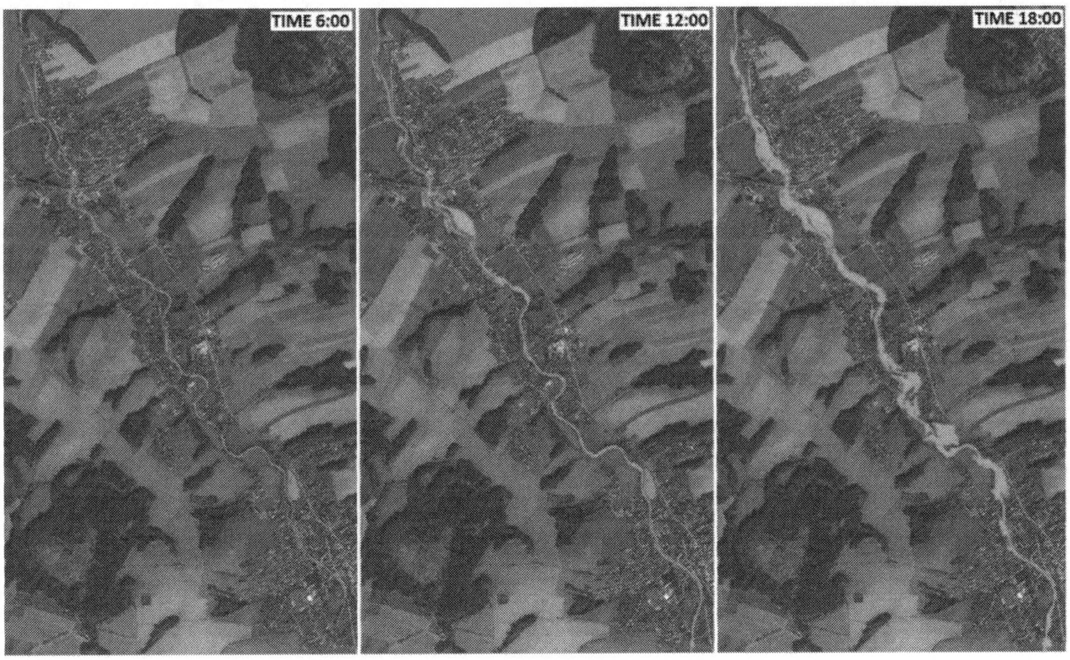

Figure 3. Inundation dynamics (6 hour step, 17th May 2010, part of Ostravice catchment, inhabited area).

A different kind of example is a situation within an uninhabited area. We recorded 44 inundation areas in uninhabited areas, but only 16 of them were confirmed by validation data (36%). In the case of unconfirmed areas, there is a problem to determine if the flooded area was created by real water conditions or it is an error of the hydrodynamic model. In figure 4 is depicted an example of the uninhabited area at the confluence of Odra river with two major tributaries (Lubina and Ondřejnice).

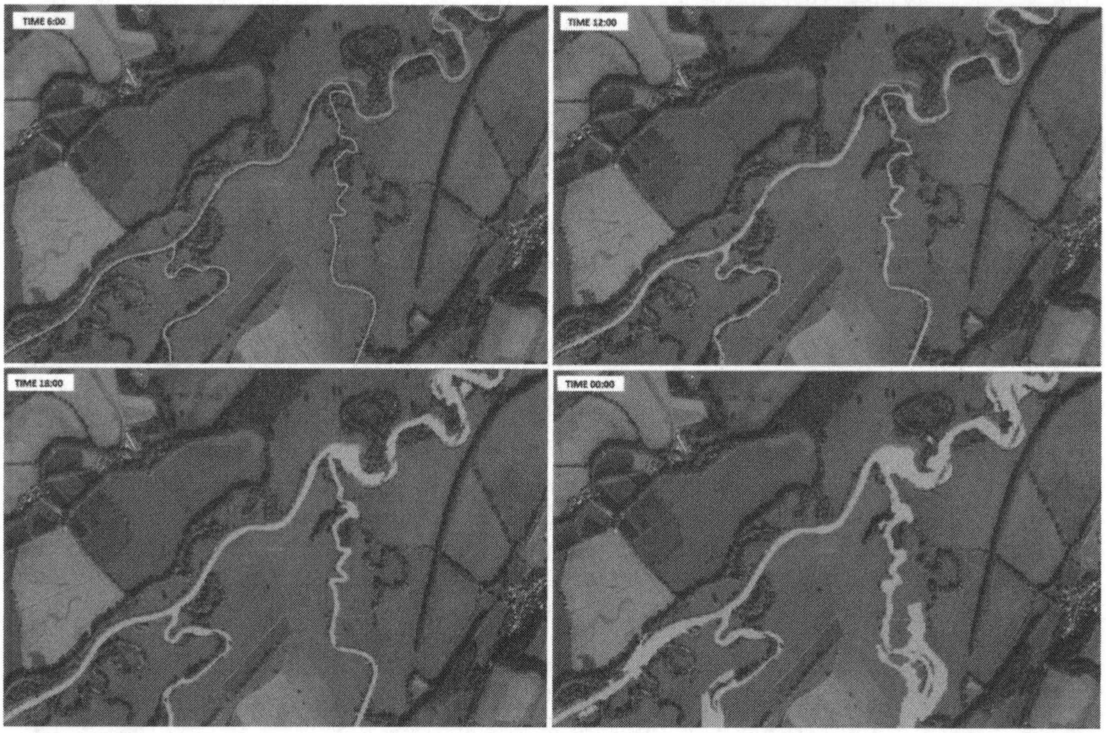

Figure 4. Inundation dynamics (6 hour step, 17th May 2010, part of Odra catchment, uninhabited area).

As a result of the stationarity of the severe storm, all inundation areas bigger than 5 km2 were created within the Olše and Ostravice catchment (north-east part). Partly affected were the Odra and Opava catchments, and thus no inundation areas with such extent were detected. Extremely important is the fact that all inundation areas bigger than 5 km2 were detected and localized correctly.

Flood areas in the MSK region were verified by comparing them with data from different sources (official reports, aerial photos and hydrological networks). However, these sources did not cover the entire study domain and therefore this study suffered from an inability to determine the presence of some inundation areas in uninhabited areas.

5. Conclusion

This study revealed the following conclusions: (1) Hydrodynamic models detected 48 inundations during the 2010 flood event within the MSK region. (2) The presence of 20 areas was confirmed by validation data. (3) 8 were fakes (model detected but no presence in the land), which was confirmed by validation data (4) 10 were placed out of the range of validation data, and therefore we cannot say whether they existed or not. (5) All inundation areas bigger than 5 km^2 were detected and estimated correctly. (6) All fakes were inundations with sizes less than 1 km^2. (7) The first flood area was detected after two days of heavy rains in the Olše catchment.

The presence and size of flood areas were evaluated based on three simple categories in this study. The study confirmed that hydraulic modeling is very useful tool for mapping and monitoring of inundation areas. This approach can bring helpful information about the size, distribution and evolution of flooded areas, and these data may therefore have high value for prevention and protection against floods in the MSK region.

Acknowledgement
This work was supported by The Ministry of Education, Youth and Sports from the National Programme of Sustainability (NPU II) project "IT4Innovations excellence in science - LQ1602".

References

[1] Bates P D, Horritt M S and Fewtrell T J 2010 A simple inertial formulation of the shallow water equations for efficient two-dimensional flood inundation modelling *J. Hydrol.* **387** pp 33–45
[2] Bissolli P, Friedrich K, Rapp J and Ziese M 2011 Flooding in eastern central Europe in May 2010 – reasons, evolution and climatological assessment *Weather* **66** pp 147-53
[3] Casas A, Lane S N, Yu D and Benito G 2010 A method for parameterising roughness and topographic sub-grid scale effects in hydraulic modelling from LiDAR data *Hydrol. Earth Syst. Sc.* **14** pp 1567-79
[4] Dutta D, Alam J, Umeda K, Hayashi M and Hironaka S 2007 A two-dimensional hydrodynamic model for flood inundation simulation: a case study in the lower Mekong river basin *Hydrol. Process.* **21** pp 1223-37
[5] Frank E, Ostan A, Coccato M and Stelling G S 2001 Use of an integrated onedimensional-two dimensional hydraulic modelling approach for flood hazardand risk mapping *Proceedings of 1st International Conference on River Basin Management* Cardiff, Wales
[6] Garcia-Pintado J, Neal J C, Mason D D, Dance S L and Bates P D 2013 Scheduling satellite-based SAR acquisition for sequential assimilation of water level observations into flood modelling *J. Hydrol.* **495** pp 252-66
[7] Gibbs M S, Clarke K and Taylor B 2016 Linking spatial inundation indicators and hydrological modelling to improve assessment of inundation extent *Ecol. Indic.* **60** pp 1298-308
[8] Giustarini L, Matgen P, Hostache R, Montanari M, Plaza D, Pauwels V R N, DeLannoy G J M, Keyser R, Pfister L, Hoffmann L and Savenije H H G 2011 Assimi-lating SAR-derived water level data into a hydraulic model: a case study *Hydrol.Earth Syst. Sci.* **15** pp 2349–65
[9] Huang C, Chen Y and Wu J 2014 Mapping spatio-temporal flood inundation dynamics at large river basin scale using time-series flow data and MODIS imagery *Int. J. Appl. Earth Obs.* **26** pp 350-62
[10] Junk W J and Wantzen K M 2004 The flood pulse concept: new aspects, approaches and applications *2nd International Symposium on the Management of Large Rivers for Fisheries* Bangkok, Thailand: FAO & Mekong River Commission
[11] Kuchar S, Golasowski M, Vavrik R, Podhoranyi M , Sir B and Martinovic J 2015 Using high performance computing for online flood monitoring and prediction *International Journal of Environmental, Chemical, Ecological, Geological and Geophysical Engineering* **9** pp 415-20
[12] Li Y, Zhang Q, Yao J, Werner A D and Li X 2014 Hydrodynamic and hydrological modeling of the Poyang Lake catchment system in China *J. Hydrol. Eng.* **19** pp 607–16
[13] Rudorff C M, Melack J M and Bates P D 2014 Flooding dynamics on the lower Amazon floodplain: 1. Hydraulic controls on water elevation, inundation extent, and river–floodplain discharge *Water Resour. Res.* **50** pp 1–16
[14] USACE 2010 *US Army Corps of Engineers., Hydrologic Engineering Center, HEC-*

IOP Conf. Series: Earth and Environmental Science **39** (2016) 012043 doi:10.1088/1755-1315/39/1/012043

RAS, River Analysis System. Hydraulic Reference manual, version 4.1

[15] USACE 2009 *US Army Corps of Engineers, Hydrologic Engineering Center, HEC-HMS, Hydrologic Modeling System HEC-HMS. User's manual, version 3.4*

[16] Ward D P, Petty A, Setterfield S A, Douglas M M, Ferdinands K, Hamilton S K and Phinn S 2014 Floodplain inundation and vegetation dynamics in the Alligator Rivers region (Kakadu) of northern Australia assessed using optical and radar remote sensing *Remote Sens. Environ.* **147** pp 43-55

[17] Yamazaki D, Kanae S, Kim H and Oki T 2011 A physically based description of floodplain inundation dynamics in a global river routing model *Water Resour.Res.* **47** W04501

[18] Zhang Q and Werner A D 2015 Hysteretic relationships in inundation dynamics for a large lake-floodplain system *J. Hydrol.* **527** pp 160-71

A method to evaluate coordination between regional economic, social development and water resources

S B Zhou[1, 2, 3, 4], W T Qi [1], A M Du[1] and H He[1]

[1] Water Problem Institute, Hohai University, Nanjing, 211100, China
[2] School of Management and Economics, Tianjin University, Tianjin, 300072, China
[3] Global Environment and Natural Resource Institute, George Mason University, US

E-mail: shenbei@hhu.edu.cn

Abstract. Coordination between regional economic, social development and water resources is the key factor for the sustainable development of regions. Scientific evaluation of the coordination and analysis of similar reasons will improve the management level of decision-makers. The Coupling Coordination Degree model (CCD) developed on synergistic theory is now considered as a better method to evaluate coordination between systems. But, there are still some deficiencies. This paper attempts to improve the method in two aspects,: (1) introduce Full Permutation Polygon Synthesis Illustration method (FPPSI) to replace the two key steps of the present CCD model. To realize the data standardization and the comprehensive evaluation of system state, and to achieve the analysis of corresponding reasons. And (2) calculate the coupling coordination degrees of systems' evolution speeds instead of comprehensive evaluation indexes, which will fully reflect the dynamic interaction between systems. To verify the feasibility of the method, Taihu Basin is taken as a case study. Results demonstrate that the improved CCD model is not only able to reflect the dynamic interaction between systems adequately, but also visually presents the specific reasons through geometrical illustration.

1. Introduction

Nowadays, with the rapid industrialization and urbanization in developing countries, the problems of environmental pollution and resource depletion exacerbate. How to coordinate social, economic development, and environmental resources protection becomes a thorny issue for decision-makers in local governance. Among various natural resources, water resource is considered as particularly important. Whether the survival of organisms or the industrial production, they are inseparable from the water. Therefore, the coordination

IOP Conf. Series: Earth and Environmental Science 39 (2016) 012044 doi:10.1088/1755-1315/39/1/012044

between social, economic development and water resource becomes the priority for governments.

Water resources system is a subsystem of the environmental resources system, so we can look for methods from the relevant theories or approaches which are adopted in the economic and ecological resources relationship studies to explore the coordination between economic development and water resources. For such kind of study, the traditional theory is Environmental Kuznets Curve (EKC) first proposed by Grossman and Krueger in 1992. The EKC argues that the relationship between the economic development and environment quality shows improvement after deterioration like the inverted letter "U." Based on this hypothesis, many scholars carry out numerous empirical research [3, 12, 18] and various improvements are made on research methods [1, 17] .Some apply EKC in the relationship between the economic development and water resources [2, 21]

Meanwhile, with the elaboration of the synergy theory [6, 8], more and more scholars try to adopt this approach in the economic and environmental resources relationship studies. In the synergy theory, Hermann Haken believes there are many different systems in natural [7]. Although they are different in the attributes, in the whole environment, the systems interact and promote, and may change between steadiness and unsteadiness. This view offers new ways to study the relationship between the economic development and environmental resources. Based on the synergy theory, some scholars draw on the capacitive coupling concept and model in physics. Developing the coupling degree template, evaluating the interactive relationship of systems or elements and apply, to the study of relationship between economic, social development and resource environment changes in different areas [4, 10, 24]. The relevant results show the coupling degree model not only can explain the interactive relationship but also can indicate the tendency of the system, which is of great significance for the regulatory of systems [20]. At the same time, some scholars point out the coupling degree is necessary for judgment on the weakness or strength of the interactions, but dependence on coupling degree may lead to the fault evaluation [15, 23]. For example, when the integrated order parameter values are close and both low in the two subsystems, the coupling degree index may regard wrongly that this state is better than others. To avoid this problem, they import the comprehensive coordinating index to improve this model, called Coupling Coordination Degree model (CCD).

Compared with the EKC and its related econometric methods, the CCD model more fits in this study. However, there are still some deficiencies in the model. For example, the model takes the geometric mean method or linear weighted sum method [14, 23], or the principle component analysis method [9, 25] in the system comprehensive evaluation index calculation. To reflect the whole states of systems, but they cannot intuitively and precisely indicate the similar reasons why the system has become the state, which makes the model stay in the shallow evaluation.

Therefore, this paper tries to make some improvements on the present CCD model. And it aims to provide a more comprehensive and scientific method for the decision-makers to evaluate the coordination between regional economic, social development and water resources and analyze the similar reasons, so that they can make appropriate policy adjustments and ultimately promote local sustainable development.

2. Improvement of Coupling Coordination Degree Model

2.1. Present Coupling Coordination Degree Model

For decades, with the development of system science and the synergy theory, the academic field has seen many studies on interaction relationship among two or more systems, with the development of network coupling theory and related models being representative. Coupling, originally a concept in physics, refers to the phenomenon that two or more systems or movements influence each other and may get together through the interaction, in which a dynamic and mutually reinforcing relationship would form under the benign interaction among the subsystems [22]. After that system coupling is introduced into the ecological environment, regional economic, social development and other fields [11]. Models to evaluate the interaction relationship between systems are also established and improved continuously in a wide application [9, 14, 23, 25]. So far, the CCD model can be regarded as a suitable method in the study of interaction relationship between systems, which comprises the following steps:

- Construct indicator system- to construct the relationship evaluation indicator system, the indicators are required to reflect the running states of the systems and the relationship among the systems, so as to effectively evaluate the coordination relationship.
- Set up efficacy function- set that system *Si* has got the parameter *ei=(ei1, ei2,…,ein)*, *n*≥1 and the value of the *eij* is no less than *αij* and no more than *βij*, the efficacy function *ui(eij)* is no less than 0 and no more than 1. Due to its ambiguity, it can determine the specific form of the function according to the method of membership degree in fuzzy mathematics. Efficacy function is divided into three types as follows according to the contribution of the parameter to the system:

Positive efficacy type: $$u_i(e_{ij}) = \begin{cases} 1, & x_{ij} < \alpha_{ij} \\ (x_{ij} - \beta_{ij})/(\beta_{ij} - \alpha_{ij}), & \alpha_{ij} \le x_{ij} \le \beta_{ij} \\ 0 & x_{ij} > \beta_{ij} \end{cases}$$

Negative efficacy type: $$u_i(e_{ij}) = \begin{cases} 1, & x_{ij} < \alpha_{ij} \\ (\beta_{ij} - x_{ij})/(\beta_{ij} - \alpha_{ij}), & \alpha_{ij} \le x_{ij} \le \beta_{ij} \\ 0 & x_{ij} > \beta_{ij} \end{cases}$$

Moderate efficacy type: $$u_i(e_{ij}) = \begin{cases} 0, & x_{ij} < \alpha_{ij} \\ (x_{ij} - \alpha_{ij})/(c_{ij} - \alpha_{ij}), & \alpha_{ij} \le x_{ij} < c_{ij} \\ 1, & c_{ij} \le x_{ij} < d_{ij} \\ (\beta_{ij} - x_{ij})/(\beta_{ij} - \alpha_{ij}), & d_{ij} \le x_{ij} < \beta_{ij} \\ 0 & x_{ij} \ge \beta_{ij} \end{cases}:$$

In the above functions, *eij* represents N.O.*j* indicator in the system *Si*, and *xij* represents the value of N.O.*j* indicator in the system *Si*.

- Calculate comprehensive evaluation index. Comprehensive evaluation index calculates the system parameters' "total contribution" to the system development. It can be achieved by the integrated method, which adopts the geometric mean method and weighted linear method in practical application:

$$U_i(u_i)=(\prod_{j=1}^{n}u_{ij})^{1/n}=\sum_{j=1}^{n}\lambda_{ij}u_{ij},0\le\lambda_{ij}\text{且}\sum_{j=1}^{n}\lambda_{ij}=1 \tag{1}$$

In (1), *Ui*(*ui*) is the comprehensive evaluation index of the system *Si*, and *λij* is the weight of N.O.*j* parameter in the system *Si*.

- Construct coupling coordination function. By the use of the capacitive coupling concept and the coupling coefficient model in physics for reference, the mutual system interaction coupling model could be set as following.

$$C_m=\left\{(U_1(u_1)\cdot U_2(u_2)\cdot\cdots\cdot U_m(u_m))/(\prod(U_i(u_i)+U_j(u_j))\right\}^{1/m},\quad (i=1,2,\cdots,m,j=1,2,\cdots,m,i\neq j) \tag{2}$$

Based on (2), it imports the comprehensive coordination index *T*, to found the CCD model.

$$T=\sum_{i=1}^{m}\delta_i U_i(u_i) \tag{3}$$

$$D=\sqrt{C_m\cdot T} \tag{4}$$

In (3) and (4), *δi* represents the weight of the system *Si*, and *D* represents the coupling coordination degree.

2.2. Improved Coupling Coordination Degree Model

As described in the introduction part, there are still some shortcomings in the present CCD model. This study attempts to improve the method in two aspects, (1) introduce the Full Permutation Polygon Synthesis Illustration method (FPPSI) to replace the two key steps of the present CCD model, realizing the data standardization, the comprehensive evaluation of system state, also and the analysis of similar reasons. And (2) calculate the coupling coordination degrees of systems' evolution speeds instead of comprehensive evaluation indexes, which will adequately reflect the dynamic interaction between systems.

The FPPSI method has been proposed by Wu et al. when he studies the indices and the evaluation process of eco-city [22]. When used in evaluation, the FPPSI method has the following advantages over other methods, (1) it provides not only precise algebraic values, but also a geometrically intuitive visualization of the status of each indicator. And (2) it alters the traditional additive approach to combining indicators by using a multidimensional approach that more appropriately reflects the integrative system principle, that the whole is more than the sum of its parts [11]. So, it is now widely used in the comprehensive evaluation of systems, especially in the social, economic system and environmental resources system [5, 19]. Although the method needs the experts to determine the critical value of indicators at first, it does not completely get rid of personal intervention, and it has been under continuous improvement in the following promotion and application [13, 11].

The basic idea of FPPSI is to first to set up standardized indicators,, and then constitute a central n-gon with the upper limit of the indicator as its radius and an irregular n-gon with the value of each index which is a full array of the indicators by the end to end. The total can form (n-1)!/2 different irregular polygons and the comprehensive index defined as the ratio of the mean area of all these irregular polygons and the area of the n-gon.

The indicator value is standardized by (5).

$$F(x) = \frac{a(x+b)}{x+c}, a \neq 0, x \geq 0 \tag{5}$$

$F(x)$ Satisfy: $F(L) = -1, F(T) = 0, F(U) = 1$

In (5), L, U, and T represent the lower limit, the upper limit, and the threshold for parameter x respectively. Thus:

$$F(x) = \frac{(U-L)(x-T)}{(U+L-2T)x+(U+L)T-2UL} \tag{6}$$

From (6), standardized function F(x) value is no less than -1 and no more than 1 when reflecting the value of the indicators from L to U. And the reflection of the increasing speed changes along the indicator value. When the indicator value is less/more than the threshold value, the growth rate of the standardized indicator will decrease/increase gradually. And the threshold value is the turning point of the growth rate of the indicator value.

Therefore, as for the indicator xi, the standardization is:

$$S_i = \frac{(U_i - L_i)(x_i - T_i)}{(U_i + L_i - 2T_i)x_i + (U_i + L_i)T_i - 2U_iL_i} \tag{7}$$

In (7), Li, Ui, and Ti represent the minimum value, maximal value and threshold value for parameter xi respectively.

A regular n-sided polygon can form by n indicators where the vertices represent Si=1, the central point represents Si=-1, and the radius from each vertex to the central point represents the value of the corresponding standardized indicator. An inner polygon, lying midway between the outer polygon and the center of the polygon, accounts for the threshold of the indicators, where Si=0 (xi= T). When inside the inner polygon, the values of the standardized indicators are negative and less than their thresholds. And when outside the inner polygon, the values are positive and greater than their threshold values [11, 19].

The number of triangles which form by the polygon center and n indicators is:

$$n(n-1)/2$$

The sum of the square is:

$$0.5\sin(\frac{2\pi}{n})\sum_{i\neq j}(S_i+1)(S_j+1)$$

Where Si is the standardized value of indicator xi, and Si+1 represents the distance from the normalized value of indicator xi to the polygon center.

When there are (n-1)!/2 indicators, there are n(n-1)!/2=n!/2 triangles, and the sum of the square is:

$$0.5\sin(\frac{2\pi}{n})\sum_{i\neq j}(S_i+1)(S_j+1)\cdot\frac{n!}{2}\cdot\frac{2}{n(n-1)}$$

Correspondingly, the sum square of $(n-1)!/2$ regular center polygons (side length=2) is:

$$0.5\sin(\frac{2\pi}{n})\cdot 4\cdot n\cdot\frac{(n-1)!}{2}$$

The ratio of the two value is the full permutation polygon comprehensive index

$$S=\frac{\sum_{i\neq j}^{i,j}(S_i+1)(S_j+1)}{2\cdot n\cdot(n-1)} \tag{8}$$

Where S represents the comprehensive index.

From the above description, it can be seen that the data standardization method in FPPSI has the same function with the efficacy function in the present CCD model; the comprehensive index may also realize the overall evaluation of the development status of the system. Thus it can adequately replace two key steps (efficacy function construction and comprehensive evaluation index calculation) in the present CCD model. Moreover, FPPSI not only has algebraic analytics, but also can make a geometrical representation. Researchers can simply find the specific indicators promoting or hindering the system development through the geometrical illustration, and that is why the system shows the present status. Therefore, compared with the current methods, the introduction of FPPSI can overcome the problems of the present methods which just stays on the surface of system evaluation, having no way to do a deeper analysis.

Also, perhaps affected by the present related research, the thinking of most scholars is limited. When they are to measure the relationship between the systems, comprehensive development level always first comes to their mind. Most scholars rarely focus on the system evolution speed. It is thought that the relationship of the systems' evolution rates may more effectively reflect the actual dynamic interaction between systems than the overall development level. Therefore, this paper puts forward another improvement to the CCD model, which is to study the system coordinated relationship based on the system evolution speed rather than the system comprehensive development level. Of course, the system development speed may be negative; then the relationship can be divided into three categories: a positive growth of both, negative growth of both and anisotropic growth (one is in negative growth; the other is in positive growth). Thus it will help researchers study the coordinated development relationship of systems more comprehensively.

3. Case Study

To verify the feasibility of the method, Taihu Basin is taken as a case study. As one of the most developed areas in China, Taihu Basin produces 21.0% national GDP with only 2.2% square and 11.65% population in 2012. With the increasing population, the rapid development of economic, industrialization and urbanization, the pressure on the water resource and environment is increasing. Thus environmental water issues have become so prominent that it has even affected the safety of the drinking water around the basin. So it is significant to coordinate the economic-social development and water-resource development for the healthy and sustainable development of the basin.

3.1. Construction of Indicator System

The construction of an appropriate evaluation indicator system is necessary for the reasonable evaluation of the coordination relationship between social, economic development and water resource. Under the principals of science, comprehensiveness, objectivity, independence, representation, and data availability, is combined with the specific circumstances of Taihu Basin and the purpose of this research. Taking the related research results for reference [13, 16, 20] , it constructs two aspects of the indicator system to reflect the state of social, economic development and water resources development in Taihu Basin (Table 1).

Table 1. dicator system of the social, economic system and the water resources system.

Type	Item	Sub-item
Social economic system	Economic development	Per capita GDP
		Per capita fiscal revenue
		Proportion of tertiary industry
	Social development	Engel coefficient
		Level of education
		Population density
		Per capita fiscal expenditure
		Level of urbanization
		Urban-rural gap
		Urban greening construction level
		Urban greening facility level
Water resources system	Water resources endowments	Total amount of water resources
		Lake water quality
		River water quality
	Water supply	Total water supply
		Total urban water supply
	Water resources protection	Rate of water resources development and utilization
		Proportion of investment in soil conservation and ecological restoration
		Ratio of the eco-water usage
		Investment in the treatment of industrial waste water
		Total industrial waste water discharged
		Qualified rate of the industrial waste water discharged
		Length of the city sewage pipes
		Daily disposal capacity of city sewage
		Non-production and non-living water proportion in cities

Note: The above indicator system mainly takes city related indicators for the comprehensive considerations of national urbanization construction background, transference from farmland water construction to city water construction, urbanization in Taihu Basin, current high-level industrialization, future development demand, and statistical indicator settings in relevant statistical yearbooks, etc.

3.2. Data Sources

The original data mostly comes from the 2001-2013 yearbooks, such as "Chinese Statistical Yearbook", "Chinese Environment Statistic", "China water resources Yearbook" and "Water resources bulletin of the Taihu Basin and southeastern rivers. Also from "Jiangsu Statistical Yearbook", "Zhejiang Statistical Yearbook", "Shanghai Statistical Yearbook", "China City Statistical Yearbook", "China Environmental Statistics Yearbook" and so on. Calculation indirectly gets some statistics and some individual years' data gets from the result of the adjacent year with the value interpolation function.

Note: geographically, Taihu Basin scatters 53% in Jiangsu province, 33.4% in Zhejiang province, 13.5% in Shanghai City, 0.1% in Anhui province. And it has been jointly governed by the governments and relevant departments of Jiangsu, Zhejiang and Shanghai. Province in China performs as the local governance unit. And data availability, statistical consistency, and data of water quality index, etc. Are taken from the lakes and rivers of Taihu Basin, while the other index data are integrated from Jiangsu, Zhejiang and Shanghai. Therefore, the Taihu Basin in this paper refers to the pan Taihu Basin.

3.3. Coupling Coordination Degree Calculation

To calculate the coupling coordination degree, the first step is to apply FPPSI to realize the data standardization and the comprehensive development index calculation.

It should be noted that, as introduced earlier, efficacy function can be divided into three types: positive efficacy, adverse efficiency, and moderate efficacy. The standardization of positive efficacy and adverse efficiency indicators are fairly straightforward,, which only requires the setting of the upper limit, lower limit, and threshold of each indicator, while that of moderate efficacy indicators are relatively complex which sees appropriate interval as the premise. As a result, many indicators are not merely positive or negative in the long run, but in the short term or according to the development present, the demand, the particular numerical situation, they can be seen as indicators of only positive or negative efficacy. Take Taihu nutrition indicator as an example. By the relevant standards, if the nutrition indicator is less than 20, then the water is poor nutrient and not conducive to the development of water resources system. If between 20 and 50, the water is moderate nutrient and supportive to the elaboration of the water resources system; if between 50 and 60, the water is mild eutrophication and not conducive to the development of water resources system. Obviously, the indicator is a moderate indicator. However, the actual data of 2001-2012 shows that the values of this indicator are all higher than 50, which indicates a negative efficacy indicator.

The application process of FPPSI is as follows:

- Calculate the parameters in equation (7) according to the processed data. And the min value, max value and average value of xi are used to replace Li, Ui, and Ti respectively, which can not only simplify the calculation method but also cut down the subjectivity to some extent.
- Standardize the parameters in equation (7). If the indicator is positive, use (7); if the indicator is negative, use the opposite of (7), that is $-Si$.

- Calculate the system comprehensive evaluation index *Uij* of the different period in equation (8) with the standardized parameters. After calculating *Uij*, calculate the evolution speed *Vij* with relative link method as (9).

$$V_{ij} = \frac{U_{ij} - U_{i(j-1)}}{U_{i(j-1)}} \quad (9)$$

Evaluate the coordination relationship of the evolution speed *Vij*.

$$C_j = (-1)^k \sqrt{\{(|V_{1j}| \cdot |V_{2j}|) / \prod (|V_{1j}| + |V_{2j}|)\}} \quad (10)$$

$$T_j = \delta_1 V_{1j} + \delta_2 V_{2j} \quad (11)$$

$$D_j = (-1)^k \sqrt{|C_j| \cdot T_j} \quad (12)$$

In (10) and (12), k represents the judgment parameter of the system coordination direction. If the two systems have the same developing direction, k=2; If the two systems have different direction developing direction, k=1. In (11), it is regarded that social, economic system development is as equally important as the water resources system development, thus δ1=δ2=0.5. Dj represents the coupling coordination degree of the social system evolution speed and the water resources system evolution speed in j period.

3.4. Results and Analysis

With the calculation method described in the previous section, the development rates of the social, economic system and water resources system in Taihu Basin from 2001 to 2014 are calculated and shown in Figure 1.

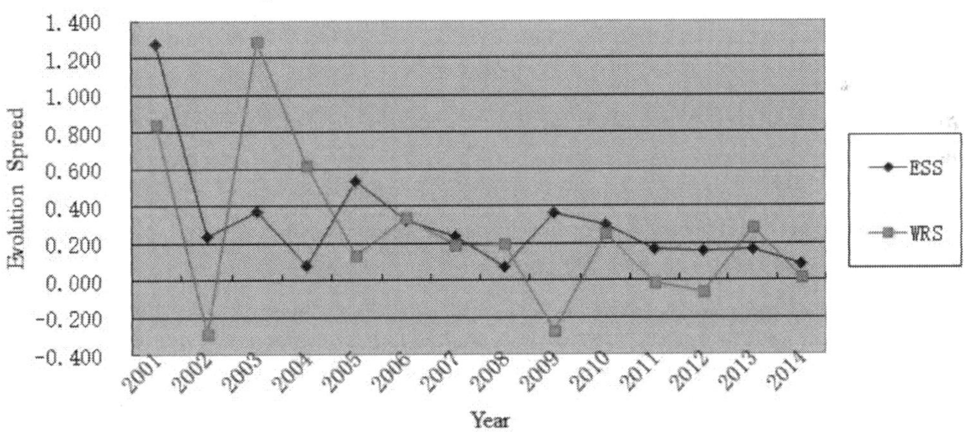

Figure 1. Evolution speed of the social, economic system (SES) and the water resources system (WRS) in Taihu Basin.

Figure 1 shows the fluctuation of evolution speed of water resources system in Taihu Basin in the 2001-2014 period, and some individual years are even with negative growth, (2002, 2009, 2011， 2012). Economic, social system fluctuates more gently since 2002 and keeps in growth. During most years, evolution speed and direction of the two systems are consistent, but in some individual years, they are inconsistent (2005, 2009, 2010); in 2001, 2002, 2005, 2007, 2009, 2010, 2011、2012、2014. Growth rate of the economic social system are greater than water resources system, and in 2003, 2004, 2006, 2008, 2013, growth rate of water resources system is higher than the economic social system. After an outbreak of blue-green algae in Taihu Lake in 2007, the water resource system growth begins to decline and appears negative growth in 2009 which shows the malpractice treatment after pollution.

Generally speaking, the change trend of the social, economic system is more stable than that of water resource system, because the adjustments effect of government on the economic, social system is more visible, while water resources system reacts slower to government adjustments.

With the improved CCD model, the coupling coordination degrees of the economic, social system and water resources system in Taihu Basin in the 2001-2014 period are calculated and shown in Figure 2.

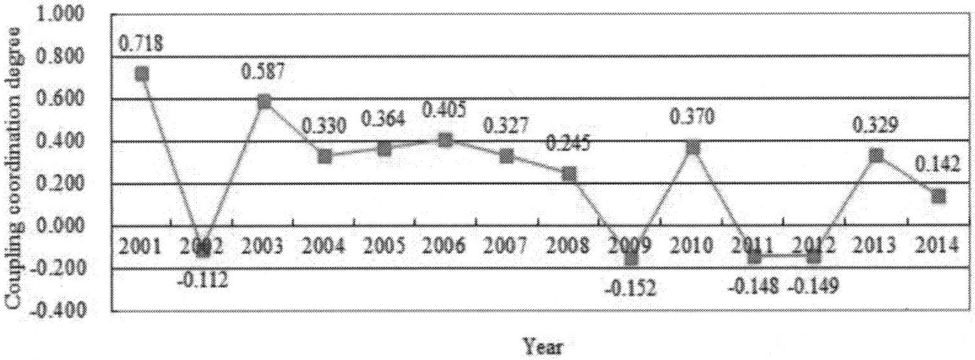

Figure 2. Coordination between the social, economic system and the water resources system in Taihu Basin.

According to Figure 2, in the 2001-2014 period, the coordination relations between the social, economic system and water resources system in Taihu Basin fluctuates. In most years, the two systems maintain a good state of positive development, but positive relationship reduces in the overall downward trend from 0.718 in 2001 to 0.142 in 2014. The two systems' coupling coordination degree is negative in 2002, 2009, 2011, 2012, but the absolute value is about 0.2 (not very serious), which means the two systems deviate from each other in the corresponding periods. Overall economic, social development has not promoted water resource system, and on the contrary, it has created a particular restriction.

Secondly, the coupling coordination degree has a slight fluctuation, and the fluctuation is not in an ideal range. The main reason is that the development of water resource system is not good, and fails to synchronize with that of the economic, social system.

Thirdly, from the perspective of annual sequence, in water resources system, the score items of the social side, such as daily sewage treatment capacity of cities, the total length of city's waterways, urban water supply, and industrial wastewater treatment success rate etc., are superior to those of the natural side,.Such as the proportion of ecological water use, water conservation, ecological investment proportion, and industrial wastewater emissions, etc. Comparatively, the score items in the economic, social system seem more balanced.

In years of 2002, 2009, 2011, 2012, coupling coordination degree of the economic, social system and water resources system in Taihu Basin is negative mainly because of the negative growth of economic social system in the corresponding period. Why does water resources system appear negative growth? This study would find the relevant reasons by applying FPPSI.

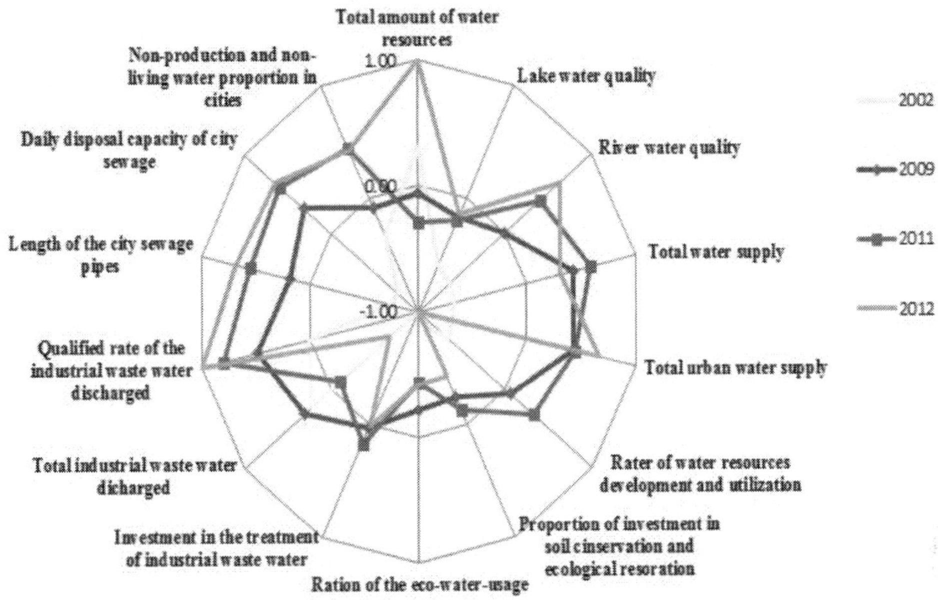

Figure 3. Comprehensive assessment of the water resources system in the water resources system negative development year.

According to Figure 3, in 2002, 2009, 2011, 2012 the main reasons for water resource system's negative development in Taihu Basin is the reduction of the efficiency value of the total amount of water resource. Lake water quality, total industrial wastewater discharged, proportion of investment in soil conservation and ecological restoration, and ratio of the eco-water usage, which have hindered the water resources system positive development (the corresponding indicator efficiency values shrink from the periphery to the center point). At the same time, it could be found that, the efficiency values of river water quality, total water supply, rate of water resources development and utilization, total urban water supply, investment in the treatment of industrial waste water, qualified rate of the industrial waste water discharged, length of the city sewage pipes, daily disposal capacity of city sewage, non-production and non-living water proportion in cities show growth trend (the corresponding indicator efficiency values expand from the center point to the periphery), which means these indicators do not hinder the development of water resources system. It indicates that relevant departments in Taihu Basin have not ignored the protection of water resource under the development, industrialization and city construction, but the overall planning of the protection, development and utilization of water resources are still inadequate, especially in soil and water conservation and ecological construction investment.

IOP Conf. Series: Earth and Environmental Science 39 (2016) 012044 doi:10.1088/1755-1315/39/1/012044

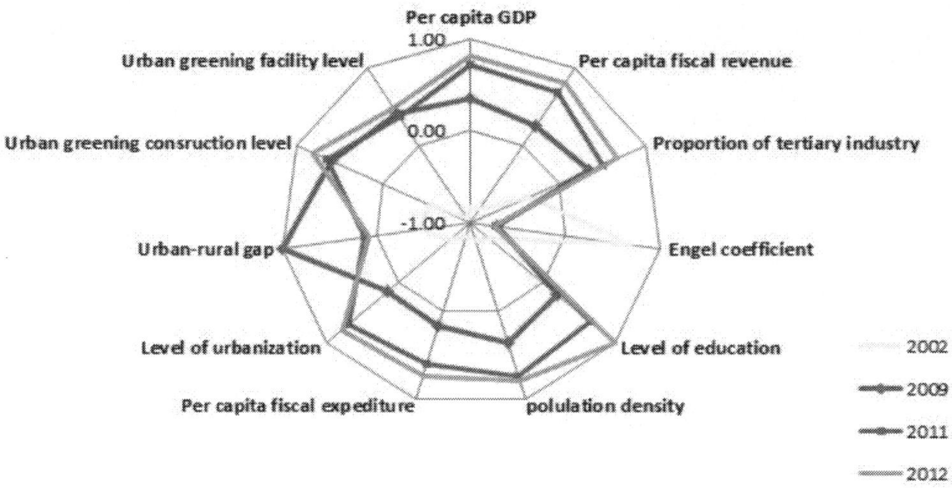

Figure 4. Comprehensive assessment of the social, economic system in the water resources system negative development year.

Figure 4 shows, in 2002, 2009, 2011, 2012, the economic social system in Taihu Basin keeps on positive growth. Efficiency values of most indicators increase (the corresponding indicator efficiency values expand from the center point to the periphery), promoting the positive development of social, economic system. However, there are also some factors hindering the development such as the gap between urban and rural areas, and Engel coefficient effect on the system (the corresponding indicator efficiency values shrink from the periphery to the center point).

From the analysis above, it is not difficult to find that there is a particular relationship between the negative growth of the water resource system and the development of the social, economic system in Taihu Basin. For example, in the corresponding periods, the efficiency value of the rate of water resources development and utilization is lower, which is the main hindering factor to the development of water supply system. And the efficiency value of the population density is low too, indicating that the increase in the population density in the economic and social system makes the rate of water resource development and utilization increase, and restricts the positive development of water resources system to some extent.

According to the results and analysis above, relevant departments can have a particular understanding of the situation. The tendency and the reasons of coordination between the regional economic, social development and water resources in Taihu Basin taking corresponding measures, increasing the proportion of ecological investment and ecological water using. Reducing the rate of water resource development and utilization through the introduction to advanced water-saving technologies, and redesigning the overall plan of water conservancy construction investment.

4. Conclusions

This paper integrates Full Permutation Polygon Synthesis Illustration method (FPPSI) and the presents coupling degree coordination model (CCD). Evaluating the coordination between systems, effectively improving the current methods of shallow evaluation to an in-depth outlook of the interaction, and uniquely illustrating the assessment results more intuitive. In addition, another improvement in this paper, calculating the coupling coordination degrees of systems' evolution speeds instead of comprehensive evaluation indexes, can adequately reflect the dynamic interaction between systems.

To verify the feasibility of the method, we apply the improved CCD model to evaluate the coordination between social, economic system and water resource system in Taihu Basin. From the evolution speed tendency analysis, it can find that whether the development direction and development speed of the two systems are consistent or not. From the coupling coordination degree calculation of the two systems' evolution speeds, it can find whether the coordination between the two systems is positive or not. From the influencing factor analysis by FPPSI, it can see the reasons why the coordination between the two systems presents above state. According to the evaluation results, decision-makers will have a clear and precise understanding of the situation, tendency and reasons of coordination between the regional economic, social development and water resources and take corresponding measures.

Therefore, the improved CCD model introduced in this paper provides a better method for the decision-makers to evaluate the coordination between regional economic, social development and water resources and to analyze the similar reasons, which will help them make appropriate policy adjustments to promote the local efficient and sustainable development.

References

[1] Agras J and Chapman D 1999 A dynamic approach to the environmental Kuznets Curve hypothesis *Ecological Economics* **28** (2): 267-77

[2] Cheng X 2012 Study on Relationship between water quality change and economic development of Taihu Lake *Environment and Sustainable Development* **5**: 73-7

[3] Dinda S 2004 Environmental Kuznets curve hypothesis: A survey *Ecological Economics* **49**: 431-55

[4] Gong S Y 2013 Study on the coupled coordination degrees between urban land intensive use system and social economic development--a case of Shandong Province *Advanced Materials Research* **726-731**(19): 4827-30

[5] Gong Y B, Zhang J G and Liang X C 2011Water quality assessment of East Route based on entire-array-polygon evaluation method *China Population, Resources and Environment* **9**: 26-31

[6] Haken H 1973 *Synergetics: cooperative phenomena in multi-component systems* Wiesbaden: Vieweg+Teubner Verlag

[7] Haken H 1981 *Synergetics-mysteries of the nature constitution* Ling FH Translation Shanghai: Shanghai Translation Publishing House **2005**: 109 -19

[8] Haken H 2004 *Synergetics: introduction and advanced topics* Berlin: Springer Berlin Heidelberg

[9] Han D J and Zhu D L 2012 Coupling relationship analysis of land use and regional economy in Guizhou province *Transactions of the Chinese Society of Agricultural Engineering* **28**(15): 1-8

[10] He Y Q, Pan X X, Wang Y, et al. 2013 Evaluation and analysis on coupling coordinated development of urban resource, environment and economy in Jiangxi Province in China *Applied Mechanics and Materials* **295-298**(14): 2457-63

[11] Jin J S, Wang R S, Li F, Huang J L, Zhou C B, Zhang H T and Yang W R 2011 Conjugate ecological restoration approach with a case study in Mentougou district, Beijing *Ecological complexity* **8:** 161-70

[12] Julianne H and Busa M 2013 Dynamite in the EKC tunnel? Inconsistencies in resource stock analysis under the environmental Kuznets curve hypothesis *Ecological Economics* **94**: 116-26

[13] Li F, Liu X S, Hu D and Wang R S 2007 Evaluation method and its application for urban sustainable development *Acta Ecologica Sinica* **11**: 4793-802

[14] Liu Y B and Song X F 2005 Coupling degree model and its forecasting model of urbanization and ecological environment *Journal of China University of Mining & Technology* **1**: 94-9

[15] Ma X D, Sun J H and Hu Z Y 2009 Co-evolutionary study on the ecological environment and the social economy multiplexed system *Advances in Water Science* **4**: 566-71

[16] Sun A J, Dong Z C and Zhang X Y 2008 Coupling degree between urban economy and technical efficiency of water use in China *Resources Science* **3**: 446-53

[17] Tsurumi T and Managi S 2010 Decomposition of the environmental Kuznets curve: scale, technique, and composition effects *Environmental Economics and Policy Studies* **11**: 19-36

[18] Vincent J R 1997 Testing for environmental Kuznets curves within a developing country *Environment and Development Economics* **2**(4): 417-31

[19] Wang S Y and Bian X M 2007 Synthesis evaluation with entire-array-polygon method to ecological economic system of Funing County *Environ. Monit Assess* **127**: 537-45

[20] Wang X N, Sun C Z and Zou W 2011 Coupling relation analysis between water poverty and economic poverty in China *China Soft Science* **12**: 180-92

[21] Wonga Y L and Lewisb L 2013 The disappearing environmental Kuznets Curve: A study of water quality in the Lower Mekong Basin (LMB) *Journal of Environmental Management* **131** (15): 415-25

[22] Wu Q, Wang R S, Li H Q and Xu X B 2005 The indices and the evaluation method of eco-city *Acta Ecologica Sinica* **8**: 2090-95

[23] Xu Y L, Wang Y D and Lin Y 2011 Research on coupling coordinated degree evaluation of Regional Sci-tech Innovation and Sci-tech Finance *Science of Science and Management of S.& T.* **12**: 116-22

[24] Xuan F H 2013 The coupling development between tourism economy and ecological environment in Harbin City *Advanced Materials Research* **726-731**(15): 4150-53

[25] Zeng M and Wang Y J 2013 Study on coupling coordination degree of China's energy,economyand environment systems based on PCA method *Journal of North China Electric Power University (Social Sciences)* **3**: 1-6

[26] Zhang J 2008 The research on synergetic development between economy and environment in Xuzhou *Contemporary Economy & Management* **9**: 59-61

[27] Zhou L, Yang S and Chen S 2011 Research on the coupling measurement of urban and rural system and the variation of coupling degree in Wuxi *Human Geography* **6**: 77-82

Acknowledgements

I would like to express my gratitude to the support by the Fund of National Social and Science of China (NO.12CGL068) & the Fundamental Research Funds for the Central Universities (No.2015B20414).

An evaluation index system of water security in China based on macroeconomic data from 2000 to 2012

X S Li[1, 2, 3], Z Y Peng[1] and T T Li[1]

[1]Economy and Management School, Wuhan University, Wuhan 430072, China
[2]Water Research Institute, Wuhan University, Wuhan 430072, China

E-mail: xuesong7401@263.net

Abstract. This paper establishes an evaluation index system of water security. The index system employs 5 subsystems (water circulation security, water environment security, water ecology security, water society security and water economy security) and has 39 indicators. Using the AHP method, each indicator is given a relative weight to integrate within the whole system. With macroeconomic data from 2000 to 2012, a model of water security evaluation is applied to assess the state of water security in China. The results show an improving trend in the overall state of China' s water security. In particular, the cycle of water security is at a high and low fluctuation. Water environment security presents an upward trend on the whole; however, this trend is unsteady and has shown a descending tendency in some years. Yet, water ecology security, water society security, and water economy security are basically on the rise. However, the degree of coordination of China's water security system remains in need of consolidation .

1. Introduction

Water is the source of life and the indispensable material basis of national economic development. Problems that are related to water resources, such as water pollution, floods, and drought can seriously restrict social and economic development. Thus, maintaining water security is the foundation and premise for achieving sustainable development. China has long been faced with serious issues of water security. The population is so large that the volume of water resources per capita is much less than the global average, which has led to a relative water shortage and an uneven distribution. In addition, problems including flooding and water pollution accidents have emerged, which have made the situation even worse. In order to realize comprehensive and sustainable development, it is necessary for China to resolve the problems that threaten water security and alleviate the water resources crisis. As the foundation of water resource management decisions, water security evaluation is aimed at estimating the state of water security in a country or region. After such evaluations have been conducted, corresponding countermeasures can be developed to reach the goal of sustainably developing the social economy and ecological environment of a country or region [1]. Therefore, a requisite premise in solving water security matters is to evaluate water security objectively and reasonably.

In recent years, many scholars and organizations have constructed evaluation index systems and made comprehensive assessments of the state of water resources in certain areas with a variety of statistical methods. The Policy Research Initiative (PRI) of Canada developed a composite water index, known as the Canadian Water Sustainability Index (CWSI), to evaluate Canadian communities' water well-being [2]. This model was widely applied to the water resource evaluation of other regions or river basins [3]. Later, the Sustainable Water Resources Roundtable (SWRR) established a water resource evaluation index system, which included 400 indicators that included three aspects: society, the economy, and the environment [4]. Constructing an evaluation index system that employed four criteria, including the economy, society, the environment and maintenance capability, Kang et al. developed a water resource sustainability evaluation model (WRSEM) [5]. In China, there were also many related studies. Han et al took several factors into consideration, such as the safety of drinking water, the contradiction between water's supply and demand, and the realization of water value, based on a multi-criteria assessment index, which was built and applied to Ningxia city for an empirical analysis [6]. Dai set personal security, economic security, social security and environmental security as four subindices under the synthetic index of water safety to constitute a water security evaluation index system, and Liupanshui City was taken as a case [7]. Generally speaking, in the existing water resource evaluation studies, assessment index systems have usually been established by dividing water security into several aspects including the economy, society, ecology and the environment. Next, corresponding indicators are selected for further calculation. However, the method of synthetic evaluation is diversified, among which the WPI [8], PCA [9] [10] , fuzzy evaluation method [11], matter-element evaluation method [12], set pair analysis [13] and AHP [14] [15] have been most commonly used. Nevertheless, in the last few years, a great number of studies on water safety evaluation have attained significant results. Most of these studies have focused on the water safety of a certain region, while countries as a whole have received minimal attention. Furthermore, those studies have been based on short time periods, focusing, for example, on a single year; thus, vertical comparisons have been lacking. In view of these elements and China's actual water safety situation, this paper develops a comprehensive water security evaluation index system and assesses the state of water security in China from 2000 to 2012 utilizing the AHP method and the water security index formula, which is expected to provide reference and foundation for China to make effective water security preserving policies.

2. Establishing an evaluation index system

2.1. *The concept of water security*

Since the idea of water security was first put forward, the concept has attracted growing attention, especially in the recent decade. In the early days, the issues that were discussed in terms of water security varied depending upon context and disciplinary perspective. For example, from a legal perspective, water security has generally been associated with allocation rules that seek to secure entitlements to desired quantities of water [16]. After 2000, water security was generally described in an integrative and interdisciplinary way, which can be sorted into four types with respect to varying priorities [17]. The first type of definition focuses on water availability, whereby water security is defined as a condition in which people have access to sufficient, safe and affordable water to satisfy their basic needs for drinking, washing, and livelihood [18]. The second type places an emphasis on human vulnerability to hazards; specifically, water security involves the protection of vulnerable water systems, protection against water related hazards such as floods and droughts, the sustainable development of water resources, and the safeguard of access to water functions and services [19]. The third type pays more attention to human needs, a representative point being that "water security is a condition where there is a sufficient quantity of water at a quality necessary, at an affordable price, to meet both the short-term and long-term needs to protect the health, safety, welfare, and productive capacity of position (households, communities, neighborhoods or nation) [20]. The last type considers sustainability as the most important element, and water security is defined as a condition in which "every person has access to enough safe water at an affordable cost to lead a clean, healthy, and productive life, while ensuring that the natural environment is protected and enhanced [21]. In China, definitions of water security that have been provided in the literature were similarly derived from the above concepts. For instance, "water security is based on an analysis of the relationship between environmental changes and security issues considering not only the situation of water resources, but the related factors of environment, ecology, society, politics, and economy [22]. Referring to the above definition, in this article, water security, based on the causes of water safety threats, is defined as a state in which the natural hydrologic cycle and human water related activities do not cause water shortages, water pollution, deluge, drought or other problems, and do not hamper the sustainable development of the social economy and the ecological environment.

2.2. The basic framework of the water security evaluation system

Obviously, with its abundant connotations, water security cannot be scaled through a few indicators. Based on the conception that is set forth in this paper, referring to existing studies, a multi-criteria water security evaluation index system is established as shown in table 1, which employs five subsystems, namely water circulation security, water environment security, water ecology security, water society security, and water economy security, which ultimately include 39 indicators.

Representing the degree of water security in the designated areas, the comprehensive water security index provides direct proof for judging the state of water security in that area. As for the five sub-indices, water circulation security is employed to describe the natural phenomena and processes that are related to water resources, mainly involving precipitation, water reserves and water-related natural hazards, recording and reflecting the law of hydrological motion. Water environment security is mainly about the situation of water quality, and it is connected with the straightforward influences that people's water-related behaviors have on their water environment. In daily life and production, the quality of the water environment is largely affected by actions such as sewage treatment and disposal. Water ecology security represents the ecological factors that affect water security, such as forests and wetlands, which are helpful for water conservation and purification, and thus play a vital role in maintaining water safety. Water social security stands for the ability to construct water security. A higher level of water security development capability can be achieved by reducing the negative impacts that natural water conditions, and the ecological environment, have on water security. Water economy security is a sub-index regarding the economic efficiency of water usage and the capacity of the economy to support water security construction. Increasingly, the utilization efficiency of water resources and guaranteed economical and sustainable development are important aspects of water security. In turn, the development of the economy can lay a foundation for water security construction.

Table 1. Water security evaluation index system.

Objective system	Subsystem	Indicators	Units	Index attributes
Comprehensive water security	Water circulation security B1	per capita volume of water resources C_1	m^3 per capita	+
		mean annual precipitation C_2	mm	+
		Groundwater quantity C_3	$10^8 m^3$	+
		Surface water quantity C_4	$10^8 m^3$	+
		The area affected by inundation C_5	10^3 hectare	-
		The area affected by drought C_6	10^3 hectare	-
	Water environment security B2	Frequency of water pollution accident C_7	times	-
		COD emission load C_8	10^4 tons	-
		Industrial waste water discharge success rate C_9	%	-
		Ratio of water quality preceding III among rivers assessed C_{10}	%	+
		Emission load of ammonia and nitrogen C_{11}	10^4 tons	-
		Total emission load of effluent C_{12}	10^4 tons	-
		Ratio of investment on dealing with environmental pollution C_{13}	%	+
		Disposing rate of urban sewage C_{14}	%	+
		Ratio of forest covering C_{15}	%	+
	Water ecology security B3	Per capita urban public greenbelt area C_{16}	m^2	+
		The proportion natural reserve area accounts for jurisdiction area C_{17}	%	+
		Soil erosion control areas C_{18}	10^3 hectare	+
		Ratio of ecological water consuming C_{19}	%	+
		The proportion wetland area accounts for country's total land area C_{20}	%	+

Water society security B4	Total amount of water supplied C_{21}	$10^8 m^3$	+
	Investment for drainage construction project C_{22}	10^8 yuan	+
	Year-end length of urban water supply pipe C_{23}	km	+
	Water popularizing rate of urban population C_{24}	%	+
	Levee protection area C_{25}	10^3 hectare	+
	The number of industrial wastewater treatment facilities C_{26}	set	+
	Effective irrigation area C_{27}	10^3 hectare	+
	Accumulated benefit rate of rural water reform C_{28}	%	+
	Ratio of loss caused by inundation C_{29}	%	-
	Ratio of economic crops loss caused by drought C_{30}	%	-
Water economy security B5	Per capita GDP C_{31}	Yuan	+
	Per capita output of grain C_{32}	kg	+
	Engel coefficient of urban households C_{33}	%	-
	Engel coefficient of rural households C_{34}	%	-
	Per capita yearly net income of rural households C_{35}	Yuan	+
	Per capita yearly real income of urban households C_{36}	Yuan	+
	Water consumption per ten thousand yuan GDP C_{37}	m^3	-
	Recycling rate of industrial water C_{38}	%	+
	Water consumption per capita C_{39}	m^3/capita	-

2.3. Determining relative weights

Water security is a complex multi-objective system, the effects of which vary among indexes. Thus, in water security assessment, it is an important step to determine the relative weights of the indicators. In this article, the AHP method is employed to determine relative weights. With respect to the principle of AHP, by consulting specialists and carrying out a questionnaire survey, a judgment matrix is structured with a scale from 1 to 9, based on a pairwise comparison between two indicators or two sub-indices. Then the relative weight of each indicator and sub-index can be confirmed through a series of calculations. The result of the relative weight determination is shown in table 2, which furnishes water security evaluations with a reliable substratum.

Table 2. Relative weights of indicators and sub-indices in water security evaluation index system.

Indicators	Weights	Indicator	Weights	Indicator	Weights	Indicator	Weights
C1	0.0750	C12	0.0207	C23	0.0051	C34	0.0050
C2	0.0205	C13	0.0369	C24	0.0274	C35	0.0120
C3	0.0142	C14	0.0680	C25	0.0137	C36	0.0071
C4	0.0518	C15	0.0497	C26	0.0073	C37	0.0262
C5	0.0378	C16	0.0101	C27	0.0137	C38	0.0254
C6	0.0378	C17	0.0151	C28	0.0218	C39	0.0178
C7	0.0100	C18	0.0328	C29	0.0177	B1	0.2372
C8	0.0184	C19	0.0084	C30	0.0177	B2	0.3354
C9	0.0499	C20	0.0248	C31	0.0155	B3	0.1410
C10	0.1116	C21	0.0385	C32	0.0060	B4	0.1677
C11	0.0201	C22	0.0050	C33	0.0038	B5	0.1186

According to table 2, among the five subsystems, water environment security, that represents the state of water quality, makes the most contribution to water safety at 0.3354, followed by water circulation security and water society security, the weights of which are 0.2372 and 0.1677, respectively. Comparatively, the influences that water ecological security and water economical security exert over water security are less, and their weights are only 0.1677 and 0.1186, respectively. The results of the relative weights determination imply that the quantity and quality of water resources are extremely vital aspects of water security, especially the quality of a water resource. In fact, the quality of a water resource determines the quantity of the available water. Further, a sufficient water supply is the fundamental premise of water security, which cannot be realized under the conditions of a water shortage. What is more, water social security has a relatively great effect on water security, which implies the inferior position of water security, which results from natural and artificial factors, can be improved with intentional corrective measures, which can increase the degree of water security. Specific to the various indicators, in view of their contribution to water security represented by the relative weights, the top ten indicators are: ratio of water quality preceding among rivers assessed, per capita volume of water resources, disposal rate of urban sewage, industrial waste water discharge success rate, surface water quantity, ratio of forest coverage, total amount of water supplied, area affected by inundation, area affected by drought, and ratio of investment in managing environmental pollution.

2.4. Consistency examination

A consistency examination is required to check the reasonableness of the relative weights, which are obtained with a judgment matrix and given a mean random consistency index (RI) that is utilized to make judgments. If the value of CR is less than 0.1, the relative weights are assigned rationally. As for the above results of the relative weights determination in the water security evaluation index system, the results of the consistency examination are displayed in table 3. Obviously, the relative weights of the indicators and sub-indices, which are attained through the AHP method, are reasonable and can be used to integrate the water security evaluation index system.

Table 3. Results of Consistency Examination.

System	A	B1	B2	B3	B4	B5
CR	0.0461	0.0754	0.0412	0.0403	0.0785	0.0680
Comparison with 0.1	<0.1	<0.1	<0.1	<0.1	<0.1	<0.1

3. A comprehensive evaluation of Chinese water security

3.1. Data sources and treatment

In this study, macroeconomic data at the national level from 2000 to 2012 is selected to evaluate the state of water security in China during that period. Data is drawn from the Chinese annual *Water Resources Bulletin, China Statistical Yearbook, China Water Conservancy Development Bulletin, China Flood and Drought Disasters Bulletin, China's Environmental Bulletin, China Environmental Statistics Yearbook,* and the CEInet statistics database. To offset the impact of prices, the data of indicators that involve value are converted to that of a level based on 2000. Additionally, some of the missing data is replaced with a linear trend analysis. Further, the range transformation method is utilized to normalize the indicator data to eliminate the differences among the indicator units.

For the positive indicators, the normalization formula is shown below:

$$D_i = \frac{X_i - X_{min}}{X_{max} - X_{min}} \quad (1)$$

For the negative indicators, the normalization formula is given by:

$$D_i = \frac{X_{max} - X_i}{X_{max} - X_{min}} \quad (2)$$

Where D_i is the value of the indicators after normalization, X_i represents the original value of the indicators, and X_{max} and X_{min} stand for the maximum and minimum values of the indicators, respectively.

3.2. The water security evaluation index formula

To calculate the index's value of each subsystem in the water security evaluation index system, the following equation is used (GC PRI, 2007; Sullivan, 2001):

$$B_j = \frac{\sum_{i=1}^{N} w_j' X_j'}{\sum_{i=1}^{N} w_j'} \quad (3)$$

Where B_j is the value of the j sub-index, X_j^i is the normalized value of i indicator that is contained in the j sub-index, and W_j^i represents the relative weights of i indicators for the j sub-index.

The comprehensive index value of water security is computed by integrating the sub-indices as follows:

$$WSI = \frac{\sum_{j=1}^{N} w_j B_j}{\sum_{j=1}^{N} w_j} \tag{4}$$

Where WSI is the comprehensive index of water security, Bj is still the value of the j sub-index and Wj stands for the relative weights of the sub-indices.

Because water security is a complicated system that is composed of several subsystems (water cycling, water environment, water ecology, water society and water economy), which are closely connected and affect each other, it is of great significance in achieving water security to coordinate the development of each subsystem. The following equation is used to scale the sub-index of the coordination degree of the water security evaluation index system [23].

$$CI = 1 - \frac{S}{\bar{B}} \tag{5}$$

Where S is the standard deviation of sub-indices, $\bar{B}$ is the mean of sub-indices and CI is the sub-index of the coordination degree of water security evaluation index system. The larger the CI is, the better the coordination of system will be.

3.3. Evaluation results and discussion

3.3.1. Water security evaluation results and analysis. Based on the water security evaluation index system that is established in this article, with the above calculation formulas and normalized data, the results of assessing Chinese water security are displayed in figure 1.

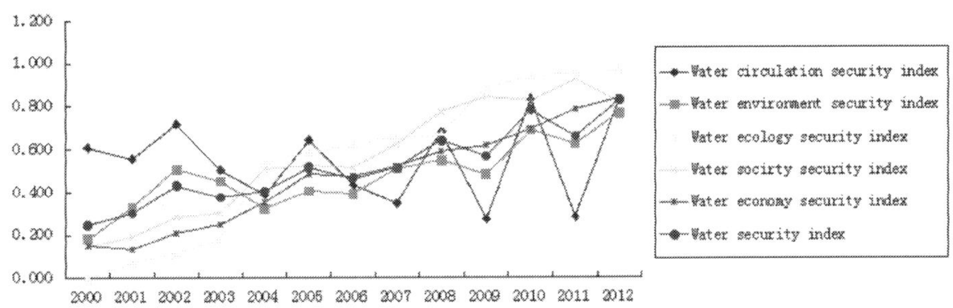

Figure 1. Variation of the water security's integrated index and sub-indices for China from 2000 to 2012.

According to the evaluation results, the water circulation security index (B1) fluctuated greatly during the studied period without fixed law. The highest value is 0.833 for 2010, while the lowest value was only 0.283 for 2011. The reason for such a consequence is that the water-related natural phenomena and the process that is described by water cycling security are often irregular and beyond human control, even though they have a close relationship with water security. The fluctuation of the water cycling security index is the very embodiment of natural hydrologic motion's randomness.

The water environment security index (B2) declined by a large margin from 0.507 to 0.323, during the period from 2002 to 2004. Moreover, it also had downward trends in the years 2003, 2006, 2009 and 2011 by 11.2%, 3.4%, 11.9% and 9.2%, respectively. A further analysis of the indicators contained in the water environment security subsystem indicates that the increase in ammonia and nitrogen emission load, and the effluent's total emission load, was accompanied by a descending ratio of water quality preceding among the rivers that were assessed, which lead to the decreases mentioned above. However, as a whole, the water security index was still on the rise from 2000 to 2012. In the last few years, accidents that threaten water security such as water pollution have occurred constantly, for example, the red river incident in Xinxiang of Henan province. As a result, governmental and public environmental awareness has grown. The central and local governments have introduced a series of laws and regulations, such as *the Prevention Plan on Water Pollution in Key River Basins from 2011 to 2015*, to control the water environment problem, and NGOs have also taken measures to actively protect the water environment. Having benefited from these efforts, there is no doubt that the problems of the water environment have been remitted to some extent, which has caused the water environment security to gradually improve.

The results show a rising trend in the water ecology security index (B3), water society security index (B4) and water economy security index (B5). In detail, the water ecology security index achieved the fastest growth in 2004 with a 69.3% jump, and the growth pace of 2008 was also comparatively large, approximately 29.3%. The water society security index dropped slightly by 11.1% and 2.4% in 2010 and 2012. Meanwhile, the water economic security index continues to steadily rise. The above consequences reflect that the ecological construction of China is in good condition, and the ability to cope with water security problems and economic strength is improving. In China, ecological safety has attracted growing attention. Ecological civilization constructions are placed in a prominent position and contribute to the beautification of China. In recent years, China's economy has been in a state of steady growth. From 2000 to 2012, the per capita GDP has risen by 52.28%. Moreover, water security has been a source of increasing concern, and as a result, investment in water security construction has increased. In 2012, water conservancy investments in fixed assets reached 411.72 billion yuan and made a gain of 19.27%. All of these facts provide support for the rising trends toward a water ecological index, a water social security index, and water economic index.

In general, China's integrated water security index still shows an overall upward trend. The year 2000 held the lowest water security index value of 0.249, while the highest value of 0.827 occurred in 2012. This period witnessed a growth rate of 232.1%. Slight drops occurred in 2003, 2006, 2009 and 2011, which is in line with the downward trend of the water environment security index. Thus, improving the water environment plays a crucial role in the future promotion of water security.

3.3.2. The water security system's coordination evaluation results and analysis. Calculated with formula (5), the coordination indexes of the Chinese water security system over the years are displayed in Table 4 and figure 2. From 2000 to 2006, the coordination index of the Chinese water security system has steadily increased from -0.049 to 0.867. Nevertheless, in the following years from 2007 to 2012, the values of the coordination index showed a trend of fluctuations, which fell badly especially in 2009 and 2011 by 32% and 28.9%, respectively. These consequences demonstrate that, in the water security system of China, a coordinating development among water circulation, water environment, water ecology, water society, and water economy has basically come to pass. However, this positive tendency has not continued in recent years, which has not had a negative impact on the overall state of water security, but positive action is still required to make the appropriate corrections.

Table 4. Coordination index of Chinese water security system from 2000 to 2012.

Year	2000	2001	2002	2003	2004	2005	2006	2007	2008	2009	2010	2011	2012
CI	-0.049	0.249	0.336	0.599	0.771	0.827	0.818	0.773	0.867	0.590	0.867	0.616	0.913
Growth rate(%)	——	-611.6	35.0	78.4	28.6	7.3	-1.1	-5.6	12.2	-32.0	47.1	-28.9	48.1

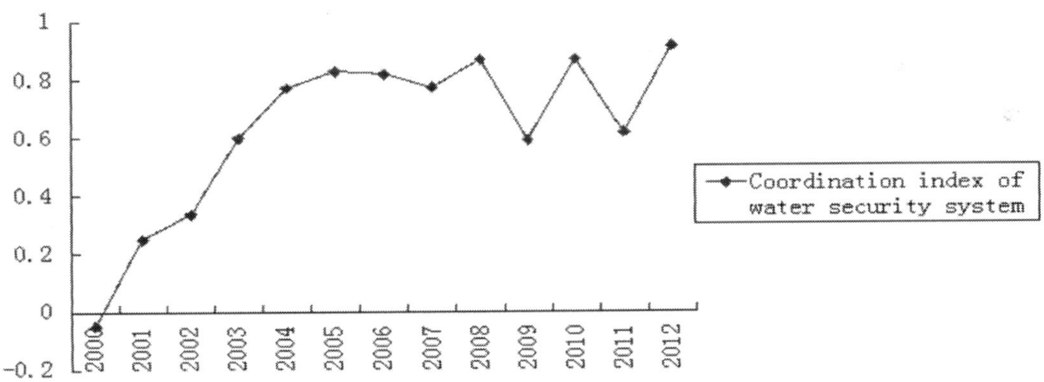

Figure 2. Variation of Chinese water security system's coordination indexes from 2000 to 2012.

4. Conclusions and Suggestions

4.1. Conclusions

This paper develops a water security model and applies it to evaluate the state of water security in China, with macroeconomic data from 2000 to 2012, and several conclusions are reached, as follows:

- As a reflection of natural hydrological motion laws, the high and low volatility of the water cycling security index reflects the inherent natural attributes as well as the randomness of the water. Although most elements of this subsystem are usually beyond human control, it can still be improved by strengthening the capability of monitoring natural incidents, and completing the mechanism of disaster protection to reduce losses and to enhance of the quality of China's water security.

- In recent years, China's water environment security has improved gradually. However, deteriorations have emerged in recent years that imply an increasing tendency toward instability. A focus on pollution control and emissions reduction is required, and the water environmental protection measures must be further implemented.
- In response to the growing economy and an increasing government awareness of environmental issues, ecological and water security construction have increased in China. As a result, in China, over the last few years, water ecological security has been well developed; moreover, the ability to preserve water security and economic support have continuously risen.
- Overall, China's water security has tended to improve, which results in an increase in vigorous regulations in the recent years. However, coordination among natural conditions, the ecology, the environment, society, and the economy still needs further consolidation.

4.2. Countermeasures and suggestions

According to the above study, some countermeasures should be taken to resolve the problems that were revealed in the empirical water security evaluation presented here, to further maintain momentum to improve Chinese water security.

- Enhancing the ability to monitor natural events and perfect the disaster protection mechanism.

The power of nature should not be underestimated; the losses that are incurred due to natural forces can be enormous. However, such possible losses can be minimized by enhancing the capability of monitoring water-related natural phenomena, and constructing a set of systematic water hazard protection mechanism. Such actions will greatly help to reduce pressures on water security. Governments should provide more support to projects that aim to develop technologies for monitoring water-related natural phenomena, introduce advanced monitoring facilities, make clear the responsibilities of monitoring sites, and release all types of hydrodynamic monitoring information in a timely manner. Moreover, non-governmental organizations that monitor natural events should be encouraged to function at full capacity. In addition, various government departments that are responsible for water security should cooperate closely with each other, construct special platforms for the collection, analysis and sharing of hydrological information, and maintain flood control and drought relief engineering facilities on a regular basis. In this way, high-level water disaster protection mechanisms can be constructed. Finally, an improved system of managing the aftermath of water hazards must be built.

- Improve the institution of water environment protection, and implement various measures to renovate the water environment

The fact that water environment security plays the most important role in a water security system makes water environment protection the key to maintaining water security. China has taken numerous actions to renovate its water environment, but they have not been fully operative due to insufficient enforcement. To resolve the problems that threaten water security to further improve the institution of water environment protection and implement various measures to renovate the water environment,, the existing laws and regulations to protect the water environment should be adjusted immediately in accordance with the requirements of the socio-economic development stage. In 2014, China passed a revised draft of *Environment protection law,* which put forward a higher standard of environmental protection. Therefore, corresponding adjustments should be made in the laws and regulations that are related to water resources, to highlight the norms of new environmental water laws. With regard to execution, governments are expected to strengthen the accountability mechanism of water environment protections. Penalties for behaviors that maliciously destroy the water environment should be strengthened, while proper awards should be given to individuals, companies, or organizations that make outstanding contributions to preserving the water environment. In addition, it is imperative to investigate the achievements of governments at all levels of water environment management, on a regular basis, to ensure that measures for water environment protection are being effectively carried out. In terms of supervision, the publicity of water-related laws and regulations should be expanded to strengthen the public's environmental awareness and cognitive abilities. Further, special feedback channels need to be explored to encourage the public to supervise others' water-related activities and government management.

- Coordinate natural environment and social economy and achieve sustainable development of water security construction.

Both the elements of the natural environment and social economy subsystems are organic components of the water security system. Achieving coordination of the development between the natural environment and the social economy will optimize the internal structure of the water safety system, which lays the foundation for water safety to achieve sustainable development. First, industrial enterprises that produce significant pollution need to be stopped, or given the option of making changes. Additionally, the agriculture industry should be properly educated in techniques of managing water pollution that result from sources such as chemical fertilizer. At the same time, it is of great significance for an evolving, environmentally friendly economy to support the development of green industries. Second, based on the rapid development of the economy, efforts should be made to intensify support for the construction of water security, bolster the construction of water conservancy projects such as laying urban pipe networks for cities and towns, and promoting the construction of rural drinking water engineering and water conservancy facilities. Third, economic and social development should positively nurture the restoration of the natural environment. Additional funds should be allocated for activities such as water resource exploitations and water environment management. Meanwhile, various modes of support, such as human and material resources, should be provided for water-related natural and environmental programs.

- Promote the innovation of water resources management systems, and consolidate the foundations for water security construction

The state of Chinese water security is affected by numerous factors such as climate, the country's social, political and economic situation, and industrialization. Moreover, water resource management should guarantee the smooth operation of China's water resource system, which should include good water security. In order to make the growing need for water security steady and continuous, it is imperative to accelerate the system's innovation of water resources management. Efforts should be made to transform the traditional governance model, with control at the core, into a modern one with good governance as its goal, converting engineering technology control into comprehensive treatment and establishing new structures and systems of water management. In the process of water resource management, the relationship among government, market, and society should be reconsidered. This relationship should be a circular and interactive one, rather than one in which governments take a one-way, leading, controlling position. Governments should shift from the direct allocation of water resources to the implementation of water rights management and the formulation of standards for water security indicators. Moreover, a unified management system of water resources and an efficient inter-departmental coordination mechanism with clear responsibilities among high authorities should be established. To improve public awareness, accurate information about water security should be published. Democratic management and broad participation should be introduced at all levels, and various online links that relate to water resources should be shared; such information will help to make decision-making more scientific, democratic, and systematized by involving the general public. As for the market, its capacity for allocation should fully exploited. It is suggested that a reasonable mechanism of water price determination should be established that will open the water market completely and achieve water's value, capitalization, and industrialization. Finally, water resource market system construction, and normative order should be completed. Through the above-described measures of marketization, the required mechanisms of water security can be established.

Acknowledgments
The study was financial supported by the CRSRI Open Research Program (Program SN：CKWV2016394/KY).

Appendix A. Process of using the AHP method to calculate weights

The rationale of the AHP method is to quantize the experiential judgments via comparisons between comparative importance of different indicators on the same level on the basis of a hierarchical index system. The concrete steps and calculation processes are as follows:

1. Build judgment matrix

Each level of principles, and its indicators, may have different weights in the measurements. The relative importance is judged by a couple of industry experts. The method of Seaty1~9 is adopted to define the judgment matrix $A=a_{ij}(n*n)$. The element a_{ij} in matrix A is the comparative importance measurement of indicators on the Row i and Column j.

Table A1. Principles of Seatyl~9 measurement method of comparative importance.

Measurement	Meaning
1	Two elements have the equal importance.
3	The former is a little more important than the latter.
5	The former is obviously more important than the latter.
7	The former is far more important than the latter.
9	The former is extremely more important than the latter.
2,4,6,8	It means the middle value of the neighboring measurements.

Reciprocal	If the relative importance measurement of indicator i and j is a_{ij}, then the relative importance measurement of indicator j and I is the reciprocal of a_{ij}

2. Calculation of weights and examination of consistency

- Calculation of weights: to calculate the geometric average of vectors in each row, conduct the normalization and consider the row vector to be the vector of weight.
- Consistency examination: Since the judger may be subjective to a certain degree, examination of consistency should be adopted to change subject one's into objective descriptions and calculate the consistency indicators.

$$CI = \frac{\lambda_{max} - n}{n - 1}$$

λ_{max} is the maximum eigenvalue of the judgment matrix. It should be compared with the values in the consistency indicators (recorded as RI) and then the CR(CR=CI/RI) can be figured out. When CR<0.1, is considered the consistency of the judgment matrix can be accepted. Otherwise the value of the judgment matrix has to be moderated.

Table A2. The Value of RI.

n	1	2	3	4	5	6	7	8	9	10	11	12	13	14	15
RI	0	0	0.52	0.89	1.12	1.26	1.36	1.41	1.46	1.49	1.52	1.54	1.56	1.58	1.59

In this paper, each principle level and weights of indicators have been calculated out according to the AHP method. In the process of building a judgment matrix are seven experts in the research field of water security—Wu Xinmu, Xia Jun, Ning Yuan, Chen jin, Fu Chun, Wang Shuyi and Wang Xianjia. They were invited to judge the relative importance of each indicator according to the principles of Seatyl~9 measurement and their judgment results are summarized. Eventually, the level of principles and weights of the indicators have been calculated using the previously described AHP method.

Appendix B. Values of water security index and sub-indices of China from 2000 to 2012

Table B1. Values of water security index and sub-indices of China from 2000 to 2012.

Year	B1	Growth rate (%)	B2	Growth rate (%)	B3	Growth rate (%)	B4	Growth rate (%)	B5	Growth rate (%)	WSI	Rank	Growth rate (%)
2000	0.608	——	0.183	——	0.149	——	0.152	——	0.000	——	0.249	13	——

2001	0.557	-8.3	0.329	80.0	0.194	29.8	0.135	-11.1	0.070	—	0.301	12	21.1
2002	0.718	28.9	0.507	54.2	0.285	46.9	0.209	54.7	0.116	65.8	0.429	9	42.7
2003	0.504	-29.8	0.450	-11.2	0.300	5.3	0.251	20.1	0.183	57.9	0.377	11	-12.3
2004	0.390	-22.7	0.323	-28.4	0.507	69.3	0.355	41.3	0.545	197.4	0.405	10	7.5
2005	0.644	65.0	0.405	25.6	0.528	4.0	0.483	36.0	0.582	6.8	0.517	6	27.6
2006	0.432	-32.8	0.391	-3.4	0.517	-2.1	0.472	-2.4	0.622	6.8	0.464	8	-10.1
2007	0.347	-19.8	0.511	30.6	0.626	21.2	0.521	10.4	0.653	5.1	0.513	7	10.4
2008	0.681	96.4	0.548	7.1	0.771	23.2	0.589	13.2	0.671	2.7	0.639	4	24.7
2009	0.274	-59.8	0.482	-11.9	0.842	9.2	0.618	4.9	0.887	32.3	0.566	5	-11.4
2010	0.833	204.2	0.687	42.5	0.828	-1.7	0.692	12.0	0.937	5.6	0.781	2	37.9
2011	0.283	-66.0	0.624	-9.2	0.922	11.4	0.787	13.7	0.954	1.8	0.659	3	-15.6
2012	0.831	193.5	0.766	22.7	0.821	-11.0	0.836	6.2	0.965	1.2	0.827	1	25.5

References

[1] Li Z Y, Wang J Y, Wang W X, et al. 2009 Comprehensive index assessment method of water safety based on weighted method with generalized contrast *Water Resources Protection* (4): 5-7

[2] GC PRI (Government of Canada, Policy Research Initiative) *Canadian Water Sustainability Index (CWSI): Project Report*

[3] Attari J and Mojahedi S A *2009* Water sustainability index: Application of CWSI for Ahwaz County *The Proc. of World Environment and Water Resources Congress* ASCE, Reston, Virginia **2009**: 1-7

[4] SWRR (Sustainable Water Resources Roundtable) *Sustainable Water Resources Roundtable Report*

[5] Kang M G and Lee G M 2011 Multicriteria evaluation of water resources sustainability in the context of watershed management *Journal of the American Water Resources Association* **47**(4): 813-27

[6] Han Y P and Ruan B Q 2003 Research on evaluation index system of water safety *Acta Scientiae Circumstantiae* **23**(2): 267-72

[7] Dai W, Chen H X and Tong S M 2012 Study on Index System for Water Resources Safety

Evaluation *Water Saving Irrigation* **2012**(3): 40-43

[8] Sullivan C 2002 Calculating a water poverty index *World Development* **30**(7): 1195-210

[9] Shi C H and Hu B 2007 Synthetically evaluating the security of urban water supply *Resources Science* **29**(3): 80-85

[10] Kang M G, Lee G M and Ko I H 2010 Evaluating watershed management within a river basin context using an integrated indicator system *Journal of Water Resources Planning and Management* **136**(2): 258-67

[11] Wang Y K, Xia Z Q and Cao S L 2008 Comprehensive evaluation method for water security *Journal of Hehai University (Natural Sciences)* **35**(6): 618-21

[12] Cao S L and Zhang L Y 2007 The Integrated Assessment of Rural Drinking Water Security *Water Resources and Power* **25**(4): 12-4

[13] Lu M, Zhang Z Y and Shi Y Z 2006 Application of set pair analysis to evaluation of water safety *Journal of Hehai University (natural sciences)* **34**(5):505-8

[14] Shao D G, Yang F S, Liu Y L, et al. 2013 Urban Water Security Index and Its Evaluation Criterion *South-to-North Water Transfers and Water Science & Technology* **2013** (1): 122-6

[15] Wang H Z, Duan Y Y and Fu C 2014 Water Security Evaluation of Anhui Province Based on Analytic Hierarchy Process *South-to-North Water Transfers and Water Science & Technology* **2014**(2): 37-41

[16] Tarlock, A.D and Wouters P 2009 Reframing the water security dialogue *Journal of Water Law* **2009**(20): 53–60

[17] Cook C and Bakker K 2012 Water security: debating an emerging paradigm *Global Environmental Change* **22**(1): 94-102

[18] Rijsberman F R 2006 Water scarcity: fact or fiction? *Agricultural Water Management* **80**(1): 5-22

[19] UNESCO-IHE 2009 *Research Themes Water Security* http:// www.unesco-ihe.org/ Research/ Research-Themes/ Water-security

[20] Witter S.G. and Whiteford S 1999 Water security: the issues and policy challenges *International Review of Comparative Public Policy* **1999**(11): 1–25

[21] *Global Water Partnership Towards Water Security: A Framework for Action* http://www.gwp.org/Global/ToolBox/References/Towards%20water%20security.%20A%20 framework%20for%20action.%20Executive%20summary%20 (GWP,%202000).pdf

[22] Ma H Q, Liu L and Chen T 2010 Water security assessment in Haihe River Basin using principal component analysis based on Kendall tau *Environmental Monitoring and Assessment* **2010**(163): 539–44

[23] Zhang W, Duan X J and Zhang W Y 2013 Evaluation and Evolution of Sustainable Development in Yangtze River Delta *Resources and Environment in the Yangtze Basin* **22**(10): 1243-49

Study on closed recycling regeneration process of $FeCl_3$ waste etching solution and recovery of Copper

X F Miao[1], X T Zhou[1, 2], R X Wang[1], D H Zhai[1] and H H Ran[1]

[1]Institute of Chemical Machinery, East China University of Science and Technology, Shanghai, China

E-mail: robert_zhou@ecust.edu.cn

Abstract. $FeCl_3$ etching solution is widely used in etching processes due to its strong oxidation and corrosion properties. However, significant amounts of waste etching solution is produced, which pollutes the environment and wastes materials. Aiming to address problems related to the use of recycled $FeCl_3$ waste etching solution containing Cu, the crystallization regeneration process in an enterprise was improved and the copper element was recovered in the form of copper chloride crystal. Based on the original process, a hydrocyclone was added to pre-separate the slurry. By investigating the change of the inlet flow rate and separation efficiency of the hydrocyclone with inlet pressure under different concentrations, the optimal inlet pressure was determined to be approximately 0.25Mpa. Experimental results indicated that the workload of the centrifuge was reduced by approximately 80%, the average specific surface area of the copper chloride crystal was decreased by 50%, the production of copper chloride crystal byproducts after washing process was increased by approximately 20% and regeneration of $FeCl_3$ waste etching solution was also realized.

1. Introduction

$FeCl_3$ etching solution has strong oxidative and corrosive effects on metals such as Fe, Cu, etc. Additionally, it is lower in price and provides a faster etching speed than other etching solutions such as cupric chloride etching solution [1]. Therefore, $FeCl_3$ etching solution is widely utilized in the etching process of Printed Circuit Boards (PCBs) [2]. However, there is a large amount of $FeCl_3$ waste etching solution produced during the etching process, which has strong corrosiveness and contains significant amounts of Fe^{3+}, Cu^{2+} and other metal ions. If directly discharged, this will cause serious pollution to the environment and waste metal resources. In recent years, with the rapid development of PCB and the increasing scarcity of metal resources in China, the use of recycled $FeCl_3$ waste etching solution has become a major topic in the field of metal resource recycling and environmental protection [3].

At present, the iron powder replacement method, the ferrous sulphide precipitation method, the crystal electrolysis method, the solvent extraction method and the ionic membrane electrolysis method are frequently used to regenerate $FeCl_3$ waste etching solution, but many problems exist in the use of these methods. For example, the purity of copper recovered by the iron powder replacement method is not high and the first reduction and second oxidation process not only wastes expensive iron powder but also increases the consumption of chlorine [4]. The ferrous sulphide precipitation method requires iron powder and sulphur powder to generate ferrous sulphide, but the reaction is violent and easily results in secondary pollution [5]. The purity of metal recovered by the crystal electrolysis method is very high, while the energy consumption of the electrolysis process is quite high and results in a large amount of chlorine generated near the anode, resulting in severe equipment corrosion [6]. The solvent extraction method achieves good separation of Fe and Cu, however the process is complicated, and the introduction of the extraction solvent reduces the purity of the regeneration liquid [7]. In the ionic membrane electrolysis method, $FeCl_3$ waste etching solution can be generated in the anode chamber, and the cathode liquid can be returned to the anode chamber after removal of Cu^{2+}. Although the ionic membrane electrolysis method is environmental friendly, it is very difficult to match the selection of the ion exchange membrane and electrolytic process [8]. Additionally, the processing cost is so high that it cannot be extended to large-scale industrial applications. Therefore, it is of great significance to determine a method of waste etching solution which possesses higher processing capacity, higher added value of recycled products, lower energy consumption and is less polluting to the environment.

Considering that the crystallization technology does not introduce additional substances, has low operating temperature and can improve the operating environment [9], this work primarily investigates the crystallization method to process $FeCl_3$ waste etching solution containing Cu, in order to achieve the regeneration of waste and the recovery of the copper element.

2. Regeneration process and equipment of $FeCl_3$ waste etching solution containing Cu

2.1. Regeneration process of $FeCl_3$ etching solution containing Cu

The common concentrations of Fe^{3+}, Fe^{2+}, Cu^{2+} are 30%, 0.5%, 2.6%, respectively, in $FeCl_3$ waste etching solution containing Cu, while in standard etching solution, the Fe^{3+} concentration is at least 38%, the Fe^{2+} concentration is less than 0.4% and the Cu^{2+} concentration is less than 2.6% [10]. Therefore, the removal of Cu^{2+} from the waste solution is the key to realizing the regeneration of $FeCl_3$ waste etching solution. This work analysed the original treatment process of waste etching solution of an enterprise. The solubility of $FeCl_3 \cdot 6H_2O$ and $CuCl_2 \cdot 2H_2O$ in water is shown in Table 1 [11]. The solubility of ferric chloride changes remarkably with temperature while the solubility of copper chloride increases slightly as temperature increases. Additionally, the crystallization temperature of ferric chloride is greater than that of copper chloride.

Table 1. The solubility of $FeCl_3 \cdot 6H_2O$ and $CuCl_2 \cdot 2H_2O$ in water.

Temperature (℃)		0	10	20	30	40	50	60	70	80	90	100
Solubility	$FeCl_3 \cdot 6H_2O$	74.4	81.8	91.9	107	—	315	—	—	526	—	536
(g)	$CuCl_2 \cdot 2H_2O$	67.7	70.8	73.7	77.3	80.8	84.2	87.6	92.3	96.1	104	110

The copper element is recovered in the form of $CuCl_2 \cdot 2H_2O$ by decompression-evaporation and concentration crystallizing. When the concentration of Cu^{2+} in solution is less than 0.5%, the remaining of Cu^{2+} is removed completely by cooling crystallization. Finally, water and chlorine are added to the waste etching solution to regenerate the $FeCl_3$ waste etching solution. The process flow chart before modification is shown in figure 1.

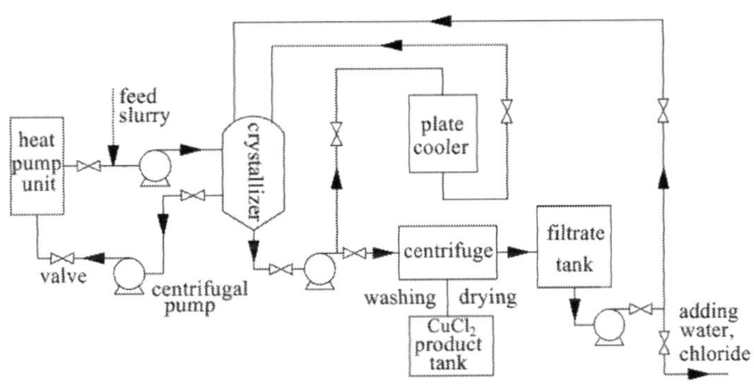

Figure 1. Process flow chart before modification.

However, when the enterprise prepared copper chloride crystals, the crystal slurry after evaporative crystallization was directly transported to a centrifuge for solid-liquid separation. The original process had many deficiencies, the main points of which are summarized as follows. First, the capacity of the centrifuge was limited, thus the evaporative crystallization system was unable to operate continuously. When the copper chloride reached saturation, the slurry became thick and sticky. If the crystals stayed in the crystallizer for a long period of time instead of being separated in a timely manner, the crystal stuck to the wall of tubes, affecting the heat transfer efficiency of the heat exchanger [12]. Secondly, grains were large enough to induce wear on the impeller of the pump, shortening its service life. In addition, the centrifuge separation process did not belong to the closed operation and small crystals often dissipated into the air with vapor, which not only increased the heat loss, but also aggravated the powder pollution. Finally, nearly half of the copper chloride crystals obtained from the centrifuge were less than 70 microns in diameter, increasing the specific surface area of the crystals, making it more possible for them to stick to ferric chloride crystals and diminish during the washing process. Based on these problems, this work improved the separation process of copper crystals and the mother liquor achieved by the centrifuge alone. The process flow chart after modification is shown in figure 2.

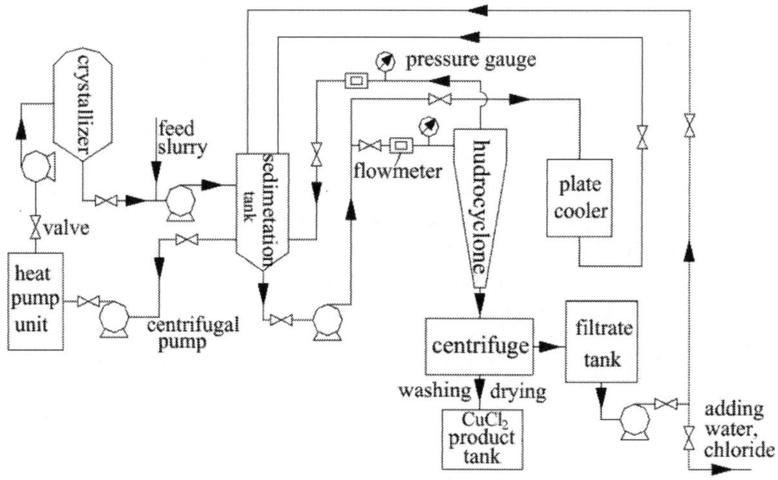

Figure 2. Process flow chart after modification.

2.2. Regeneration equipment of $FeCl_3$ waste etching solution

Based on the original equipment, a sedimentation tank and a hydrocyclone were added to pre-separate the crystal slurry, and then the centrifuge was employed to separate the thick slurry from the underflow. The crystal slurry was separated in a hydrocyclone [13] in a closed environment. Most of the fine slurry passed through the overflow tubes and returned to the setting tank, and approximately 20% of the thick slurry passed through the underflow tube into the centrifuge. This could reduce the processing capacity of the centrifuge and relieve heat loss and powder pollution. Furthermore, there were many crystals in the overflow liquid of the hydrocyclone. These were returned to the setting tank as crystal seeds to promote the secondary nucleation [14], which could result in a more uniform particle size.

This equipment primarily included one water source heat pump unit with a heat supply of 455KW and a power of 144.4KW, one vortex air cooled water chilling unit with a cooling capacity of 119.1KW and a power of 41.7KW, one root water jet vacuum pump unit with a power of 14KW, a hydrocyclone with a nominal diameter of 75mm, one three-foot centrifuge with a rotating speed of 1200r/min and a motor power of 5.5KW, five centrifugal pumps with 11KW of power, one crystallizer, regulating valves, pressure gauges, flow meters, pipelines, etc.

3. Industrial trial operation of regeneration of $FeCl_3$ waste etching solution

3.1. Materials of industrial trial operation

IOP Conf. Series: Earth and Environmental Science **39** (2016) 012046 doi:10.1088/1755-1315/39/1/012046

Before the $FeCl_3$ etching solution was treated, the concentration of Fe^{3+}, Fe^{2+} and Cu^{2+} were 31.5%, 0.6%, and 3%, respectively. The copper chloride crystal size distribution before process modification is shown in table 2. The average size of the crystals was approximately 70 microns; therefore, it was necessary to separate particles smaller than 70 microns and return them to the setting tank to continue their growth in order to obtain high quality crystals.

Table 2. The size distribution of copper chloride crystal separated directly by the Centrifuge.

Size (μm)	Distribution (%)	Size (μm)	Distribution (%)	Size (μm)	Distribution (%)	Size (μm)	Distribution (%)
1.19	0.1	20	1.38	63.25	5.82	200	2.87
1.74	0.27	22.44	1.71	70.96	5.74	224.40	2.32
2.69	0.3	25.18	2.13	79.62	5.69	251.79	2.01
4.26	0.59	28.25	2.73	89.34	5.34	282.51	1.75
6.02	0.96	31.70	3.21	100.24	4.73	316.98	1.06
8.45	0.95	35.57	3.59	112.47	4.38	355.66	0.74
10.62	1.22	39.91	4.12	126.19	4.19	399.05	0.49
13.39	1.54	44.77	4.63	141.59	3.65	447.74	0.36
15.00	0.83	50.24	4.85	158.87	3.44	533.03	0.57
17.83	1.22	56.37	4.97	178.25	3.25	671.04	0.3

3.2. Process of industrial operation

During system operation, 0.65 tons of waste etching solution was treated per hour with a vacuum pressure of 0.08Mpa and a heating temperature of 55℃. If the temperature was too low, ferric chloride would crystallize and reduce the purity of the copper chloride crystals; if the temperature was too high, copper chloride crystals would dissolve and production would be reduced. Regulating the valve opening allowed the inlet flow rate to be controlled at 6-10m^3/h and the inlet pressure ranged from 0.15-0.30Mpa.

As the structural parameters of the hydrocyclone had been determined, its operating parameters primarily influenced the quality of the products. The change of inlet flow rate and separation efficiency with inlet pressure were investigated under three different inlet concentrations. As shown in figure 3 and figure 4, the inlet flow rate increased continuously while the separation efficiency first increased and then decreased as inlet pressure increased. However, the influence of inlet concentration on the inlet flow rate and separation efficiency was very complicated. When the inlet pressure was approximately 0.25Mpa and the concentration was 10%, the separation efficiency reached its maximum at 72.4%. Meanwhile, the inlet flow rate was 6.926m3/h with a split ratio of 20.6% and the concentration of the underflow liquid was 28.6%. The average output of crystals from the underflow reached 350kg/h, significantly satisfying the requirements of copper crystal production.

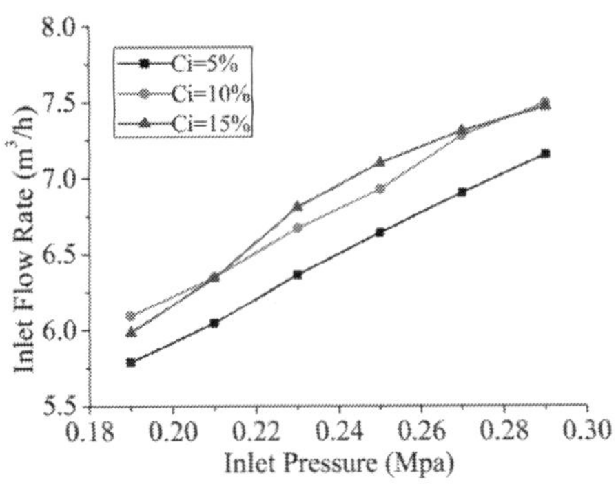

Figure 3. Relationship between inlet flow rate and inlet pressure.

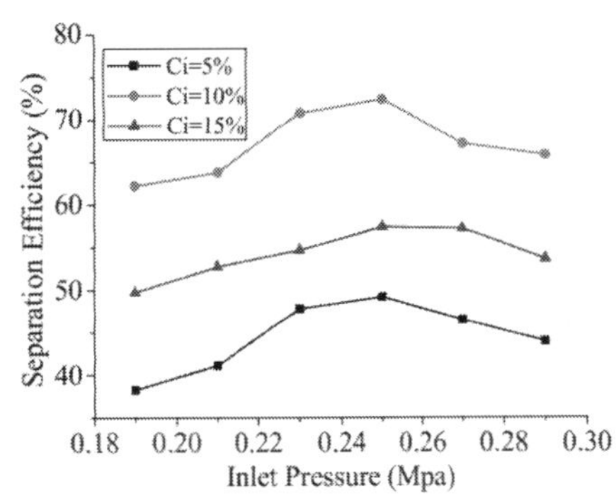

Figure 4. Relationship between separation efficiency and inlet pressure.

4. Data analysis of trial operation

In the experiments, 16 tons of waste etching solution were treated in 25 hours to obtain two tons of copper chloride crystals and approximately nine tons of regenerated $FeCl_3$ etching solution. Under the inlet concentration of 10% and an inlet pressure of 0.25Mpa, the size distribution of crystals in overflow slurry and underflow slurry were conducted by Mastersizer 2000. The results are depicted in Figures 5(a) and 5(b). The particle size in the overflow slurry was generally approximate to 40 microns and particles greater than 90 microns only accounted for 10%, which effectively decreased the possibility that circulating mother liquor would become stuck on the heat exchange tubes. In addition, the particle size in the underflow slurry generally consisted of 105 microns, and particles less than 40 microns only accounted for 8%. Therefore, the slurry after hydrocyclonic separation achieved an obvious thickening effect, which helped to improve the separation efficiency of the centrifuge and decreased the possibility that fine crystals could block the filter cloth. The following section primarily compares the cost and the quality of products before and after process optimization.

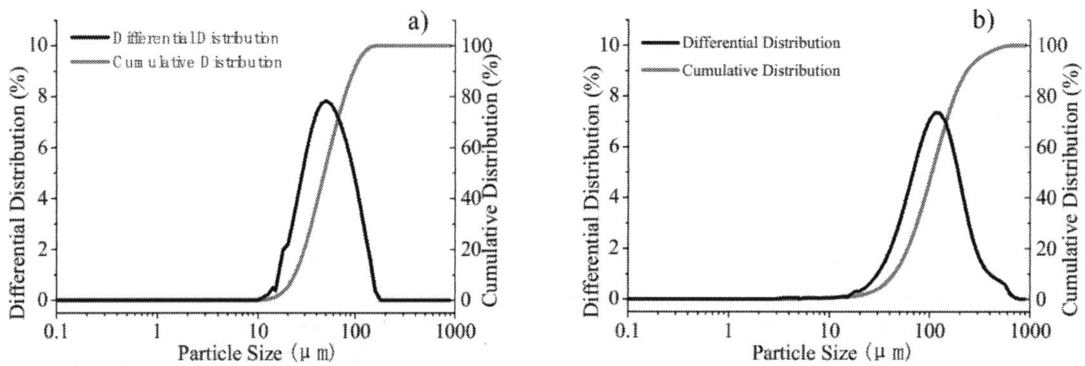

Figure 5. Particle distribution under the inlet concentration of 10% and inlet pressure of 0.25Mpa: (a) overflow particle; (b) underflow particle.

4.1. Analysis of product quality

After the slurry was concentrated by the hydrocyclone and then separated by the centrifuge, the particle mainly consisted of 105 microns and the surface area of the copper chloride crystals decreased by 50%. Therefore, after the washing treatment, production increased by 20% and its concentration increased from 95% to 98%. In the regenerated $FeCl_3$ etching solution, the concentrations of Fe^{3+}, Fe^{2+}, Cu^{2+} were 44.27%, 0.07%, 0.13%, respectively, and its density was 1454kg/m3, which meets the concentration requirements regenerated etching solution. Consequently, the closed recycling regeneration process of $FeCl_3$ waste etching solution and copper extraction successfully achieved the resource utilization.

4.2. Analysis of operating cost

After the process was modified, the waste etching solution was treated as a capacity of 0.65 tons per hour. According to the actual situation, the average utilization rate of the water source heat pump unit, the vortex air cooled water chilling unit and the root water jet vacuum pump unit reached nearly 90% per hour. The utilization rate of the electric motor connected to the hydrocyclone and the centrifuge was approximately 50%. Therefore, the total power per hour of the system is approximately equal to $(144.4+41.7+14+3\times11)\times90\%+(11+5.5)\times50\%=218$kW/h.

Since the capacity of the centrifuge and the heat loss of separation was reduced, the treatment of the waste etching solution per ton could save approximately10% of electric energy as compared to the unmodified process.

4.3. Analysis of resources consumption

In terms of resource consumption, if one ton of waste etching solution was treated, considering the instability of Fe^{2+}, hydrolysis of Cu^{2+}, density and oxidation of recycled etching solution, the process required only three cubic meters of chlorine, 2.5 kilograms of hydrochloric acid and 100 kilograms of water.

The reaction of iron powder replacement of sponge copper not only consumes a large amount of expensive iron powder, but also emits significant heat, requiring high heat resistance of equipment. Additionally, the price of copper chloride crystals is approximately two times that of copper powder. Taking the consumption of chemical resources and the benefits brought from by-products into consideration, the process proposed in this work is a better choice for the processing of $FeCl_3$ waste etching solution containing Cu.

5. Conclusion

The significance of the closed recycling regeneration process of $FeCl_3$ waste etching solution is primarily the result of the following conclusions.

- After the process was modified, the evaporation and crystallization system can be run continuously, the crystal slurry flows fluently and the faults from heat pipe fouling and impeller erosion are greatly reduced.
- Crystal slurry concentration and classification are conducted in the closed environment of the hydrocyclone. Less than 20% of the concentrated slurry is separated by the centrifuge, which greatly reduces the workload of the centrifuge and thus decreases the heat loss and crystal loss caused by vapour dissipation. Simultaneously, because the average surface area of copper chloride is reduced, the surface adhesion of ferric chloride crystals and loss during the washing process are also decreased. As a result, the production of copper chloride crystals is increased by 20% and the concentration after washing is increased by 3%.
- Adopting the combination method of evaporation crystallization and cooling crystallization has many advantages compared to other methods. The production process is conducted in a closed environment and no irritating gas leakage and toxic liquid discharge is produced, which achieves zero discharge of the waste etching solution and increases the recovery of metal resources.

Acknowledgments

The authors are grateful for the financial support from the Shanghai Shiyu Solid Waste Disposal Co.,Ltd and the technical support from the East China University of Science and Technology, the Institute of Chemical Machinery.

References

[1] Cakir O, Temel H and Kiyak M 2005 Chemical etching of Cu-ETP copper *Journal of Materials Processing Technology* **162** 275–9

[2] Liu P, Du Y G, Zhang W J, et al. 2006 A study of the Regeneration of $FeCl_3$ *Etching Solution Electroplating & Pollution Control* **26** 36–9

[3] Tong X X, Xu J, Ge T, et al. 2014 Research Process in the Recycling of $FeCl_3$Waste *Etching Solution Plating & Finishing* **36** 17–20

[4] Wang B, Zhou J W and Yu X Z 2009 *An regeneration recovery method of waste etching solution containing copper chloride and ferric chloride* CN101462803A

[5] Xie D F and Tan Z G 2005 Comprehensive utilization of waste ferric chloride etchant containing nickel *Energy Environmental Protection* **19** 37–9

[6] Koene L and Janssen L J J 2001 Removal of nickel from industrial process liquids *Electrochimica Acta.* **4** 695–703

[7] Lee M S and Lee K J 2005 Separation of iron and nickel from a spent $FeCl_3$ etching solution by solvent extraction *Hydrometallurgy* **80** 163–9

[8] Zheng W F, Xiong Z D and Sun S P 2008 *Recovery technique and device for waste etching solution* CN101307446A

[9] Femandez-Torres M J , Randall D G, Melamu R, et al. 2012 A comparative life cycle assessment of eutectic freeze crystallization and evaporative crystallization for the treatment of saline wastewater *Desalination* **306** 17–23
[10] Fu P Q 2014 *Study on Copper Chloride waste liquid containing resources utilization* Guangdong Chemical Industry **41** 162–5
[11] Liu G Q, Ma L X and Liu J 2002 *Handbook of Physical Property Data for Chemical Industry-organic volume* (Beijing: Chemical Industry Press) pp 476–7
[12] Zheng P Y, Yu J S, Zhang S P, et al. 2006 Study of Heat and Mass Transfer Patterns in Evaporative Crystallization Systems *Science Technology and Engineering* **6** 1002–06
[13] Zhao Q G and Zhang M X 2003 *Hydrocyclone Separation Technology* (Beijing: Chemical Industry Press) pp 241–2
[14] Wu C, Yue Y P, Yang H Q, et al. 2003 *Research progress on crystallization from solution* Jiangxi Chemical Industry **32** 33–5

Landfall tropical cyclone rainstorms on the north slope of the Dabie Mountains

Z Y Huang[1, 2, 3], J Y Wang[1, 4] and C Lee[1]

[1] Institute of Heavy Rain, Wuhan, CMA 430205, China
[2] China University of Geosciences, Wuhan 430074, China
[3] State Key Laboratory of Severe Weather, Beijing 100089, China

E-mail: amywang060@163.com

Abstract. The formation and development mechanism of landfall cyclone rainstorms that occur on the north slope of the Dabie Mountains were investigated by the determination of typical occurrences. Interaction between the tropical cyclone and the westerly trough was characterized by the favorable circulation backgrounds of landfall tropical cyclone rainstorms on the north slope of the Dabie Mountains. A conveyor belt was created between the easterly jet flow of the tropical cyclone and the subtropical high pressure of the western equatorial Pacific Ocean and the southerly jet flow of the westerly trough front, creating a huge amount of energy and vapor from the landfall tropical cyclone in the rainstorm area and destabilizing the stratification. These conditions were advantageous to the frontogenesis of a warm front and the development of Mesoscale convective systems (MCS) in the westerly cold air that met the inverted trough located at the northern portion of the tropical cyclone. The existence and development of the mesoscale front area in the ground provide a trigger mechanism for the rainstorm. The MCS occurred and developed in the equivalent potential temperature theta se (θse) frontal zone, which is located between the low pressure area of the typhoon and the cold air, which is located at the rear of the westerly trough. The terrain block slowed or stopped the motion of the low pressure system formed by the landfall tropical cyclone, which was conducive to the enhancement of the rainstorm.

1. Introduction

The area on the north slope of the Dabie Mountains primarily includes the upper and middle reaches of the HuaiHe River, which is one of multiple rainstorm areas in China [1].The track movement of Tropical Cyclones after landfall are of three types: towards the west, towards the north, and towards the northwest. When the tropical cyclone moves toward the northwest, it is prone to the enhancement of rainstorms on the north slope of the Dabie Mountains, which can produce extreme precipitation [1]. For example, the extreme Henan "75.8" rainstorm was enhanced by the interaction between a tropical cyclone and a mid-altitude weather system after the cyclone landed, resulting in extreme precipitation values that reached 189.5mm within one hour and 680mm within six hours, and resulted in a total precipitation of 1631mm [1, 2].

This type of rainstorm occurs outside of tropical cyclone circulation; however, a physical connection with the tropical cyclone is inherent to this event [3, 4]. The characteristics of these rainstorms include a sudden, strong, relatively concentrated precipitation period, local precipitation intensity and an obvious mesoscale system with prominent mesoscale features [3, 4, 5]. The interaction between weather systems of different scales and at different latitudes provides a favorable large-scale circulation background for the occurrence and development of the MCS of the landfall cyclone rainstorm. When southwest monsoons and tropical cyclones interact, the abundant water vapor induced by the southwest monsoon contributes significantly to the formation and development of MCS, which results in the occurence of rainstorms [6, 7].

The occurrence and development of tropical cyclone circulation potential vorticity, water vapor and energy transport provide favorable conditions for the development of the MCS. As a powerful source of disturbance, in a suitable atmospheric node, the effect of a tropical cyclone on the environment is characterized by nonlinear dynamics similar to a Rossby disturbance wave form, which stimulates the occurrence and development of MCS. A large amount of potential energy is converted into kinetic energy, which causes the MCS and promotes the occurrence of heavy rain [8, 9, 10].

Based on special rainstorm occurrences, there are obvious geographical characteristics and limitations that affect the formation mechanisms of rainstorms caused by tropical cyclones, which must be investigated in greater depth. This study investigates the mechanism of tropical rainstorm cyclones in the northern Dabie Mountains, including the background characteristics of large-scale circulation, water vapor sources and transport processes, as well as MCS triggering mechanisms [11, 12, 13, 14, 15].

2. Date and methodology

A total of 57 stations within the 113-117^{0}E and 33-38^{0}N region were identified. A rainstorm occurrence was defined as an event that affected no fewer than five stations and resulted in precipitation exceeding 50mm over 24h of constant rain. From 1996 to 2015, there were a total of seven documented tropical cyclone rainstorms that occurred on the northern slope of the Dabie Mountains.

Precipitation data was obtained from the Atmospheric Detection Center of China Meteorological Administration. The diagnosis analysis data was the NCEP final analysis data.

3. Results and discussion

3.1. Circulation Characteristics

On July 19, 2005 (UTC), the "Haitang" typhoon reached land in Lianjiang County of Fujian Province. After touching land, the storm system moved northwest on July 20th (UTC) and was reduced to a low pressure system. On July 22, a wide area of heavy rainfall occurred on the northern slope of the Dabie Mountains, delivering more than 50mm of precipitation at 38 rainfall stations, more than 100mm of precipitation at 13 stations, and a maximum recorded precipitation of 295mm.

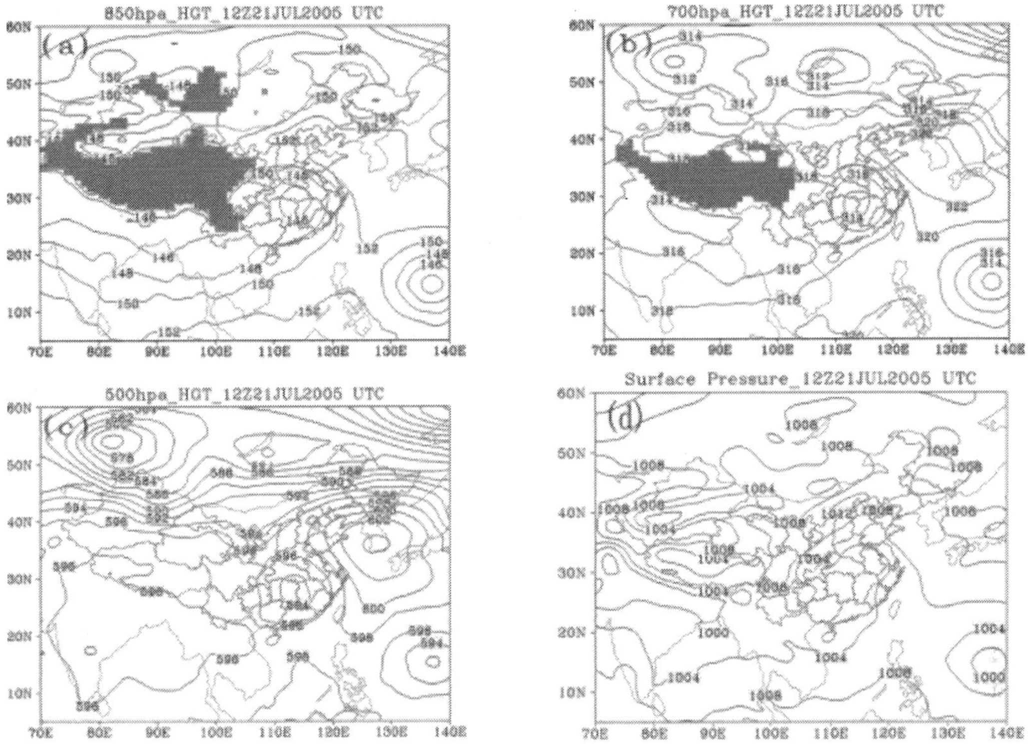

Figure 1. Circulation characteristics before the rainstorm at 12:00 (UTC) on 21 July 2005. (a) 850 hPa; (b) 700 hPa; (c) 500 hPa; (d) surface.

As shown in figure 1, prior to the rainstorm on July 21, at 12:00 (UTC), in 500hPa, the low-pressure center of the "HaiTang" typhoon was located in the Poyang Lake area, and the trough of "HaiTang" stretched northward to the north slope of the Dabie Mountains. The sub-tropical high pressure area in eastern and northern China created a trough that moved eastward into the Hetao area. Additionally, a tropical storm was forming to the east of the Philippines, as shown in figure 1(c). As shown in figures 1(a) and (b), respectively, in the Huaihe area at 700hPa and 850hPa, there was a low level of easterly jet flow, with the greatest wind speed reaching 18m/s. The atmospheric circulation evolved from 00:00 to 12:00 on July 22 in the "HaiTang" low-pressure center, positioned in the Poyand Lake region, though the subtropical intensity and scope did not change significantly. However, the low trough in the Hetao area moved toward the northern Dabie Mountains and began to weaken and contract in the north. On July 12, at 21:00 (UTC), the weak cold air located at the surface accumulated in the Hetao area, and as of 22:18 (UTC), it began to approach the north slope of the Dabie Mountains, as shown in Figure 1(d). The rainstorm on the north slope of the Dabie Mountains in the 500hPa trough then moved to the trough located to the north of the "HaiTang" typhoon. This occurred as a result of the interactions between the tropical storm, the subtropical high, the westerly trough, the low-level jet and the cold air at the surface.

4. Water vapor transport and convergence

Adequate water vapor supply is one of the basic criteria for the formation of heavy rainfall. Tropical cyclones travel over the warm ocean and thus serve as heat engines, and they transfer water vapor and energy to the area of heavy rainfall. On July 22, 2005 at 12:00 (UTC), before the heavy rain began, abundant moisture had gathered in the troposphere of the Dabie Mountains, nearly saturating the atmosphere with a large amount of convective available potential energy (CAPE). As shown in figure 2, when "HaiTang" travelled to the northwest, the southwest monsoon from the Indian Ocean advanced, and the monsoon and the "HaiTang" typhoon merged to form a southerly jet in eastern China. Additionally, the easy flow of the subtropical high combined with the southerly airflow of the eastern portion of "HaiTang" to form a southeastern low-level jet in the region of eastern China, as shown in figure 2. Finally, J-1 and J-2 combined to form a strong partial low-level easterly jet flow in the HuaiHe area, as shown in figure 2. This transported water vapor and energy to the Dabie Mountains area and led to the strong water vapor convergence in the area. Additionally, the southerly airflow on the western side led to water vapor convergence on the north slope of the Dabie Mountains. As a result, the low-level jet flow stream on the eastern side of the "HaiTang" monsoon provided a link through which the middle latitude and low latitude weather systems could interact.

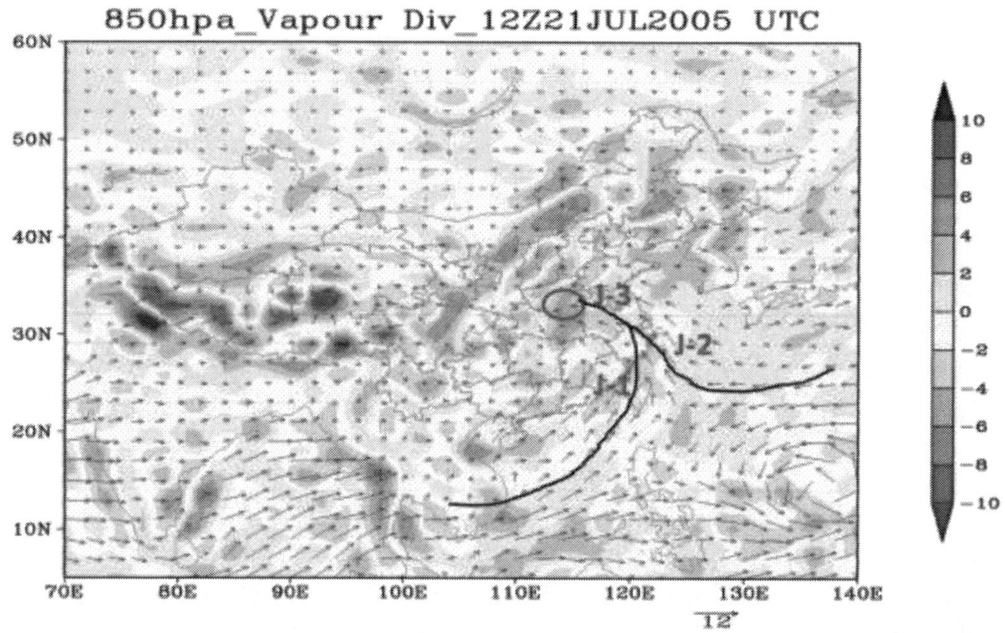

Figure 2. 850hPa water vapor flux divergence and wind vector at 12:00 (UTC) on 21 July 2005. Water vapor flux divergence (color-filled, $g{\cdot}hPa^{-1}{\cdot}cm^{-2}{\cdot}s^{-1}$) and wind vector (black arrows, $m{\cdot}s^{-1}$).

According to the water vapor transport data provided in figure 3, the transport of water vapor occurred primarily in the 700 hPa layer, while the transport of water vapor was very low at 500 hPa and above. The 850 hPa layer and the 950 hPa layer included two vapor transport belts. One belt was consistent with the direction of the southwest monsoon, which transported water vapor from the Indian Ocean and the South China Sea to the area of heavy rainfall. The other belt was located to the southwest of the eastern portion of the "HaiTang" typhoon and the subtropical high, and transported water vapor from the East China Sea to the area of heavy rainfall. These two conveyor belts were in contact with J-2 and J-1, respectively, and finally merged in the Huaihe area. In addition, the terrain of the Dabie Mountains blocked the easterly jet, forcing the jet to climb in altitude, which was therefore conducive to the further development of upward movement.

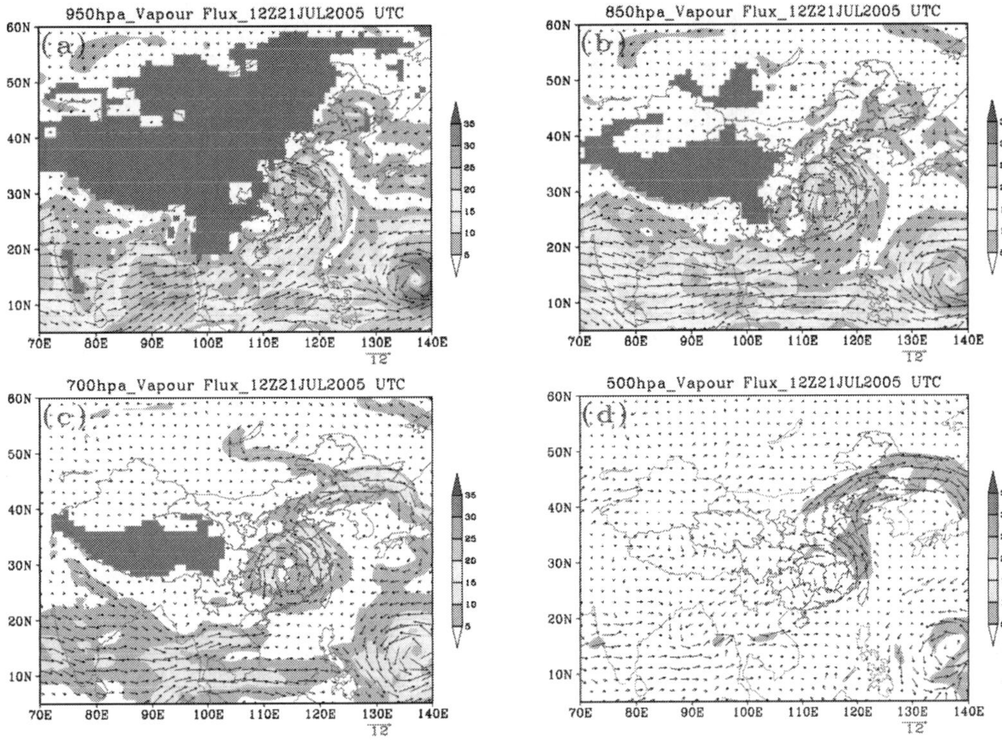

Figure 3. Water vapor flux and wind vector at 12:00 (UTC) on 21 July 2005. Water vapor flux (color-filled, $g \cdot hPa^{-1} \cdot cm^{-1} \cdot s^{-1}$) and wind vector (black arrows, $m \cdot s^{-1}$).

5. Unstable features and dry intrusion

As shown in figure 1, the rainstorm on the north slope of the Dabie Mountains occurred in the westerly "HaiTang" trough on the northern front, as a result of the interaction between the cold and warm air on the north slope of the Dabie Mountains. Figure 3 depicts the evolution of θse. As the westerly trough moved eastward and "HaiTang" moved to the northwest, a θse frontal zone formed between the cold air from the westerly trough and the warm air from the "HaiTang" typhoon. From 950 hPa to 500 hPa, the θse frontal zone tilted toward the west. The mesoscale clouds developed a pattern in the 850-700 hPa θse frontal zone, resulting in heavy rainfall. Additionally, the θse face steeped, easily leading to the development of moist slantwise vorticity that was conducive to the enhancement of ascending motion.

IOP Conf. Series: Earth and Environmental Science **39** (2016) 012047 doi:10.1088/1755-1315/39/1/012047

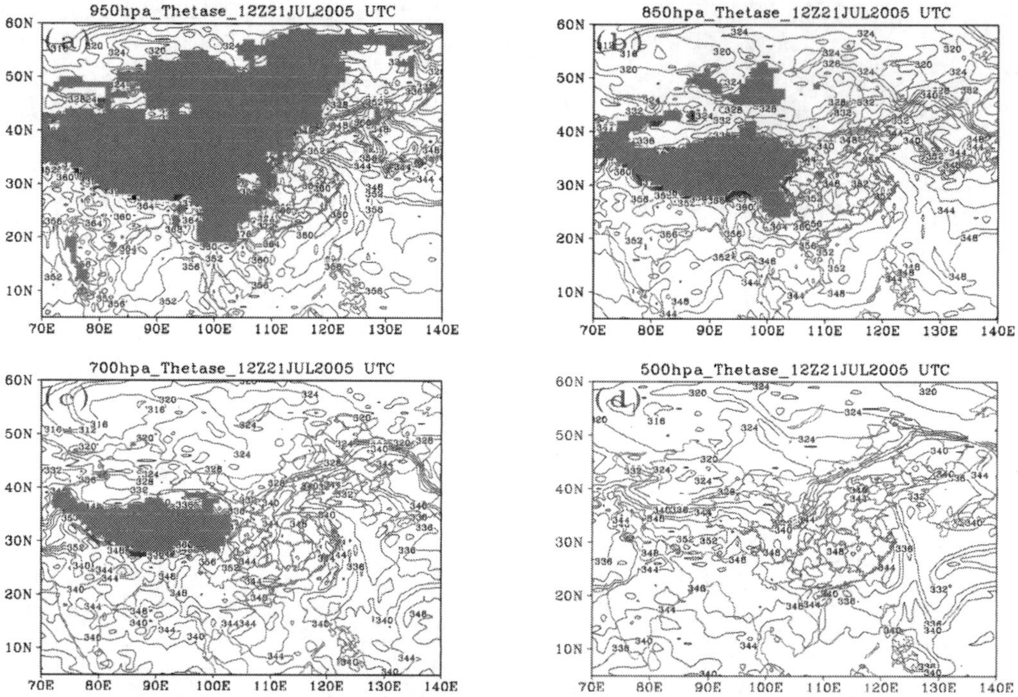

Figure 4. The distribution of θse (contour, ℃) at 12:00 (UTC) on 21 July 2005.

Figure 4 depicts the low level dry cold air invasion process of tropical cyclone troughs. The intersection between the westerly flow and the easterly flow occurred on the west side of the storm zone. In the 925hPa layer, cold air extended into the heavy rain area, and in the 850-700 hPa layer, the partial easterly jet at the bottom of the cold air began an upward climb. The cold air from the 700-500 hPa layers were divided into two portions, as depicted in figure 5: the invasive air in the trough of "HaiTang" was forced over the storm and to the western side of the rainstorm zone. This indicates that the cold air in the middle and lower layers was continuously increasing as a result of the westerly trough. The middle level of the cold air intrusion under the "HaiTang" trough consisted of warm moist air flow at the bottom, which forced the cold air to climb. These circulation conditions were very conducive to the development of MCSs. Additionally, the storm area was located to the upper right front of the exit area of the upper air jet stream and the left front of the low-level wind jet. The coupling between the upper level westerly jet and the low-level jet strengthened the low level convergence and the upper level divergence in order to promote and maintain the vertical movement of the developed storm.

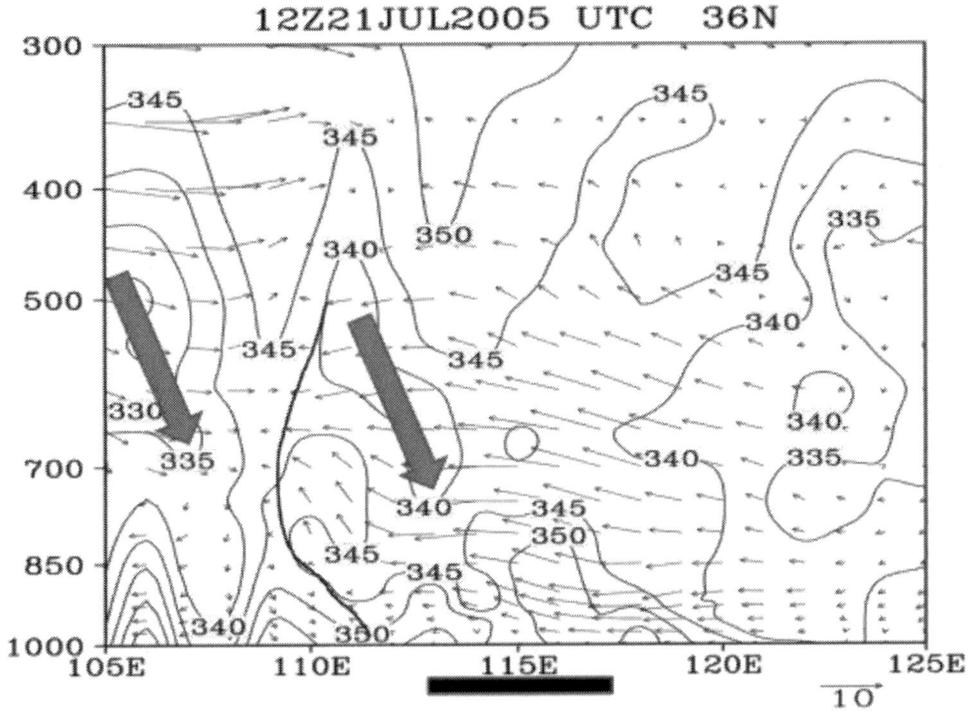

Figure 5. Zonal section of the rainstorm cente r("▬"shown) along 36^{0}N at 12:00 (UTC) on 21 July 2005. θse (contour, ℃), wind vector (black arrows, m•s^{-1}).

5.1. Upward movement mechanism

As shown in figure 6, the 850hPa layer demonstrated strong convergence over the area of heavy rainfall, the 500-400hPa layers demonstrated strong radial scattering, and the 200hPa layer demonstrated strong radial scattering. The convergence of the 850hPa layer was caused by the easterly flow. The divergence of the 500-400hPa layers was a result of the cold advection of the westerly trough. The convergence of the 200hPa layer was caused by the upper level jet stream. As demonstrated by the vertical distribution of the vertical velocity, a rising movement occurred from the boundary layer to the middle troposphere in the rainstorm area. These results clearly indicate that the high-level divergence and low-level convergence of the atmospheric circulation conditions were conducive to the development of upward movement. In addition, the vertical velocity of the zonal profile (figure 6a) demonstrates that both sides of the storm, to the east and west, exhibited sinking movement. The sinking movement on the edges and the rising area were induced by secondary circulation rings. The two secondary circulation rings were located to the east and west of the rainstorm area, which served to maintain the rainstorm.

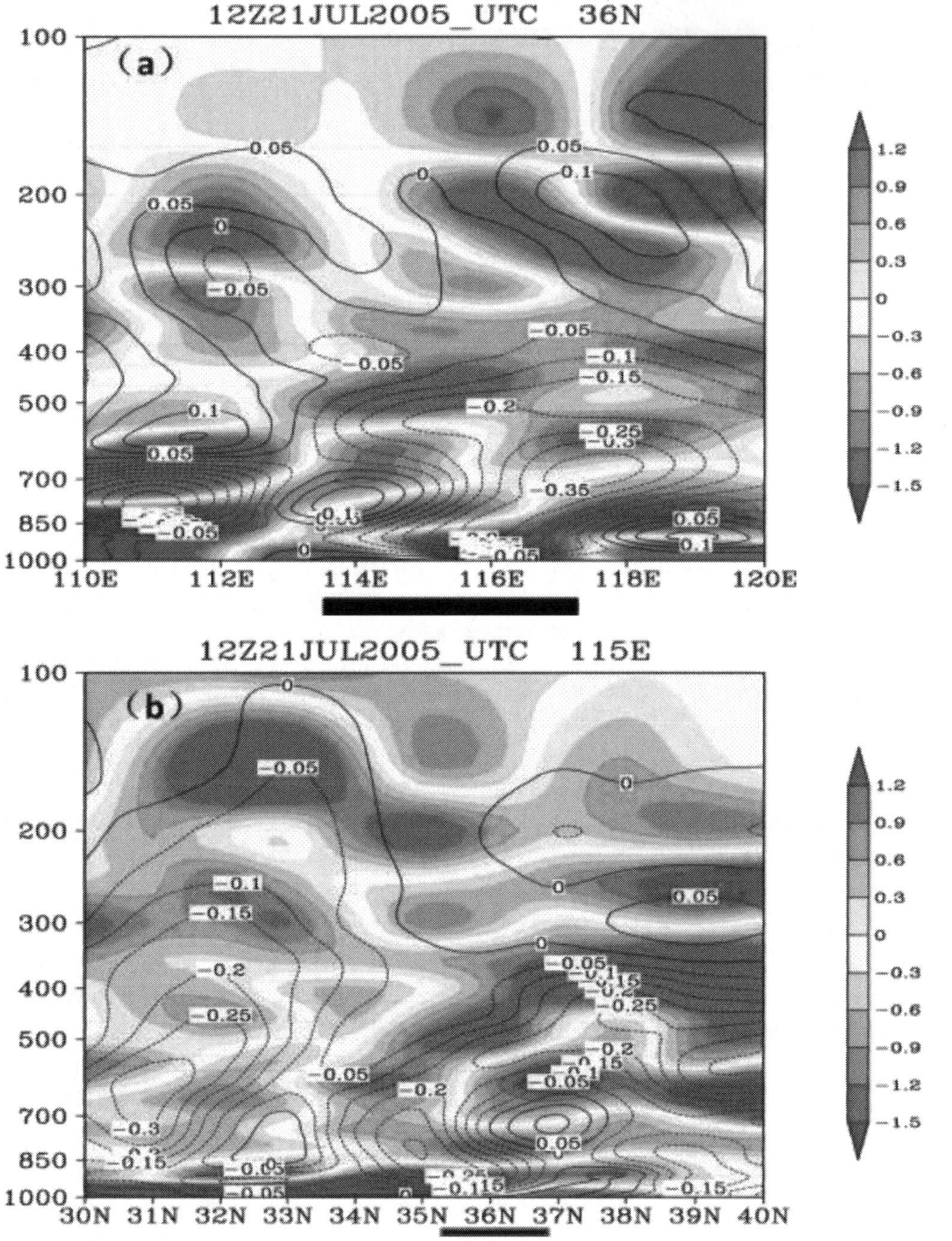

Figure 6. Zonal section of the rainstorm center ("■"shown) along 36^{0}N. (a) Meridional cross-section of the rainstorm center along 115^{0}E (b) at 12:00 (UTC) on 21 July 2005. Vertical velocity (color-filled, Pa•s^{-1}) , wind vector (black contour, m•s^{-1}).

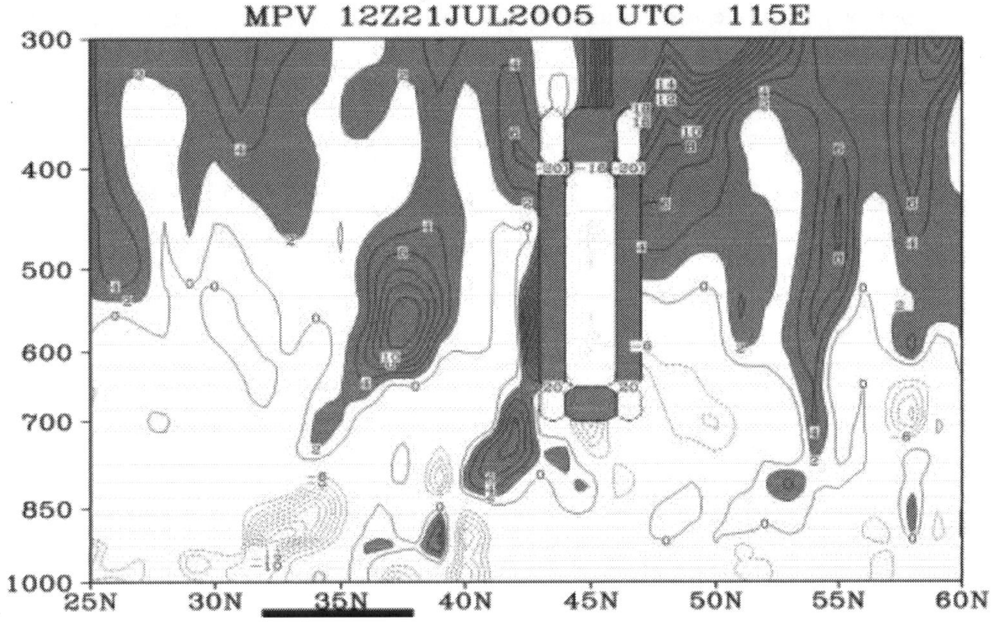

Figure 7. Meridional cross-section of the rainstorm center ("▬"shown) along 115^{0}E at 12:00 (UTC) on 21 July 2005. Moist potential vorticity (contour line, $10^{-5}s^{-1}$).

Under moist potential vorticity conservation conditions, the increase in the vertical shear of the horizontal wind or the wet compression of the horizontal wind would lead to significant development of the vertical vorticity as well as enhancement of the upward movement. The rainstorm area was located in the center of the negative center of the 850hPa moist potential vorticity and in the center of the 500hPa moist potential vorticity (figure 7). The vertical distribution of the moist potential vorticity strengthened and maintained the rainstorm.

6. Conclusions

Based on the NECP reanalysis data, the occurrence and development mechanisms of the tropical cyclone rainstorm that occurred on the north slope of Dabie Mountains were investigated. The study suggests the following conclusions.

- As the tropical cyclone moved toward the northwest, the subtropical high was strengthened and extended westward. A low level easterly jet was formed between the two systems in the east, and warm, wet air flow was transported to the north by the easterly jet. Induced by the movement of the middle-latitude westerly trough, cold air moved to the south and the cold and warm air on the northern slope of the Dabie Mountains experienced confluence, resulting in torrential rain.
- A southerly low-level jet was formed between the east side of the tropical cyclone, the west side of the subtropical high and the front of the westerly trough, which conveyed steam and energy to the rainstorm area. Between the jet and the jet stream, positive feedback made the circumstances conducive to the enhancement of low level convergence and the development of vertical motion.
- The cold air at the surface was the trigger mechanism for the development of MCSs. Among the low level dry air intrusion of the low pressure trough to the north, the

atmospheric stratification became extremely unstable. Mesoscale convective clouds developed in the θse frontal zone, which led to an increase in the heavy rainfall.

Acknowledgements

This work was supported by the Public welfare industry special (meteorological) funds (grant GYHY201306012) and Research and Operation Project for IHR (IHRKYYW201604).

References

[1] Chen L S, Luo H B and Duan Y H 2004 An overview of tropical cyclone and tropical meteorology research progress *Advances in Atmospheric Sciences* **21**(3): 505-14

[2] Meng Z Y, Chen L S and Xu X D 2002 Recent progress on tropical cyclone research in China *Advances in Atmospheric Sciences* **19(**2): 103-10

[3] Kepert J D 2006 Tropical cyclone inner core dynamics *Report on Topic 1.2 of 6th International Workshop on Tropical Cyclones. San Jose, Costa Rica,WMO* 79-103

[4] Chen L S 2007 Forecast on landfall tropical cyclone rainfall *International Training Workshop on Tropical Cyclone Disaster Reduction (Lecture Note Volume)* Guangzhou: WMO **8-19**

[5] Teixeira M D Sand Satyamurty P 2007 Dynamical and synoptic characteristics of heavy rainfall episodes in Southern Brazil *Mon.. Wea. Rev.* **135**(2): 598-617

[6] Elsberry R L 2002 Predicting hurricane landfall precipitation: Optimistic and pessimistic views from the symposium on precipitation extremes *Bull Amer. Meteor. Soc.* **83**(9): 1333-39

[7] Carvalho L M V, Jones C and Liebmann B 2002 Extreme precipitation events in southeastern South America and large—scale convective patterns in the South Atlantic convergence zone *J. Climate* **15**(17): 2377-94

[8] Atallah E, Bosart L F and Aiyyer A R 2007 Precipitation distribution associated with landing falling tropical cyclones over the eastern United States *Mon. Wea. Rev.* **135**(6): 2185-06

[9] Franklin C N, Holland G J and May P T 2006Mechanisms for the generation of mesoscale vorticity features in tropical cyclone rain bands *Mon. Wea. Rev.* **134**(10): 2649- 69

[10] Lin Y L, Chiao S, Wang T A, et al. 2001 Some common ingredients of heavy orographic rainfall *Wea. Forecasting* **16**(6): 633-60

[11] Jones S C 2004 On the ability of dry tropical-cyclone-like vortices to withstand vertical shear *J. Atmos. Sic.* **61**(1):114-19

[12] Corbosiero K L and Molinari J 2002 The effects of vertical wind shear on the distribution of convection in tropical cyclones *Mon Wea. Rev.* **130**(8):2110-23

[13] Corbosiero K L and Molinari J 2003 The relationship between storm motion, vertical wind shear and convective asymmetries in tropical cyclones *J. Atmos. Sic.* **60(**2):366-76

[14] Chen S S, Knaff J A and Marks F D Jr. 2006 Effects of vertical wind shear and storm motion on tropical cyclone rainfall asymmetries deduced from TRMM *Mon. Wea.. Rev.* **134**(11): 3190- 208.

[15] Chow K C, Chan K L and Laua K H 2002 Generation of moving spiral bands in the tropical cyclones *J. Atmos. Sci.* **59**(20): 2930-50

Effect of manure under different nitrogen application rates on winter wheat production and soil fertility in dryland

H Q Zhang [1,2], X Y Yu [1,2], B N Zhai [1,2,3], Z Y Jin [1,2] and Z H Wang [1,2]

[1] College of Natural Resources and Environment, Northwest A&F University, Yangling, Shaanxi 712100, China
[2] Key Laboratory of Plant Nutrition and the Agro-environment in Northwest China, Ministry of Agriculture, Yangling, Shaanxi 712100, China

E-mail: bingnianzhaitg@126.com

Abstract. Exploring an effective fertilization practice is crucial for achieving a sustainable dryland winter wheat cropping system. Following a split-plot design, this study was conducted to investigate the combined effect of manure (-M or +M; main plot) and various rates of nitrogen (N) fertilizer (0, 75, 150, 225, and 300 kg N ha-1; sub plot) on grain yield, water and N use efficiencies of winter wheat, and soil nutrients. The results showed that the treatments with manure improved the grain yield by 8%, and WUE by 10% relative to that without manure throughout the study years. The highest winter wheat yield and WUE were both recorded in the M+N225 treatment, which were not significantly different from those for M+N75 and M+N150 treatment. In contrast, high levels of N fertilizer (> 150 kg N ha-1) combined with manure not only caused a reduction in the N use efficiency (NUE), but it also caused an increase in the soil residual nitrate-N (from 43.7 to 188.9 kg ha-1) relative to without manure. After three years of continuous cropping, the treatment combining manure with 150 kg N ha-1 fertilizer had the highest SOM, available P and available K, which was 24%, 379% and 102% higher than that for unfertilized treatment (CK), and 10%, 267%, and 55% higher than that for without manure, respectively. Thus, the combination of manure (17.5 t ha-1 poultry or 30 t ha-1 pig manure) with 75-150 kg N ha-1 N fertilizer is recommended for improving winter wheat yield, water and N use efficiencies, and reducing soil nitrate-N residue as well.

1. Introduction

Water deficit and soil infertility are two major factors that limit cereal production in rainfed drylands that do not receive any supplemental water. The Loess Plateau in China had an average annual precipitation between 200-600 mm and a strong evaporation as a result of low infiltration and precipitation per event [1].Therefore, it is characterized as a typical dryland. Furthermore, greater than 90% of the croplands in this region are rainfed. Winter wheat (*Triticum aestivum L.*) is one of the main cereal crops cultivated on the Loess Plateau, and it has a planting area of about 5 million hectares, which accounts for 40% of the total food crops cultivation area [2]. However, the winter wheat grain yields have been low and unstable for a long time, only averaging from 2.5 t ha^{-1} to 3.7 t ha^{-1} [3], which is much lower than the average yields in China. On the one hand, the irrational local cultivation and fertilization practices accelerates the degradation of soil quality, which results in a low soil organic matter content (SOM). Currently, two-thirds of the croplands in this region have an SOM below 1.0%. On the other hand, the rainfall distribution in this region is temporally uneven. For example, 60% to 70% of the precipitation falls during July-September [4], which is when the winter wheat fields stay fallow. Drought often occurs during the growing season of winter wheat, and this water deficiency hastens maturation, shortens the grain-filling period, and reduces dry matter production, thereby resulting in low wheat yields on the Loess Plateau [5].

To combat this, fertilization is becoming crucial method to enrich soil fertility and "muster water" for meeting the crop growth demands and improving the crop production in the Loess Plateau [6]. However, the unadvised and excessive application of chemical fertilizer (especially N) by local farmers has brought a series of environmental issues, including eutrophication, groundwater contamination, and greenhouse gas emissions [7]. With the rapid development of agriculture, there is an increasing emphasis on soil fertility build-up and environmental protection to achieve the sustainability of dryland farming, rather than simply improving crop yields. Several studies have reported that the combined application of manure with inorganic fertilizers played an important role in improving soil nutrients, crop yields, fertilizer, and water use efficiency [8-11]. However, an unreasonable combination ratio of manure-N and inorganic-N does not result in high yields or water and N use efficiencies, but it instead caused a large amount of soil nitrate-N residue accumulation, which migrates down to deeper soil [12] and causes pollution in the underground water. Thus, the present study aimed to: (1) investigating the combined effect of manure with different rates of N fertilizer on grain yield, water and fertilizer use efficiencies of winter wheat, and soil nutrients; (2) exploring a locally optimal combination of manure and chemical N fertilizer for winter wheat production and environmental conservation. The results of this study will be theoretical basis for winter wheat production and soil fertility build-up in dryland.

2. Materials and methods

2.1. Experimental site

A 3-year winter wheat monoculture experiment was conducted from October 2011 to June 2014 in Bai Shui (109°58'N 35°18'E, 850 m asl) Shaanxi Province, China. Bai Shui County is located in the Loess Plateau, with an average annual temperature of 11.4°C and precipitation of 577.8 mm. The precipitation amounts during the 3 study years were 710.1 mm, 391.4 mm, and 603.8 mm in 2011-2012, 2012-2013, and 2013-2014, respectively. Based on the Generalized Precipitation Classification Scheme [13], 2011-2012 was categorized as a wet year, 2012-2013 was a dry year, and 2013-2014 was a normal year.

The soil at the experimental site is classified as Calcic Cambisols (FAO taxonomy), with a bulk density of ~1.30 g cm-3. The background soil sample collected from the 0–20 cm layer at the experimental site had 14.4 g kg^{-1} of organic matter, 0.8 g kg^{-1} of total N, 9.2 mg kg^{-1} of Olsen-P, and 177.2 mg kg^{-1} of available K.

2.2. Experimental design

A locally widely used winter wheat cultivar, 'Jinmai - 47', was used as the test crop. Two factors were included in this study: with manure and without manure (+M and -M), and five N application rates, including 0, 75, 150, 225, and 300 kg N ha^{-1} (N0, N75, N150, N225, and N300). These 10 treatments were arranged in a split plot design with 4 replications, with the manure treatments as the main plot and the five N fertilizer rates as the sub plot. The non-fertilizer treatment was defined as the control (CK). The plot size was 6-m wide and 10-m long, and all the plots were arranged randomly in the field.

The manure-added areas (+M) were fertilized with a combination of manure and N fertilizer according to the five different application rates. Well-composted poultry manure and pig manure were collected from local farmers. Poultry manure was applied in the first year (2011-2012). In the latter two years (2012-2014) of the experiment, due to the scarcity of poultry manure resources, pig manure was applied. The application rates and nutrient contents of the poultry and pig manure are given in Table 1. The non-manure areas (-M) were only fertilized with the five rates of chemical N fertilizer. In addition, 90 kg P2O5 ha^{-1} and 60 kg K2O ha^{-1} were applied in all plots. The N fertilizer was applied as urea: 70% of which was applied at sowing, and 30% was at the elongation stage in all treatments. Phosphorus (P) fertilizer was superphosphate, and potassium (K) fertilizer was sulfate of potash. All the fertilizers were evenly applied and incorporated into the top 0-20 cm soil layers before seeding.

Winter wheat was sown on 12th October, 17th October, and 15th October of each respective year by a seeding machine at a rate of 150 kg ha-1 and with 20 cm-wide row spacing. The individual plots were plowed before sowing for seedbed preparation, and no irrigation was provided. All other field management practices were consistent with locally adopted practices during the three years study. The field was left fallow with low stubble after the winter wheat harvest, and the stubble and weeds were crushed and incorporated into the soil in late September before seedbed preparation.

Table 1. Nutrient contents and application rates of the manures.

Manure type	Application rates (t ha^{-1})	pH	Organic matter (g kg^{-1})	Total N (g kg^{-1})	Total P (g kg^{-1})	Total K (g kg^{-1})
Poultry manure (2011-2012)	17.5	8.7	47.5	18.3	21.3	20.0
Pig manure (2012-2013)	30.0	7.1	28.8	8.0	12.5	14.5
Pig manure (2013-2014)	30.0	7.6	25.3	7.1	10.6	12.4

2.3. *Sampling and sample analysis*

In early June, Three quadrats covering a 4 m^2 area were selected randomly and manually harvested from the respective plot every study year. The crop was tied into bundles and then left in mesh bags for drying for about a week. After threshing, the air-dried grain was weighed to determine the grain yield. Additionally, crops which were randomly collected from the 3 by 1 m long rows were taken to the laboratory, and dried at 65 ℃ for 24 h. Then the oven-dried plant material were ground separately (grain and straw) with a grinder for nutrient analysis. The N contents in the samples were determined by the Kjeldahl method [14].

Three cores of soil samples (0-200 cm depth) in each plot were collected randomly by using an auger (inner diameter of 4.0 cm) before wheat sowing (early October) and after harvest (late June) every year. (This occurred in every year except for 2011, when only 7 cores were randomly selected in the whole experimental area before sowing). The soil from each 20 cm interval over the 0–200 cm layer was separated, and the soil from the same layer was mixed. The fresh soil samples were brought into the laboratory, and 10 g of each was used for estimating soil water content, by dryng in an oven at 105℃ for 12 hours for estimating the soil water content. Additionally, 5 g of each soil sample was extracted by 50 mL of 1 mol L-1 KCl and shaking for 1 h for the soil nitrate-N estimation [15]. Soil samples for soil nutrient determination were collected at a depth of 20 cm at the end of the experiment (June 2014). The SOM was determined by oxidizing SOM in the soil samples with K2Cr2O7 in concentrated sulfuric acid for 30 min followed by titration of the excess K2Cr2O7 with ferrous-ammonium sulfate [16]. The Kjeldahl method was used for soil total N determination, the Olsen extraction method was used for available P estimation[17], and the available K was determined by a flame photometer [18].

2.4. *Calculation methods*

The soil water storage was calculated as:

$$W = h \times \rho \times \theta \times 1000 \quad (1)$$

where h is the soil depth (cm); ρ is the soil bulk density (g cm^{-3}), and θ is the soil gravimetric water content (%).

The WUE was calculated as:

$$WUE = \frac{Y}{ET} = Y/(W1 - W2 + P) \quad (2)$$

where Y is the grain yield (kg ha^{-1}); ET is the evapotranspiration (mm); W1 is the soil water storage before the winter wheat planting (mm); W2 is the soil water storage after the winter wheat harvest (mm); and P is the precipitation during the winter wheat growing season (mm).

The NUE was calculated as:

$$NUE = \frac{N\ uptake\ by\ the\ fertilized\ treatment - N\ uptake\ in\ the\ control}{total\ N\ applied} \times 100 \quad (3)$$

Analysis of variance (ANOVA) was conducted to detect differences using SPSS 19.0 for Windows (SPSS Inc., Chicago, IL, USA). A one-way ANOVA was used to test for differences between the N application rates in the with manure group and without manure group, respectively. A two-way ANOVA was used to test for differences between groups with and without manure. Comparisons of means were performed using Duncan's test.

3. Results

3.1. Grain yield

From table 2, the grain yield varied greatly among the study years, and its variation was consistent with the annual precipitation. More specifically, the wet year (2011-2012) produced the highest winter wheat yield, whereas the dry year (2012-2013) produced the lowest. Single inorganic fertilizer application improved the winter wheat yield by 16%, 26% and 13%, whereas the combination of manure and chemical fertilizer resulted in an increase of yield by 18%, 34% and 16% as compared to that of the CK treatment in 2011-2012, 2012-2013, and 2013-2014, respectively. In general, additional manure increased grain yield by a mean of 8% relative to that of single inorganic fertilizer application throughout the three experimental years.

Table 2. The grain yield of winter wheat affected by manure and N rate.

Treatment	2011-2012 (kg ha^{-1})		2012-2013 (kg ha^{-1})		2013-2014 (kg ha^{-1})	
	-M	+M	-M	+M	-M	+M
N0	5156b	5456b	1906b	2414b	3805b	4145b
N75	5722ab	6004ab	2147b	2684a	4164ab	4410ab
N150	6300a	6325a	2504a	2760a	4459a	4595a
N225	6267a	6345a	2532a	2854a	4520a	4679a
N300	5656ab	5950ab	2489a	2595ab	4192ab	4264b
Average	5820B	6016A	2316B	2661A	4228B	4419A

Different lowercase letters (one-way ANOVA) and uppercase (two-way ANOVA) letters indicate significant differences at $p < 0.05$ (Duncan's test).

The grain yield increased first and declined afterwards with the increasing N rate, and the highest grain yield was always observed in the M+N225 treatment during the study years, which was 6345 kg ha^{-1} in 2011-2012, 2854 kg ha^{-1} in 2012-2013, and 4679 kg ha^{-1} in 2013-2014. However, no significant differences were observe between M+N75, M+N150 and M+N225 treatments for grain yield. When the N rate was exceeded 225 kg N ha^{-1}, increasing N input would not lead to a further increase in the winter wheat yield. The regression analysis indicated a parabolic relationship between the N application rate and the winter wheat grain yield (Figure 1). Maximum yield values in the –M group were 6,294, 2,559, and 4,500 kg ha^{-1} corresponding to N rates of 176, 204, and 201 kg N ha^{-1}, and in the +M group, the maximum yield values were 6,331, 2,825, and 4,648 kg ha^{-1} corresponding to N rates of 187, 179, and 173 kg N ha^{-1}. The results revealed that, compared to that without manure, less N fertilizer is needed to achieve a higher maximum winter wheat yield after combining the fertilizer with manure. This effect was more obvious in the dry year when 70 kg N ha^{-1} fertilizer was reduced, but a 10% higher grain yield was achieved by additional manure.

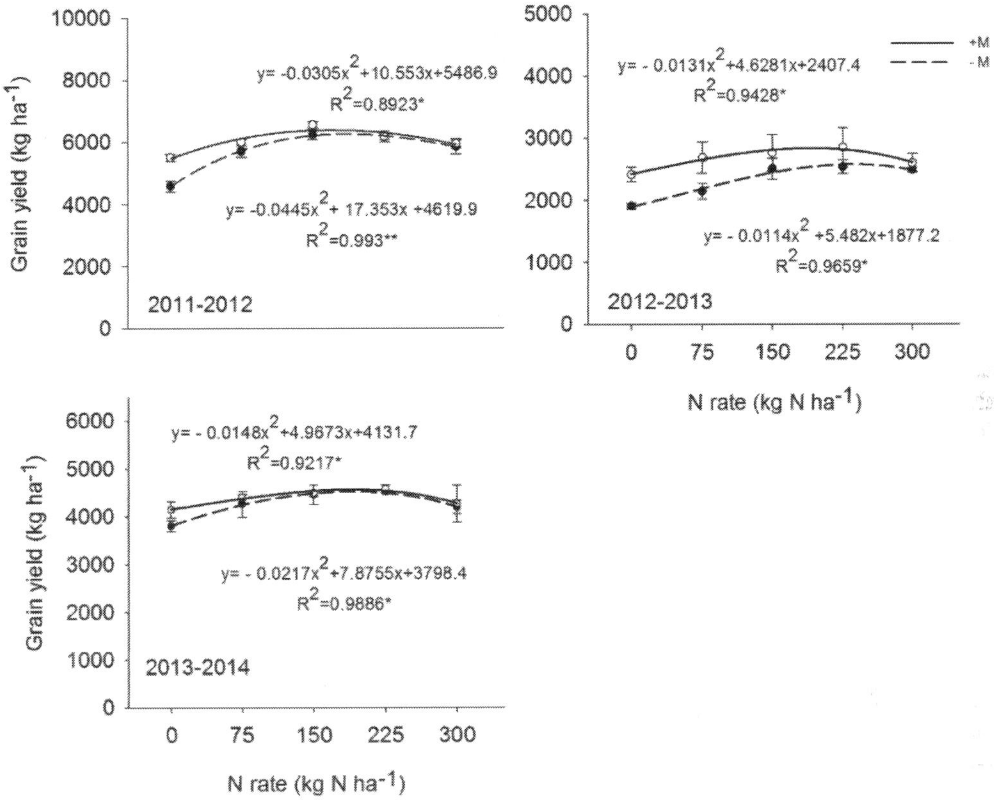

Figure 1. The variation of grain yield along with increasing N rate.

3.2. Water use efficiency

Due to the scarcity of available soil water on drylands, improving the efficiency of water uptake by plants is crucial in promoting crop yields. "Mustering water with fertilizer" is an effective and important method for increasing the soil water uptake by plants and therefore improving water use efficiency (WUE).

Table 3 reveals that the ET was greatly affected by the precipitation, as the wet year resulted in the highest ET, whereas the dry year lead to the lowest. However, the effects of N rate and additional manure on ET were not significant. Compared to that of CK treatment, single inorganic fertilizer application increased WUE by 9%, 42%, and 20%; the combination of manure and inorganic fertilizer increased WUE by 14%, 62%, and 36% in 2011-2012, 2012-2013 and 2013-2014, respectively. In general, the WUE of the winter wheat increased in the order of: CK < single manure application < single N fertilizer application < combination of manure and N fertilizer. Additional manure increased WUE by 17% in 2012-2013, and 11% in 2013-2014. The highest WUE was recorded in the M+N225 treatment for 2012-2013 and 2013-2014, and it was 73% and 43% higher than that for CK and 29% and 16% higher than that for the N225 treatment in 2012-2013 and 2013-2014, respectively. However, the differences between M+N75, M+N150 and M+N225 treatments were not significant for WUE.

Table 3. The evapotranspiration (ET) and water use efficiency (WUE) affected by manure and N rate.

Treatment	ET (mm)						WUE ($kg\ ha^{-1}\ mm^{-1}$)					
	2011-2012		2012-2013		2013-2014		2011-2012		2012-2013		2013-2014	
	-M	+M	-M	+M	-M	+M	-M	+M	-M	+M	-M	+M
N0	542a	580a	255a	245a	344a	336a	9.5b	9.4b	7.9b	10.4b	10.1b	11.3b
N75	550a	573a	247 a	251a	323a	336a	10.0b	10.5ab	9.2ab	11.3ab	11.5ab	13.1a
N150	524a	560a	238 a	251a	326a	348a	12.0a	11.3a	11.2a	11.6ab	12.3a	13.8a
N225	553a	561a	251a	222a	337a	332a	11.3ab	11.3a	10.6a	13.7a	12.5a	14.5a
N300	565a	583a	228b	228a	335b	343a	8.9b	10.2ab	11.6a	12.1a	11.9ab	12.4b
Average	547A	571A	244A	239A	333A	339A	10.3A	10.5A	10.1B	11.8A	11.7B	13.0A

Note: Different lowercase letters (one-way ANOVA) and uppercase (two-way ANOVA) letters indicate significant differences at $p < 0.05$ (Duncan's test).

3.3. N use efficiency (NUE) of winter wheat and soil nitrate-N accumulation

After three years consecutive cultivation, the average N uptake of the winter wheat increased following the order of: CK < single manure application < single N fertilizer application < combination of manure and N fertilizer. Among all the treatments, M+N225 produced the highest N uptake, which was 59.4% higher than that for CK and 14% higher than that for N225 treatment. Combining manure with 75 and 150 kg N ha^{-1} N fertilizer had higher NUE than those for single 75 and 150 kg N ha^{-1} N fertilizer application, but when the N rate was exceeded 150 kg N ha^{-1}, additional manure would cause a decrease of NUE.

Table 4. N uptake and NUE affected by manure and N rate.

Treatment	Shoot N uptake ($kg\ ha^{-1}$)		0-200 cm soil nitrate-N residue kg · hm^{-2})		NUE (%)	
	-M	+M	-M	+M	-M	+M
N0	206.1b	226.8b	79.6e	109.2e	-	-
N75	261.3ab	289.5ab	165.7d	192.1d	24.5	27.9
N150	287.1ab	328.5a	317.0c	360.7c	18.0	22.6
N225	312.0a	325.2a	475.4b	586.8b	15.7	14.6
N300	269.1ab	277.5ab	769.7a	958.6a	7.0	5.6
Average	267.1B	289.5A	361.5B	441.5A	16.3A	17.7A

Note: Different lowercase letters (one-way ANOVA) and uppercase (two-way ANOVA) letters indicate significant differences at $p < 0.05$ (Duncan's test).

The increasing N rate caused the NUE to gradually decline whereas the 0-200 cm soil nitrate-N residue to increase in both –M and +M groups. The differences in soil nitrate-N residue between –M and +M groups became more significant with the increasing N input. More specifically, when the N input was 150 kg N ha^{-1}, the difference in soil nitrate between –M and +M group was only 43.7 kg ha^{-1}, and when the N input increased to 300 kg N ha^{-1}, the difference came to 188.9 kg ha^{-1}.

3.4. Soil nutrients in 0-20 cm layer

As shown in table 5, after three years of consecutive planting, the soil nutrients (organic matter, total N, available P, and available K) decreased by different extents in the CK treatment, whereas the fertilization practices generally improved the soil nutrients as compared to the initial values before the study: The single N fertilization improved the SOM by 3%, total N by 6%, available P by 32%; The combination of manure and inorganic fertilizer improved the SOM by 18.9%, total N by 20%, available P by 268%, and available K by 9%. Among all the treatments, M+N150 resulted in the highest SOM, available P and available K, which was 24%, 379% and 102% higher than that for the CK treatment, and 10%, 267%, and 55% higher than that for N150 treatment.

Table 5. Effects of manure and different rates of N fertilizer on soil nutrients.

Treatment	SOM (g · kg^{-1})		Total N (g · kg^{-1})		Available P (mg · kg^{-1})		Available K (mg · kg^{-1})	
	-M	+M	-M	+M	-M	+M	-M	+M
Before sowing	14.3		0.81		9.23		177.2	
N0	14.0b	14.6b	0.81a	0.89a	8.2b	28.7b	109.9bc	153.8b
N75	14.5ab	16.9a	0.84a	0.95a	13.7a	32.0a	172.0a	206.0a
N150	15.7a	17.3a	0.87a	0.96a	10.7b	39.3a	143.5ab	222.2a
N225	14.7ab	17.0a	0.84a	0.99a	11.6ab	29.7b	89.9c	188.2ab
N300	13.9b	16.8a	0.88a	1.06a	12.7a	40.1a	118.9bc	199.1ab
Average	14.6B	16.5A	0.85B	0.97A	11.4B	34.0A	126.8B	193.9A

Note: Different lowercase letters (one-way ANOVA) and uppercase (two-way ANOVA) letters indicate significant differences at $p < 0.05$ (Duncan's test).

4. Discussion

Previous studies have reported that the combination of manure and chemical fertilizer increased winter wheat yield relative to applying inorganic fertilizer alone [19-21]. In the present study, only rational combinations of manure and N fertilizer could achieve a high yield of winter wheat. Excessive N (>225 kg N ha^{-1}) input would not lead to a further increase in the winter wheat yield, which was consistent with previous studies [22-23]. Moreover, the present results revealed that the combination of manure and inorganic fertilizer was more effective for increasing the winter wheat yield and WUE during the dry year (15%) as compared to those of the wet (3%) and average year (5%). Wang et al. [24] reported similar results in a long-term experiment conducted in Gansu province, China (on the Loess Plateau). However, different results were reported by Hao et al., [25] who observed better yield improvement through additional manure during wet years. The opposite results may have manifested from varied soil conditions, nutrient contents in the manure, and the experimental period of different studies.

Water is one of the key factors that limits dryland farming development. Understanding the effect of fertilization practices on water use by crops and WUE is crucial for overcoming these limitations in dryland farming. In the present study, there were not any significant differences for ET between –M and +M groups, and this result is consistent with that of both Wang et al. [26] and Liu et al. [10]. Considering that there would be more water lost from transpiration to obtain higher yields, this observation indicates that combining manure with chemical fertilizer may enhance the transpiration of plants and decrease the water loss from evaporation through the soil surface. On the one hand, additional manure increased the numbers of ears, the aboveground biomass, and the canopies of winter wheat, which were effective barriers to prevent direct radiation, thus reducing water loss through evaporation [27]. On the other hand, the application of manure increased the soil infiltration rate, total porosity, water field saturated hydraulic conductivity, and soil water stable macro-aggregates (>0.25 mm) significantly, thereby enhancing the soil water holding capacity [10, 28]. However, the differences in the WUE between –M and +M groups were not significant in the first year; this is likely due to the prolonged salutary effect of manure on soil quality and productivity, and the differences could not be significant in just one year.

As agriculture has developed, the soil degradation and environmental pollution caused by N fertilizer has become increasingly serious over the years. As a result, the concerns for sustainable soil productivity and efficient use of N fertilizers have emerged as important issues. The present study showed that the combination of manure with N fertilizer increased the shoot N uptake as compared to applying chemical fertilizer alone. Liang et al. [29] found that the combined application of manure and chemical fertilizers significantly improved the immobilization of N fertilizer during the early stages of wheat growth when the N demand of the crop was low. Then, mineral N was released from the microbial N pool during the key process of wheat growth. This mechanism indicated that the combined application of manure and inorganic fertilizer increased the synchrony between the N supply and the crop demand, thus reducing N losses from agriculture. In addition, Li et al. [30] also reported an increase in the root activity of winter wheat after manure application, which improved the N uptake by the crop.

Fertilization practices offers a great opportunity to soil nitrate-N residue. In the present study, the NUE declined and the residual nitrate-N in the 0-200 cm soil layer increased gradually with increasing N fertilizer rate, which was consistent with some previous studies [31-32]. Furthermore, there would be more nitrate-N which accumulated in the soil in +M group compared to that in –M group as the N rate increased. However, M+N75 and M+N150 resulted in a comparable value of winter wheat yield and a higher NUE compared to that for N150 and N225, whereas the amount of residual nitrate-N in the soil profile was decreased, thus indicating the potential of manure to replace a portion of N fertilizer.

As a direct indicator of soil fertility, an improved SOM is important for achieving a sustainable dryland farming system. Several studies have reported the beneficial effect of a combined application of manure and inorganic fertilizer on SOM improvement [33-35]. The present results showed that, after three years of continuous application, the combined application of manure and chemical fertilizer improved the SOM by 18.9% relative to the initial value before the experiment, and by 15.6% relative to the single N fertilizer application. Additionally, inorganic N fertilization caused the SOM to increase, which is primarily because the larger biomass and root area stimulated by the N fertilizer left more crop residues in the soil, thereby increasing the SOM. Moreover, the manure group improved the soil available P and available K, which is primarily due to manure was rich in P and K, and enriched the soil fertility significantly.

5. Conclusions

Additional manure did not cause more soil water depletion than the without manure group did, but achieved a higher grain yield, which resulted in a higher WUE. However, only rational combinations of manure and chemical fertilizers was able to improve the N use efficiency of winter wheat and preventing more soil nitrate-N accumulation. The excessive N fertilizer (> 150 kg N ha^{-1}) combined with manure would cause a decrease in the NUE and the accumulation of soil residual nitrate-N as compared without manure. After three years of consecutive planting, the combination of manure with chemical N fertilizer significantly increased the SOM, total N, Olsen-P, and available K as compared to the initial values and the single N fertilizer application. The M+N150 treatment had a better effect on the soil fertility improvement than the other treatments.

Thus, considering the issue of environmental protection and soil fertility build-up, rather than simply improving crop yields, it can be concluded that a combination of 75-150 kg N ha^{-1} of N fertilizer with manure (17.5 t ha^{-1} poultry manure or 30 t ha^{-1}pig manure) should be recommended in order to achieve a sustainable winter wheat cropping system in a dryland.

References

[1] Li S X, Wang Z H, Malhi S S, Li S Q, Gao Y J and Tian X H 2009 Nutrient and water management effects on crop production, and nutrient and water use efficiency in dryland areas of China *Agronomy Journal* **102** 223-65

[2] Li, S X, Wang Z H, Hu T T, Gao Y J and Stewart B A 2009 Nitrogen in dryland soils of China and its management Advanced Agronomy **101** 123-81

[3] Zhang D B, Yao P W, Zhao N, Wang Z, Yu C W, Cao Q H, Cao W D and Gao Y J 2015 Responses of winter wheat production to green manure and nitrogen fertilizer on the Loess Plateau *Agronomy Journal* **107** 361-74

[4] Shi H and Shao M G 2000 Soil and water loss from the Loess Plateau in China *Journal of Arid Environments* **45** 9–20

[5] Angus J F and Herwaarden van A F 2001 Increasing water use and water use efficiency in dryland wheat *Agronomy Journal* **93** 290–8

[6] Deng X P, Shan L, Zhang H P and Turner N C 2006 Improving agricultural water use efficiency in arid and semiarid areas of China *Agricultural Water Management* **80** 23–40

[7] Ju X T, Xing G X, Chen X P, Zhang S L, Zhang L J, Liu X J, Chen Z L, Yin B, Christie P, Zhu Z L and Zhang F S 2009 Reducing environmental risk by improving N management in intensive Chinese agricultural systems *PNAS* **106** 3041-46

[8] Duan Y H, Xu M G, Gao S D, Yang X Y, Huang S M, Liu H B and Wang B R 2014 Nitrogen use efficiency in a wheat－Corn cropping system from 15 years of manure and fertilizer applications *Field Crop Research* **157** 47-56

[9] Jiang D, Hengsdijk H, Dai T B, Boer de W, Jing Q and Cao W X 2006 Long-term effects of manure and inorganic fertilizers on yield and soil fertility for a winter wheat-maize system in Jiangsu, China *Pedosphere* **16** 25-32

[10] Liu C A, Li F R, Zhou L M, Zhang R H, Jia Y, Lin S L, Wang L J, Siddique K H M and Li F M 2013 Effect of organic manure and fertilizer on soil water and crop yields in newly-built terraces with loess soils in a semi-arid environment *Agricultural Water Management* **117** 12-32

[11] Meade G, Lalor S T J and Mc Cabe T 2011 An evaluation of the combined usage of separated liquid pig manure and inorganic fertilizer in nutrient programmes for winter wheat production *European Journal of Agronomy* **34** 62-70

[12] Fan J, Hao M D and Shao M A 2003 Nitrate accumulation in soil profile of dryland farming in Northwest China *Pedosphere* **13** 367-74

[13] Sun H Y, Shen Y J, Yu Q, Flerchinger G N, Zhang Y Q, Liu C M and Zhang X Y 2010 Effect of precipitation change on water balance and WUE of the winter wheat-summer maize rotation in the North China Plain *Agricultural Water Management* **97** 1139-45

[14] Jackson M L 1956 Soil *Chemical Analysis-Advanced Course* Chapter 3 (Wisconsin: University of Washington) pp 101-68

[15] Dai J, Wang Z H, Li F C, He G, Wang S, Li Q, Cao H B, Luo L C, Zan Y L, Meng X Y, Zhang W W, Wang R H, and Malhi S S 2015 Optimizing nitrogen input by balancing winter wheat yield and residual nitrate-N in soil in a long-term dryland field experiment in the Loess Plateau of China *Field Crop Research* **181** 32–41

[16] Nelson D W and Sommers L E 1982 *Methods of Soil Analysis* **2** (Wisconsin: Soil Sciene Society of America) pp 539-79

[17] Olsen S R and Sommers L E 1982 *Methods of soil analysis* part 2 (Wisconsin: Soil Science Society of America) pp 403-30

[18] Abbasi M K and Tahir M M 2012 Economizing nitrogen fertilizer in wheat through combinations with organic manures in Kashmir, Pakistan *Agronomy Journal* **104** 169-77

[19] Kato Y and Yamagishi J 2011 Long-term effects of organic manure application on the productivity of winter wheat grown in a crop rotation with maize in Japan *Field Crops Research* **120** 387-95
[20] Shah S A, Shah S M, Mohammad W, Shafi M and Nawaz H 2009 N uptake and yield of wheat as influenced by integrated use of organic and mineral nitrogen *International Journal of Plant Production* **3** 45-56
[21] Zhong Y Q W and Shangguan Z P 2014 Water consumption characteristics and water use efficiency of winter wheat under long-term nitrogen fertilization regimes in Northwest China *PLoS ONE* **9**
[22] Zhou J B, Wang C Y, Zhang H, Dong F, Zheng X F, Gale W and Li S X 2011 Effect of water saving management practices and nitrogen fertilizer rate on crop yield and water use efficiency in a winter wheat - summer maize cropping system *Field Crops Research* **122** 157-63
[23] Wang S Y, Fan T L, Ding N P and Zhao G 2010 Crop yield and water use efficiency under long-term fertilization in Loess Plateau *Journal of Nuclear Agricultural Sciences* **24** 1044-50
[24] Hao M D, Lai L, Wang G L and Dang T H 2003 Effects of long-term fertilization on wheat yield on Loess Plateau *Chinese journal of Applied Ecology* **14** 1893-96
[25] Wang J, Liu W Z and Dang T H 2011 Responses of soil water balance and precipitation storage efficiency to increased fertilizer application in winter wheat *Plant Soil* **347** 41–51
[26] Rehman S, Khalil S, Muhammad F, Rehman A, Khan A Z, Rehman A, Zubair M and Khalil I F 2010 Phenology, leaf area index and grain yield of rainfed wheat influenced by organic and inorganic fertilizer *Pakistan Journal of Botany* **42** 3671-85
[27] Rasool R, Kukal S S and Hira G S 2008 Soil organic carbon and physical properties as affected by long-term application of FYM and inorganic fertilizers in maize-wheat system *Soil and Tillage Research* **101** 31-6
[28] Liang B, Zhao W, Yang X Y and Zhou J B 2013 Fate of nitrogen-15 as influenced by soil and nutrient management history in a 19-year wheat–maize experiment *Field Crops Research* **144** 126-34
[29] Li X H, Yang S X, Yu Z W and Yu S L 2005 Effects of organic manure application on growth and senescence of root in winter wheat *Plant Nutrition and Fertilizer Science* **11** 467-72
[30] Iqbal M, Idowu O J, Hassan A, Khan A G and Amjad M 2014 Dairy manure and nitrogen fertilizer effects on residual nitrate and phosphate, and wheat yield in a sandy clay loam soil *Journal of Plant Nutrition* **37** 562-74
[31] Liu H B, Li Z H, Zhang W L and Lin B 2004 Study on N use efficiency of Chinese cabbage and nitrate leaching under open field cultivation *Plant Nutrition and Fertilizer Science* **10** 286-91
[32] Hati K M, Mandal K G, Misra A K, Ghosh P K and Bandyopadhyay K K 2006 Effect of inorganic fertilizer and farmyard manure on soil physical properties, root distribution, and water-use efficiency of soybean in Vertisols of central India *Bioresource Technology* **97** 2182-88
[33] Manna M C, Swarup A, Wanjari R H, Ravankar H N, Mishra B, Saha M N, Singh Y V, Sahi D K and P A Sarap 2005 Long-term effect of fertilizer and manure application on soil organic carbon storage, soil quality and yield sustainability under sub-humid and semi-arid tropical India *Field Crops Research* **93** 264-80
[34] He Y T, Zhang W J, Xu M G, Tong X G, Sun F X, Wang J Z, Huang S M, Zhu P and He X H 2015 Long-term combined chemical and manure fertilizations increase soil organic carbon and total nitrogen in aggregate fractions at three typical cropland soils in China *Science of the Total Environment* **532** 635-44

IOP Conf. Series: Earth and Environmental Science **39** (2016) 012049 doi:10.1088/1755-1315/39/1/012049

Screening and identification of aerobic denitrifiers

K Shao[1], H M Deng[1], Y T Chen[1], H J Zhou[1] and G X Yan[1, 2]

[1]State Key Laboratory of Heavy Oil Processing, Beijing Key Laboratory of Oil and Gas Pollution Control, China University of Petroleum, Beijing 102249, China

E-mail: yangx0919@163.com

Abstract. With the standards of the effluent quality more stringent, it becomes a quite serious problem for municipalities and industries to remove nitrogen from wastewater. Bioremediation is a potential method for the removal of nitrogen and other pollutants because of its high efficiency and low cost. Seven predominant aerobic denitrifiers were screened and characterized from the activated sludge in the CAST unit. Some of these strains removed 87% nitrate nitrogen at least. Based on their phenotypic and phylogenetic characteristics, the isolates were identified as the genera of Ralstonia, Achromobacter, Aeromonas and Enterobacter.

1. Introduction

For a long time, removing COD (chemical oxygen demand) and suspended solids was the main goal for the treatment of wastewater, while nitrogenous compounds removal was not taken into full consideration. In recent years, the pollution of nitrogenous compounds has been seriously increasing, because of the fast development of economy and urbanization, which have endangered the safety of environmental water in China, particularly in drinking water. For nitrogen removal from wastewater, nitrification and denitrification processes by microbes are the most easily available on account of their low maintenance cost, as well as the effective contaminant removal performance [1][2]. However, nitrification demands an aerobic condition for successful oxidation of ammonium and nitrite, and denitrification uses electron donors in the anoxic environment conversely. Thus conventional theories have assumed that nitrification and denitrification can only be achieved in different levels of oxygen, which were complicated operations.

In 1984, aerobic denitrification was first proposed by Robertson and Kuenen [3]. Since then, many reports describing aerobic denitrification have been published [4][5][6]. Aerobic denitrification by microbes offers a new way for biological degradation of nitrogen.That is to say, aerobic denitrification, which is conducted by aerobic denitrifiers, may provide the potential to eliminate the requirement for two separate tanks and reduce sludge yield[7]. The alkalinity consumption during the nitrification process can be partly compensated by the accumulation of the alkalinity in denitrification, so that pH of the biosystem keeps relatively stable. In addition, some aerobic denitrifiers are able to conduct combined heterotrophic nitrification and aerobic denitrification [8][9][10].

The purpose of this writing is to screen aerobic denitrifiers from activated sludge in CAST unit, and evaluate the characterization of these denitrifiers.

2. Materials and methods

2.1. Sample sites

The sample was obtained from the CAST unit in the Liaohe petrochemical sewage treatment plant and then stored in a sealed bottle. The sample was resuspended in the enrichment medium to obtain a homogeneous suspension.

2.2. Media

We used a denitrification medium (DM), a R_2A medium, a screening medium and denitrifying performance detecting cultures in this research.

The constituents of the DM were as follows: 7.9 gL^{-1} $Na_2HPO_4{\cdot}7H_2O$; 1.5 gL^{-1} KH_2PO_4; 0.3 gL^{-1} $(NH_4)_2SO_4$; 0.5 gL^{-1} $MgSO_4{\cdot}7H_2O$; 4.7 gL^{-1} sodium succinate; 1.5 gL^{-1} KNO_3, and 2mL trace element solution. Adjusting the pH of the cultures to 7.5 is necessary [11]. The components of trace element solution were as follows: 50 gL^{-1} EDTA; 2.2 gL^{-1} $ZnSO_4$; 5.5 gL^{-1} $CaCl_2$; 5.1 gL^{-1} $MnCl_2{\cdot}4H_2O$; 1.1 gL^{-1} $(NH_4)_6Mo_7O_2{\cdot}4H_2O$; 5.0 gL^{-1} $FeSO_4{\cdot}7H_2O$; 1.6 gL^{-1} $CuSO_4{\cdot}5H_2O$; 1.6 gL^{-1} $CoCl_2{\cdot}6H_2O$.

The constituents of the R_2A medium were as follows: 0.5 gL^{-1} yeast extract; 0.3 gL^{-1} K_2HPO_4; 0.5 gL^{-1} peptone-D; 0.5 gL^{-1} casamino acid; 0.5 gL^{-1} D-glucose; 0.5 gL^{-1} starch soluble; 0.05 gL^{-1} $MgSO_4{\cdot}7H_2O$; 0.3 gL^{-1} $C_3H_3NaO_3$. 15 g agarose was added to the R_2A mixture as needed for plating.

The screening medium was comprised of following reagents: 1 gL^{-1} KNO_3; 0.2 gL^{-1} $CaCl_2$; 1 gL^{-1} KH_2PO_4; 0.5 gL^{-1} $FeCl_3{\cdot}6H_2O$; 1 gL^{-1} $MgSO_4{\cdot}7H_2O$; 8.6 gL^{-1} sodium succinate; 21 gL^{-1} agarose; 1mL bromothymol blue. The final pH of the medium was adjusted to 7.0±0.2.

Denitrifying performance detecting cultures were comprised of following reagents: 1.31 gL^{-1} $C_6H_5Na_3O_2{\cdot}2H_2O$; 1.10 gL^{-1} CH_3COONa; 0.361 gL^{-1} KNO_3; 0.2 gL^{-1} $MgSO_4{\cdot}7H_2O$; 1.0 gL^{-1} KH_2PO_4; 5.0 gL^{-1} K_2HPO_4; 0.5 gL^{-1} NaCl; and 1mL trace element solution. Adjusting the pH of the cultures to 7.4 is necessary.

2.3. Accumulation culture and screening of aerobic denitrifiers

To cultivate the aerobic denitrifiers, 10mL of the sample was transferred into the Erlenmeyer flask with 250 mL of the denitrification medium. The Erlenmeyer flask was sealed with a breathable sealing membrane, and cultivated in a shaker at 30℃ and 175 rpm for two days. After enrichment, the resultant bacterial suspensions with different concentration gradient were plated on screening agar media and incubated aerobically. Colonies with different apparent characteristics covered the agar plates, and each colony was selected to fresh screening mediums until purified isolates were obtained. In order to detect nitrogen removal performance, isolates were incubated in the denitrifying performance detecting cultures, using KNO_3 as the sole nitrogen source. In this study, seven single strains owning high nitrogen removal efficiency were selected, and the strains were stored on R_2A medium in the fridge at 4℃ for further use.

2.4. Identification of the aerobic denitrifiers

The experiments to examine the physiological and biochemical characteristics of the strains were conducted as previous description in *Bergey's manual of determination bacteriology* and *identification and guidance of bacteria* [12]. To put it simply, to measure the oxidase activity, the oxidation of tetramethyl-p-phenylenediamine reacting on filter paper was monitored. The activity of catalase was determined via examining the bubble formation of the fresh colonies that had been added in 3% hydrogen peroxide (V/V). Moreover, the microscopic characteristics of the isolates were obtained by Gram staining and agar-stab culture.

Genomic DNA of the single isolate was extracted from bacterial suspension using the Bacterial DNA Kit (50) (Omega Bio-TEK), which was following the manufacturer's instructions. Strains were identified by PCR (polymerase chain reaction amplification) of the 16S rRNA gene of the isolated strain by bacterial universal primers, 27f (AGAGTTTGATCATGGCTCAG), and 1492r (TACGGTTACCTTGTTACGACTT) [13]. The PCR was conducted under the following conditions: 94℃ for 5 min; 35 cycles at 94℃ for 30s; 55℃ for 30s; 72℃ for 1 min, and extension at 72℃ for 7 min. The PCR products were analyzed with agarose gel electrophoresis using 0.8% agarose gels which were stained with ethidium bromide. Then, the amplified bands were visualized on a UV light box (Molecular Imager Gel Doc EX System by Newbio industry). The PCR products were then sequenced by Sangon Biotech Co. Ltd. (Shanghai, China). Finally, there were multiple alignments for the sequences, in comparison with available sequences in the GenBank database, using the Basic Local Alignment Search Tool[14]. A phylogenetic tree figure was then built by MEGA software (version 7.0) as well as the neighbor-joining algorithm.

2.5. Nitrogen removal performance of the aerobic denitrifiers

The isolates were pre-cultured in R_2A medium at 30℃ and incubated in a shaker with a speed of 175 rpm and centrifuged at 4℃ and 10000 rpm for 10 minutes during their logarithmic phase. Bacteria were washed by buffer solution three times, and then inoculated into denitrifying performance detecting cultures (5% inoculum). After seven days' cultivation, the supernatants which were used to detect the content of nitrite (NO^{2-}), nitrate (NO^{3-}), total nitrogen (TN) and chemical oxygen demand (COD) were obtained by centrifuging the cultures.

2.6. Analytical methods

The OD600 value was tested at a wavelength of 600 nm by a spectrophotometric assay (SHIMADZU UVmini^{-1}240). NO^{2-}, NO^{3-}, TN and COD were tested by procedures described in the standard methods [15]. In a brief, NO^{2-} was examined by the N-(1-naphthalene)-diaminoethane photometry method. TN and nitrate were analyzed using the hydrochloric acid photometry method. COD was tested through the potassium dichromate method (CHEMetrics). All indexes were tested in triplicate.

3. Results and discussion

3.1. Isolation of the aerobic denitrifiers

In the initial screening work, a total of thirteen strains were isolated from the activated sludge sample from Liaohe petrochemical sewage treatment plant under aerobic conditions, and then seven strains that could remove nitrate nitrogen were selected according to their BTB plate's reaction.

3.2. Identification of the aerobic denitrifiers

According to the morphological observation, colonies of strain 1 and strain 4 were gray white, moist and transparent; strain 6 and strain 7 were cream white, smooth and non-transparent. The colonies of strain 2, strain 3 and strain 5 showed light yellow with moist and transparent surface. Gram staining indicated that all the isolates belonged to rod-shaped, or short rod-shaped, gram-negative bacteria. Strain 2 and stain 3 were oxidase-negative and catalase-positive, and other strains showed positive characteristics in both catalase and oxidase activities.

The high similarities to the genera were listed in table 1 and BLAST was used in the homology searching of the sequences in GenBank. Through partial 16S rRNA sequences analysis in comparison with each other, or with some of the reported bacteria with nitrogen removal performance, a phylogenetic tree figure of the isolates using neighbor-joining method was built by MEGA7.0 shown in figure 1, and several formerly recorded[16] representative aerobic denitrifiers were also included like *Achromobacter* sp.

Table 1. Similarities of 16SrRNA sequence between the strains and related affiliations.

Isolate No.	Affiliation	GenBank No.	Similarity %
1	*Ralstonia* sp.	KM823929.1	100
2	*Achromobacter* sp.	HM246520.1	100
3	*Enterobacter cloacae*	JX188069.1	100
4	*Aeromonas hydrophila*	KC252599.1	100
5	*Aeromonas aquariorum*	KC202271.1	100
6	*Enterobacter cloacae*	KP313043.1	100
7	*Aeromonas* sp.	KJ817212.1	99

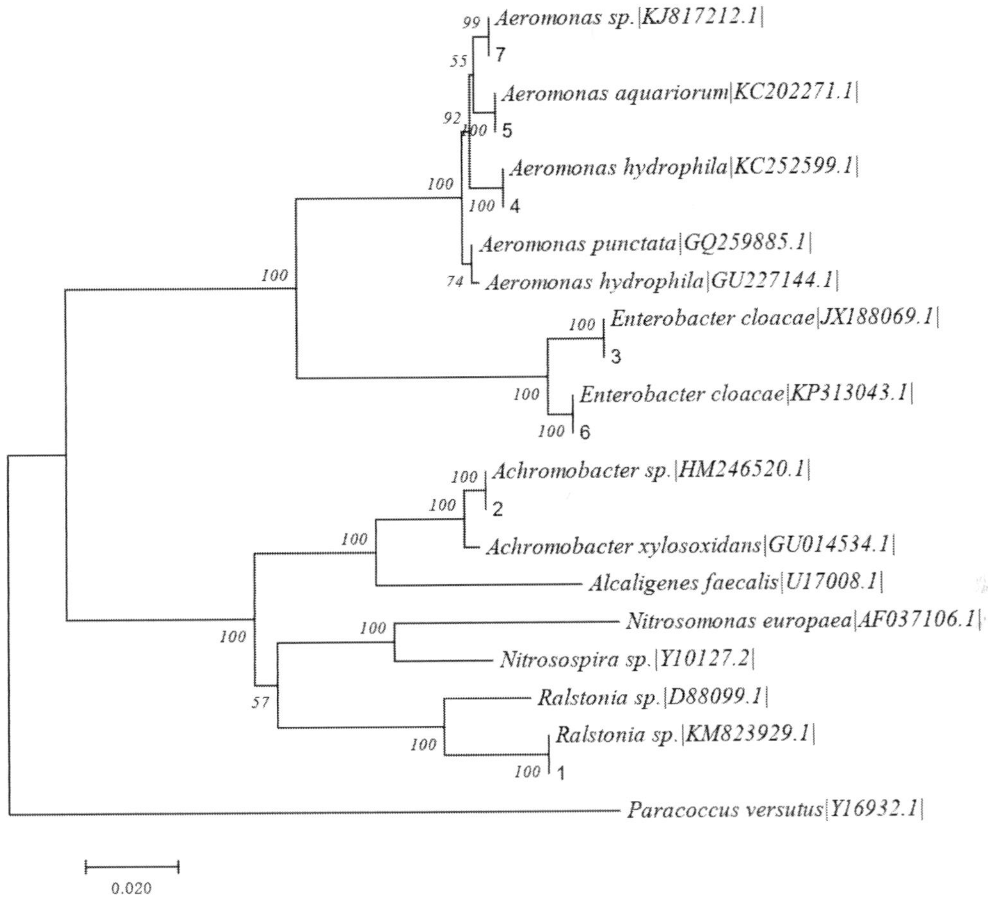

Figure 1. The phylogenetic tree figure according to partial 16S rRNA sequences revealing phylogenetic relationships between the isolates and some other typical nitrogen-removing aerobic denitrifiers from other groups.

3.3. Nitrogen removal performance of the aerobic denitrifiers

The nitrate-removal rate of each isolate in denitrifying performance detecting cultures under aerobic conditions was shown in figure 2. It revealed that strain 2, strain 3 and strain 7 tended to be good at degrading nitrate nitrogen, while the removal rate of strain 5 was unsatisfactory.

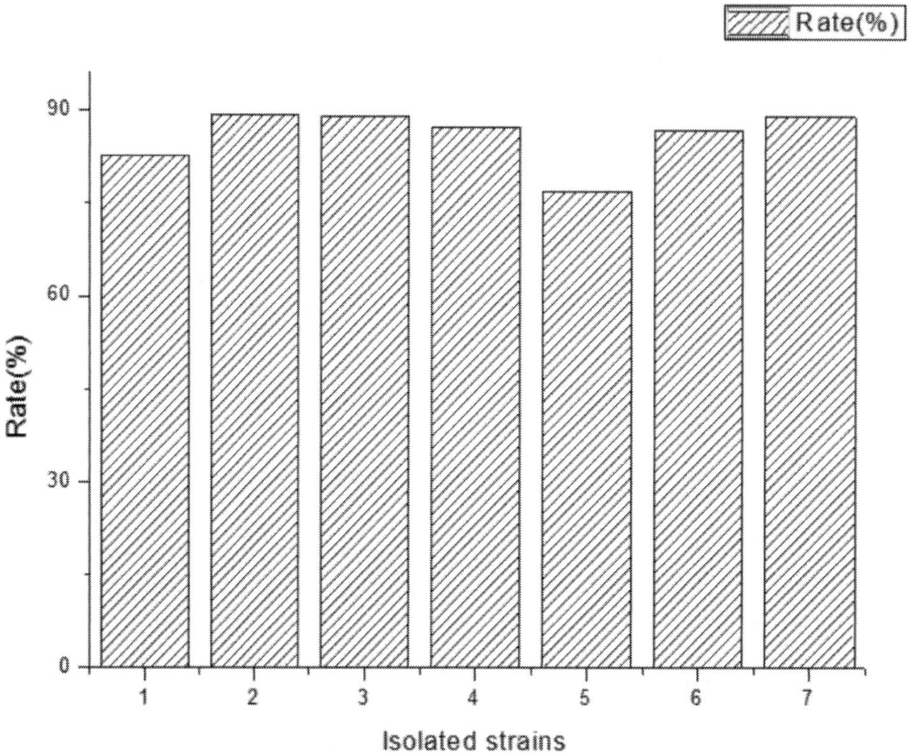

Figure 2. Nitrate nitrogen removal rate of the isolates.

4. Conclusion

In this paper, 7 predominant aerobic denitrfiers were screened and characterized from the activated sludge in the CAST unit in Liaohe petrochemical sewage treatment plant, and identified as the genera of *Ralstonia*, *Achromobacter*, *Aeromonas* and *Enterobacter*. Five of them can remove more than 86% nitrate nitrogen.

The aerobic denitrification process rebuts the traditional nitrogen removal theory, which believes that denitrification is proceeded by heterotrophs only in anaerobic environments. Discovery of aerobic denitrifiers made it possible for nitrification and denitrification processes to be conducted simultaneously, which would greatly diminish operating costs. Therefore, aerobic denitrifiers have an outstanding potential in the application of contaminated source water bioremediation.

In my succeeding work, I will optimize the conditions of the culture (including DO concentration, C/N load ratio, carbon source, temperature and pH) for the isolated strains, discuss the nitrogen removal efficiency of the combination of the isolated strains, and try to make a practical application in industrial effluent.

References

[1] Jechalke S, Vogt C, Reiche N, Franchini A G, Borsdorf H, Neu T R and Richnow H H 2010 Aerated treatment pond technology with biofilm promoting mats for the bioremediation of benzene, MTBE and ammonium contaminated groundwater *Water Res.* **44**: 1785–96

[2] Perelo L W 2010 Review: in situ and bioremediation of organic pollutants in aquatic sediments *J. Hazard. Mater* **177** 81–89

[3] Robertson L A and Kuenen J G 1984 *Aerobic denitrification--old wine in new bottles? Antonie Van Leeuwenhoek* **50** 525-44

[4] Krul J M 1976 Dissimilatory nitrate and nitrite reductuin under aerobic conditions by an aerobically and anaerobically grown Alcaligines sp. and by active sludge *J. Appl. Bacteriol* **40** 245–60

[5] Marshall R O, Dishburger H J, Mac Vicar R and Hallmark G D 1953 Studies on the effect of aeration on nitrate reduction by pseudomonad dprvies using N15 *J. Bacteriol* **66**: 254–58

[6] Meschner K and Wuhrmann K 1963 Beitrag zur kenntnis der microbiellen denitrifying *Pathol Microbiol* **26**: 579–91

[7] Seifi M and Fazaelipoor M H 2012 Modeling simultaneous nitrification and denitrification (SND) in a fluidized bed biofilm reactor *Appl. Math. Model.* **36:** 5603‑13

[8] Joo H M H and Shoda M 2005 Characteristics of ammonium removal by heterotrophic nitrification-aerobic denitrification by Alcaligenes faecalis No. 4 *J. Biosci. Bioeng* **100** 184-91

[9] Padhi S K, Tripathy S, Sen R, Mahapatra A S, Mohanty S and Maiti N K 2013 Characterisation of heterotrophic nitrifying and aerobic denitrifying Klebsiella pneumoniae CF-S9 strain for bioremediation of wastewater *Int. Biodeter. Biodegr.* **78** 67-73

[10] Zhang J, Wu P, Hao B, and Yu Z 2011 Heterotrophic nitrification and aerobic denitrification by the bacterium Pseudomonas stutzeri YZN-001 *Bioresour. Technol.* **102** 9866-69

[11] Robertson L A and Kuenen J G 1988 Heterotrophic nitrification in Thiosphaera pantotropha: oxygen uptake and enzyme studies *J. Gen. Microbiol.* **134** 857–63

[12] Bergey D H amd Holt J G 1994 *Bergey's Manual of Determinative Bacteriology, nineth* ed. Williams & Wilkins, Baltimore

[13] Suzuki M T, Giovannoni S J 1996 Bias caused by template annealing in the amplification of mixtures of 16S rRNA genes by PCR *Appl. Environ. Microbiol.* **62** 625

[14] http://www.ncbi.nlm.nih.gov/BLAST/bast.cgi (accessed May 2016)

[15] National Environmental Protection Bureau 2002 *Guide of Analytical Method on Monitoring Water and Wastewater* China Environmental Science Press, Beijing

[16] Liang Z, Wei D, Feng L J, et al. 2012 Isolation of aerobic denitrifiers and characterization for their potential application in the bioremediation of oligotrophic ecosystem *Bioresource Technology* **108**(4) 1-7

Real-time simulation of large-scale floods

Q Liu[1], Y Qin[1, 3], G D Li[1], Z Liu[2], D J Cheng[2] and Y H Zhao[2]

[1] Key Laboratory of the North-West Water Resources and Ecology Environment, Ministry of Education, Xi'an University of Technology, Xi'an 710048, China;
[2] Tianjin Water Authority, 210 Weidi Road, Hexi District, Tianjin 300074, China

E-mail: 13571991500@126.com

Abstract. According to the complex real-time water situation, the real-time simulation of large-scale floods is very important for flood prevention practice. Model robustness and running efficiency are two critical factors in successful real-time flood simulation. This paper proposed a robust, two-dimensional, shallow water model based on the unstructured Godunov-type finite volume method. A robust wet/dry front method is used to enhance the numerical stability. An adaptive method is proposed to improve the running efficiency. The proposed model is used for large-scale flood simulation on real topography. Results compared to those of MIKE21 show the strong performance of the proposed model.

1. Introduction

In recent years, flood inundation mapping projects have been widely implemented in China. The flood inundation maps of the maximum water depth, the flood arrival time, and the flood duration are drawn according to many typical cases of hydrology condition and dyke burst. Those maps should be very useful for flood prevention, especially for emergency planning. However, the real-time water situation and hydraulic engineering states are very complex and changeable, so the flood inundation maps of typical cases might be quite different with the real flood situation, and the real-time simulation of large-scale floods is very important for flood prevention practice.

Many flood simulation models have been developed in recent years, including the 1D [1], 2D [2-4], and 1D-2D coupled [5,6] hydrodynamic models. A number of 2D flood simulation models used the 2D shallow water equations as the governing equations and adopted the Godunov-type finite-volume methods for numerical resolution. Due to the well conservation property, the Godunov-type finite-volume method is more suitable than the finite-difference method for flood simulation with wetting and drying on complex topography. Although finite-volume based flood simulation models have been researched for many years, methods published were mainly focused on the model accuracy and numerical stability [2-4]. Some parallel models [7] were proposed for faster computation. Parallel models should achieve real-time simulation of large-scale floods, but they need an expensive computation server, which also requires great power consumption.

According to the two critical factors of model robustness and running efficiency, this paper proposed a robust, two-dimensional, shallow water model based on the unstructured Godunov-type finite volume method. A robust wet/dry front method is used to enhance the numerical stability. An adaptive method is proposed to improve the running efficiency.

2. Numerical model

The 2D shallow water equations are used as the governing equations [6],

$$\frac{\partial \mathbf{U}}{\partial t}+\frac{\partial \mathbf{E}}{\partial x}+\frac{\partial \mathbf{G}}{\partial y}=\mathbf{S} \tag{1}$$

in which,

$$\mathbf{U}=\begin{bmatrix} h \\ hu \\ hv \end{bmatrix},\quad \mathbf{E}=\begin{bmatrix} hu \\ hu^2+g(h^2-b^2)/2 \\ huv \end{bmatrix},\quad \mathbf{G}=\begin{bmatrix} hv \\ huv \\ hv^2+g(h^2-b^2)/2 \end{bmatrix},$$

$$\mathbf{S}=\mathbf{S}_0+\mathbf{S}_f=\begin{bmatrix} 0 \\ g(h+b)S_{0x} \\ g(h+b)S_{0y} \end{bmatrix}+\begin{bmatrix} 0 \\ -ghS_{fx} \\ -ghS_{fy} \end{bmatrix}$$

where h is the depth; u and v are the x- and y- velocity; b is the bottom elevation; S_{0x} and S_{0y} are the x- and y- bed slopes, $S_{0x}=-\partial b/\partial x$ and $S_{0y}=-\partial b/\partial y$; g is the gravity acceleration; S_{fx} and S_{fy} are the friction terms estimated by Manning formulae:

$$S_{fx}=\frac{n^2u\sqrt{u^2+v^2}}{h^{4/3}} \qquad S_{fy}=\frac{n^2v\sqrt{u^2+v^2}}{h^{4/3}} \tag{2}$$

The discretization equation can be obtained by using Green's theorem to integrate equation (1), which is given as

$$\mathbf{U}_i^{n+1}=\mathbf{U}_i^n-\frac{\Delta t}{\Omega_i}\sum_{k=1}^{3}\mathbf{F}_k(\mathbf{U}_\mathrm{L},\mathbf{U}_\mathrm{R})\cdot\mathbf{n}_kL_k+\frac{\Delta t}{\Omega_i}\mathbf{S}_i \tag{3}$$

where the subscript i is the cell number; $\mathbf{U}_i$ is the cell-averaged conservative vector; n is the time step; k and L is the edge number and length respectively; $\mathbf{n}$ is the unit outward vector normal to the edge; Ω is the cell area; $\mathbf{S}_i$ is the source approximation; $\mathbf{F}$ is the outward normal flux vector; $\mathbf{U}_\mathrm{L}$ and $\mathbf{U}_\mathrm{R}$ are the local left and right states for Riemann problem respectively; Δt is the time step.

The MUSCL-Hancock scheme presented in [6] is also used in this paper for numerical computation. Model details can be found in [6], and this paper only presents improvement on numerical stability and running efficiency.

In [6], the VFRs are used to calculate water level, and cells are classified as wet-cell and dry-cell, based on the nodal water depth. Only the continuity equation is updated for dry cells, whereas the continuity and momentum equations are simultaneously updated for wet cells. Those should compose a robust method for wetting/drying treatment. However, since the momentum equation is not updated for dry cells, the flow velocities are set to zero to the cells at the wet/dry fronts, which would be inconsistent with real flow condition. To deal with this problem, a new wet/dry front method is proposed for reconstruction to the cells at the wet/dry fronts, given as,

$$h_{i,k}^{\mathrm{Rec}}=\begin{cases} 0 & \text{if} \quad \eta_i\le b_{i,k}^{\min} \\ \dfrac{(\eta_i-b_{i,k}^{\min})^2}{2(b_{i,k}^{\max}-b_{i,k}^{\min})} & \text{if} \quad b_{i,k}^{\min}<\eta_i\le b_{i,k}^{\max} \\ \eta_i-\dfrac{b_{i,k}^{\min}+b_{i,k}^{\max}}{2} & \text{if} \quad \eta_i>b_{i,k}^{\max} \end{cases} \qquad (k=1,2,3) \tag{4}$$

$$u_{i,k}^{\mathrm{Rec}}=u_i \quad v_{i,k}^{\mathrm{Rec}}=v_i \qquad (k=1,2,3) \tag{5}$$

where η_i is the water surface elevation; $b_{i,k}^{\min}$ and $b_{i,k}^{\max}$ are the minimum and maximum bed elevations of the edge. When the numerical fluxes are computed using the reconstructed values, the continuity and momentum equations are simultaneously updated for wet-cells, as well as the cells at the wet/dry fronts.

In [6], a fixed time step is used for 1D-2D coupled computation, since the 1D hydrodynamic model, based on the finite-difference method, should adopt a fixed time step for computation. However, the fixed time step should be very small, to preserve the stability condition during the whole simulation. This would markedly decrease the running efficiency of the 2D model. To deal with this problem, an adaptive method is proposed to improve the running efficiency, given as,

$$\Delta t = C_r \cdot \min_{i,k}\left[\left(\frac{\Omega}{(|u_\perp| + \sqrt{gh})_k L_k}\right)_i\right] \quad i = 1,2,\cdots,N;\ k = 1,2,3 \tag{6}$$

where Δt is the time step; C_r is the Courant number, and $C_r = 0.95$; $u_\perp$ and h are the average of the reconstructed value; N is the total number of cells. The adaptive time step should improve the running efficiency for dyke break flood simulation, since the time step would be very small when dyke breach occurs, but would be larger when the flow velocity is small.

3. Case study: a real flood simulation in Wei River Basin

3.1. Computational mesh

The study area is the Wei River Basin, which is triangulated by 179296 grids, and the length of edges is around 200m-250m. The average area of grids is 0.053km2. The area of the computation domain is about 9503 km2, which is a large-scale field. Figure 1 shows the local computational mesh. Figure 2 shows the bed elevation.

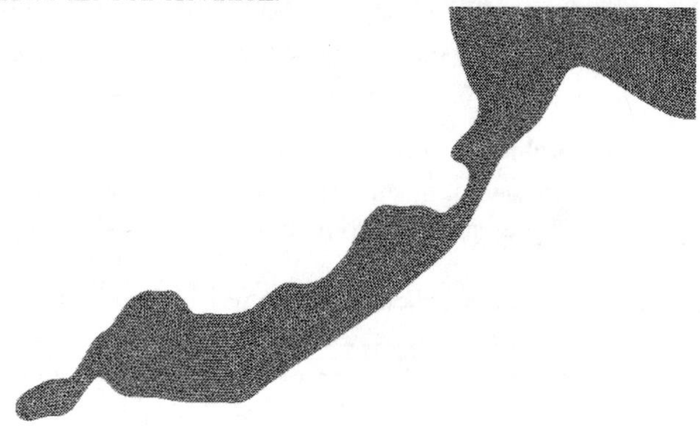

Figure 1. The local computational mesh.

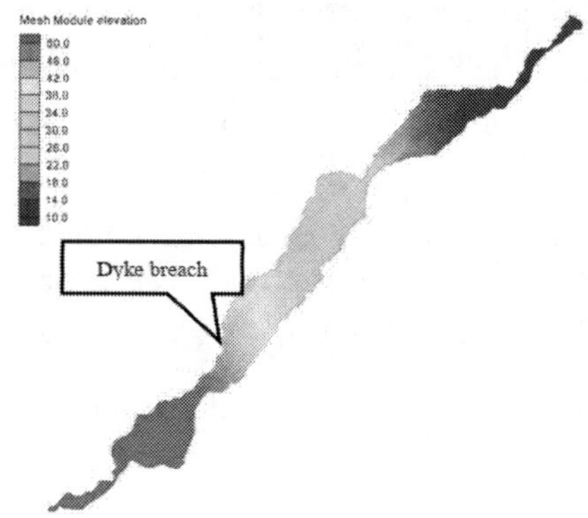

Figure 2. The bed elevation.

3.2. Boundary condition

A hypothetical dyke breach and the position are shown in Figure 2. The discharge across the breach is shown in Figure3. The peak discharge is 496m^3/s, and the flooding time is about 263h. To simulate the whole process of flooding and recession, the computation time is set to 878h, and the discharge across the breach is 0 during 264h-878h.

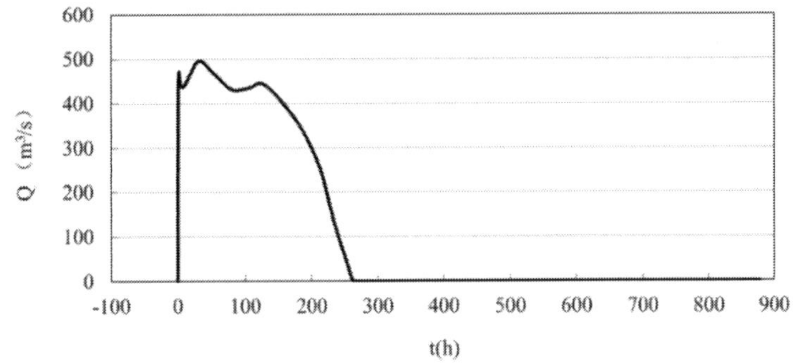

Figure 3. The discharge across the breach.

3.3. Simulated results
Figure 4 shows the distribution of water depth at different times, which gives a reasonable process of flood routine.

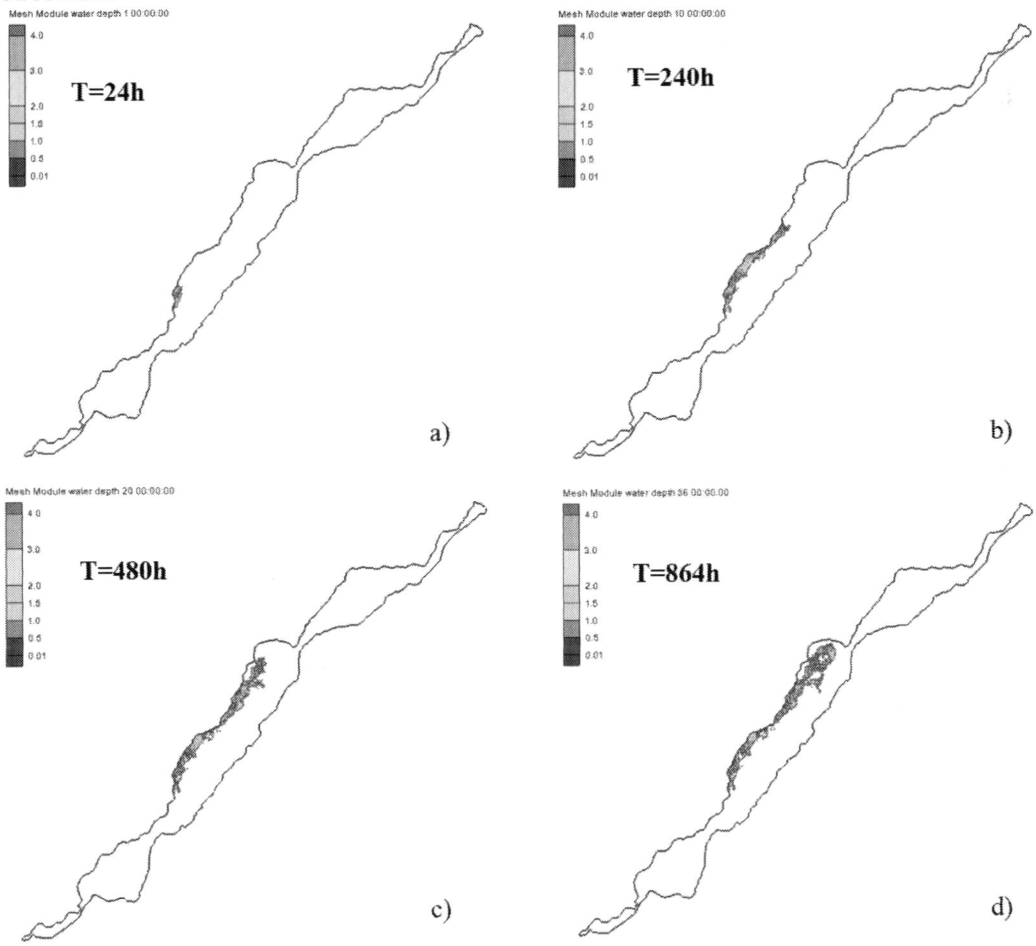

Figure 4. The distribution of water depth at different times.

Figure 5 shows the mass conservation result, which validates the well conservation property of the proposed model. During the simulation, the water volume from breach is almost equal to the water volume of cells, and the relative error is about 0.017% at the end. The mass error is very small, and it should not induce a falsity flow.

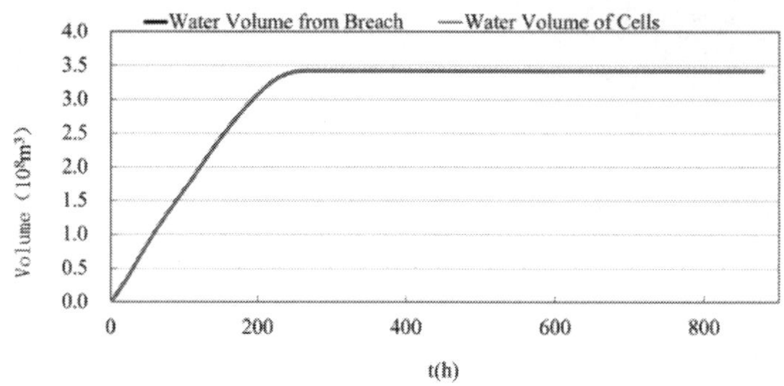

Figure 5. The mass conservation result.

Table 1 gives the comparison of the final flooded area between the proposed model and the MIKE 21 model. Results show the good agreement with MIKE 21.

Table 1. The comparison of the final flooded area (km^2).

Model	Flooded area	0.05-0.5m	0.5-1m	1-2m	2-3m	>=3m
MIKE 21	864.26	370.35	234.5	200.34	50.14	8.93
The proposed model	856.28	364.62	241.19	193.46	47.03	9.98
Relative error(%)	-0.92%	-1.55%	2.85%	-3.43%	-6.20%	11.76%

The flood inundation maps of the maximum water depth, and the flood arrival time are also comprised in MIKE 21 results, see figure 6 and figure 7. Results also show the good agreement with MIKE 21.

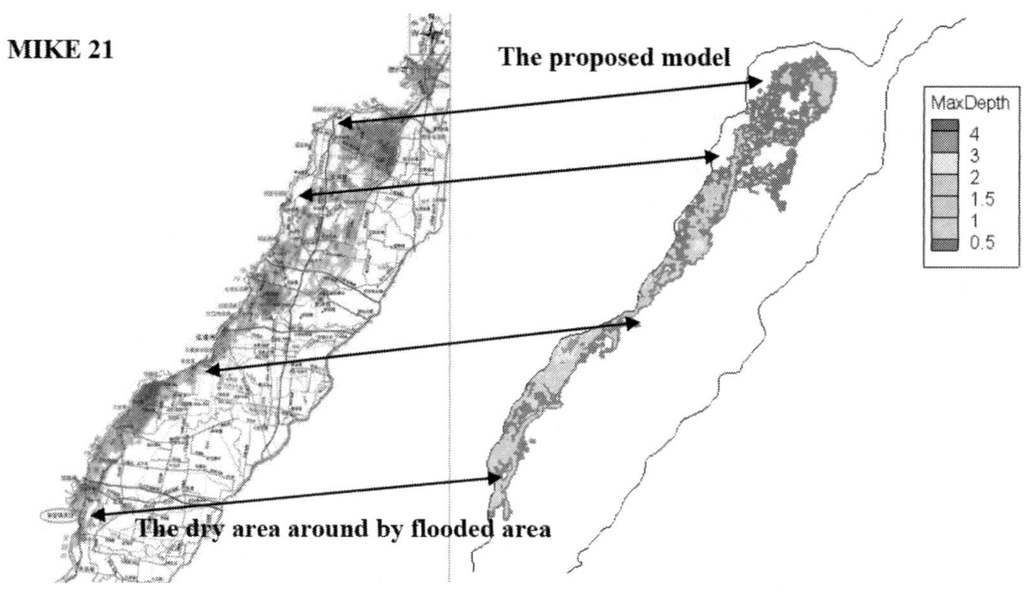

Figure 6. The flood inundation maps of the maximum water depth.

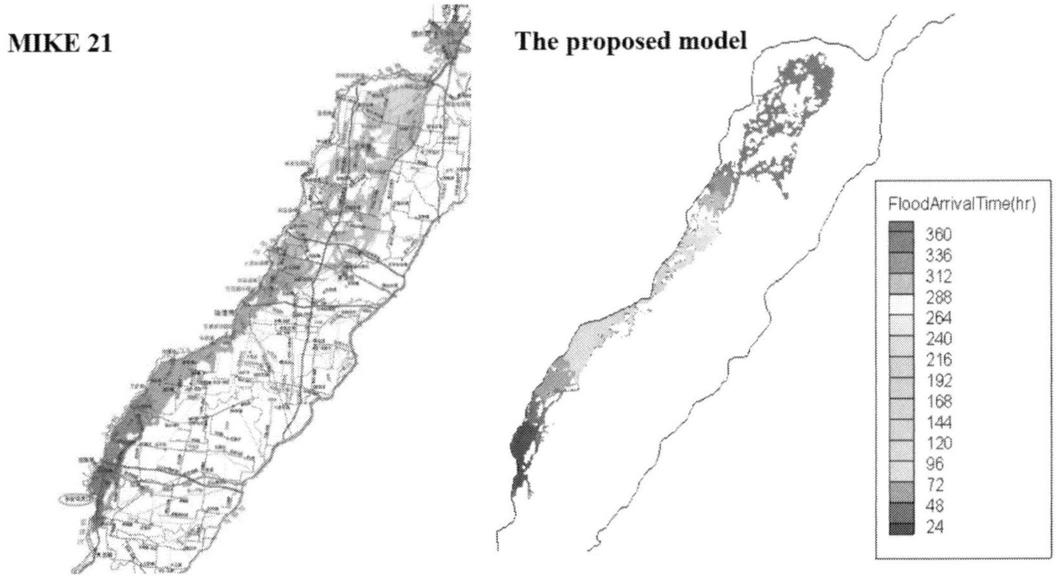

Figure 7. The flood inundation maps of the flood arrival time.

The running time of the proposed model is 1.7h, which shows the computational efficiency of the proposed model. In this study, a 22-day flooding process can be simulated in one hour, so it can be concluded that the proposed model is useful for real-time simulation of large-scale floods

4. Conclusions

According to the two critical factors of model robustness and running efficiency, this paper proposed a robust, two-dimensional, shallow water model based on the unstructured Godunov-type finite volume method. A robust wet/dry front method is used to enhance the numerical stability. An adaptive method is proposed to improve the running efficiency. The proposed model is used for large-scale flood simulation on real topography. Results compared to those of MIKE21 show the strong performance of the proposed model.

References

[1] Zhu D, Chen Y, Wang Z and Liu Z 2011 Simple, robust, and efficient algorithm for gradually varied subcritical flow simulation in general channel networks *ASCE Journal of Hydraulic Engineering* **137**(7): 766-74

[2] Song L, Zhou J, Guo J, Zou Q and Liu Y 2011 A robust well-balanced finite volume model for shallow water flows with wetting and drying over irregular terrain *Advances in Water Resources* **34**(7): 915-32

[3] Song L, Zhou J, Li Q, Yang X and Zhang Y 2011 An unstructured finite volume model for dam- break floods with wet/dry fronts over complex topography *International Journal for Numerical Methods in Fluids* **67**(8): 960-80

[4] Liu Y, Zhou J, Song L, Zou Q, Liao L and Wang Y 2013 Numerical modelling of free-surface shallow flows over irregular topography with complex geometry *Applied Mathematical Modelling* **37**(23): 9482-98

[5] Blade E, Gomez-Valentin M, Dolz J, et al. 2012 Integration of 1D and 2D finite volume schemes for computations of water flow in natural channels *Advances in Water Resources* **42**: 17-29

[6] Liu Q, Qin Y, Zhang Y, et al. 2015 A coupled 1D–2D hydrodynamic model for flood simulation in flood detention basin *Natural Hazards* **75**(2):1303-25

[7] Sanders B F, Schubert J E and Detwiler R L 2010 ParBreZo: A parallel, unstructured grid,

Godunov-type, shallow-water code for high-resolution flood inundation modeling at the regional scale *Advances in Water Resources* **33**(12): 1456-67

IOP Conf. Series: Earth and Environmental Science **39** (2016) 012051 doi:10.1088/1755-1315/39/1/012051

Hydrogeochemical and isotopic composition of groundwater in the Horgos River Watershed in the North-Western Ili Basin in Sinkiang Province, China

S H Jin[1, 2], J Y Jiang[1, 3] and J P Rao[1]

[1]Institute of Disaster Prevention, Beijing, 101601, China

E-mail : jiangjiyi@cidp.edu.cn

Abstract. In this study, hydrogeochemical and stable environmental isotopic (oxygen-18 and deuterium) data of groundwater in the Horgos River watershed in the north-western Ili Basin were examined to determine the groundwater chemical characteristics, as well as the origins of the groundwater and surface water. The quaternary aquifer's structure and spatial distribution is complex. Hydrogeochemical data shows that the groundwater is alkaline due to the presence of bicarbonate as the dominant anion. In the north, the dominant groundwater cation was Ca^{2+} while Mg^{2+} ions were highest in the south. In addition, water types varied from HCO_3-SO_4-Ca and HCO_3-SO_4-Ca-Mg to HCO_3-SO_4-Mg-Ca and SO_4-HCO_3-Mg-Ca. Overall, the groundwater in this study showed high total hardness. The dissolving and evaporation inspissation were considered the main controlling factors. Isotopic data (oxygen-18 and deuterium) indicated that the Horgos River was the main recharge source for the groundwater, which rapidly infiltrated across sandstone macropores. Furthermore, surface water and shallow groundwater were significantly affected by evaporation.

1. Introduction

The climate is relatively dry in northwest China. Annual evaporation exceeds annual precipitation significantly. Moreover, due to surface water scarcity, groundwater is particularly important for water resources.

As an emerging border city (figure 1), Horgos City has become a focus area for economic development and ecological environmental protection to the west of Sinkiang. Water consumption in the city in agriculture, industry, and domestic use has grown rapidly. Because of the complex groundwater environment of the Horgos River watershed and the fragility of its ecological environment, groundwater exploitation, management, and protection is critical from socio-economic and environmental points of view [1].

Studying the geochemistry of groundwater contributes to a better understanding of the main hydrogeochemical processes that control the groundwater environment and its relationship with groundwater migration. A goal of this study was to assess groundwater quality; thus, parameters such as pH, electrical conductivity (EC), total dissolved solids (TDS), total hardness (TH), and major ions (cations and anions) were evaluated. Analyses of environmentally stable isotopes (oxygen-18 and deuterium) were used to define groundwater sources and origins, as well as the main recharge sources feeding the groundwater aquifers [2-5].

In this study, groundwater resources and environment problems of the Ili River Basin were investigated. Also, groundwater distribution characteristics and the process of controlling groundwater geochemistry formation were analyzed using hydrogeochemistry and stable environmental isotope (oxygen-18 and deuterium) data.

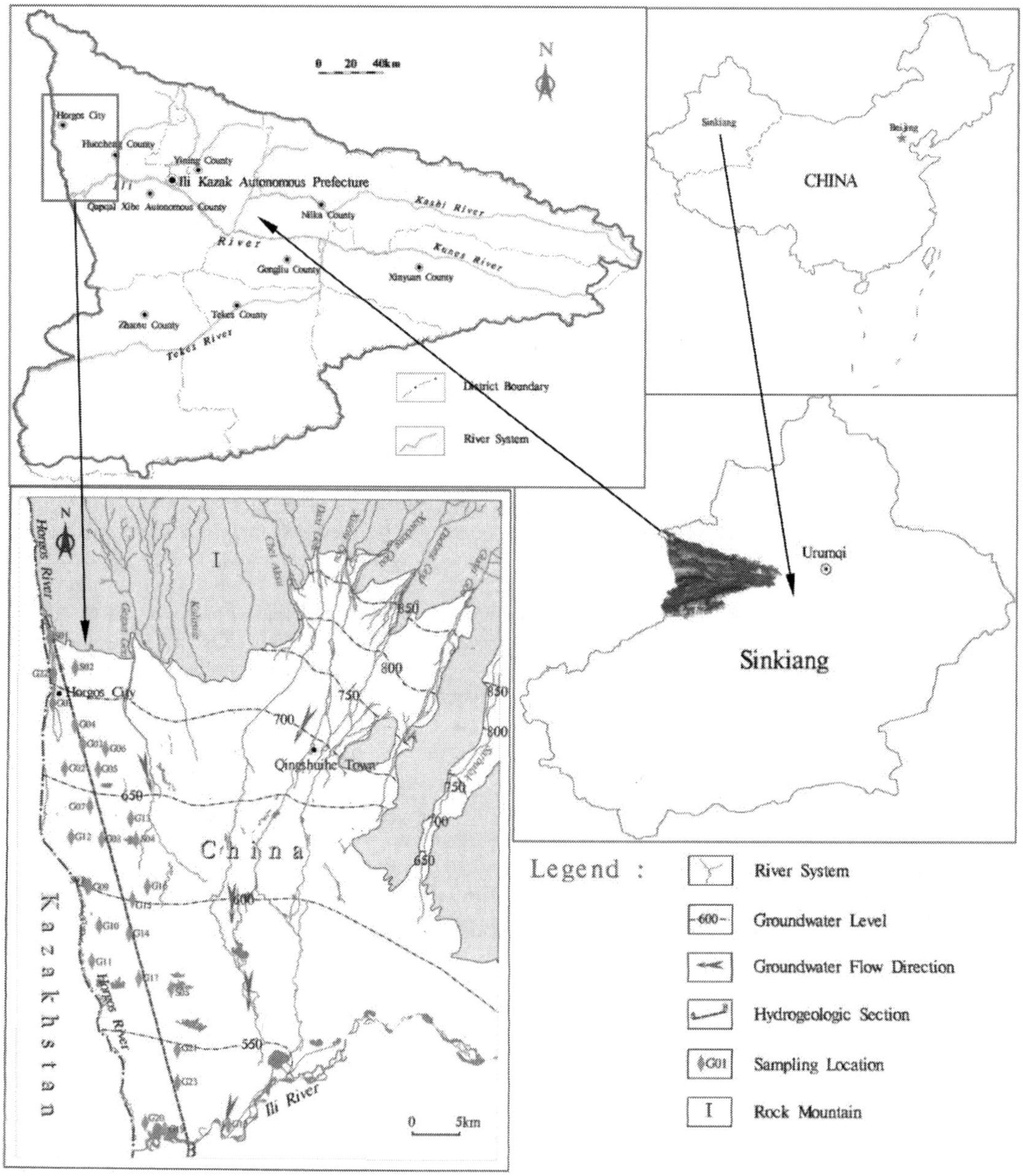

Figure 1. Location map of studied area.

2. Studied area

2.1. Physiography and climate of the study area

Horgos River is located in the northwest of the Ili Basin in Sinkiang, at the border of China and Kazakhstan (figure 1). The Horgos River originates in the north Tianshan mountain, flows through the plains from north to south, and then ends in the Ili River. The portion of the Horgos River within

China's border is 69 kilometers long, its watershed area is 2,736 square kilometers, and it has a natural drop of 2,500 meters. It covers topographical features of rock mountain, piedmont desert belt, fine soil plain, and then desert [6]. This study focused on the plain portion, which is about 50 kilometers long, where the area is relatively flat and the altitude of the terrain gradually descends from north to south.

The study area is a typical arid or semi-arid climate region of North-West China. According to statistical data from the Huocheng meteorological station, the average yearly temperature is 9.3° C and the multi-year average precipitation and evaporation are 218.9 mm (mainly between July and August) and 1401.1 mm, respectively.

2.2. *Hydrogeologic setting*

Aquifers in the plain of the Horgos River watershed are mainly composed of quaternary aquifer. There are three types of aquifer matrices and structural features (figure 2). In the north, gravel is the main aquifer matrix; toward the center of the plain, the aquifer matrix is comprised of gravel, sand, and clay, and the southern end of the plain features a coverage of silt, fine sand, and clay. Therefore, this section is mainly an alluvial aquifer of a fine soil plain in the lower reaches of the Ili River, and lacks high-quality groundwater resources.

According to the hole structure in this area, the quaternary aquifer is more than 500 meters thick. The groundwater level line on the plane indicates that groundwater migration is from north to south and is eventually emptied into the Ili River.

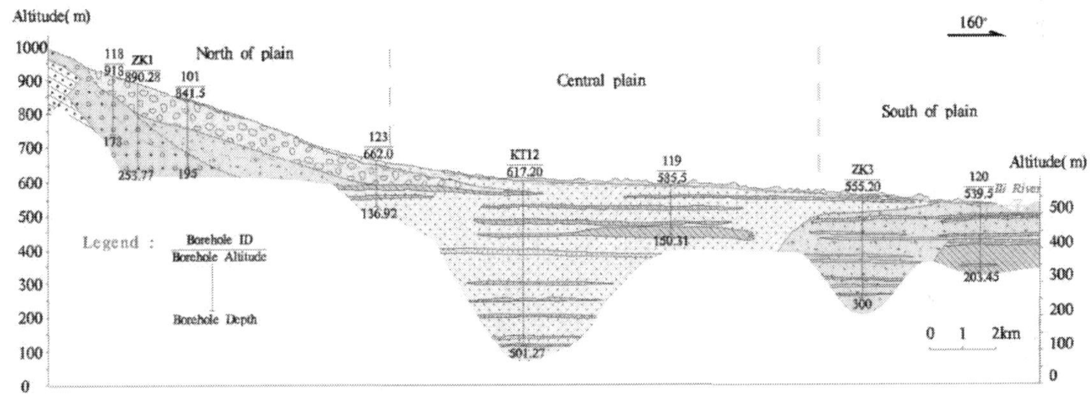

Figure 2. Hydrogeologic section of A—B in studied area.

3. Sampling and analysis

This study was carried out on 23 groundwater samples and 5 surface water samples (figure 1). The samples were collected in August 2014 while investigating groundwater resources and environmental geological problems in the Ili River Valley; all sampling sources were picked after long-term mining.

The samples were collected into 500 ml PET plastic bottles that had been washed at least 3 times. No preservatives were added to the samples. After collecting each sample, the bottle was sealed with a membrane. Samples for oxygen-18 and deuterium (δ18O and δD) concentration determinations were collected in 30 ml PET brown plastic bottles that had been washed at least 3 times; these samples were filtered by a 0.02 μm membrane before sampling. Its treatment was the same as the other samples during the complete analysis. Coordinates of the sampling position were determined by a Venture HC handheld GPS. The sample situ parameters (temperature, pH, oxidation-reduction potential, dissolved oxygen, electrical conductivity, and total dissolved solids) were determined in situ using a HACH-CEL900 portable water analyzer. Stable isotope (oxygen-18 and deuterium) concentrations were determined at Chang'an University using an IWA-35EP laser liquid water isotope analyzer via the axis integral cavity absorption spectrum technique with a testing accuracy of ±1‰ to ±2‰. Complete hydrogeochemical analysis samples were measured at the Geological Environment Monitoring

Institute of Sinkiang by an atomic absorption spectrophotometer (WFX-110 type); its measuring accuracy was ±2%.

4. Results and discussion

4.1. Hydrogeochemical characteristics of groundwater

Table 1 shows the results of the hydrogeochemical characteristics of the water samples. The pH of the groundwater samples varied from 7.12 to 8.28; the pH of surface water varied from 8.34 to 8.91, indicating that it was weakly alkaline. The surface water had significantly higher pH levels than the groundwater; and the pH of shallow groundwater was higher than that of deep groundwater.

Table 1. Hydrogeochemical parameters and major constituents (in mg/L) of the studied water.

Samples	Cations mg/L				Anions mg/L			pH	Cond μS/cm	TDS mg/L	Total hardness
	K^+	Na^+	Ca^{2+}	Mg^{2+}	Cl^-	SO_4^{2-}	HCO_3^-				
S01	0.5	5.5	36.1	7.3	3.5	38.4	103.7	8.51	218.8	151.3	120.1
S02	0.3	0.0	32.1	7.3	2.1	38.4	91.5	8.34	214.1	147.8	110.1
S03	4.1	34.9	56.1	41.3	31.9	167.1	231.9	8.41	702.0	473.1	310.2
S04	2.0	9.4	32.1	19.4	5.0	76.8	112.3	8.9	251.0	209.4	160.1
S05	4.6	54.1	36.1	36.4	28.4	180.6	183.1	8.91	598.0	440.0	240.2
G01	0.6	6.2	56.1	2.4	3.5	67.2	97.6	8.28	290.0	197.7	150.1
G02	0.7	6.6	49.7	7.8	2.8	50.0	134.2	8.09	317.0	202.8	156.1
G03	1.8	8.8	68.1	24.3	10.6	76.8	219.7	7.83	482.0	342.6	270.2
G04	1.0	7.2	36.1	19.4	3.5	76.8	128.1	8.25	263.0	227.2	170.1
G05	2.5	25.3	100.2	29.2	12.8	124.9	292.9	7.59	712.0	497.7	370.3
G06	1.9	34.3	68.1	17.0	26.9	115.3	207.5	7.97	587.0	390.8	240.2
G07	1.9	9.9	80.2	21.9	14.2	67.2	262.4	7.8	601.0	374.9	290.2
G08	18.1	106.5	192.4	70.5	67.4	547.5	396.6	7.28	1668.0	1308.6	770.6
G09	2.1	27.5	60.1	21.9	21.3	103.7	207.5	7.74	579.0	372.0	240.2
G10	1.3	9.0	32.1	19.4	7.1	38.4	158.7	7.12	296.0	214.6	160.1
G11	2.5	49.6	60.1	34.0	42.5	192.1	195.3	8.10	703.0	504.4	290.2
G12	9.8	41.6	92.2	41.3	24.8	201.7	292.9	7.57	849.0	585.8	400.3
G13	1.6	20.2	101.8	25.8	16.3	98.0	353.9	7.31	676.0	478.6	360.3
G14	2.4	19.7	80.2	24.3	14.2	86.5	278.3	7.48	574.0	399.4	300.2
G15	1.9	26.5	70.5	25.3	10.6	76.8	288.0	7.43	553.0	379.4	280.2
G16	4.3	139.0	222.8	79.7	156.0	551.4	476.0	7.15	1927.0	1414.8	884.7
G17	3.1	29.0	72.1	38.9	24.8	96.1	329.5	7.67	674.0	464.8	340.3
G18	2.2	25.3	56.1	35.0	28.4	99.9	246.5	7.68	546.0	389.0	284.2
G19	--	--	--	--	--	--	--	7.59	2910.0	--	--
G20	2.0	36.2	41.7	28.2	10.6	57.6	268.5	7.60	2350.0	332.2	220.2
G21	4.0	29.8	41.7	26.7	17.7	109.5	195.3	8.19	622.0	349.9	214.2

Electrical conductivity (EC) is used to characterize groundwater circulation [7]. Electrical conductivity of the collected samples varied from 263 μS/cm in sample G04 up to 2910 μS/cm in sample G19; the surface water samples varied from 214.1 μS/cm to 702 μS/cm. As was the case with pH, the EC of shallow groundwater was significantly higher than that of deep groundwater. This indicated that shallow groundwater was obviously influenced by evaporation and concentration.

The total dissolved solids (TDS) of the studied groundwater samples ranged from 197.7 mg/L to 585.8 mg/L, except for samples G08 (1308.6 mg/L) and G16 (1414.8 mg/L). For surface water samples, the total dissolved solids ranged from 147.8 to 473.1 mg/L. In the horizontal direction, the total dissolved solids of the groundwater samples increased from north to south, and the same changing trend was demonstrated as the groundwater depth increased. This phenomenon indicated that groundwater of the Horgos River watershed traveled from north to the south, which was consistent with the research results of Yin X X, et al. on the zoning of groundwater chemical types and the hydrochemistry action in the same area [8].

The total hardness (TH) was mainly determined by the concentration of calcium and magnesium ions [9]. The ground water samples of the Horgos River watershed plain ranged from 150.1 mg/L in sample G01 up to 884.7 mg/L in sample G16 for total hardness. Accordingly, 65% of the groundwater samples (13 samples) were considered hard, and the remaining samples (7 samples) were considered very hard. Samples G19, G22, and G23 were not detected. Samples S01 and S02 were considered moderately hard at levels of 120.1 mg/L and 110.1 mg/L, respectively. The TH of 2 of the 5 surface water samples (S04 and S05) were considered hard (160.1 mg/L and 240.2 mg/L) and sample G03 was considered very hard at 310.2 mg/L.

Table 1 provides concentrations of the major constituents (cations and anions). The Na+ concentrations ranged from 6.2 to 49.6 mg/L, except for samples G08 and G16, which were more than 100 mg/L. The Na+ ion concentrations were 5.5 mg/L in the upper reaches of the Horgos River. K+ concentrations were quite low, varying from 0.6 to 18.1 mg/L in the groundwater samples, and was detected at about 0.5 mg/L in sample S01. Groundwater Ca2+ concentrations ranged from 32.1 mg/L in sample G10 to 222.8 mg/L in sample G16, and the water of Horgos River was 36.1 mg/L. The Mg2+ ion concentrations ranged from 2.4 mg/L to 79.7 mg/L in groundwater samples, and 7.3 mg/L in sample S01. The Cl- ion concentrations in the groundwater samples were high, up to 156 mg/L; however, it was only 3.5 mg/L in sample S01. The bicarbonate HCO3- content varied from 97.6 mg/L to 396.6 mg/L in the groundwater samples; it was 103.7 mg/L in sample S01. The sulfate ion SO42- concentrations in the groundwater samples ranged from 50 mg/L to 201.7 mg/L, except for samples G08 and G16, which recorded the highest values at 547.5 mg/L and 551.4 mg/L, respectively. Sample S01 was recorded at 38.4 mg/L. Higher Ca2+ and bicarbonate concentrations were observed in samples G08 and G16 because of the abundance of clay minerals and sulfates (such as gypsum) in the aquifer. The Ca2+ concentrations were greater than those of Mg2+ in all samples. The ion concentrations in sample S01 were very low, but very high in samples G08 and G16. The ion concentrations of other samples decreased as the depth increased; they also gradually reduced from the north to the south.

From the calculation results, it can be seen that bicarbonate was the dominant anion in 84% of the samples. Sulfate ions SO42- were the dominant anion in only 4 samples (G08, G11, G16, and S05). Ca2+ was the dominant cation in the groundwater in the northern portion of the plain, while Mg2+ ions were dominant in the southern portion. In this change process, ion concentration order varied from Ca2+ >Mg2+ to Mg2+> Ca2+ or Mg2+>Na+ >Ca2+. Water types varied from HCO3-SO4-Ca or HCO3-SO4-Ca-Mg to HCO3-SO4-Mg-Ca or even SO4-HCO3-Mg-Ca.

4.2. Environmental isotopes

Some evidence indicates that the stable environment isotopes of oxygen-18 ($\delta^{18}O$) and hydrogen ($\delta^2H/\delta D$) are excellent tracers for determining groundwater origins [10]. Table 2 gives the environmental isotopic composition results. The studied groundwater samples ranged from -12.47‰

to -3.07‰ for $\delta^{18}O$, and from -86.92‰ to -40.69‰ for δD, and -11.79‰ for $\delta^{18}O$ and -78.20‰ for δD in sample S01.

Table 2. Isotopic composition of water samples.

Samples	δD ‰	$\delta^{18}O$ ‰	Samples	δD ‰	$\delta^{18}O$ ‰	Samples	δD ‰	$\delta^{18}O$ ‰
S01	-78.02	-11.79	G060	-76.92	-11.67	G16	-79.73	-11.68
S02	-79.15	-11.45	G07	-83.05	-12.20	G17	-83.37	-11.98
S03	-66.57	-8.14	G08	-86.92	-12.23	G18	-81.42	-11.60
S04	-73.23	-9.70	G09	-84.91	-11.86	G19	-79.66	-11.28
S05	-40.69	-3.07	G10	-84.82	-12.47	G20	-77.79	-11.54
G01	-81.02	-11.60	G11	-81.37	-11.84	G21	-49.00	-5.42
G02	-81.87	-12.34	G12	-84.08	-12.41	G22	-82.15	-12.15
G03	-83.81	-11.84	G13	-81.08	-11.69	G23	-83.77	-12.15
G04	-83.17	-12.42	G14	-78.45	-11.30			
G05	-79.34	-11.56	G15	-81.55	-11.63			

It is important to study the isotope characteristics of the water in the study area, especially with respect to the local meteoric water line (LMWL). A given year's average concentration of environmental isotopes of atmospheric precipitation in Urumqi are -86.25‰ for δD and -12.42‰ for $\delta^{18}O$; this was used as the background value of precipitation in the study area. The $\delta^{18}O$ and δD of the majority of the samples were higher than the background value. Furthermore, both the maximum $\delta^{18}O$ and δD occurred in the plain area of the surface water, which indicated that significant evaporation occurred when precipitation translated into groundwater. Evaporation was more intense in the plain compared to the mountain area.

According to IAEA data from 1986-2002, the equation of the local meteoric water line of Urumqi is as follows:

$$\delta D = 7.21\delta^{18}O + 4.50 \quad (1)$$

This diagram shows that some surface water and groundwater samples fell between the LMWL of Urumqi (Figure 3). Additionally, most groundwater samples were characterized by relatively depleted oxygen-18 and deuterium contents, indicating that they were not significantly affected by evaporated open water or soil water, which implied rapid infiltration of Horgos River waters across sandstone macropores. As a result, the environmental isotope concentrations of groundwater were diluted, while surface samples were characterized by a relative enrichment of oxygen-18 and deuterium contents. This indicated that they were significantly affected by evaporation and could not be quickly replenished by fresh water.

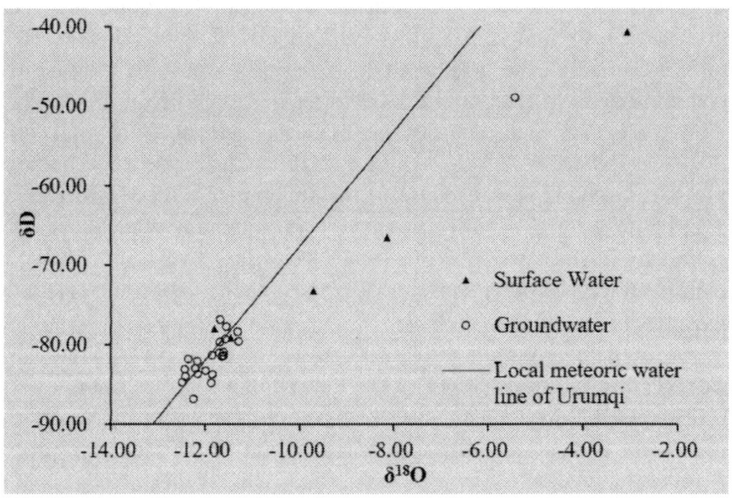

Figure 3. Relationship between $\delta^{18}O$ and δD for the studied groundwater samples.

In the vertical direction, oxygen-18 and deuterium concentrations in groundwater presented an obvious depletion tendency as depth increased.

5. Conclusions

The studied groundwater hydrogeochemical compositions were strongly influenced by dissolving and evaporation inspissation. The groundwater in the study area originated from atmospheric precipitation, which is mainly replenished by the Horgos River. This was evident from the following results:

- The pH measurements revealed an alkaline chemical environment of the groundwater in the Horgos River watershed plain because bicarbonate was the dominant anion for the majority of groundwater samples.
- The pH and EC of shallow groundwater was higher than groundwater occurrence in deep water; total dissolved solids and total hardness of groundwater samples presented a changing trend of increasing with an increase of depth and the same from north to south;
- Except for samples G19, G22, and G23, 65% of groundwater samples were considered hard; the remaining samples were considered very hard.
- As depth increased and the plain moved from north to south, oxygen-18 and deuterium depletion became more apparent.

Acknowledgments

This study was funded by The Graduate Student Innovation Fund of the Basic Scientific Research Service Fee of the Central University (ZY20160310, ZY20110103), “1:50,000 Hydrogeological Investigation in the Western Plain of Ili River Valley” (121201011000150021) carried out by the Xi'an geological survey center of the China Geological Survey. The authors of this paper are grateful to all members of the research team for their hard work.

References

[1] Xia Y B, Xie H L, Wang B, Hu Y Z, Wang X D, Bai Y N and Zhang J Q 2012 Review on regional groundwater cycle and evolvement *Geological Survey and Research* vol **35** ed Tianjin Institute of Geology and mineral resources pp 299-303

[2] Wang Z, Xaymurt A and Hu X G 2012 Analysis on water environment change and its influence factors in Huocheng county of Sinkiang *Water Conservancy Science and Technology and Economy* **21** pp 46-50

[3] Mamattursun E, Hamid Y and Anwar M 2011 Geochemical characteristics of groundwater in Ili River irrigated area, western Tianshan Mountains *Geochimica* vol **40** ed Science Press pp 179-87

[4] Hamid Y, Tursunjan I, Mihrigul M, Gulnar T and Huang Z F 2011 Investigation on dynamical laws of salt in groundwater in Ili River basin on Sinkiang *Agricultural Research in the Arid Areas* **21** pp 168-73

[5] Mamattursun E, Hamid Y, Zulpiya M and Xu N 2012 The response of soil salinization to characters of groundwater in Ili River basin *Journal of China Hydrology* **32** pp 14-20

[6] Li H and Zhou H F 2007 Variation characteristics of δD and δ^{18}O stable isotopes in the precipitation of Urumqi *Joumal of Arid Land Resources and Environment* vol **21** ed Science Press pp 46-50

[7] Song X F, Li F D and Yu J J 2007 Characteristics of groundwater cycle using deuterium, oxygen-18 and hydrochemistry in Chaobai River basin *Geographical Research* **26** pp 11-21

[8] Yin X X, Chi B M, Jiang J Y, Kong H M and Li B Y 2014 The study of groundwater encironment characteristicsof Huocheng county plain located in the western part of Ili～Künes basin of Sinkiang *Quaternary Sciences* vol **2** ed Science Press

[9] Salem W M and Ei-Sayed M 2015 Hydro-geochemical and isotopic composition of ground water in Helwan area *Egyptian Journal of Petroleum* **24** pp 411-21

[10] Hamed Y and Dhahri F 2013 Hydro-geochemical and isotopic composition of groundwater,with emphasis on sources of salinity, in the aquifer system in Northwestern Tunisia *Journal of African Earth Sciences* **83** pp 10-24

Research on evaluation methods for water regulation ability of dams in the Huai River Basin

G H Shan[1, 2], S F Lv[1] and K Ma[1]

[1]University of Science and Technology of China, Hefei, China

E-mail: jackshan@mail.ustc.edu.cn

Abstract. Water environment protection is a global and urgent problem that requires correct and precise evaluation. Evaluation methods have been studied for many years; however, there is a lack of research on the methods of assessing the water regulation ability of dams. Currently, evaluating the ability of dams has become a practical and significant research orientation because of the global water crisis, and the lack of effective ways to manage a dam's regulation ability has only compounded this. This paper firstly constructs seven evaluation factors and then develops two evaluation approaches to implement the factors according to the features of the problem. Dams of the Yin Shang ecological control section in the Huai He River basin are selected as an example to demonstrate the method. The results show that the evaluation approaches can produce better and more practical suggestions for dam managers.

1. Introduction

Recent studies on evaluating the water regulation abilities of dams have responded to the wide concerns about the global water resource crisis [6, 8]. In contrast, previous research mainly focused on water scheduling for flood control, drainage, water supply, pollution prevention, control, and so on. [7]. According to Abbas et al. [1], research on related water schedules in the past did not pay enough attention to the evaluation of the specific water regulating ability of dams that can be useful for making water schedules. With the development of social economy and the need for the efficient usage of water resources, the precise evaluation of water regulation ability has become an urgent need [11]. Although there are many evaluation methods [2, 3], research on evaluating dams' regulation abilities are rare. Furthermore, considering that managers are in need of water regulating suggestions, this study has crucial practical significance.

This paper selects dams in the Yin Shang ecological control section as examples to implement the idea. The multi-dam regulation of the Huai He River basin has been developed for many years [9, 10]. Recent research has mainly focused on the dams' regulation of multiple indexes of water quantity, water quality, and water ecology. The evaluation model and related evaluation methods presented in this paper can produce a more precise evaluation of dams' regulation abilities when considering water ecology and water quality.

This paper adapts the Technique for Order Preference by Similarity to Ideal Solution (TOPSIS) and Fuzzy Comprehensive Evaluation approaches to implement the evaluation process based on the adopted evaluation model. The results can provide technical support for the connection regulations for dams and suggestions for related decision-makers.

2. Evaluation factors

For evaluating the specific ability of dams to regulate water schedules, this paper adopts an assessment model that includes seven influential factors: water quantity that a dam is able to use (V1), water transmitting time of the schedule (V2), status of getting water along the river of a dam (V3), environmental remains of different contaminants (V4), dam management status (V5), construction status of dams (V6), and emergency handling capability of dams (V7). These factors can be classified into three aspects, which are included in the model as shown in Figure 1.

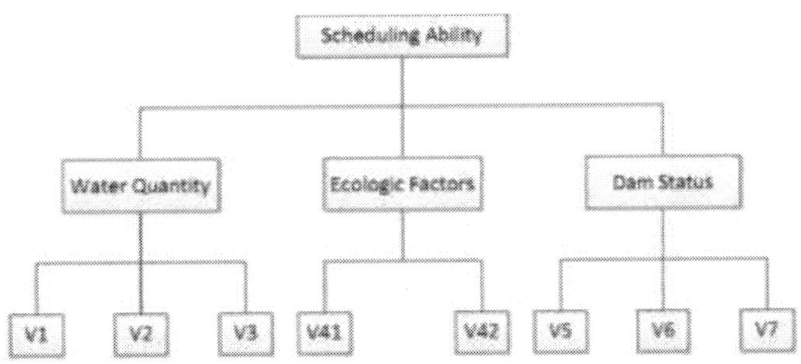

Figure 1. Classification of influential factors for evaluating the dam ability.

From Figure 1, the factors affecting the water regulating ability for evaluation of a dam can be divided into three types. This model depicts how the factors are composed and how they influence the regulating ability of dams.

3. Methods

Using these evaluation factors, this paper presents two approaches to implement the model and compares their results. Determining the regulating ability of dams is, in fact, the process of making an evaluation. There are many methods of evaluation. This paper uses the methods TOPSIS and Fuzzy Comprehensive Evaluation because of their suitability for this evaluation problem.

3.1. TOPSIS

Technique for Order Preference by Similarity to an Ideal Solution (TOPSIS) is an evaluation solution that assesses the target according to the distance between attributes producing the best and the worst composite solutions [5].

This paper sets every dam as a target and its evaluation indexes as its attributes. For N dams, the best scheduling composite dam is listed according to all the best attributes, and the worst scheduling dam is obtained by combining all its worst attributes. The specific notation for the process is shown in Table 1.

Table 1. TOPSIS Evaluation Process for Dams.

Dams	V1	V2	V3	V4	V5	V6	V7
Dam 1	v_{11}	v_{12}	v_{13}	v_{14}	v_{15}	v_{16}	v_{17}
Dam 2	v_{21}	v_{22}	v_{23}	v_{24}	v_{25}	v_{26}	v_{27}
…	…	…	…	…	…	…	…
Dam n	v_{n1}	v_{n2}	v_{n3}	v_{n4}	v_{n5}	v_{n6}	v_{n7}

Table 1 lists all the dams and their relevant weights. In the table, v_{ij} is the value of dam i's j^{th} index. The results of the TOPSIS method can be calculated by the following steps.

- Step 1: Data normalization

Any positive index can be standardized by the following formula:

$$\overline{v_{ij}} = \frac{v_{ij} - v_i^{mi}}{v_i^{ma} - v_i^{mi}} \tag{1}$$

Any negative index can be standardized by the following formula:

$$\overline{v_{ij}} = \frac{v_i^{ma} - v_{ij}}{v_i^{ma} - v_i^{mi}} \tag{2}$$

Using equations (1) and (2), all values of indexes can be standardized into the range [0-1]. In the equations, vima is the best (maximum) performance value of Vi among all the dams, and vimi indicates the worst (minimum) performance value of Vi among all the dams. vij is the jth value of dam i , and $\overline{v_{ij}}$ is the standardized value of vij.

- Step 2: Select the best and worst performance indexes among all dams to arrive at two solutions.

Solution 1: $MaDam = (v_1^{ma}, v_2^{ma}, v_3^{ma}, v_4^{ma}, v_5^{ma}, v_6^{ma}, v_7^{ma})$

Solution 2: $MiDam = (v_1^{mi}, v_2^{mi}, v_3^{mi}, v_4^{mi}, v_5^{mi}, v_6^{mi}, v_7^{mi})$

MaDam represents the best possible performance of a dam, whereas *MiDam* shows the worst possible performance of a dam, given the current data.

- Step 3: Calculate the distance.

Distance 1: $$Dma_i = \begin{bmatrix} \left(v_{i1} - v_1^{ma}\right)^2 + \left(v_{i2} - v_2^{ma}\right)^2 + \left(v_{i3} - v_3^{ma}\right)^2 + \left(v_{i4} - v_4^{ma}\right)^2 \\ + \left(v_{i5} - v_5^{ma}\right)^2 + \left(v_{i6} - v_6^{ma}\right)^2 + \left(v_{i7} - v_7^{ma}\right)^2 \end{bmatrix}^{1/2}$$

Distance 2: $$Dmi_i = \begin{bmatrix} \left(v_{i1} - v_1^{mi}\right)^2 + \left(v_{i2} - v_2^{mi}\right)^2 + \left(v_{i3} - v_3^{mi}\right)^2 + \left(v_{i4} - v_4^{mi}\right)^2 \\ + \left(v_{i5} - v_5^{mi}\right)^2 + \left(v_{i6} - v_6^{mi}\right)^2 + \left(v_{i7} - v_7^{mi}\right)^2 \end{bmatrix}^{1/2}$$

The two distances above tell how far any dam i is away from the best and worst results. In these two solutions, Dma_i shows how far the water scheduling ability of dam i is away from the best performing composite dam, while Dmi_i indicates how far the scheduling ability of dam i is away from the worst performing composite dam. Furthermore, v_{ij} is the jth evaluation index of dam i.

- Step 4: Obtain the schedule ability.

$$SA_i = \frac{Dmi_i}{Dma_i + Dmi_i}$$

Step 4 gives the final evaluation result of dam i's water scheduling ability. The SA_i represents the water resource scheduling capability of dam i evaluated by TOPSIS.

3.2. Fuzzy Comprehensive Evaluation (FCE)

Fuzzy Comprehensive Evaluation (FCE) was developed from the concept of the Fuzzy set [4]. FCE quantifies the fuzzy indexes of evaluation objects through establishing grade fuzzy subsets and utilizes the fuzzy variable principle to integrate each index so as to better solve the fuzzy problem.

FCE involves the following factors: Factor set U; Comment set V; Membership Function F; Fuzzy relation matrix R; Weight distribution vector W; and Fuzzy Comprehensive Evaluation Model M.

- Establish factor set $U= \{u_1, u_2, \dots, u_7\}$.

u_1 is the water quantity that a dam is able to handle; u_2 is the water transmitting time of the schedule; u_3 is the status for using water along the river of a dam; u_4 indicates the remains of two pollutants (*NH3-N* and *COD*) after water has passed through the dam; u_5 indicates the dam management status; and u_6 is the construction status of a dam. Finally, a dam's emergency handling capability is represented by u_7.

- Establish Comment set $V= \{v_1, v_2, v_3\}$.

The measure v_1 shows that the actual value of the factor is in a bad condition, whereas v_2 represents that the actual value of the factor is in a middle condition, and v_3 means that the actual value of the factor is good.

- Membership Function

The key to evaluating the scheduling ability of dams using the fuzzy mathematical method is selecting the Membership Function F. There are many methods that can be selected. The Trichotomy method is used in this paper to identity the membership function of all factors. Although different factors have different physical quantities, data standardization of all factors can eliminate the differences among them.

- Evaluation matrix of a single factor

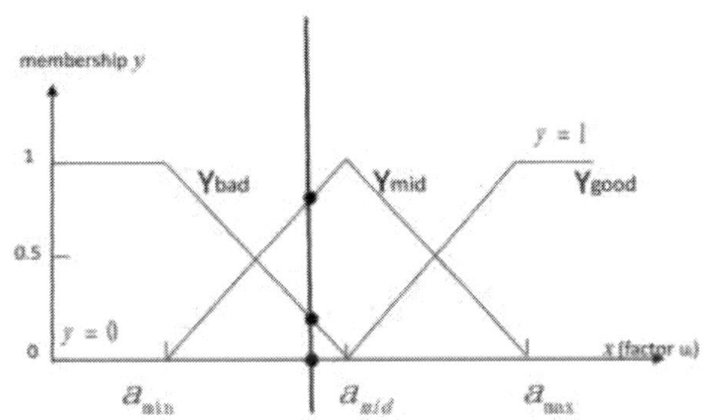

Figure 2. Evaluation matrix of a single factor.

IOP Conf. Series: Earth and Environmental Science **39** (2016) 012052 doi:10.1088/1755-1315/39/1/012052

In this part, the membership of each factor is computed, according to the value of the factor. The values of each factor for all dams are listed. For example, for calculating the factor u_1, a list of numbers is obtained: a_1, a_2, a_3, ..., a_n. Then, the maximum value (a_{max}) of that factor is found as well as the minimum value (a_{min}) of that factor. a_{mid} is set to equal (a_{max} +a_{min})/2. The fuzzy distribution is shown as follows:

$$\begin{cases} y_{good} = 0, & x \in [0, a_{mid}) \\ y_{good} = \dfrac{x - a_{mid}}{a_{max} - a_{mid}}, & x \in [a_{mid}, a_{max}] \\ y_{good} = 1, & x \in (a_{max}, +\infty) \end{cases}$$

$$\begin{cases} y_{mid} = 0, & x \in [0, a_{min}) \\ y_{mid} = \dfrac{x - a_{min}}{a_{mid} - a_{min}}, & x \in [a_{min}, a_{mid}] \\ y_{mid} = \dfrac{a_{max} - x}{a_{max} - a_{mid}}, & x \in [a_{mid}, a_{max}] \\ y_{mid} = 0, & x \in (a_{max}, +\infty) \end{cases}$$

$$\begin{cases} y_{bad} = 1, & x \in [0, a_{min}) \\ y_{bad} = \dfrac{a_{mid} - x}{a_{mid} - a_{min}}, & x \in [a_{min}, a_{mid}] \\ y_{bad} = 0, & x \in (a_{mid}, +\infty) \end{cases}$$

Common fuzzy distributions include y_{good}, y_{mid}, y_{bad}. When a factor is getting a value, there will be three membership values associated with it.

- Fuzzy relation matrix R

$$R = \begin{bmatrix} R_1 \\ R_2 \\ \cdots \\ R_7 \end{bmatrix} = \begin{bmatrix} r_{11} & r_{12} & r_{13} \\ r_{21} & r_{22} & r_{23} \\ \cdots & \cdots & \cdots \\ r_{71} & r_{72} & r_{73} \end{bmatrix}$$

In this paper, seven factors that influence the scheduling ability of dams are considered. According to the membership function of each factor, the membership degree can be found by using number normalization. All the membership degrees consist of an evaluation matrix R.

- Weight distribution vector W

For each influence factor, the weighing vector $W=(w_1, w_2, \ldots, w_m)$ indicates the respective weight, and the sum of all factor weights in this set is 1, which is consistent with the normalization principle. In this work, the expert inquiry method is used. The results are as follows:

Table 2. Weight of factors.

Factor	V_1	V_2	V_3	V_4	V_5	V_6	V_7
Weight	0.20	0.15	0.11	0.13	0.12	0.16	0.13

- Fuzzy Comprehensive Evaluation Model M

The composite operator is selected. Synthesis of W and R will be S, which means the comprehensive evaluation results are in comment set V. The Fuzzy operator $M(\bullet,\otimes)$ is employed.

$$S=W^{\circ}R$$

3.3. Experiment

This paper takes the Yin Shang ecological control section on the Huai He River basin as an example to illustrate the method. All of the data for this study came from the Key State Science and Technology Project (2014ZX07204-006-05). From the Yin Shang control section, this paper selects five dams as experimental objects: Zhou Kou, Huai Dian, Gen Lou, Fu Yang, and Yin Shang. For each of them, the data for the seven indexes are listed in Table 3.

Table 3. TOPSIS Evaluation Data of Dams.

Dams	V1	V2	V3	V4	V5	V6	V7
Zhou Kou	445	259.3	3	1.00	2	3	3
Huai Dian	82	133.1	2	0.74	2	3	2
Gen Lou	3970.61	143.7	2	0.24	2	2	2
Fu Yang	2875	78.4	1	0.41	3	2	2
Yin Shang	5431.46	0	1	0.07	5	4	5

Firstly, the authors set up the experiment environment. Secondly, a database was devised to store the necessary information. Finally, the evaluation process was conducted by performing the experiment on the environment.

The database was constructed by an SQL Server, and the experiment environment was built by the MyEclipse tool. SQL and MyEclipse are suitable to perform the evaluation process of dams.

4. Results and discussion

The results calculated by the TOPSIS approach are shown in Table 4.

Table 4. Evaluation Results of TOPSIS.

Dams	Yin Shang	Fu Yang	Huai Dian	Zhou Kou	Gen Lou
SA[a]	0.68	0.42	0.42	0.41	0.40

[a] SA means scheduling ability of dams.

The results calculated by the FCE approach are summarized in Table 5.

Table 5. Evaluation Results of FCE.

Dams	Bad	Middle	Good
Zhou Kou	0.44	0.41	0.15
Huai Dian	0.44	0.46	0.10
Gen Lou	0.32	0.45	0.23
Fu Yang	0.39	0.36	0.25
Yin Shang	0.13	0.02	0.85

The results show that the Yin Shang dam is the best of the five dams because of its top score in the V1, V2, V3, V5, V6, and V7 indexes. Although the scores of the dams may be different, the tendency of the dams' scores is almost the same, which means that different evaluation approaches could obtain the same results based on the common data. Furthermore, the different results of the specific regulating ability of the dams are caused by the two different evaluation approaches. In addition, it is easy for administrators to choose the dam with the highest water regulating ability to address the urgent problems that are occurring in the ecological control section after ranking the dams.

5. Conclusion

This paper introduces a new model to evaluate the water regulating ability of dams. The paper uses two evaluation methods, TOPSIS and FCE, to implement the model. In the experiment stage, five dams of the Sha Ying ecological control section were selected, and their differences were compared using the assessment results from the TOPSIS and FCE approaches. From the results, it is clear that the Yin Shang dam is the best capable dam for water environment protection in terms of regulating ability.

The next step to extend this study should be to use these evaluation methods to find the best way to conduct water regulation using dams.

Acknowledgements

The authors would like to thank Chuang Liu at USTC for his comments on the methods and paper.

This work was supported by the Major Science and Technology Program for Water Pollution Control and Treatment of the Key State Science and Technology Project under Grant No. 2014ZX07204-006-05.

References

[1] Abbas S, Alireza M, Aminzadeh A H, Hamed F A and Piltan T S 2012 *A review of the sand castle test for assessing collapsibility of filters in dams.*

[2] Paul R A and Stockwell L T 1983 *A method for wetland functional assessment:* Volume **1** *Critical review and evaluation concepts* Technical report

[3] Chumley-Jones H S, Dobbie A and Alford C L 2002 Web-based learning: Sound educational method or hype? a review of the evaluation literature *Academic Medicine* **77**(10):S86–93

[4] Shao C Y 2009 The implication of fuzzy comprehensive evaluation method in evaluating

internal financial control of enterprise *International Business Research* **2**(1):210

[5] Su C H, Ken T-K Chen and Fan K K 2013 Rough set theory based fuzzy topsis on serious game design evaluation framework *Mathematical Problems in Engineering*

[6] Vleck L D V and Murphy P A 1983 Comparison of dams' estimated transmitting abilities from first lactation herdmates or all records of herdmates to predict sons' milk evaluations *Faculty Papers and Publications in Animal Science* p391

[7] Wang Z L, Gao S C and Zhang H Y 2011 Research on influential factors of dams' regulating on river water quality *Engineering Journal of Wuhan University / Wuhan Daxue Xuebao* **44**(5):571–5

[8] Wieland M 2012 Seismic design and performance criteria for large storage dams *Proc. 15th World Conf. on Earthquake Engineering* Lisbon, Portugal

[9] Zhan C S, Zhang R R, Song X M, and Liu B L 2015 An enhanced environmental multimedia modeling system based on fuzzy-set approach: I. theoretical framework and model development *Frontiers of Environmental Science & Engineering* **9**(3):494–505

[10] Zhan C S, Zhang R R, Song X M, and Liu B L 2015 An enhanced environmental multimedia modeling system based on fuzzy-set approach: Ii. model validation and application *Frontiers of Environmental Science & Engineering* **9**(6):1025–35

[11] Zuo Q T, Liu Z H, Dou M and Gao J S 2011 Research framework of assessment of dams' impact on water quality and quantity and identification of regulation ability *South-to-North Water Diversion and Water Science & Technology* **2**:009

Research on the water resources regulation ability model of dams in the Huai He River Basin considering ecological and management factors

Y Shui[1], H C Liu[1], L H Li[1, 2], G G Yu[1] and J Liu[1]

[1]Scientific Research Department of Water Resource Preservation on Huai He, Bengbu, China

E-mail: 982386590@qq.com

Abstract. Research that assesses the scheduling ability of dams garners a great deal of attention due to the global water resource crisis. These studies can provide useful and practical suggestions for scheduling the water resources of dams to solve problems, such as addressing ecological water needs and so on. Recent studies have primarily evaluated the schedule ability of dams according to their quantifiable attributes, such as water quantity, flow velocity, etc. However, the ecological and management status can directly determine the possibility and efficiency of a dam's water resource scheduling. This paper presents an evaluation model to assess the scheduling capacity of dams that takes into consideration ecological and management factors. In the experiment stage, this paper takes the Sha Ying river of the Huai He River Basin as an example to evaluate the scheduling ability of its dams. The results indicate that the proposed evaluation model can provide more precise and practical suggestions.

1. Introduction

Recent studies evaluating the scheduling abilities of dams have increased amidst wide concerns about the global water resource crisis [8]. However, past research has primarily focused on water schedule on flood control, drainage, water supply, pollution prevention, control, and so on [7]. With the development of social economy and the increased efficiency of water resources, water ecology has been attracting a great deal of attention [11]. Past studies assessing the water schedule ability of dams were more concerned about water quantity and water quality but ignored the importance of the management status of dams [2]. To the best of our knowledge, related researchers avoided taking both management and ecological factors into consideration during their modeling for assessing water schedule ability [5]. However, the ecological and management factors can have a serious effect on the water schedule ability of dams [3]. Thus, some researchers adopted these qualified factors to obtain an imprecise assessment of dam scheduling ability [4]. Nonetheless, these indexes can be quantified and considered in order to make precise evaluations of dam scheduling abilities with scientific approaches [6]. This paper investigates fifty experienced experts who have been working on managing related dams for many years and obtains their rates on the management status of the dams. As for the ecological factors, this paper utilizes water quantity and the density of NH3-N and COD elements, which are critical influences on the ecological status of water around dams, to represent the water ecology variable.

2. Materials and Methods

2.1. Dams information

In this paper, the dams selected are primarily located in the Huai He River, which is one of the most important and largest rivers in China.

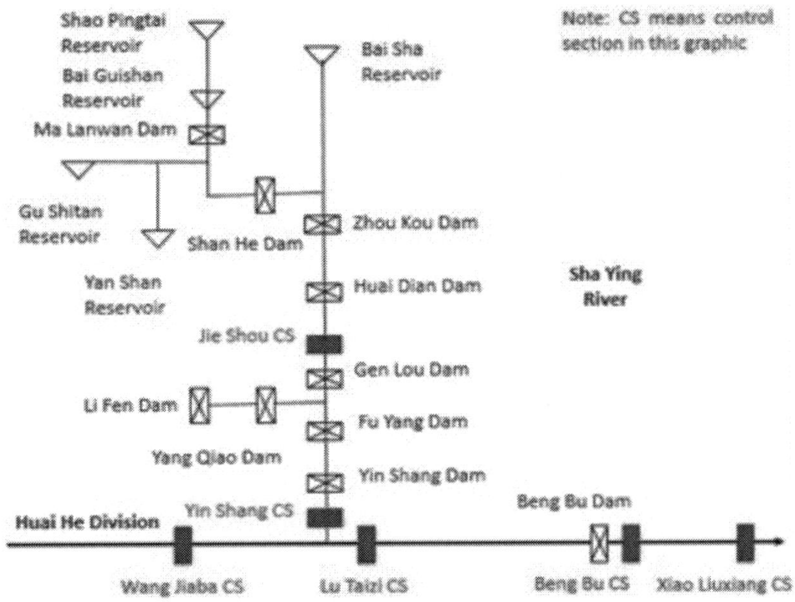

Figure 1. Dam layout in the Yin Shang control section.

Figure 1 illustrates the layout of the Yin Shang control section. From this figure, it can be seen that many control sections reside along the Huai He River division, and there are many dams and reservoirs along with the Yin Shang control section.

2.2. Evaluation model

For evaluating the specific schedule ability of dams, this paper constructs an assessment model that includes seven influential factors: water quantity that dams are able to use (V1), the water transmitting time of the schedule (V2), status of getting water along the river of a dam (V3), environmental remains of different contaminants (V4), dam management status (V5), construction status of dams (V6), and emergency handling capability of dams (V7). These factors can be classified into three aspects, which are included into the model as figure 2.

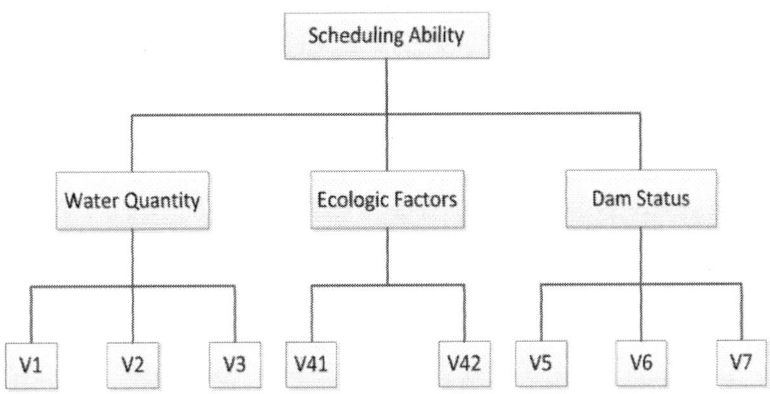

Figure 2. Dam Layout of the Yin Shang Control Section.

From figure 2, the factors affecting the scheduling ability evaluation of a dam can be divided into three categories. This model depicts how the factors are composed and how they influence the scheduling ability assessment of dams.

2.3. Evaluation Indexes

Based on past research, the evaluation indexes for assessing the schedule ability of dams mainly considered water quantity and water ecological factors. However, management factors are also critical when assessing dams' schedule abilities. This paper optimizes the factors considered in the past and presents new indexes according to practical conditions. The meanings of the indexes are shown in the following.

2.3.1. Water quantity that dams are able to use. The water quantity that a dam is able to schedule represents the amount of water that a dam is free to use. This index indicates the scheduling possibility and ability rather than the total water amount of a dam. The specific calculation process can be depicted by formula 1.

$$W = \eta * (w_1 - w_2) \quad (1)$$

In Equation 1, W is the water quantity that a dam is capable of using, w_1 is the total amount of water in a dam, and w_2 is the water quantity that a dam is unable to utilize. The ratio η is a water scheduling attribute of a dam; when η equals 1, the total water can be used for scheduling, and when η equals 0, no water can be scheduled. In reality, the water of a dam cannot be used up due to the existence of restrictions imposed by the dam and other urgent emergency concerns, such as shipping. Thus, the ratio η can make the water quantity factor more reasonable and practical.

2.3.2. Water transmitting time of the schedule. When a place requires water for ecological problems, a dam requires time to transmit water to the place. In the Huai He River, this dam index is mainly decided by the dam's distance from the control section of the river. This can be explained by figure 1. A hybrid model combining hydrodynamic effects has been developed to describe the water flow and its time [9] [10]. According to the model, this factor can be calculated. Thus, taking this factor into consideration can express the urgent aspect of the schedule ability of dams.

2.3.3. Status of getting water along the river of a dam. The status of getting water along the river of a dam describes the status of the scheduling water being stopped and used during the process of transmitting it to the destination. This parameter can influence the amount of water that reaches an ecological control section, which affects the water scheduling of a dam.

This paper utilizes fifty experienced experts who previously worked in this area for a long time. The experts rate this factor for each dam, and two rating criterions are established for them. The rating standards are the water number used and the number of dams that exists during the process when the water was transmitted from the dams to the ecological control section, and the results of the control section are listed in Table 1.

From the table, the specific evaluation score for an expert to a dam on V3 can be obtained. Thus, removing the largest and smallest score of a particular dam, obtain the value of V3 can be obtained as equation 2.

$$S_j = \frac{\sum_{i=1}^{n} RD_{ij}}{n = 2} (i \neq ima, imi) \quad (2)$$

In equation 2, S_j is the V3 value of Dam j; RD_{ij} means Rater i rates the ability of Dam j, which ranges from 0-5; m represents the number of dams along the control section; and n is the effective expert's rating. Furthermore, ima and imi are the maximum and minimum raters for rating the ability of dam j, respectively. In this paper, five dams above the Yingshang control section were selected as research objects. The five dams include Zhou Kou, Huai Dian, Gen Lou, Fu Yang, and Yin Shang. The five dams are marked as Dam 1, Dam 2,..., Dam 5, respectively.

Table 1. Score table for V3.

Raters	Dam 1	Dam 2	...	Dam m
Rater 1	RD_{11}	RD_{12}	...	RD_{1m}
Rater 2	RD_{21}	RD_{22}	...	RD_{2m}
...	...	...	...	....
Rater n	RD_{n1}	RD_{n2}	...	RD_{nm}

2.3.4. Environmental remains of different water quality parameters. The environmental remains of different contaminants are taken into consideration for evaluating the dams' purifying water ability. The better this factor, the higher the purifying ability of the water in the dam is, which also means there is a higher water quality. This factor can be calculated as follows.

$$R = 100 *(C_0 - C_1) \quad (3)$$

From equation 3, this factor is represented by R. C_0 is the standard water quality in water function zones, which can be defined according to [1]. The number 100 is the unit exchanging ratio, and C_1 ia the water quality of the dams. In this case, it can be concluded that the higher the R value, the more purified the water is in the relevant dams.

As for water quality, this paper calculates two water quality parameters, NH_3-N and COD, as representatives for assessing the dams' water quality. In this case, equation 3 can be modified into Equation 4.

$$R = 100 * \frac{\alpha * (C_{0NH3-N} - C_{1NH3-N}) + \beta * (C_{0COD} - C_{1COD})}{\alpha + \beta} \quad (4)$$

From equation 4, α represents the weight of NH_3-N, while β is the weight of COD. Thus, by Equation 4, the value of the environmental remains of the water quality parameters can be obtained.

2.3.5. Dam management status. Dam management status is an index explaining how the dams are utilized, and whether it is easy to schedule the water. This paper adopts the same approach for obtaining its values as was done for V3. Thus, this evaluating index can be defined.

2.3.6. Construction status of dams. The construction status of a dam is the current status of a dam that can affect its scheduling ability. This paper rates a dam's construction status as a 0-5 score according to its build time, maintenance time, dam level, and research results. For a specific dam, a rating table can generated, such as Table 2.

Table 2 gives the four indexes for calculating the construction status of a dam. The first three indexes, the dam build time, dam maintenance time, and dam level can be defined by looking up relevant literatures. As for the fourth index, the research results score can be generated by assessing whether the dam has a flow monitoring facility and a security facility for ecological flow. Using all of the scores in Table 2, this paper calculates the construction status of dams according to Equation 5.

$$C = \frac{\varepsilon *RB + \phi *RM + \varphi *RL + \gamma * RR}{\varepsilon + \phi + \varphi + \gamma} \quad (5)$$

Table 2. Rating table for V6.

Raters	Build time	Maintenance time	Dam level	Research results
Rater 1	RB_{11}	RM_{12}	RL_{13}	RR_{14}
Rater 2	RB_{21}	RM_{22}	RL_{23}	RR_{24}
...	...	...	...	....
Rater n	RB_{n1}	RM_{n2}	RL_{N3}	RR_{n4}

C represents the final value for this factor. The coefficients ε, ϕ, φ, and γ of the indexes show the weights of the four indexes, respectively. Moreover, the value of the indexes in Equation 5 can be calculated according to the method for determining V3.

2.3.7. Emergency handling capability of dams. The emergency handling capability of a dam is an index evaluating the ability of a dam to cope with emergencies, which is a factor that can affect the water scheduling ability of a dam. This paper calculates this value in the same method by which V3 is calculated. Thus, the value of this factor can be defined in a scientific approach.

3. Experiment

This paper employs the Yin Shang control section of the Huaihe River as an example for performing the experiment. In this experiment, this paper adopts the Technique for Order Preference by Similarity to Ideal Solution (TOPSIS) method to implement the model, and all the data comes from the Key State Science and Technology Project (2014ZX07204-006-05). TOPSIS is a commonly used approach for evaluating regulation ability [12]. The results are shown in table 3.

Table 3. TOPSIS evaluation process for dams.

Dams	V1	V2	V3	V4	V5	V6	V7
Zhou Kou	445	259.3	3	1.00	2	3	3
Huai Dian	82	133.1	2	0.74	2	3	2
Geng Lou	3970.61	143.7	2	0.24	2	2	2
Fu Yang	2875	78.4	1	0.41	3	2	2
Ying Shang	5431.46	0	1	0.07	5	4	5

From the results, it can be determined that the Yin Shang dam is the best dam of the five dams because it has the best scores for the V1, V2, V3, V5, V6, and V7 indexes. The new factors model for evaluating dams' regulation abilities combined with the adopted assessing approach can give a specifically quantifiable number of the ability. The model is useful and practical for managers to regulate dams.

4. Discussion

This paper presents a new model for evaluating the water scheduling ability of dams, which can provide practical suggestions to related managers to make the best choice to maintain ecological water needs. In a follow-up study, further research can be conducted to determine the best algorithm to simulate the scheduling process based on the evaluation results. Additionally, based on this new model, researches on finding the best way to implement this model can be performed in detail.

Acknowledge

The authors would like to thank Chuang Liu at USTC for his comments on the methods and paper.

This work was supported by the Major Science and Technology Program for Water Pollution Control and the Treatment of the Key State Science and Technology Project under Grant No. 2014ZX07204-006-05.

Reference

[1] Maite M A, Ashok K, Chapagain, Arjen Y H, and Mesfin M M 2012 *The water footprint assessment manual: Setting the global standard. Routledge*

[2] Brandt S A 2000 *Classification of geomorphological effects downstream of dams Catena* **40**(4):375–401

[3] Richard T K 2000 Ecological impacts of dams, water diversions and river management on floodplain wetlands in Australia *Austral Ecology* **25**(2):109–27

[4] Krawczyk J B 1993 A decision support tool for controlling a dam to environmentally acceptable standards *In Proceedings of the 29* Operational Research Society of New Zealand Conference, Auckland* pp 219–26

[5] Newham L T H, Letcher R A, Jakeman A J, Heathwaite A L, Smith C J, and Large D 2002 Integrated water quality modelling: Ben chifley dam catchment, Australia *International Environmental Modelling and Software Society* **1**:275–80

[6] Palmieri A, Shah F and Dinar A 2001 Economics of reservoir sedimentation and sustainable

management of dams *Journal of Environmental Management* **61**(2):149–63

[7] Wang Z L, Gao S C, and Zhang H Y 2011 Research on influential factors of dams' regulating on river water quality Engineering *Journal of Wuhan University / Wuhan Daxue Xuebao* **44**(5):571–5

[8] Wieland M 2012 Seismic design and performance criteria for large storage dams *In Proc. 15th World Conf. on Earthquake Engineering, Lisbon, Portugal*

[9] Zhan C H, Zhang R R, Song X M and Liu B L 2015 An enhanced environmental multimedia modeling system based on fuzzy-set approach: I. theoretical framework and model development *Frontiers of Environmental Science & Engineering* **9**(3):494–505

[10] Zhang R G, Zhan C G, Song X M and Liu B L 2015 An enhanced environmental multimedia modeling system based on fuzzy-set approach: Ii. model validation and application *Frontiers of Environmental Science & Engineering* **9**(6):1025–35

[11] Zuo Q T, Liu Z H, Dou M and Gao J S 2011 Research framework of assessment of dams' impact on water quality and quantity and identification of regulation ability *South-to-North Water Diversion and Water Science & Technology* **2**:009

[12] Su C H, Ken T-K Chen, and Fan K K 2013 Rough set theory based fuzzy topsis on serious game design evaluation framework *Mathematical Problems in Engineering*

Evaluation system of water ecological civilization of irrigation area in China

Z Liu[1], J Chen[1], D Chen[1, 2], S Zhang[1], X C Li[1], Y Zhu[1] and Y Li[1]

[1]Key Laboratory of Efficient Irrigation-Drainage and Agricultural Soil-Water Environment in Southern China (Ministry of Education), College of Water Conservancy and Hydropower Engineering, Hohai University, Nanjing 210098, China

E-mail: 23543458@qq.com

Abstract. Irrigation area is an important carrier, and also has a pivotal role in the construction of water ecological civilization in China, as well as worldwide. This work extracted the five basic characteristics of water ecological civilization of irrigated area, namely "resource saving, efficient production, ecological nature, beautiful environment, and civilized consciousness". Further, based on the frequency analysis of indicators related to the evaluation of irrigation area, we proposed the evaluation system of water ecological civilization of irrigated area. Taking an irrigation district of Huaian City, Jiangsu Province, China as an example, we carried out the case evaluation in use of the fuzzy comprehensive evaluation method. Thus, we provide the theoretical and technical reference for the construction and assessment of water ecological civilization of irrigation district to both China and abroad.

1. Introduction

As the source of life, water resource is also the control element of the ecosystem. Water conservancy construction, in some way, can be regarded as the core of ecological civilization. Early in 2012, China vigorously put forward that, from the aspects of resource conservation and ecological environmental protection, the country should speed up water conservancy construction and strengthen water management. In 2014, China again stressed the need to develop and improve the laws and regulations of ecological compensation, water pollution control, and marine environmental protection. Later, the new water management concept "water-saving priority, spatial equilibrium, system governance, handed force" was determined, so as to pointing out the direction of how to accelerate the development of water reform.

As an important grain-producing district, irrigation area becomes a vital carrier of water ecological civilization-due to its large proportion of water use in the whole society. In recent years, considering the modes such as "irrigation modernization", "ecological irrigation", "water-saving and ecology irrigation area", some scholars and managers analyzed the content, features and theoretical basis of the corresponding irrigated area. Further, they put forward the main technical systems and evaluation system. Some also carried out the relevant practice and pilot studies for ecological construction of irrigation area[2-8]. All of the above research can be used to guide and reference the construction of water ecological civilization of irrigated area. However, at present, the theory, ideas, and practice are still relatively weak. It is very hard to make an interpretation or evaluation of water ecological civilization of irrigation area. Therefore, from both a theoretical and practical point of view, the construction of water ecological civilization of irrigation area in China is still in its infancy, and needs to strengthen the relevant research and practice.

2. Construction of evaluation system

2.1. The first-level index

Index, as the core of evaluation system, is the basis for the evaluation of construction of water ecological civilization in irrigation area. It should follow the principles of comprehensive, representative, and operational feasibility. Based on the background process of water ecological civilization, and related guidelines and objectives from "suggestions on accelerating the construction of water ecological civilization" by Ministry of Water Resources in China, the paper summarized the characteristic of water ecological civilization in irrigation area, as well as the basic evaluation criteria layer—resource saving, efficient production, ecological nature, beautiful environment and civilized consciousness. The specific meanings are as follows:

Resource saving-Corollary, water-saving transformation, universal resource conservation awareness, and efficient technology should be widely promoted. Many aspects, such as irrigation planning, design, construction and management processes, should all focus on high efficient use on soil and water resources and energy conservation. Efficient production-Irrigation area should set goals of both high yield and low cost, focusing on maximizing the overall efficiency. Ecological nature-Water conservancy project in irrigation area should be in harmony with the ecological environment. Ecological engineering and near-natural construction methods should be applied. Irrigated ecosystem can get the restoration and maintain homeostasis. Soil erosion and water environment have been effectively controlled. Agricultural non-point source pollution has been effectively regulated, so as to achieve restoration and protection of biodiversity. Beautiful environment—Environment in the irrigation district should be nice and clean. Irrigation and drainage canals, and rivers, should be kept in good condition. Hydraulic engineering should be in harmony with the surrounding environment and have a certain aesthetic value within the landscape. It would be better if they can reflect the local cultural characteristics of water. Civilized consciousness-awareness of resource conservation has been widely popularized. Concept of water ecological civilization is deeply rooted among people. Irrigated farmers are happy and satisfied with the participation in the irrigation area planning, design, construction and management.

2.2. The secondary-level index

In reference to a number of relevant pieces of literature on irrigated area assessment, ecological civilization, modernization and regulations about water, agriculture, environment, ecology in China, the frequency analysis of related indicators were carried out. Indicators which received relatively high frequency were then selected from the five aspects—"resource saving, efficient production, ecological nature, beautiful environment and civilized consciousness". The Specific indicators system is shown in Table 1, including 5 first-level and 18 secondary-level indicators, in which 15 are quantitative and 3 are qualitative.

3. Evaluation method

3.1. Choice of method

Evaluation system of water ecological civilization irrigation has the following comprehensive characteristics: layered, fuzzy, gray and a certain degree of subjectivity. This paper, considering the adaption, recommends selecting the fuzzy comprehensive evaluation method [22] for the current work.

3.2. Level delineated

In the fuzzy evaluation, it is necessary to select the appropriate fuzzy operator. Evaluation process of water ecological civilization in irrigation area needs to consider the overall factors comprehensively. It will be relatively accurate to calculate through weighted average. In this paper, we take the operator type of weighted average.

The evaluation sets are respectively given a score of 5, 4, 3, 2 or 1, from good to bad, thus building a model in view of the fuzzy analytic hierarchy process. The evaluation results are represented by a number from 1 to 5—Score>4(Category Ⅰ), 3<Score<4(Category Ⅱ), Score<3(Category Ⅲ).

4. Case evaluation of water ecological civilization in irrigation area

4.1. Overview of irrigation area

The irrigation area is located in Huai'an City, Jiangsu Province, China, mainly relying on the water of Yangtze River and Huai River for irrigation. Due to the large water leakage, ecological lining through the backbone water channel was carried out in irrigation area. Sprinkling, pipes and other water-saving projects and technical measures were promoted. Some arable land also needs pumping for irrigation. With a relatively good infrastructure and natural condition, the agricultural production in this irrigation area maintained well, as well as the overall environment. Ditches and roads were mostly green. Irrigation management organizations, which were divided into four agencies, have a strict system.

4.2. Index weight and situation evaluation

Using AHP[1], index weights in each level are fingered out. Considering the core research subject and the length restrictions of the article, interviews, questions, matrix and weight calculation process are not shown in this paper. It should be noted that the index weight to all the levels of the evaluation system is not the same. They will make a corresponding change with different irrigation districts in different natural and geographical conditions. Specific weights were calculated in the AHP process and are shown in table 1. The evaluation involves a wide range of factors, most of which can be quantified through surveys, measurements and calculations. Some indicators are difficult to express mathematically; qualitative description is necessary at this time. Evaluation situations were primarily from the planning report, water-saving transformation planning report, Huaian yearbooks and administration statistics in irrigation areas. Concrete results are shown in table 1.

Table 1. Evaluation system of water ecological civilization in irrigation area.

First-level index	Weights	Secondary-level index	Weights	Value	Comment
		water use efficiency of irrigation[9]	0.303	0.6	excellent
		area ratio of water-saving irrigation project	0.233	0.58	excellent
		ratio of high standard farmland[10]	0.280	50.2%	good
resource saving	0.141	average device efficiency of pumping station[11]	0.184	56%	medium
		water productivity[12]	0.357	1.53	good
		grain yield per mu	0.386	468.2	medium
efficient production	0.283	agricultural input-output ràtio[13]	0.257	0.17	medium

		roads and canals greening rate[15]	0.253	83%	good
		ecological revetment rate	0.266	79%	good
		loss control rate[16]	0.195	74%	medium
		quality compliance rate of irrigation water[17]	0.178	69%	medium
ecological nature[14]	0.283	intensity of chemical fertilizer[18]	0.108	267	bad
		compliance rate of village environment	0.388	86%	medium
		irrigation building ornamental[19]	0.313	/	good
beautiful environment	0.166	comprehensive utilization of crop stalks[20]	0.299	58%	inferior
		awareness penetration of water management and ecological civilization	0.363	/	medium
		resource conservation awareness of irrigation farmers	0.298	68%	medium
civilized consciousness[21]	0.126	integral satisfaction of Irrigation farmers	0.339	/	medium

According to the steps of the fuzzy comprehensive evaluation method, evaluation matrix of index was calculated.

For example, calculation process of resource saving evaluation matrix is as follows:

$$B_1 = A_1 \bullet R_1 = \begin{bmatrix} 0.303 & 0.233 & 0.280 & 0.184 \end{bmatrix} \bullet \begin{bmatrix} 0.8 & 0.2 & 0 & 0 & 0 \\ 0.8 & 0.2 & 0 & 0 & 0 \\ 0.15 & 0.7 & 0.15 & 0 & 0 \\ 0 & 0.15 & 0.7 & 0.15 & 0 \end{bmatrix}$$

$$= \begin{bmatrix} 0.471 & 0.331 & 0.129 & 0.028 & 0 \end{bmatrix}$$

Similarly, the comprehensive evaluation matrix of irrigation area can be drawn:

$$R=\begin{bmatrix}B_1\\B_2\\B_3\\B_4\\B_5\end{bmatrix}=\begin{bmatrix}0.471&0.331&0.129&0.028&0\\0.054&0.346&0.504&0.096&0\\0.078&0.419&0.355&0.132&0.016\\0.047&0.277&0.319&0.118&0.239\\0&0.150&0.700&0.150&0\end{bmatrix}$$

$$B=A\bullet R=\begin{bmatrix}0.141&0.283&0.283&0.166&0.126\end{bmatrix}\bullet$$

$$\begin{bmatrix}0.624&0.274&0.088&0.014&0\\0.045&0.313&0.537&0.105&0\\0.096&0.493&0.311&0.091&0.010\\0.035&0.258&0.506&0.183&0.018\\0&0.150&0.700&0.150&0\end{bmatrix}$$

$$=\begin{bmatrix}0.112&0.329&0.403&0.107&0.044\end{bmatrix}$$

To quantify the results of the evaluation,

$$W=B\bullet Q=\begin{bmatrix}0.112&0.329&0.403&0.107&0.044\end{bmatrix}\bullet\begin{bmatrix}5&4&3&2&1\end{bmatrix}^{T}=3.34$$

4.3. Results analysis

Comprehensive evaluation score of the irrigation area is 3.34, which belongs within Category II. Essentially, it basically met the evaluation criteria-the construction and management substantially satisfied the requirements of water ecological civilization in irrigation area, but still needs to be improved in some aspects.

However, the evaluation result does not reflect the grade of the first-level index, nor explain the shortcomings of the status quo in the irrigation area. Membership, namely the level of related special subjection, can offer a more accurate way to the evaluation results. Thus, this work further analysed the degree of membership of the first-level indicator, reaching the evaluation results of the five first-level indicators. According to the principle of membership matrix, the maximum membership degree is the evaluation situation of the index. Memberships of the first-level index are shown in figure 1. The evaluation levels are excellent, medium, good, medium and medium, respectively. Analysis results show that in this irrigation area, resource saving rates the best, followed by the ecological nature. Efficient production, beautiful environment and civilized consciousness are relatively weak, and still need some targeted measures to improve the building standard, thus fully realizing the water ecological civilization in irrigation area.

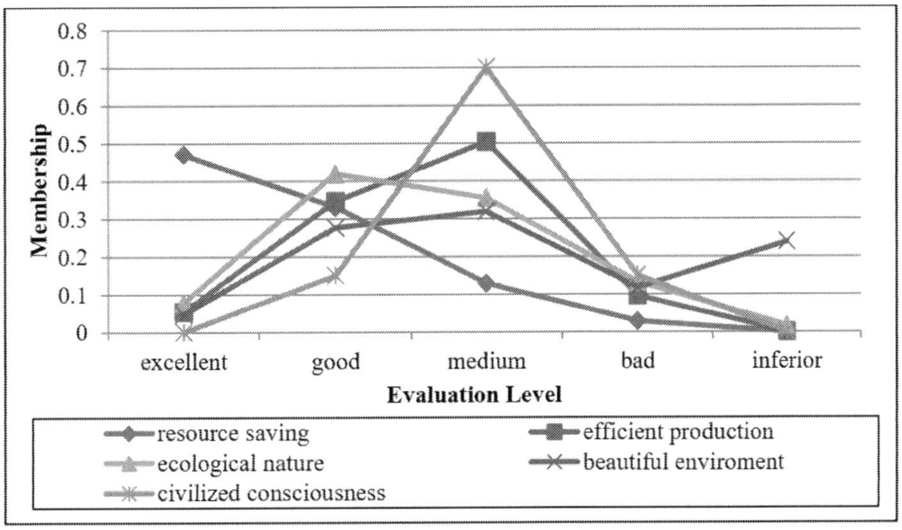

Figure 1. Membership of the first-level index.

5. Conclusion

Construction evaluation is the most important content of water ecological civilization in irrigation area in China. This paper presented an appropriate evaluation system, as well as the comprehensive method. Further, an irrigation area in Huaian City was taken as an example to be evaluated, providing a specific and feasible theoretical basis for the assessment of water ecological civilization in irrigation area. It should be noted that the weights of indexes at all levels are not the same. With constant development, irrigated conditions and infrastructure are also changing, as is the evaluation system. Only through change will they meet the new requirements of different periods. In addition, the construction of water ecological civilization in irrigation district is a research blend of many disciplines including ecological economics, ecological ethics, landscape ecology and water conservancy projects. Therefore, it needs interdisciplinary study of thinking in order to achieve better research results.

Acknowledgments

This study has been supported by the Science and Technology Project on Water Conservancy in Jiangsu Province (2015074), the Fundamental Research Funds for the Central Universities (2015B20614), and Training Programs of Innovation and Entrepreneurship for Undergraduates in Hohai University (201610294023), the Program for Distinguished Talents in Hohai University, and a Project Funded by the Priority Academic Program Development of Jiangsu Higher Education Institutions (PAPD).

References

[1] Jiang G 2005 *Fuzzy AHP synthetical appraise method and application* Tianjin University

[2] Jiang K P 2004 Construction thinking of ecological irrigation area-in use of ecological civilization to promote the sustainable development of irrigation district *China Rural Water Conservancy and Hydropower* 4-10

[3] Tang X J 2007 Water resources situation, current problems and countermeasures of Dujiangyan irrigation area *Proc. of the 3rd Youth Science and Technology Forum of China Water Conservancy Institute*

[4] Gu B J 2006 *Construction principle and key technology of ecological irrigation district* Hohai University

[5] Zhang Z Z 2010 Study on ecological irrigation area and its healthy development *Modern High*

Efficiency Water Saving Agriculture and Construction of Ecological Irrigation District

[6] Zhang Z Y 2006 Brief analysis on construction of ecological irrigation district in China *Academic annual meeting of China Water Conservancy Society*

[7] Fang Y X 2011 Fuzzy comprehensive evaluation model and its application of eco-system's health in irrigation district *Journal of Agricultural Engineering* **27** 199-205

[8] Yang P L 2009 Theory and technology of construction on ecological irrigation district *China Water Resources* 32-35

[9] Li Y N 2003 Study on water use efficiency in irrigation *China Rural Water Conservancy and Hydropower* 23-26

[10] Shen M 2012 Research on classfication of provincial high standard farmland—Taking Guangdong Province as an example *China Land Science* **26** 28-33

[11] Zhang R T 1996 Research on conversion method of device efficiency of pumping station *Pump Technology* 27-31

[12] Li Y H 2001 Calulation method and application of water productivity *China Water Resources* 65-66

[13] Xie K J 1992 Investigation on agricultural input-output ratio of Xigangzi Town *Journal of Hei River* 40-42

[14] Jin D G 2004 Theory and practice of natural ecological engineering *Proc. of International Ecological City Construction*

[15] Xie X H 2008 Research of levee greening and manangement in Weishan Irrigation District *Shandong Water Resources* 52-53

[16] Liu J G 2008 Research on comprehensive evalution system of loss control rate *Journal of Yangtze River Academy* **25** 82-85

[17] Zhao J R 2008 *Influence of irrigation water quality to Soil* Northwest A&F University

[18] Yu Y X 2009 Trends and countermeasurs of ferterlizer pollution in China *Proc. of China Sustainal Development Academy*

[19] Zhong S M 2004 Research on combination between practicality and appreciation of buildings *Building Development Orientation*

[20] Zhang K 2012 Comprehensive utilization of crop stalks will be increased—Interpretation of the implementation of comprehensive utilization of crops in 12th Five-Year *New Rural Technology* 49-50

[21] Zhuo Y 2007 Critical thinking of how to promote civil's sense of ecological civilization *Marx Doctrine and Reality* 106-111

[22] Jin J L 2004 Fuzzy comprehensive evalution model based on AHP *Journal of Hydraulic Engineering* 65-70

Algae inhibition experiment and load characteristics of the algae solution

L Xiong[1], J X Gao[1, 2], Y X Zhang[1], Z K Yang[1], D Q Zhang[1] and W He[1]

[1]State Key Laboratory of Power Transmission Equipments & System Security and New Technology Chongqing University, Chongqing 400044, China

E-mail: 1551849658@qq.com

Abstract: It is necessary to inhibit microbial growth in an industrial cooling water system. This paper has developed a Monopolar/Bipolar polarity high voltage pulser with load adaptability for an algal experimental study. The load characteristics of the *Chlorella pyrenoidosa* solution were examined, and it was found that the solution load is resistive. The resistance is related to the plate area, concentration, and temperature of the solution. Furthermore, the pulser's treatment actually inhibits the algae cell growth. This article also explores the influence of various parameters of electric pulses on the algal effect. After the experiment, the optimum pulse parameters were determined to be an electric field intensity of 750 V/cm, a pulse width per second of 120μs, and monopolar polarity.

1. Introduction

Due to the large reproduction of bacteria, numerous microbes are present in industrial circulating cooling water, and as a result, many problems occur, such as pipe blockage and severe corrosion [1]. Among the methods of bacteriostasis and algal inhibition, the applicable scope of biological methods is primarily limited for a particular microorganism, and the chemical methods can easily corrode the equipment and cause secondary pollution [2]. Therefore, the physical methods of electromagnetic pulse processing technology have become a popular research topic [3].

However, there is a lack of experimental data about the load characteristics of the algae solution. There are many contrary opinions about the different effects of monopolar and bipolar pulses [4]-[6]. There is also little information about the major impact parameter and the most superior treatment parameter among electric field intensity, pulse width, etc.

This paper aims at identifying the optimum combination of electric parameters for the best algae inhibition effectiveness within the parameter scope of the pulse generator, and the load properties of the algal solution are analyzed.

IOP Conf. Series: Earth and Environmental Science **39** (2016) 012055 doi:10.1088/1755-1315/39/1/012055

2. Experimental devices

2.1. Power supply design

This paper developed a novel monopolar/bipolar polarity high-voltage pulser based on the 9 - stage Marx pulse generator and H-bridge. Load Z_L is the algae fluid. The main structure is shown in Figure 1. In order to change the pulse width and frequency, the Pulse Width Modulation (PMW) of the Digital Signal Processor (DSP) is adjusted. The DSP can make the output DC voltage continuous and adjustable by communicating with the high-voltage DC source using the serial port. In addition, it can also yield monopolar or bipolar polarity by controlling the switch of the H Bridge. The hardware platform for the High Voltage Pulser is shown in Figure 2.

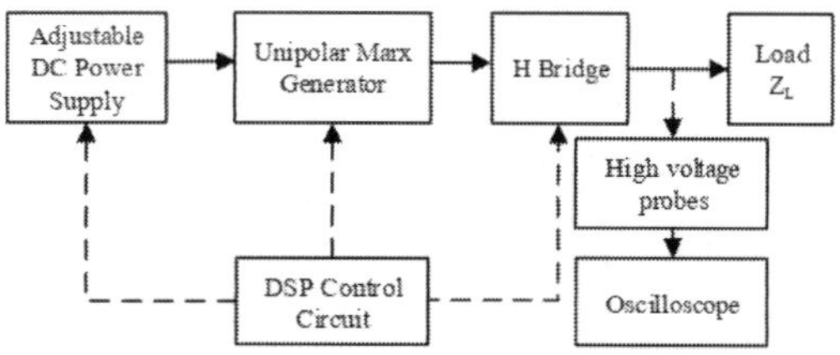

Figure 1. Polarity Tunable High Voltage Pulser's system structure diagram.

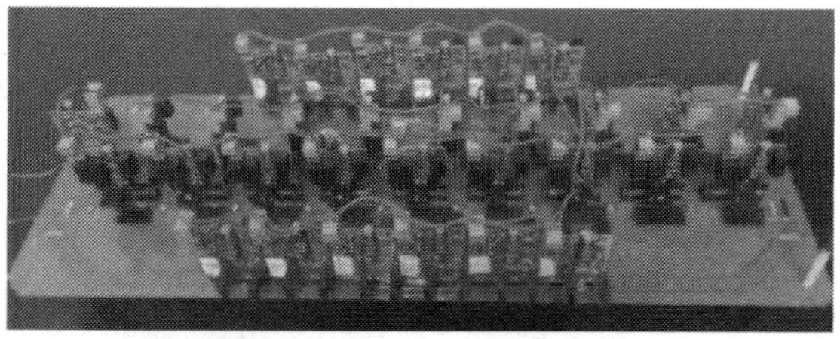

Figure 2. Hardware platform for the High Voltage Pulser.

2.2. Treatment chamber

In this paper, a small rectangular chamber is designed with an interior volume of 50 mm*80 mm*100 mm (ignoring the electrode size). We repeatedly compared both the advantages and disadvantages of a copper electrode, platinum electrode, and graphite electrode. Taking into consideration the high price of platinum electrodes and the electrolytic effect of copper electrode, we selected the graphite electrode, which is shown in Figure 3.

Figure 3. The experimental treatment chamber.

3. Load characteristics of the algal solution

The study employs the *Chlorella pyrenoidosa* algae as the experimental subject. Three identical micro-algal solutions with different concentration were made and were named as x solution, y solution, and z solution ($x=4.76y$, $x>z>y$). The parameters for the pulser were set as the following: bipolar pulse, a pulse frequency of 1 kpps, and a pulse width of 6 μs. The waveforms of measured load under x concentration with a graphite electrode chamber or copper electrode chamber are shown in Figure 4a and 4b.

IOP Conf. Series: Earth and Environmental Science **39** (2016) 012055 doi:10.1088/1755-1315/39/1/012055

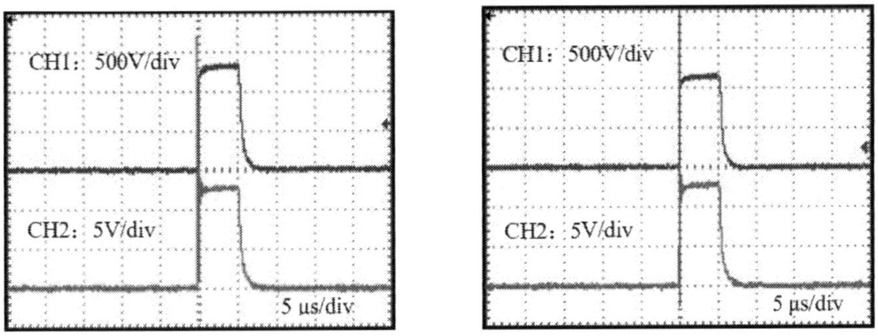

(**a**) with graphite electrode plates (**b**) with copper electrode plates

Figure 4. The load waveform when treating the algal liquid.

It was found that the algal solution load is highly resistive. Furthermore, the load characteristics exhibit no difference under the two systems, and the other parameters, including the liquor concentration, are also the same.

In summary, we produced a list of the equivalent resistance of the algae solution, which was connected to the concentration, temperature, and plate area, as shown in Table 1.

Table 1. The resistance value of the algae liquid.

	Concentration /x=4.76y		Temperature /℃		Volume/mL	
	x	y	23.5	66	60	130
Equivalent resistance	552	604	550	275	768.7	314.5

The results showed that the load characteristic of the algae solution was resistance. With increasing temperature and concentration, the equivalent resistance of the algae fluid is reduced. Additionally in this process chamber, as the plate area of the algae solution is increased, its equivalent resistance is also reduced.

4. The algae inhibition experiment

4.1. Experimental methodologies

We divided the pulse electric field intensity and pulse width per second (, multiplied by the number of pulse per second and the single pulse width) into three levels, and employed the monopole or bipolar pulse condition. Then, in this paper, we conducted several pre-experiments in order to determine the proper values of three electric parameters, shown in Table 2. The distance between the electrode plates is 40mm so that the pulse voltage can directly reflect the electric field intensity.

Table 2. The electrical parameters

Parameters	Level 1	Level 2	Level 3
Voltage (kV)	1	2	3
Pulse width per second (μs)	180	120	60
Polarity	Bipolar	Monopolar	

Therefore, there are 18 experimental groups with different parameters, as shown in Table 3. In the table, Bipolar-2 kV-180 μs indicates a bipolar pulser, the voltage is 3 kV, the pulse width time per second is 180 μs, and so on.

Table 3. The pulser parameter group.

Group	Pulser Parameters	Group	Pulser Parameters	Group	Pulser Parameters
1	Bipolar-3 kV-180 μs	7	Bipolar -3 kV-60 μs	13	Monopolar-3kV-120μs
2	Bipolar -2 kV-180 μs	8	Bipolar -2 kV-60 μs	14	Monopolar-2kV-120μs
3	Bipolar -1 kV-180 μs	9	Bipolar -1 kV-60 μs	15	Monopolar-1kV-120μs
4	Bipolar -3 kV-120 μs	10	Monopolar-3kV-180μs	16	Monopolar-3kV-60μs
5	Bipolar -2 kV-120 μs	11	Monopolar-2kV-180μs	17	Monopolar-2kV-60μs
6	Bipolar -1 kV-120 μs	12	Monopolar-1kV-180μs	18	Monopolar-1kV-60μs

4.2. Experimental procedure

- The algal solution was divided into 19 groups, one blank group, and 18 treatment groups. Then, different pulse parameters were used to treat the algal solution for 5 minutes.
- Each time we finished one experimental group, the algal solutions were then put into the thermostatic water tank for growth. All experiments were finished in the first day. As for the blank group, the measuring time, concentration value, and temperature were recorded when the algal solution had been shaken up, sampled, and diluted five times.
- From the second day to the seventh day, we tested the concentration value and temperature of the algal solution in all of the experimental groups. The detection was done every 24 hours after the first detection time. The experiment platform is shown in Figure 5.

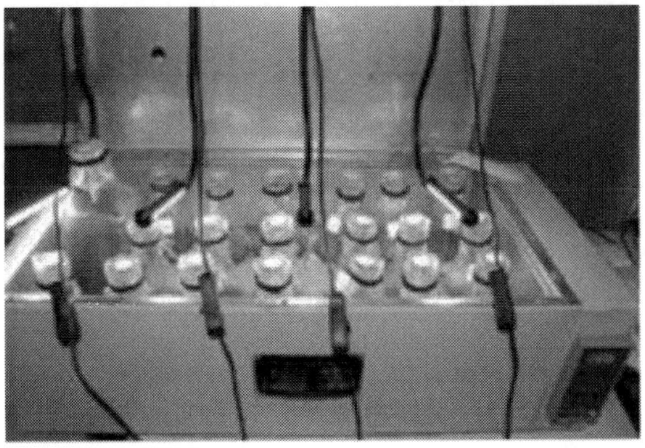

Figure 5. The test platform of culturing the *Chlorella pyrenoidosa*.

4.3. Results and discussion

After a week, we obtained the growth rate of the algal cells according to formula (1), by which we evaluated the inhibition of algal cell growth by high-voltage pulse. The lower the value, the better the suppressing effect is.

$$P = \frac{C_n - C_1}{C_1} \tag{1}$$

In formula (1), P refers to the growth rate of the algal cells, C_n is the concentration of the algae cells of one group in the nth day, and C_1 is the concentration that was measured before the experiment. The growth rates that were recorded over seven continuous days are shown in Table 4.

Table 4. The algal cell growth rate.

The growth rates after the experiment																
	Initial value /%	Training 24 hours /%	Training 48 hours /%	Training 72 hours /%	Training 96 hours /%	Training 120 hours /%	Training 144 hours /%		Initial value /%	Training 24 hours /%	Training 48 hours /%	Training 72 hours /%	Training 96 hours /%	Training 120 hours /%	Training 144 hours /%
1	-2.13	-3.04	15.19	33.12	25.22	38.28	44.06	10	1.51	9.66	17.52	45.90	45.00	43.19	51.34
2	4.63	3.39	27.46	39.18	45.97	56.15	67.87	11	4.42	5.90	9.73%	25.95	33.32	48.06	63.40
3	-3.77	0.58	19.12	30.13	29.55	38.54	55.05	12	2.09	8.35	7.16%	41.48	49.23	54.61	79.07
4	-2.65	0.59	15.88	17.93	28.22	37.93	36.46	13	-1.48	4.45	7.71%	23.73	30.85	39.45	34.41
5	-5.54	-5.54	14.28	18.65	26.52	40.80	41.09	14	4.66	7.58	5.54%	29.43	37.30	43.72	73.73
6	-1.83	3.96	22.86	34.13	31.69	42.06	58.82	15	-0.87	9.30	16.27	30.22	35.16	27.90	59.28
7	0.61	0.31	15.94	34.96	30.66	44.46	43.54	16	2.55	1.28	8.94%	22.67	30.98	24.91	36.73
8	-5.57	4.40	10.55	22.87	26.38	44.56	42.80	17	1.62	9.74	24.99	38.30	44.46	59.07	71.72
9	0.63	10.44	20.88	37.33	37.64	59.47	72.44	18	-0.60	4.19	12.87	23.64	32.32	34.72	45.79
Blank		13.17	20.05	34.72	52.37	55.67	68.24								

According to the data, group 13 has the best inhibitory effect. Next, we used a single variable method to analyze the data; that is, two random variables in Table 4 were kept unchanged, and the third variable was changed. By analyzing the first variable (voltage), we found that all groups produce good inhibitory effects with 3kV voltage. Thus, it may be concluded that when the voltage is 3kV, the main factor of inhibition is voltage. By analyzing the second variable (pulse width time per second), we found that the pulse width time has a negative influence on the algae growth, and it has nothing to do with polarity when the voltage is 3 kV. The inhibition effect is 120>60180. As for the third variable (polarity of pulse), monopole has a greater effect than bipolar when the parameters are 3kV-120.

In summary, the best inhibitory effect takes place when the pulser parameters are monopole-3 kV-120In this condition, the growth rate of the algal cells is 34.41%, which is a 33.83% decrease compared to 68.24% in the blank group. When the electric field intensity is 750 V/cm, the high-voltage pulser has a common inhibitory effect. The influence of pulse width per second follows the following rule: 120, and the monopole effect is superior to that of bipolar. However, when the electric field intensity is too low, there is no obvious regularity to determine how all parameters affect the growth of the algal cells.

5 Conclusions

- This paper applied a graphite electrode to replace the copper electrode in an algae removal with high-voltage pulser experiment, which can avoid the electrolytic effect of the copper electrode.
- This paper found that the *Chlorella pyrenoidosa* load manifests resistor characteristics. Its equivalent resistance is closely related to the plate area, temperature, and concentration of the algae solution.
- The data proved that the pulse voltage or electric field intensity was the major impact parameter of algae inhibition. The optimal parameters that may lead to the best inhibitory effect were monopole-3 kV-120.

Acknowledgments
This work is supported by the National Natural Science Foundation of China (51077139) and the National "111" Project (B08036).

References

[1] Xiong L and Ma L 2015 *IEEE Transactions on Dielectrics and Electrical Insulation* **22** 1887
[2] Wei X L, Li J H and Yan E 2010 *Electric Machines and Control* **5** 84-90
[3] Fang X D and Guan Z C 2002 *High Voltage Engineering* **26** 29-32
[4] Zabihi S, Zare F, Ledwich G and Ghosh A 2011 *IEEE Transactions on Dielectrics and Electrical Insulation* **18** 1181-88
[5] Li C X and Yao C G 2006 *High Voltage Engineering* **1** 54-56
[6] Li H T, Hong-Je R and Jong-Soo K 2009 *IEEE Transactions on Plasma Science* **37**190-4

Estimating groundwater velocity using apparent resistivity tomography: A sandbox experiment

J L Chen[1,2], C H Chen[1], C L Kuo[1], C S Fen[1] and C C Wu[1]

[1]Dept. of Environmental Engineering and Science, Feng Chia University, 100 Wenhwa Rd., Taichung City, Taichung 40724, Taiwan

E-mail: chenjl@fcu.edu.tw

Abstract. The electrical resistivity tomography (ERT) technique can estimate groundwater velocity to within 5% of the pre-set groundwater velocity. The apparent conductivity obtained by the ERT technique is linearly related to the groundwater conductivity, as described by Archie's law. Gaussian-like profiles of the tracer concentration were demonstrated with the ERT technique, and the estimated dispersion coefficient was between 0.0015 and 0.0051 cm^2/sec. In terms of monitoring changes in groundwater conductivity, the ERT technique has two major advantages over monitoring wells: (1) it measures a larger area and provides more representative results; and, (2) it does not withdraw groundwater samples, and therefore does not affect the groundwater flow. The objective of this research is to measure groundwater velocity with the ERT technique using only one well. The experiments in this research were divided into two parts. The first part evaluated the accuracy and repeatability of the ERT technique using a dipole-dipole array, and the second part estimated the groundwater velocity in a sandbox using the ERT technique. The length, width, and height of the sandbox, which was made of acrylic, were 1.5, 0.5, and 1.0 m, respectively. The ERT sandbox was sequentially filled with 5-cm layers of the silica sand to a total height of 70 cm. A total of 32 electrodes spaced every 5-cm were installed in the center of the sandbox. Three monitoring wells were installed along the line of the electrodes. Both no-flow and constant flow (NaCl solution with electrical conductivity and concentration of 5,000 μs/cm and 2.456 g/L, respectively) tracer experiments were conducted.

1. Introduction

The technique of electrical resistivity tomography (ERT) for estimating the electrical conductivity of soil matrix has two major advantages over traditional methods such as monitoring wells: it is cost-effective and able to measure a large area in a short period of time[1], including mapping of leachate contamination[2, 3], soil remediation[4], and evaluating the moisture content of the soil[5].

The theory of the ERT technique is to apply a current through a soil matrix and measure the induced electrical potential, which depends on the resistivity of the soil matrix and the geometry of the electrodes. A schematic diagram of the electrode arrangement is shown in Figure 1, which depicts the current electrodes (C1 and C2) and the electrical potential electrodes (P1 and P2). The relationship between the applied current, I, and the measured electrical potential difference, ΔV, can be written as equation (1).

$$\rho_a = K \, \Delta V / I \quad (1)$$

where ρ_a is the apparent soil resistivity and K is the geometric factor. The geometric factor is a function of the arrangement of the four electrodes (i.e., C1, C2, P1, and P2). Assuming the soil apparent resistivity to be homogenous, the geometric factor of the electrode arrangement shown in Figure 1 can be written as equation (2).

$$K=2\pi\left[\frac{1}{\overline{C1P1}}-\frac{1}{\overline{C1P2}}-\frac{1}{\overline{C2P1}}+\frac{1}{\overline{C2P2}}\right]^{-1} \tag{2}$$

The electrode arrangement shown in Figure 1 measures one value of apparent resistivity. The ERT technique installs a series of linearly-aligned electrodes into the surface soil and applies a constant current through two of the electrodes (C electrodes) while measuring the electrical potential between two of the electrodes (P electrodes). The above process is repeated by using different arrangements of C and P electrodes, typically resulting in more than one hundred measurements of ρ_a in one sequence of measurement. The sequence of measurement is generally called the electrode array. Three of the most used electrode arrays are the Wenner array, the dipole-dipole array, and the pole-pole array. The dipole-dipole array, which is used in this research, places the C electrodes at one side and the P electrodes at the other side such that the distance between the C and P electrodes is larger than 10 times the distance between the C electrodes.

Referring to Figure 2, the sequence of measurements starts with C1 and P1 next to each other and continues as P1 moves away from C1. The distance between the C1 and P1 electrodes determines the depth (i.e., location) of the measured ρ_a. As the P1 electrode moves away from the C1 electrode, the depth increases, resulting in a 2 dimensional (2D) depiction of ρ_a.

The 2D measurement of soil resistivity enables the ERT technique to detect any changes in ρ_a at depths below the groundwater table without the use of monitoring wells or drilling. Furthermore, because the soil resistivity is a function of the groundwater resistivity, detection of changes in ρ_a can be used to estimate the changes in groundwater resistivity. When the soil is composed of sand and filled with brine water, the Archie's law equation can be used to describe the relationship between the soil resistivity, ρ_0, and brine resistivity, ρ_w, as shown in equation (3)[6].

$$\rho_0=F\rho_w \tag{3}$$

where F is the formation factor. As the conductivity is the reciprocal of the resistivity, equation (3) can be used to estimate the resistivity of groundwater when the soil resistivity is available. The soil resistivity obtained by the ERT technique can be applied to estimate the groundwater resistivity using equation (3). One application of this technique is to measure the groundwater velocity by conducting a tracer study using brine water with high electrical conductivity as the tracer. The major advantage of this method is that it requires only one injection well, whereas conventional tracer studies require at least one injection and several monitoring wells. In addition, the ERT results represent a spatial distribution of the tracer concentration, whereas a monitoring well provides only a point measurement. Therefore, the groundwater velocity can be estimated using the spatial distribution of the tracer concentration measured at different times. Thus, the objective of this research is to evaluate the ERT technique for measuring groundwater velocity. The other objective is to evaluate the sensitivity of the ERT technique in detecting changes in groundwater conductivity.

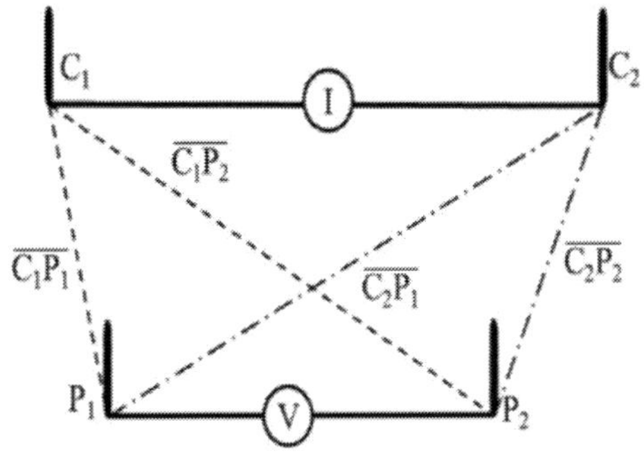

Figure 1. Schematic diagram of electrode arrangement for measuring soil resistivity. *C1* and *C2* are the current electrodes; *P1* and *P2* are the potential electrodes.

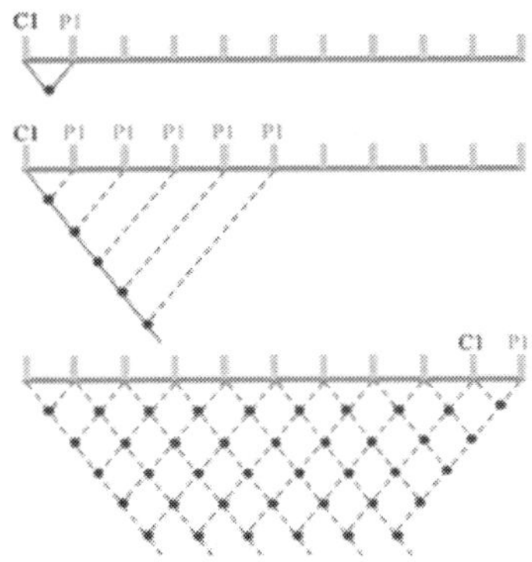

Figure 2. Schematic diagram showing the relationship between electrode spacing and soil depth during an ERT test. *C1* denotes the electrode where current is applied, whereas *P1* denotes the electrode where potential is measured.

2. Material and methods

The ERT experiments were conducted in a sandbox filled with silica sand saturated with sodium chloride (NaCl) solution. The length, width, and height of the sandbox, which was made of acrylic, were 1.5, 0.5, and 1.0 m, respectively. The diameter of the silica sand ranged between 0.6 and 1.18 mm. A smaller sandbox was constructed to evaluate the relationship between the apparent electrical conductivity and solution electrical conductivity using Ohm's law. The length, width, and height of the smaller sandbox were 25, 15, and 10 cm, respectively.

The ERT sandbox was sequentially filled with 5-cm layers of silica sand until a total height of 70 cm was reached. After laying down each layer of sand, deionized water (DI) water was added to overflow the sand layer by 1 to 2 cm. This packing procedure was intended to minimize the amount of air bubbles trapped within the sand and maximize the uniform distribution of sand particles of different sizes[7]. The electrodes for the ERT experiments were stainless steel rods with a length of 60 cm and a diameter of 0.6 cm. To ensure that the electrodes behave as point sources/sinks during the ERT experiments, each electrode was insulted along the length with a plastic sleeve so that only a pointed tip of 0.5 cm in length contacted the sand. The layout of the electrodes was centered in the sandbox with 5 cm of distance between adjacent electrodes (Figure 3). As shown in Figure 3, three wells were installed along the line of the electrodes, with well 1 serving as the tracer injection well and wells 2 and 3 serving as monitoring wells. All the electrodes were connected to ERT equipment (McOHM profiler 4XP 32) manufactured by OYO. With a maximum output of 120 mA and 400 V, the McOHM profiler could be connected to 32 electrodes and simultaneously collect 4 sets of current and voltage data. During an ERT experiment, the main output is the apparent resistivity of the soil matrix at different locations.

The sandbox experiments were divided into two parts: accuracy and repeatability characterization, and a tracer study for estimating groundwater velocity. The accuracy and repeatability test was conducted under no-flow conditions. Before each no-flow ERT experiment, NaCl solution with electrical conductivity between 600 and 1,200 S/cm was injected as a plane front from the inlet of the sandbox to purge the DI water until the electrical conductivity of the outflow was the same as the inflow, after which both the inlet and outlet of the sandbox were shut off. The accuracy test was performed by comparing the ERT results to the measured apparent resistivity, which was obtained by first placing the sand and salt water in a small rectangular box and then applying a constant current across the soil matrix and measuring the voltage difference between two adjacent points, and finally using Ohm's law to calculate the apparent resistivity.

IOP Conf. Series: Earth and Environmental Science **39** (2016) 012056 doi:10.1088/1755-1315/39/1/012056

The tracer study was conducted under controlled flow conditions. In a controlled flow experiment, the hydraulic heads on both ends of the sandbox were maintained at a constant level. The electrical conductivity of the inflow solution was first set at 600 S /cm. Once the flow velocity was steady, the solution of 5,000 S /cm was injected at a flow rate of 30 mL/min as a point source 5 cm upstream from the electrode matrix. ERT measurements were then conducted at different time intervals. Three flow velocities were used and compared to the ERT results: 0.25, 0.6, and 0.8 cm/min. During the controlled flow experiments, the resistivity of the solution in both monitoring wells was measured and compared to the ERT results.

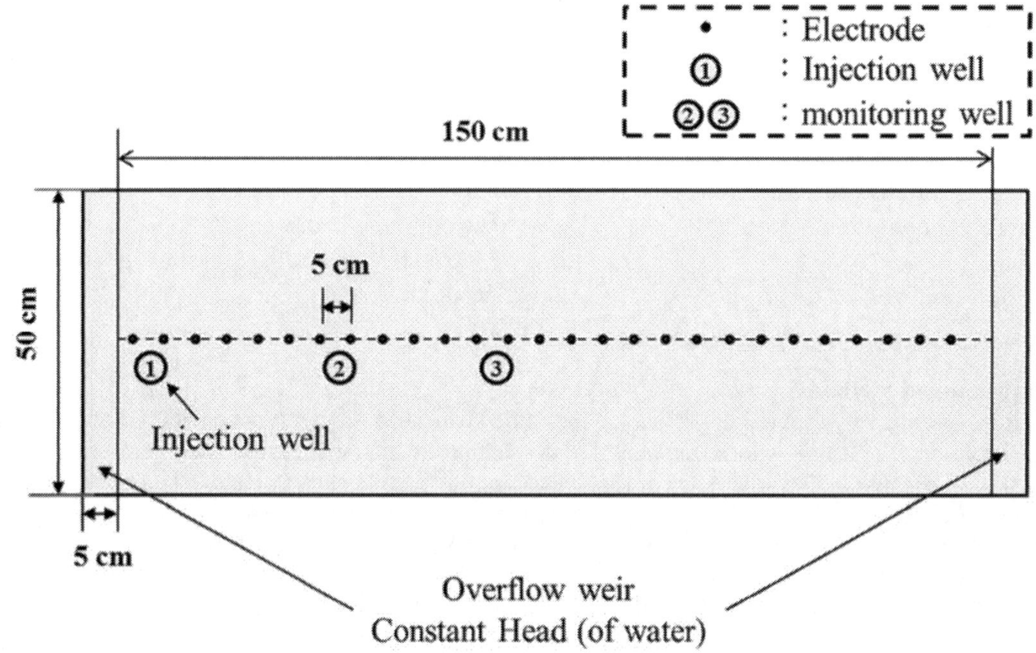

Figure 3. Schematic description of the sandbox ERT experiment.

3. Results and discussion

3.1. Accuracy and repeatability of the ERT technique

The baseline resistivity with which the ERT results were compared is the resistivity obtained from the rectangular box tests. As the conductivity is commonly used to indicate the property of the soil matrix and the soil solution, the resistivity data is converted to conductivity in this section. The relationship between the solution conductivity and the soil conductivity is linear, indicating that Archie's law (equation 3) is valid for this experiment (Figure 4). The formation factor estimated from Figure 4 is 3.1, which is comparable to values found in literature[6].

The results obtained from the ERT experiments also indicate a linear relationship between the apparent conductivity and solution conductivity (Figure 5), with slopes similar to that of Figure 4. This suggests that the ERT technique can detect changes to solution conductivity. The results from the ERT technique demonstrate the measured apparent conductivity is a spatial function. This is unexpected, as the whole sandbox was carefully prepared and can be assumed to be homogeneous and have uniform conductivity. Therefore, the ERT measurements suggest that the soil matrix in the sandbox is heterogeneous. The degree of non-uniformity can be characterized by the formation factor, which can be calculated with equation (3). The formation factor in the first layer ranges between 3.3 and 5.7 with an average of 4.7, whereas it ranges between 2.4 and 4.2 with an average of 3.3 in the second layer.

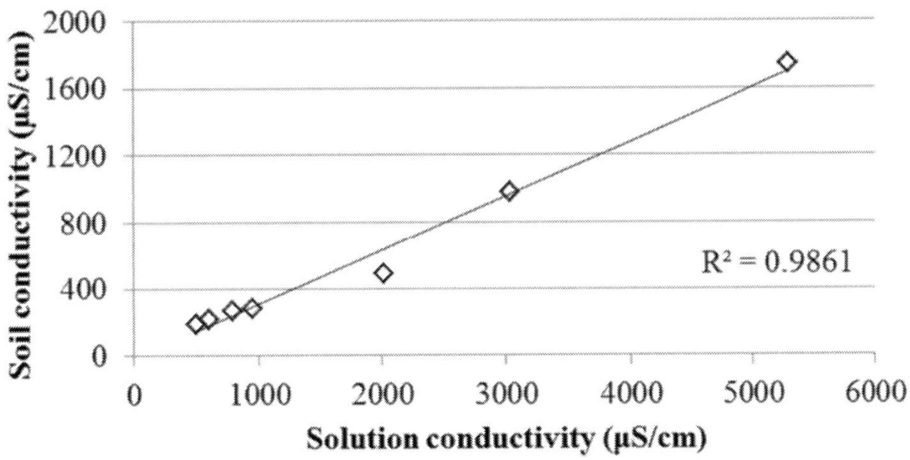

Figure 4. Soil conductivity as a function of solution conductivity.

In addition, the solution conductivity in the monitoring wells was measured and used in equation (3) to evaluate the accuracy of the ERT technique. The results show that the solution conductivities in the three monitoring wells were almost the same, indicating that the solution conductivity in the sandbox was uniformly distributed (Figure 6). It also shows a linear relationship between the solution conductivity and the apparent conductivity. The variation shown in Figure 6 is smaller than that in Figure 5, with the formation factor ranging between 1.4 and 2.5 and an average of 1.2. From the above result, it can be concluded that the formation factor is non-uniform in the sandbox and the change in solution conductivity can be detected by the ERT technique.

The repeatability of the ERT technique was evaluated by measuring the apparent resistivity at 3 different times when the sandbox was filled with a solution of uniform conductivity: initial, 16 h, and 22 h. The results show little changes of apparent resistivity at different measuring times (Figure 7). The coefficients of variation for the data shown in Figure 7 a-c are 4.72%, 3.99%, and 3.75%, respectively. With the coefficients of variation being less than 5% at different measuring times, the repeatability of the ERT technique is acceptable. The ERT experiments were conducted in a sandbox filled with silica sand saturated with saline water.

3.2. Estimation of groundwater velocity

The ERT technique estimates the groundwater velocity by monitoring the changes in apparent resistivity during the tracer study. As shown in the previous section, the apparent resistivity is linearly related to solution conductivity. Therefore, the larger the change in the apparent resistivity is, the larger the solution conductivity is. A typical profile of change in the apparent resistivity is shown in Figure 8. The change in apparent resistivity is calculated by subtracting the background apparent resistivity from the current measurements. It should be noted that each point has a unique background apparent resistivity, which needs to be measured before tracer injection. This characterization of background apparent resistivity is critical for detecting changes during tracer studies. After tracer injection, the plume of the tracer advanced down the gradient as reversed Gaussian-like curves with well-defined maximum changes in apparent resistivity. As the conductivity of the tracer is larger than that of the background solution, the apparent resistivity is expected to decrease as the tracer plume advances. As the plume advances, the spread of the curve increases as a result of dispersion. The shape of the curve suggests that the transport of the tracer follows the advection-dispersion mechanism (Figure 8). The curves shown in Figure 8a were used to estimate the dispersion coefficients in the direction of the groundwater flow. The results range between 0.0015 and 0.0051 cm^2/sec with an average of 0.0018 and standard deviation of 0.0015 cm^2/sec.

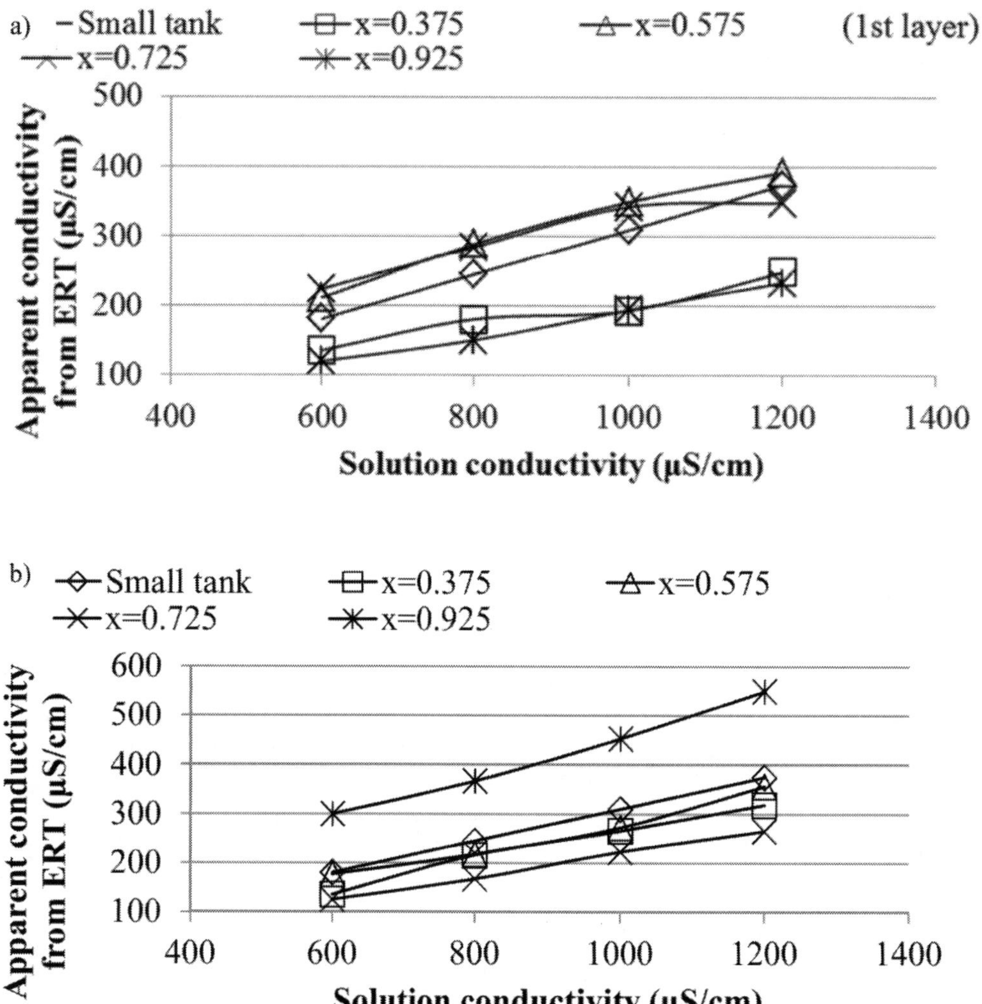

Figure 5. Apparent conductivity from ERT as a function of solution conductivity. (a) 1st layer, depth = 5 cm; (b) 9th layer, depth = 14 cm. x values indicate distance from the left end along the electrode array. Small tank indicates apparent conductivity obtained from rectangular box tests

The point where the maximum change in apparent resistivity occurred can be considered as the center of the mass of the plume and can be used to estimate the velocity of the plume. As the tracer is inert, the velocity of the plume equals the velocity of the groundwater. To estimate the groundwater velocity, the center of the mass of each plume shown in Figure 8 is plotted as a function of the time after tracer injection (Figure 9). The slope of the curve in Figure 9 is taken as the groundwater velocity. The ERT technique used the first 3 layers of apparent resistivity profiles to estimate the groundwater velocity with %Difference less than 5% (Table 1). This indicates that the ERT technique is able to estimate the velocity of the groundwater by monitoring the advance of the tracer plumes and confirms the feasibility of the main concept of this research.

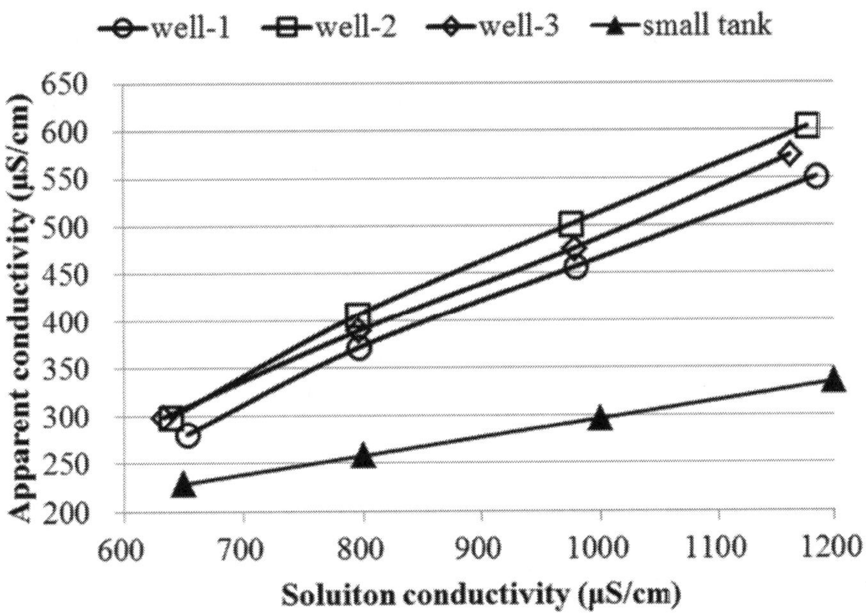

Figure 6. Apparent conductivity at the locations of monitoring wells as a function of measured solution conductivity

Table 1. Summary results of groundwater velocity estimation from ERT results.

Controlled Velocity (cm/min)		0.25	0.60	0.81
Velocity estimated from ERT (cm/min)	1st layer	0.26	0.55	0.86
	2nd layer	0.25	0.62	0.86
	3rd layer	0.27	0.58	0.73
	Average ERT velocity (1st-3rd layers)	0.26	0.58	0.82
	% Difference	4.0	3.3	1.2

% Difference is calculated using the following equation.

$$\%\text{Difference} = \frac{|\text{Controlled velocity} - \text{Average ERT velocity}|}{\text{Controlled velocity}} \times 100$$

3.3. Groundwater conductivity measured by ERT under constant flow conditions

The groundwater resistivity in wells 1 to 3 (Figure 3) during the tracer study with a controlled velocity of 0.8 cm/min was measured in order to monitor the advance of the plume. In the injection well (well 1), the groundwater resistivity and the ERT results had the same pattern: both reached a minimum at 16 min after injection and went back to background values when the tracer plume passed (Figure 10a). The groundwater conductivity in well 2, which was located 30 cm from well 1, had the same pattern as well 1 (Figure 10b). This is unexpected, as the groundwater velocity was only 0.8 cm/min. The reason for this abnormality can be attributed to possible preferential flow paths between the two wells. The ERT results, which reached a minimum value at 65 min after tracer injection, show a more reasonable pattern than the groundwater resistivity (Figure 10b). The above comparison clearly demonstrates one of the main advantages of the ERT technique over monitoring wells: the ERT technique measures the resistivity over a larger area than that of a monitoring well, and hence provides more representative observations than a monitoring well.

The groundwater conductivity in well 3, which is located 60 cm from well 1, has a similar pattern to that of the ERT measurements (Figure 10c). The minimum groundwater resistivity was observed at 66 min after tracer injection, whereas the apparent resistivity from ERT technique reached a minimum at 90 min after tracer injection. This indicates that well 3 was in the path of the groundwater flow and was able to detect the change in groundwater resistivity. The results from the ERT measurements also suggest that changes in groundwater resistivity can be detected by the ERT technique. The reason for the lag between the peaks of groundwater conductivity and ERT measurements can be attributed to the fact that it could take time for the plume to enter the monitoring well. Therefore, to be able to detect real time groundwater resistivity change, forced withdrawal of groundwater sample is necessary.

However, forced withdrawal of groundwater could alter the flow patterns in the vicinity of the monitoring well and cost more time and effort to collect groundwater samples. Therefore, it can be concluded that the ERT technique can detect real time changes in apparent resistivity without disturbing the flow regime and provide a more representative measurement of groundwater resistivity than a monitoring well.

4. Conclusions

The ERT technique is able to detect the changes in apparent resistivity of soil matrix caused by the changes in solution resistivity. The relationship between the apparent conductivity measured with the ERT technique and solution conductivity is linear and can be described with Archie's law. The formation factors described in Archie's law are non-uniform within the sandbox. The ERT technique can be used to estimate groundwater velocity with reasonable accuracy and monitor the advance of a conductive plume. The dispersion coefficient of a sandbox can be estimated with the ERT technique. The ERT technique has two major advantages over monitoring wells in detecting changes in groundwater resistivity: results are more representative, and procedures do not disturb the existing flow regime. The ERT has the ability to estimate groundwater velocity with just one injection well.

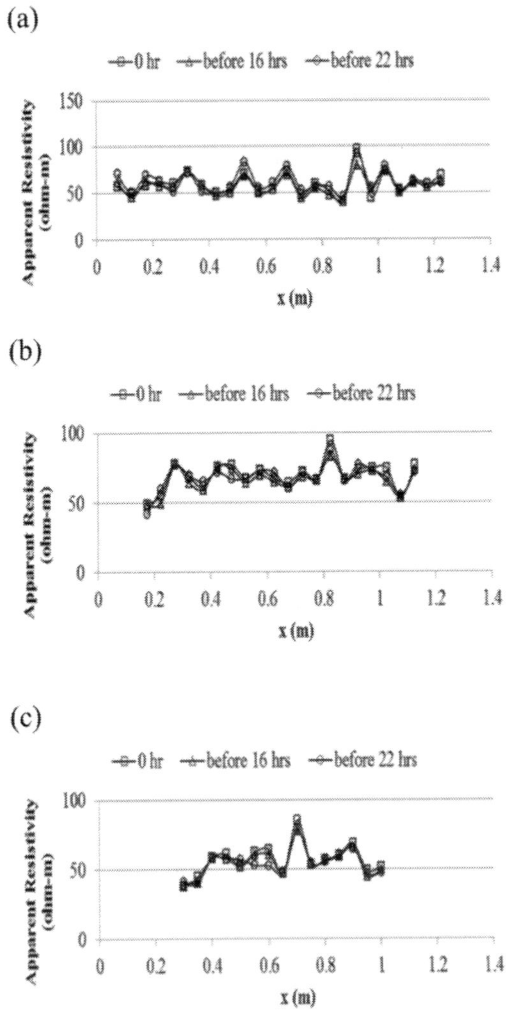

Figure 7. Apparent resistivity obtained by ERT as a function of distance from the left end of the sandbox at different layers: (a) 1st layer, (b) 5th layer, and (c) 10th layer.

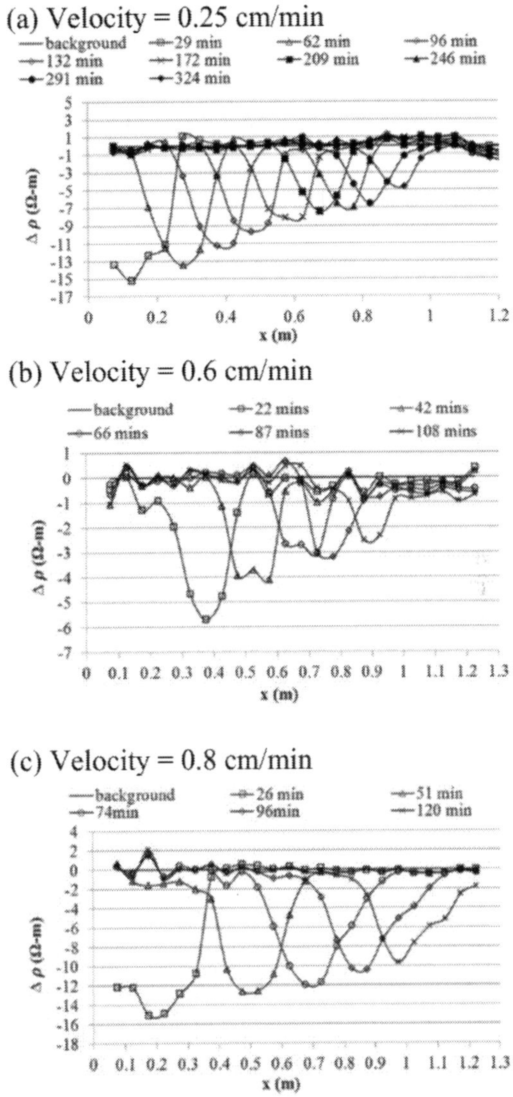

Figure 8. Change of apparent resistivity as a function of distance from the tracer injection point at 3 different controlled groundwater velocities: (a) 0.25, (b) 0.6, and (c) 0.8 cm/min.

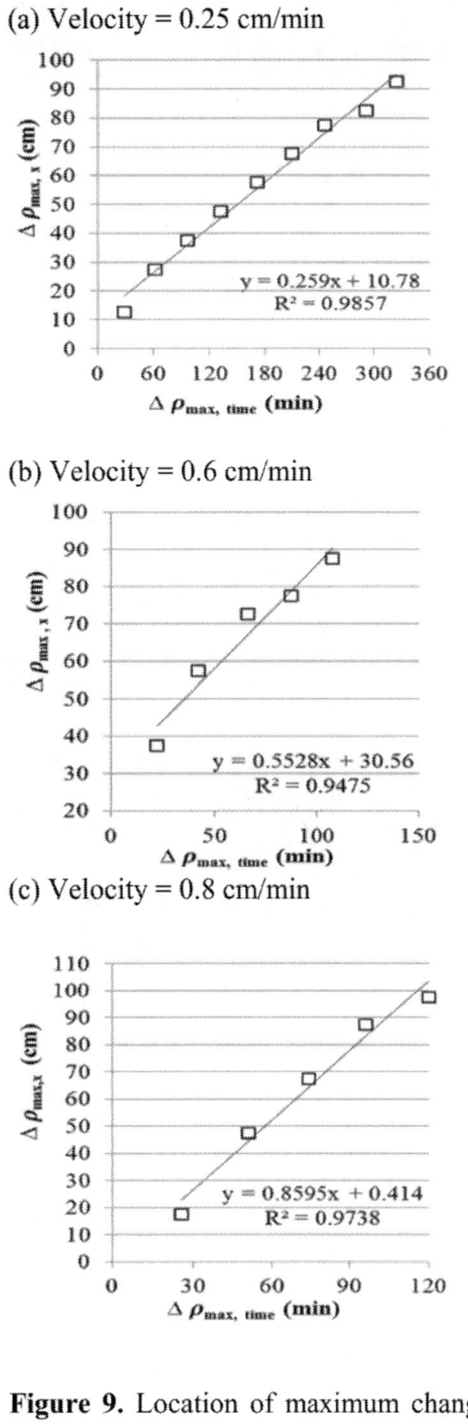

Figure 9. Location of maximum change in apparent resistivity as a function of time at 3 different controller groundwater velocities: (a) 0.25, (b) 0.6, and (c) 0.8 cm/min

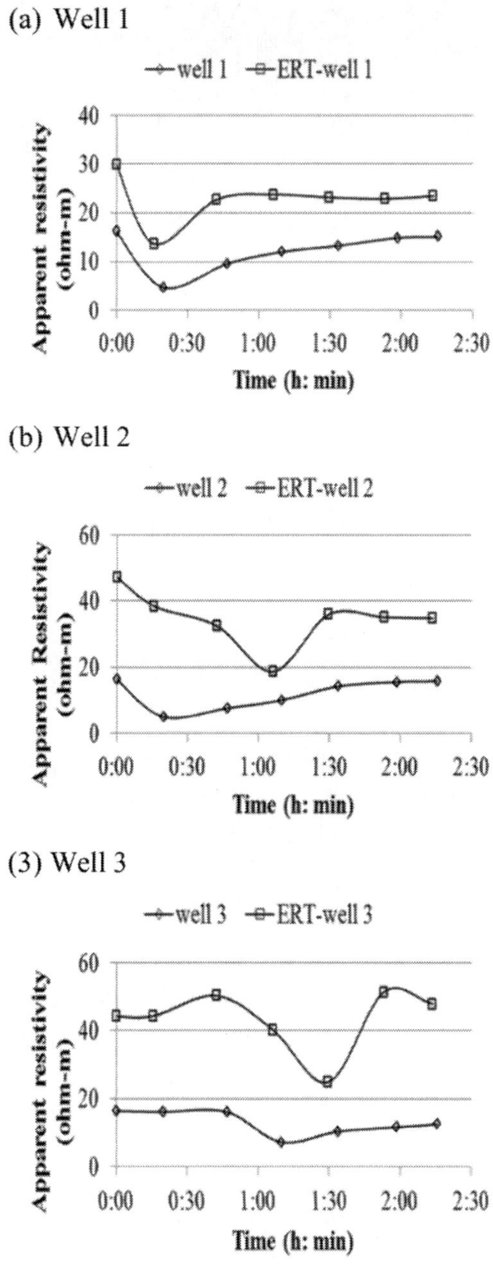

Figure 10. Apparent resistivity (open square) and groundwater resistivity (open diamond) as a function of time after injection at 3 different wells: (a) well 1, (b) well 2, and (c) well 3.

Acknowledgement

This study was funded by a research project supported by the Taiwan EPA. The views or opinions expressed in this article are those of the writers and should not be construed as opinions of the Taiwan EPA. Mention of trade names, vendor names, or commercial products does not constitute endorsement or recommendation by Taiwan EPA.

References

[1] Samouëlian A, Cousin I, Tabbagh A, Bruand A and Richard G 2005 Electrical resistivity survey in soil science: A review *Soil and Tillage Research* **83** 173-93

[2] Samsudin A R, Rahim B-d E A, Yaacob W Z W and Hamzah U 2006 Mapping of contamination plumes at municipal solid waste disposal sites using geoelectric imaging technique: Case studies in Malaysia *J. Spat Hydro.* **6** 13-22

[3] Michot D, Benderitter Y, Dorigny A, Nicoullaud B, King D and Tabbagh A 2003 Spatial and temporal monitoring of soil water content with an irrigated corn crop cover using surface electrical resistivity tomography *Water Resour. Res.* **39** 1138

[4] Slater L, Binley A, Versteeg R, Cassiani G, Birken R and Sandberg S 2002 A 3D ERT study of solute transport in a large experimental tank *J. Appl. Geophys.* **49** 211-29

[5] Merritt A J, Chambers J E, Wilkinson P B, West L J, Murphy W, Gunn D and Uhlemann S 2016 Measurement and modelling of moisture—electrical resistivity relationship of fine-grained unsaturated soils and electrical anisotropy *J. Appl. Geophys.* **124** 155-65.

[6] Archie G E 1942 The electrical resistivity log as an aid in determining some reservoir characteristics *Transactions of the American Institute of Mining, Metallurgical, and Petroleum Engineers* **146** 54-62

[7] Bowen I R, Devlin J F and Schillig P C 2012 Design and Testing of a Convenient Benchtop Sandbox for Controlled Flow Experiments *Ground Water Monit. Remediat* **32** 87-91

IOP Conf. Series: Earth and Environmental Science **39** (2016) 012057 doi:10.1088/1755-1315/39/1/012057

A critical review and database of biomass and volume allometric equation for trees and shrubs of Bangladesh

H Mahmood[1, 4], M R H Siddique[1] and M Akhter[2, 3]

[1]Forestry and Wood Technology Discipline, Khulna University, Khulna - 9208, Bangladesh
[2]Assistant Conservator of Forests, Forest Department, Ministry of Environment and Forest, Dhaka, Bangladesh
[3]Food and Agricultural Organization of the United Nations, Bangladesh

E-mail: mahmoodhossain@hotmail.com

Abstract. Estimations of biomass, volume and carbon stock are important in the decision making process for the sustainable management of a forest. These estimations can be conducted by using available allometric equations of biomass and volume. Present study aims to: i. develop a compilation with verified allometric equations of biomass, volume, and carbon for trees and shrubs of Bangladesh, ii. find out the gaps and scope for further development of allometric equations for different trees and shrubs of Bangladesh. Key stakeholders (government departments, research organizations, academic institutions, and potential individual researchers) were identified considering their involvement in use and development of allometric equations. A list of documents containing allometric equations was prepared from secondary sources. The documents were collected, examined, and sorted to avoid repetition, yielding 50 documents. These equations were tested through a quality control scheme involving operational verification, conceptual verification, applicability, and statistical credibility. A total of 517 allometric equations for 80 species of trees, shrubs, palm, and bamboo were recorded. In addition, 222 allometric equations for 39 species were validated through the quality control scheme. Among the verified equations, 20%, 12% and 62% of equations were for green-biomass, oven-dried biomass, and volume respectively and 4 tree species contributed 37% of the total verified equations. Five gaps have been pinpointed for the existing allometric equations of Bangladesh: a. little work on allometric equation of common tree and shrub species, b. most of the works were concentrated on certain species, c. very little proportion of allometric equations for biomass estimation, d. no allometric equation for below-ground biomass and carbon estimation, and d. lower proportion of valid allometric equations. It is recommended that site and species specific allometric equations should be developed and consistency in field sampling, sample processing, data recording and selection of allometric equations should be maintained to ensure accuracy in estimation of biomass, volume, and carbon stock in different forest types of Bangladesh.

1. Introduction

Bangladesh is one of the most densely populated countries in the world with only 0.06% of the world's forest [1] and 0.017 ha of per capita forest land [2]. The huge population (166.3 million) exerts immense pressure on her forest resources for the demand of timber, fuel wood, and other forest products. Measurement of forest biomass and volume are important for estimating forest productivity. Appropriate estimation of forest stocking, productivity, nutrient cycling, nutrient budget, amount of carbon stock, and prediction of future status are important considerations for the sustainable management of forest resources [3, 4]. Tree biomass and volume can be measured from both destructive (clear-cut) and non-destructive (allometric equation) methods [3, 5]. Allometric method is frequently used for estimating the biomass and volume of forest plant species, which is the most powerful tool of measurement [6, 7, 8, 9]. The use of appropriate equations for estimating biomass and volume will contribute in improving the accuracy of forest resource assessment, and also guide the forest policies and management interventions [10, 11]. Considering this importance, different research and academic institutions and individual researchers have developed biomass and volume allometric equations for estimating biomass and volume stock of a particular forest of Bangladesh.

Choice of allometric equations is one of the key sources of uncertainty in forest biomass estimation. About one-fourth of the published equations contain blunders, oversights, or forecast unrealistic values [12]. Therefore, verification and validation of allometric equation is imperative, before its application in estimation of biomass and volume [13]. Validating and verifying an allometric equation is vast in extent, and diverse in regards to methods, assumptions and conclusions. Recently, [13] proposed a systematic transparent quality control method to quantify the degree of confidence for the allometric equations without involving equation development. Quantitative reviews of available allometric equations have been implemented at a regional scale, but few countries have developed a national database for biomass and volume allometric equations [12]. In South Asia, the first regional database was developed in 2014 [14], while [15] have developed the first database for Bangladesh in 2013. However, these databases were not comprehensive and quality controlled. Repetition of some allometric equations was also observed. The use and development of allometric equations should be specific to the conditions of the country, which is crucial for three reasons. The first is that the assessment of tree and forest resources of a country should be accurate. The second is that country specific models are more efficient, and the third reason is the amount of uncertainty of the models used. However, the overall objective is to improve the quality of estimates for a multitude of purposes including timber volume, wood energy biomass, carbon stocks, etc. in trees and forest resources of Bangladesh. To achieve this overall objective, we have to have the status of the existing tree allometric equations, forest resource estimation methods, and a verified database of tree allometric equations. Therefore, a quality controlled comprehensive database of allometric equations is needed to assess the gaps and scope for the development of new allometric equations with important tree species of different forest types of Bangladesh. The specific objectives of the present study were to: i. develop a compilation with valid allometric equations of biomass, volume and carbon for trees and shrubs of Bangladesh, ii. find out the gaps and scope for further development of allometric equations with different trees and shrubs.

2. Material and Methods

2.1. Collection of Information

Key stakeholders (government departments, research organizations, academic institutions, and potential individual researchers) were identified considering their involvement and experiences in the use and development of allometric equations. A draft list of documents containing allometric equations was prepared through consultation with key stakeholders and online search (Google Scholar) results. This list was updated by taking time and identifying unlisted documents. The listed documents were collected from bibliographic databases such as: Science Direct, Springer Link, CABI, AGRIS, AGRICOLA, JSTOR, ResearchGate. Sometimes personal and official communications were established with identified key stakeholders to obtain their research articles, reports, theses, bulletins, monographs, inventory reports and proceedings papers. Hard and soft copies of the collected documents were maintained for references.

2.2. Compilation of Information

The collected documents were sorted considering relevance and repetition. The information of the allometric equations in the sorted documents was grouped and recorded into 8 different categories. The categories were plant ecology (Population), geographical location, ecoregions (FAO, Udvardy, Bailey, WWF, Holridge and IUCN Bioecological zones-Bangladesh), equation parameters (coefficients, constants, variables and ranges), plant components (leaves, Branches, Stump, Stems, Bark, Root, etc.), taxonomical description (Family, Genus, Species), fit statistical information (R2, adjusted R2, RMSE, sample number, bias correction, Akaike information criterion, furnival index) and Bibliography.

2.3. Quality control of the allometric equations

All the allometric equations were tested through a quality control scheme following four types of verification (operational verification, conceptual verification, applicability, statistical credibility) according to [13]. The detail process of verification has been given below:

- Operational verification: Too large or too small predicted biomass or volume values
- Conceptual verification: Predicted biomass or volume are lower than "0" or have negative values
- Applicability: Under which condition the model can be applied (Population ecology, environmental condition of the site where the equation was developed, tree component measured, Taxonomic reference, Range of applicability)
- Statistical credibility: Sample size should be at least 30 trees and the coefficient of determination should be higher than 0.85

3. Results

3.1. Documents of allometric equations

A total of 53 documents were identified that contained the allometric equations for plants of Bangladesh. The collected documents were sorted, considering relevance and repetition, and 50 documents were found (i.e. 96% of the total document of allometric equations). Most of the documents were Journals (52%) followed by reports (24%), bulletins and theses (figure 1). The list of documents containing allometric equations of Bangladesh has been presented as Appendix 1.

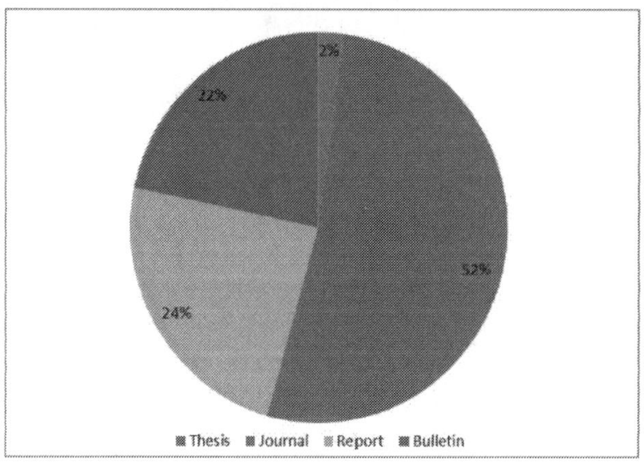

Figure 1. Document type of available literature in percentage.

Most of the documents were prepared during the last decade of the twentieth century (1991-2000). However, the number of studies on the development of allometic equations has increased rapidly during 2011 to 2016, which contributed 26% of the total allometric documents of Bangladesh (figure 2).

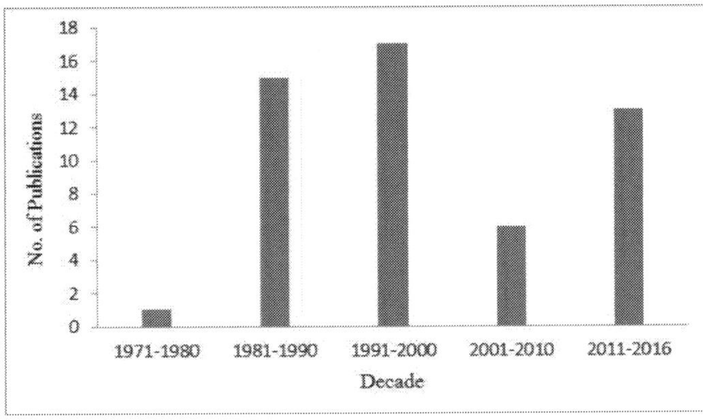

Figure 2. Year-wise publication of the documents.

3.1. Total number of allometric equations

This study recorded a total of 517 allometric equations on volume, biomass, carbon and nutrients for 80 species of tree, shrub, palm and bamboo. Higher preference, about 92% of the total allometric equations, was recorded for trees and 6% for shrubs. Among these equations, 70% equations were for volume estimation and only 6% equations were for oven-dry biomass estimation. Unfortunately, there is not a single equation for below-ground biomass estimation (table 1).

Table 1. Total number of allometric equations according to plant and equation types in Bangladesh.

Category	Volume	Green biomass	Oven-dried biomass	Air-dried biomass	Carbon	Nutrients	Length of split leaf	Total
Tree	360	78	11	0	25	3	0	477
Shrub	1	1	20	0	3	6	0	31
Palm	0	2	0	0	0	0	1	3
Bamboo	0	3	0	3	0	0	0	6
Total	361	84	31	3	28	9	1	517

3.2. Verified allometric equations

Considering operational and conceptual verification, applicability and statistical credibility, the total number of allometric equations was reduced to 222, which was 43% of the total allometric equations. Most of the equations (45%) failed to meet the requirements of statistical credibility and conceptual verification (24%) (table 2). About 97% of the valid allometric equations were for individual species, while only 3% equations for mixed species. Irrespectively, about 77% of allometric equations were for plantation followed by natural forest (15%) and home garden (7%).

Table 2. Number of allometric equations in each category of verification.

Category	Operational verification	Conceptual verification	Applicability	Statistical credibility	Final validation
Valid	473	394	517	285	222
Not valid	44	123	0	232	295
Total equation	517	517	517	517	517

Trees contained 196 verified equations which were 41% of the total allometric equations under the tree category. Shrubs contained 26 verified equations which were 84% of the total allometric equations of the shrub category (table 1 and 3). Surprisingly, not a single verified allometric equation was found for palm and bamboo. Under valid allometric equations, about 62% and 12% of the equations were observed for volume and oven-dry biomass respectively (figure 3).

Table 3. Number of verified allometric equations according to plant and equation types in Bangladesh.

Category	Volume	Green biomass	Oven-dried biomass	Carbon	Nutrients	Total
Tree	138	44	10	1	3	196
Shrub	0	0	17	3	6	26
Palm	0	0	0	0	0	0
Bamboo	0	0	0	0	0	0
Grand total	138	44	27	4	9	222

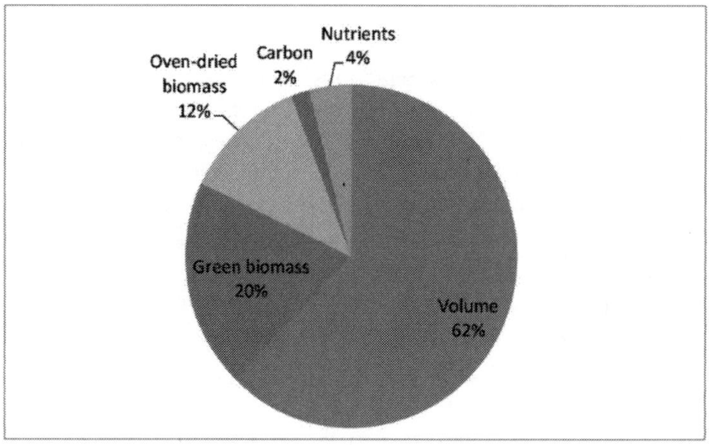

Figure 3. Percentage of valid allometric equations in different categories.

Thirty-nine species of 18 families and 31 genera have a total of 222 verified allometric equations. But, *Sonneratia apetala, Acacia mangium* and *Acacia auriculiformis* have each of 12 volume equations. *Senna siamea* has 23 allometric equations for green biomass, but other species have few equations under each category. These four species contributed 37% of the total verified allometric equations (Table 4). Species wide valid allometric equations in Bangladesh have been prepared and presented in Appendix 2.

Table 4. List of species with verified allometric equations in Bangladesh.

Genus	Species	Local name	Volume	Green biomass	Oven-dried biomass	Carbon	Nutrients	Remark
Acacia	*mangium*	Mangium	12	12				
Acacia	*auriculiformis*	Akashmoni	12	5	1			
Acacia	*nilotica*	Babla	2					
Aegialitis	*rotundifolia*	Nuniya			2			
Aegiceras	*corniculatum*	Khulshi			4	1	3	
Albizia	*procera*	Koroi	2					
Albizia	*spp*	Koroi	6					Mixed species
Albizia	*saman*	Rain tree	6					
Albizia	*richardiana*	Rajkoroi	2					
Aphanamixis	*polystachya*	Pitraj	2					
Artocarpus	*chaplasha*	Chapalish	2		2			
Artocarpus	*heterophyllus*	Kathal	2					
Avicennia	*officinalis*	Baen	2					
Azadirachta	*indica*	Neem	2					
Breonia	*chinensis*	Kadam	1					
Ceriops	*decandra*	Goran			5			
Dalbergia	*sissoo*	Sissoo	5					
Dipterocarpus	*turbinatus*	Telya garjan	3					
Eucalyptus	*camaldulensis*	Eucalyptus	7					
Eucalyptus	*tereticornis*	Eucalyptus	1					
Eucalyptus	*brassiana*	Eucalyptus	1					
Excoecaria	*agallocha*	Gewa			5	1	3	
Falcataria	*moluccana*	Moluccna	2					
Gmelina	*arborea*	Gamar	2					
Hevea	*brasiliensis*	Rubber	10					
Kandelia	*candel*	Goria			5	1	3	
Lagerstroemia	*speciosa*	Jarul	1		1			
Lannea	*coromandelica*	Badi	1					
Mangifera	*indica*	Am	2					
Melia	*azadarach*	Bokain	2					
Mixed			1	4	1	1		
Pinus	*caribaea*	Pine	8					
Senna	*siamea*	Minjiri	8	23				
Shorea	*robusta*	Sal	7					
Sonneratia	*apetala*	Keora	12					
Swietenia	*macrophylla*	Mahogany	6					
Syzygium	*cumini*	Kalojam	1					
Tectona	*grandis*	Teak	2		1			
Terminalia	*arjuna*	Arjun	2					

Xylia	*xylocarpa*	Pyinkado	1

A total of 58 verified allometric equations were observed for all over Bangladesh for different species. These allometric equations were developed from sample trees that were collected from different locations of Bangladesh. Therefore, these equations overlap different ecoregions of Bangladesh. Conversely, the other 164 verified allometric equations were found under all categories of ecoregions in Bangladesh as described by FAO, Udvardy, WWF and Bailey (table 5).

Table 5. Numbers of allometric equations and species in different ecoregions of Bangladesh.

Ecoregion	Zones	Equation number	Species number
FAO	Tropical moist Deciduous Forest	38	12
	Tropical rain forest	26	12
Udvardy	Tropical humid forests	76	14
	Tropical dry forests / Woodlands	12	9
WWF	Tropical humid forest	12	6
	Tropical and subtropical moist broadleaf forests	10	1
	Tropical and subtropical dry broadleaf forests	77	20
	Mangrove	33	5
Bailey	Rainforest Division	101	21
	Rainforest Regime Mountain	12	9
Holdridge	Subtropical moist	13	4
	Subtropical wet	13	10
	Tropical moist	10	2
Bangladesh IUCN	Brahmaputra-Jamuna flood plain	4	2
	Chittagong Hills and the CHTs	6	2
	Ganges flood plain	2	1
	Offshore island	14	2
	Sundarbans	33	5
	Surma-Kushiara flood plain	1	1
	Sylhet hills	11	8

* Overlapped ecoregions have not considered in this table.

4. Discussion

Allometric equations for trees and shrubs are fundamental to assess standing volume, biomass or carbon stock, bioenergy, nutrient cycling, payment for environmental services etc. [7, 10, 11]. Inappropriate use and development of allometric equations may give an inaccurate estimate of forest resources that may lead to inappropriate decisions on forest management issues and initiatives [16]. In Bangladesh, 5% species of trees and shrubs have allometric equations for estimating biomass and volume. However, this percentage was reduced to 2.5% considering only the verified equations. Four tree species contributed about 37% of the total valid allometric equations, and 12% verified equations were for oven-dried biomass of 10 species. This scenario pinpointed five gaps in the existing allometric equations of Bangladesh. This situation indicates: a. little work on allometric equations of common tree and shrub species, b. most of the works were concentrated on certain species, c. very little proportion of allometric equations for biomass estimation, d. no allometric equation for below-ground biomass and carbon estimation, and d. lower proportion of valid allometric equations.

Bangladesh Forest Research Institute (BFRI) was usually responsible for the development of allometric equations for estimating volume and biomass since 1971. Forest Department also developed some volume allometric equations in cooperation with BFRI under specific forest inventory from 1971 to 2000. The contribution of individual researchers in developing allometric equations was quite low during that period. Recently (2001-2016), the contribution from the individual researchers has increased. Previously, almost all efforts were given to derived volume equations for different tree species in the natural forest and plantation, to estimate the commercial volume stock of timber in particular forest and species as well. This could be the reason a very small number of biomass allometric equations were found in Bangladesh during 1971 to 2000. Few fuel wood species (*Acacia mangium, Acacia auriculiformis, Senna siamea* and *Sonneratia apetala*) have gotten more emphasis to derive allometric equations.

Development of biomass and volume allometric equations requires extensive planning, field works, sample analysis in the laboratory, and data analysis. These activities are mostly destructive, difficult, and expensive to repeat. Therefore, these activities require consistency in field work and equation selection process. Most of the allometric equations, 45% and 24% of the total allometric equations, failed to meet the requirements of statistical credibility and conceptual verification respectively under quality control scheme. This large proportion of invalid equations (57%) may be from lack of awareness on the quality control scheme of the derived allometric equations. So, it is suggested to include interval of calibration, residual standard deviation, coefficient of regression value, number of sample trees and location of sample tree or data collection in the document to meet the requirements of quality control scheme.

During the last National Forest Assessment in Bangladesh (2005-07), globally available equations and factors were used to calculate the above-ground biomass of forest. The Sundarban Carbon inventory in 2009-10 used the globally available equations for the mangrove species to calculate the carbon stock in the Sundarban Reserved Forest [17]. However, he accuracy can be questionable as they were using some general allometric equations [16, 18]. Therefore, development of allometric equations for local species, considering various factors for different forest types, is essential to ensure accuracy in volume, biomass and carbon estimation [12]. Major species of bamboo natural/ homestead, coastal afforestation, fresh water swamp forest, inland chars, mango plantation, associate species of Sal forest, major species of the Sundarbans and tree species outside the forest (table 6) should be given more emphasis during the development of biomass and volume allometric equations in Bangladesh.

Table 6. List of recommended species for further development of allometric equations in Bangladesh.

Acacia catechu	*Bombax ceiba*	*Dendrocalamus longispathus*
Adina cordifolia	*Borassus flabellifer*	*Dillenia indica*
Albizia lebbeck	*Bruguiera gymnorrhiza*	*Dillinia pentagyna*
Albizia odoratissimus	*Bruguiera sexangula*	*Diospyros peregrina*
Anacardium occidentale	*Butea monosperma*	*Disopyros philippensis*
Areca catechu	*Calophyllum inophyllum*	*Duabanga grandiflora*
Avicennia alba	*Cassia fistula*	*Dysoxylum binectariferum*
Avicennia marina	*Cerbera manghas*	*Erythrina orientalis*
Avicennia officinalis	*Chickrassia tabularis*	*Excoecaria indica*
Bambusa arundinacea	*Clerodendrum inerne*	*Feronia limonia*
Bambusa balcooa	*Cocos nucifera*	*Ficus bengalensis*
Bambusa longispiculata	*Cynometra ramiflora*	*Ficus hispida*
Bambusa polymorpha	*Dalbergia sisoo*	*Ficus religiosa*
Bambusa tulda	*Dalbergia spinosa*	*Heritiera fomes*
Bambusa vulgaris	*Dellinia pentagyna*	*Hibiscus tiliaceus*
Barringtonia acutangula	*Delonix regia*	*Khaya anthotheca*
Leucaena leucocephala	*Pongamia pinnata*	*Pithecellobium dulce*
Litchi chinensis	*Psidium guajava*	*Tamarix indica*
Lumnitzera racemosa	*Rhizophora apiculata*	*Terminalia belerica*
Melocanna baccifera	*Rhizophora mucronata*	*Terminalia catappa*

Michelia champca	*Schima wallichii*	*Toona ciliata*
Mimosops elengi	*Sonneratia apetala*	*Trema orientalis*
Moringa oleifera	*Sonneratia caseolaris*	*Xylocarpus granatum*
Nypa fruticans	*Spondias dulce*	*Xylocarpus mekongensis*
Phoenix paludosa	*Syzygium grandis*	*Zizyphus mauritiana*
Phoenix sylvestris	*Tamarindus indica*	

Acknowledgements
We greatly acknowledge the financial support of FAO through GCP/BGD/058/USA (LOA Code: FAOBGDLOA 2015-026) to accomplish this work. The authors also thank FAO for providing financial support (Travel and Accommodation) to present the paper at the International conference WRE2016, Shanhai, China. We would like to thank Bangladesh Forest Department, Bangladesh Forest Research Institute, Institute of Forestry and Environmental Sciences, University of Chittagong, Department of Forestry and Environmental Science, Shahjalal University of Science & Technology, and Forestry and Wood Technology Discipline, Khulna University for their cordial logistic supports during the collection of the extensive literature.

Appendix A

Sl no	Source
1	Alamgir, M., Al-Amin, M. 2008. Allometric models to estimate biomass organic carbon stock in forest vegetation. Journal of Forestry Research 19 (2): 101-106
2	Chaffey, D.R., Miller, F.R., Sandom, J.H. 1985. A forest inventory of the Sundarbans, Bangladesh. Project report 140, Overseas Development Administration, Land Resources Development Centre, England
3	Cox, F.Z. 1984. Volume functions for plantation species and elements for growth models for Teak. Field Document no 2, Assistance to the Forestry Sector of Bangladesh, Food and Agricultural Organization of the United Nations, FAO/UNDP Project BGD/79/017
4	Das, N. 2014. Modeling Develops to Estimate Leaf Area and Leaf Biomass of Lagerstroemia speciosa in West Vanugach Reserve Forest of Bangladesh. ISRN Forestry, doi. org/10.1155/2014/ 486478
5	Das, S., Davidson, J., Latif, M.A., Rahman, F., Das, S. 1985. Tree volume tables for Moluccana (Paraserianthes falcataria syn. Albizia falcataria syn. A. moluccana) in Bangladesh. Bulletin no 4, Inventory Division, Bangladesh Forest Research Institute, Chittagong.
6	Das, S., Rahman, M.F., Reza, N.A., Latif, M.A. 1992. Tree volume tables for Sal (Shorea robusta Gaertn. f.) in the plantations of Bangladesh. Bulletin 7, Forest Inventory Series, Bangladesh Forest Research Institute, Chittagong, 1-11 pp.
7	Davidson, J., Das, S., Khan, S.A., Latif, M.A., Zashimuddin, M. 1985. Tree volume tables for small Eucalypt round wood in Bangladesh. Bulletin no 4, Silviculturer Research Division, Bangladesh Forest research Institute, Chittagong
8	Deb, J.C., Halim, M.A., Ahmed, E. 2012. An allometric equation for estimating stem biomass of Acacia auriculiformis in the north-eastern region of Bangladesh. Southern Forests 74(2): 103–113

9	Drigo, R., Latif, M.A., Chowdhury, J.A., Shaheduzzaman, M. 1987. The maturing mangrove plantations of the coastal afforestation project. Food and Agricultural Organization of the United Nations, FAO/UNDP Project BDG/85/085, Assistant to the Forestry Sector.

Sl no	Source
10	Drigo, R., Shaheduzzaman, M., Chowdhury, J.A. 1988. Inventory of forest resources of Southern Sylhet Forest Division. Assistance to Forestry Sector - Phase II, Field Document no 3, Food and Agriculture Organisation of the United Nations, FAO/UNDP Project BGD/85/085
11	Hossain, S.M.Y., Martin, A.R. 2013. Merchantable timber production in Dalbergia sissoo plantations across Bangladesh: regional patterns, management practices and edaphic factors. Journal of Tropical Forest Science 25(3): 299-309
12	Islam, S.M.Z., Ahmed, K.U., Khan, M.I. 2014. Mathematical models for estimating stem volume and volume tables of Rubber tree. Bulletin 10, Forest Inventory Series, Bangladesh Forest Reearch Institute, Chittagong.
13	Islam, S.M.Z., Khan, M.I. Ahmed, K.U. 2012. Volume equations and tables for Rajkoroi (Albizia richardiana King and Prain) planted in the southern part of Bangladesh. Bangladesh Journal of Forest Science 32 (1): 28-39
14	Islam, S.S., 1988. Commercial volume table for teak (Tectona grandis) in Bangladesh by regression technique. Bano Biggyan Patrika 17 (1&2): 55-67
15	Islam, S.S., Kabir, J., Masum, A.K.M. 2012. Volume Table of Raintree (Samanea saman) in Bangladesh by Regression Technique. Open Journal of Statistics 2: 115-119
16	Islam, S.S., Reza, N.A., Hasnin, M., Khan, M.A.S., Islam, M.R., Siddiqi, N.A. 1992. Volume table of young Keora (Sonneratia apetala) trees for the western coastal belt of Bangladesh. Bulletin 1, Plantation Trial Unit Series, Bangladesh Forest Research Institute, Chittagong, 1-23 pp.
17	Khan, M.N.I., Faruque, O. 2010. Allometric relationships for predicting the stem volume in a Dalbergia sissoo Roxb. plantation in Bangladesh. iForest 3: 153-158
18	Kingston, B. 1979. A collation of tree and bamboo volume tables of Bangladesh. Field Document no 15, Food and Agricultural Organization of the United Nations, UNDP/FAO Project BGD/72/005, Forest Research Institute, Chittagong
19	Latif M.A., Habib, M.A. 1994. Biomass tables for Acacia mangium grown in the plantations in Bangladesh. Journal of Tropical Forest Science 7(2): 296- 302
20	Latif, M.A., 1988. Biomass tables for young Eucalyptus grown in Bangladesh. Bano Biggyan Patrika 17 (1 & 2): 46-54

Sl no	Source
21	Latif, M.A., Das, S., Rahman, M.F., Chowdhury, J.A. 1994. Tree volume tables for Baen (Avicennia officinalis L.) in the coastal plantations of Bangladesh. In: Latif, M.A. (ed.), Tree volume table for keora (Sonneratia apetala) and Baen (Avicennia officinalis) in the coastal plantation of Banglaesh. Bulletin 8, Forest Inventory Division, Bangladesh Forest Research Institute, Chittagong, 21-23 pp.
22	Latif, M.A., Habib, M.A. 1993. Biomass table for Acacia auriculiformis grown in the plantation in Bangladesh. Indian Journal of Forestry 16 (4): 323-327
23	Latif, M.A., Habib, M.A. 1994. Biomass tables for minjiri (Cassia Siamea Lam.) grown in the plantations in Bangladesh. Bangladesh Journal of Forest Science 23 (1): 59-64
24	Latif, M.A., Habib, M.A., Das, S., 1993. Tree volume tables for Acacia mangium in the plantations of Bangladesh. Bangladesh Journal of Forest Science 22 (1 & 2): 23-29
25	Latif, M.A., Islam, M.N. 1984a. Tree volume volume tables for Syzygium grande (Wt.) Wald (Dhakijam). In: Choudhury, J.H., and Davidson, J. (eds.), Tree volume tables for four species grown in plantation in Bangladesh. Bulletin 2, Inventory Division, Bangladesh Forest Research institute. Chittagong, Bangladesh, pp. 25-57.
26	Latif, M.A., Islam, M.N. 1984b. Tree volume tables for Artocarpus chaplasha Roxb. (Chapalish). In: Choudhury, J.H., and Davidson, J. (eds.), Tree volume tables for four species grown in plantation in Bangladesh. Bulletin 2, Inventory Division, Bangladesh Forest Research institute. Chittagong, Bangladesh, pp. 58-92.
27	Latif, M.A., Islam, M.N. 1984c. Tree volume tables for Dipterocarpus turbintus Gaertn. F. (Tali Garjan). In: Choudhury, J.H., and Davidson, J. (eds.), Tree volume tables for four species grown in plantation in Bangladesh. Bulletin 2, Inventory Division, Bangladesh Forest Research institute. Chittagong, Bangladesh, pp. 122-128.
28	Latif, M.A., Islam, M.N. Choudhury, J.H. 1984. Tree volume tables for Gmelina arborea Roxb. (Gamar). In: Choudhury, J.H., and Davidson, J. (eds.), Tree volume tables for four species grown in plantation in Bangladesh. Bulletin 2, Inventory Division, Bangladesh Forest Research institute. Chittagong, Bangladesh, pp. 93-121.
29	Latif, M.A., Islam, M.N., Islam, S.S. 1985. Tree volume tables for Teak (Tectona grandis) in Bangladesh. Bulletin no 5, Inventory Division, Bangladesh Forest Research Institute, Chittagong.

Sl no	Source
30	Latif, M.A., Islam, M.S., Islam, S.M.Z. 1999. Volume tables for sissoo, koroi, mahogany, eucalyptus and bokain planted on croplands in the western part of Bangladesh. Bangladesh Forest Research Institute, Chittagong.
31	Latif, M.A., Islam, M.S., Islam, S.M.Z., 2000. Volume tables for Sissoo, Koroi, Akashmoni, Babla, Mahogany, and Rain tree planted on embankments and road sides in the coastal areas of Bangladesh. Bulletin 9, Forest Inventory Series, Bangladesh Forest Research Institute, Chittagong.
32	Latif, M.A., Islam, S.M.Z. 2000. Volume tables for 11 important tree species grown in the home gardens of Bangladesh. Forest Inventory Division, Bangladesh Forest Research Institute, Chittagong.
33	Latif, M.A., Islam, S.M.Z. 2004. Timber and fuelwood volume tables for Acacia auriculiformis, A. mangium, Eucalyptus camaldulensis and Dalbergia sissoo in plantations in Bangladesh. Forestry Sector Project, Bangladesh Forest Department and Bangladesh Forest Research Institute, Chittagong.
34	Latif, M.A., Islam, S.S., Davidson, J. 1986. Metric volume tables for some tree species found in the natural forests of Bangladesh. Bulletin 6, Inventory Division, Bangladesh Forest Resarch Institute, Chittagong.
35	Latif, M.A., Khan, A.F.M.K., Hossain, M.M. 1998. Stump diameter -DBH - volume relationships for Teli Garjan (Dipterocarpus turbinatus), Dhakijam (Sygyzium grande) and Teak (Tectona grandis) in Bangladesh. Bangladesh Journal of Forest Science 27 (1): 16-24
36	Latif, M.A., Rahman, M.F., Das, S. 1995. Volume table for Acacia auriculiformis, Cassia siamea and Pinus caribaea in Bangladesh. Bangladesh Journal of Forest Science 24 (2): 22-30
37	Mahmood, H., Saha, C., Abdullah, S.M.R., Saha, S., Siddique, M.R.H. 2015b. Allometric biomass, nutrient and carbon stock models for Kandelia candel of the Sundarbans, Bangladesh. Trees DOI: 10.1007/s00468-015-1314-0
38	Mahmood, H., Shaikh, M.A.A., Saha, C., Abdullah, S.M.R., Saha, S., Siddique, M.R.H. 2016. Above-ground biomass, nutrients and carbon in *Aegiceras corniculatum* of the Sundarbans. *Open Journal of Forestry* 6 (2): 72-89
39	Mahmood, H., Siddique, M.R.H., Bose, A., Limon, S.H., Chowdhury, M.R.K. Saha, S. 2012. Allometry, above-ground biomass and nutrient distribution in Ceriops decandra (Griffith) Ding Hou dominated forest types of the Sundarbans mangrove forest, Bangladesh. Wetlands Ecology and Management 20: 539-548

Sl no	Source
40	Mahmood, H., Siddique, M.R.H., Saha, S., Abdullah, S.M.R. 2015a. Allometric models for biomass, nutrients and carbon stock in Excoecaria agallocha of the Sundarbans, Bangladesh. Wetlands Ecology and Management 23 (4): 765-774
41	Rahman, M.F., Das, S., Latif, M.A. 2001. Volume table for Koroi (albizia procera) and Arjun (Terminalia arjuna) trees planted in the central part of Bangladesh. Bangladesh Journal of Forest Science 30 (1): 39-46.
42	Rahman, M.F., Das, S., Reza, N.A., Chowdhury, J.A., Latif, M.A. 1994. Tree volume table for Keora (Sonneratia apetala Buch.-Ham) in the coastal plantation of Bangladesh. In: Latif, M.A. (ed.), Tree volume table for keora (Sonneratia apetala) and Baen (Avicennia officinalis) in the coastal plantation of Banglaesh. Bulletin 8, Forest Inventory Division, Bangladesh Forest Research Institute, Chittagong, 1-20 pp.
43	Rahman, M.M., Kamaluddin, M. 1996. Volume table for natural hybrid trees of Acacia mangium X Acacia auriculiformis in plantations of Bangladesh. Chittagong University Studies, 20 (1): 89-94
44	Revilla, J.A.V., Ahmed, I.U., Hossain, A. 1998a. Forest Inventory of the Sundarbans Reserved Forest, Final Report, Volume 1. Mandala Agricultural Development Corporation and Forest Department, Ministry of Environment and Forests, Dhaka, Bangladesh
45	Revilla, J.A.V., Ahmed, I.U.., Saha, U.K. 1998b. Forest Inventory of the Natural Forest and Forest Plantations (Sylhet Forest Division), Final Report. Gob/Wb, Forest Resources Management Project, Technical Assistance Component. (Mandala Agricultural Development Corporation and Forest Department, Ministry of Environment and Forests, Dhaka, Bangladesh).
46	Roy, B. 2012. Species-specific allometric models for estimation of aboveground stem biomass of three dominant tree species at Satchari National park. Unpublished MS thesis. Department of Forestry and Environmental Science, ShahJalal University of Science and Technology, Sylhet, Bangladesh
47	Sarker, S.K., Das, N., Chowdhury, M.Q., Haque, M.M. 2013. Developing allometric equations for estimating leaf area and leaf biomass of Artocarpus chaplasha in Raghunandan Hill Reserve, Bangladesh. Southern Forests 75(1): 51-57
48	Siddique, M.R.H., Mahmood, H., Chowdhury, M.R.K. 2012. Allometric relationship for estimating above-ground biomass of Aegialitis rotundifolia Roxb. of Sundarbans mangrove forest, in Bangladesh. Journal of Forestry Research 23 (1): 23-28.
49	Sylvander, R., Latif, M.A., Karlsson, A. 2001. Forest inventory of the Sal forest of Bangladesh, Volume 1, Technical Report, Forestry Sector Project, Forest Department, Ministry of Environment and Forest, Dhaka

50 Ullah, M.R., Banik, G.R., RajibBanik. 2014. Developing Allometric Models for Carbon Stock Estimation in Eighteen Year Old Plantation Forests of Bangladesh. Jacobs Journal of Microbiology and Pathology 1(1): 006

Appendix B

Genus	Species	Local name	Family	Unite of Y	Vegetation Component	Equation	Sample size	R2	MSE	Ref. No
Senna	*siamea*	Minjiri	Leguminosae	kg	Total above-ground green biomass	Log (Green biomass) = -1.5851 + 2.4855 * Log (DBH)	120	0.9728	0.0232	23
Senna	*siamea*	Minjiri	Leguminosae	kg	Total above-ground green biomass	Log (Green biomass) = -4.4303 + 2.4855 * Log (GBH)	120	0.9728	0.0232	23
Senna	*siamea*	Minjiri	Leguminosae	kg	Total above-ground green biomass	Log (Green biomass) = -2.0847 + 2.1723 * Log (DBH) + 0.5141 * Log (Height)	120	0.9777	0.0189	23
Senna	*siamea*	Minjiri	Leguminosae	kg	Total above-ground green biomass	Log (Green biomass) = -4.5714 + 2.1723 * Log (GBH) + 0.5141 * Log (Height)	120	0.9777	0.0189	23
Senna	*siamea*	Minjiri	Leguminosae	kg	Stem green biomass	Log (Green biomass) = -2.1442 + 2.5917 * Log (DBH)	120	0.9665	0.0313	23
Senna	*siamea*	Minjiri	Leguminosae	kg	Stem green biomass	Log (Green biomass) = -5.1110 + 2.5917 * Log (GBH)	120	0.9665	0.0313	23
Senna	*siamea*	Minjiri	Leguminosae	kg	Stem green biomass	Log (Green biomass) = -2.7095 + 2.2372 * Log (DBH) + 0.5817 * Log (Height)	120	0.9722	0.0257	23
Senna	*siamea*	Minjiri	Leguminosae	kg	Stem green biomass	Log (Green biomass) = -5.2705 + 2.2372 * Log (GBH) + 0.5817 * Log (Height)	120	0.9722	0.0257	23
Senna	*siamea*	Minjiri	Leguminosae	kg	Branch green Biomass	Log (Green biomass) = -2.2732 + 1.9752 * Log (DBH)	120	0.5707	0.3528	23
Senna	*siamea*	Minjiri	Leguminosae	kg	Branch green Biomass	Log (Green biomass) = -4.5343 + 1.9752 * Log (GBH)	120	0.5707	0.3528	23
Senna	*siamea*	Minjiri	Leguminosae	kg	Branch green Biomass	Log (Green biomass) = -3.2955 + 1.3142 * Log (DBH) + 1.0521 * Log (Height)	120	0.5855	0.3355	23
Senna	*siamea*	Minjiri	Leguminosae	kg	Branch green Biomass	Log (Green biomass) = -4.7999 + 1.3142 * Log (GBH) + 1.0521 * Log (Height)	120	0.5855	0.3355	23
Senna	*siamea*	Minjiri	Leguminosae	kg	Leaves and twigs green biomass	Log (Green biomass) = -2.1219 + 1.9299 * Log (DBH)	120	0.7614	0.1568	23
Senna	*siamea*	Minjiri	Leguminosae	kg	Leaves and twigs green biomass	Log (Green biomass) = -4.3311 + 1.9299 * Log (GBH)	120	0.7614	0.1568	23
Senna	*siamea*	Minjiri	Leguminosae	kg	Leaves and twigs green biomass	Log (Green biomass) = -0.6183 + 2.8726 * Log (DBH) - 1.5471 * Log (Height)	120	0.8209	0.098	23
Senna	*siamea*	Minjiri	Leguminosae	kg	Leaves and twigs green biomass	Log (Green biomass) = -3.9067+ 2.8726 * Log (GBH) - 1.5471 * Log (Height)	120	0.8209	0.098	23
Senna	*siamea*	Minjiri	Leguminosae	kg	Stem and branch green biomass	Log (Green biomass) = -2.0512 + 2.6006 * Log (DBH)	120	0.9641	0.0339	23
Senna	*siamea*	Minjiri	Leguminosae	kg	Stem and branch green biomass	Log (Green biomass) = -5.0282 + 2.6006 * Log (GBH)	120	0.9641	0.0339	23
Senna	*siamea*	Minjiri	Leguminosae	kg	Stem and branch green biomass	Log (Green biomass) = -2.9256 + 2.0525 * Log (DBH) + 0.8996 * Log (Height)	120	0.9785	0.0201	23

Senna	*siamea*	Minjiri	Leguminosae	kg	Stem and branch green biomass	Log (Green biomass) = -5.2752 + 2.0525 * Log (GBH) + 0.8996 * Log (Height)	120	0.9785	0.0201	23
Senna	*siamea*	Minjiri	Leguminosae	kg	Branch, leaves and twigs green biomass	Log (Green biomass) = -2.5173 + 2.281 * Log (DBH)	120	0.8098	0.1641	23
Senna	*siamea*	Minjiri	Leguminosae	kg	Branch, leaves and twigs green biomass	Log (Green biomass) = -2.9974 + 1.98 * Log (DBH) + 0.494 * Log (Height)	120	0.8115	0.1613	23
Senna	*siamea*	Minjiri	Leguminosae	kg	Branch, leaves and twigs green biomass	Log (Green biomass) = -5.264 + 1.98 * Log (GBH) + 0.494 * Log (Height)	120	0.8115	0.1613	23

Genus	**Species**	**Local name**	**Family**	**Unite of Y**	**Vegetation Component**	**Equation**	**Sample size**	**R2**	**MSE**	**FI**	**Ref. No**
Sonneratia	*apetala*	Keora	Lythraceae	m3	Total volume over bark	Volume = 0.0117 + 0.0000280056 * DBH^(2) * Height	713	0.912			42
Sonneratia	*apetala*	Keora	Lythraceae	m3	Total volume over bark	Volume = 0.0117 + 0.00000283756 * GBH^(2) * Height	713	0.912			42
Sonneratia	*apetala*	Keora	Lythraceae	m3	Volume under bark	Volume = 0.0041 + 0.0000246325 * DBH^(2) * Height	713	0.937			42
Sonneratia	*apetala*	Keora	Lythraceae	m3	Volume under bark	Volume = 0.0041 + 0.00000249579 * GBH^(2) * Height	713	0.937			42
Avicennia	*officinalis*	Baen	Aviceniaceae	m3	Total volume over bark	Volume = 0.0089 + 0.0000264 * DBH^(2) * Height	308	0.859	0.00026		21
Avicennia	*officinalis*	Baen	Aviceniaceae	m3	Total volume over bark	Volume = 0.0089 + 0.00000267 * GBH^(2) * Height	308	0.859	0.00026		21
Albizia	*procera*	Koroi	Mimosaceae	m3	Total volume over bark	Log (Volume) = -12.0901 + 2.502194 * Log (GBH)	221	0.983			41
Albizia	*procera*	Koroi	Mimosaceae	m3	Total volume over bark	Log (Volume) = -11.6632 + 1.941989 * Log (GBH) + 0.754839 * Log (Height)	221	0.991			41
Terminalia	*arjuna*	Arjun	Combretacear	m3	Total volume over bark	Log (Volume) = -11.1885 + 2.222144 * Log (GBH)	177	0.986			41
Terminalia	*arjuna*	Arjun	Combretacear	m3	Total volume over bark	Log (Volume) = -11.3794 + 1.896423 * Log (GBH) + 0.653558 * Log (Height)	177	0.997			41
Shorea	*robusta*	Sal	Dipterocarpaceae	m3	Total volume over bark	Log (Volume) = -9.1727759 + 2.5178944 * DBH	499	0.9666	0.03856	0.0575	38
Shorea	*robusta*	Sal	Dipterocarpaceae	m3	Total volume over bark	Log (Volume) = -12.0554 + 2.5178944 * Log (GBH)	499	0.9666	0.03856	0.0575	38
Shorea	*robusta*	Sal	Dipterocarpaceae	m3	Total volume over bark	Log (Volume) = -9.615639 + 2.033071 * Log (DBH) + 0.7361229 * Log (Height)	499	0.9551	0.00776	0.0881	38
Shorea	*robusta*	Sal	Dipterocarpaceae	m3	Total volume over bark	Log (Volume) = -11.938881 + 2.033071 * Log (GBH) + 0.7361229 * Log (Height)	499	0.9551	0.00776	0.0881	38

Shorea	*robusta*	Sal	Dipterocarpaceae	m3	Total volume under bark	Volume = 0.0032556 + 0.0000269 * DBH^(2) * Height	499	0.9622	0.00652	0.0808	38
Shorea	*robusta*	Sal	Dipterocarpaceae	m3	Total volume under bark	Volume = 0.003255 + 0.0000027255 * GBH^(2) * Height	499	0.9622	0.00652	0.0808	38
Sonneratia	*apetala*	Keora	Lythraceae	m3	Total volume over bark	Volume = 0.0052 - 0.0022 * X + 0.0005 * DBH^(2)	461	0.86	0.00008	0.009	16
Sonneratia	*apetala*	Keora	Lythraceae	m3	Total volume over bark	Log (Volume) = -9.1937 + 1.7683 * Log (DBH) + 0.7358 * Log (Height)	461	0.95	0.0547	0.003	16
Sonneratia	*apetala*	Keora	Lythraceae	m3	Total volume over bark	Volume = 0.0042 - 0.0017 * DBH + 0.0005 * DBH^(2)	464	0.85	0.0001	0.0109	16
Sonneratia	*apetala*	Keora	Lythraceae	m3	Total volume over bark	Log (Volume) = -9.2587 + 1.6463 * Log (DBH) + 0.9138 * Log (Height)	464	0.94	0.0744	0.0051	16
Acacia	*mangium*	Mangium	Fabaceae	m3	Total volume over bark	Log (Volume) = -8.209 + 2.2178 * Log (DBH)	132	0.98	0.0152	0.0062	24
Acacia	*mangium*	Mangium	Fabaceae	m3	Total volume over bark	Log (Volume) = -10.7488 + 2.2178 * Log (GBH)	132	0.98	0.0152	0.0062	24

Genus	Species	Local name	Family	Unite of Y	Vegetation Component	Equation	Sample size	R2	MSE	FI	Ref. No
Acacia	*mangium*	Mangium	Fabaceae	m3	Total volume over bark	Log (Volume) = -9.1426 + 1.7612 * Log (DBH) + 0.83335 * Log (Height)	132	0.989	0.0091	0.0048	24
Acacia	*mangium*	Mangium	Fabaceae	m3	Total volume over bark	Log (Volume) = -11.1587 + 1.7612 * Log (GBH) + 0.83335 * Log (Height)	132	0.989	0.0091	0.0048	24
Acacia	*mangium*	Mangium	Fabaceae	m3	Total volume under bark	Log (Volume) = -9.00226 + 2.3246 * Log (DBH)	132	0.969	0.0299	0.0056	24
Acacia	*mangium*	Mangium	Fabaceae	m3	Total volume under bark	Log (Volume) = -11.6633 + 2.3246 * Log (GBH)	132	0.969	0.0299	0.0056	24
Acacia	*mangium*	Mangium	Fabaceae	m3	Total volume under bark	Log (Volume) = -10.2221 + 1.74054 * Log (DBH) + 1.07596 * Log (Height)	132	0.977	0.0219	0.0048	24
Mixed				kg	Total above-ground green biomass	Log (Green biomass) = -1.3933 + 2.39602 * Log (DBH)	294	0.948			20
Mixed				kg	Total above-ground green biomass	Log (Green biomass) = -4.136 + 2.39602 * Log (GBH)	294	0.948			20
Mixed				kg	Total above-ground green biomass	Log (Green biomass) = -2.228 + 1.81492 * Log (DBH) + 0.85007 * Log (Height)	294	0.9575			20
Mixed				kg	Total above-ground green biomass	Log (Green biomass) = -4.306 + 1.81492 * Log (GBH) + 0.85007 * Log (Height)	294	0.9575			20
Tectona	*grandis*	Teak	Lamiaceae	cft	Total volume over bark	Volume = 0.084 * DBH^(2.263)	635	0.88	0.23	40.79	17
Tectona	*grandis*	Teak	Lamiaceae	cft	Total volume over bark	Volume = 0.000465 * DBH^(1.58) * Height^(1.603)	645	0.92	0.16	33.36	17
Acacia	*auriculiformis*	Akashmoni	Leguminosae	kg	Total above-ground green biomass	Log (Green biomass) = -1.3577 + 2.4177 * Log (DBH)	139	0.9674			22
Acacia	*auriculiformis*	Akashmoni	Leguminosae	kg	Total above-ground green biomass	Log (Green biomass) = -2.2782 + 1.9736 * Log (DBH) + 0.8113 * Log (Height)	139	0.986	0.0084		22
Acacia	*auriculiformis*	Akashmoni	Leguminosae	kg	Stem green biomass	Log (Green biomass) = -2.3176 + 2.6075 * Log (DBH)	139	0.9698	0.0218		22
Acacia	*auriculiformis*	Akashmoni	Leguminosae	kg	Stem green biomass	Log (Green biomass) = -3.1661 + 2.1982 * Log (DBH) + 0.7477 * Log (Height)	139	0.983	0.0119		22
Acacia	*auriculiformis*	Akashmoni	Leguminosae	m3	Total volume over bark	Log (Volume) = -8.208 + 2.2389 * Log (DBH)	139	0.959		0.0061	36
Acacia	*auriculiformis*	Akashmoni	Leguminosae	m3	Total volume over bark	Log (Volume) = -10.7709 + 2.2389 * Log (GBH)	139	0.959		0.0061	36
Acacia	*auriculiformis*	Akashmoni	Leguminosae	m3	Total volume over bark	Log (Volume) = -9.125 + 1.918 * Log (DBH) + 0.67988 * Log (Height)	139	0.988		0.0648	36
Acacia	*auriculiformis*	Akashmoni	Leguminosae	m3	Total volume over bark	Log (Volume) = -11.3205 + 1.918 * Log (GBH) + 0.67988 * Log (Height)	139	0.988		0.0648	36
Acacia	*auriculiformis*	Akashmoni	Leguminosae	m3	Total volume under bark	Log (Volume) = -9.187 + 2.468 * Log (DBH)	139	0.9688		0.0059	36

IOP Conf. Series: Earth and Environmental Science **39** (2016) 012057 doi:10.1088/1755-1315/39/1/012057

Acacia	*auriculiformis*	Akashmoni	Leguminosae	m3	Total volume under bark	Log (Volume) = -12.0121 + 2.468 * Log (GBH)	139	0.9688	0.0059	36
Acacia	*auriculiformis*	Akashmoni	Leguminosae	m3	Total volume under bark	Log (Volume) = -10.2398 + 2.100244 * Log (DBH) + 0.780214 * Log (Height)	139	0.9773	0.0048	36

Genus	Species	Local name	Family	Unite of Y	Vegetation Component	Equation	Sample size	R2	MSE	FI	Ref. No
Acacia	auriculiformis	Akashmoni	Leguminosae	m3	Total volume under bark	Log (Volume) = -12.6440 + 2.100244 * Log (GBH) + 0.780214 * Log (Height)	139	0.9773		0.0048	36
Senna	siamea	Minjiri	Leguminosae	m3	Total volume over bark	Log (Volume) = -8.602 + 2.4038 * Log (DBH)	120	0.9796		0.0098	36
Senna	siamea	Minjiri	Leguminosae	m3	Total volume over bark	Log (Volume) = -11.3536 + 2.4038 * Log (GBH)	120	0.9796		0.0098	36
Senna	siamea	Minjiri	Leguminosae	m3	Total volume over bark	Log (Volume) = -9.514 + 1.871 * Log (DBH) + 0.897 * Log (Height)	120	0.9898		0.0054	36
Senna	siamea	Minjiri	Leguminosae	m3	Total volume over bark	Log (Volume) = -11.6557 + 1.871 * Log (GBH) + 0.897 * Log (Height)	120	0.9898		0.0054	36
Senna	siamea	Minjiri	Leguminosae	m3	Total volume under bark	Log (Volume) = -9.334 + 2.55686 * Log (DBH)	120	0.976		0.0081	36
Senna	siamea	Minjiri	Leguminosae	m3	Total volume under bark	Log (Volume) = -12.2632 + 2.55686 * Log (GBH)	120	0.976		0.0081	36
Senna	siamea	Minjiri	Leguminosae	m3	Total volume under bark	Log (Volume) = -10.1766698 + 2.0641847 * Log (DBH) + 0.8290937 * Log (Height)	120	0.986		0.0049	36
Senna	siamea	Minjiri	Leguminosae	m3	Total volume under bark	Log (Volume) = -12.5396 + 2.064187 * Log (GBH) + 0.8290937 * Log (Height)	120	0.986		0.0049	36
Pinus	caribaea	Pine	Pinaceae	m3	Total volume over bark	Log (Volume) = -8.7854 + 2.410755 * Log (DBH)	122	0.986		0.002	36
Pinus	caribaea	Pine	Pinaceae	m3	Total volume over bark	Log (Volume) = -11.545 + 2.410755 * Log (GBH)	122	0.986		0.002	36
Pinus	caribaea	Pine	Pinaceae	m3	Total volume over bark	Log (Volume) = -9.39412 + 1.867386 * Log (DBH) + 0.839034 * Log (Height)	122	0.9945		0.0052	36
Pinus	caribaea	Pine	Pinaceae	m3	Total volume over bark	Log (Volume) = -11.5317 + 1.867386 * Log (GBH) + 0.839034 * Log (Height)	122	0.9945		0.0052	36
Pinus	caribaea	Pine	Pinaceae	m3	Total volume under bark	Log (Volume) = -9.11552 + 2.483187 * Log (DBH)	122	0.9858		0.0084	36
Pinus	caribaea	Pine	Pinaceae	m3	Total volume under bark	Log (Volume) = -11.9580 + 2.483187 * Log (GBH)	122	0.9858		0.0084	36
Pinus	caribaea	Pine	Pinaceae	m3	Total volume under bark	Log (Volume) = -9.7505 + 1.935397 * Log (DBH) + 0.851715 * Log (Height)	122	0.9933		0.0058	36
Pinus	caribaea	Pine	Pinaceae	m3	Total volume under bark	Log (Volume)= -11.9660 + 1.935397 * Log (GBH) + 0.851715 * Log (Height)	122	0.9933		0.0058	36
Acacia	auriculiformis	Akashmoni	Leguminosae	m3	Total volume over bark	Log (Volume) = -11.839665 + 2.404568 * Log (GBH)	124	0.973			31
Acacia	auriculiformis	Akashmoni	Leguminosae	m3	Total volume over bark	Log (Volume) = -11.506528 + 1.973377 * Log (GBH) + 0.623823 * Log (Height)	124	0.979			31
Swietenia	macrophylla	Mahogany	Meliaceae	m3	Total volume over bark	Log (Volume) = -12.52620808 + 2.5653795 * Log (GBH)	245	0.942			31

Swietenia	macrophylla	Mahogany	Meliaceae	m3	Total volume over bark	Log (Volume) = -12.4361459 + 1.8661846 * Log (GBH) + 1.2282822 * Log (Height)	245	0.96	31
Albizia	spp	Koroi	Mimosaceae	m3	Total volume over bark	Log (Volume) = -12.8715358 + 2.6994968 * Log (GBH)	178	0.929	31

Genus	Species	Local name	Family	Unite of Y	Vegetation Component	Equation	Sample size	R2	MSE	FI	Ref. No
Albizia	spp	Koroi	Mimosaceae	m3	Total volume over bark	Log (Volume) = -12.4 + 1.7131 * Log (GBH) + 1.58245 * Log (Height	178	0.967			31
Dalbergia	*sissoo*	Sissoo	Fabaceae	m3	Total volume over bark	Log (Volume) = -12.427775 + 2.6056676 * Log (GBH)	202	0.902			31
Dalbergia	*sissoo*	Sissoo	Fabaceae	m3	Total volume over bark	Log (Volume) = -12.5189939 + 1.9800535 * Log (GBH) + 1.0775148 * Log (Height)	202	0.934			31
Acacia	*nilotica*	Babla	Mimosaceae	m3	Total volume over bark	Log (Volume) = -11.2782859 + 2.34743 * Log (GBH)	128	0.91			31
Acacia	*nilotica*	Babla	Mimosaceae	m3	Total volume over bark	Log (Volume) = -11.875835 + 1.8823999 * Log (GBH) + 1.0819988 * Log (Height)	128	0.91			31
Albizia	*saman*	Rain tree	Mimosaceae	m3	Total volume over bark	Log (Volume) = -12.287524 + 2.5086408 * Log (GBH)	190	0.952			31
Albizia	*saman*	Rain tree	Mimosaceae	m3	Total volume over bark	Log (Volume) = -12.3213818 + 1.8912934 * Log (GBH) + 1.183443 * Log (Height)	190	0.974			31
Ceriops	*decandra*	Goran	Rhizophoraceae	g	Leaf	Oven-dried biomass = 2.99 * (Collar girth)^(1.95)	48	0.89	0.02		39
Ceriops	*decandra*	Goran	Rhizophoraceae	g	Branch	Oven-dried biomass = 0.23 * (Collar girth)^(3.09)	48	0.94	0.03		39
Ceriops	*decandra*	Goran	Rhizophoraceae	g	Bark	Oven-dried biomass = 0.77 * (Collar girth)^(2.23)	48	0.97	0.01		39
Ceriops	*decandra*	Goran	Rhizophoraceae	g	Stem with bark	Oven-dried biomass = 3.22 * (Collar girth)^(2.27)	48	0.97	0.01		39
Ceriops	*decandra*	Goran	Rhizophoraceae	g	Total above-ground	Oven-dried biomass = 4.70 * (Collar girth)^(2.41)	48	0.97	0.01		39

Genus	Species	Local name	Family	Unite of Y	Vegetation Component	Equation	Sample size	R2	MSE	RMSE	AIC	Bias correction	Ref. No
Aegialitis	*rotundifolia*	Nuniya	Plumbaginaceae	g	Leaf	Oven-dried biomass = 13.96 * (Collar girth) - 12.38 * (Height)^(2)- 0.01 * (Height at girth measurement point)^(2) + 0.08 * (Collar girth) * (Height) * (Height at girth measurement point)	50	0.88	1392.78	37.32			48
Aegialitis	*rotundifolia*	Nuniya	Plumbaginaceae	g	Branch	Oven-dried biomass = 3.09 * (Collar girth)^(2) - 22.887 * (Height)^(2) + 0.13 * (Collar girth) * (Height) * (Height at girth measurement point)	50	0.92	8626.98	92.882			48
Excoecaria	*agallocha*	Gewa	Euphorbiaceae	kg	Leaf	Log 10 (Oven-dried biomass) = 0.9256 * Log 10 (DBH^(2)) - 2.133	30	0.8499	0.051	0.226	-86.652	1.146	40
Excoecaria	*agallocha*	Gewa	Euphorbiaceae	kg	Branch	Log 10 (Oven-dried biomass) = 1.1656 * Log 10 (DBH^(2)) - 1.7047	30	0.9669	0.016	0.126	-122.159	1.043	40
Excoecaria	*agallocha*	Gewa	Euphorbiaceae	kg	Bark	Log 10 (Oven-dried biomass) = 1.0824 * Log 10 (DBH^(2)) - 1.7568	30	0.9933	0.003	0.052	-175.484	1.007	40
Excoecaria	*agallocha*	Gewa	Euphorbiaceae	kg	Stem without bark	Log 10 (Oven-dried biomass) = 1.0927 * Log 10 (DBH^(2)) - 1.0275	30	0.9937	0.003	0.051	-176.616	1.007	40
Excoecaria	*agallocha*	Gewa	Euphorbiaceae	kg	Total above-ground	Log 10 (Oven-dried biomass) = 1.0996 * Log 10 (DBH^(2)) - 0.8572	30	0.9953	0.002	0.044	-185.005	1.005	40
Excoecaria	*agallocha*	Gewa	Euphorbiaceae	kg	Total above-ground	Log 10 (Nitrogen) = 1.0972 * Log 10 (DBH^(2)) - 3.0845	30	0.9922	0.0032	0.05677		1.008583	40
Excoecaria	*agallocha*	Gewa	Euphorbiaceae	kg	Total above-ground	Log 10 (Phosphorus) = 1.0947 * Log 10 (DBH^(2)) - 5.6790	30	0.9905	0.0039	0.0623		1.010333	40
Excoecaria	*agallocha*	Gewa	Euphorbiaceae	kg	Total above-ground	Log 10 (Potassium) = 1.0990 * Log 10 (DBH^(2)) - 3.0370	30	0.9929	0.0029	0.054		1.007774	40

Genus	Species	Local name	Family	Unite of Y	Vegetation Component	Equation	Sample size	R2	MSE	RMSE	AIC	FI	Bias correction	Ref. No
Excoecaria	*agallocha*	Gewa	Euphorbiaceae	kg	Total above-ground	Log 10 (Carbon) = 1.1 * Log 10 (DBH^(2)) - 1.1937	30	0.9953	0.0019	0.044			1.005136	40
Kandelia	*candel*	Goria	Rhizophoraceae	kg	Leaf	Oven-dried biomass = 0.014 * (DBH)^(2) + 0.03	25	0.89	0.004	0.06	-133.53	0.063		37
Kandelia	*candel*	Goria	Rhizophoraceae	kg	Branch	Sqrt (Oven-dried biomass) = 0.29 * (DBH) - 0.21	25	0.87	0.03	0.16	-86.947	0.23		37
Kandelia	*candel*	Goria	Rhizophoraceae	kg	Bark	Sqrt (Oven-dried biomass) = 0.66 * sqrt (DBH) - 0.57	25	0.86	0.01	0.1	-110.07	0.111		37
Kandelia	*candel*	Goria	Rhizophoraceae	kg	Stem without bark	Sqrt (Oven-dried biomass) = 1.19 * Sqrt (DBH) - 1.02	25	0.86	0.03	0.18	-80.521	0.345		37
Kandelia	*candel*	Goria	Rhizophoraceae	kg	Total above-ground	Oven-dried biomas = 0.21 * (DBH)^(2) + 0.12	25	0.94	0.38	0.62	-18.875	0.616		37
Kandelia	*candel*	Goria	Rhizophoraceae	kg	Total above-ground	Nitrogen = 0.39 * (DBH)^(2) + 0.49	25	0.95						37
Kandelia	*candel*	Goria	Rhizophoraceae	kg	Total above-ground	Phosphorus = 0.77 * (DBH)^(2) + 0.14	25	0.95						37
Kandelia	*candel*	Goria	Rhizophoraceae	kg	Total above-ground	Potassium = 0.87 * (DBH)^(2) + 0.07	25	0.95						37
Kandelia	*candel*	Goria	Rhizophoraceae	kg	Total above-ground	Carbon = 0.09 * (DBH)^(2) + 0.05	25	0.96						37
Acacia	*auriculiformis*	Akashmoni	Leguminosae	kg	Stem biomass	Oven-dried biomass = 0.092486 * ((DBH) * (Height))^(1.4765)	600	0.9674			-600.02		1.01066	8
Albizia	*saman*	Rain tree	Mimosaceae	m3	Stem volume	Log (Volume) = -8.3013 + 2.1746 * Log (DBH)	205	0.86	0.07			0.21		15
Albizia	*saman*	Rain tree	Mimosaceae	m3	Stem volume	Log (Volume) = -9.1864 + 1.8502 * Log (DBH) + 0.8234 * Log (Height)	205	0.9	0.05			0.18		15

Mixed Shrub				g	Above-ground biomass	Oven-dried biomass = 0.696735 + 0.536662 * (Biomass)		0.87201					1
Mixed Shrub				g	Above-ground biomass	Carbon = -0.379625 + 0.500132 * (Biomass)		0.89708					1

Genus	**Species**	**Local name**	**Family**	**Unite of Y**	**Vegetation Component**	**Equation**	**Sample size**	**R2**	**MSE**	**AIC**	**FI**	**Bias correction**	**Ref. No**
Acacia	*mangium*	Mangium	Fabaceae	kg	Total above-ground green biomass	Log (Green biomass) = -1.4659 + 2.3256 * Log (DBH)	132	0.9795	0.0193				19
Acacia	*mangium*	Mangium	Fabaceae	kg	Total above-ground green biomass	Log (Green biomass) = -4.1281 + 2.3256 * Log (GBH)	132	0.9795	0.0193				19
Acacia	*mangium*	Mangium	Fabaceae	kg	Total above-ground green biomass	Log (Green biomass) = -1.7073 + 2.1922 * Log (DBH) + 0.2331 * Log (Height)	132	0.9772	0.0185				19
Acacia	*mangium*	Mangium	Fabaceae	kg	Total above-ground green biomass	Log (Green biomass) = -4.2168 + 2.1922 * Log (GBH) + 0.2331 * Log (Height)	132	0.9772	0.0185				19
Acacia	*mangium*	Mangium	Fabaceae	kg	Stem green biomass	Log (Green biomass) = -2.2782 + 2.5213 * Log (DBH)	132	0.9553	0.0431				19
Acacia	*mangium*	Mangium	Fabaceae	kg	Stem green biomass	Log (Green biomass) = -5.1644 + 2.5213 * Log (GBH)	132	0.9553	0.0431				19
Acacia	*mangium*	Mangium	Fabaceae	kg	Stem green biomass	Log (Green biomass) = -2.7344 + 2.2692 * Log (DBH) + 0.4406 * Log (Height)	132	0.9584	0.0397				19
Acacia	*mangium*	Mangium	Fabaceae	kg	Stem green biomass	Log (Green biomass) = -5.3320 + 2.2692 * Log (GBH) + 0.4406 * Log (Height)	132	0.9584	0.0397				19
Acacia	*mangium*	Mangium	Fabaceae	kg	Stem and branch green biomass	Log (Green biomass) = -1.8493 + 2.3906 * Log (DBH)	132	0.9751	0.0211				19
Acacia	*mangium*	Mangium	Fabaceae	kg	Stem and branch green biomass	Log (Green biomass) = -4.5859 + 2.3906 * Log (GBH)	132	0.9751	0.0211				19
Acacia	*mangium*	Mangium	Fabaceae	kg	Stem and branch green biomass	Log (Green biomass) = -2.4276 + 2.0709 * Log (DBH) + 0.5586 * Log (Height)	132	0.9818	0.0153				19
Acacia	*mangium*	Mangium	Fabaceae	kg	Stem and branch green biomass	Log (Green biomass) = -4.7982 + 2.0709 * Log (GBH) + 0.5586 * Log (Height)	132	0.9818	0.0153				19

Artocarpus	*chaplasha*	Chapalish	Moraceeae	kg	Leaf	Log (Oven-dried biomass) = -4.44814 + 2.0483 * Log (DBH)	200		-58.298		1.019	47
Lagerstroemia	*speciosa*	Jarul	Lythraceae	kg	Leaf	Log (Oven-dried biomass) = -1.34008 + 0.83123 * Log (DBH) + 0.47969 * Log (Height)	312	0.9638	-309.79	0.47	1.0066	4
Artocarpus	*chaplasha*	Chapalish	Moraceeae	kg	Stem biomass	Log (Oven-dried biomass) = -0.53361 + 0.988759 * Log ((DBH)^(2) * (Height) * (Wood density))	101	0.99	-427.62			46

Genus	Species	Local name	Family	Unit e of Y	Vegetation Component	Equation	Sample size	R2	RMSE	AIC	FI	Ref. No
Tectona	*grandis*	Teak	Lamiaceae	kg	Stem biomass	Log (Oven-dried biomass) = 0.07908 + 0.89315 * Log ((DBH)^(2) * (Height) * (Wood density))	101	0.94		-76.89		46
Albizia	*richardiana*	Rajkoroi	Leguminosae	m3	Total volume over bark	Log (Volume) = -10.996396 + 2.247808 * Log (GBH)	511	0.98				13
Albizia	*richardiana*	Rajkoroi	Leguminosae	m3	Total volume over bark	Log (Volume) = -10.831293 + 1.699319 * Log (GBH) + 0.813706 * Log (Height)	511	0.98				13
Mixed				m3	Total volume over bark	Log (Volume) = -9.4209 + 1.7480 * Log (DBH) + 0.9310 * Log (Height)	954	0.984				7
Eucalyptus	*brassiana*	Eucalyptus	Myrtaceae	m3	Total volume over bark	Log (Volume) = -9.5783 + 1.6783 * Log (DBH) + 1.0483 * Log (Height)	164	0.985				7
Eucalyptus	*tereticornis*	Eucalyptus	Myrtaceae	m3	Total volume over bark	Log (Volume) = -9.4264 + 1.6850 * Log (DBH) + 0.9840 * Log (Height)	279	0.981				7
Eucalyptus	*camaldulensis*	Eucalyptus	Myrtaceae	m3	Total volume over bark	Log (Volume) = -9.3520 + 1.8055 * Log (DBH) + 0.8590 * Log (height)	511	0.986				7
Artocarpus	*chaplasha*	Chapalish	Moraceeae	m3	Total volume over bark	Log (Volume) = -8.179774 + 2.24074 * Log (DBH)	427	0.976	0.03		3.27	26
Artocarpus	*chaplasha*	Chapalish	Moraceeae	m3	Total volume over bark	Log (Volume) = -8.9449526 + 1.82851 * Log (DBH) + 0.735381 * Log (Height)	427	0.984	0.02		2.67	26
Gmelina	*arborea*	Gamar	Lamiaceae	m3	Total volume over bark	Log (Volume) = -7.9022697 + 2.1472 * Log (DBH)	486	0.933	0.08		3.68	28
Gmelina	*arborea*	Gamar	Lamiaceae	m3	Total volume over bark	Log (Volume) = -8.4687076 + 1.63502 * Log (DBH) + 0.784847 * Log (Height)	486	0.966	0.04		2.65	28
Dipterocarpus	*turbinatus*	Telya garjan	Dipterocarpaceae	m3	Total volume over bark	Log (Volume) = -8.5116354 + 2.35556 * Log (DBH)	436	0.979	0.034		1.82	27
Dipterocarpus	*turbinatus*	Telya garjan	Dipterocarpaceae	m3	Total volume over bark	Volume = 0.000390878 + 0.00064549776 * (DBH)^(2) + 0.0001478277 * (DBH) * (Height) + 0.00002407 * (DBH)^(2) * (Height)	436	0.69			1.47	27
Falcataria	*moluccana*	Moluccna	Leguminosae	m3	Stem Volume over bark	Log (Volume) = -8.9942 + 1.4963 * Log (DBH) + 1.1461 * Log (Height)	343	0.932			0.0141	5
Falcataria	*moluccana*	Moluccna	Leguminosae	m3	Stem Volume over bark	Log (Volume) = -10.707106 + 1.4963 * Log (GBH) + 1.1461 * Log (Height)	343	0.932				5
Hevea	*brasiliensis*	Rubber	Euphorbiaceae	m3	Total volume over bark	Log (Volume) = -10.5628 + 2.1502 * Log (GBH)	583	0.95				12
Hevea	*brasiliensis*	Rubber	Euphorbiaceae	m3	Total volume over bark	Log (Volume) = -11.2768 + 1.8795 * Log (GBH) + 0.6928 * Log (Height)	583	0.97				12
Hevea	*brasiliensis*	Rubber	Euphorbiaceae	m3	Total volume under bark	Log (Volume) = -10.6451 + 2.1607 * Log (GBH)	583	0.95				12

Hevea	*brasiliensis*	Rubber	Euphorbiaceae	m3	Total volume under bark	Log (Volume) = -11.3509 + 1.8930 * Log (GBH) + 0.6848 * Log (Height)	583	0.97	12
Hevea	*brasiliensis*	Rubber	Euphorbiaceae	m3	Total volume over bark	Log (Volume) = -10.4946 + 2.1365 * Log (GBH)	388	0.93	12

Genus	Species	Local name	Family	Unit e of Y	Vegetation Component	Equation	Sampl e size	R2	Ref . No
Hevea	*brasiliensis*	Rubber	Euphorbiaceae	m3	Total volume over bark	Log (Volume) = -11.355075 + 1.90505 * Log (GBH) + 0.67956 * Log (Height)	388	0.96	12
Hevea	*brasiliensis*	Rubber	Euphorbiaceae	m3	Total volume under bark	Log (Volume) = -10.58495 + 2.14861 * Log (GBH)	388	0.93	12
Hevea	*brasiliensis*	Rubber	Euphorbiaceae	m3	Total volume under bark	Log (Volume) = -11.43443 + 1.92013 * Log (GBH) + 0.670876 * Log (Height)	388	0.96	12
Hevea	*brasiliensis*	Rubber	Euphorbiaceae	m3	Total volume over bark	Volume = 0.01097 - 0.00064 * (GBH) + 0.000055 * (GDB)^(2)	195	0.96	12
Hevea	*brasiliensis*	Rubber	Euphorbiaceae	m3	Total volume under bark	Volume = 0.016931 - 0.00085 * (GBH) + 0.000055 * (GBH)^(2)	195	0.96	12
Acacia	*auriculiformis*	Akashmoni	Leguminosae	m3	Timber volume over bark	Volume = 0.027119694 + 0.00000240953 * (GBH)^(2) * (Height)	141	0.942	33
Acacia	*auriculiformis*	Akashmoni	Leguminosae	m3	Timber volume over bark	Volume = 0.02059085 + 0.00000257258 * (GBH)^(2) * (Height)	68	0.929	33
Acacia	*auriculiformis*	Akashmoni	Leguminosae	kg	Branch and stem less than 30 cm girth to 10 cm girth green biomass	Green biomass = 17.17526 + 0.011026 * (GBH)^(2)	68	0.857	33
Eucalyptus	*camaldulensis*	Eucalyptus	Myrtaceae	m3	Timber volume over bark	Volume = 0.003083594 + 0.00000291538 * (GBH)^(2) * (Height)	117	0.974	33
Eucalyptus	*camaldulensis*	Eucalyptus	Myrtaceae	m3	Timber volume over bark	Volume = 0.005034521 + 0.00000269095 * (GBH)^(2) * (Height)	60	0.96	33
Eucalyptus	*camaldulensis*	Eucalyptus	Myrtaceae	m3	Total volume over bark	Volume = 0.076339 - 0.00058066 * (Height) + 0.000016216 * (GBH)^(2) + 0.0000032565 * (GBH)^(2) * (Height)	94	0.978	33
Eucalyptus	*camaldulensis*	Eucalyptus	Myrtaceae	m3	Timber volume over bark	Volume = 0.00444242 + 0.00000274348 * (GBH)^(2) * (Height)	94	0.975	33
Acacia	*mangium*	Mangium	Fabaceae	m3	Total volume over bark	Volume = 0.0379401 - 0.0027469 * (GBH) + 0.000099945 * (GBH)^(2)	44	0.935	33
Acacia	*mangium*	Mangium	Fabaceae	m3	Total volume over bark	Volume = 0.01368013 - 0.00018226 * (Height) + 0.000005503 * (GBH)^(2) + 0.00000352188 * (GBH)^(2) * (Height)	44	0.971	33
Acacia	*mangium*	Mangium	Fabaceae	m3	Timber volume over bark	Volume = 0.047423 - 0.00387 * (GBH) + 0.000109 * (GBH)^(2)	37	0.908	33
Acacia	*mangium*	Mangium	Fabaceae	m3	Total volume over bark	Volume = -0.04085 + 0.00437656 * (Height) + 0.0000627199 * (GBH)^(2) + 0.00000248335 * (GBH)^(2) * (Height)	159	0.965	33
Acacia	*mangium*	Mangium	Fabaceae	m3	Timber volume over bark	Volume = 0.010632025 + 0.00000289124 * (GBH)^(2) * (Height)	133	0.95	33
Dalbergia	*sissoo*	Sissoo	Fabaceae	m3	Total volume over bark	Volume = 0.012282107 + 0.00168945 * (Height) - 0.000019455 * (GBH)^(2) + 0.00000392037 * (GBH)^(2) * (Height)	80	0.972	33

Genus	Species	Local name	Family	Unite of Y	Vegetation Component	Equation	Sample size	R2	Ref. No
Dalbergia	*sissoo*	Sissoo	Fabaceae	m3	Total volume under bark	Log (Volume) = -12.14678171 + 2.49978991 * Log (GBH)	181	0.973	30
Dalbergia	*sissoo*	Sissoo	Fabaceae	m3	Total volume under bark	Log (Volume)= -11.8405276 + 2.07000287 * Log (GBH) + 0.6152993 * Log (Height)	181	0.982	30
Swietenia	*macrophylla*	Mahogany	Meliaceae	m3	Total volume under bark	Log (Volume) = -12.045383 + 2.460647 * Log (GBH)	120	0.979	30
Swietenia	*macrophylla*	Mahogany	Meliaceae	m3	Total volume under bark	Log (Volume) = -11.716535 + 2.084968 * Log (GBH) + 0.534389 * Log (Height)	120	0.99	30
Albizia	spp	Koroi	Mimosaceae	m3	Total volume under bark	Log (Volume) = -12.093533 + 2.463398 * Log (GBH)	103	0.931	30
Albizia	spp	Koroi	Mimosaceae	m3	Total volume under bark	Log (Volume) = -11.961135 + 1.967741 * Log (GBH) + 0.907724 * Log (Height)	103	0.947	30
Eucalyptus	*camaldulensis*	Eucalyptus	Myrtaceae	m3	Total volume under bark	Log (Volume) = -11.177929 + 2.297689 * Log (GBH)	151	0.94	30
Eucalyptus	*camaldulensis*	Eucalyptus	Myrtaceae	m3	Total volume under bark	Log (Volume) = -11.523307 + 1.911628 * Log (GBH) + 0.738982 * Log (Height)	151	0.955	30
Melia	*azadarach*	Bokain	Meliaceae	m3	Total volume under bark	Log (Volume) = -11.041653 + 2.1705 * Log (GBH)	143	0.935	30
Melia	*azadarach*	Bokain	Meliaceae	m3	Total volume under bark	Log (Volume) = -10.962743 + 1.888957 * Log (GBH) + 0.505435 * Log (Height)	143	0.951	30
Mangifera	*indica*	Am	Anacardiaceae	m3	Total over bark volume	Log (Volume) = -11.27269 + 2.24506 * Log (GBH)	343	0.975	32
Mangifera	*indica*	Am	Anacardiaceae	m3	Total over bark volume	Log (Volume) = -11.25377 + 1.96697 * Log (GBH) + 0.52237 * Log (Height)	343	0.981	32
Lannea	*coromandelica*	Badi	Anacardiaceae	m3	Total over bark volume	Log (Volume) = -11.519102 +2.01724 * Log (GBH) + 0.56356 * Log (Height)	87	0.971	32
Syzygium	*cumini*	Kalojam	Myrtaceae	m3	Total over bark volume	Log (Volume) = -11.24854 + 2.24804 * Log (GBH)	99	0.966	32
Artocarpus	*heterophyllus*	Kathal	Moraceeae	m3	Total over bark volume	Log (Volume) = -11.06320 + 2.18203 * Log (GBH)	119	0.97	32
Artocarpus	*heterophyllus*	Kathal	Moraceeae	m3	Total over bark volume	Log (Volume) = -10.99533 + 1.80823 * Log (GBH) + 0.68951 * Log (Height)	119	0.983	32
Albizia	spp	Koroi	Mimosaceae	m3	Total over bark volume	Log (Volume)= -11.50692 + 2.31757 * Log (GBH)	140	0.968	32
Albizia	spp	Koroi	Mimosaceae	m3	Total over bark volume	Log (Volume) = -11.19651 + 1.85690 * Log (GBH) + 0.67878 * Log (Height)	140	0.979	32
Swietenia	*macrophylla*	Mahogany	Meliaceae	m3	Total over bark volume	Log (Volume) = -11.46122 + 2.29592 * Log (GBH)	105	0.981	32
Swietenia	*macrophylla*	Mahogany	Meliaceae	m3	Total over bark volume	Log (Volume) = -11.27102 + 1.88064 * Log (GBH) + 0.64629 * Log (Height)	105	0.99	32
Azadirachta	*indica*	Neem	Meliaceae	m3	Total over bark volume	Log (Volume) = -11.33340 + 2.25814 * Log (GBH)	36	0.974	32
Azadirachta	*indica*	Neem	Meliaceae	m3	Total over bark volume	Log (Volume) = -11.42823 + 1.89235 * Log (GBH) + 0.71493 * Log (Height)	36	0.985	32
Aphanamixis	*polystachya*	Pitraj	Meliaceae	m3	Total over bark volume	Log (Volume) = -11.25645 + 2.25821 * Log (GBH)	105	0.973	32

Aphanamixis	*polystachya*	Pitraj	Meliaceae	m3	Total over bark volume	Log (Volume) = -11.25528 + 1.98544 * Log (GBH) + 0.47163 * Log (Height)	105	0.987	32
Albizia	*saman*	Rain tree	Mimosaceae	m3	Total over bark volume	Log (Volume) = -11.37623 + 2.26924 * Log (GBH)	153	0.981	32

Genus	Species	Local name	Family	Unit e of Y	Vegetation Component	Equation	Sampl e size	R2	MSE	RMS E	AIC	FI	Ref . No
Albizia	*saman*	Rain tree	Mimosaceae	m3	Total over bark volume	Log (Volume) = -11.31983 + 1.91118 * Log (GBH) + 0.63606 * Log (Height)	153	0.99					32
Breonia	*chinensis*	Kadam	Rubiaceae	m3	Total volume over bark	Log (Volume) = -10.4647 + 2.3911 * Log (DBH) + 0.6373 * Log (Height)	51	0.9906 5					10
Dipterocarpu s	*turbinatus*	Telya garjan	Dipterocarpacea e	m3	Total volume over bark	Log (Volume) = -9.5258 + 2.1229 * Log (DBH) + 0.5993 * Log (Height)	49	0.9666 7					10
Lagerstroemi a	*speciosa*	Jarul	Lythraceae	m3	Total volume over bark	Log (Volume) = -9.6744 + 2.1065 * Log (DBH) + 0.6675 * Log (Height)	74	0.9862 8					10
Xylia	*xylocarpa*	Lohakat	Leguminosae	m3	Total volume over bark	Log (Volume) = -9.4303 + 2.0988 * Log (DBH) + 0.6042 * Log (Height)	94	0.9872 3					10
Shorea	*robusta*	Sal	Dipterocarpacea e	m3	Total volume over bark	Log (Volume) = -10.0253 + 2.1163 * Log (DBH) + 0.7588 * Log (Height)	79	0.9878 7					10
Sonneratia	*apetala*	Keora	Lythraceae	m3	Total volume over bark	Log (Volume) = -8.66152 + 1.5856 * Log (DBH) + 0.77152 * Log (Height)	91	0.98					9
Sonneratia	*apetala*	Keora	Lythraceae	m3	Total volume over bark	Log (Volume) = -9.29715 + 1.70514 * Log (DBH) + 0.95088 * Log (Height)	236	0.98					9
Sonneratia	*apetala*	Keora	Lythraceae	m3	Total volume over bark	Log (Volume) = -9.23507 + 1.69673 * Log (DBH) + 0.92309 * Log (Height)	133	0.98					9
Sonneratia	*apetala*	Keora	Lythraceae	m3	Total volume over bark	Log (Volume) = -8.75215 + 1.75034 * Log (DBH) + 0.64233 * Log (Height)	214	0.92					9
Aegiceras	*corniculatu m*	Khulshi	Myrsinaceae	kg	Leaf	Log 10 (Oven-dried biomass) = 0.76 * Log 10 ((DBH^(2)) - 1.39	29	0.93	0.02	0.12	-119.05	0.0 5	38

Aegiceras	*corniculatum*	Khulshi	Myrsinaceae	kg	Bark	Log 10 (Oven-dried biomass) = 1.04 * Log 10 ((DBH^(2)) - 1.80	29	0.99	0.004	0.07	-154.68	0.02	38
Aegiceras	*corniculatum*	Khulshi	Myrsinaceae	kg	Stem without bark	Log 10 (Oven-dried biomass) = 1.04 * Log 10 ((DBH^(2)) - 0.99	29	0.99	0.004	0.07	-154.68	0.14	38
Aegiceras	*corniculatum*	Khulshi	Myrsinaceae	kg	Total above-ground	Sqrt (Oven-dried biomass) = 0.48 * DBH - 0.13	29	0.99	0.03	0.18	-96.57	0.68	38
Aegiceras	*corniculatum*	Khulshi	Myrsinaceae	g	Total above-ground	Sqrt (Nitrogen) = 0.67 * DBH + 0.11	29	0.99	0.09	0.31	-66.27	1.85	38
Aegiceras	*corniculatum*	Khulshi	Myrsinaceae	g	Total above-ground	Sqrt (Phosphorus) = 0.94 * DBH + 0.08	29	0.98	0.19	0.43	-45.94	3.69	38
Aegiceras	*corniculatum*	Khulshi	Myrsinaceae	g	Total above-ground	Sqrt (Potassium) = 1.06 * DBH - 0.18	29	0.99	0.17	0.41	-49.25	3.66	38
Aegiceras	*corniculatum*	Khulshi	Myrsinaceae	kg	Total above-ground	Sqrt (Carbon)= 0.33 * DBH - 0.09	29	0.99	0.02	0.13	-177.67	0.39	38

References

[1] F 2010 The Food and Agriculture Organization of the United Nations *Global Forests Resource Assessment*

[2] Altrell D, et al. 2007 National forest and tree resources assessment: 2005 Bangladesh Forest Department, Ministry of Environment and Forests *Bangladesh Space Research and Remote Sensing Organization, Ministry of Defense and Food and Agriculture Organization of the United Nations* **5** 116

[3] Golley B F, et al. 1975 *Mineral Cycling in a Tropical Moist Forest Ecosystem* University of Georgia Press. Athens

[4] Mahmood H 2014 Carbon pools and fluxes in Bruguiera parviflora dominated naturally growing mangrove forest of Peninsular Malaysia *Wet. Ecol. Magt.* **22**(1) 15

[5] Ketterings Q M, et al. 2001 Reducing uncertainty in the use of allometric biomass equations for predicting above-ground tree biomass in mixed secondary forest *For. Ecol. Magt.* **146** 199

[6] Mahmood H, et al. 2004 Allometric relationships for estimating above and below-ground biomass of saplings and trees of Bruguiera parviflora (Wight and Arnold) *Malaysia App. Biol.* **33**(1) 37

[7] Mahmood H, et al. 2012 Allometry, above-ground biomass and nutrient distribution in *Ceriops decandra* (Griffith) Ding Hou dominated forest types of the Sundarbans mangrove forest, Bangladesh *Wet. Ecol. Magt.* **20** 539

[8] Komiyama A, et al. 2005 Common allometric equations for estimating the tree weight of mangroves *J. Trop. Ecol.* **21** 471

[9] Komiyama A, et al. 2008 Allometry, biomass, and productivity of mangrove forest: a review *Aqua. Bot.* **89** 128

[10] Morgan W B and Moss P A 1985 Biomass energy and urbanisation: commercial factors in the production and use of biomass fuels in tropical Africa *Biomass* **6** 285

[11] Bombelli A, et al. 2009 Assessment of the status of the development of the standards for the Terrestrial Essential Climate Variables: Biomass *Food and Agriculture Organization – Global Terrestrial Observation System* **18**

[12] Henry M, et al. 2011 Estimating tree biomass of sub-Saharan African forests: a review of available allometric equations *Silva Fennica* **45**(3B) 477

[13] Birigazzi L, et al. 2015 Toward a transparent and consistent quality control procedure for tree biomass allometric equations *Xiv World Forestry Congress, Durban, South Africa* **7** 11

[14] Sandeep S, et al. 2014 *Inventory of volume and biomass tree allometric equations for South Asia* Food & Agriculture Organization of the United Nations, Rome, Italy

[15] Akhter M, et al. 2013 *Tree volume and biomass allometric equations of Bangladesh* FD and FAO, Dhaka, Bangladesh

[16] Mahmood H, et al. 2015 Allometric models for biomass, nutrients and carbon stock in Excoecaria agallocha of the Sundarbans, Bangladesh *Wet. Ecol. Magt.* **23** (4) 765

[17] Rahman M M, et al. 2015 Carbon stock in the Sundarbans mangrove forest: spatial variations in vegetation types and salinity zones *Wet. Ecol. Magt.* **23** (2) 269

[18] Mahmood H, et al. 2016 Above-ground biomass, nutrients and carbon in Aegiceras corniculatum of the Sundarbans *Open J. For.* **6** (2) 72

Influences of increased daily repeated upstream releases and varying meteorological conditions on temperature distributions in a river-reservoir system

G Chen [1] and X Fang [1, 2, 3]

[1] Department of Civil Engineering, Auburn University, Auburn, Alabama, USA
[2] Beijing Cooperative Innovation Research Center on Architectural Energy Saving and Emission Reduction, Beijing University of Civil Engineering and Architecture, Beijing, China

E-mail: xzf0001@auburn.edu

Abstract. Temperature distribution in a river-reservoir system was simulated using a calibrated three-dimensional Environmental Fluid Dynamics Code model under various hypothetical weather conditions and daily repeated large releases (DRLRs) from the upstream boundary. Both DRLRs and weather conditions affect and control the formation and spread of density currents and then affect the bottom-layer temperatures. The DRLRs with longer durations (e.g., 6 or 8 hours) can relatively quickly push cooler release water to the Gorgas upstream monitoring station (GOUS) and the river intake. With the air temperature drops in the first 6 days, simulated bottom temperatures at GOUS for 6- and 8-hr DRLRs are lower than one under 4-hr DRLR, but relatively larger bottom-layer temperature drops only primarily occur during the air-temperature drop and rise period. The release with larger flow rate can also maintain the cooler water temperature downstream. Releasing the same amounts of water, with different release durations and flow rates, has a very similar effect on the downstream water temperatures.

1. Introduction

Water temperature is one of the significant and important drivers of stream ecosystems. Water temperature affects all biological and chemical reactions and the density of water that influences the transport of water and pollutants in aquatic systems [1]. Human civilization and domination (through dam operation, industrial production, cropping, deforestation, etc.) would alter water temperatures in rivers [2]. Under the natural processes, many climate parameters can also possibly affect water temperature in streams/rivers, for example, solar radiation, relative humidity, wind speed [3]. Weather condition is one of the most important physical parameters that affects water temperature in aquatic systems. Harmeson and Schnepper [4] indicated graphically that water temperatures in rivers follow closely the pattern of variations in daily mean air temperatures.

The alteration of flow in the river can also be responsible for changes in river water temperature [5]. The reservoir release is a common operating procedure for regulated rivers to manage the water resources. The releases to meet hydro-electric power demands can affect the flow dynamics in downstream river. For the summer period or warm season, minimum release from upstream reservoir is typically required to maintain lower downstream temperature in relatively shallow rivers for environmental protection [6]. It is common that the upstream release results in rapid stage changes (such as more than 1 m) in the downstream river [7, 8]. The large release from a large dam results in wave propagations that exist more than 100 km downstream [9]. Besides the momentum and mixing effect from the upstream large release, atmospheric conditions will play an important role in the river temperature variations after the reservoir release.

The released temperature from stored water is much colder than the temperature in the shallow river downstream. Hence, many river-reservoir systems develop complex density currents in the downstream river/reservoir that has the large diurnal variations in atmospheric heating rates. A density current occurs when the density of the water flowing into a water body is greater than the density of the ambient water. For the traditional or classic gravity current, it's driven by the density difference. In the natural rivers and reservoirs, a density current is kept in motion by the force of gravity acting on differences in density, the slope of river and reservoir bottom (gravity), and the momentum effect from a large release that plays a role in density current propagation.

Several studies about density currents have been studied in situ [10]. The density currents are also investigated in the laboratory [11]. Using various simplifying assumptions, researchers dealt with sloping channels with rectangular cross sections to develop analytical models with laboratory data [12]. Recently, three-dimensional (3D) Environmental Fluid Dynamics Code (EFDC) [13] has been widely used in modelling hydrodynamics and transport processes in rivers, lakes, reservoirs, and estuaries. Hamrick and Mills [14] developed and used the EFDC model to simulate thermal transport and water temperature distributions in the Conowingo Pond that was influenced by thermal discharges from the Peach Bottom atomic power plant.

In this study, we focused on combined or integrated effects of variable weather conditions, and larger discharge of daily repeated large releases (DRLRs), on the downstream water temperatures. We examined simulated temperature distributions under different types of weather scenarios to evaluate site-specific correlation with weather conditions. The results of this study are important with regard to water quality modelling and management, and habitat assessment in rivers.

2. Study area

The study deals a river-reservoir system in Alabama (AL), USA, which is 124.2 river km long. For the modelling study of water temperature distributions in the system, the upstream boundary is the Smith Dam Tailrace (SDT), and the downstream boundary is Bankhead Lock& Dam (BLD). The system includes Sipsey Fork, the lower Mulberry Fork, and a reservoir segment of Bankhead Lake. Since Sipsey Fork and the lower Mulberry Fork are the riverine portions of Bankhead Lake, the study area is referred to as the Bankhead river-reservoir system (BRRS) in the paper. BRRS receives inflows from four tributaries: Upper Mulberry, Locust Fork, Lost Creek, and Blackwater Creek. A power plant is located at the bank of the upstream portion of BRRS and has a river intake that withdraws cooler water from two bottom layers in the Black Warrior River. One of the monitoring stations is called GOUS, 5.58 km upstream of the power plant or 64 km downstream from SDT. The average bottom slope of BRRS is 0.014% with bottom elevations changing from 65.8 to 77.2 m. The water surface elevation in BRRS depends on SDT's large water releases, flows from its tributaries, and the water surface elevation in BLD. The BLD elevation is typically influenced by outflow through hydro turbines, spillage through gates of BLD, and loss of water through the Bankhead navigation lock.

3. Model development and boundary conditions

The EFDC model is a general purpose modelling package that can be configured to simulate one-, two- and three-dimensional flow, transport, and biogeochemical processes in various surface water systems including rivers, lakes, estuaries, reservoirs, wetlands, and coastal regions [15-17]. Details of governing equations and numerical schemes for EFDC hydrodynamic models are given by Hamrick [13]. The EFDC model for BRRS was designed to model or simulate temperature distributions along the Black Warrior River and along depth. Therefore, temperature model component in EFDC was activated to model thermal or heat budget for each computational cell. The heat input is from the upstream release water. It's the source of the lower water temperature which can make the downstream temperature lower. The measured upstream release temperatures varying with time directly link to the upstream computation cells. At the downstream boundary of the study area, water (heat) flux leaves the control volume. The model has the surface heat exchange between the atmosphere and the water surface. The heat exchange with the sediment is also applied around the river bed.

Using hydrographic bathymetry data, the EFDC model for the BRRS is developed. The simulation domain of the model has a total of 6974 horizontal grids and 10 horizontal layers along the depth direction for each grid. The grid size DX in transverse direction along the river ranged from 9.5 m to 189.8 m and DY in longitudinal flow direction ranged from 10.0 m to 277.1 m. Average grid sizes DX and DY are 25.7 m and 100.9 m, respectively.

The model of BRRS used the upstream boundary at the tailrace of Smith Dam and the downstream boundary at Bankhead L&D. There are two types of water releases from Smith Dam: (1) more or less constant continuous release (2.83 m^3/s) with almost 9.6 °C to support the downstream environment and ecosystem, (2) intermittent large releases from hydro-turbine units of Smith Dam in order to meet peak electric generation demand, which were about 4–5 °C higher than constant release temperature. For the downstream boundary, 15-minute water surface elevations are used for the BRRS. The typical water surface elevation at downstream is 77.6 m.

The atmospheric boundary condition was meteorological data from the Birmingham regional airport obtained from NOAA's Southeast Regional Climate Center (SERCC), which is about 40 km east of the study area. The data included hourly air temperature, atmospheric pressure, wind speed, wind direction, rainfall, and cloud cover. Required solar radiation data were obtained from Cleveland, AL, which is the closest Auburn Mesonet station from the study area.

4. Previous model application results

In the first study [7], a three-dimensional hydrodynamic and temperature EFDC model was calibrated with 2011 observed data to understand flow dynamics and temperature variations in BRRS under observed irregular upstream releases and varying meteorological conditions recorded in Birmingham. Comparing observed data, the simulation results show good performance for simulated water surface elevations, discharge, velocity, and temperatures at different locations. Especially for bottom temperature, the EFDC model predicted the magnitude and duration of these temperature drops with reasonable accuracy. It is essential that the EFDC model can predict the bottom temperatures because they are directly related to how the model simulates the density currents along the river bottom, which is the foundation for the next step.

Each flow release from SDT promoted and enhanced the movement of density currents moving from upstream towards downstream. In order to understand the cause/effect of upstream release on downstream bottom temperature in BRRS, all boundary conditions except the upstream release were fixed as constant in the second and third phases of the study [18]. Hourly varying climate variables under hypothetical constant weather condition were used as atmospheric boundary and based on data from a relatively warmer day in 2011, which have no cooling or warming trend, called as constant weather in this study, but have hourly variations in each day (Figure 1). Simulation results show that the density currents in BRRS are more complex than classic density currents and have discontinuous propagations. The density currents form at different reaches, and are destroyed at upstream locations due to the flow momentum of the releases, then form again due to solar heating. The propagation of density current is affected by the different durations of large releases and atmospheric conditions. Overall average surface and bottom temperatures are lower for longer duration of DRLRs [18].

5. Model scenarios

5.1. Durations of DRLR

After analyzing intermittent releases of 2010 and 2011 from SDT, and the statistical summary of the releases from SDT shows that both average and median releases were about 140 m^3/s and the releases typically started around 1:00 pm [7], additional data analysis was also performed to determine what possible durations of DRLRs from SDT could be. Table 1 shows calculated release durations of DRLRs from 2010 to 2014 if the release discharge at SDT is set at 140 m^3/s for each release. The available release data are about 184 days, mostly from May 1 to October 31 in each year. Calculated daily release duration ranges from 3.48 to 7.91 hours. Most release durations are from 4 to 6 hours, which will be used for the model scenario runs in this study. A few scenario runs will have the durations of 8-hr to generate comparative results, and the 8-hr DRLRs are highly possible in a relative warmer year, e.g., 2013 (table 1).

Table 1. Calculated daily upstream release durations of 140 m^3/s discharge based on 2010 to 2014 release data.

Year	Data period (Julian Day)	Recorded Total Release Volume (m^3)	Calculated Duration for DRLR (hr)
2010	151 to 333	359,557,358	3.48
2011	120 to 303	370,758,545	3.59
2012	121 to 304	443,569,903	4.39
2013	120 to 303	763,621,151	7.91
2014	120 to 303	631,369,260	6.45

5.2. Climate scenarios

Two types of meteorological scenarios were used in this study in order to identify weather impacts on density current movements: (1) constant weather (figure 1), and (2) hypothetical cooling and then warming weather (figure 2). For the constant weather scenario, the daily maximum and minimum air temperatures are 33.3 and 21.1 °C (figure 1). The hourly air temperatures were calculated from the daily maximum and minimum temperatures applying the sinusoidal wave function model [19]

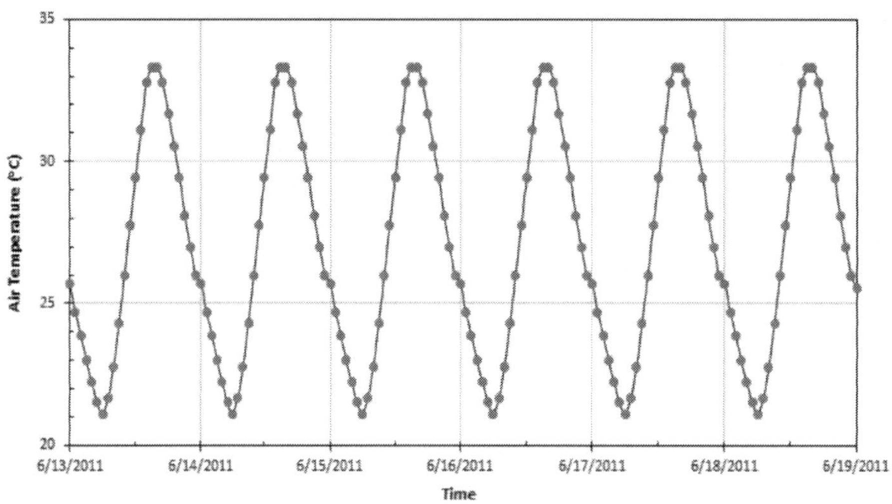

Figure 1. Six-day (June 13–18) example time-series of hourly varying air temperatures under hypothetical constant weather condition.

The calibrated EFDC model provided simulated water surface elevation in different cross sections and temperature, velocity and discharge in 10 layers (depths) for all grids. How do surface and bottom water temperatures at the downstream locations of BRRS drop or increase with the changes of weather conditions such as cooling and warming trend over a few days? To mimic typical weather variations in the Birmingham area [20], a new hypothetical weather scenario was proposed and developed: air temperature and solar radiation have a 6-day drop (June 13 - 18) and then a 6-day rise (June 18 - 23), and the constant weather (figure 1) was used before and after the drop and rise period (figure 2). This is an 11-day drop and rise period (June 13 - 23). During the drop and rise period, the air temperature has a drop or rise of 2 °C each day. The solar radiation has the same drop or rise pattern (trend) as the air temperature does, and it is assumed to be 10% change of the solar radiation under the constant weather condition each day.

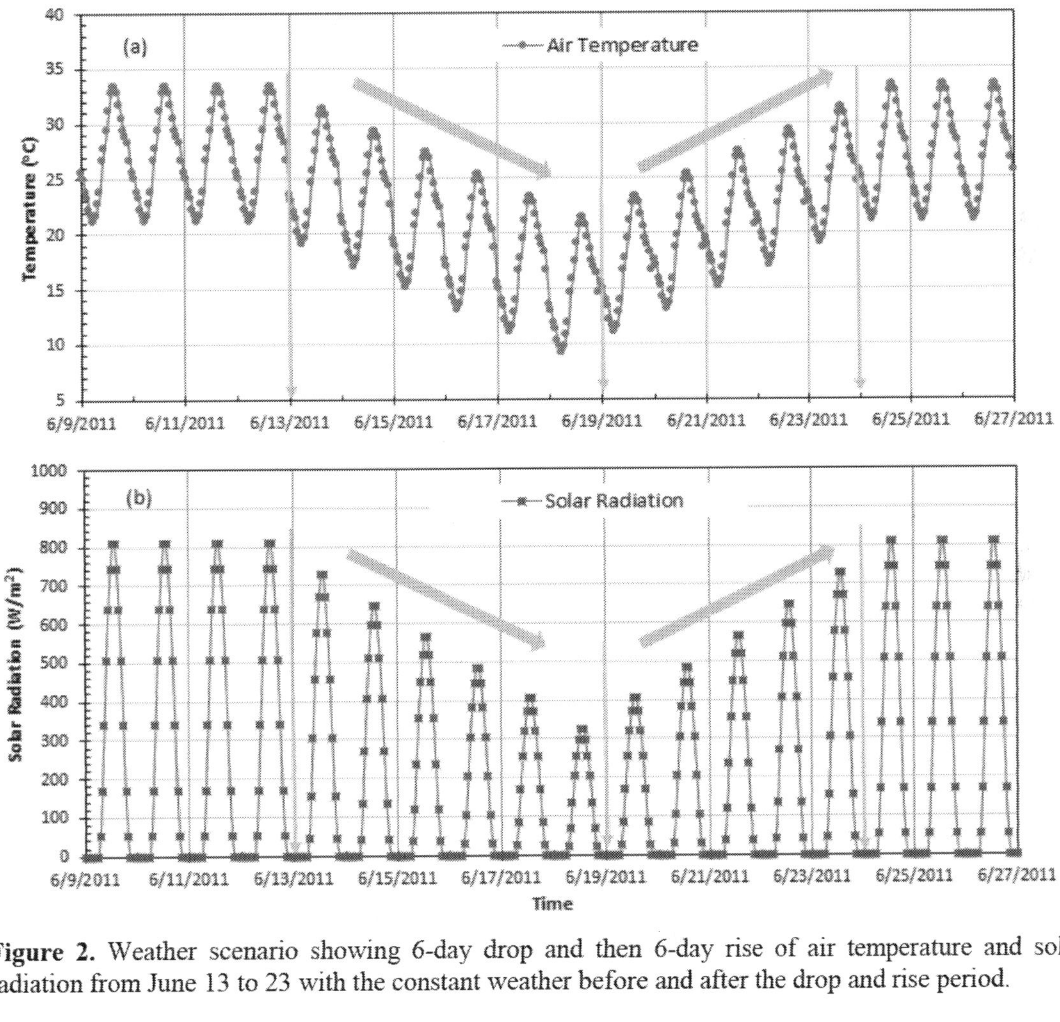

Figure 2. Weather scenario showing 6-day drop and then 6-day rise of air temperature and solar radiation from June 13 to 23 with the constant weather before and after the drop and rise period.

6. Results and discussion

6.1. Under two weather scenarios

Railsback (1997) concluded that significant saving can be realized using release patterns adapted to changing atmospheric conditions when the downstream river of the release is relatively shallow. Figure 3 shows time series of simulated surface- and bottom-layer water temperatures for 4-hr DRLR under two weather scenarios: the constant weather and the 11-day drop and rise. Under the constant weather, daily mean bottom and surface temperatures are almost constant for 4-hr DRLR from June 4 to July 18, and surface temperatures have 4.2 °C diurnal fluctuation when there is 12.2 °C variation of daily air temperature (figure 1).

For the first 6-day drop period, the surface temperature begins to drop quickly when the air temperature begins to drop from June 13. However, the bottom temperature shows a 3-day delay. After air temperature drops 4 days, the surface and bottom temperatures are well mixed for about 8 hours in each day (about 11 hr after the release) and continuously drop for 5 days. The surface temperature has much large diurnal variations that range from 3.8 to 9.0 °C (average of 6.9 °C and standard deviation of 1.8 °C) under the same 12.2 °C variation of daily air temperature. Under the 6-day rise period, when the air temperature starts to increase on June 19, the surface and bottom temperatures begin to increase with a 3-day delay. The surface temperature at GOUS rises quickly from 21.0 °C to 26.9 °C (constant daily-mean surface temperature under the constant weather condition) in 15 days. However, the bottom temperature takes much longer time to increase to the stable temperature simulated under the constant weather condition. Because of air temperature dropping and rising in 11 days, the bottom temperature can maintain a cooler temperature for about 32 (June 17 to July 19) days. Overall, water temperature at GOUS increases or drops corresponding to the air temperature change, but it can maintain the cooler condition for a longer period, especially for bottom layers.

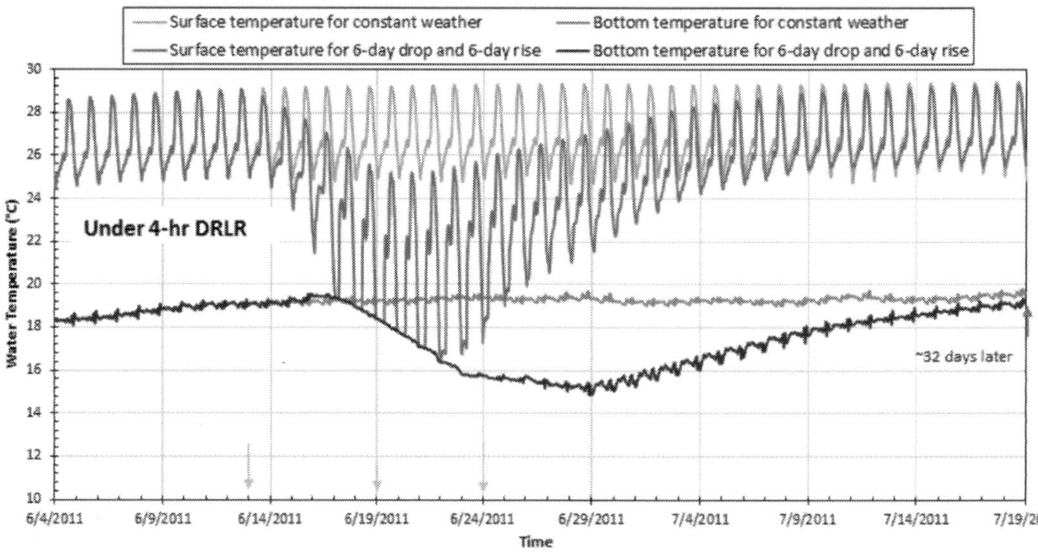

Figure 3. Time series of simulated surface and bottom water temperatures at GOUS for 4-hr DRLR under two weather scenarios: constant weather and the 11-day drop and rise.

6.2. DRLR duration changes during the drop and rise period

If more water is released during the cooler weather condition, could downstream temperatures become lower? Figure 4 shows time series of simulated bottom-layer water temperatures at the river intake under the 6-day drop and 6-day rise weather scenario when DRLRs with 140 m^3/s at SDT over the 11 days last for 4, 6, and 8 hours. The bottom-layer water temperatures at the river intake on June 13 (just before air temperature drop starts) are 19.3 °C. There are 4-hr DRLRs and the constant weather before and after the 11-day (June 13 to 23) air temperature drop and rise period. Comparing with the 4-hr DRLR for the whole period under constant weather, the scenario for 4-hr DRLR during 11-day drop and rise period has 12 °C air temperature drop, and solar radiation drops from June 13 to 19 (figure 2), and then increase on both parameters. The maximum and average daily mean differences during the cooler period (June 15 to July 9) between these two scenarios are 4.0 and 1.6 °C, respectively, which shows the benefit of air temperature drop on reducing the bottom-layer temperatures. For the 4-hr, 6-hr and 8-hr DRLRs during the drop and rise period, the bottom-layer temperature at the river intake begins to drop after 2 days of delay when it compares temperature time-series under the constant weather. The bottom-layer temperature drop is larger for longer DRLR. The lowest bottom-layer water temperatures at the river intake during the 11-day drop and rise period under 4-hr, 6-hr and 8-hr DRLRs are 15.4 °C, 13.9 °C and 13.5 °C on June 24, 22 and 21 with the cooler duration 25 days (June 15 – July 9), 31 days (June 15 – July 15), and 31 days (June 15 – July 15), respectively.

To see the increased duration effects with cooler weather condition, the differences of daily mean temperatures for 6-hr, 8-hr DRLRs comparing with 4-hr DRLR were also calculated. The maximum daily mean differences for 6-hr and 8-hr DRLR scenarios are 2.23 °C and 3.13 °C, respectively. Over the cooler duration (June 15 – July 15), the average differences under 6-hr and 8-hr DRLRs are 0.6 °C and 0.9 °C in comparison to the 4-hr DRLR, respectively. Therefore, the large releases with longer durations (e.g., 6 or 8 hours) can relatively quickly push cooler release water to the river intake. With the air temperature drop in the first 6 days, simulated bottom temperatures at the river intake for 6-hr and 8-hr DRLRs are lower than ones under 4-hr DRLR, but relatively larger bottom-layer temperature drops primarily occur during the air-temperature drop and rise period. During or after the rise period, the bottom-layer temperatures are almost the same for 6-hr and 8-hr DRLRs and show similar patterns.

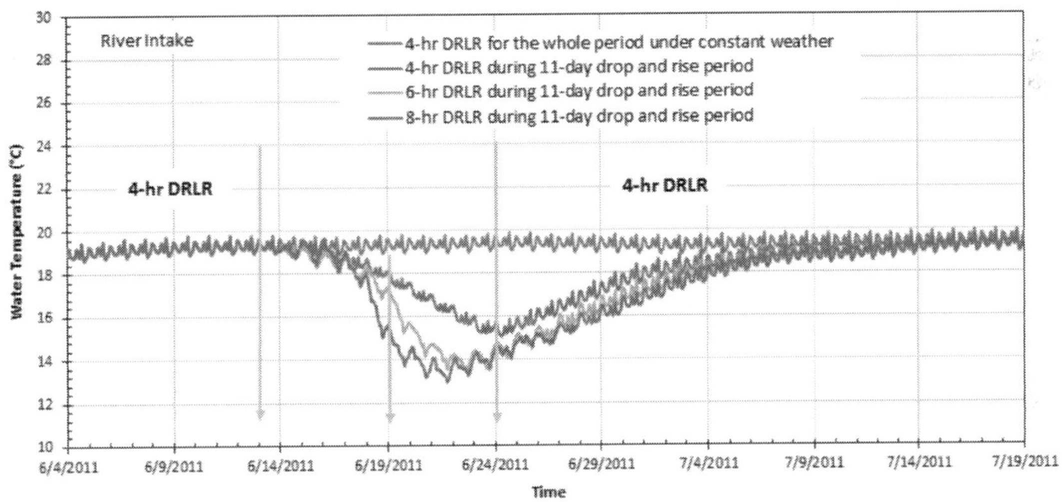

Figure 4. Time series of simulated bottom temperatures at the river intake when the DRLR release lasts for 4-hr, 6-hr and 8- from June 13 to 23 under 11-day drop and rise weather scenarios. There are 4-hr DRLRs and the constant weather before and after the air temperature drop and rise period.

6.3. Increase release flow rate scenario

6.3.1. Under constant weather. For the above scenario runs, the release flow rate from SDT is fixed as 140 m^3/s with different durations because, for example, 50% of 2011 recorded releases ranged from 137.3 to 142.1 m^3/s with a median of 138.8 m^3/s [7]. Under the actual release, there are some releases up to about 280 m^3/s in 2011. Figure 5 shows the time series of simulated bottom temperature at GOUS with increased release of 280 m^3/s starting from June 13 and lasting 2, 4 and 6 days under constant weather. The release duration is 4 hours each day. Due to the 280 m^3/s DRLR for 2, 4 and 6 days, simulated bottom temperature at GOUS increases from 19.1 °C to 20.2 °C in one day and keeps more or less constant for about 3 days due to the large flow momentum. Then, the bottom temperature decreases and maintains the cooler temperature from June 17 to July 13. It seems these 2, 4, and 6 days of larger release (280 m^3/s), push more cooler water to GOUS and have the cooler bottom temperatures for 26 days in comparison to temperatures for 4-hr 140 m^3/s DRLR. Although there are 4 days with higher bottom temperature immediately after the larger release (figure 5). The lowest bottom temperatures resulted from 280 m^3/s 4-hr DRLR for 2, 4, and 6 days are 16.8, 16.0, and 15.9 °C, respectively. The daily mean bottom temperatures from June 18 to July 8 for 2 days of 280 m^3/s 4-hr DRLRs are 0.2 to 1.4 °C higher than temperatures for 4 days of 280 m^3/s 4-hr DRLRs. Overall, simulated bottom temperatures at GOUS for 4 and 6 days of 280 m^3/s 4-hr DRLRs are very similar (figure 5).

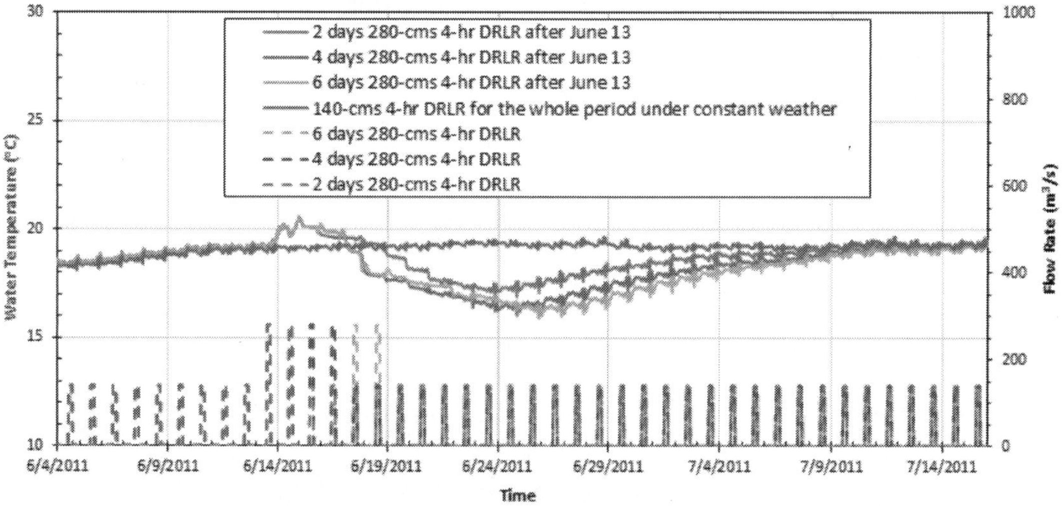

Figure 5. Time series of simulated bottom temperature at GOUS for 4-hr 140 m3/s DRLR with 4-hr 280 m^3/s DRLR starting from June 13 and lasting 2, 4 and 6 days under constant weather, including time series of release flow rate from upstream.

6.3.2. Simulation of release scenarios under drop and rise period. Can increased duration, or the flow rate, or both of the large releases make the downstream bottom temperature lower? Figure 6(a) shows time series of simulated bottom temperatures at GOUS under the 6-day drop and 6-day rise weather scenario under 4 release patterns, which are 4 combinations of DRLRs from two release flows (140 or 280 m^3/s) and two release durations (4 or 8 hours). There are 4-hr DRLRs and the constant weather before and after the air temperature drop and rise period (June 13 to 23). Same as Figure 4, longer and larger release will maintain the bottom temperature much longer and cooler. One interesting phenomenon is that the pattern of bottom temperatures is almost the same for 4-hr DRLR with 280 m^3/s and 8-hr DRLR with 140 m^3/s. It may be due to the same amount of water that was released for the two scenario runs. Due to the flow momentum, there are two days of the water temperature going up, which is an increase from about 19 to 20 °C. Afterwards, the bottom temperature decreases and maintains cooler temperatures for about 40 days. The scenario 1, 140 cms 4-hr DRLR, is a reference run because there is no change on the duration and the flow rate of the release before and after the drop and rise period. The minimum bottom temperature at GOUS is 15.2 °C occurring on June 29 under the scenario 1, and the bottom temperature increases to 19.2 °C on July 19.

For the other 3 scenarios, the minimum bottom temperatures at GOUS are 14.0, 13.8, and 12.9 °C occurring on June 20, June 20, and June 19, respectively. The bottom temperature gradually increases to 19.1, 19.1, and 18.9 °C on July 19, respectively. Figure 6(b) shows time series of differences of daily mean simulated bottom-layer temperatures between the reference scenario 1 and the other 3 scenarios. Figure 6(b) is derived from the data in figure 6(a). The positive differences indicate that the other 3 scenarios make the bottom-layer water temperatures at GOUS cool further. Although the bottom temperatures for 8-hr DRLR with 140 m^3/s are a little bit colder (on average -0.1 °C) than temperatures under the 4-hr DRLR with 280 m^3/s, overall, releasing the same amounts of water may have almost the same impact to maintain lower bottom-layer temperature downstream. Mean differences with standard deviations were calculated over the lower temperature periods (~35 days from June 15 to July 19), which are 0.7 (± 1.1) °C, 0.9 (± 1.0) °C and 2.0 (± 1.5) °C for the scenarios of 280 cms 4-hr, 140 cms 8-hr, and 280 cms 8-hr DRLRs. It should be mentioned that the scenario of 280 cms 8-hr DRLRs releases two times the amount of water in comparison to two scenarios of 280 cms 4-hr and 140 cms 8-hr DRLRs, and four times the amount of water in comparison to the reference scenario of 140 cms 4-hr DRLRs.

What is the optimal combination of release duration and flow rate? There is a maximum temperature at the river intake that we need to determine in order to meet downstream temperature constraints set by a regulatory agency. The releasing of more water during the drop and rise period has created a lower-bottom-temperature effect for ~35 days (figure 6(b)); it is still an efficient approach to make the bottom temperature at the river intake lower to compensate for potential temperature increase due to warmer weather after the air temperature drop period. More scenario runs and result analyses are still necessary to identify the optimal release pattern. Using weather forecasts and adjusted release schedules may be useful to the reservoir release management.

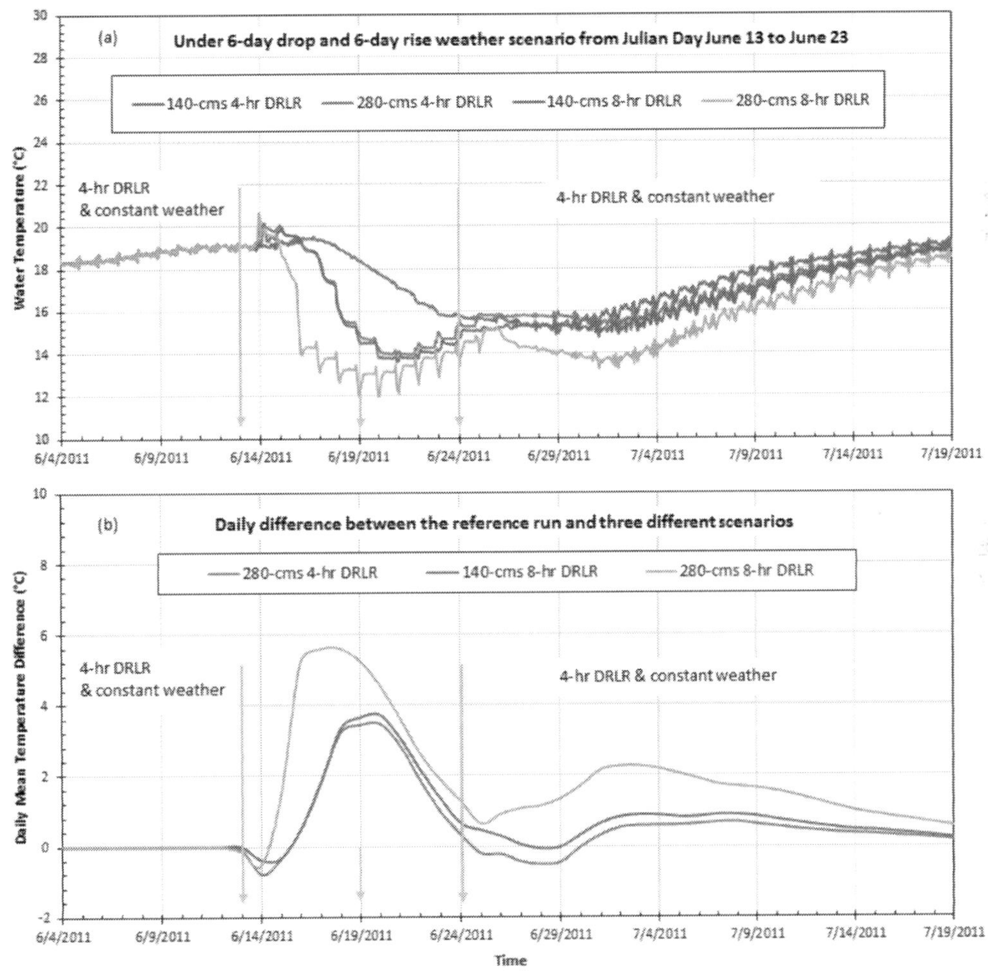

Figure 6. (a) Time series of simulated bottom temperatures at GOUS under the 6-day drop and 6-day rise weather scenario when 140-cms or 280-cms DRLRs last for 4 or 8 hrs (4 combinations). There are 4-hr DRLRs and the constant weather before and after the air temperature drop and rise period (June 13 to 23). (b) Difference of daily mean simulated bottom-layer temperatures between the reference scenario 1 and other 3 scenarios.

7. Conclusions

In this paper we applied a previously calibrated 3D EFDC model for a river-reservoir system to simulate temperature distributions under various hypothetical weather conditions and increased daily repeated large releases or DRLRs from upstream boundary. A series of model scenario runs were performed to further understand the bottom-layer water temperature changes at downstream locations of BRRS (e.g., GOUS and the river intake) corresponding to hypothetical meteorological changes under different release scenarios. All hypothetical weather and release scenarios were based on the data analysis of local weather conditions and actual release patterns.

Both the large releases and weather conditions affect the formation and spread of density currents and then bottom temperature distribution in BRRS. The large releases with longer durations (e.g., 6 or 8 hours) can relatively quickly push cooler release water to GOUS and the river intake. The air temperature drop of 12 °C in the first 6 days simulated maximum and average daily-mean differences during the cooler period (June 15 to July 9) in comparison to ones under the constant weather scenario are 4.0 and 1.6 °C, respectively. Daily mean simulated bottom temperatures at GOUS for 6- and 8-hr DRLRs are also on average 0.6 and 0.9 °C lower than ones under 4-hr DRLR, but relatively larger temperature drops only primarily occur during the air-temperature drop and rise period (figure 4). The release with larger flow rate can also maintain the cooler water temperature downstream. Releasing the same amounts of water have very similar effects to the downstream water temperatures (figure 6). More research is still necessary to identify the optimal release pattern.

References

[1] Thomann R V and Mueller J A 1987 *Principles of surface water quality modeling and control* (New York: Harper & Row)
[2] Caissie D 2006 The thermal regime of rivers: a review *Freshw. Biol.* **51** 1389-406
[3] Johnson A C, Acreman M C, Dunbar M J, Feist S W, Giacomello A M, Gozlan R E, Hinsley S A, Ibbotson A T, Jarvie H P and Jones J I 2009 The British river of the future: how climate change and human activity might affect two contrasting river ecosystems in England *Sci. Total Environ.* **407** 4787-98
[4] Harmeson R H and Schnepper V M 1965 Temperatures of surface waters in Illinois. In: *Illinois State Water Survey Report of Investigation* **49** p 45
[5] Sinokrot B A and Gulliver J S 2000 In-stream flow impact on river water temperatures *Journal of Hydraulic Research* **38** 339-49
[6] Consultants W-C 1986 *Stanislaus project instream temperature study* Pacific Gas and Electric Company, San Ramon, California
[7] Chen G, Fang X and Devkota J 2015 Understanding flow dynamics and density currents in a river-reservoir system under upstream reservoir releases *Hydrological Sciences Journal*
[8] Fang X, Weems T B, Devkota J and Chen G 2013 *Watershed modeling, water balance analysis, three-dimensional flow and thermal discharge modeling for the William C.* Gorgas Plant. Department of Civil Engineering, Auburn AL **368** p 124
[9] Petts G E 1984 Impounded rivers: perspectives for ecological management. Wiley, Chichester, UK **329** pp
[10] Dallimore C J, Imberger J and Ishikawa T 2001 Entrainment and turbulence in saline underflow in Lake Ogawara *Journal of Hydraulic Engineering* **127** 937-48
[11] Akiyama J and Stefan H G 1984 Plunging flow into a reservoir - Theory *Journal of Hydraulic Engineering ASCE* **110** 484-99
[12] Fernandez R L and Imberger J 2008 Time-varying underflow into a continuous stratification with bottom slope *Journal of Hydraulic Engineering* **134** 1191-8
[13] Hamrick J M 1992 A three-dimensional environmental fluid dynamics computer code: Theoretical and computational aspects. In: *Speical Report* **317** (Gloucester Point, VA: The College of William and Mary, Virginia Institute of Marine Science) p 63
[14] Hamrick J M and Mills W B 2000 Analysis of water temperatures in Conowingo Pond as

influenced by the Peach Bottom atomic power plant thermal discharge *Environmental Science & Policy* **3** 197-209

[15] Shen J and Lin J 2006 Modeling study of the influences of tide and stratification on age of water in the tidal James River *Estuarine, Coastal and Shelf Science* **68** 101-12

[16] Kim C-K and Park K 2012 A modeling study of water and salt exchange for a micro-tidal, stratified northern Gulf of Mexico estuary *Journal of Marine Systems* **96–97** 103-15

[17] Devkota J and Fang X 2014 Numerical simulation of flow dynamics in a tidal river under various upstream hydrologic conditions *Hydrological Sciences Journal*

[18] Chen G and Fang X 2015 Sensitivity analysis of flow and temperature distributions of density currents in a river-reservoir system under upstream releases with different durations *Water* **7** 6244-68

[19] Chen G and Fang X 2015b Accuracy of hourly water temperatures in rivers calculated from air temperatures *Water* **7** 1068-87

[20] Chen G and Fang X 2016 Influences of meteorological conditions and daily repeated upstream release on temperature distributions in a river-reservoir system *Journal of Hydro-environment Research*

IOP Conf. Series: Earth and Environmental Science 39 (2016) 012059 doi:10.1088/1755-1315/39/1/012059

Distribution selection for hydrologic frequency analysis using subsampling method

S Das[1, 2]

[1]College of Hydrometeorology, Nanjing University of Information Science and Technology (NUIST), Nanjing, China

E-mail: samirandas@gmail.com

Abstract. This paper investigates the potential utility of subsampling, a resampling technique with the aid of a goodness of fit test to select the best distribution for frequency analysis. Subsampling draws samples (of smaller size) from the original sample without replacement. The performance of the methodology is assessed by applying the methodology to an observed annual maximum (AM) hydrologic data series. Several AM discharge series of different record lengths are used as case studies to determine the performance. Overall, it is found that the methodology is suitable for a longer data series and a good performance can be obtained when the subsample size is around half of the underlying data sample. The methodology has also outperformed the standard AD test in terms of effectively discriminating between distributions. All results indicate that the subsampling technique can be a promising tool in discriminating between distributions.

1. Introduction

Hydrologic frequency analysis is concerned with the assessment of magnitudes (such as flood or heavy rainfall) of stated frequency or stated degree of rarity for use as input to the process of hydrologic risk assessment and management. The application of probability theory in a standard frequency approach requires the determination of a probability distribution and the estimation of its parameters. While several procedures related to parameter estimation, such as L-moments [1], provide satisfactory results in dealing with parameter estimation of a wide range of probability distributions, that has not been the case about the methods that select an appropriate probability distribution function. The selection of a suitable probability distribution for a particular gauged site is still a major problem in frequency analysis. The selection of a suitable distribution leads to a reduction in the error of quantile estimates. In hydrologic frequency analysis, different types of distributions have been considered in different contexts, and a list of such can be obtained from the WMO report [2]. The choice of such distributions is generally selected on the basis of graphical methods (such as a probability plot or L-moment ratio diagram) or by goodness of fit tests such as a chi-square test [3], Kolmogorov–Smirnov (KS) test [4], Anderson–Darling (AD) test [5,6], or probability plot correlation coefficient test [7]. These tests, however, have often been criticized for their inability to discriminate between statistical distributions for the same application [2,6]. At the same time, it has been acknowledged that GOF tests can sensibly be used in frequency analysis as the basis for rejecting some distributions, but not to select a best population distribution.

Thus, there is a need for alternative approaches to attempt further distinguishing between flood frequency distribution selections. In a more recent time, [8] presented an objective model selection criterion based on Akaike information, Bayesian Information and Anderson–Darling Criterion. However, as shown by the authors, the performance in the context of three-parameter distribution selection is not satisfactory. Other notable developments in the context of flood frequency analysis are the Bayesian hierarchical model and climate-informed methods [9,10].

The objective of this paper is to assess how subsampling [11], a data intensive resampling technique [12], can be used as a tool to overcome difficulties in selecting a best population distribution for a site with relatively long record length where the discrimination between statistical distributions is challenging. Subsampling draws samples (of smaller size) from the original sample without replacement. The reason the subsampling can be used in distribution selection is due to the fact that the subsamples are actually samples from the true unknown distribution, as was the case with the original sample. So, if the traditional methods failed to discriminate between distributions for the original sample, it is expected that a huge number of subsamples drawn from the original sample would provide a pool of data series that can effectively be used to determine an outright single choice for a gauged site. The study assesses the performance by applying the methodology to observed annual maximum (AM) flow data series. Several AM flow series of different record lengths pertaining to catchments located in different regions of U.K. are used as case studies.

2. Methodology

This section details the methodology applied in the study, which includes subsampling technique, statistical models and the proposed test procedure.

2.1. Subsampling

Subsampling falls within the broad area of resampling technique. Popular resampling techniques such as bootstrap [13] and subsampling [11] are data intensive simulation methods for statistical inference. They create an ensemble of datasets, each of which is drawn from the original sample of size n. In classical bootstrap, samples of size n are drawn with replacement while in subsampling, and samples of size b (b<n) are drawn without replacement. The reason the subsampling can be used in distribution selection is that the subsamples are actually samples (of smaller size) from the true distribution F. On the other hand, resamples (e.g. bootstrap) are samples from an estimator of F. Here, F is a probability (or cumulative) distribution function, $F(q) = prob(Q_i \le q), for \quad i = 1, \cdots, N$, where q is any real number. A number of B i.i.d. samples $\mathbf{Q}^{*(1)}, \cdots, \mathbf{Q}^{*(B)}$ (each of size b) are drawn from the sample population consisting of the observations $\{Q_1, \cdots, Q_N\}$. In subsampling terminology, these B i.i.d. samples are called subsamples. A good summary of subsampling can be obtained from [11].

2.2. Statistical techniques

The statistical approach associated with frequency analysis is focused on finding an appropriate form to model the underlying distribution of flood data and then estimating the parameters of this underlying distribution. In this study, two-parameter distributions, such as Extreme value I (EV1) or Gumbel (G), and three-parameter distributions, such as Generalized Extreme Value (GEV), Generalized Logistic (GLO), Pearson Type 3 (PE3), Generalized Normal (GNO) and Generalized Pareto (GPA), are considered. These distributions have been widely used in flood frequency analysis [14]. This study uses the method of L-moments to estimate distribution parameters. The theoretical expressions and the formulae for parameter estimation using L-moments of the candidate distributions are given by [14]. In this study, a goodness of fit test (GOF) based on the Anderson-Darling (AD) test statistic [6,15,16] is used to evaluate the suitability of different probability distributions. The AD test is regarded as the most powerful and preferred test available at present [6].

2.3. *Test procedure*

The proposed subsampling methodology can be applied to an at-site annual maximum (AM) hydrologic data series to select the most appropriate distribution for frequency analysis. The test procedure has the following steps:

- A large number of subsamples (i.e. 1000) of size b (b < n) are drawn from the original data series of size n without replacement.
- The Anderson-Darling (AD) test is then applied to each subsample drawn from the original sample. The AD test statistic is computed for each candidate distribution.
- The distribution is chosen that is accepted most frequently by the test statistic at the 5% significance level

By the above description, one can understand that the methodology is aimed to develop a hydrologic decision tool for a longer data series. Generally, the recommended size (record length) for performing at-site frequency analysis is 20 years [17]. In this procedure, the AD test is applied to each subsample to determine their frequency distribution. Thus, the record length 20 years is considered as minimum subsample size (b). It then requires the sample size (n) to be greater than 20 in the subsampling procedure. However, the selection of an appropriate subsample size with respect to a sample is an important aspect of this methodology. This is also regarded as an important issue when determining a confidence interval with the subsampling approach [18]. This will be examined in detail in the following sections.

3. Application and results

Four gauged stations are selected as case studies to apply the subsampling methodology. They (Stations 39001, 54002, 21006, 12001) pertain to catchments located in different regions of U.K. They are different in record lengths, and the records vary from 46 to 130 years. Table 1 shows the necessary information about the data series such as location, record length and statistical characteristics in terms of L-moments. These datasets are taken from the Centre for Ecology & Hydrology (CEH), U.K.

Table 1. Site information and statistical characteristics of flood data.

Station No.	Station Name	River	Record Length (yrs)	AREA (km2)	Statistical characteristics			
					Mean/L1	L-CV	L-Skew	L-kur
39001	Kingston	Thames	130	9948	325.4	0.195	0.138	0.168
54002	Evesham	Avon	76	2210	154.2	0.272	0.164	0.206
21006	Boleside	Tweed	46	1500	428.7	0.189	0.174	0.140
12001	Woodend	Dee	70	1370	427.6	0.207	0.075	0.197

Popular tests, namely the Anderson-Darling (AD) goodness of fit test and L-moment ratio (LMR) diagram, are applied to identify the appropriate distributions for each of the selected data series. Table 2 summarizes the AD test result for each of the AM data series. By 'Reject' (or 'Accept'), it means that the distributions are rejected (or accepted) by the test statistic at the 5% significance level. For stations 39001, 54002 and 21006, the AD test rejected only one distribution, namely GPA, out of six candidate distributions, while for station 12001 the test rejected all the considered distributions. Overall, for each AM data series, the test statistic has failed to discriminate among the candidate distributions. Plots of dimensionless L-moment ratios (LMR) are shown in figure 1, which compares the observed and the theoretical relations between L-skewness and L-kurtosis for each of the AM data series. The LMR values of stations 12001 and 54002 lie above the theoretical GLO line, while for station 39001 it is positioned between the GEV and GLO line. In the case of 21006, the value lies between the GNO and PE3 line, and the value is very close to the GEV line and the EV1 (G) point. In all cases, there is enough doubt to get an outright choice for respective data series. Thus, in this context, the L-moment ratio diagrams cannot be regarded as suitably reliable.

Table 2. Anderson-Darling goodness of fit test results at the 5% significance level applied to the selected stations.

	NOR	EV1	GEV	GLO	PE3	GNO	GPA
ST39001	Reject	Accept	Accept	Accept	Accept	Accept	Reject
ST54002	Reject	Accept	Accept	Accept	Accept	Accept	Reject
ST21006	Accept	Accept	Accept	Accept	Accept	Accept	Reject
ST12001	Reject	Reject	Reject	Reject	Reject	Reject	Reject

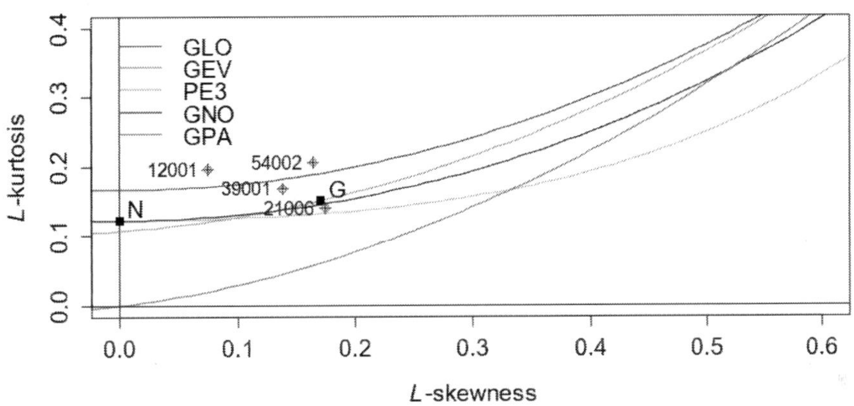

Figure 1. L-moment ratio diagram showing the point values for selected AM series.

The proposed subsampling methodology is applied to each of the AM data series. A large number of subsamples (e.g. 10000) were drawn from the original data series, taking a set of different subsample size (b) without replacement. The set of different b was chosen for particular AM series according to their record lengths. This has been indicated in figure 2. The algorithm in [19] was used to draw subsamples from the original data series without replacement. The Anderson-Darling (AD) test was then applied to each subsample, and a summary result has been produced for each distribution. Figure 2 shows the results in terms of percentages of accepted distributions for each station. Different values are obtained for different distributions and for different subsample sizes. It is noted that for the distributions that are accepted at the 5% significance level for the original data series, the acceptances of those distributions increase with the increase of b. However, at certain subsample sizes, discrimination is difficult to achieve. For example, at b= 100 for station 39001, an 100% acceptance rate was observed for four distributions: GEV, GLO, PE3 and GNO. However, for other subsample sizes such as b= 20, 40 and 60, the discrimination was good enough to choose the most appropriate distribution. A similar kind of conclusion can be obtained for station 54002. Due to its short record length (46), the results of station 21006 are reported for b= 20, 25, 30, 35 and 40. The discrimination is difficult in this context, as it was also shown by the AD test and LMR diagram with the original data. However, at b=20 the GNO was accepted slightly more than the GEV and EV1. A similar observation was also noted for b=25. Although none of the distributions were accepted by the AD test for station 12001, the test procedure allows the identification of an appropriate distribution. The results so obtained were then used to produce frequency curves. Figure 3 shows four probability plots: the GLO fit for stations 39001, 54002 and 12001 and the GNO fit for station 21006. The flood data are plotted against the return periods corresponding to the logistic variate for GLO and the normal reduced variate for GNO. The observed data are presented as small circles and the flood quantiles are presented as lines. There are two relatively high events (see the probability plot for station 12001 in Figure 3), which might have caused the distributions to perform so poorly for station 12001, but with this test procedure, the discrimination among distributions was achieved. The relatively poor performance of the test method in differentiating between distributions for station 21006 may be owing to reasons such as short record length and the characteristics of L-moment statistics.

One of the key observations from the case studies is that the finding of an appropriate size of b is an important task. It is found that the discrimination was recognized when the size of the subsample was around half of the sample size. Thus, taking b= ½ of n is a good choice to start a subsampling experiment in this context. The methodology also provides a degree of acceptance that is absent for a GOF test. This can allow ranking of the distributions or selection of an appropriate distribution in case none of the considered distributions are accepted by a GOF test.

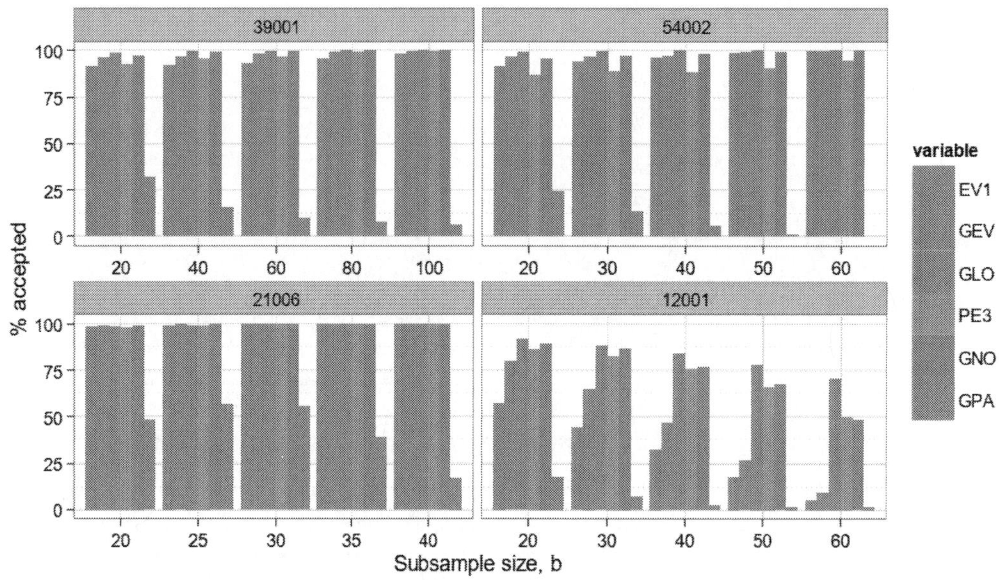

Figure 2. Bar charts showing the results of the Anderson- Darling goodness of fit test applied to the subsamples of different sizes drawn from the selected AM data series. Each panel represents the station number, and different colors indicate different types of distribution considered in this study.

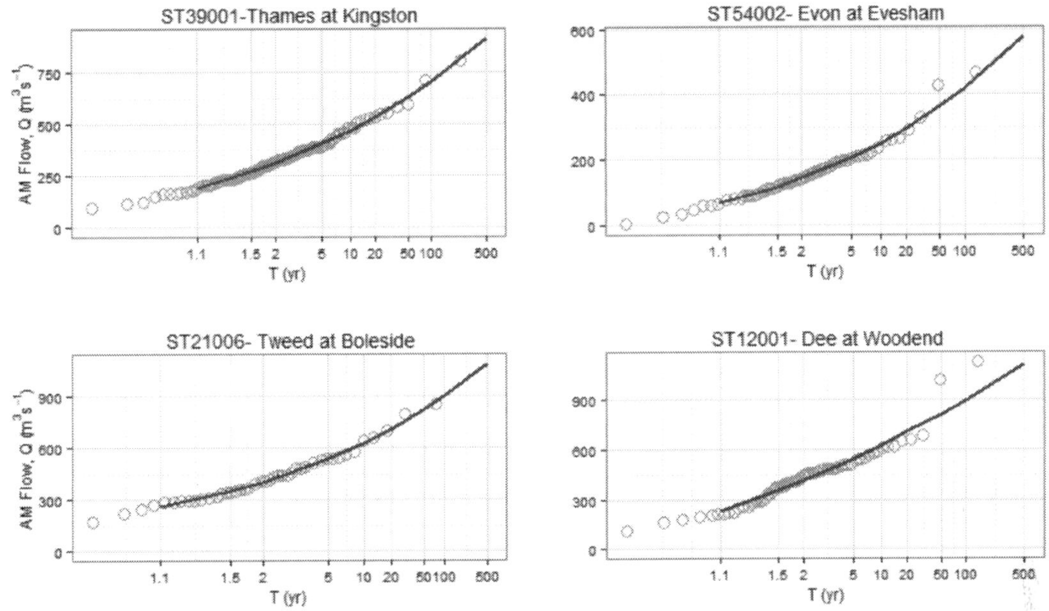

Figure 3. Probability plot: GLO fit for stations 39001, 54002, and 12001 and GNO fit for station 21006. The observed data are presented as small circles and the flood quantiles are presented as lines. The flood data are plotted against the return periods (T).

4. Discussion and conclusion

This study was conducted to investigate the potential utility of a subsampling technique to discriminate between distributions for use in hydrologic frequency analysis. The concept is applicable to the fact that the subsamples are actually samples from the true unknown distribution, as was the case with the original sample. The idea is to analyse a huge number of subsamples obtained from the original sample to better discriminate between distributions that otherwise may not be possible with the original sample. The methodology can be thought of as the framework opposite to the regional analysis [20, 21]. In a regional analysis framework, the flood data from different sites in a homogeneous region are assumed to come from a population distribution, and the distribution that is accepted in the highest number of cases or a similar sort of arrangement is chosen for frequency analysis. In this context, a large number of subsamples are drawn from a sample, and the distribution that is accepted by a goodness of fit test in the highest number of cases is chosen for frequency analysis.

To demonstrate the performance, the methodology was applied to observed data. Four gauged stations pertaining to catchments in U.K. were selected as case studies to apply the subsampling methodology. When applied to observed data, it was found that with the proper choice of subsample size, the methodology performed better in discriminating among distributions. Generally, the performance of the methodology decreased as the subsample size approached the original sample size. The methodology performed better when the subsample was around half of the sample size. Thus for practical purposes, the use of n/2 can be endorsed, but it is recommended to try a range of sizes and see what the overall behaviour is. It was also found that a longer data series produced better results compared to a shorter record length. The method also provides a degree of acceptance, which can be useful for practitioners in decision making.

However, it is recognized that in some cases it may not perform as a full proof tool because it depends on the power of AD testing, which has its own limitations. Then again, in comparison with the standard AD tests, the proposed methodology performs considerably better, and a sensible use of the methodology can give good results. A good operational strategy could be to use the AD and LMR tests in combination with the subsampling technique. The first two methods can be used to eliminate inappropriate distributions from the candidate distributions. If two or more are selected by the tests, then the proposed methodology can be applied to select the best population distribution. Overall, the methodology can be regarded as an interesting complement to standard statistical testing because it presents the advantage of identifying the best probability distribution with a degree of acceptance.

Acknowledgement

The author would like to acknowledge the financial support in the form of a faculty start-up fund made available by the College of Hydrometeorology, Nanjing University of Information Science and Technology (NUIST), Nanjing, China. The observed annual maximum flood data series of selected stations was obtained freely from the Centre for Ecology & Hydrology (CEH), UK website: http://www.ceh.ac.uk/data/nrfa/data.

References

[1] Hosking J R M 1990 L-moments: analysis and estimation of distributions using linear combinations of order statistics *Journal of the Royal Statistical Society* Series C (Applied Statistics) **52** 105–24

[2] Cunnane C 1989 Statistical distributions for flood frequency analysis (Geneva: Operational *Hydrology Report* (WMO))

[3] Das S, Millington N and Simonovic S P 2013 Distribution choice for the assessment of design rainfall for the city of London (Ontario, Canada) under climate change *Can. J. Civ. Eng.* **40** 121–9

[4] Chowdhury J U, Stedinger J R and Lu L 1991 Goodness-of-fit tests for regional generalized extreme value flood distributions *Water Resources Research* **27** 1765–76

[5] Ahmad M I, Sinclair C D and Spurr B D 1988 Assessment of flood frequency models using empirical distribution function statistics *Water Resources Research* **24** 1323

[6] Laio F 2004 Cramer-von Mises and Anderson-Darling goodness of fit tests for extreme value distributions with unknown parameters *Water Resources Research* **40** 1–10

[7] Vogel R W and McMartin D E 1991 Probability plot goodness-of-fit and skewness estimation procedures for the Pearson type 3 distribution *Water Resources Research* **27** 3149–58

[8] Laio F, Di Baldassarre G and Montanari A 2009 Model selection techniques for the frequency analysis of hydrological extremes *Water Resources Research* **45** 1–11

[9] Yan H and Moradkhani H 2014 A regional Bayesian hierarchical model for flood frequency analysis *Stochastic Environmental Research and Risk Assessment*

[10] Renard B and Lall U 2014 Regional frequency analysis conditioned on large-scale atmospheric or oceanic fields *Water Resources Research* **50** 9536–54

[11] Anon (Springer Series in Statistics) Dimitris N P, Joseph P R and Michael W 1999 *Subsampling-Springer*

[12] Politis D N 1998 Computer-intensive methods in statistical analysis Ieee Signal Processing Magazine **15** 39–55

[13] Efron B 1979 Bootstrap methods: another look at the jackknife *The Annals of Statistics* **7** 1–26

[14] Hosking J R M and Wallis J R 1997 *Regional frequency analysis: an approach based on L-moments* (Cambridge University Press)

[15] Stephens M 1986 *Tests based on EDF statistics. In Goodness-of-fit techniques* ed R.B. D'Agostino and M A Stephens (New York: Marcel Dekker, Inc.)

[16] Das S and Simonovic S P 2012 Assessment of uncertainty in flood flows under climate change impacts in the upper Thames River Basin, Canada *British Journal of Environment & Climate Change* **2** 318–38

[17] Institute of Hydrology 1999 *Flood Estimation Handbook* (Wallingford, U.K.: Institute of Hydrology) **5**

[18] Bickel P J and Sakov A 2008 On the choice of m in the m out of n bootstrap and confidence bounds for extrema *Statistica Sinica* **18** 967–85
[19] Green B F 1977 Fortran subroutines for random sampling without replacement *Behavior Research Methods & Instrumentation* **9** 559–59
[20] Das S and Cunnane C 2011 Examination of homogeneity of selected Irish pooling groups *Hydrology and Earth System Sciences* **15** 819–30
[21] Das S and Cunnane C 2012 Performance of flood frequency pooling analysis in a low CV context *Hydrological Sciences Journal* **57** 433–44

Spatiotemporal variation of precipitation based on three indicators in the flood season in Beijing

Y Chen[1], J H Wang[1], W H Xiao[1, 2], Y Y Zhou[1], L T Wang[1] and X Y Song[1]

[1] State Key Laboratory of Simulation and Regulation of Water Cycle in River Basin, China Institute of Water Resources and Hydropower Research, China

E-mail: xiaoweihua@iwhr.com

Abstract. This paper attempts to analyze the spatiotemporal variation of precipitation in Beijing in terms of three indicators, i.e., the total precipitation, the maximum daily precipitation, and the maximum 5h precipitation in the flood season, with data collected from 20 meteorological stations from 1975 to 2012 (June to September). The Mann-Kendall non-parametric test, the linear regression, and five-year moving average were used. The results showed that: (i) the three indicators showed a decreasing trend; however, from 1975 to 1995 and from 1995 to 2012, the precipitation in the flood season showed an increasing trend with an increasing rate of 3.36 and 2.90mm/a, respectively. (ii) The indicators demonstrated a similar spatial variation. The high value precipitation areas were mainly in the northern and central parts of the city. The central districts had increasing centers with maximum increasing rates of 0.35, 0.65, and 0.49mm/a, respectively. However, the northern part exhibited different degrees of decreasing trends. (iii) The increase in the precipitation in the city center was mainly due to the urban heat island effect and the change in the underlying surface. The temperature gradient was intensified due to the topography in the city center and the northern suburb, causing a decreasing trend of precipitation.

1. Introduction

In Beijing, 80% of the precipitation falls in the flood season, when short-range strong rainfall often occurs. The rapid urbanization has led to an obvious change of precipitation in the downtown city. Precipitation is the key indicator in the water resource spatiotemporal pattern influenced by climate change. An extreme precipitation situation has a negative effect on the urban environment and the social economy. It is important to analyze the spatiotemporal distribution of precipitation in order to study the effect climate change has on water resource planning and management [1]. Therefore, this paper will study the extreme situations of precipitation, including the spatiotemporal variation of the maximum daily precipitation and the maximum 5h precipitation based on the study of the spatiotemporal distribution of the total precipitation in the flood season. It attempts to reveal the characteristics of the concentrated rainfall from three different time scales, which can help aid in understanding the characteristics of the extreme situation of precipitation and its future trend, to better allocate water resources, and to use the hydrological forecast to prevent flood and drought.

Currently, studies on the precipitation of Beijing mainly concentrate on the following aspects: (i) The first is studies focusing on the water distribution characteristics in the annual time scale [2]. For example, Xu Zongxue *et al.* used the non-parameter test method to analyze the spatiotemporal variation of the annual and interannual precipitation in Beijing. The result showed that the annual precipitation decreased over time in Beijing. (ii) The second is studies focusing on urbanization's impact on precipitation [3, 4, 5, 6]. For example, Sun Jisong *et al.* analyzed the influence of the city island effect on the winter and summer precipitation in Beijing [7]. It was proposed that there is a clear difference in the relative variation trend of the precipitation days and the amount of city and suburbs in winter and summer. (iii) The third is studies focusing on the variation characteristics of the extreme precipitation [8, 9]. For example, You Huanling *et al.* analyzed the spatial distribution of the extreme precipitation characteristics. Their results indicated that from 1981 to 2001, the amount, days, and intensity of extreme precipitation showed a declining trend in most of the stations in Beijing [8]. The precipitation of Beijing mainly occurs in the flood season, which is usually accompanied with several great rainstorms that very likely result in flood disasters. In recent years, the precipitation has undergone distinct changes. However, the previous studies have mainly focused on the spatiotemporal variation of a long sequence of the annual precipitation in Beijing or only on the maximum daily precipitation, which fails to reflect the spatiotemporal variation of the concentrated precipitation of Beijing. Therefore, based on the precipitation data from 1975 to 2012, this paper attempts to analyze the spatiotemporal variation from three different times, i.e., the total precipitation, the maximum daily precipitation, and the maximum 5h precipitation in the flood season with the Mann-Kendall non-parametric test method, linear regression method, and the five-year moving average method. In terms of time, a comparison of long-sequence precipitation indicators and the variation of urbanization in different phases are analyzed. In terms of space, the distribution disparity and variation of the central urban area, the northern urban area, the southern plain area, and the western mountain area are analyzed, and potential explanations are given based on the analysis.

2. Materials and method

2.1. *Study area*

Beijing is located in the northern part of the North China Plain, surrounded by the Yanshan Mountains to the north and west and the Taihang Mountains in the northern section (figure 1). The mountainous area accounts for 62% of the entire region [10, 11]. It is a city of typical sub-humid continental monsoon climate in the North Temperate Zone, with great variation in annual precipitation and uneven precipitation distribution in the seasons. In summer, it is hot and rainy, whereas in winter, it is cold and dry. The multi-year average annual precipitation of Beijing is 585mm, among which the mountainous area is 591 mm, and the plain is 600mm. Approximately 80% of the precipitation falls in the flood season, which is 488 mm.

Song Xiaomeng *et al.* provided a potential method to divide Beijing into different zones [10]. Here, a similar approach is followed; Beijing is divided into six zones: the central urban area, northern urban area, suburb area, northern mountain area, western mountain area, and southern plain area (figure 1).

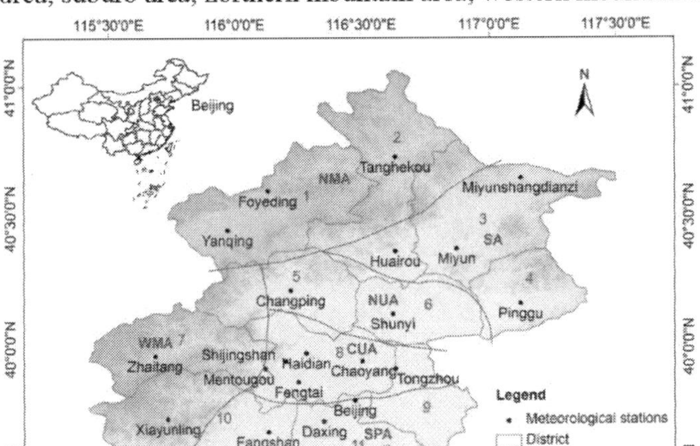

Figure 1. Distribution of the meteor ological stations in Beijing. Blue soli d lines denote the boundaries of six a reas. CUA, NUA, SA, NMA, WMA, and SPA refer to the central urban a rea, northern urban area, suburb are a, northern mountain area, western m ountain area, and southern plain are a, respectively. Numbers indicate dif ferent districts: 1. Yanqing; 2. Huair ou; 3. Miyun; 4. Pinggu; 5. Changpi ng; 6. Shunyi; 7. Mentougou; 8. Cent ral city (including six districts: Dong cheng, Xicheng, Haidian, Chaoyang, Fengtai, and Shijingshan); 9. Tongz hou; 10. Fangshan; and 11. Daxing.

2.2. Data sources

The data of the paper were collected from the hourly precipitation records of 20 meteorological stations in Beijing from 1975 to 2012 as provided by the National Meteorological Administration. The stations include six from the urban area, eight from suburbs, and six from the mountain area, which covers the city center, suburb, and mountain area. The specific distribution is illustrated in figure 1. Both the Foyeding station and Xiayunling station have missing data for certain years, but this is corrected by interpolation and extension. Statistically speaking, the data is adequate to acquire a reliable result. This paper ascertains the monthly precipitation, the maximum daily precipitation, and the maximum 5h precipitation in the flood season based on the hourly precipitation data.

2.3. Method

In this paper, linear regression and the five-year moving average method are used to analyze the variation trend of the precipitation in the long-sequence and in twenty years in the flood season. The Mann-Kendall method was used together with the professional mapping software of Surfer 8.0, adapting the kriging spatial interpolation method to interpolate all the precipitation of all stations and the Mann-Kendall tendency rate to obtain the spatial distribution for analysis. The Mann-Kendall test sets up two series: a forward one (UF) and a backward one (UB). If the UF and UB curves intersect and then diverge and acquire specific threshold values, then a statistically significant trend exists. The point of intersection shows the approximate change point at which the trend begins [12-14].

3. Results and discussion

3.1. The temporal variation of the precipitation in the flood season

In order to eliminate the periodical variation of the data, this paper adopts the five-year moving average method to analyze the precipitation anomaly in Beijing and the five-year moving average variation trend (Figure 2). The precipitation time series below are based on the average of all 20 meteorological stations in each year.

From figure 2, it can be observed that: (i) The precipitation in the flood season is declining with time, yet the decline is not that obvious. The decrease rate for the multi-years average precipitation is 1.24mm/a, and the decrease rate for the 5-year moving average precipitation is 1.60mm/a. (ii). The precipitation anomalies before 1995 were mainly positive anomalies; afterwards, they were mainly negative. A distinct disparity exists before and after 1995. In figure 4, the UF and the UB curves indicating the average precipitation in the flood season also show a sudden change in 1995.

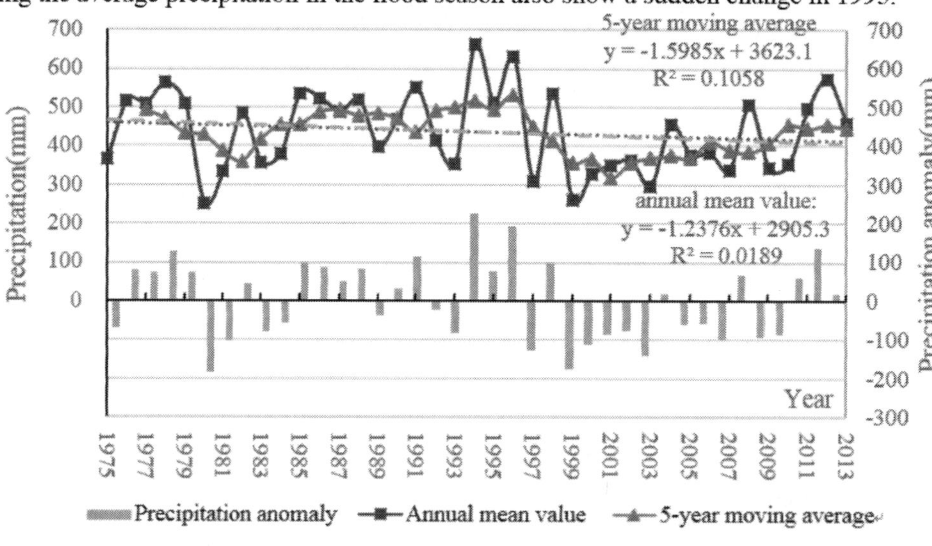

Figure 2. Temporal distribution and trends of rainfall in the flood season in Beijing.

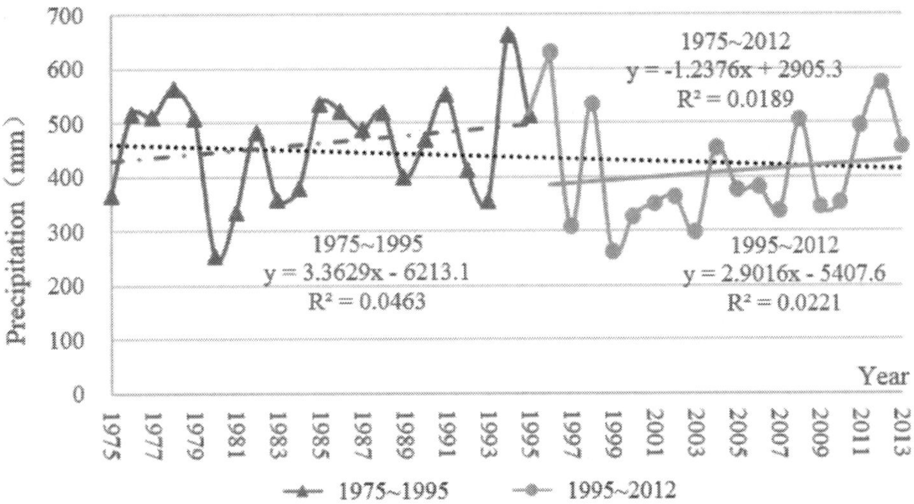

Figure 3. The linear variation trend of the precipitation in different periods in the flood season in Beijing.

Figure 3 illustrates the variation trends of the precipitation in different time periods. It can be found that in different time periods, the precipitation trends varied. Although the precipitation from 1995 to 2012 is less than that from 1975 to 1995, thus showing a decreasing trend in the long-sequence variation, there is clearly an increasing trend in these two periods at a rate of 3.36mm/a and 2.90mm/a, respectively. This reflects that the city island effect that emerged with rapid urbanization has little influence on the long-sequence precipitation trend, but it has a significant influence on the precipitation of certain areas in the most recent 20 years.

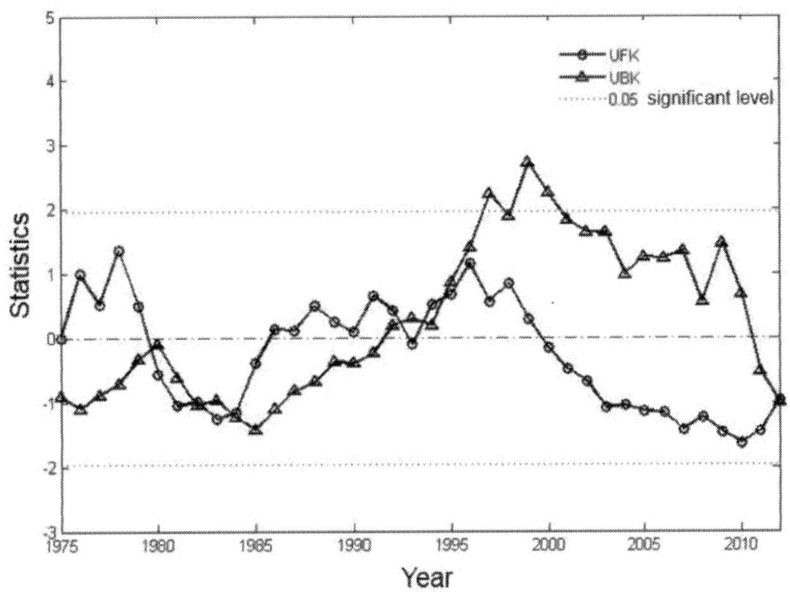

Figure 4. Mann-Kendall values of precipitation data series of flood season from 1975 to 2012 in Beijing.

In order to understand the variation trend of the monthly precipitation, a calculation is made on the statistical magnitude Z of Mann-Kendall and gradient β of the average precipitation month by month in the flood season. The result is shown in table 1. It can be concluded that: (i) the precipitation of the flood season occurs mainly in July and August, which is when 38.6% and 31.1% of the flood season precipitation falls, and (ii) among the four monthly series, only August passes the significance level test (α=0.05) and shows a decreasing trend. This indicates that August is the month that contributes the most to the decrease in precipitation in the flood season.

Table 1. The Mann-Kendall trend analysis of each month in the flood season.

Time sequence	The average precipitation (mm)	The parameter value of statistical characteristics		Trend
		Z	β	
Flood season	437.5	-0.895	-1	Tends to decrease
June	81.3	-0.251	-0.155	Tends to decrease
July	168.5	-0.226	-0.339	Tends to decrease
August	136.0	-2.816	-2.062	Distinct decrease
September	51.1	1.861	0.812	Tends to increase

3.2. The temporal variation of the extreme precipitation in the flood season

Due to the climate change, the intensity and the frequency of the hydrologic extremum have undergone changes. In recent years, Beijing has encountered many rainstorms, among which the "7.21" rainstorm produced the maximum precipitation in recorded meteorological history since 1951. To give context to the magnitude of the rainstorm, the average precipitation is 170 mm. However, during the rainstorm, the precipitation in the Fangshan district that was at the center of the rainstorm reached 541 mm, resulting in huge losses. Figures 5 and figure 6 show the variation trend of the maximum daily precipitation and the maximum 5h precipitation in the last 38 years in Beijing by using the linear regression method. Because 2012 was a year of extreme precipitation, it has a great impact on the analysis of the precipitation variation trend from 1995 to 2012. Therefore, it is excluded from the analysis. From the variation trend of the maximum daily precipitation in Figure 5, it can be observed that there is a decreasing trend either in the long-time sequence or in the 20-year time sequence in maximum daily precipitation, among which the decrease rate of 1995 to 2011 is 1.24mm/a. Its maximum daily precipitation is 7mm, which is less than that of 1975 to 1995. From the variation trend of the maximum 5h precipitation in Figure 6, it can be found that its variation trend is similar to that of the maximum daily precipitation. This is particularly distinct from 1975 to 2011 with a decreasing rate of 0.56mm/a.

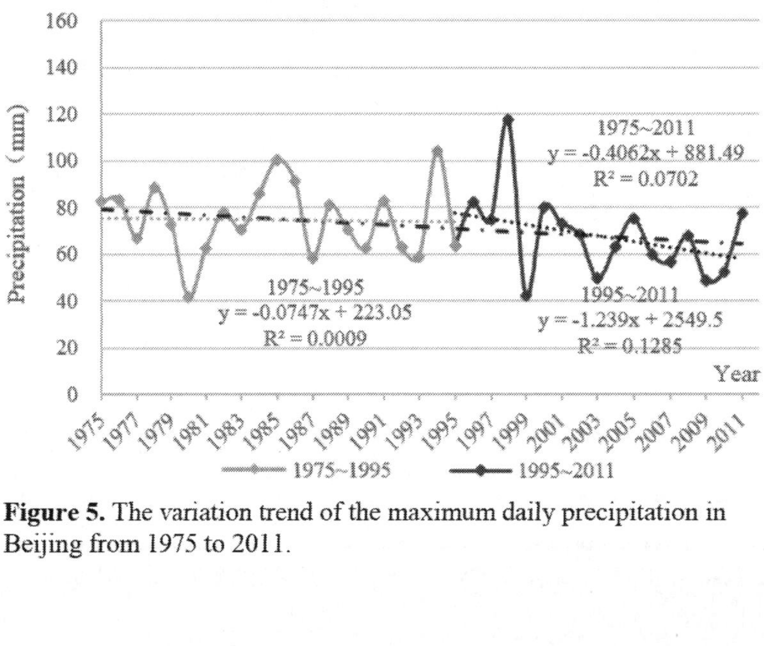

Figure 5. The variation trend of the maximum daily precipitation in Beijing from 1975 to 2011.

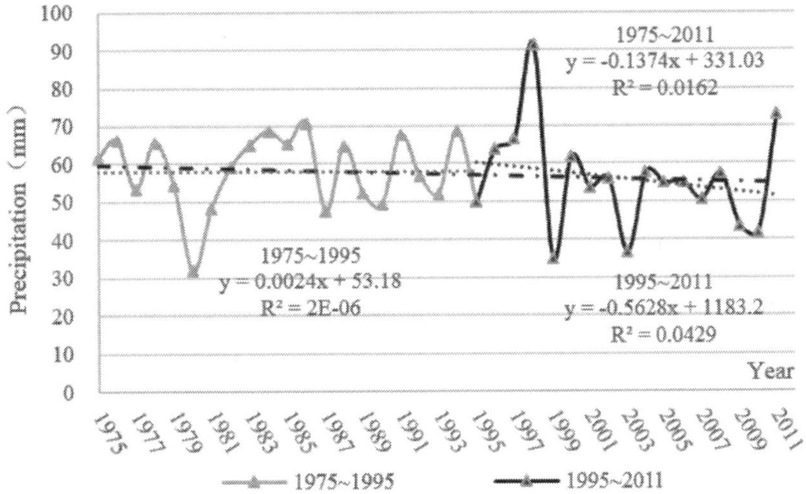

Figure 6. The variation trend of the maximum 5h precipitation in Beijing from 1975 to 2011.

3.3. The spatial variation of the precipitation in the flood season

From the long-time average precipitation contour map of Beijing in the flood season in figure 7, it can be observed that the precipitation of Beijing in the flood season is unevenly distributed. The high value districts of precipitation are mainly in the city center and the windward slope in front of the mountain in the northern part of the city. Generally speaking, the precipitation is decreasing from east to west. The precipitation in the southeast area is normally higher than 400mm, while the precipitation of the remote northwest area in the flood season is less than 400mm, among which the precipitation of the surrounding area of the Miyun district reaches 500mm.

From the Mann-Kendall gradient contour map of the long-term average precipitation in the flood season in figure 8, it can be observed that since the 1980s, different districts of Beijing have shown different degrees of a decreasing trend, among which the northern part of the city is the most obvious, forming a decreasing center in the Huairou district with a decrease rate of 4.8mm/a. However, in the city center, it exhibits a slight increasing trend, forming an increasing center in Haidian with an increase rate of 0.35mm/a. The west mountain area is quite stable.

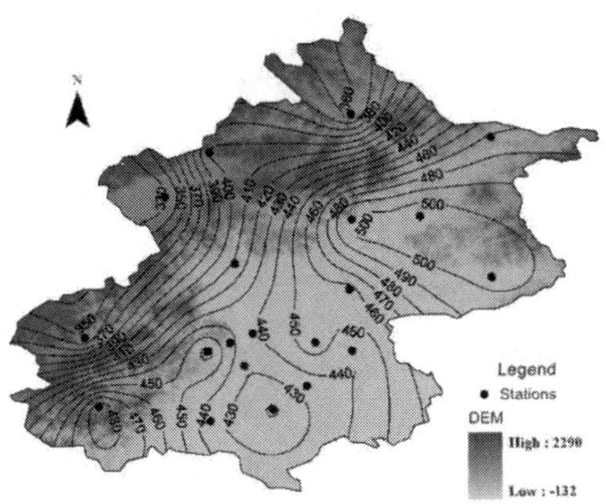

Figure 7. The long-time average precipitation contour map of Beijing in the flood season (mm).

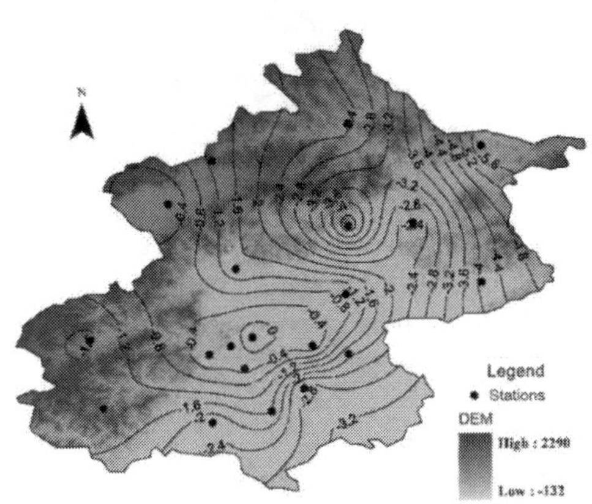

Figure 8. The Mann-Kendall gradient contour map of the long-term average precipitation in th e flood season (mm/a).

3.4. The spatial variation of the extreme precipitation in the flood season

3.4.1. The spatial variation of the maximum daily precipitation. From figure 9, it can be found that the spatial distribution of the maximum daily precipitation is almost the same as the total precipitation in the flood season. The high value district is mainly in the city center, the northern part of the city. and the southern plain area, while the maximum daily precipitation is less in the western mountain area. From Figure 10, it can be found that the there is a clear increasing trend in the city center, forming an increasing center in Mentougou with a maximum increasing rate of 0.65mm/a. The northern district and the mountainous area exhibit a clear decreasing trend, forming decreasing areas in Huairou and Miyun, with decreasing rates of 1.12mm/a and 0.03mm/a, respectively. The southern plain has no significant change. This result is in accordance with that proposed by You Huanling *et al.* [8]: the extreme precipitation intensity of the daily precipitation in the northern part of Beijing is decreasing, while the extreme precipitation intensity of the city center is increasing.

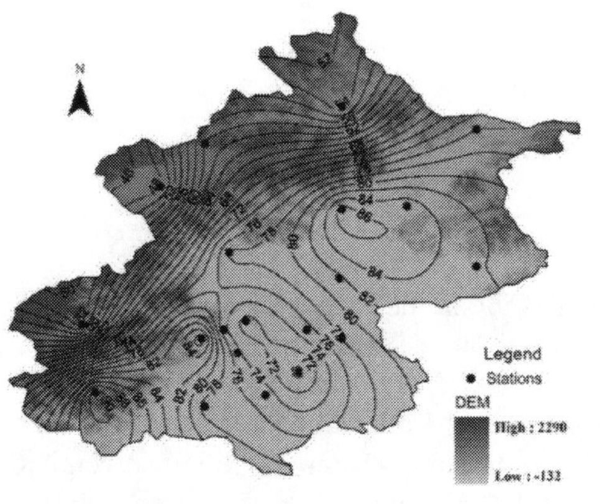

Figure 9. The contour map of the maximum daily precipitation of Beijing in the flood season (mm).

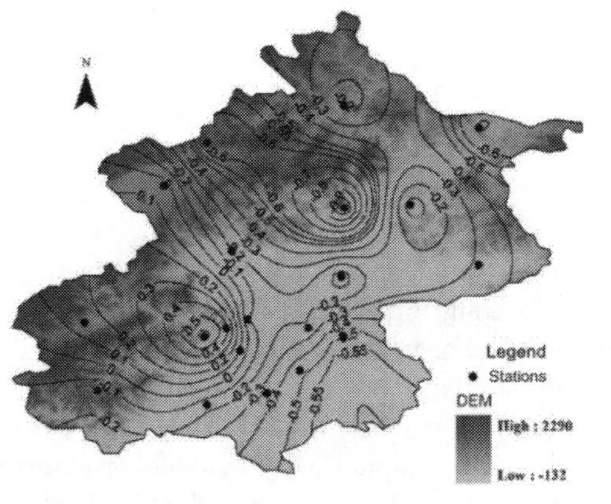

Figure 10. The Mann-Kendall gradient contour map of the maximum daily precipitation of Beij ing in the flood season (mm/a).

3.4.2. The spatial variation of the maximum 5h precipitation

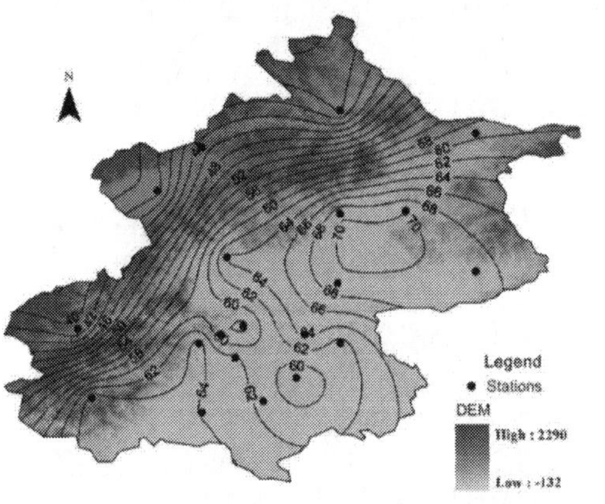

Figure 11. The contour map of the maximum 5h precipitation of Beijing in the flood season (mm).

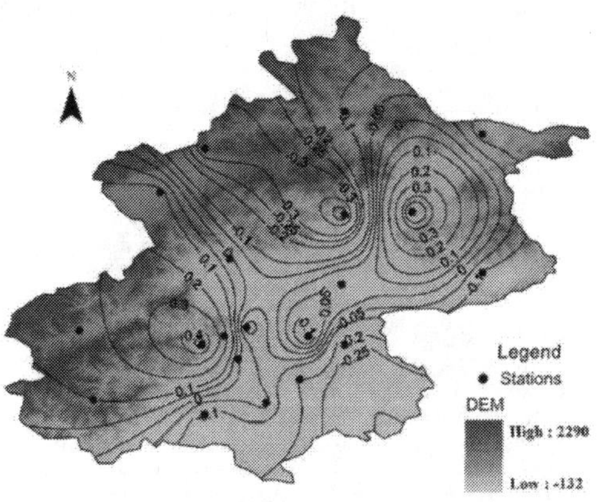

Figure 12. The Mann-Kendall gradient contour map of the maximum 5h precipitation of Beijing in the flood season (mm/a).

From figure 11, it can be found that there are two short-term rainstorm centers in the city center and its northern part, among which the maximum 5h precipitation in the north is 70mm/5h, and the precipitation of the west mountainous area is 45mm/5h, which is relatively small. The precipitation decreases from east to west. From figure 12, it can be found that the city center and the northern part exhibit a clear increasing trend, forming two increasing centers in the Mentou Gou and Chaoyang districts with increasing rates of 0.49mm/a and 0.20mm/a, respectively. There is also an increasing center in Miyun with a rate of 0.56mm/a and a decreasing center in Huairou with a rate of 0.51mm/a in the northern part of the city. The variation trend is not obvious in the western mountainous area and the southern plain.

3.5. The preliminary analysis of the reasons of the spatiotemporal variation in the flood season

The spatiotemporal distribution of Beijing is influenced by various meteorological factors, such as the general atmospheric circulation, temperature, wind, and so on, as well as by underlying surface factors, such as the landscape, vegetation cover, and types of land utilization [4, 6, 15].

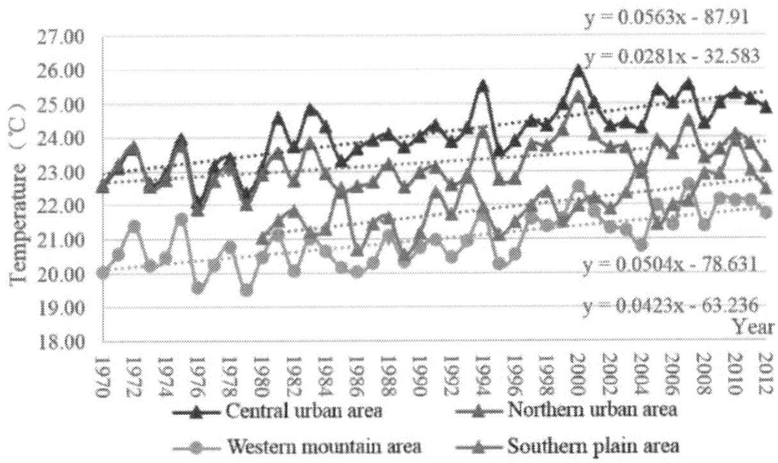

Figure 13. The temperature variation trend of different areas in Beijing.

Before 1980, the city of Beijing developed slowly. However, since 1980, Beijing has undergone rapid development [15]. The condition of the underlying surface became quite different from the suburb and the mountain areas due to the rapid expansion and development of the city. It can be observed from figure 13 that the temperature of Beijing increased differently in different zones. Since 1980, out of all the areas, the temperature of the city center increased the fastest with a rate of 0.56℃/10a and had the highest temperature. The temperature in the urban area was higher than that in the western mountain area and southern plain area. Sun *et al.* also found a variation in the temperature between the urban area and suburb, which explained the commonly observed 'urban heat island effect' [16]. This indicates that the urban heat island effect brought about by the urbanization and the variation of the underlying surface condition are probably the major factors that affect the precipitation variation trend and the long-term sequence variation in the last 20-years in the city center. The precipitation of the northern part of the city has remained at a high level. One possible cause of this phenomenon is that the topography has enhanced the temperature gradient between the city and the northern part. It would result in an increase in weak precipitation regions in the north [4]. In addition, the extreme precipitation intensity and the number of days decreased in the northern part of the city [4], which results in the decreased precipitation in the north. Because the southern part of the city is upstream of the city wind flow field, it is less influenced by the urban heat island effect, with no significant change in precipitation. The western mountain area is precipitous and is less influenced by the urban heat island circulation and monsoons. Therefore, the precipitation variation is not obvious.

4. Conclusions

This paper conducted the Mann-Kendall non-parametric test method, the linear regression method, and the five-year moving average method to analyze the spatiotemporal distribution features and the variation trend of the total precipitation, the maximum daily precipitation, and the maximum 5h precipitation in the last 38 years from 20 meteorological stations in Beijing. The following conclusions can be made based on the results:

- The precipitation in the flood season decreased with time in Beijing, with a rate of 1.43mm/a. This decrease resulted from the reduction of precipitation in August. However, 1975-1995 and 1995-2012 showed a clear increasing trend, with increase rates of 3.36 mm/a and 2.90mm/a, respectively. The maximum daily precipitation and the maximum 5h precipitation exhibited decreasing trends in three time periods. The decreasing rates of 1995-2011 were 1.24mm/a and 0.56mm/a, respectively.
- The spatiotemporal distribution of the precipitation in the flood season was unevenly distributed. The high value precipitation districts were mainly in the north and the center of the city, forming increasing centers with maximum increasing rates of 0.35mm/a, 0.65mm/a, and 0.49mm/a. In addition, the three time indicators in the north all had decreasing centers in the Huairou station, with decreasing rates of 4.8mm/a, 1.12mm/a, and 0.51mm/a. However, the maximum 5h precipitation resulted in increasing centers in the Miyun district, with an increasing rate of 0.56mm/a.
- The increasing trend of precipitation in the last 20 years in the city center was mainly caused by the heat island effect and the change to the underlying surface brought about by the rapid development of the city over the last 20 years. At the same time, this also increased the short-term strong precipitation in certain areas. Due to the topography, the northern part of the city has intensified the temperature gradient in the center city and the northern suburb. Meanwhile, because of the monsoons, the weak precipitation process in the northern part of the city increased, and the extreme precipitation decreased, resulting in a decreasing trend in the northern part of the city. Because the western mountain area was less influenced by the urban heat island circulation and the monsoon, the precipitation was in accordance with the decreasing trend with the long sequence precipitation.

References

[1] Kampata J M, Parida B P and Moalafhi D B 2008 Trend analysis of rainfall in the headstreams of the Zambezi River Basin in Zambia *Phys. Chem. Earth* **33**:621–5
[2] Xu Z X, Zhang L and Ruan B Q 2006 Analysis on the spatiotemporal distribution of precipitation in the Beijing *Region Arid Land Geogr.* **29**(2):186–92
[3] Liu X M, Hu F, Li L and Wang Y 2006 Summer urban climate trends and environment effect in the Beijing area *Chinese J Geophys.* **49**(3):689–97
[4] Yu S Q 2007 The annual variation of precipitation in Beijing and its effects on urban city *Prog. Nat. Sci.* **17**(5):632–8
[5] Zhang L, Xu Z X and Ruan B Q 2006 The influence of the urban heat island effect on temperature and precipitation in Beijing *Prog. Nat. Sci.* **21**(5):746–55
[6] Zheng Z F, G H, Wang Z W and Li Q C 2014 Analysis on Spatial Distribution of Precipitation in Beijing and Its City Effect *Plateau Meteorol.* **33**(2):522–9
[7] Sun J S and Shu W J 2007 The effect of urban heat island on winter and summer precipitation in Beijing region Chinese *Atmos. Sci.* **31**(2):311–20
[8] You H L, Liu W D and Ren G Y 2014 Changes in precipitation extremes in Beijing during 1981-2010 *Climatic Environ. Res.* **19**(1):69–77
[9] Zhong Y D, Jia Y W and Li Z W 2013 Spatial and Temporal Changes of Maximum 1h Precipitation Intensity in Beijing Region in Last 53 Years *China Hydro.* **33**(1):32–7
[10] Song X M, Zhang J Y, Amir A, Shouraseni S R, Xuan Y Q, Wang G Q, He R M,Wang X J and Liu C S 2014 Rapid urbanization and changes in spatiotemporal characteristics of

precipitation in Beijing metropolitan area. *Geophys. Res. Atmos.* **119** :1–22
[11] Zhou Y Y, Xiao W H, Wang J H, Zhao Y, Huang Y, Tian J Y and Chen Y 2016 Evaluating spatiotemporal variation of groundwater depth/level in Beijing Plain, a groundwater-fed area from 2001 to 2010 *Adv. Meteorol.*
[12] Kendall M G 1975 *Rank Correlation Measures* London: Charles Griffin
[13] Khaled H H 2007 Trend detection in hydrologic data: The Mann–Kendall trend test under the scaling hypothesis *Hydro.* **349** (3):350–63
[14] Mann H B 1945 Non-parametric tests against trend *Econometrica* **13**(3): 245–9
[15] Zhao N, Liu S H and Yu H Y 2011 Urbanization effects on local climate in Beijing in recent 48 years *Chinese Atmos. Sci.* **35**(2):373–85
[16] Li W L, Liu H L and Zhou X J 2003 Analysis of influence of Taihu lake and urban heat island on local circulation in Yangtze Delta *Sci. China D.* **46**(4): 405–15
[17] Gao S F, Ye Y L, Zhou Y 2012 Regional rainfall characteristic analysis for the Kaoping River Basin *Hydraul. Eng.* **43**(11):1341–8
[18] Jiang Y, Peng Q D, Luo H H and Ma W 2011 Analysis of spatial and temporal variation of water quality in Huaihe River Basin *Hydraul. Eng.* **42**(11):1283–8
[19] Li P C, Li Q F, Cai T, Yu M X, Xie W, Bai X and Li H Y 2010 Feature analysis of 51 years precipitation change in Beijing and Tianjin area *Water Res. Power* **28**(10):6–9
[20] Luo L, Wang Z J, Liu X Y and Zhang X C 2013 Changes in characteristics of precipitation in flood season over five typical basins of middle reaches of the Yellow River in China *Hydraul. Eng.* **44**(7):848–55
[21] Santosh M P, Deepak K, Mahesh K J and Jan A 2014 Spatial and temporal trends of mean and extreme rainfall and temperature for the 33 urban centers of the arid and semi-arid state of Rajasthan, India *Atmos. Res.* **138**:73–90
[22] Shi W L, Yu X Z, Liao W G, Wang Y and Jia B Z 2013 Spatial and temporal variability of daily precipitation concentration in the Lancang River basin, China. *Hydro.* **495**:197–207

IOP Conf. Series: Earth and Environmental Science 39 (2016) 012061 doi:10.1088/1755-1315/39/1/012061

Dynamic computing resource allocation in online flood monitoring and prediction

S Kuchar[1, 2], M Podhoranyi[1], R Vavrik[1] and A Portero[1]

[1] IT4Innovations, VSB - Technical University of Ostrava, 17. listopadu 15/2172, 708 33 Ostrava, Czech Republic

E-mail: stepan.kuchar@vsb.cz

Abstract. This paper presents tools and methodologies for dynamic allocation of high performance computing resources during operation of the Floreon+ online flood monitoring and prediction system. The resource allocation is done throughout the execution of supported simulations to meet the required service quality levels for system operation. It also ensures flexible reactions to changing weather and flood situations, as it is not economically feasible to operate online flood monitoring systems in the full performance mode during non-flood seasons. Different service quality levels are therefore described for different flooding scenarios, and the runtime manager controls them by allocating only minimal resources currently expected to meet the deadlines. Finally, an experiment covering all presented aspects of computing resource allocation in rainfall-runoff and Monte Carlo uncertainty simulation is performed for the area of the Moravian-Silesian region in the Czech Republic.

1. Introduction

Europe has suffered a rising number of all kinds of extreme flood events [7] from coastal and fluvial to pluvial and reservoir floods. These events do not have natural borders of influence, and therefore many scientists and researchers study them. Extreme flood events have increased rapidly in frequency and magnitude during last decades [4], and flood monitoring systems were thus developed in order to analyze and model the current flood situation and provide short-term prediction to support emergency decision processes related to the threat of a given coming flood. River flood monitoring and control requires measurement and notification of the water level, velocity and precipitation [10].

Several types of monitoring systems have been developed to provide information about prediction of extreme flood events and monitoring of their current behavior [13]. These systems provide similar functionality, but on different levels of complexity [3]. There are systems, which offer information only about the current state and mainly serve as dispatching systems, that coordinate usage of different kinds of human and material resources, preparing emergency plans and deploying experts and equipment. A different type of system can be more focused on prediction of the upcoming situation based on the current and historical situation as well as the meteorological forecast. These systems execute various hydrologic model implementations and analyze their results to provide flood warnings before the flooding situation escalates. As these systems often work automatically in near-real time, they need appropriate computing resources to provide accurate up-to-date results and timely warnings. Allocating these resources is not a trivial task because the executed models can be computationally intensive, and often the results must be computed quickly to be of any use for the emergency services that base their decisions on timely and precise information. The execution time of simulations during emergency scenarios should be in the order of tens of minutes, and the first warning should come at least several hours or even days before the emergency as far as the atmospheric forecast data are available. Lately, high performance computing (HPC) infrastructures and methods have piqued the interest of researchers and practitioners in the field of flood monitoring to increase the spatial and temporal scale and resolution of flood inundation models [16], improve the accuracy of modelling results with calibration [17], work with the probabilistic nature of the numerical weather prediction in ensemble flood forecasting [6], provide urgent computing in flood decision support [2] and tackle other computation intensive tasks while ensuring short execution times.

Nevertheless, effective flood monitoring systems are not only about providing the most precise results as soon as possible, but they also have to react to the seasonality of floods. In general, floods have seasonal character and there are periods that are not affected by this phenomenon [8]. This means that it is not economically effective to operate a flood monitoring system in the full performance mode during its whole lifetime (i.e., using a large amount of computing resources to provide results as quickly as possible even during non-emergency scenarios). On the other hand, it is important to keep the system running even in non-flood seasons with a low number of computing resources (and therefore a lower energy consumption) because the risk of flood is still present. Therefore, proper distribution of various kinds of resources based on the current flood situation may decrease cost of the operation.

In this paper, an HPC based online flood prediction system is presented that is able to dynamically allocate appropriate computational resources in hydrological modelling to increase the energy efficiency during non-flood situations and ensure more precise results with shorter execution time during emergencies.

2. Floreon$^+$ system

Floreon$^+$ is a system for monitoring, modeling, prediction and support in disaster management that was primarily developed for the Moravian-Silesian region in the Czech Republic. As it is a web-based platform backed by an HPC infrastructure, its modularity, flexibility and responsiveness allows easy integration of various thematic domains, regions and data. Its main goal is to support operational and tactical disaster management and emergency processes by providing and integrating simulations from several domains in near-real time.

The system currently focuses on the domains of flood monitoring and prediction, monitoring of the real time traffic situation and modeling of toxic cloud spread.

In the area of flood monitoring and prediction, Floreon$^+$ computes an ensemble of rainfall-runoff (RR) and hydrodynamic (HD) models. The system currently supports two semi-distributed event RR models – HEC-HMS 3.4 (developed by USACE) [14] and Math1D (experimental internally developed model) - and executes them both for all monitored catchments. The measured input data to the models are provided every 10 minutes from a network of water and meteorological gauges managed by the Odra catchment management office. A meteorological forecast for 72 hours is acquired every 6 hours from the numerical weather forecast model MEDARD provided by the Academy of Sciences of the Czech Republic. Results from the RR models are passed to the HD models that compute water levels, velocities and discharge, which are then visualized on the map interface. Currently, the only supported HD model in the system is HEC-RAS 4.1 (developed by USACE) [15], but we are planning to extend this support with additional models (focusing on 2D models, such as MIKE21FM by DHI).

Standard simulations of the supported RR models in the Floreon$^+$ system can also be enhanced by running automatized calibrations of their parameters that use optimization and inverse modelling methods. These methods estimate schematization parameter values that minimize the error between the measured and simulated values based on the current state of the catchment and meteorological information. Some input data cannot be calibrated due to its uncertain nature (e.g. weather forecasts), but this data significantly influences the accuracy of the results. Fortunately, it is possible to model the distribution of errors from the comparison of historical results with real measured values [5]. These distributions are then used by the Floreon$^+$ system to run Monte Carlo (MC) uncertainty simulations that provide a probabilistic view on possible scenarios in the near future. Both automatized calibration and uncertainty modelling execute a high number of simulations with changing parameters, so their computation time is long when executed sequentially. Parallelization and HPC infrastructure is used to decrease the execution time to meet the required quality of service. (Additional information about the complete online flood monitoring and prediction process in the Floreon$^+$ system can be found in [9]).

In the domain of traffic monitoring, Floreon$^+$ uses statistical methods to provide an overview of the current road traffic situation based on the velocity and location data from on-board GPS devices coming to the system every minute.

3D computational fluid dynamics solvers are also used in the system to simulate the extent and concentration of a possible toxic cloud spread for accidents occurring in places with known risks or contamination from mobile sources.

All outputs of the system are provided through a web interface with a map component (see Figure 1). This solution consolidates all of the integrated sections into one complex geographical framework and therefore delivers a detailed overview of the current situation as well as historical events. The combination of three simulation domains puts additional pressure on the computing resource management, sharing and balancing because available resources have to be shared according to the current priority in various domains (e.g., during a flood emergency, the majority of the computing resources should be used for flood simulations and not expended on toxic cloud modelling). Even though we currently focus only on resource allocation in the online flood monitoring process, the proposed methodology is also usable for other domains.

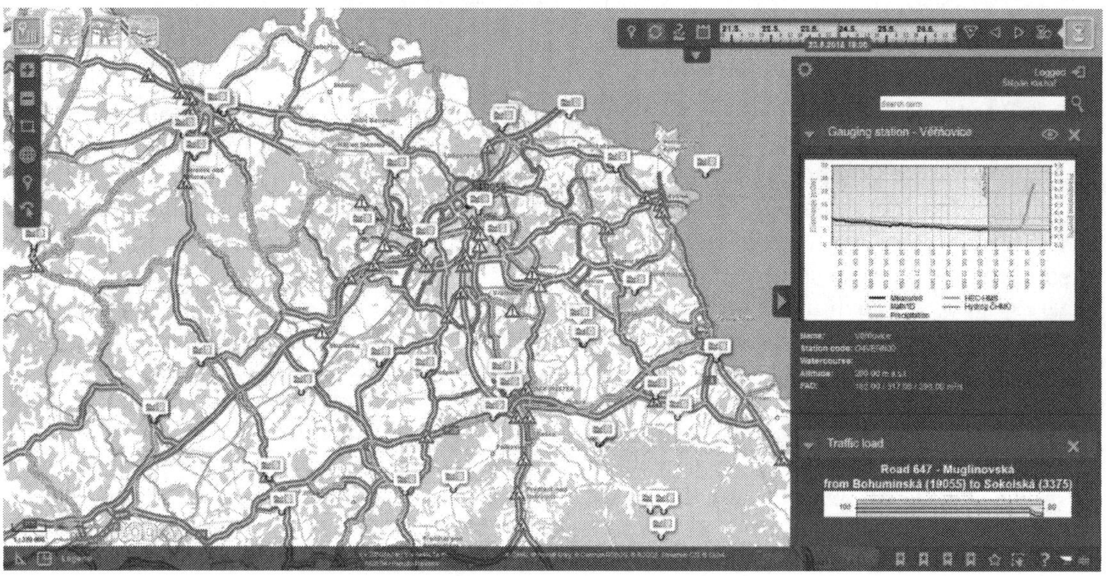

Figure 1. Screenshot of the Floreon+ web interface.

3. Runtime resource management

Applications running simulations in any domain have to satisfy certain runtime constraints. Generally, they need to provide defined precision of results in a given time limit. To ensure fulfilment of the constraints, when several applications run concurrently and share resources of the same machine, some sort of automatic resource management has to be applied. In our system, the HARPA-OS run-time resource manager [1], which is the topmost layer of the HARPA platform [12], is used for dynamic resource allocation. It is able to monitor the status of system resources and assign or reallocate resources between all managed applications during their execution. The main aim of the resource allocation strategy is to provide a sufficient amount of resources for the application to fulfill its constraints, but the objective is also to keep power consumption as low as possible and save energy

[11]. All managed applications have to implement a specific interface, which enables runtime reconfiguration of the application in cycles and contains the following methods:

- onSetup() – this method is executed only once at the start of the application. It contains initialization of the application, loading of input data and configuration of parameters as well as necessary pre-processing. For example, for RR modelling, a number of MC samples is set based on the expected precision, and an initial simulation without uncertainty is executed to provide referential results. For HD modelling, a time step for the HD simulation is set and appropriate data are extracted from input hydrographs.
- onConfigure(int awmId) – this method is executed when the HARPA-OS assigns resources to the application for the first time or changes the allocated resources to meet the expected service quality. HARPA-OS selects the most appropriate set of resources (passed to the method by the awmId parameter; see section 4. for more information) based on the current scenario, resource allocation strategy, state of the application and required service quality, and passes it to the onConfigure method. The application then has to reconfigure its parameters to properly run on the allocated resources. RR models reconfigure the number of MC samples processed in the current cycle. HD models change the number of time steps executed concurrently in the current cycle (a steady flow approach is used for HD modelling, so each time step of the simulation is independent and can be executed in parallel).
- onRun() – this method performs the main computation of the application. The configured number of MC samples and HD time step simulations for the current cycle is processed, and the average time required to simulate one sample or time step is measured.
- onMonitor() – this method monitors the current performance and evaluates the ongoing progress of the application execution and its constraints. After all MC samples and HD time steps in the current cycle are processed, the time required to process the remaining samples/time steps is computed, and a total time for the whole simulation is estimated. If the estimated time would exceed the time constraint specified in the service quality level, more resources are requested.
- onRelease() – this method performs all required post-processing steps and cleanup operations after the whole computation is finished. In RR modelling, percentiles are extracted from the MC sample results and saved as final results. In HD modelling, the results are combined with a digital elevation model of the area to extract the inundated areas.

Figure 2 illustrates a simplified example for execution of three independently managed applications running concurrently on the same machine and sharing its 16 CPU cores. These can represent several uncertainty modelling instances, each processing a different modelled area or a combination of uncertainty and hydrodynamic simulations. The first application has a high priority and a short execution time limit (e.g. a flood emergency occurring in the processed catchment), so it is clearly preferred in the first cycle with 8 core allocation, but needs only 6 cores in the second cycle to finish its computation. The resource manager can therefore reallocate the 2 released cores to the second application, while resources of the third application remain unchanged (note that no onConfigure method is called in the third application as its resources did not change). Finally, 6 cores released by the end of the first application are reallocated to the remaining applications in the third cycle. The resource manager compares the monitoring results from their onMonitor methods and decides to allocate 4 cores to the third application and the 2 remaining cores to the second application.

In real simulations, the duration of each cycle depends on the implementation, but it should be short enough to provide enough possibilities to re-allocate the resources during the whole runtime. In our implementation of RR uncertainty modelling, each cycle is just a few seconds long, and for HD simulations, it is in the order of tens of seconds. This means that during a standard 10-minute RR uncertainty simulation execution, the re-allocation can occur hundreds of times. The re-allocation can therefore be highly dynamic and it can quickly adapt to the changing load of the computing resources and required service quality levels of all running applications.

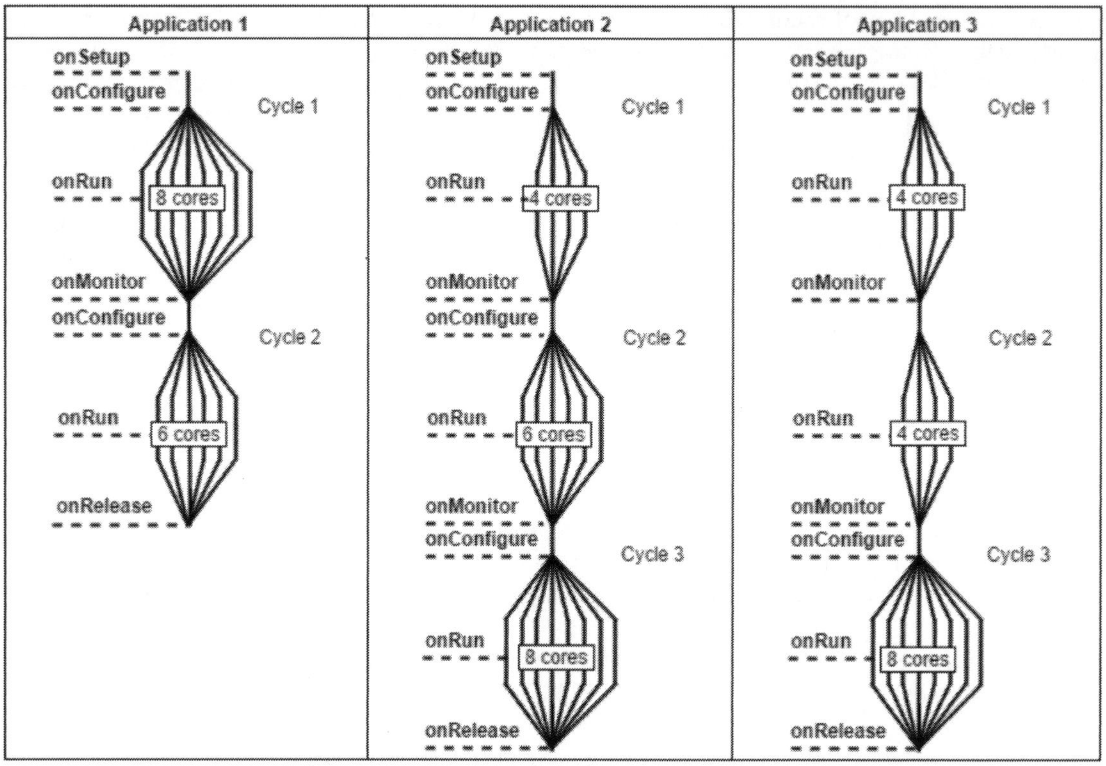

Figure 2. Example of runtime resource management.

4. Quality of provided services

All allowed resource configurations (e.g. number of allocated x86-64 cores and cluster nodes, amount of required memory, allocation of accelerators, etc.) that the managed application is able to work with have to be pre-specified by a set of application working modes (AWM). The applications also have to set their quality of service (QoS) levels that specify their runtime requirements, such as the priority of each task or its requested execution time limit. The HARPA-OS then allocates computing resources to each running task based on the best fitting AWM for the current quality expectations in each cycle.

We have identified three major QoS levels during the Floreon+ system operation that are based on the requirements of the system users (integrated emergency center in the Moravian-Silesian region and expert users) and periodicity of the input data coming to the system from third party providers. The switching between these QoS levels is triggered by exceeding specific flood activity degrees (FAD). FAD describes the state of the water level during flood events. Usually, there are three levels of FAD quantified by a water level and defined for all major water gauges in the Czech Republic; the first degree is alertness, the second degree is preparedness and the third degree is flood hazard. These flood activity degrees are based on the local water and river channel conditions.

The QoS levels used in our system are:

- Standard operation – a standard scenario for non-flood situations. The computation can be relaxed with lower precision, and the system should only use a minimal number of resources for flood monitoring. Rainfall-runoff simulations are executed every hour along with their uncertainty simulations generating 1000 MC samples. Hydrodynamic simulations are started once a day with a 6-hour execution time.
- Alertness – when measured or simulated discharge on any station exceeds the first flood activity degree, the system switches to an alerted mode. Simulations are started more often with a higher focus on precision to more carefully monitor the development in rivers. Rainfall-

runoff simulations are executed every 30 minutes with a 10 minute time step, along with their uncertainty simulations generating different numbers for different modelled catchments – 1250 samples for the Opava catchment, 2000 for the Odra (upper part) catchment, 2500 for the Olše catchment and 3000 for the Ostravice catchment (the modelled catchments are described in section 5.1.). Hydrodynamic simulations are started every 6 hours with a 2-hour execution time.

- Emergency – when the measured or simulated discharge on any station exceeds the second flood activity degree, the system switches to an emergency mode. Excess resources are allocated to provide the simulation results as soon as possible with high precision demands, and the flood monitoring domain has the highest priority. Rainfall-runoff simulations are executed every 10 minutes, along with their uncertainty simulations, generating 3000 MC samples. Hydrodynamic simulations are started every 30 minutes with a 30-minute execution time.

Application working modes have to be specified separately for each type of simulation (i.e. one set of AWMs for simple rainfall-runoff, different set for uncertainties and another one for hydrodynamics) to support the identified QoS. One AWM for each type has to cover the maximum allocable resources needed to support the Emergency mode. One AWM has to contain only minimal resources for economical operation in standard mode. Several AWMs have to exist in between to ensure proper dynamic reallocation of resources during execution when simulation is delayed because of more prioritized requests or errors in the infrastructure.

5. Experiment

To show the functionality of the HARPA-OS runtime resource manager, we have conducted an experiment that executes and manages rainfall-runoff simulations and their uncertainty simulations for the modelled area of the Floreon^{+} system.

5.1. Area of interest

The Moravian-Silesian region is situated in the northeastern part of the Czech Republic in central Europe on the borders with Poland and Slovakia. The western and eastern parts of the region are mountainous. Hydrologically, the territory of the Moravian-Silesian region lies within the river Odra and Morava catchments. The Floreon^{+} system currently only covers the Odra catchment, which can be divided into four delineated sub-catchments – the Ostravice, Olše (excluding the part of the catchment in Poland) and Opava catchments and the upper part of the Odra catchment up to the confluence with the Opava river (see Figure 3). Basic characteristics of the modeled catchments are depicted in **Table 1**.

5.2. Experimental infrastructure

The experiment has been conducted on a machine with a SuperMicro X10DAI Motherboard with two sockets. Each socket arms an Intel(R) with 8 cores Core(TM) i7-4870HQ CPU @ 2.50GHz, providing 16 cores in total, as well as 32 threads with hyper-threading that is disabled for these experiments. The machine is equipped with 256 GB DDR3 1866 MHz without ECC memory.

5.3. Runtime resource management settings

To retain large flexibility and show even small changes in resource allocations, we opted to define finely grained AWMs in this experiment. In total, there are 16 AWMs starting with one CPU core and going through by one core to the whole machine. Numbers of Monte Carlo samples and related execution time limits were set to the QoS levels identified in section 4.

IOP Conf. Series: Earth and Environmental Science 39 (2016) 012061 doi:10.1088/1755-1315/39/1/012061

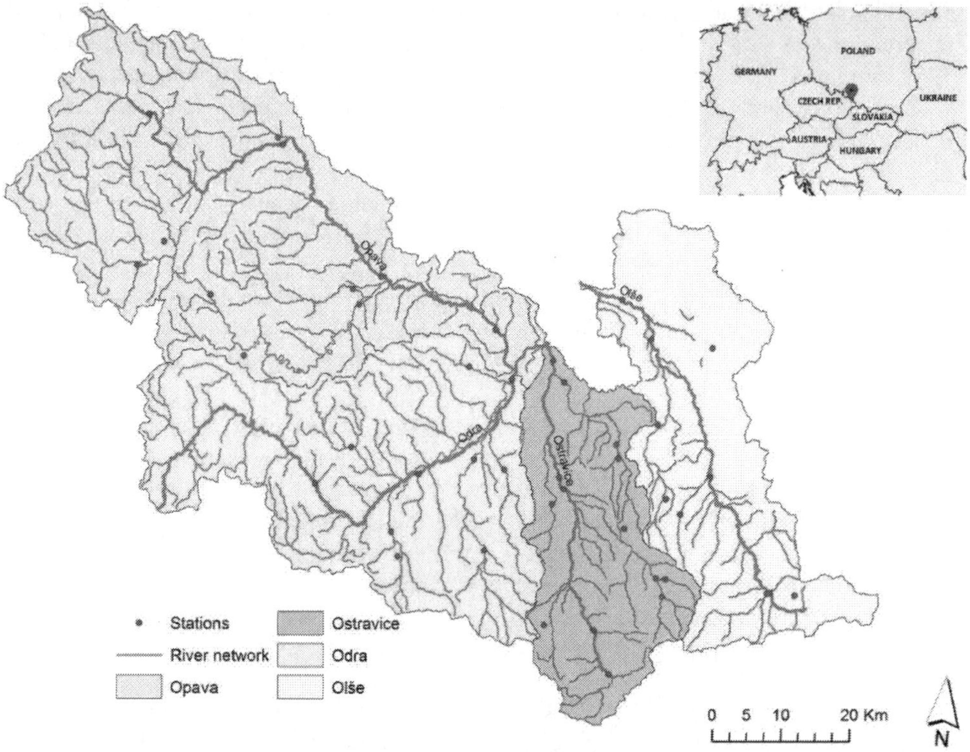

Figure 3. Area of interest.

Table 1. Basic characteristics of studied catchments (in order of importance).

Catchment	Area (km^2)	Length (km)	Spring (m.s.l.)	Related risks
Ostravice	828	64	720	High population density Industrialized zones Major cities included Mountains in the upper part can cause significant runoff
Olše	914	72.8	909	Probability of fluvial floods Narrow channel Cross border catchment Problematic monitoring of Polish tributaries
Odra (upper part)	1615	112	633	Flash flood occurrence High density of streams in the upper part No dams Missing water stations at tributaries
Opava	2089	110.7	529*	Dense river network Extensive area Lack of water monitoring stations Agricultural areas

* confluence of Bíla, Střední and Černá Opava

5.4. Results

Based on the four modelled catchments, four instances of rainfall-runoff and uncertainty simulations were started on the experimental infrastructure at the same time and had to compete for resources. We performed three experimental runs, one for each identified emergency level. All instances started with the same amount of resources, as the resource manager divides all resources in the machine equally.

Figure 4 shows the Standard operation mode where the resource manager quickly decreases equally divided allocated resources of all instances as it receives information that the simulations do not need so much computing power to meet their specified execution time limits. All instances are then kept at the lowest possible resource levels until the end of the computation, leaving the remaining resources free to be used by other applications. The gap with ideal performance grows exponentially, as one core is more than enough to compute simulations in this mode.

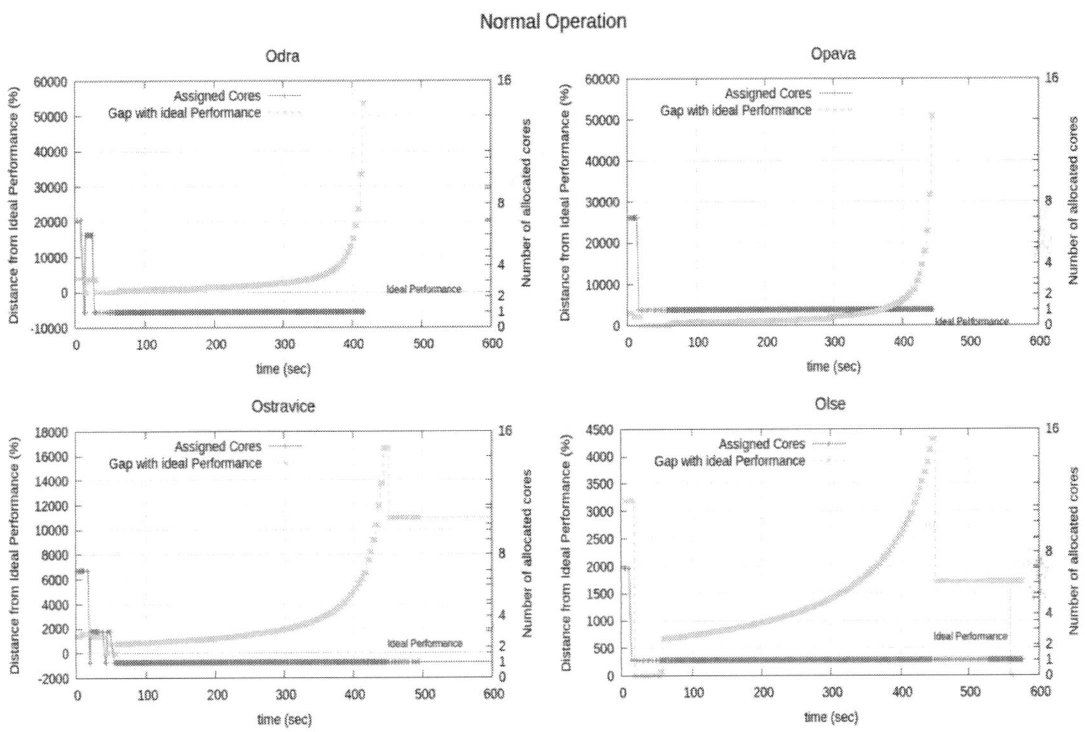

Figure 4. Experimental results for the Standard operation mode.

Figure 5 shows the Alertness mode, in which the Opava (top right) catchment has the least MC samples to compute. Therefore, it only needs minimal resources and even finishes before its time limit to release its allocated core for use by other catchments. The Odra catchment (top left) has the second lowest priority and MC sample count, but it already needs more than one core to finish its computation in the specified time limit for the Alertness mode. The results show how it switches between a 2 and 3-core allocation during its execution as the total execution time estimate reacts to the current allocation. The Olše catchment with the second highest importance (bottom right) has an even higher allocation of 3 or 4 cores and even 5 near the end of the simulation. Allocated resources also increase near the end of the simulation for the most important Ostravice catchment (bottom left) that shows the most dynamic behavior. This is caused by the fact that the total execution time estimate is becoming more accurate near the end and the resource manager tries to finish the simulation right on time.

Figure 6 shows the Emergency mode, where all simulations finally fill the entire machine and the resource manager should pause all other non-flood applications or decrease their resources to the absolute minimum. All instances balance on a 4-core allocation, competing only when any of them become significantly delayed.

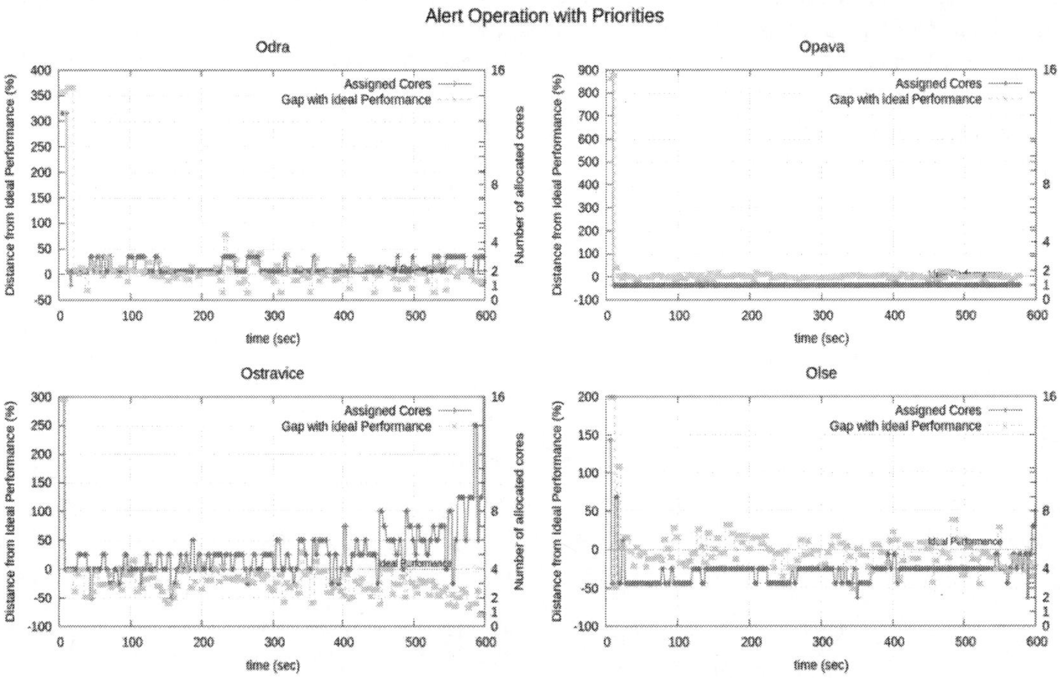

Figure 5. Experimental results for the Alertness mode.

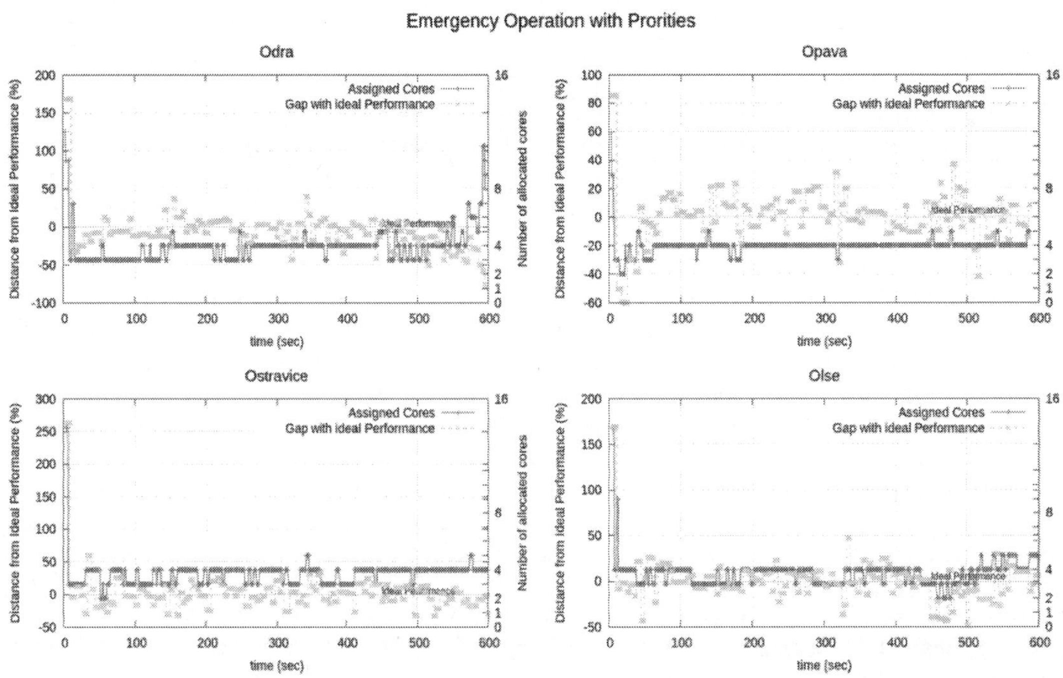

Figure 6. Experimental results for the Emergency mode.

6. Conclusion and future work

The main aim of the Floreon$^+$ project is to provide detailed information about the possibility of any incoming flood and its extent. In the future, Floreon$^+$ should be able to estimate possible property damage before it happens, provide recommendations about flood measures and online decision support for emergency committees to minimize the number of casualties and property damage in the case of extreme weather events. HARPA-OS integration will significantly increase the quality of the provided services, mainly by being able to adapt the computation demand to the changing environment and weather situation, conserving energy and costs of continuous operation and ensuring availability of the system.

The main advantages that HARPA-OS integration can bring to online flood monitoring systems are:

- More cost-efficient continuous operation in non-emergency scenarios, heightened alertness and prioritized execution in warning scenarios, and urgent computation in emergency scenarios
- Optimal reallocation of computational resources during runtime for concurrent execution of several different simulations with different service quality levels
- Faster reactions to errors in the infrastructure – the runtime resource manager can easily reallocate computation from the faulty components

The experiment in this paper focused only on a single-node operation, but to effectively utilize HPC resources, a multi-node approach has to be covered as well. We are currently developing an experimental version of an MPI-based solution that will be able to manage resources over multiple computing nodes. Our future research will also focus on enhancing the proposed methodology by adding priorities for individual instances that can dynamically change during execution based on the state of the catchment and meteorological forecast to better react to local changes (e.g. local flash floods). We also plan to extend the methodology further to other problem domains in the Floreon$^+$ system.

Acknowledgement

This work was supported by The Ministry of Education, Youth and Sports from the National Programme of Sustainability (NPU II) project "IT4Innovations excellence in science - LQ1602", by the Large Infrastructures for Research, Experimental Development and Innovations project "IT4Innovations National Supercomputing Center – LM2015070" and by the 7th EU framework programme project no. FP7-612069 HARPA.

References

[1] Bellasi P, Massari G and Fornaciari W 2015 Effective runtime resource management using Linux Control Groups with the BarbequeRTRM Framework *ACM Trans. Embed. Comput. Syst.* **14**:2 pp 39:1–39:17

[2] Brzoza-Woch R, Konieczny M, Kwolek B, Nawrocki P, Szydło T and Zieliński K 2015 Holistic approach to urgent computing for flood decision support *Procedia Computer Science* **51** pp 2387–96

[3] Cools J, Innocenti D and O'Brien S 2016 Lessons from flood early warning systems *Environmental Science and Policy* **58** pp 117-22

[4] Dobrovicova S, Dobrovic R and Dobrovic J 2015 The Economic impact of floods and their importance in different regions of the world with emphasis on Europe *Procedia Economics and Finance* **34** pp 649–55

[5] Golasowski M, Litschmannova M, Kuchar S, Podhoranyi M and Martinovic J 2015 Uncertainty modelling in rainfall-runoff simulations based on Parallel Monte Carlo Method *Neural Network World* **25**:3 pp 267–86

[6] Haiyun S, Li T, Liu R, Chen J, Li J, Zhang A and Wang G 2015 A service-oriented architecture for ensemble flood forecast from numerical weather prediction *Journal of Hydrology* **527** pp 933–42

[7] Kjeldsen T R, Macdonald N, Lang M, Mediero L et al. 2014 Documentary evidence of past floods in Europe and their utility in flood frequency estimation *Journal of Hydrology* **517** pp 963-73

[8] Koutroulis A G, Tsanis I K and Daliakopoulos I N 2010 Seasonality of floods and their hydrometeorologic characteristics in the island of Crete *Journal of Hydrology* **394** pp 90-100

[9] Kuchar S, Golasowski M, Vavrik R, Podhoranyi M, Sir B and Martinovic J 2015 Using high performance computing for online flood monitoring and prediction *International Journal of Environmental, Chemical, Ecological, Geological and Geophysical Engineering* **9**:5 pp 11-16

[10] Merkuyeva G, Merkuyev Y, Sokolov B V, Potryasaev S, Zelentsov V A and Lektauers 2015 Advanced river flood monitoring, modelling and forecasting *Journal of Computational Science* **10** pp 77-85

[11] Portero A, Kuchar S, Vavrik R, Golasowski M and Vondrak V 2014 System and application scenarios for disaster management processes, the rainfall-runoff model case study *Lecture Notes in Computer Science* **8838** pp 315–26

[12] Rodopoulos D, Corbetta S, Massari G, Libutti S, Catthoor H, Sazeidesz Y, Nicopoulos C, Portero A, Cappe E, Vavrik R, Vondrak V, Soudris D, Sassi F, Fritsch A and Fornaciari W 2015 HARPA: Solutions for dependable performance under physically induced performance variability *Proceedings of International Conference on Embedded Computer Systems: Architectures, MOdeling, and Simulation SAMOS XV Conference* July 20-23, 2015.

[13] Sunkpho J and Ootamakorn Ch 2011 Real-time flood monitoring and warning system *J. Sci.Technol.* **33** pp 227–35

[14] USACE 2009 *US Army Corps of Engineers, Hydrologic Engineering Center, HEC-HMS, Hydrologic Modeling System HEC-HMS. User's manual, version 3.4*

[15] USACE 2010 *US Army Corps of Engineers., Hydrologic Engineering Center, HEC-RAS, River Analysis System. Hydraulic Reference manual, version 4.1*

[16] Yu D 2010 Parallelization of a two-dimensional flood inundation model based on domain decomposition *Environmental Modelling & Software* **25**:8 pp 935–45

[17] Zhang A, Li T, Si Y, Liu R, Shi H, Li X, Li J and Wu X 2016 Double-layer parallelization for hydrological model calibration on HPC systems *Journal of Hydrology* **535** pp 33–45

Hydrogeochemical analysis of volcanic and geothermal fluids in the Andes from Ecuador using hydrochemical plots (Stiff, Piper and Schoeller-Berkaloff diagrams)

D Carrera-Villacrés[1, 2], A Hidalgo[1], P Guevara-García[1], M T Vivero[1] and V Delgado-Rodríguez[1]

[1]Universidad de las Fuerzas Armadas-ESPE, Departamento de Ciencias de la Tierra y la Construcción, Grupo de Investigación en Contaminación ambiental (GICA), Av.,Gral. Rumiñahui S/N, Sangolquí, Ecuador

E-mail: dvcarrera@espe.edu.ec

Abstract. The formation of several sources of hot springs in the Andes from Ecuador was the result of intense volcanic activity due to the subduction of the Nazca oceanic plate under the South American continental plate. The aims of this study include the presentation of chemical analysis in graphical form in order to describe the hidrogeochemistry water geothermal origins, their chemical classification and their relationship to the complex geology of Ecuador using different hydro chemical plots such as Stiff's polygonal diagram, Piper's trilinear diagram and Schoeller-Berkaloff's logarithmic vertical columns diagram. Geothermal waters can be divided into two groups. The first group was associated with an extinct volcanic activity produced in the Cenozoic and were qualified based on the type of water Na+-Cl- , while the second group was associated with young Quaternary volcanic activity, and the types of water were Mg^{2+}-HCO_3^-, Na^+-HCO_3^-, Na^+-SO_4^{2-},Mg^{2+}-SO_4^{2-}.

1. Introduction

Ecuador has various sources of geothermal water as a result of the strong magmatic activity produced in the Quaternary by the subduction of the Nazca oceanic plate under the South American continental plate [1]. These sources are located in the volcanic arc in Ecuador. The aims of this study include the presentation of chemical analysis in graphical form in order to describe the hydro geochemistry from hot springs used in spas located in the Andes of Ecuador, including their behavior and their predominant ions, their chemical classifications and their relationship to the complex geology of Ecuador according to different hydrochemical diagrams such Stiff's polygonal diagram, Piper's trilinear diagram and Schoeller-Berkaloff's logarithmic vertical columns diagram.

2. Materials and methods

The samplings were taken from 34 locations along the Andes in Ecuador. The distance between sampling sites was approximately 3215 kilometers. The sampling points are shown in figure 1.

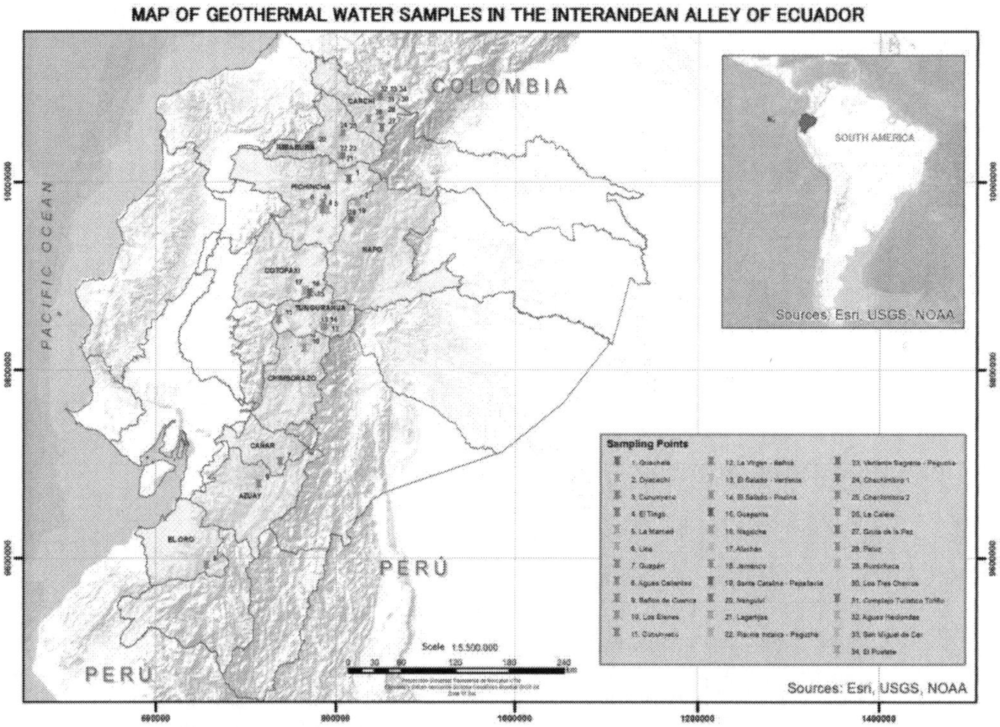

Figure 1. Sampling points of hot springs in the Andes of Ecuador.

2.1. Determination of physico-chemical parameters

Any physic-chemical parameters were measured for all water samples. The first parameter was water temperature as measured in-situ by a mercury thermometer. Sample pH values were analyzed at the laboratory of the Universidad de las Fuerzas Armadas ESPE with a pH meter, Thermo Scientific Orion 3-Star model, and the electrical conductivity (EC) was measured with a conductivity meter, model HACH HQ14d. Finally, Total Dissolved Solids (TDS) and Waste Dry Calcined (WDC) were measured with APHA support [2]. Anions and cations of the first 24 samples were sent Havoc laboratory in Ecuador, accredited by the Accreditation Service Ecuadorian. The concentrations of Ca^{2+} and Mg^{2+} were obtained by colorimeter methods, the concentrations of Na^{+} and K^{+} were measured with an Ion Meter inoLab® pH / ION 7320, Cl^{-} content was determined by colorimetry according to NTE INEN 0976 [3], $HCO3^{-}$ was measured according to the volumetric method and SO_4^{2-}, NO_3^{-} and PO_4^{3-} content were determined spectrophotometrically according to EPA guidelines [4].

Determination of the error rate in the variables was measured in the laboratory.

After obtaining the results of the analysis, a control is necessary to check any result discrepancies according to the APHA [2] equation 1.

$$\%\,Mistake = \left[\frac{\sum\frac{meq}{L}_{cation} - \sum\frac{meq}{L}_{anion}}{\sum\frac{meq}{L}_{cation} + \sum\frac{meq}{L}_{anion}}\right] * 100 \qquad (1)$$

Table 1 shows the percentages of anions and cations admissible for each range.

Table 1. Percentage of allowable difference in analytical results.

Sum of cations (meq/L)	% Acceptable difference
0.0 – 3.0	± 0.2 %
3.0 – 10.0	± 2 %
10 – 800	± 2.5 %

Source:[2]

It is necessary to check the results with the information the Table 1 to review the accuracy of parameters.

2.2. Development of hydrochemical diagrams

The hydrochemical diagrams were developed with Diagrammes software versión 6.5, developed by the laboratory of hydrogeology at the Université d'Avignon.

2.3. Mapping

The maps were produced using Arc GIS software with shape files obtained by the Instituto Geográfico Militar of Ecuador (IGM) and geological mapping accomplished by the Instituto Nacional de Investigación Geológico Minero Metalúrgico in Ecuador (INIGEMM). The maps were added to the polygonal Stiff diagrams to analyze the variation between the cations and anions of the samples and the spatial arrangement of the hydro geochemical families.

3. Results and discussions

To achieve better analysis and interpretation, the hydrogeochemical data was divided into three zones: the northern zone (Carchi, Imbabura and Pichincha), central zone (Cotopaxi, Napo, Tungurahua and Chimborazo) and the southern zone (Cañar, Azuay and El Oro).

In the northern zone, the greatest numbers of samples (20) were obtained due to the presence of volcanoes with their bases in the Cordillera Occidental and the Andes Valley. The recorded temperature values ranged from 18 to 60 ° C, with pH values between 4.96 to 7.65 and electrical conductivity values ranging from 166 to 6795 μS / cm. Relative pH values <6 (Lloa and Aguas Hediondas) indicated a possible interaction with acidic gases [5]. Table 2 shows the parameters measured in Ecuador.

Table 2. Temperature, pH, EC, WDC, cation and anion in water samples in the Andes of Ecuador.

No.	Location	T°C	pH	EC (dS/m)	TDS (ppm)	WDC (ppm)	CATION (mg/l)				ANION (mg/l)			
							K^+	Na^+	Ca^{+2}	Mg^{+2}	$SO4^{-2}$	Cl^-	HCO_3^-	CO_3^{-2}
1	Guachalá	39	6,4	2,6	1424,0	260,0	43,5	335,4	36,5	96,4	<2	340,0	980,0	0,0
2	Oyacachi	50	6,6	5,5	3184,0	2056,0	32,1	893,7	96,2	93,8	101,8	810,0	1701,9	0,0
3	Cununyacu	26	7,0	1,2	660,0	172,0	12,2	105,2	29,7	72,0	8,0	170,0	460,0	0,0
4	El Tingo	40	7,0	3,5	3848,0	972,0	22,7	445,0	24,8	135,9	10,0	380,0	1340,0	0,0
5	La Merced	35	6,5	1,2	1912,0	196,0	11,0	103,1	38,9	71,2	6,0	200,0	440,0	0,0
6	Lloa	29	6,0	1,9	1076,0	152,0	13,8	182,2	76,0	87,0	<2	250,0	680,0	0,0
7	Guapán	60	7,1	20,2	11308,0	10124,0	91,3	4462,0	76,2	95,7	<2	5857,5	2140,0	0,0
8	Agua Caliente - Portovelo	52	8,1	3,4	4012,0	1840,0	34,0	459,5	172,0	47,1	78,0	1099,1	120,0	0,0
9	Baños de Cuenca	62	7,2	4,6	4512,0	2028,0	54,3	637,5	85,0	41,1	72,0	1050,0	440,0	0,0
10	Los Elenes	21	6,9	2,3	1576,0	952,0	7,0	158,8	91,8	168,2	878,0	50,0	320,0	0,0
11	Cununyacu (Tungurahua)	47	8,3	5,1	3408,0	1816,0	4,7	690,0	253,3	94,0	96,0	1633,0	140,0	2,7
12	La Virgen (Baños)	53	6,5	5,9	4232,0	2712,0	59,8	409,9	14,6	481,9	965,0	790,0	920,0	0,0
13	El Salado VT (Baños)	40	6,3	7,2	6128,0	4156,0	56,0	361,3	281,4	490,1	856,8	1001,1	1421,9	0,0
14	El Salado Piscina (Baños)	45	6,6	9,0	7932,0	5476,0	69,1	478,2	306,2	641,3	3234,2	652,1	396,5	0,0
15	Guapante	25	7,2	1,2	1008,0	540,0	10,1	84,0	29,7	71,8	93,0	41,0	520,0	0,0
16	Nachehe	27	6,8	4,0	2616,0	1480,0	38,4	338,0	11,2	303,5	16,0	380,6	1781,0	0,0
17	Aluchán	43	7,0	1,5	1008,0	732,0	6,8	187,1	26,5	46,3	38,0	214,8	420,9	0,0
18	Jamanco	61	6,7	7,7	4772,0	4060,0	138,9	1239,0	256,8	38,9	297,0	2272,5	330,0	0,0
19	Santa Catalina (Papallacta)	56	7,1	2,1	1376,0	980,0	5,7	235,3	192,3	9,0	375,0	450,0	66,0	0,0
20	Nangulví	50	7,4	4,9	3016,0	2624,0	7,6	641,7	229,9	90,9	335,0	1466,9	72,0	0,0
21	Lagartijas	18	6,1	0,5	284,0	116,0	4,5	22,0	14,0	33,9	2,0	29,8	226,0	0,0
22	Peguche (Piscina Incaica)	23	6,5	1,9	1132,0	712,0	13,6	137,4	61,7	101,6	13,4	230,0	630,0	0,0
23	Peguche (Vertiente Sagrada)	23	6,6	2,2	1332,0	900,0	23,3	165,0	46,1	130,5	5,5	290,0	750,0	0,0
24	Chachimbiro 1	52	7,6	6,6	4128,0	3528,0	210,7	968,2	98,0	84,2	57,0	1880,5	500,0	0,0
25	Chachimbiro 2	60	6,4	6,8	4076,0	3432,0	183,2	964,0	83,8	82,2	56,0	1731,7	572,4	0,0
26	La Calera	34	6,3	1,3	852,0	552,0	29,1	73,4	74,2	85,5	31,0	64,3	737,4	0,0
27	Gruta de la Paz	40	6,9	2,6	1748,0	1232,0	74,6	156,1	27,7	215,1	3,5	159,1	1416,5	0,0
28	Paluz	21	6,2	1,1	764,0	508,0	4,7	32,9	68,1	69,9	3,5	38,9	576,5	0,0
29	Rumichaca	34	7,1	3,1	2032,0	1444,0	96,3	267,5	29,9	179,4	6,5	259,0	1490,8	0,0
30	Los 3 Chorros (Neptuno)	25	6,0	0,8	588,0	456,0	27,4	57,3	40,3	45,3	18,0	42,3	435,0	0,0
31	Complejo Turístico Tufiño	25	6,6	0,9	640,0	484,0	24,4	69,6	45,3	44,5	140,8	90,9	270,5	0,0
32	Aguas Hediondas	58	5,0	1,8	1572,0	1288,0	56,6	149,5	87,6	70,2	702,7	118,9	24,4	0,0
33	San Miguel de Car	23	6,2	0,4	376,0	240,0	10,6	26,5	18,0	24,2	37,4	32,0	151,3	0,0
34	El Puetate	18	6,7	0,2	208,0	128,0	1,3	7,1	8,6	12,1	4,4	11,4	83,5	0,0
	Minimun	18	4,96	0,1669	208	116	1,329	7,13	8,6	9,0425	2	11,36	24,4	0
	Average	38,7	6,7	3,7	2610,0	1716,1	43,5	457,2	89,2	128,1	275,5	708,5	663,4	0,1
	Maximum	62	8,3	20,2	11308,0	10124,0	210,7	4462,0	306,2	641,3	3234,2	5857,5	2140,0	2,7

Total dissolved solid (TDS) content ranged from 208 to 4128 mg l^{-1}. Figure 2a represents the associated Piper Diagram, and Map 1 depicts the related Stiff Diagram. Four water families were determined in the Piper and Stiff diagrams: sodium bicarbonate, (Guachalá and El Tingo), sodium Chloride (Chachimbiro y Nangulví), sulfated sodium (Aguas Hediondas) and bicarbonate magnesium, while the remaining samples were characterized by their low salinity. The hot springs from Pichincha Province are associated with the Caldera Chacana located in the volcanic center of the largest rhyolite volcanic quaternary in Ecuador [6]. Carchi Province is associated with the Chiles volcanoes (Aguas Hediondas) and the pyroclastic deposits of extinct volcanoes in Imbabura to the volcanic deposits of the Yanahurco volcanoes (Chachimbiro) and Imbabura (Peguche).

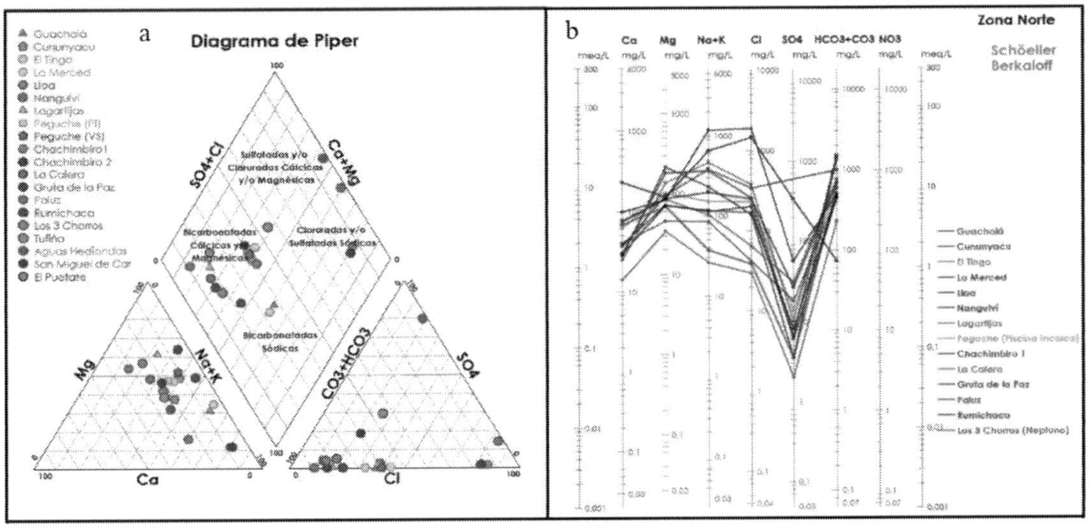

Figure 2. (a) Piper diagram of sample points in the northern zone.

(b) Schoeller – Berkaloff diagram of sample points In the northern zone.

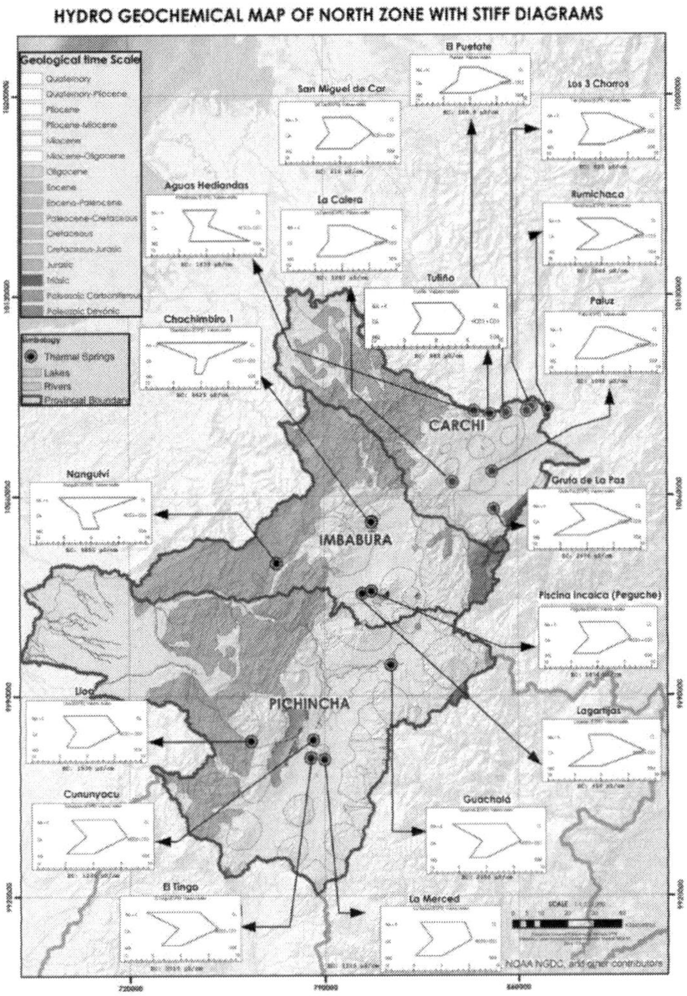

Figure 3. Hydrogeochemical map of central zone via Stiff diagrams.

IOP Conf. Series: Earth and Environmental Science **39** (2016) 012062 doi:10.1088/1755-1315/39/1/012062

In the central zone, recorded temperature values ranged from 21 to 61 ° C, with pH values between 6.2 to 8.3 and EC values between 1150 to 8970 µS / cm.

The values from 1008 to 7932 mg l^{-1} were measured in TDS (El Salado). As shown in the Piper diagram (figure 3a) and the Stiff diagrams (Map 2), four family types were determined. The first family was bicarbonate magnesium (Guapante and Nagsiche), likely the result of shallow water areas where sediment can accumulate [7] and matching its low mineralization and temperature. The second family was sulphate magnesium (El Salado, La Virgen and Los Elenes). The third family was sodium chloride (Papallacta, Cununyaku and Jamanco). The fourth family was sodium bicarbonate (Aluchán and Oyacachi). The hot springs in this area are located within the Cordillera Real and associated with volcanic activity produced in the Quaternary period by existing strata volcanoes as characterized by their lava flows and pyroclastic basaltic to rhyolitic composition [8]).

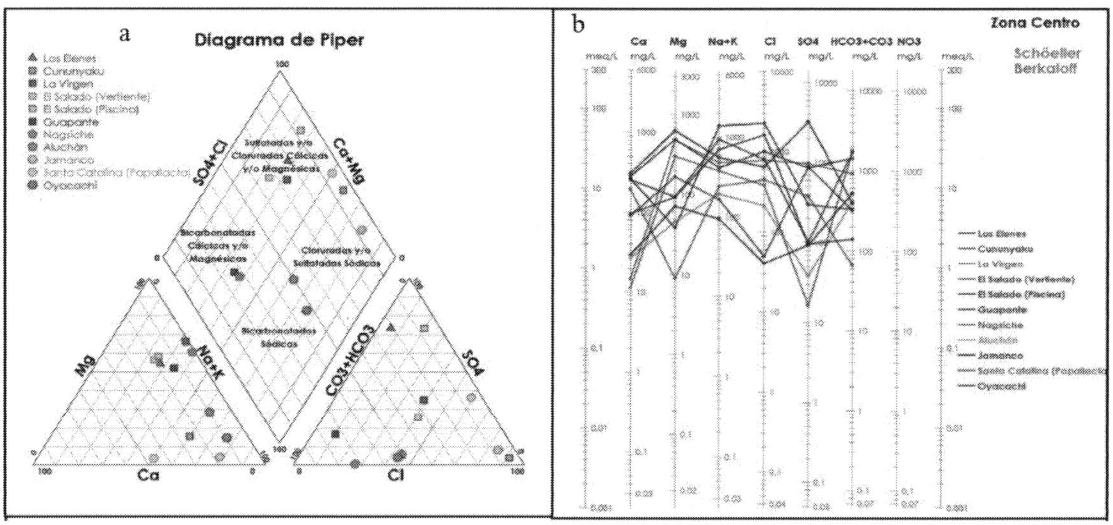

Figure 4. (a) Piper diagram of samples points in central zone. (b) Schoeller – Berkaloff diagram of sample points.

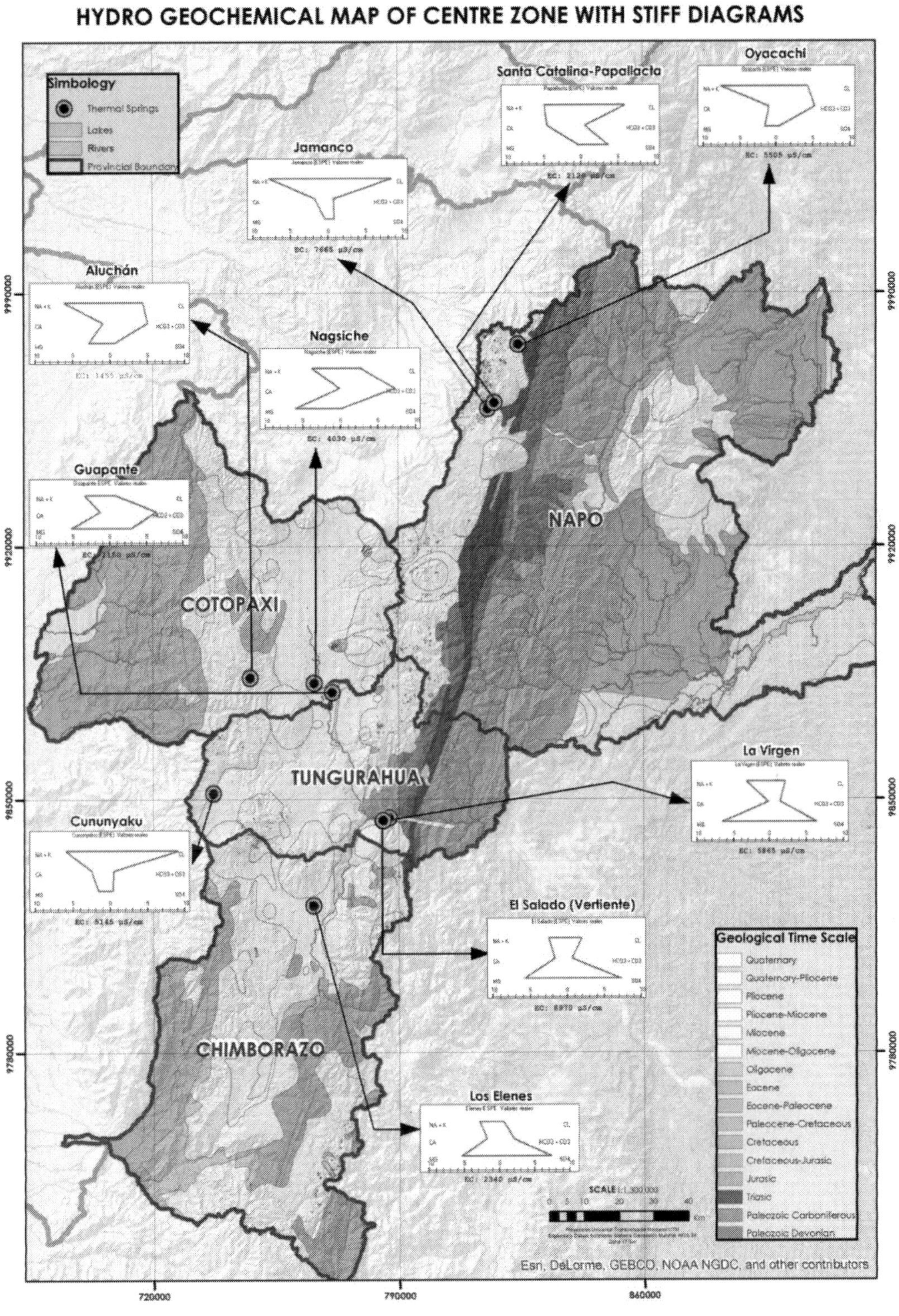

Figure 5. Hydro geochemical map of central zone Stiff diagrams.

In the southern zone, the highest temperature and EC values were measured as follows: Baños de Cuenca (62 °C) and Guapán (20,220 μS cm^{-1}). The pH values were neutral except for Portovelo (8.09) which indicated a slightly basic value. Portovelo also indicated high TDS values greater than 4000 mg l^{-1}, except for Guapán (11308 mg l^{-1}). This high salinity is the result of prolonged interaction with rock-water and processes of evaporation [5]. As shown in the related Stiff diagrams (Map 3) and Scholler-Berkaloff graph (figure 4b), the most abundant ions in these samples are Na^+ and Cl^-.

The Piper diagram (figure 4a) identified a water family: sodium chloride characterized by deep ancient aquifers and remaining geology of the extinct volcanic activity in the Miocene [9] in this area of Ecuador. These waters are likely near or pass through salt diapirs [10].

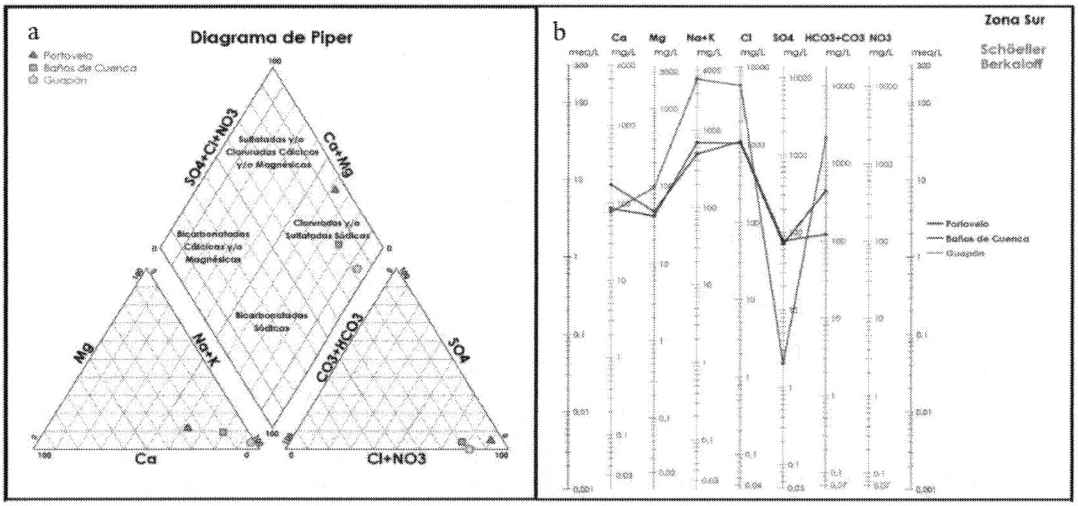

Figure 6. (a) Piper diagram of sample points in the southern zone. (b) Schoeller – Berkaloff diagram of sample points in the southern zone.

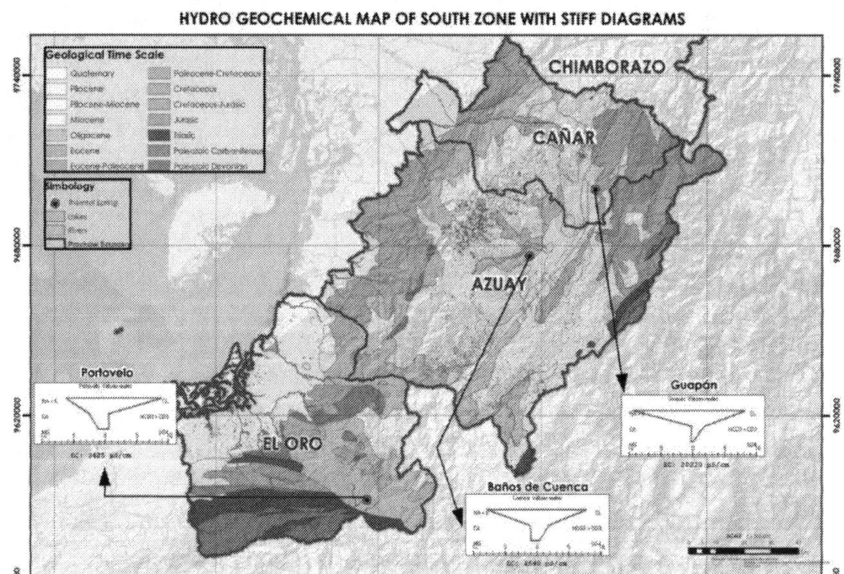

Figure 7. Hydrogeochemical map of south zone via Stiff diagrams.

IOP Conf. Series: Earth and Environmental Science **39** (2016) 012062 doi:10.1088/1755-1315/39/1/012062

4. Conclusions

The hot springs in Ecuador can be divided into two groups. The first group was associated with extinct volcanic activity produced in the Cenozoic (Oligocene, Miocene and Pliocene) and the second group was associated with young volcanic activity in the Quaternary. The first group was characterized by sodium chloride water (Guapán, Los Elenes, Baños de Cuenca, Cununyaku and Nagulví); these samples were associated with higher recorded temperatures and higher mineralization than other groups, and Stiff diagrams in "T" shapes. The water samples from the Quaternary were magnesium bicarbonate and sodium bicarbonate, sodium sulfate and magnesium sulfate. These types of waters are related to the complex geology of characteristic volcano deposits of igneous rocks (basalts and rhyolites) and metamorphic rocks (schists) of the Cordillera Real and Inter-Andes Valley where they were found, and characterized by Stiff diagrams shaped like arrowheads (Mg^{2+}- $HCO_3^-Na^+$ - HCO_3^-) and irregular polygons (Mg^{2+}-SO_4^{2-}, Na^+ -SO_4^{2-}).

References

[1] Lonsdale P and Klitgord 1978 Structure and tectonic history of the Eastern Panama Basin *Geological Society of America Bulletin* **89**: 981-99

[2] American public health association 1995 *Standard methods for the examination of water and wastewater 19th edition* Publication office American public health association Washington D.C. p 199 and 211

[3] NTE INEN 1984 Agua potable *Determinación de Cloruros*

[4] Environmental Protection Agency 1983 *Methods for Chemical Analysis of Water and Wastes* Washington D.C.

[5] Inguagggiato S, Hidalgo S, Beate B and Bourquin J 2010 Geochemical and isotopic characterization of volcanic and geothermal fluids discharged from the Ecuadorian volcanic arc *Geofluids* 1-17

[6] Villares F 2010 Estudio geovulcanológico de la Zona sur de la Caldera Chacana *Escuela Politécnica Nacional*

[7] Custodio E and Llamas M 2001 *Hidrología Subterránea* Tomo I. Ediciones Ortega. España

[8] Aspden J A and Litherland M 1992 The geology and Mesozoic Collisional history of the Cordillera Real , Ecuador *Tectonophysics* **205**: 187-204

[9] Beate B, Monzier M, Spikings R, Cotton J, Silva J, Bourdon E and Eissen J P 2001 Mio-Pliocene adakite generation related to flat subduction in southern Ecuador: Quimsacocha volcanic center *Earth and Planetary Science Letters* **192**: 561-70

[10] Carrera D, Guevara P, Hidalgo A, Vivero M T and Maya M 2015 Removal of physical information chemistry of spa that is utilizing geothermal water in Ecuador *Procedia Earth and Planetary Sciences* **15**: 367-73

IOP Conf. Series: Earth and Environmental Science **39** (2016) 012063 doi:10.1088/1755-1315/39/1/012063

Optimization strategy integrity for watershed agricultural non-point source pollution control based on Monte Carlo simulation

Y Gong [1], Y J Yu [1,2] and W Y Zhang [1]

[1]School of Materials Science & Engineering,Beijing Institute of Technology, Beijing 100000, China

Email: 04575@bit.edu.cn

Abstract. This study has established a set of methodological systems by simulating loads and analyzing optimization strategy integrity for the optimization of watershed non-point source pollution control. First, the source of watershed agricultural non-point source pollution is divided into four aspects, including agricultural land, natural land, livestock breeding, and rural residential land. Secondly, different pollution control measures at the source, midway and ending stages are chosen. Thirdly, the optimization effect of pollution load control in three stages are simulated, based on the Monte Carlo simulation. The method described above is applied to the Ashi River watershed in Heilongjiang Province of China. Case study results indicate that the combined three types of control measures can be implemented only if the government promotes the optimized plan and gradually improves implementation efficiency. This method for the optimization strategy integrity for watershed non-point source pollution control has significant reference value.

1. Introduction

Agricultural non-point source pollution refers to the surface and groundwater environmental pollution caused when sediment, nutrients, pesticides and other pollutants in farmland are integrated into water through surface runoff, interflow, farmland drainage and underground seepage during the precipitation or irrigation processes[1]. These pollutants are derived primarily from farmland fertilization, pesticides, livestock, aquaculture and rural residents. With the acceleration of modernization in rural areas, farm production and lifestyles have changed gradually, and the contaminant components of waste have become increasingly complex. Agricultural non-point source pollution is the most significant and most widely-distributed non-point source pollution, and is becoming one of the primary causes of the deterioration of the ecological environment in rural areas[2]. This has severely restricted the sustainable development of agriculture and the rural economic environment[3]. In order to promote the stable and rapid development of the rural economy and realize harmonious development between man and nature, an integrated study of agricultural non-point source pollution is therefore immediately necessary[4]. Implementing the optimization strategy integrity for pollution control can improve governance efficiency[5]. Additionally, it has significant value in both theoretical and practical applications[6].

2. Research method

Based on actual conditions, a field study was conducted in the northeast area of China. The survey primarily included the basic conditions of agricultural land such as position, topography, range, elevation, soil type, planting type and area, etc. A survey of nitrogen and phosphorus loss was also conducted. Based on the characteristics of the area including concentrated rainfall, high soil nutrient content, strong soil freezing and thawing and serious loss of nitrogen and phosphorus, the Ashi River watershed was chosen as a case study, and various proposed measures for the governance of agricultural non-point source pollution were investigated. A software model was then constructed to simulate the optimization scheme.

The general idea is as follows: First, the source of watershed agricultural non-point source pollution is divided into four aspects, including agricultural land, natural land, livestock breeding and rural residential land. Secondly, different pollution control measures at the source, midway and ending stages are chosen. Thirdly, the optimization effects of pollution load control in three stages are determined based on Monte Carlo simulation.

The overall technical route is shown in figure 1. The research procedures are as follows. In figure 1, different “X” values indicate different scenarios. Rlb and Rub indicate the upper and lower bounds of reduction rates.

- Collate each optimized measure of control efficiency as the input parameters of the software model.
- Select six optimization measures according to specific geographic, topographic and climatic conditions of the case study area.
- Simulate the corresponding optimization results of each scenario with MATLAB software based on Monte Carlo simulation and adjust the scenario model input parameters. Obtain the corresponding output parameters of each optimization scheme.
- Integrate output parameters. Finally, select the best optimization schemes by comparison to provide technical support for the formulation of watershed agricultural non-point source pollution control measures.

The primary collected control measures are as follows; each measure has upper and lower bounds of the rate of nitrogen reduction:

- Source control measures. These primarily include N_{11} centralized treatment of rural sewage, N_{12} centralized treatment of rural life garbage, N_{13} scale of livestock and poultry, N_{14} grain for green, N_{15} soil improvement technology, N_{16} reasonable application of livestock manure, N_{17} testing soil for formulated fertilization, N_{18} water-saving irrigation farmland, and N_{19} agricultural structure optimization.
- Path control measures. These primarily include N_{21} technique of artificial pond, N_{22} oxidation pond technology, N_{23} artificial wetlands technology, and N_{24} vegetation buffer.
- River bank control measures. This primarily include N_{31} water shore buffer.

A simulation of an optimization scheme of total nitrogen pollution was chosen as an example. The reduction plan for total nitrogen includes the following six scenarios.

- Choose common optimization measures and build optimization scenario 1——TN_SA1: N_{13}+ N_{15}+ N_{17}+ N_{23}+ N_{31};
- Choose policy guidance optimization measures and build optimization scenario 2 —— TN_SA2: N_{13}+ N_{14}+ N_{16}+ N_{23}+ N_{24}+ N_{31};
- Choose optimization measures in which farmers can actively participate and build optimization scenario 3——TN_SA3: N_{12}+ N_{16}+ N_{18}+ N_{21};
- Choose more scientific optimization measures and build optimization scenario 4——TN_SA4: N_{13}+ N_{16}+ N_{13}+ N_{19}+ N_{22}+ N_{23}+ N_{31};
- Choose efficient cutting optimization measures and build optimization scenario 5 —— TN_SA5: N_{12}+ N_{15}+ N_{23}+ N_{31};

- Choose the source control optimization measures and build optimization scenario 6——TN_SA6: N_{12}+ N_{13}+ N_{14}+ N_{15}+ N_{16}+ N_{17}+ N_{18}+ N_{19}.

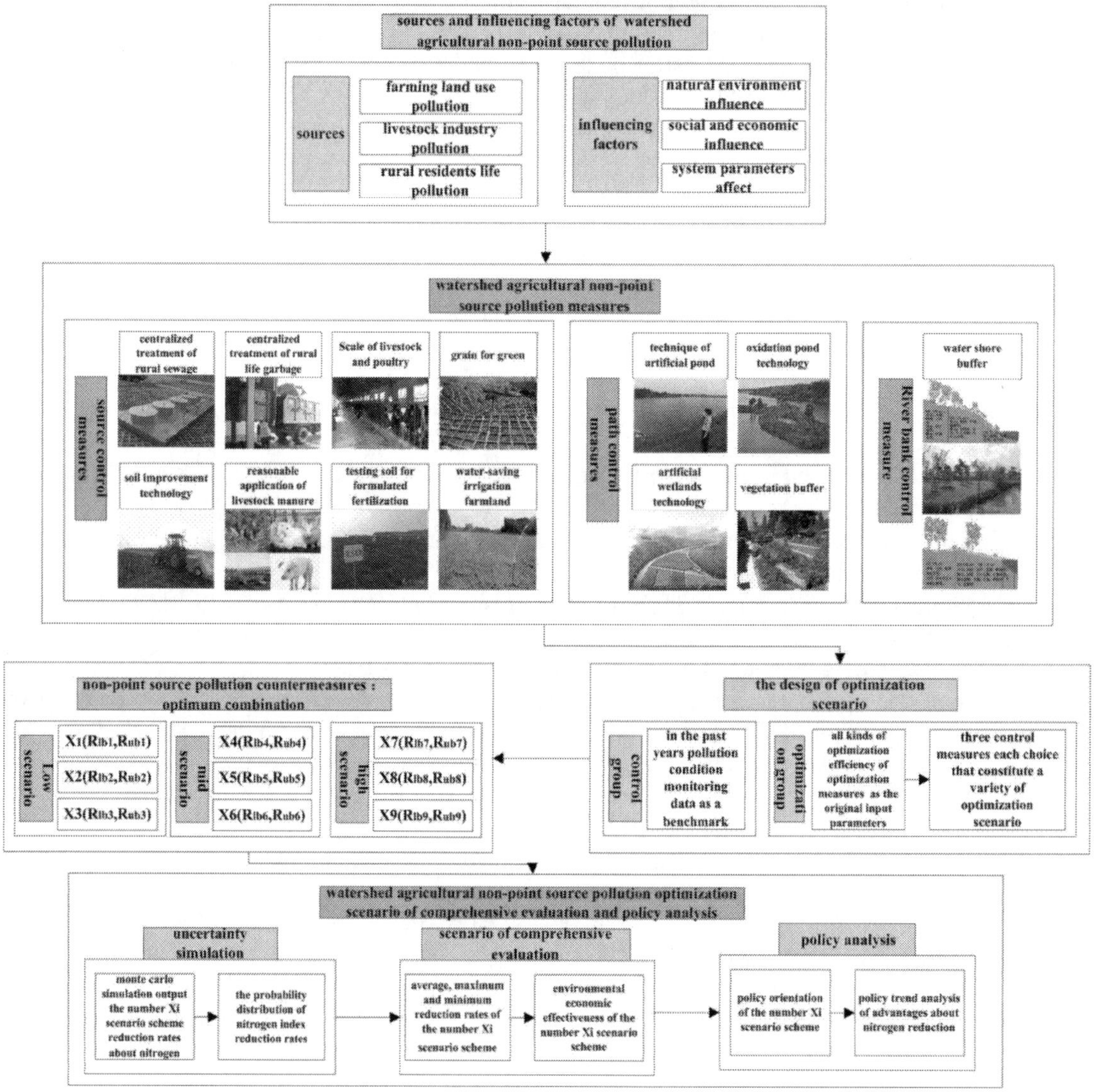

Figure 1. Research diagram for agricultural non-point source pollution control.

3. Results and discussion

The optimization results of total nitrogen pollution were respectively simulated based on Monte Carlo simulation. The probability distribution is shown in figure 2 and the optimization effect of the distribution box is shown in figure 3. The optimization effects of the total nitrogen pollution situations are respectively distributed as numbers 1 through 6 in 67, 58, 65, 68, 52, and 68 which can be seen in figure 2 and figure 3. The optimization effects of scenario 2 and Ccenario 5 are relatively good. The optimization effects of Scenario 6 are relatively poor. The optimization effects of scenarios 1, 3 and 4 are similar and fall in the middle of the observed efficacies.

The optimization effects of efficiency and policy-oriented optimization schemes are relatively good. The optimization effect of source control optimization is relatively poor. Due to the existence of a certain reading error, errors exist between the average value displayed in the box diagram and the normal distribution, but overall optimization results similar.

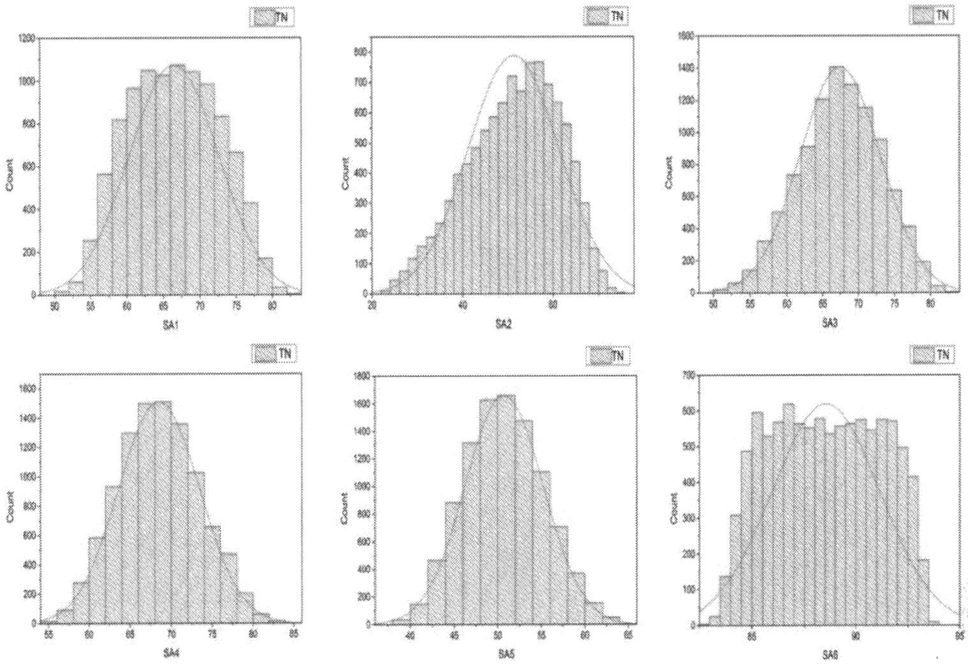

Figure 2. Normal distribution of the optimization effect of watershed agricultural non-point source total nitrogen pollution.

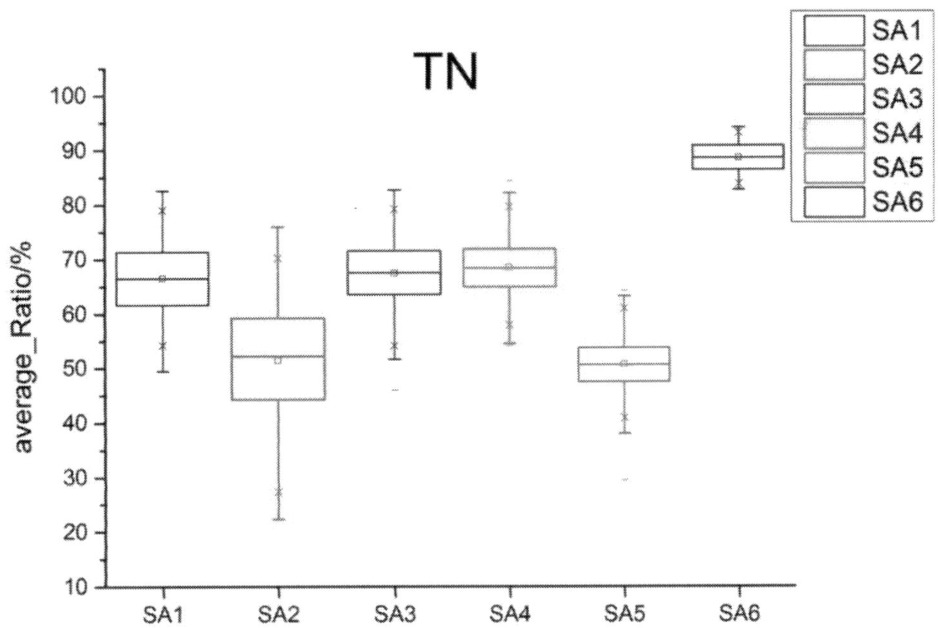

Figure 3. Box figure of optimization effect of watershed agricultural non-point source total nitrogen pollution.

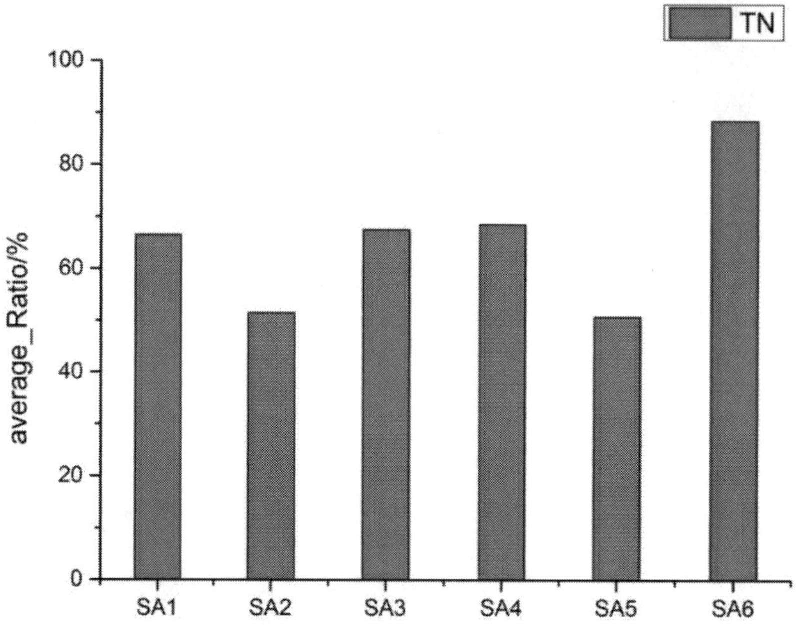

Figure 4. Histogram of optimization effect of watershed agricultural non-point source total nitrogen pollution.

The optimization results of each nitrogen pollution condition was respectively simulated based on Monte Carlo simulation. The computed optimization effect of each total nitrogen pollution scenario is shown in figure 4. Results indicate average optimization reduction. The optimization effects of senarios 1 through 6 are 66.47, 51.45, 67.46, 68.54, 50.78 and 88.55, respectively. The results are similar to analytical results presented in the normal distributions, box figures and histograms.

Analysis of figure 2, figure 3 and figure 4 indicates that the order of optimization effects induced by the six scenarios is as follows: S2, S5>S1, S3, S4>S6. In scenario 2 and scenario 5, source control measures, path control measures and riverbank control measures combined. Therefore, in the governance process of watershed agricultural non-point source pollution, three types of control measures should be combined. Source control measures require relatively low cost; however, it is difficult to control the agricultural non-point source pollution in watersheds with source control measures alone. scenario 3 indicates that farmers are actively involved in the optimization measures, source control measures and path control measures combined. The execution rate of this scenario is relatively high but each optimization measure is not sufficient and lacks scientific and policy guidance. scenario 1 and scenario 4 represent the combination of three types of control measures. scenario 1 represents a common optimization measure, and lacks scientific and policy guidance. scenario 4 provides a more scientific optimization measure, but the execution rate of this scenario is relatively low and cannot be widely used across a large area.

4. Conclusion

Based on the analysis of optimization scenarios regarding total nitrogen pollution, the governance of watershed agricultural non-point source pollution requires policy guidance and more efficient optimization measures. The proposed plan can be implemented only if the government promotes the optimized plan gradually improves implementation efficiency. This simulation scheme is feasible and has significant reference value.

Acknowledgements
The authors would like to express a deep sense of gratitude for the following contributors: (1) the National Natural Science Foundation of China (No. 51474033, No. 41301636); (2) the National Key Technology Research and Development Program of the Ministry of Science and Technology of China (No. 2012BAD15B05).

References

[1] Huang J H, *et al.* 2015 *Journal of Hydrology* **524** 100-10
[2] Liu R M, *et al.* 2016 *Journal of Hydrology* **533** 379-88
[3] Crespo R. de J, *et al.* 2016 *Ecological indicators* **66** 47-54
[4] Lu L, *et al.* 2016 *Agriculture, Ecosystems and Environment* **222** 48-59
[5] Yeboah F K, *et al.* 2015 *Land Use Policy* **49** 75-85
[6] Liu X Y, *et al.* 2015 *International Soil and Water Conservation Research* **3** 130-40

Research framework of integrated simulation on bilateral interaction between water cycle and socio-economic development

C F Hao[1, 2, 4], X L Yang[3], C W Niu[2] and Y W Jia[2]

[1] Department of Hydraulic Engineering, Tsinghua University, Beijing 100084, China
[2] State Key Laboratory of Simulation and Regulation of Water Cycle in River Basin, China Institute of Water Resources and Hydropower Research, 1 Yu-Yuan-Tan South Road, Beijing 100038, China
[3] Beijing Institute of Water, 21 Chegongzhuang West Road, Beijing 100044, China

E-mail: andychao2004@126.com

Abstract. The mechanism of bilateral interaction between natural water cycle evolution and socio-economic development has been obscured in current research due to the complexity of the hydrological process and the socio-economic system. The coupling of economic model CGE (Computable General Equilibrium) and distributed hydrological model WEP (Water and Energy transfer Processes) provides a model-based tool for research on response and feedback of water cycle and social development, as well as economic prospects under the constraint of water resources. On one hand, water policies, such as water use limitation and water price adjustment under different levels of socio-economic development, are to be evaluated by CGE model as assumed conditions and corresponding results of water demand could be put into WEP model to simulate corresponding response during the whole process of water cycle. On the other hand, variation of available water resources quantity under different scenarios simulated by WEP model may provide proper limitation for water demand in CGE model, and corresponding change of economic factors could indicate the influence of water resources constraints on socio-economic development. The research is believed to be helpful for better understanding of bilateral interaction between water and society.

1. Introduction

The mechanism and characteristics of the water cycle has been one of the most essential problems in the field of hydrological research. With rapid socio-economic development, the evolution of the water cycle is becoming increasingly complex due to ever expanding human activities. Solar radiation and gravity are primary driving forces for a natural water system with little human influence, but it has been disturbed to a large extent by water supply and drainage. Dualistic characteristics of hydrological processes have attracted more and more attention in water resources research. Hydrological interactions between human society and the nature system are considered to be the key issue of water resources assessment and management under the changing environment especially for countries facing water scarcity [1].

In view of the high impacts of socio-economic water use on the natural water cycle, hydrological researchers turn to integrated simulation of natural hydrological processes and water-using processes. The theory of dualistic water cycles was proposed and distributed hydrological model developed on the basis of a natural-artificial water cycle [2]. However, the quantity and process of water demand is not yet well regarded by the current dualistic hydrological model in which water use is considered simply by sectoral quotas and total quantity control. The mechanism of bilateral interaction between natural water cycle evolution and socio-economic development has not been clearly illuminated due to complexity of the hydrological process and the socio-economic system. As a fact of vital relation between water consumption and socio-economic development, elements of water and socio-economic system should be taken into consideration, including water resources distribution and industrial structure.

2. Methodology

To discover the coordinated variation of water demand to socio-economic development, the CGE (Computable General Equilibrium) model is recommended, which is able to explain every aspect of the socio-economic system and provide the quantity and the price of all factors in the market, such as labor, land and capital, by quantitative description of the input-output relationship among each account in SAM (Social Accounting Matrix) on the basis of Walras general equilibrium theory [3]. Water resources as one of those factors could also be appropriately considered. Regional water demand of each sector and effects of water resources constraint could be predicted. Meanwhile, WEP model, short for Water and Energy transfer Processes, can well reflect the change of runoff and convergence under different conditions of land surface and water use by simulation of the whole water cycle process based on physical mechanism, in which the spatial varieties inside specific region are fully considered, such as hydrological processes, input variables, boundary conditions and geometric basin characteristics. The quantity of available water resources and corresponding hydrological processes driven by natural factors and human activities play an intermediate role in the connection of human society and the water system. The integration of the economic model CGE and the distributed hydrological model WEP as a model-based tool would be helpful for better understandings on interactive relationships between hydrological evolution and socio-economic development [4], by quantitative analysis of the water consumption and hydrological process variations under preset scenarios.

2.1. Economic model CGE

The CGE model originated in Walras' general equilibrium theory for research on complicated causality in socio-economic system, and has been widely used in policy analysis [5-9]. Different sectors, commodities and production factors are involved in the CGE model. It describes intricate interdependence of production activities and trading behaviors through simultaneous mathematical equations defining the behavior of the different actors. The CGE model is able to indicate the direct and indirect impacts of different sectoral policies in the socio-economic system with chain reactions under various disturbances, by providing the quantity and the price of all factors in the market by numerical simulation of the input-output relationship among each account in the social accounting matrix. It follows the SAM disaggregation for comprehensive actions in socio-economic system. Production activities in CGE are examined according to profits maximization. Consumption behaviors are decided on the basis of utility maximization. There are several essential constraints for overall systems to be satisfied in the model. The modelling approaches and functions in the CGE model are referred to in [3].

Environment and resources can be incorporated into the model to reflect corresponding responses of economic factors, as a result of preset decisions on resource allocation principle and income distribution mechanism, considering the balancing requirement and feedback system of environment, resources and economic activities [10]. For the research of economic impacts of water policy and water constraint under various situations, the key problem is how to embody water resources in the model to realize the dynamic estimation of water supply and demand. There are two major types of methods incorporating water resources into the traditional CGE model. One is to build relationships between water resources and correlative accounts that already exist in SAM structure by subdivision of those accounts according to the diverse characteristics associated with water supply and demand. Functional expressions and new equations are usually added and relationships between water use and economic departments are defined in terms of input and output benefits, which is easy to be achieved particularly for the circumstance lack of enough data required by SAM. The other way is to transform the social accounting matrix by embedding water resources in the model. The factors of water resources are introduced into the new version of SAM as one of its basic factors. The price and demand of water could be calculated as an endogenous variable. It's essential for the model to illustrate domestic and productive water uses and corresponding economic variations under different scenarios [11].

2.2. Hydrologic model WEP

The WEP model [12] was designed using Physically-Based Spatially Distributed (PBSD) models and Soil Vegetation Atmosphere Transfer (SVAT) models for references. It has demonstrated powerful capability for integrated hydrological simulation under changing environment in Japan and China. There are several technological advantages in the model: (1) organic combination of physically-based hydrologic simulation and energy balance in water cycle, (2) fine description on spatially uneven distribution of underlying surface on small scale by the mosaic method, and (3) self-adaptive operation of different runoff generation methodologies by simultaneously numerical simulation of surface runoff and groundwater on basis of topography and soil characteristics and precipitation features. The modelling structure and corresponding approaches adopted in the WEP model are referred to in [13].

Refinement of simulation grids on micro-scale and physically-based numeric calculation of hydrological components in the WEP model brings tremendous computational work which prevents its application in large basins. Besides, the hydrological processes of artificial branch of the water cycle are not well considered in the original model structure. The WEP-L model is developed on the basis of the theory of the dualistic water cycle by significant improvements of the WEP model [13]. The WEP-L model takes the subdivisions of contour bands intersecting small sub-watersheds as basic calculation units to realize the balance between proper computational work and satisfactory representation of underlying surface. The subdivisions in the model are coded according to the Pfafstetter rule. The vegetation and land use types in the WEP-L model are grouped into three classifications including irrigated fields, non-irrigated farmland, and other land cover to reflect the impacts of artificial irrigation on the water cycle. Moreover, artificial interferences on hydrologic processes by water dispatching of reservoirs, as well as water use and drainage, are simulated by newly the developed water allocation and regulation model (WARM). The WARM model is coupled with the WEP-L model for integrated simulation of the water cycle, in which the processes and distributions of water use and drainage are interpolated by statistics data. Besides, the quantities and processes of snow melting are calculated by improved snow melt module on the basis of temperature-index approach.

The improved WEP model is able to accomplish water cycle simulation and general water resources assessment for preset conditions with and without water use, and fully illustrate water cycle components including both of surface runoff and groundwater. In addition, hydrological variations under a changing environment can be carried out by updating meteorological data or land covers.

2.3. Coupling of CGE and WEP

Along with the rapid socio-economic development and the constant advance of industrialization and urbanization, water scarcity has become a bottleneck which restricts sustainable economy especially for arid regions. Traditional hydrological research focusing on natural water cycle has encountered great challenges derived from the influence of the increasingly powerful artificial water cycle concerning whole procedure from water in-taking and water consumption to water drainage. The hydrological model's incapability on water use simulation and prediction is amplified when dealing with the response and feedback between human system and water system, while the CGE model gives a chance for people to understand the relationship of socio-economic development and corresponding water demand. The integration of the quantitative economic model CGE and distributed hydrological model WEP opens up a new possibility to study the evolution mechanism of coupled socio-economic and water system on basis of dualistic water cycle.

On one hand, water policies, including water use limitation and water price adjustment, are evaluated by CGE as preset scenarios under different levels of socio-economic development. Sectoral water demand of concerned regions, according with either current situation or planning upgrading and readjustment of industrial structure, will be calculated to meet the principle of general equilibrium in the CGE model. Moreover, the spatial distribution of water use for each sector will be available by multi-area and dynamic improvement of traditional CGE. As multi-area and dynamic CGE model for research unit is usually hard to achieve due to shortage of data, water resources rational allocation model ROWAS (Rules-driven Object-oriented simulation model for Water resources System) is eligible for downscaling the global results of CGE model to a reasonable distribution and corresponding process of water use [14]. The ROWAS model translates complex water system to quantitative framework by mathematical description based on generalization of hierarchical network of the water system. The water sources in the ROWAS model consist of surface water, groundwater and recycled wastewater, while the water users include agriculture, industry, services, urban and rural domestic water, off-stream eco-environment, hydropower, shipping and inner-river ecological water. The confluence and conversion of different water sources are considered to describe water balance of water supply. Subsequently, the results of water use could be input into the WEP model for the simulation of water resources responses with consideration of artificial water cycle. Comprehensive analysis of water use in the socio-economic system and its impacts on regional water resources conditions, including evolution of hydrological processes and water flux transported, will be carried out for the reference of decision-making. The mechanism of interactions between the water system and the socio-economic system would be a rational conclusion by means of coupling the CGE and the WEP model.

On the other hand, variability of available water resources under climate change and land use adjustment may cause enormous impacts on socio-economic development. It's predicted that available water resources will decrease by half by 2050. The distributed hydrological model WEP possesses powerful capabilities on hydrological simulation based on physical mechanism, by which variation of available water resources under different conditions could be well simulated. The results of the WEP model about distribution and evolution rule of regional water resources may provide proper constraints for water demand in the CGE, which is intended to indicate the influence of water resources on socio-economic development especially for changing environment in the future.

3. Scenarios

Different policy orientations of national and regional programming may be tested by coupled CGE and WEP. The reform of water price, such as raising industrial water price and public subsidies for irrigation water, could be treated as an exogenous variable of the CGE model so that the water demand of different sectors is endogenously calculated in terms of general equilibrium theory. Within this context, the response of hydrological process on preset water price policy would be simulated by the WEP and the reasonability and consequence of water policy can be assessed. While the water policy is aimed at water use control, the water price and industrial structure would be endogenous variables. In addition, virtual water trade is also an appreciable subject for water-stressed countries. In China, regime of the most stringent water management has been carried out for each province and main river basin. Application of the coupled CGE and WEP would provide valuable information for decision-makers to evaluate and improve water policy.

Global warming has already drawn much attention of the international communities as an inevitable fact. The Intergovernmental Panel on Climate Change (IPCC for short) has released five assessment reports on global climate change. Scientists studied the possibility of various circumstances on greenhouse gas emission and proposed several scenarios of global climate change denoted by precipitation and temperature on the basis of ensemble forecast. More research on downscaling the global climate change to certain regions has also been accomplished. The impacts of global warming on regional water resources and the corresponding ripple effects on socio-economic development would be an interesting theme of research by the coupling of CGE and WEP. Besides, land use planning is also a probable element worth studying as a vital factor for the both hydrological cycle and the socio-economic system.

4. Conclusions

For the purpose of reasonable simulation on the whole water cycle of natural-artificial dualistic characteristic, the integration of the economic model CGE and the distributed hydrological model WEP is proposed. The coupled model would be a powerful tool to simulate the response of hydrological processes in water system under different socio-economic policies, as well as the feedback of water constraints to socio-economic development of human system. Water demand for each sector of the socio-economic system in the CGE model serves as a key element by which an artificial branch of global water cycle in the WEP model is driven, and the natural water cycle is greatly changed. Water demand in the CGE could be considered by either simply building relationships between water resources and correlative accounts that already exist in SAM structure, or adding the water resources account into SAM of traditional CGE model as an essential factor. Although there are a number of problems to implement this framework, including insufficient data for multi-area and dynamic CGE model, different scales of temporal and spatial water cycle in CGE and WEP, parameter sensitivity and uncertainty, and inconclusive scenes under changing environment, the coupling of the CGE and WEP provides an idea for inspection of the global water cycle and examination of water management. The intensive study, of coupled model and interdisciplinary research involving hydrology and socioeconomics, exhibits a great potential for the overall interpretation of the water cycle which would be helpful for better understanding on bilateral interaction between water system and socio-economic system.

Acknowledgements

This study was financially supported by the NSFC (50939006, 51379215) and the MOST of China (2015CB452701).

References

[1] Liu J, Qin D, Wang H, Wang M and Yang Z 2010 Dualistic water cycle pattern and its evolution in Haihe River basin *Chin. Sci. B.* **55** 1688
[2] Wang H, Wang J H, Qin D Y and Jia Y W 2006 Theory and methodology of water resources assessment based on dualistic water cycle model *J. Hydro. Eng.* **12** 15
[3] Lofgren H, Harris R L and Robinson S 2002 *A standard computable general equilibrium (CGE) model in GAMS* **5** (Intl Food Policy Res Inst)
[4] Baldassarre G D, Viglione A, Carr G and Kuil L 2014 Socio-hydrology: conceptualising human-flood interactions *Hydrol. Earth. Syst. Sc.* **10** 4515
[5] Britz W and Hertel T W 2011 Impacts of eu biofuels directives on global markets and eu environmental quality: an integrated PE, global CGE analysis *Agr. Eco. Env.* **142** 102
[6] Sassi M and Cardaci A 2013 Impact of rainfall pattern on cereal market and food security in sudan: stochastic approach and CGE model *Food Policy* **43** 321
[7] Wissema W and Dellink R 2014 CGE analysis of the impact of a carbon energy tax on the irish economy *Ecol. Econ.* **61** 671
[8] Yu H W and Shen D J 2014 Application and outlook of CGE model in water resources *J. Nat. Res.* **29** 1626
[9] Li N, Wang X, Shi M and Yang H 2015 Economic impacts of total water use control in the heihe river basin in northwestern china - an integrated CGE-BEM modeling approach *Sust.* **7** 3460
[10] Zhao Y, Wang J F and Cai H J 2008 Review of CGE models on water resources *Adv. Water Sc.* **19** 756

[11] Feng S, Li L X, Duan Z G and Zhang J L 2007 Assessing the impacts of South-to-North Water Transfer Project with decision support systems *Decis. Sup. S.* **42** 1989

[12] Jia Y, Ni G, Kawahara Y and Suetsugi T 2001 Development of WEP model and its application to an urban watershed *Hydrol. Proc.* **15** 2175

[13] Jia Y, Wang H, Zhou Z, Qiu Y, Luo X, Wang J, Yan D and Qin D 2006 Development of the WEP-L distributed hydrological model and dynamic assessment of water resources in the Yellow River Basin *J. Hydrol.* **331** 606

[14] You J, Gan H, Wang H and Wang L 2005 A rules-based object-oriented simulation model for water resources system *J. Hydro. Eng.* **36** 1043

IOP Conf. Series: Earth and Environmental Science 39 (2016) 012065 doi:10.1088/1755-1315/39/1/012065

Advances in preparation of modified activated carbon and its applications in the removal of chromium (VI) from aqueous solutions

Z L Deng[1, 2], M N Liang[1, 2, 3], H H Li[1, 2] and Z J Zhu[1, 2]

[1]College of Environmental Science and Engineering, Guilin University of Technology, Guilin 541004, P. R. China
[2]Guangxi Collaborative Innovation Center for Water Pollution Control and Water Safety in Karst Area, Guilin University of Technology, Guilin 541004, P. R. China

E-mail: liangmeinaa@163.com

Abstract. The wastewater in which Cr(VI) is not fully treated has drawn environment researchers' attention increasingly, due to its environmental pollution and harms to human health. Thus a high efficiency of modified activated carbon (MAC) to remove Cr(VI) has become one of the hot topics among environmental material research. This paper introduces the modification methods from the physical structure features and chemical properties of the activated carbon (AC) surface. At the same time, it briefly analyses the chemical characteristics of Cr(VI) in aqueous solutions, and on the basis of the aforementioned introduces the modification methods of the surface chemical characteristics of AC, such as: oxidation modification, reduction modification, loaded metal modification, and microwave modification. Combining studies on removing Cr(VI) from aqueous solutions by MAC in recent years, this paper anticipates the new trends of preparing MAC and the points in absorption research, offering some suggestions for future studies.

1. Introduction

Along with industrial development, the contaminants arising from human activities also increased significantly. Serious environmental problems, caused by heavy metals, have drawn extensive attention from environmental researchers [1], especially some heavy metals with high toxicity, which can cause serious problems to the ecosystem even at low concentrations [2]. Chromium is one of the most toxic heavy metals, coming from the milling machine and sheet metal, paint, tannery, and fertilizer industries [3]. Usually, Cr has two major oxidation states in water, Cr(VI) and Cr(III) [4]. However, due to its carcinogenicity, high toxicity and high aqueous solubility [5], Cr(VI) is 300 times more toxic than Cr(III) [6]. Most of the present domestic and foreign studies about treating wastewater containing Cr(VI), focus on ion exchange [7], electrolysis reduction [8], chemical precipitation [9] and adsorption [10]. At present, adsorption receives increasing attention from people because of its advantages such as high efficiency, simple operation, and low costs.

With great adsorption capacity, fast adsorption, and renewability after saturation, activated carbon (AC) is widely used in processing wastewater containing Cr(VI) [11,12,13]. For instance, Shashikant et al. [14] used $CaCl_2$ and NaCl solution to activate mango skin. Ramakrishna et al. [15] used $ZnCl_2$ to activate orange skin to prepare AC to remove Cr(VI) from solution; oth of them received good effects. However, traditional AC, with many ashes, performed poorer in adsorptive selectivity, and was limited in removing pollutants in water. Therefore, in order to further improve adsorption and removal efficiency of AC towards Cr(VI), many researchers at home, and abroad, manage to apply some modification methods.. Modified activated carbon (MAC) for absorbing and removing Cr(VI) in solution efficiently has become one of the hot research topics at present. This paper reviewed several kinds of preparations for MAC and their performances of removing Cr(VI) from the water, so as to offer some references for future studies on MAC to remove Cr(VI).

2. Chemical characteristics of Cr(VI) in water

Generally, Cr(VI) exists water in the following four forms: H_2CrO_4, $HCrO_4^-$, CrO_4^{2-} and $Cr_2O_7^{2-}$. The proportion of four kinds of ions in aqueous solutions is determined by its concentration and pH value (namely, the H^+ concentration) [16]. Research shows that when the mass concentration of Cr(VI) is greater than 1g L^{-1}, the pH value of the solution influences the major forms of Cr(VI) as table 1 demonstrated [16].

Table 1. Major forms of Cr(VI) in aqueous solutions determined by its pH value.

pH value in aqueous	Major forms of Cr (VI)
pH<1	H_2CrO_4
1<pH<7	$HCrO_4^-$ $Cr_2O_7^{2-}$
pH>7	CrO_4^{2-}

Equilibrium equations among all the forms of Cr(VI) are as follows.

$$H_2CrO_4 + H_2O \leftrightarrow H_3O^+ + HCrO_4^- \quad lgK_1=0.2 \tag{1}$$

$$HCrO_4^- + H_2O \leftrightarrow H_3O^+ + CrO_4^{2-} \quad lgK_2=-5.9 \tag{2}$$

$$2HCrO_4^- \leftrightarrow Cr_2O_7^{2-} + H_2O \quad lgK_3=1.53 \tag{3}$$

Cr(VI) has a strong oxidability. The oxidation of Cr(VI) is related to its concentration and pH value (namely, the concentration of H+). The higher the Cr(VI) concentration, the lower the pH value (the higher the concentration of H+), the greater the redox potential, and then the stronger the oxidation of Cr (VI). When reacting with a reducing agent, Cr(VI) will be reduced to Cr3+ ions [17]. Under acid condition, the electrode reaction equation of Cr(VI) reacting with reducing agents in redox reaction can be expressed by equation (4).

$$Cr_2O_7^{2-} + 14H^+ + 6e \rightarrow 2Cr^{3+} + 7H_2O \tag{4}$$

3. Preparation of modified activated carbon

Activated carbon (AC) is a kind of absorption material with well-developed pore structure and huge specific surface area, thus it enjoys a stronger adsorptive performance. Because the adsorptive performance of AC is mainly determined by its surface physical structures and chemical properties, it is necessary to modify the physical structure and chemical properties of AC to a certain degree for improving its performance in adsorbing and removing specific pollutants.

3.1. Modification methods towards the surface physical structure of the activated carbon

The modification of surface physical structure of AC refers to utilizing physical or chemical methods to increase the specific surface area and changing the pore size (larger or smaller) or pore size distribution, in order to change the physical absorption performance of AC during its preparation, and to change the adsorptive structure of AC's surface.

The modification of the surface physical structure of AC generally falls into two steps, a) carbonizing AC to remove volatile components, b) and using some oxidative gas to enrich the pore structure of AC. This can develop the pore structure of AC, and uniform the pore size distribution, through a good control of some carbonization conditions, such as: the carbonized time, carbonized temperature, activator types, and the activation time during activation, and addition of some oxidants [18]. Generally, the pore size can be reduced by adopting the thermal shrinkage method, gas phase pyrolysis for blocked pores, and impregnation and covering method [19].

3.2. Modification of surface chemical properties of activated carbon and research on its performance for removing Cr(VI) from water

The modification of surface chemical properties of MAC refers to using physical or chemical methods to change the surface functional groups on AC and the ions and compounds loaded on the surface, making it the active point in specific adsorption, to control the hydrophilic/hydrophobic properties of AC and enhance its binding capacity with metal (or metal oxide).The modification methods of activated carbon's surface chemical properties usually include: surface oxidation method, surface reduction method, load material modification method, microwave processing modification method, and alkali method [20]. According to the available reports, the modification methods to prepare AC which can remove Cr(VI) from water efficiently mainly include: oxidation modification, reduction modification, loaded metal modification and microwave processing modification.

3.2.1. Activated carbon modified by oxidation. After treating the surface of activated carbon by strong oxidant, under certain experimental conditions, the acidic groups are increased and the adsorption ability of activated carbon towards polar material is also improved. The types and quantities of groups on the surface of activated carbon, modified by different oxidants are also different. In general, the stronger oxidability of oxidant leads to a greater quantity of the oxygen-containing functional groups on the surface of the AC. At present, commonly used oxidants are nitric acid, potassium permanganate, sulfuric acid, and phosphoric acid [21].

Using sulfuric acid, phosphoric acid, nitric acid, Yao et al. [22] held a modified experiment on granular activated carbon after it was sieved by 10 to 16 mesh screens. The results showed that, under the optimum adsorption conditions, the highest removal rates for Cr(VI) of three kinds of acid-modified activated carbons were 95.79%, 97.50% and 99.2% respectively, and were increased by more than 30% compared with the virgin ones. Ryszard et al. [23] applied HNO_3 to the modification, and adopted argon gas to protect it under 1100 °C for degassing and modifying activated carbon. Making a comparison of the adsorptive performance for Cr(VI) of these two kinds of modified activated carbons, the results showed that when pH was 5, temperature was 25 °C, adsorption time was 3 hours. The maximum adsorption capacity for Cr(VI) of HNO_3 modified activated carbon was 6.92 mg/g, while degassing modified activated carbon was 2.07 mg/g, and the virgin activated carbon (before modification) was only 1.28 mg/g. This proved that the NO_3^- and Cl^- can reduce the adsorption performance of degassing modified activated carbon, while HNO_3 modified activated carbon was not affected by them. This further proved that the HNO_3 modified activated carbon demonstrated a better removal performance for Cr(VI). Jin et al. [24] used potassium permanganate to modify activated carbon, and the results showed that the optimum preparation condition was when potassium permanganate concentration was 0.01 mol/L, the solution pH was 3, the modification time was 5 hours, and the temperature of the solution was 35 °C. At the moment, the adsorption capacity of modified activated carbon for Cr(VI) in water reached 0.489 mg/g. Compared with the virgin activated carbon, adsorption capacity of modified activated carbon for Cr(VI) was increased by 85.89%. Ma et al. [25] used phosphoric acid modified activated carbon as raw material, modified it by nitric acid with 10% concentration for 4 hours under 40 °C, and obtained nitric acid modified activated carbon. At room temperature, 0.2 g of such MAC was used to treat 50 mL Cr(VI) solution with a concentration of 100 mg/L, and the adsorption rate of Cr(VI) was increased from 49.58% to 79.21%. The above studies showed that adsorption capacity for Cr(VI) of oxidation modified activated carbon was improved greatly.

3.2.2. Reduction-modified activated carbon. Surface reduction modification mainly refers to, under proper temperature, when suitable reducing agents are used to reduce the functional groups on the surface of the AC, increase the number of oxygenous alkaline groups on the surface of AC, enhance its surface nonpolar property, and enable a stronger adsorption capacity for nonpolar substances [26].

Common reducing agents are H_2, N_2, NaOH, KOH and ammonia. In the process of water treatment, the number of the surface oxygenous alkaline groups on AC after reduction treatment increased significantly. This helps AC to absorb some pollutants (particularly organic pollutants) [27]. Under acid condition, through the electrode equation of the redox reaction between Cr(VI) and reducing matters, the increasing of alkaline groups would undermine the adsorption of Cr(VI). However, studies of Park et al. [28] suggested that the removal rate of AC after being treated by NaOH increased by two times, after adsorbing Cr(VI) for 170 minutes. The results showed that this may be attributed to the specific surface area and decrease of pore volume. At the same time, Chiang et al. [29] proved that under the alkaline condition, OH- will react with functional groups on the surface of the activated carbon.

3.2.3. Loaded metal modified activated carbon. The modification of loaded metal mainly takes advantage of the reducibility and adsorption characteristics of AC. During that process, the metal ions are adsorbed on the surface of AC, and then the reducing property of AC is used to reduce metal ions into ions in forms of simple substance or low valence. Due to the strong binding capacity of metal or metal ions towards absorbed substances, the adsorption capacity of AC towards absorbed substance is greatly enhanced. At present, loaded metals are commonly used, such as iron, manganese and silver [30]. Zuo et al. [31] took advantage of ferric chloride solution to modify AC, making it loaded iron oxide by the method of impregnation and roasting, and studied adsorption performance of AC towards Cr(VI) in water. The results showed that under the same conditions, virgin AC possesses a 38.6% removal rate towards Cr(VI) ion, while the removal rate of MAC was increased to 91.4%. Therefore, iron-MAC can increase the adsorption capacity towards chromium ions and improve adsorption efficiency. Leng et al. [32] studied the modification of AC through manganese ion solution with different concentrations. They thought that under the same conditions, manganese ion modified activated carbon can increase the adsorption capacity towards Cr(VI), and follow the Langmuir adsorption model. Qin et al. [33] loaded Fe_2O_3 and MnO_2 onto mulberry stem to study the adsorption performance of composite adsorbent towards Cr(VI). The results showed that under the same conditions, compared with mulberry stem activated carbon, the maximum adsorption capacity of mulberry stem/iron manganese oxide composite adsorbent towards Cr(VI) was increased by 33.02%. And, after 3 times of desorption and regeneration, the regeneration and utilization efficient of composite adsorbent was 86.80%. Kakavandi et al. [34] utilized zero-valent iron and silver bimetallic nanoparticles to modify AC, and studied the Fe^0/Ag activated carbon's performance for removing Cr(VI). The results showed that when pH value was 3, Cr(VI) adsorption effect was best, and could reach adsorption equilibrium within 60 minutes, which was in line with the Langmuir adsorption model. The maximum adsorption capacity then was 100 mg/g, and Fe^0/Ag activated carbon possesses magnetism, which helped the adsorbent regeneration. Therefore, it has great potential in application, demonstrating the trends in future.

3.2.4. Modified activated carbon by microwave treatment. Microwave modification mainly indicates that the activated carbon absorbs the energy of microwave irradiation, rises the temperature of the whole system rapidly, causes a chemical reaction between the functional groups on the surface of AC and modifier, produces chemical reaction, and further changes the structure of the functional groups on the pore surface. At the same time, the rising system temperature will promote carbonization, and lead to certain changes in the pore structure [35, 36]. Ma et al. [37] adopted nitric acid-microwave modified AC and microwave directly modified AC into their studies. The results showed that the adsorption capacity of Cr(VI) was significantly increased, with a removal rate towards Cr(VI) of 78.62%, increased by 47.49% compared with the virgin activated carbon. The adsorption capacity towards Cr(VI) of microwave directly modified AC was 22.01%, which was 2.83% less than gas-free modified carbon (removal efficiency was 24.84%). However, its removal rate towards aniline was enhanced. This showed that the microwave directly modified AC favored the adsorption of hydrophobic organic, and did not favor the adsorption and removal of metal ions in water.

4. Conclusion and prospect

Either changing the surface physical structure, or surface chemical properties of the AC, can improve the adsorption performance of AC. Activated carbons applied to remove heavy metals are mainly chemical modified ones. It is generally acknowledged that changing the surface chemical properties of AC can better the effects of removing Cr(VI) from water. Although AC enjoys extensive sources, using waste biomass to prepare AC, and improving its adsorption performance through modification, such a recycle mode of "treating wastes by wastes" is still undoubtedly one of the research trends in future.

Current methods of studying MAC to removing Cr(VI) in water are mainly oxidation modification, reduction modification, loaded metal modification, and microwave modification. Among these methods, oxidation modified activated carbon and loaded metal modified activated carbon perform better in removing Cr(VI), and thus are studied more. Through the magnetic metal nanoparticles loaded in AC, not only does it improve the adsorption performance of AC towards Cr(VI), but it is also conducive to recovery. Therefore, it has a great application value, and is one of the development trends of activated carbon.

Removal effect of MAC towards Cr(VI) is often affected by the chemical characteristics of Cr(VI), and a variety of other factors. Meanwhile, the added modified material and AC also react with each other during the process of adsorption. At present, a lot of researchers adopt artificially configured Cr(VI) solution, while the factory chromium wastewaterconsists of complicated composition. Hence the competition mechanism of Cr(VI), and other ions, and their effects on the adsorption performance of modified activated carbon may be an important trend in future studies.

Acknowledgements

This research was financially assisted by the National Natural Science Foundation of China (NSFC21367010), the Guangxi Science and Technology Development Project (GuiKeGong14124004-3-3) and the Provincial Natural Science Foundation of Guangxi (2014GXNSFBA118054).

References

[1] Ye S H, Zhang M P and Yang H 2014 Biosorption of Cu^{2+}, Pb^{2+} and Cr^{6+} by a novel exopolysaccharide from Arthrobacter *Carbohydrate Polymers* **101** 50-56

[2] Luo C, Tian Z, Yang B and Zhang L 2013 Manganese dioxide/iron oxide/acid oxidized multi-walled carbon nanotube magnetic nanocomposite for enhanced hexavalent chromium removal *Chem. Eng. J.* **234** 256-65

[3] Lv X, Xu J, Jiang G, Tang J and Xu X 2012 Highly active nanoscale zero-valent iron (nZVI)–Fe_3O_4 nanocomposites for the removal of chromium (VI) from aqueous solutions *J. Colloid Interface Sci.* **369** 460-69

[4] Zheng J Y, Zhao Q L and Luo Y H 2014 Treatment of Cr(VI) micro-polluted water by ACF prepared by cotton woven waste *J. Chinese Journal of Environmental Engineering* **8** 4333-39

[5] Lv X, Xu J, Jiang G and Xu X 2011 Removal of chromium (VI) from wastewater by nanoscale zero-valent iron particles supported on multiwalled carbon nanotubes *Chemosphere.J.* **85** 1204-09

[6] Gatto N, Kelsh M and Mai D 2010 Occupational exposure to hexavalent chromium and cancers of the gastrointestinal tract: a meta-analysis *J. Cancer Epidemiol.* **34** 388

[7] Tiravanti G, Petruzzelli D and Passino R 1997 Pretreatment of tannery wastewaters by an ion exchange process for Cr(III) removal and recovery *Water Science and Technology* **36** 197-207

[8] Seaman J C, Bertsch P M and Schwallie L 1999 In situ Cr(VI) reduction within coarse-textured, oxide-coated soil and aquifer systems using Fe(II) solutions *Environmental Science & Technology* **33** 938-44

[9] Zhou X J and Korenaga T 1993 A process monitoring/controlling system for the treatment of wastewater containing chromium (VI) *Water Research* **27** 1049-54

[10] Gosh P K 2009 Hexavalent chromium (VI) removal by acid modified waste activated carbons

Journal of Hazardous Materials **171** 116-22
[11] Acharya J, Sahu J N, Sahoo B K , Mohanty C R and Meikap B C 2009 *Chem. Eng. J.* **150** 25
[12] Hamadi N K, Chen X D, Farid M M and Lu M G 2001 *Chem. Eng. J.* **84** 95
[13] Demirbas E, Kobya M, Senturk E and Ozkan T 2004 *Water S.A.* **30** 533
[14] Shashikant M and Trupti N P 2015 Adsorption studies of chromium (VI) on activated carbon derived from mangifera indica (mango) seed shell *J. Inst. Eng. India Ser.* **96** 237-47
[15] Ramakrishna G and Susmita M 2016 Preparation of microporous activated carbon from Aegle Marmelos fruit shell and its application in removal of chromium (VI) from aqueous phase *Journal of Industrial and Engineering Chemistry* **36** 355-63
[16] Sengupta A and Cllfford D 1986 Important process variables in chromate ion exchange *Environmental Science and Technology* **20** 149
[17] Zhang F, Ma L and Zhang Z L 2012 Application of chemical reduction technology to remediation of Cr contaminated soil *Environmental Protection of Chemical Industry* **32** 419
[18] Yang T and Lua A C 2003 Characteristics of activated carbons prepared from pistachio-nut shells by physical activation *J. Colloid Interface Sci.* **267** 408-17
[19] Jiang X, Jiang W J and Jin Y 2003 Modified activated carbon and its application in environmental protection *Environmental Science & Technology* **26** 55-57
[20] Liang X and Wang X J 2011 The technology of activated carbon modification and its application in wastewater treatment *Technology of Water Treatment* **37** 1-6
[21] Chen X Y, Lin X L and Wei Q H 2008 Development of the surface chemical modification of the activated carbon and its applications *Science Technology and Engineering* **8** 5463-67
[22] Yao Y 2014 *Study on the preparation of acid modified activated carbon and its removal of Cr(VI)* (Changsha, Hunan University)
[23] Dobrowolski R and Otto M 2010 Study of chromium (VI) adsorption onto modified activated carbons with respect to analytical application *Adsorption* **16** 279-86
[24] Jiang Z H 2013 *Research on removal of Cr(VI) in water by modified activated carbon* (Kunming, Kunming University of Science and Technology)
[25] Ma Y 2015 *Adsorption of Cr(VI) ions by modified activated carbon* (Nanjin, Nanjing Forestry University)
[26] Jiang Y X, Huan X and Jiang W J 2012 Research and application of the treatment of heavy metals wastewater by activated carbon *Journal of Anhui Agricultural Sciences* **7** 4156-58
[27] Sahin O and Saka C 2013 Preparation and characterization of activated carbon from acorn shell by physical activation with H_2O-CO_2 in two-step pretreatment *Bioresource Technology* **136** 163-68
[28] Park S J and Jang Y S 2002 Pore structure and surface properties of chemically modified activated carbons for adsorption mechanism and rate of Cr(IV) *J. Colloid Interface Sci.* **249** 458-63
[29] Chiang H L, Huang C P and Chiang P C 2002 The surface characteristics of activated carbon as affected by ozone and alkaline treatment *Chemosphere* **47** 257-65
[30] Chen J P and Wu S N 2004 Simultaneous adsorption of copper ions and humic acid onto an activated carbon *Journal of Colloid and Interface Science* **280** 334-42
[31] Zuo W Y, Tong H J and Shi B F 2015 Cr(VI) adsorption from aqueous solutions using activated carbon modified Fe *Chinese Journal of Environmental Engineering* **9** 45-50
[32] Leng Y, Chen L and Zhang X Y 2013 Study on Cr(VI) adsorption using activated carbon modified with Mn *Sichuan Environment* **32** 6-11
[33] Qin Y M., Wang D Q and Liang M N 2016 Preparation of mulberry stem activated carbon/Fe-Mn oxide composite sorbent and its effects on the adsorption of Cr(VI) *Environmental Chemistry* **35** 783-92
[34] Kakavandi B and Kalantary R R 2014 Enhanced chromium (VI) removal using activated carbon modified by zero-valent iron and silver bimetallic nanoparticles *Journal of Environmental Health Science & Engineering* **12** 115
[35] Hienendze J A 1996 On the modification of chemical surface properties of active carbon: In the

IOP Conf. Series: Earth and Environmental Science **39** (2016) 012065 doi:10.1088/1755-1315/39/1/012065

search of carbon with stable basic properties *Langrnuir* **12** 4404

[36] Radovic L R 1997 An experimental and theoretical study of adsorption of aromatics processing electron-withdrawing and electron-donating functional groups by chemical modified activated carbon *J.Carbon.* **35** 1339

[37] Ma Y N 2009 *Study on microwave modified activated carbon and its removal of aniline and Cr(VI)*(Changsha, Hunan University)

Influence of reaction period on a new packing SBBR process treating high-salt and phosphorus deficiency

J X Fu[1], M J Zhou[1, 2], P F Yu[1], M Sun[1], X Q Ji[1] and J Zhang[1]

[1]Municipal and environmental school of Shenyang Jianzhu University, Shenyang 110168, China

E-mail: 724427990@qq.com

Abstract. In order to solve the problem of high-salt ballast wastewater treatment, Sequencing Biofilm Batch Reactor of new packing activated sludge process used to simulate an experimental study. When Chloride ion concentration is 20±2g/L, the impacts of reaction period and anoxic time (Ta) / aerobic time (To) on the effect of the treatment process and sludge activity were investigated. The results show the salt acclimation SBBR process can effectively remove organic contaminants; when the reaction period was 49h, the removal rates of Chemical Oxygen Demand (COD) reached more than 91.0%, and the removal rate of NH+ 4 –N was 83.3%, the removal rate of Total Nitrogen (TN) was 67.7%, and the effluent concentration of COD, NH+ 4 –N and TN were respectively 45.7mg/L, 7.8mg/L and 18.6mg/L. At this time, TF reached 43.6μg/ml. With the Ta / To increase, the degree of denitrification increased and the nitrification rate reduced. When Ta/To 1:3, the optimal nitrogen removal appeared, ammonia removal efficiency was 83.8%, the effluent concentration was 7.8mg / L, TN removal rate was 67.7%, and the effluent concentration was 18.6mg / L, to reach effluent standards. Technical support was also provided to solve the problem of coastal salt waste low phosphorus wastewater.

1. Introduction

Sequencing batch biofilm reactor (SBBR), completely mixed in the space, with the time to completely push on streaming, has a high response speed. Under the same processing efficiency, SBBR pool body is less than the volume of a continuous flow process, saving floor space, and the same reactor by changing the operation parameters can perform different functions of wastewater treatment, with flexible operation mode[1]. Chlorine (water) for high concentrations of chlorine ions in waste water raw water fluctuation, the characteristics of the strong role of microbe poisoning[2-5], was combined with the advantage of SBBR craft, in order to obtain a stable biological treatment effect[6-9]. Pans wadden et al.[10] found that the treatment of wastewater with high salinity by salinity-domesticated activated sludge can improve the removal efficiency of organic matter and ammonia nitrogen. When SBBR crafts a period after the end of the water, raw water from the reactor, the change of quantity of inlet water quality within the reactor on biological treatment system no longer has any effect, which gives it high resistance to impact load capacity[11]. Therefore, this research through the study of a new type of packing SBBR craft obtained correct and reasonable experimental data, so as to find out feasible process design parameters and operation parameters for actual high chloride wastewater treatment engineering design, debugging, operation and management. This data will provide a strong technical basis.

2. Materials and methods

2.1. Test device

The test apparatus is shown in figure 1: SBBR reactor made of organic glass 200 mm in diameter, 1 m high, effective volume 14 L, water from the bottom in, and drainage in the middle. The bottom of the device is equipped with an aeration disc, and the internal temperature is kept constant with heating instrument control. Internally, add polyester filter material, cylindrical, size of φ×h = 5 mm x 10 mm, nonpoisonous and harmless, with good chemical stability. Fill more than 50% with polyester fiber filter material with a porosity of 95%. The reactor temperature should be controlled at 25 ℃. Dissolve oxygen at 3 to 5 mg/L, pH of 7-8. After a period of time, the cultivation of the activated sludge is domesticated, make advantageous salt resistant bacteria strains, and adjust the parameters to achieve smoothness.

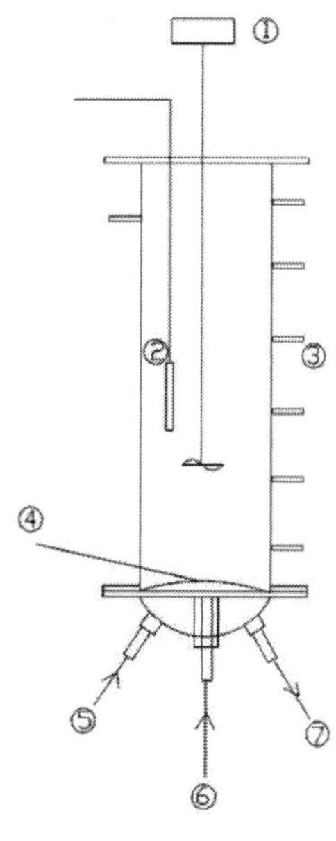

Figure 1. SBBR test apparatus.
Note: ①Agitator; ②Temperature controller; ③Water intake; ④Aeration disc; ⑤Water inlet; ⑥Air inlet; ⑦ All landowners mud mouth

2.2. Test raw water

This test was of Dalian new port sewage treatment plant water quality for the study of reference. This paper studies conditions after dealing with the oil removal of separation tank water. Based on a test, the water COD concentration is 450-550 mg/L, ammonia nitrogen, 35 to 45 mg/L, TN50-60 mg/L, pH at 7.5-8.5, aerobic phase DO for 3 to 5 mg/L, the phosphate content is 0.05, the C/P for 10000:1, and the phosphorus source has an extreme lack of water.

2.3. Test methods

In the test for early domestication salt-tolerance of aerobic sludge, sludge concentration of 5 g/L [12]: The SBBR reaction period consists of five stages, namely the water phase, oxygen mixing stage, aerobic aeration stage, sedimentation stage, and drainage stage. Initial conditions: SBBR operation periods are selected for 13 h, 25 h, 37 h, 49 h, 61 h and 73 h, in which the phases are: water phase 5 min, anoxic and aerobic time ratio of 1:3, sludge sedimentation 50 min, drainage 5 min, and decantation pen 1:1. The temperature should be in the 20 ~ 24 ℃ range. Investigate the treatment effect of the SBBR reactor in different reaction periods, and measure the different sludge dehydrogenase activity, and the optimization of process conditions. Adjust the SBBR run aerobic and anaerobic period. It will conclude the best Ta/To.

2.4. Analytical methods

In this study, all methods are in accordance with the method of determination of "Water and Wastewater Monitoring Analysis Method" (Fourth Edition) (National Environmental Protection Agency 2002) stipulations. The specific indicators and test methods are shown in table 1.

Table 1. Analytical methods.

Test items	Analytical methods	Use of instruments
NH_4^+-N	Nessler reagent spectrophotometric	WFJ2100 type visible spectrophotometer
NO_3^--N	UV spectrophotometry	UV-Vis spectrophotometer
TN	Persulfate oxidation - UV spectrophotometry	UV-Vis spectrophotometer
COD	Fast hermetic digestion method (photometric)	WFJ2100 type visible spectrophotometer
		Electric oven thermostat blast
TTC	Detection of enzyme activity spectrophotometric [13]	WFJ2100 type visible spectrophotometer

Note: COD (sealed digestion using nitric acid, due to high chlorine ions, is contained in water samples, so the appropriate dilution water samples, and according to Hg: Cl = 10:1 join HgSO4 masking).

3. Results

In high salinity (Cl - concentration of 20000 mg/L) conditions, through the successful domestication of activated sludge, the reactor treatment effect stabilized, DO for 5 ~ 6 mg/L, temperature of 12 ℃, PH 7.5 ~ 8.5, water filling ratio of 0.5, change reaction period.

3.1. The influence of the reaction period on COD removal

During the test, the influent water COD concentration control was between 450-550 mg/L, the reaction period was selected from 13h, 25h, 37h, 49h, 61h and 73h. In turn, run 3 cycles per phase, monitoring data of three times the average of the analysis results, comparing the different reaction periods on the degradation of organic matter.

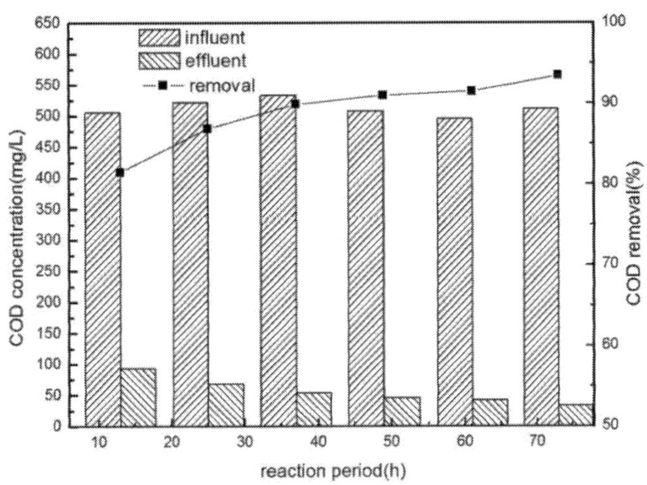

Figure 2. The efficiency of different reaction periods on COD removal.

As can be seen from figure 2, the longer reaction period, the better COD removal efficiency. The salinity influences the COD degradation. The activated sludge salt-tolerant bacteria in the high salt environment of the SBBR reactor, has the ability of degradating organic matter less than ordinary microbial slightly in the sludge, but has a strong effect on the degradation of organic matter. When the reaction cycle was 13 h, 25 h and 24 h, the effluent COD concentrations of 93.6 mg/L, 68.4 mg/L and 53.7 mg/L, did not meet the level of the A discharge standards. When the reaction period was 49 h, 61 h and 73 h, the effluent COD concentration was less than 50 mg/L, and the COD removal rate was over 90%, which meets the national and local regulations of COD emission levels of A standard. Therefore, the best operation period is 49h.

3.2. The influence of reaction period on ammonia nitrogen removal

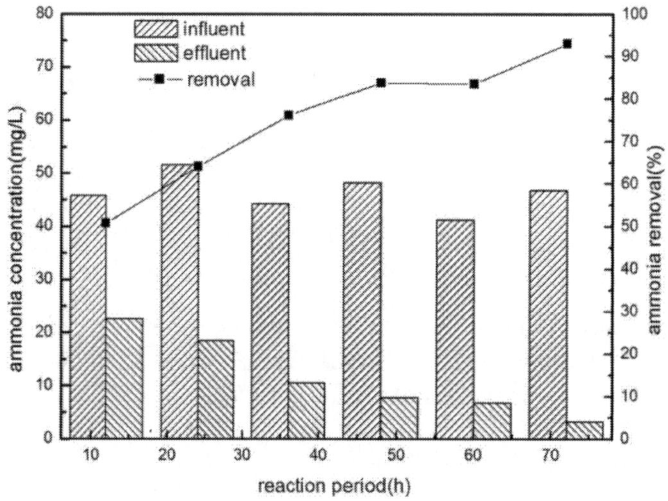

Figure 3. The efficiency of different reaction periods on ammonia nitrogen removal.

The ammonia nitrogen concentration of effluent water is 41.2 ~ 51.6 mg/L. Ammonia nitrogen effluent water concentration is significantly affected by reaction period changes. When the reaction period was 49h, 61h and 73h, the average water ammonia nitrogen concentration was 7.8 mg/L, 6.8 mg/L and 3.3 mg/L, the average removal rate of ammonia nitrogen was 83.8%, 83.5% and 92.9%, respectively. Therefore, under the condition of high salinity, ammonia nitrogen was still able to achieve better removal, and nitrosation bacteria can still work effectively. The longer the reaction period, the better treatment effect of the system. When the reaction period exceeded 37h, the effluent water concentration of ammonia nitrogen was under 8 mg/L, which can meet the national and local regulations for ammonia nitrogen emission levels of A standard.

3.3. The influence of reaction period of TN removal

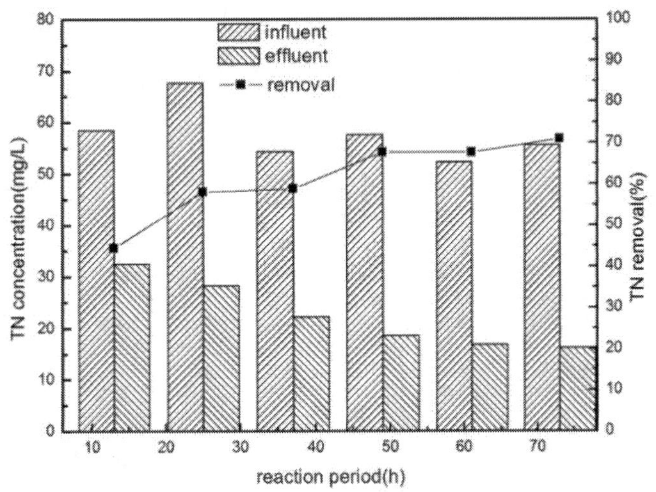

Figure 4. The efficiency of different reaction periods on total nitrogen removal.

As can be seen from figure 4, with an influent water total nitrogen concentration of 52.3 ~ 67.69 mg/L, the total nitrogen was affected by the reaction period significantly. When the reaction period was 49h, the average discharge total nitrogen concentration was 18.6 mg/L, and the total nitrogen removal rate was 67.7%. Visibly, under high salinity conditions, as the reaction period increased, the total nitrogen removal can get a better treatment effect, significantly. Biological denitrification needs two processes, include nitrification and denitrification, but the total nitrogen removal was affected by those two processes. First ammonia nitrogen nitration was under aerobic conditions, then the aerobic phase generated nitrate as an electron acceptor in the anoxic environment, and the organic carbon source as an electron donor for the denitrification reaction, so as to realize the removal of nitrogen. Therefore, the denitrification stage really removes nitrogen, and nitrification is the necessary condition to realize nitrogen nitration[14]. Figure 3 and figure 4 show that with a less than 25h reaction period, owing to the bad effect on nitrification, it was unable to provide a sufficient electron acceptor for the denitrification reaction, which causes the system total nitrogen removal rate to be low. The reaction period, by contrast, over 37h is good for removal of total nitrogen, and effluent water was less than 20 mg/L, which reaches level 1 B emission standards.

3.4. The influence of reaction period on microbial activity

Sludge dehydrogenase activity is the core factor affecting the system organic matter degradation. Microbial growth and metabolism requires energy, and as the key step in the biological oxidative decomposition of organic matter dehydrogenation in cells, molecular oxygen is the ultimate hydrogen donor, and the more hydrogen comes through the oxidative respiratory chain, the higher the dehydrogenase activity. The level of dehydrogenase directly reflects the activity of the microbial metabolism. Sludge samples are taken at the end of each reaction period; an elevated TF value of the sludge indicates that the more hydrogen oxygen in the process of oxidation respiratory chain is transferred to the oxygen in the process of microbial metabolism, as growth and metabolism of microorganisms is fast, so the biochemical effect of sludge utilization is good. Different reaction periods of dehydrogenase activity of activated sludge were compared to instruct the sludge biochemical ability of the SBBR system.

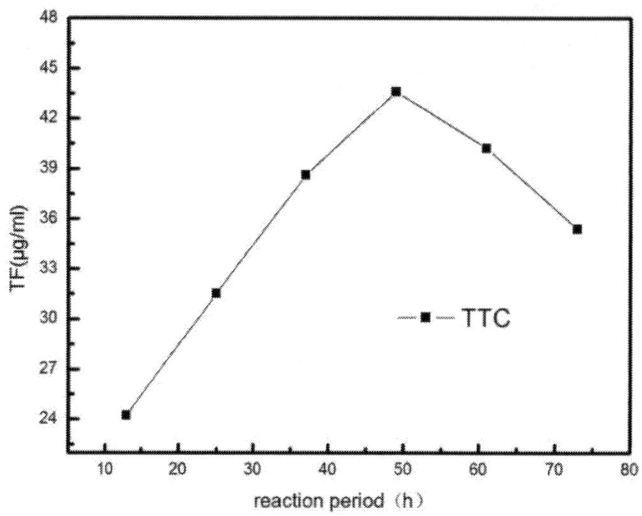

Figure 5. The influence of reaction period on microbial activity.

As can be seen from figure 5, when the reaction period was less than 49h, the dehydrogenase activity of microorganisms in the SBBR reactor increases with the extension of the reaction period. When the reaction period was 13 h, TF was 24.2 mu g/ml; when the reaction period increased to 49h, the maximum TF was 43.6 mu g/ml. Then, as the reaction period continued to extend, the microbial activity began to fall; when the reaction period was 73h, TF dropped to 35.4 mu g/ml. The results show that when the reaction period was short, the microorganism in the reactor had just recovered from the last period. The microorganism was in a slow growth phase, and metabolic system needed to adapt to a new environment at the same time as synthetase, coenzyme, and other metabolic intermediate metabolites, so that the metabolic activity in this period is not strong. As the shorter reaction period did not reach the vigorous stage of microbial growth, so the dehydrogenase activity was relatively low; when the reaction cycle was longer, the microorganism in the reactor had been prepared for the period of adjustment, which had reached the exponential phase of growth strength, and the period of the microbial growth had enough material base. At the same time, the external environment was also best, along with the microbial metabolism and enzyme activity. Thus, the dehydrogenase activity is higher. When the reaction period continued to increase and reached a certain length, microorganisms in the reactor necessary to maintain the survival of the material basis were exhausted, nutrient ratios imbalanced, microbial growth entered the decline phase, and the external environment continued to grow worse. As harmful metabolites increased, cellular catabolism is greater than synthetic metabolism, which led millions of bacteria to death, and dehydrogenase activity dropped with a long reaction period.

3.5. The influence of reaction period on operation time A and O paragraphs

On the basis of the experimental study above, select the best reaction cycle as 49 h, control temperature 12 ℃, pH 7.5 ~ 8.5, water filling ratio 0.5, fixed water, precipitation and drainage length of time constant, adjust single cycle of oxygen session with the aerobic session run length, and set Ta (A period of time) and To the period of time (O) ratio on the SBBR reactor effluent indicators. Specifically of oxygen in the test of time and operation of aerobic sessions as table 2:

Table 2. Single cycle anoxic and aerobic running time.

	Anaerobic time T_A	Aerobic time T_O	TA/TO
Plan A	24	24	1:1
Plan B	16	32	1:2
Plan C	12	36	1:3
Plan D	8	40	1:5
Plan E	6	42	1:9

The SBBR reactor anoxic and aerobic section running time of ammonia nitrogen and total nitrogen removal had some influence, as figure 6 shows, so that the Ta/To decreased, and the ammonia effluent water concentration decreased. When the Ta/To was 1:3, the ammonia effluent water concentration was 7.8 mg/L and the removal rate was 83.8%, which reaches the level of A effluent water standard. The greater the Ta/To, the smaller the effluent water concentration of TN. When Ta/To was 1:3, the TN effluent concentration was 18.6 mg/L, with a removal rate of 67.7%. In terms of the reason for this, when Ta/Ta is greater, the aerobic period is relatively shorter, causing incompleteness of nitrification, and the ammonia nitrogen removal rate is low, resulting in a smaller aerobic phase of nitrate and less nitrite[15]. When the lack of oxygen time is relatively longer, the anoxic period has enough time for denitrification. In addition, due to the short aerobic period, the removal rate of carbonaceous organic is reduced, which makes the anoxic period of the carbon source sufficient. Therefore, as the Ta/To becomes higher, the SBBR has higher denitrification, and the denitrification effect is better. Also, whe Ta/To is reduced, the aerobic period is prolonged, and the ammonia nitrogen removal rate increases, but the lack of oxygen time is short and the carbon reduces, causing the denitrifying effect become poor and the TN removal rate to reduce. Based on comprehensive consideration, the best Ta/To is 1:3.

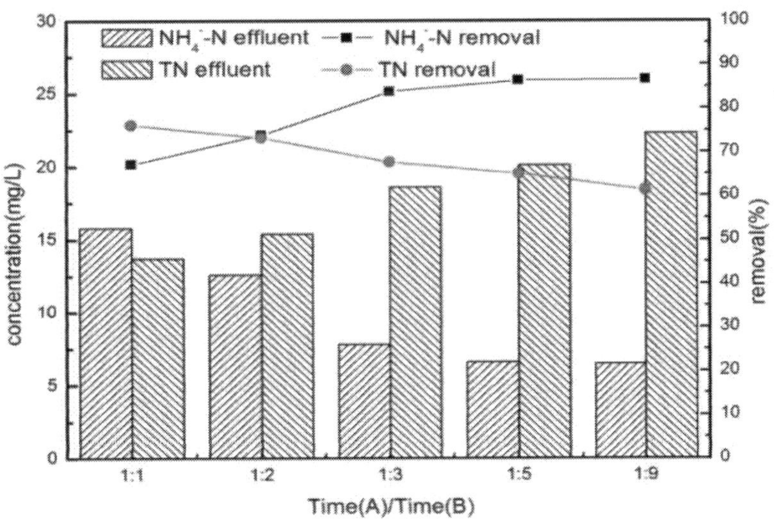

Figure 6. TA:TO effect on denitrification SBBR reactor.

4. Conclusion

Under the condition of high salinity, new type packing SBBR craft adopted through domestication of the success of the activated sludge can effectively remove organic pollutants. When the reaction period was more than 36 h, the removal rate of COD was over 90%, and the effluent COD concentration was less than 50 mg/L; the ammonia nitrogen removal rate was over 80%, and the ammonia water concentration was under 8 mg/L, which meets the national and local regulations of the COD emission level of A standard. Therefore, the best operation period was 48 h. The total nitrogen also had a better removal rate. With the extension of the reaction period, the treatment effect was obviously better. When the reaction period was more than 37h, the reactor for removal of total nitrogen was better, and the total nitrogen was less than 20 mg/L, which reaches level 1 B emission standards. In high salt sludge in new packing SBBR system activity, the microbial growth rule is closely related with the reaction cycle. When the reaction cycle was 49 h, microbes reached the exponential phase of growth strength, microbial metabolism, and enzyme activity, so the dehydrogenase activity reached maximum, TF is 43.6 mu g/ml. At the same time, the sludge dehydrogenase activity was high in salt SBBR system, which is one of the core factors of organic matter degradation ability. For a 49 h reaction period, the system of COD, ammonia nitrogen and total nitrogen removal efficiency to achieve optimization, an additional reaction cycle treatment effect is not obvious. The Ta/To (the ratio of anoxic and aerobic time) influenced the denitrification effects obviously; with an increase of Ta/To, the denitrification rates increased, and the nitrification rates decreased. When the Ta/To was 1:3, the denitrification effect achieve optimization, the ammonia nitrogen removal rate was 83.8%, the water concentration was 7.8 mg/L, the TN removal rate was 67.7%, the effluent concentration was 18.6 mg/L, and the primary standard level of the water was reached.

Acknowledgments
This research was support by the national water pollution control and management of science and technology major special zx07202 (2012-2012). The authors thank assistance and encouragement from colleagues, special work by technical staff with field and laboratory analyses, and two anonymous reviewers for helpful suggestions for revision.

References

[1] Xu Z Y, Yang Z H, Zeng G M, *et al.* 2006 Sequencing batch biofilm reactor type processing high ammonia leachate denitrification mechanism research *Journal of environmental science* **26**(1) 55-60

[2] Hamoda M F and Alattar I M S 1995 The effects of high sodiumchloride concentrations on activated sludge treatment *Wat. Sci. Tech.* **31**(9) 61-72

[3] Panswad T and Anan C 1999 Impact of high chioride wastewater on the an anaerobic/anoxic/aerobic with and without inoculation of chloride acclimated *Water Research* (5) 1165-72

[4] Zhang Y Q and Shen X N 2003 Seawater desalination after drainage impact on biological treatment of municipal sewage treatment plant *Urban Environment and Urban Ecology* (6) 126-27

[5] Qun K, Ma W C, Xu J C, *et al.* 2005 The influence of high salt concentration of industrial wastewater biochemical treatment study *Journal of Environmental Pollution Treatment Technology and Equipment* **6**(8) 42-45

[6] Feng Y H, Zhan X M and Wen X H 2000 Activated sludge processing system for saline wastewater impact load resistance *Journal of Environmental Science* **1**(1) 106-08

[7] Zhang Y S, Wang J and Jiang L D 2000 Salinity effect on the properties of the second pond sludge settling *China Water Supply and Drainage* (2) 18-19

[8] Cui Y W, *et al.* 2004 NaCl salinity effect on activated sludge treatment system of integrated *Journal of Environmental Engineering* **22**(1) 19-21

[9] Zou S Y, Zhang J P, Wu J R, *et al.* 2008 Of biotechnology research progress of high salt wastewater treatment *Industrial Water Treatment* **28**(11) 1-4

[10] Pans W T and Anan C 1997 Salt how of Carbon and Nitro - gen Bacteria in an Anaerobic/Anoxic/Aerobic Process *Proceedings of the Third Australian Conference on Biologi - CAL Nutrient Removal from Wastewater, Australia in Water and Wastewater Association, Victoria, another awarding*

[11] Ahmet U and Fikret K 2004 Alt inhibition on in nutrient removal from saline wastewater in a

sequencing batch reactor *Journal of Enzyme and Microbial Technology* (3/4) 313-18

[12] Wang Y and Han L 2011 Quick start and operation of SBR technology and saline wastewater *Journal of Environmental Science and Technology*(12) 287-91

[13] Zhou C S and Yin J 1996 Of TTC - dehydrogenase activity detection method research *Journal of Environmental Science* (4) 400-05

[14] Fu J X, Wang B H and Liu D Y 2014 SBR process in beer wastewater treatment of high organic nitrogen denitrification study *Water Treatment Technology* **40** (2) 65-9

[15] Zhang L H, Fu Y X, Tao Y, *et al.* 2013 NaCl salinity shortcut nitrification denitrification effect of SBR technology *Journal of Beijing University of Technology* **33**(2) 280-87

IOP Conf. Series: Earth and Environmental Science **39** (2016) 012067 doi:10.1088/1755-1315/39/1/012067

A review of environmental flow assessment: methodologies and application in the Qianhe River

C F Hao[1, 2, 4], L M He[3], C W Niu[2] and Y W Jia[2]

[1] Department of Hydraulic Engineering, Tsinghua University, Beijing 100084, China
[2] State Key Laboratory of Simulation and Regulation of Water Cycle in River Basin, China Institute of Water Resources and Hydropower Research, 1 Yu-Yuan-Tan South Road, Beijing 100038, China
[3] Power Construction Corporation of China, 22 Chegongzhuang West Road, Beijing 100048, China

E-mail: andychao2004@126.com

Abstract. Environmental flow is of great significance for maintenance of ecological services in the riverine ecosystem. The paper reviews present methodologies for environmental flow assessment on the basis of three classifications, including the Hydrological Index Methodologies, the Hydraulic Rating Methodologies and the Habitat Simulation Methodologies. Both advantages and disadvantages of each classification are fully analysed, as well as applicable conditions. Moreover, representative methods of different classifications are applied to prescribe environmental flow in the Qianhe River of north China with consideration of hydrological series, hydraulic characteristics and habitat suitability of targeted species. The results of environment flow by the Montana Method, Wetted Perimeter Method and Physical Habitat Simulation System are 1.23 m^3/s, 2.07 m^3/s and 0.52 m^3/s for the Qianyang section in middle reach of the Qianhe River. In view of seasonal variation of the Qianhe River, the paper recommends 1.23 m^3/s as the minimum runoff in the dry season and 2.07 m^3/s in the wet season. For further improvement of environmental flow assessment, studies of quantitative correspondence relationship between each component of instream flow and its ecological functions of riverine ecosystem, and the development of Holistic Methodologies by combination of various methodologies and multi-disciplinary information have great potential.

1. Introduction

In recent years, rapid socio-economic development and accompanying intensive interruption of the local water cycle have brought about severe ecological environmental problems in many places. The protection and restoration of regional ecological environments draw more and more attention, and environmental flow assessment comes to a focus of study both in China and abroad. Generally, the ecological environment system consists of rivers, lakes, wetlands, forests and so on. The paper focuses on the environmental flow of riverine ecosystem, by investigating and summarizing different ways to work out how much water a river needs to maintain its ecological functions.

Environmental flow, the quantity and corresponding processes of instream flow for preset target of ecological functions in the context of specific hydrological and ecological conditions, serves as the most fundamental element in riverine ecosystem for its balances of heat, sediment, salinity and water. The environmental flow assessment requires multidisciplinary study and integrated application of hydrology, hydraulics, river geomorphology, environmental science, biology, and so on. Flow regime, altered by socio-economic water use and dam constructions in many rivers, is believed to be the key point of environmental flow as the master variable of fluvial geomorphology and aquatic community evolution, as well as mass and energy exchanging between streamway and floodplain area [1-2], and advanced hydraulic models provide powerful tools for numeric simulation of hydrological processes in regional water cycle [3].

Current methodologies of environmental flow assessment could be grouped into three categories, including Hydrological Index Methodologies, Hydraulic Rating Methodologies and Habitat Simulation Methodologies. Different methods of environmental flow assessment should be adopted in water resources management of given basins according to data collection, technical skills and research budgets. The paper completes environmental flow assessment of Qianhe River in China on the basis of application and comparison of different methodologies.

2. Methodologies

Hydrological Index Methodologies, which generally assume that riverine ecosystem goes well under natural status, simply set some percentage of simulated or monitor runoff as necessary eco-environment water for the studied river. Hydraulic Rating Methodologies take selected hydraulic indexes as reasonable representation of integrated functions in the riverine ecosystem, so corresponding runoff in accordance with appropriate hydraulic indexes needs to be satisfied. Habitat Simulation Methodologies focus on the specific requirement of species of concern in the riverine ecosystem and evaluate environmental flow by establishing a quantitative relationship between biological suitability of certain species and hydrological and hydraulic factors.

2.1. Hydrological Index Methodologies

The basis of Hydrological Index Methodologies is historical runoff series. Environmental flow assessment of the studied river is carried out by selected characteristic parameters of hydrological data and proper ratio. The historical hydrological runoff of a specific river section, either simulated or monitored data, is the only necessary input data, which means widely applicable capacity and very few calculations. Since lacking of clear mechanism linkage on hydrological variation and ecological response, Hydrological Index Methodologies are adopted only for rivers with lower priority or very rough approximations in river planning and management. Several common methods of this classification include Montana Method, Range of Variability Approach, etc.

2.1.1. Montana Method. Montana Method, also known as Tennant Method named after its presenter, proposed by American scientist Don Tennant in 1976 and first applied in the middle and western parts of the United States [4], has been widely used in many places in the world since the early stages of environmental flow assessment. Minimum environmental flow of the river section concerned is converted on the basis of annual average natural runoff by different ratios according to wet season and dry season, varying from 10% to 200%. The advantage of the Montana Method is its simplicity, but it ignores the daily or seasonal variation characteristics of a runoff data series. In some cases, recommended minimum environmental flow is even larger than natural instream runoff, especially in a dry season. Hence, it's not suitable for seasonal rivers and other rivers with apparent temporal fluctuation of runoff. The method has been improved by ecologists and hydrologists, who set more detailed options of ratios according to a requirement of key protected targets in different stages [5]. Environmental flow assessment by the Montana Method needs a natural runoff series of more than 30 years by numerical simulations or restored data based on observations.

2.1.2. Range of Variability Approach. Range of Variability Approach, or RVA for short, deriving from aquatic ecology theory concerning the critical value of hydrological variability, is a method for environmental flow assessment and eco-hydrological variation evaluation on the basis of Indicators of Hydrologic Alteration (IHA for short) index system [6-7]. The IHA index system consists of five classifications of runoff characteristics to represent ecologically related hydrological processes, including: scale, timing, frequency, duration and rates of change [8]. In general, each hydrologic indicator values between one standard deviation from the mean, or range of values variation between 25% and 75% (or 33% and 67%) frequency according to natural runoff series, are considered to be acceptable for conservation of local biological diversity, following the principle of Intermediate Disturbance Hypothesis [9]. The comparisons between hydraulic indicators of a natural runoff series before intensive disturbance and those of natural-artificial water cycle [10] are carried out to evaluate the variation of hydrological characteristics and corresponding impacts on the riverine ecosystem at different stages. RVA just provides possible integrated impacts on the riverine ecosystem by flow alternation, instead of detailed responses of hydro-ecological factors and processes.

2.2. Hydraulic Rating Methodologies

Hydraulic Rating Methodologies believe that aquatic habitat availability is closely related with certain hydraulic indexes, such as: river width, average depth, flow velocity, wetted perimeter and hydraulic radius. Correlative curves between runoff and selected hydraulic indexes are obtained from long-term data series of a hydrologic monitoring system or a hydraulic computation on the basis of theoretical formulae. The turning points or specific characteristic values of studied rivers are regarded as environmental flow for the local riverine ecosystem. Although river channel characteristics, transection shapes and longitudinal slope for instance, are considered by Hydraulic Rating Methodologies, direct and well-defined effects of river runoff on specific ecosystem functions and seasonal variations are still not reflected. The representative method of this classification is Wetted Perimeter Method.

The Wetted Perimeter Method [11] evaluates environmental flow on the basis of one-to-one correspondence between wetted perimeter of river transection and runoff. The roughness of river channel, transection shapes and riverbed gradient, obtained from field surveys or hydrological yearbook, are necessary parameters. The wetted perimeter is the line length of the wetted part of selected river cross section under water level. According to the relation curve of wetted perimeter and corresponding runoff, the inflection point is considered to be representative for the basic riverine ecological state. The ecological functions of the riverine ecosystem for both of instream living beings and riparian regions would be significantly damaged if stream flow falls below critical runoff. There are various ways to recognize the critical point on wetted perimeter-runoff curve. Maximum curvature principle, and slope coefficient of 1, is the most popular method. When the Wetted Perimeter Method is used for environmental flow assessment, fitting functions are often introduced. Fitting functions of wetted perimeter-runoff curves vary with different transection shapes. Power function is suitable for fitting of triangle, U shaped or parabola transections, while exponential function is used for rectangular or trapezoid transections.

2.3. Habitat Simulation Methodologies

Habitat Simulation Methodologies, the most complex way to compute environmental flow, set up experiential or physical relationships between the stream flow and the habitat requirement of aquatic organisms in a certain living period [12]. Hydraulic data and biological information are collectively considered to establish the direct and indirect response of available habitat for protected species on the depth and velocity of river flow, channel slope and section configuration of river channel, roughness and overburden of riverbed, as well as water temperature and water quality. Integrated applications of Habitat Simulation Methodologies need both quantitative correspondence of steam flow and habitat suitability, and explicit relationship between habitat suitability and species individual amount. Hydraulic model, habitat simulation model and hydrological series analysis model are essential components to support comprehensive environmental flow assessment. Moreover, not only single recommended environmental flow, but also dynamic habitat improvement or degradation for targeted species with environmental flow variation, are realized by Habitat Simulation Methodologies to provide meaningful decision-making reference for a river administrator, as long as protected species are decided. Since the relationship curve between habitat suitability and hydraulic indexes are empirical at present, it's still not transplantable in different rivers, even for similar species. A large number of surveys and experiments are required to establish the explicit relationship curve. In addition, complicated technique and high budget prevent the wide application of the Habitat Simulation Methodologies. The most representative one of this classification is Instream Flow Incremental Methodology.

Instream Flow Incremental Methodology (IFIM for short), involving hydrodynamic, water quality, hydrological and ecological models, is a theoretical system of environmental flow assessment to provide alternative analysis and decision support in integrated planning and scientific management of river basins [13]. Generally, fish are selected as targeted species by IFIM as the top of the food chain in freshwater biomes. Macrohabitat and microhabitat are all considered by IFIM to demonstrate available habitat area in different periods of fish life, such as the spawning period, the nursing period and the mature period. Factors of macrohabitat include river morphology, water quality, temperature, turbidity and transparency, while those of microhabitat include water depth, velocity, riverbed base and surface coverage. One of the representative methods of microhabitat simulation is Physical Habitat Simulation System (PHABSIM for short) developed by US Fish and Wildlife Service [14]. River channel characteristics and relationship curve of habitat suitability index for specific fish are input into the hydraulic model in microhabitat simulation, and responsive value of weighted usable area (short for WUA) in accordance with instream runoff are calculated. The inflection point of WUA-runoff curve is considered as minimum environmental flow, and the peak point represents the best status of the targeted species. While in macrohabitat simulation, SSTEMP model and QUAL2E model are popularly used concerning water temperature, dissolved oxygen, ammonia nitrogen and other harmful materials.

3. Study area

Qianhe River, a tributary of the Weihe River in the Yellow River Basin in north China, originates from Shimiaoliang of Gansu province and flows from the northwest to the southeast. The length of Qianhe River is 129.6 km with the catchment area of 3493 km^2. The altitude of the upstream basin is between 1500m and 2000m, 1000m-1500m in the middle reach and below 1000m in the downstream basin. There are three reservoirs located at the main stream of the Qianhe River, Duanjiaxia reservoir, Fengjiashan reservoir and Wangjiaya reservoir from upstream to downstream, and one national aquatic germ plasm resources conservation zone along the middle and lower reach, as is shown in figure 1. The study of environmental flow of Qianhe River is essential for regional ecological environment conservation.

Qianyang hydrometric station, located in the middle reach of Qianhe River near the end of conservation zone, is selected as a representative section for environmental flow assessment. Data collections and preparations for environmental flow assessment of the Qianyang section are completed as following: (1) Data series of meteorological stations including precipitation, air temperature, humidity, wind velocity, and sunshine duration, underlying surface information including DEM, agrotype, land use and vegetation cover, water resources system networks including river network and diversion projects, water use and drainage data, as well as monthly hydrological series of instream flow of Qianyang section from 1973 to 2010, are collected and processed for the establishment and validation of the hydraulic model to simulate natural runoff of study area [15]. (2) River channel characteristics including transection shape, riverbed roughness and longitudinal slope are collected from hydrologic yearbooks of the Yellow River Basin in China for hydraulic indexes computation of selected section. The transection shape of Qianyang section is shown in figure 2.

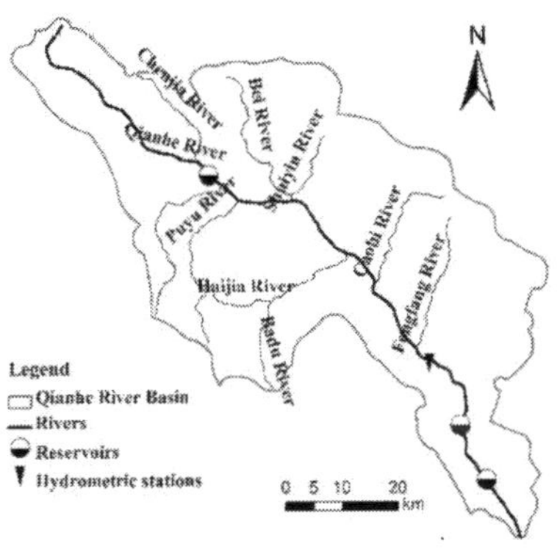

Figure 1. Overview of Qianhe River basin.

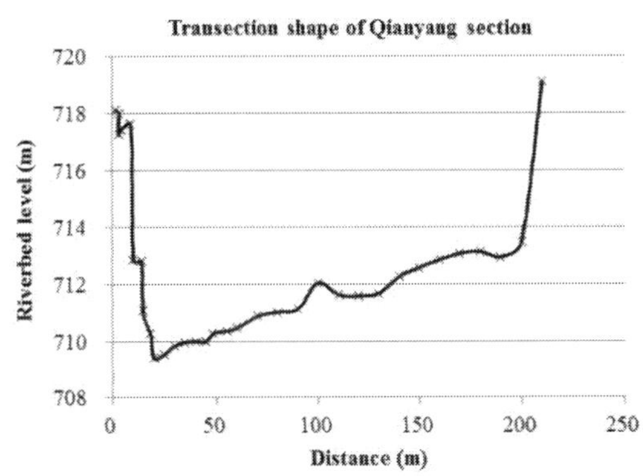

Figure 2. Transection shape of Qianyang section.

4. Results and discussions

4.1. Environmental flow assessment by Hydrological Index Methodologies
For natural runoff process simulation of the Qianhe River, distributed hydrological model WEP (Water and Energy transfer Processes) based on physical mechanism is adopted to analyze variation of runoff yield and convergence. The WEP model was developed on the basis of Physically-Based Spatially Distributed (PBSD) models and Soil Vegetation Atmosphere Transfer (SVAT) models. It is capable of hydrological processes simulation under changing environment. The model was encoded in Fortran Language. It's suitable for both a personal computer and a giant server to run the model. The structure of the model consists of horizontal subdivisions of contour bands, intersecting small sub-watersheds and vertical composition of 9 layers. The modelling approaches and establishment of the WEP model are referred to in published papers and reports [3, 15, 16].

The comparison of simulated and measured average annual runoff in Qianyang section is shown in figure 3. The relative error of runoff simulation is 4.5%, and the Nash Efficiency Coefficient of yearly runoff simulation is 0.78, with monthly runoff simulation of 0.56, indicating a satisfactory accuracy. On this basis, monthly and annual natural runoff of Qianyang section are simulated. The average annual runoff of long series in Qianyang section is 12.28 m^3/s. According to the Montana Method, 10 percent of average annual runoff regarded as base flow is the minimum water requirement for the riverine ecosystem. Hence, the basic line of environmental flow for Qianyang section of the Qianhe River is 1.23 m^3/s by Hydrological Index Methodologies.

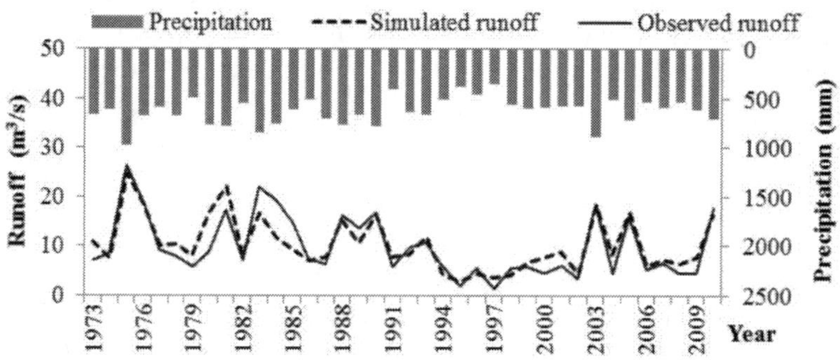

Figure 3. Validation of runoff simulation in Qianyang section.

4.2. Environmental flow assessment by Hydraulic Rating Methodologies
The paper collects hydraulic indexes of Qianyang section by field research and literature investigation, including river channel roughness, transection shape and longitudinal slope. The relationship of wetted perimeter and water depth is obtained according to the transection shape of Qianyang section. Fitting curve of wetted perimeter and runoff by power function is shown in Figure 4. To eliminate the influence of coordinate scale, relative values of each factor, derived from ratio of actual value and maximum value of overall series, are presented in the figure.

Runoff of Qianyang section is 2.07 m^3/s in accordance with slope coefficient of fitting curve as 1. The calculated runoff is supposed to be the critical point of the riverine ecosystem functions. Basic ecological services may be interrupted when instream flow falls below this value.

4.3. Environmental flow assessment by Habitat Simulation Methodologies
The weighted usable area (WUA) is a key factor of Habitat Simulation Methodologies to represent suitability of stream communities. The paper selects ephemeroptera benthos, one of the most common riverine benthos in Qianhe River, as an aquatic indicator species. Velocity and depth of instream flow are considered to be the key hydraulic indexes. The relationship curve between habitat suitability of ephemeroptera benthos and hydraulic indexes are referred to in former achievements [17]. Manning's

formula is adopted to demonstrate the correspondence relationship between WUA and runoff according to river channel roughness, transection shape and longitudinal slope of the Qianyang section. Variation of WUA with runoff of Qianyang section is shown in figure 5.

As is shown in the figure, the value of WUA rises up with increasing instream flow at the beginning. The peak of curve represents the maximum usable habitat for ephemeroptera benthos, and corresponding runoff of the Qianyang section is 0.52 m^3/s. After that, the WUA falls down and becomes zero when the runoff is about 0.9 m^3/s.

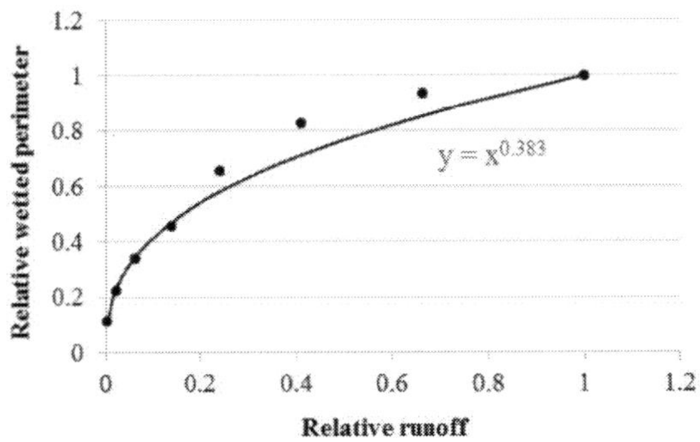

Figure 4. Fitting curve of wetted perimeter and runoff of Qianyang section.

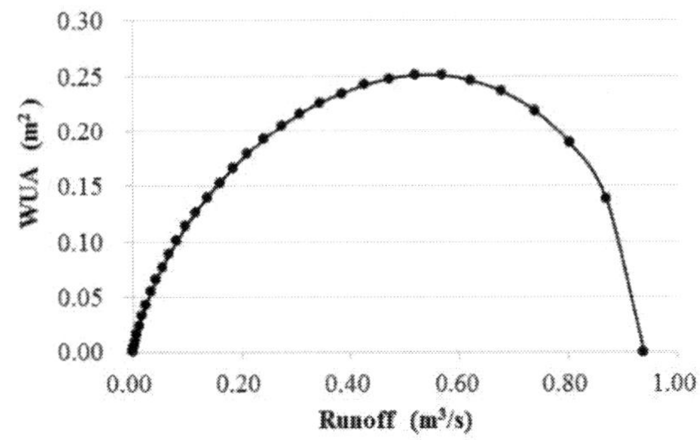

Figure 5. Variation of WUA with runoff of Qianyang section.

5. Conclusions

There are three kinds of methodologies for environmental flow assessment: the Hydrological Index Methodologies, Hydraulic Rating Methodologies and Habitat Simulation Methodologies. The paper summarizes the advantages and disadvantages of different methodologies, as well as their applicable conditions. Generally, the Hydrological Index Methodologies are based on hydrological statistics without consideration of specific characteristics of different sections and rivers. The Hydraulic Rating Methodologies reflects the influence of river channel characteristics, but detailed correspondences of ecological functions and each component of environmental flow are not embodied. The Habitat Simulation Methodologies quantitatively presents the habitat suitability of targeted species by weighted usable area, but ignores other ecological services in integrated management of the river basin.

Besides, the applications of the representative methods of each classification are carried out in the Qianyang section of the Qianhe River in north China. The results of environment flow by the Hydrological Index Methodologies, Hydraulic Rating Methodologies and Habitat Simulation Methodologies are 1.23 m^3/s, 2.07 m^3/s and 0.52 m^3/s. It's worth noting that the minimum result of 0.52 m^3/s by the Habitat Simulation Methodologies can just meet the requirement of the targeted species, ephemeroptera benthos. The maximum result of 2.07 m^3/s by the Hydraulic Rating Methodologies is regarded as the critical point of the instream flow for effective survival environment of the riverine ecosystem. The mediate result of 1.23 m^3/s by the Hydrological Index Methodologies represents the base flow of the Qianhe River for fundamental requirements of river continuum and self-purification. In view of seasonal variation of the Qianhe River, like most rivers in north China, the paper suggests that the base flow of 1.23 m^3/s is set to be the minimum runoff for a dry season while the critical flow of 2.07 m^3/s for a wet season.

The results of this study put forward a preliminary evaluation of environmental flow for the Qianhe River. There are still several key issues to be improved. The relationship curve between habitat suitability of targeted species and hydraulic indexes introduced from field studies of other places may not be precisely suitable for the studies river, and water temperature and overburden of the riverbed have not been taken into consideration yet. The combination of present methodologies, named as Holistic Methodologies [18], may bring significant improvement for environmental flow assessment. To make the environmental flow assessment more practical for the integrated ecological management of the Qianhe River, quantitative correspondence relationship between each component of instream flow and ecological function is to be clarified, and discharge processes for different periods should be carried out.

Acknowledgements
This work was financially supported by the National Natural Science Foundation of China (50939006, 51379215) and the National Key Basic Research Program of China (2015CB452701).

References

[1] Poff N L and Zimmerman J K H 2010 Ecological responses to altered flow regimes: a literature review to inform the science and management of environmental flows *Freshwater Biol.* **55** 194
[2] Mcmanamay R A and Frimpong E A 2015 Hydrologic filtering of fish life history strategies across the United States: implications for stream flow alteration *Ecol. Appl.* **25** 243
[3] Jia Y, Wang H, Zhou Z, Qiu Y, Luo X, Wang J, Yan D and Qin D 2006 Development of the WEP-L distributed hydrological model and dynamic assessment of water resources in the Yellow River Basin *J. Hydrol.* **331** 606
[4] Tennant D L 1976 Instream flow regimes for fish, wildlife, recreation and related environmental resources *Fish. Manage. & Ecol.* **1** 6
[5] Zheng Z H, Zhang Z Z, Huang Q, Qiu L, Li Q and Wei M H 2010 Improvement and application on ecological water requirement Tennant method *J. Sichuan Univ.: Eng. Sci. Ed.* **42** 34
[6] Brian R, Jeffrey B, Robert W and David B 1997 How much water does a river need? *Freshwater Biol.* **37** 231
[7] Yang P, Yin X A, Yang Z F and Tang J 2013 A revised range of variability approach considering the periodicity of hydrological indicators *Hydrol. Processes* **28** 6222
[8] Leach J M, Kornelsen K C, Samuel J and Coulibaly P 2015 Hydrometric network design using streamflow signatures and indicators of hydrologic alteration *J. Hydrol.* **529** 1350
[9] Connell J H 1978 Diversity in tropical rain forests and coral reef *Science* **199** 1302
[10] Liu J, Qin D, Wang H, Wang M and Yang Z 2010 Dualistic water cycle pattern and its evolution in Haihe River basin *Chin. Sci. Bull.* **55** 1688
[11] Gippel C J and Stewardson M J 1998 Use of wetted perimeter in defining minimum environmental flows *Regul. Rivers: Res. & Manage.* **14** 53
[12] Acreman M C and Dunbar M J 2004 Defining environmental river flow requirements – A review *Hydrol. Earth Syst. Sci.* **8** 861
[13] Bovee K D 1986 Development and evaluation of habitat suitability criteria for use in the instream flow incremental methodology (Washington: National Ecology Center)
[14] Gore J A, Crawford D J and Addison D S 1998 An analysis of artificial riffles and enhancement of benthic community diversity by physical habitat simulation (PHABSIM) and direct observation *Regul. Rivers: Res. & Manage.* **14** 69
[15] Liu J J 2013 Distributed simulation and evolution law of water cycle in the Weihe river basin under changing environment (Beijing: Institute of Water Resources and Hydropower Research) p 82
[16] Jia Y, Ni G, Kawahara Y and Suetsugi T 2001 Development of WEP model and its application to an urban watershed *Hydrol. Processes* **15** 2175
[17] Gore J A, Layzer J B and Mead J 2001 Macroinvertebrate instream flow studies after 20 years: a role in stream management and restoration *Regul. Rivers: Res. & Manage.* **17** 527
[18] Tharme R E 2003 A global perspective on environmental flow assessment: emerging trends in the development and application of environmental flow methodologies for rivers *River Res. & Appl.* **19** 397

Synthesis and photocatalytic activity of sepiolite supportednano-TiO_2 composites prepared by a mild solid-state sintering process

L M Liao[1, 2, 3, 5], Z Q Wang[1, 2, 3], H Liang[1, 4], J Feng[1, 2, 3] and D Zhang[1, 2, 3]

[1]Changjiang River Scientific Research Institute, Wuhan 430010, China
[2]Research Centerof National Dam Safety Engineering Technology, Wuhan 430010, China
[3]Collaborative Innovation Center for Geo-hazards and Eco-Environment in Three Gorges Area, Hubei Province, Yichang 443002, China
[4]College of Water Conservancy and Hydroelectric Engineering,Hohai University, Nanjing 210098, China

E-mail: stone142@163.com

Abstract. Supported nano-TiO_2photocatalysts play an important role in water environment restoration because of their potential application to photocatalytic degradation of organic contaminants in waste water. With sepiolite as the support, the nano-TiO_2/sepiolite composite photocatalysts were synthesized by an easily operated and mild solid-state sintering process.The microstructureand photocatalytic property of the sepiolite supportednano-TiO_2 composites were characterized and analyzed by X-ray diffraction spectroscopy, UV-Visible spectroscopy and fluorescence spectroscopy. In addition, the influences of calcination temperature and load ratios on the photocatalytic activity of sepiolite supported nano-TiO_2 composites were studied.The results indicated that appropriate ratios of sepiolite supports to nano-TiO_2contributed to uniform dispersion of nanoparticles, and enhanced the absorption ability within the UV-Vis range, and consequently increased the photocatalytic activity of the composites.Under the preparation conditions of 90 wt. % TiO_2 loading and calcinated at 400 °C, a maximum in photocatalytic activity ofnano-TiO_2 sepiolite composite was obtained.

1. Introduction

Owing to stable photochemical performance, high activity, and non-toxity, nano-TiO_2 has good prospects in the photocatalytic degradation of organic pollutants in water [1-3]. However, its practical application is limited by the difficult separation and recycling in water. Supported and immobilized nano-TiO_2on a high surface area carrieris an effective method to solve the problem [4-5]. Currently, the porous structure materials are considered as favorable load carrier of nano-TiO_2, which is beneficial for the effective utilization of photocatalytic activity of nano-TiO_2. As a kind of hydrous magnesium-rich silicate mineral with chain structure, sepiolite has the largest surface areain non-metallic minerals (up to about 900m^2/g), the most unique pore structure, andthe strongest adsorptive capacity of clay minerals, as well as stable chemical property, rich resource and low cost .Thus, sepiolite is a good carrier candidate for supporting nano-TiO_2.

In the past few years, many different approaches of preparing mineral supported nano-TiO_2 composites have been developed, including: sol-gel method, settling method, cross-coupling method, powder sintering method and so on [6-9]. Powder sintering or a solid diffusion method can build strong binding force between each phase because of the thermal diffusion effect. However, synthesis of sepiolite supported nano-TiO_2 composites with this method is rarely reported.

In this paper, sepiolite supported nano-TiO_2 composites were synthesized by an easily operated and mild solid-state sintering process as high performance photocatalysts. The influences of calcination temperature and load ratios on the photocatalytic activity of sepiolite supported nano-TiO_2 composites were studied, which provided a theoretical basis for its application in degradation of organic contaminants in waste water.

2. Materials and methods

The main raw materials were: sepiolite ($Mg_4Si_6O_{15}(OH)_2 \cdot 6H_2O$, SiO_2content about 50 wt.%, MgO content about 21 wt.%); nano-TiO_2 particles P25 (Degussa, Germany, TiO_2 content > 98 wt.%); ethanol (AR); NaOH (AR); terephthalic acid (AR); distilled water. At first, nano-TiO_2particles and different weight of sepiolite powder were mixed and wet milled for 30 min and calcinated at a certain temperature for 6 h. After cooling off to room temperature slowly, the samples were grinded until the particle size was well-distributed. A series of sepiolite supported nano-TiO_2 composites of different loadratios were obtained. According to TiO_2 weight percentage, these materials were respectively marked as 40%T/S (TiO_2/sepiolite), 70%T/S, 80%T/S, 90%T/S, 95%T/S. The percentages involved in this paper are all mass ones.

Micro-structural observations and light absorption characteristic analysis were carried out with an X-ray diffractometer (XRD, Bruker D8 Advanced, German) and a UV-Visible spectrophotometer (UV-Vis, Shimadzu UV-2550, Japan).The •OH concentration produced during photocatalytic reaction could effectively characterize the photocatalytic activity of photocatalysts. According to the capture effect of terephthalic acid on •OH, the as-produced hydroxylate 2-hydroxyterephthalic acid(TAOH)could transmit unique signals near the spectral peak of the fluorescence at 426 nm, and thereby indirectly indicated •OH concentration formed in the solution. In this experiment, •OH relative concentration generated in a 90 min reaction time was used to characterize photocatalytic activity of composite photocatalytic materials.

3. Results and discussion

3.1. Effect of preparation conditions on the structure of sepiolite supportednano-TiO_2 composites. Figure 1 (a) shows XRD spectra of 70%T/S composites prepared at different calcination temperatures. While the calcination temperature increases from 300 °C to 400 °C,the ratio of peak intensities of characteristic peaks of anatase phase at $2\theta = 25.3°$ and rutile phase at $2\theta = 27.4$ °,54.5° increased slightly, indicating that crystalline phase of anatase TiO_2 is more intact. As the calcination temperature rises, the ratio of peak intensities of characteristic peaks of anatase phase and rutile phase decreased, indicating that the anatase to rutile phase transformation gradually occurs in the composites. Generally, anatase is considered more active than rutile in photocatalysis, so the calcination temperature at 400 °C is preferred for preparation of the composite photocatalysts. Figure 1 (b) gives XRD spectra of T/S composites with different load ratios (wt. %) calcinated at 400 °C. There was no apparent change in TiO_2 characteristic peak position and FWHM, which proved the uniform dispersion of nanoparticles on the sepiolite carrier. With loading content of TiO_2 increases, anatase diffraction peak intensity gradually increased, while the peak intensity of sepiolite decreased.

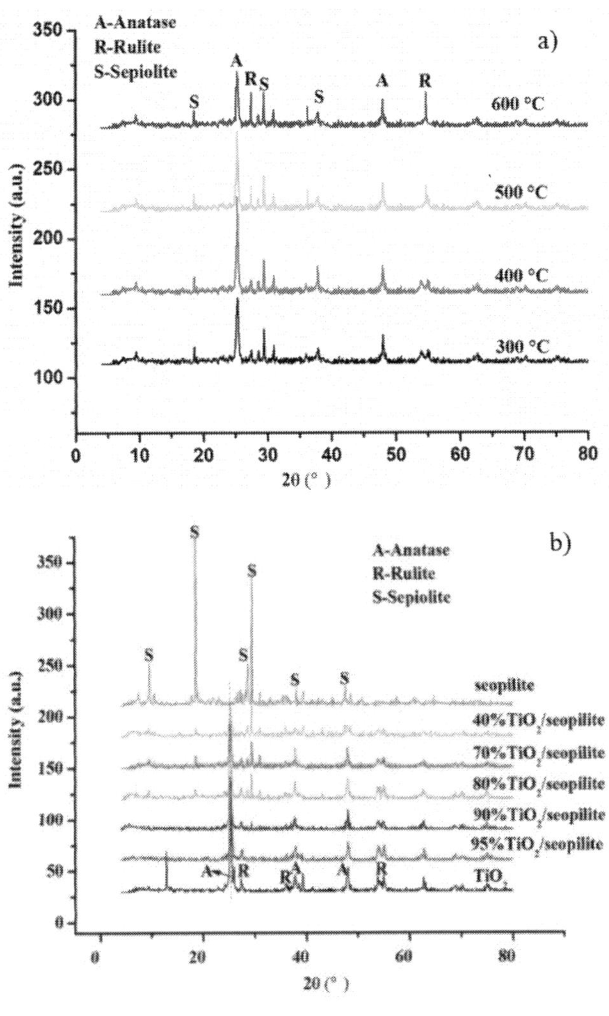

Figure 1. X-ray diffraction spectra (XRD) of (a) 70%T/S composite samples obtained from different calcination temperatures and (b) T/S composite samples with different load ratios (wt. %) calcinated at 400 °C.

3.2. Photocatalytic property of sepiolite supportednano-TiO_2 composites

UV-Visible light diffuse reflectance spectra (DRS) of T/S composites with different load ratios are shown in figure 2. In general, compared with pure TiO_2, visible light (390~780 nm) absorption ability of the composites are all enhanced, which is induced by relatively strong visible light absorption ability of sepiolite. That is to say, the composites of sepiolite and nano-TiO_2 can improve the visible light utilization for photocatalysis. Furthermore, 40%T/S composites have the strongest visible light absorption intensity, 90%T/S composites take second place. However, the UV light (200~380 nm) absorption intensity of 40%T/S composites are apparently depressed, which might be caused by UV-shielding effect of high content sepiolite. Thus, in accordance with the utilization efficiency of both ultraviolet and visible light for these photocatalysts, 90% T/S composites represent better photocatalytic response ability in UV-Vis range, which is more beneficial for the photocatalytic activity.

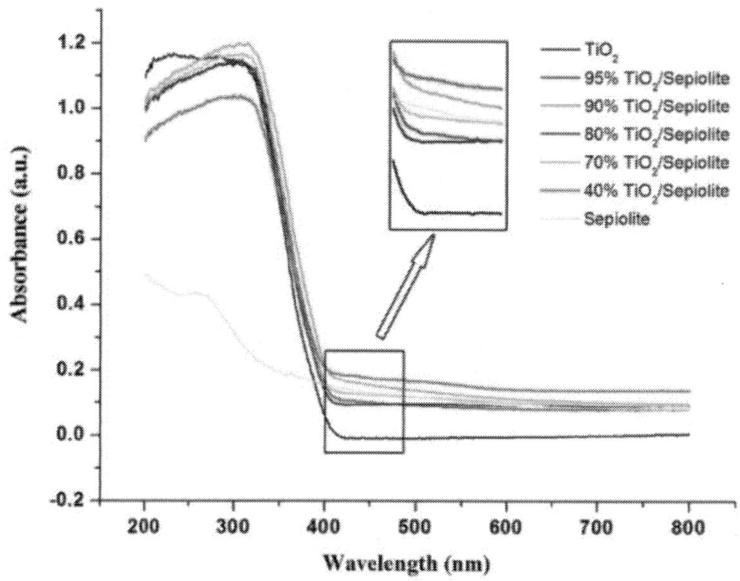

Figure 2. Diffuse reflectance spectra (DRS) of raw TiO_2, sepiolite and T/S composite samples with different load ratios (wt. %).

Figure 3 shows time dependences of fluorescence intensity at 426 nm for T/S composites in terephthalic acid solution irradiated by UV-visible light. The "intensity-time" slopes k_{I-T} of the composites could be used to estimate the generation speed of •OH in the solution. The greater the k_{I-T} is, the higher the generation speed of •OH is, and then the stronger the photocatalytic activity. As shown in figure 3(a), the curve slope k_{I-T} for the composites increases first and then decreases with the increment of loading content of TiO_2. The k_{I-T} for 90%T/S composites is the highest one, which implies the optimalphotocatalytic activity. It should be attributed tothe uniform dispersion of nanoparticles and favorable photocatalytic response ability in UV-Vis range, proved in figure 1(b) and figure 2. Excessive TiO_2 load on sepiolite would result in nanoparticle accumulation and blocking in sepiolite pores, which hinders improvement on the photocatalytic activityon the contrary. In addition, as shown in figure 3(b), the curve slope k_{I-T} for 90% T/S composite prepared by different calcination temperatures increases first and then decreases with the rise of the temperature. The photocatalytic activity for the composites calcinated at 400 °C is preferred, which is consistent with the results shown in figure 1(a).

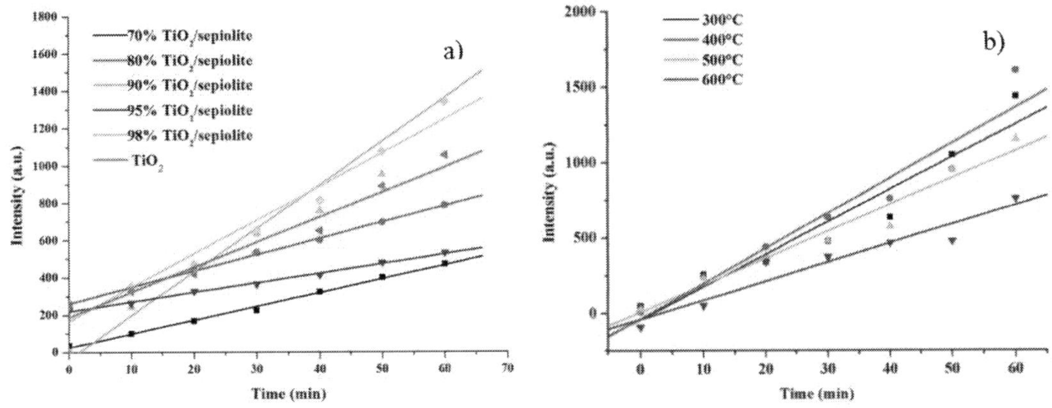

Figure 3. Time dependences of fluorescence intensity at 426 nm for T/S composites in terephthalic acid solution irradiated by UV-visible light: (a) T/S composites with different load ratios calcinated at 400°C; (b) 90%T/S composites prepared by different calcination temperatures.

4. Conclusions

Sepiolite supported nano-TiO_2 composites were synthesized by an easily operated and mild solid-state sintering process. The existence of sepiolite carrier could significantly contribute to uniform dispersion of nanoparticles, enhance their photocatalytic response ability in UV-Vis range, and subsequently improve photocatalytic activity of the composites. The photocatalytic activity of the composites was influenced by the preparation conditions includingload ratios and calcination temperature. 90%T/S composites prepared at 400°C were preferred to achieve optimal photocatalytic activity.

Acknowledgment

National Nature Science Foundation of China (No. 51209023, 51509015, 51309018), Non-profit Industry Financial Program of MWR (No. 201301023) and Grants of National Dam Safety Engineering Technology Research Center (No. Ggfky15043)

References

[1] Siddhapara K S and Shah D V2014 *J. Nanosci. Nanotechnol.***14** 6337
[2] Jiang X D, Wang Y Q and Pan C X 2011 *J. Am. Ceram. Soc.***94** 4078
[3] Dai K, Zhang X H, Fan K, Peng T Y and Wei B Q 2013*Appl. Surf. Sci.* **270** 238
[4] Wang W D, Serp P, Kalck P and Faria J L 2005 *J. Mol. Cata.A: Chem.* **235** 194
[5] Corma A and Garcia H 2004 *Chem. Comm.***13** 1443
[6] Anandan S and Yoon M 2003 *J. Photochem. andPhotobiol. A: Chem.***160** 181
[7] Trujillo M E, Hirales D, Rincon M E, Hinojosa J F, Leyva G L and Castillon F F 2013 *J. Mater. Sci.* **48** 6778
[8] Yan G Y, Wang X X, Fu X Z and Li D Z 2004 *Catal. Today* **93-95** 851
[9] Li S P, Wu Q S, Cui C, Lu G S, Zhang C Sand Yan Z Y 2013 *Mater. Sci. Poland* **31** 372

IOP Conf. Series: Earth and Environmental Science **39** (2016) 012069 doi:10.1088/1755-1315/39/1/012069

Enhancement of dewatering performance of digested paper mill sludge by chemical pretreatment

Y Q Lin[1], C Zeng[1, 2], H H Wu[1] and B X Zeng[1]

[1] College of Natural Resources and Environment, South China Agricultural University, Guangzhou 510640, China

[2] The State Key Laboratory of Pollution Control and Resource Reuse, School of Environmental Science and Engineering, Tongji University, Shanghai 200092, China

E-mail address: yqlin@scau.edu.cn

Abstract. The wide application of anaerobic digestion (AD) for waste sludge results in a huge amount of digested sludge, while the appropriate reuse of digested sludge depends on effective solid-liquid separation. Thus, chemical (acid/alkali) pretreatment effects on dewaterability of digested paper mill sludge (DPMS) for better downstream reuse based on enhanced solid-liquid separation were investigated in this research. The dewatering properties of paper mill sludge (PMS) were also investigated to elucidate the impact of AD on sludge dewaterability. The results indicated that a higher DPMS dewaterability was noted with acid pretreatment (pH5). A 41.37% moisture content and 74.41% dewatering efficiency were determined for DPMS after acid (pH5) pretreatment within 25 min. In addition, a 7.13 mg•g-1 VSS of extracellular polymeric substances (EPS) and 101.50 μm of average particle size were observed. It was also observed that both EPS concentrations and particle sizes were key parameters influencing DPMS dewaterability. Lower EPS concentrations with larger average particle sizes contributed to enhanced sludge dewaterability. Moreover, dewaterability of PMS was higher than that of DPMS, which illustrated that AD would decrease the sludge dewaterability.

1. Introduction

As a solid by-product of the papermaking industry, paper mill sludge (PMS) is continuously produced by wastewater treatment, and originates from pulping and paper-making processes. It is reported that nearly 400 million tons of paper production around the world annually will produce approximately 18 million dry PMS [1]. The majority of them have been directly delivered to landfills or discarded carelessly. Otherwise, PMS is a type of significant biosolid waste containing plenty of organic components. Thus, biological treatments of PMS, such as composting and anaerobic digestion, are popular. Anaerobic digestion has been, and continues to be, widely applied in the stabilization process of sludge management, providing effective pathogen destruction, reduction of volatile solids (VS) and odor potential and an energy source in the form of biogas [2]. Anaerobic digestion of PMS for methane production has been reported in China, which indicates the potential benefits in the resource utilization of this industrial organic waste as well as in the solving of energy shortage problems [3].

However, as a result of the wide application and utilization of the anaerobic digestion (AD) of PMS, digested paper mill sludge (DPMS) presents another disposal problem with its high moisture content. It contains plenty of secondary fiber fines with a lower total solid (TS) content (generally less than 8%) [4]. Discharging DPMS to sewage sludge plants causes excessively high organics loads for the wastewater treatment process. Application of raw DPMS as a liquid fertilizer in agriculture or horticulture is also unfeasible due to the low fertility and possible pathogens. Proper reusage of DPMS is always based on efficient solid-liquid separation. Otherwise, the dewaterability of sewage sludge has been focused on for a long time, instead of paying much attention to the dewaterability of anaerobic digestrate [2, 5, 6, 7]. On the other hand, based on the few current results for the dewaterability of anaerobic digestrate, different results about the impact of anaerobic digestion on sludge dewaterability have been reported, owing to various sludge. Some indicated that anaerobic digestion generally impeded sludge dewaterability, while the other reported that sludge dewaterability was enhanced after anaerobic digestion [2, 5, 8]. According to authors' knowledge, no related report on DPMS dewaterablity is available.

To determine sludge dewaterability, moisture content and dewatering efficiency are used as common indexes. Besides, extracellular polymeric substance (EPS) is regarded as one of the most important factors that influence the dewatering characteristics of sludge [9, 10]. As metabolic products accumulate on the bacterial cell surface, EPS consists of organic polymers, proteins, polysaccharides and DNA, which are highly hydrated and may contain up to 99% water by weight, thereby retaining a high level of water in the sludge [11, 12]. Due to its physico-chemical properties, EPS is thought to have considerable impact not only on the charge of sludge surface, bioflocculation process, settling and dewatering [13], but also serves as the protective cover for cells against the harsh external environment or as the carbon and energy reserves during starvation [14].

On the other hand, sludge dewaterablity is related to its composition and structure. Various pretreatments have been applied to enhance sludge dewaterability, including the common chemical pretreatments. The influence of acid pretreatment on sludge dewaterability, EPS concentrations and settling performance of activated sludge has been investigated, and the results suggested that the moisture contents of dewatered sludge varied with pH values, and the optimum pH value for filtration dewatering was 2.5 [15]. He et al. [16] observed that there were no comprehensive changes in the sludge dewatering efficiency when acid pretreatment was applied, but the moisture content of dewatered sludge decreased evidently. This phenomenon indicated that the degree of dehydration improved. However, no research has reported on the effect of acid/alkali pretreatment on DPMS dewaterablity.

The objective of this study was to evaluate the dewaterability of DPMS by pretreatment with acid or alkali at different pH values. Moisture contents, dewatering efficiencies and EPS concentrations were analyzed in the dewatering process. It is proposed to provide a feasible reusage method for anaerobic digestate via enhanced solid-liquid separation for downstream application.

2. Materials and methods

2.1. Preparation of sludge

PMS was collected directly from the secondary clarifiers (normally settling tanks) of the Guangzhou Pulp & Paper Plant, China. There are two processes in this plant: one is bleaching chemi-thermo-mechanical pulp (BCTMP) made from Masson pine; the other is papermaking from waste paper after de-inking. Waste water arises from three sections – pulping, papermaking and de-inking – and is usually dewatered to 60% - 70% moisture content at the end of wastewater treatment. In order to get the optimal C/N ratio, monosodium glutamate waste liquor was applied, which was collected from Ao-Sang Monosodium Glutamate Factory (Guangzhou, China).

DPMS was obtained from anaerobic digestion of PMS at (37 ± 1)°C, TS of 10% for 40 days, inoculating with 10% seed sludge (based on TS) that was obtained from the sewage tank (near Building 4 in the South China Agricultural University, Guangzhou, China) and was acclimatized with PMS step by step in the laboratory for 3 months. The anaerobic digestate was settled for 24 hours. The clear supernatant was removed to obtain the tested sludge. The tested sludge was stored in the refrigerator (0 - 4℃) prior to the experiment.

The target sludge of PMS was charactered as pH 7.63, TS 35.68%, VS 40.74% (based on TS), and EPS concentration 19.34 mg•g^{-1} VSS. The tested sludge of DPMS was charactered as pH 8.00, TS 24.96%, VS 27.54% (based on TS), and EPS concentration 28.62 mg•g^{-1} VSS.

2.2. Experimental procedure

2.2.1. A subsubsection. Pretreatment: A series of bench-scale experiments were conducted in 500 mL beakers loaded with 250 mL tested sludge. 20% (w/v) NaOH solution and 20% (v/v) HCl solution were added until the final pH value of each beaker reached 5, 7, 9 and 11, respectively. The tested sludge was continuously stirred slowly for 5 min while adding chemicals, till the target pH value was obtained. Then, the mixture was allowed to react for 20 min. As for the control treatment, no chemicals were applied. All experiments were carried out at room temperature.

2.2.2. Suction filtration test. Suction filtration was conducted in a Buchner funnel with a quantitative filter paper under a 0.04 MPa vacuum. 20 mL sludge was poured into the Buchner funnel to filter for 40 min. The moisture content of dewatered sludge was determined according to the Standard Methods [17]. Dewatering efficiency was determined according to the following equation:

$$Dewatering\,efficiency = \frac{Qr - Qc}{Qr} \times 100(\%)$$

where Qr is the weight of the water in sludge before dewatering and Qc is the weight of the water in sludge after dewatering.

2.2.3. Settling test. 100 mL tested sludge was poured into a measuring cylinder and kept subsiding for 90 min to investigate the sludge settling performance. The volume of settled sludge was recorded once every 10 min.

2.3. Analytical methods

TS, VS, volatile suspended solids (VSS), moisture content and pH were determined according to the Standard Methods [17]. Particle size was determined by the screen method. Tested sludge was screened step by step with tap water on a sieve with a diameter of 1.25, 0. 9, 0.8, 0.63, 0.45 and 0.2mm, respectively. The sludge that remained on each sieve was collected and dried at 105℃ for 24 h. The weight percentage of particles for each size was equal to the sludge weight on each sieve divided by the total sludge weight, based on dry mass. Proteins were analyzed using the ultraviolet absorption spectrum method with bovine serum albumin as the standard, polysaccharides were measured according to the anthrone-H_2SO_4 method, and DNA analysis was carried out using the ultraviolet absorption spectrum method [18]. As the main components of EPS were proteins, polysaccharides and DNA, a EPS concentration could be determined by the sum of its components' concentrations. In order to know the configurations of tested sludge with different pretreatments, a scanning electron microscope (SEM) (XL30, FEI, USA) was applied for sludge samples.

EPS was extracted based on formaldehyde-NaOH method. Firstly, 50 mL sludge was centrifuged at 6000G for 5 min. Then, sludge pellets were resuspended and diluted to the original volume by adding buffer solution (2 mmol/L Na3PO4, 4 mmol/L NaH2PO4 and 1 mmol/L KCl, pH=7), followed by adding 0.3 ml formaldehyde solution and stirring for 30 min. Afterwards, 50 mL NaOH solution (0.04 mol/L) was added to the sludge and the mixture was agitated at 300 r/min for 1 h, then centrifuged at 6000G for 10 min. The supernatant filtered with 0.2 μm microfiltration membrane was used to test the concentrations of proteins, polysaccharides and DNA.

3. Results and discussion

3.1. Moisture content and dewatering efficiency

As presented in table 1, the lowest moisture content of 41.37% was obtained in the pretreated DPMS (pH 5), while the highest moisture content of 46.59% was obtained in the original DPMS (pH 8). The moisture content in the dewatered sludge dropped down by 11.2% when HCl pretreatment with pH 5 was applied. The moisture contents of the pretreated DPMS (pH 7), pretreated DPMS (pH 9) and pretreated DPMS (pH 11) were 41.66%, 44.65% and 44.45%, respectively, all of which were lower than those of the original DPMS (pH 8). The results of moisture contents indicated that pretreatment with HCl (pH 5) provided a better condition for DPMS to obtain a higher dewaterability.

Besides, the highest dewatering efficiency of 76.04% was obtained in the pretreated DPMS (pH 7), the second highest of 74.41% was obtained in the pretreated DPMS (pH 5), and both of them were higher than that which was obtained in the original DPMS (pH 8). The dewatering efficiencies of the pretreated DPMS (pH 9) and the pretreated DPMS (pH 11) were 72.34% and 65.71%, respectively. The results indicated that acid pretreatment with pH 5 and pH 7 showed a higher dewatering efficiency than alkali pretreatment with pH 9 and pH 11. Also, the lowest dewatering efficiency was noted in the treatment with pH 11.

In this test, the moisture contents of dewatered sludge for different treatments ranked in the order of pretreated DPMS (pH 5) < pretreated DPMS (pH 7) < pretreated DPMS (pH 11) ≈ pretreated DPMS (pH 9) < original DPMS (pH 8). Meanwhile, the dewatering efficiency observed in each treatment followed the order of pretreated DPMS (pH 7) > pretreated DPMS (pH 5) > pretreated DPMS (pH 9) > original DPMS (pH 8) > pretreated DPMS (pH 11). It could be inferred that a strong alkaline environment is not favorable to enhance DPMS dewaterability, while a neutral or weakly acidic condition behaves in the opposite way. The possible main reason is that the sludge surface is negatively charged, and the acid changes its surface properties by neutralizing the negative charges and compressing the double electrode layer, resulting in more flocculation of flocs, which probably favors the dewatering process. What's more, it was reported that the hydrophobic fraction of carboxyl and other polar groups was enhanced with the decrease of pH value, which reduced their adsorption ability for water molecules, resulting in a transformation of water distribution in sludge [19]. Based on those results, the higher content of bound water in sludge reduced the mechanical dewatering ability. An excess of alkali could possibly increase the negative charges of sludge particles, and then disperse the sludge flocs, resulting in a worse dewaterability. Overall, taking the moisture content of dewatered sludge and dewatering efficiency into consideration, pretreatment with pH 5 was optimal to enhance the DPMS dewaterability compared to other acid/alkali pretreatments.

Table 1. Moisture contents of dewatered sludge and dewatering efficiencies of DPMS.

	Original DPMS (pH8)	**Pretreated DPMS (pH5)**	**Pretreated DPMS (pH7)**	**Pretreated DPMS (pH9)**	**Pretreated DPMS (pH11)**
Moisture Content (%)	46.59±0.37	41.37±0.34	41.66±0.57	44.65±0.27	44.45±0.17
Dewatering Efficiency (%)	69.28±0.24	74.41±0.21	76.04±0.33	72.16±0.09	65.71±0.13

DPMS: Digested paper mill sludge

3.2. EPS concentrations and compositions

According to figure 1, the lowest EPS concentration of 7.13 mg•g^{-1} was noted in the pretreated DPMS (pH 5), while the highest of 54.82 mg•g^{-1} was obtained in the pretreated DPMS (pH 11). EPS concentrations in the original PMS (pH 8), original DPMS (pH 8), pretreated DPMS (pH 7) and pretreated DPMS (pH 9) were 19.34 mg•g^{-1}, 28.62 mg•g^{-1}, 18.10 mg•g^{-1} and 37.20 mg•g^{-1}, respectively. Therefore, the EPS concentrations in the reactors with different pretreatments followed the trend of pretreated DPMS (pH 5) < pretreated DPMS (pH 7) < original PMS (pH 8) < original DPMS (pH 8) < pretreated DPMS (pH 9) < pretreated DPMS (pH 11). It was clearly observed that EPS concentrations increased with the increase of pH values in reactors. Additionally, it was found that the main component of EPS in each treatment was proteins, amounting to 60-85%, no matter whether pretreatments were applied or not. The concentration of polysaccharides was the second most in the sludge before pretreatments. However, the concentration of DNA increased to be the second most component in sludge after pretreatments, amounting to 11-14%, which was followed by polysaccharides, amounting to 2-5%. Generally, a high concentration of DNA in EPS could be an alarming indication that the cells were lysed in the harsh extraction process [13]. However, it could be considered that the extraction process for EPS was effective in this study because the percentage of DNA concentration was less than 15% [12]. With regard to reducing the polysaccharide concentration after acid/alkali pretreatments, it was possibly due to the disintegration of polysaccharides, which caused less extracted polysaccharides.

An excess of alkali possibly led to the dissolution of proteins and polysaccharides, and then the EPS solubility increased. Thus, abundant EPS were released from the cellular surface [12]. In this study, the EPS concentration of DPMS was higher than that of PMS. One possible reason might be related to the fact that the pH value of PMS increased after anaerobic digestion [3, 4], and then the solubility of EPS increased.

As shown in figure 2, less EPS was extracted, and a lower moisture content was noted. Meanwhile, the dewatering efficiency decreased with the increase of EPS concentration. Therefore, the correlation between dewaterability of pretreated DPMS, evaluation with moisture content and dewatering efficiency, and EPS concentration are presented (see figure 2). Based on this fact, it could be inferred that an increase of EPS concentration would reduce DPMS dewaterability. This is possibly due to the highly hydrated nature of EPS that retains water within the sludge matrix, which counteracted the formation of flocculation [9]. Meanwhile, acid pretreatment removed some EPS from the sludge surface, and then a majority of bound water was released. Consequently, the proportion of free water that could be removed by mechanical force increased, making pretreated DPMS easy to pack the sludge aggregates and reduce the moisture content of dewatered sludge [16]. Therefore, it can be concluded that EPS was important for DPMS dewaterability, and acid pretreatment could remove EPS, resulting in an enhanced degree of DPMS dehydration. Otherwise, as the main component of EPS, proteins were reported to likely play a significant role in its properties [20]. Further research on the influence of EPS components on DPMS dewaterability could be carried out in the future.

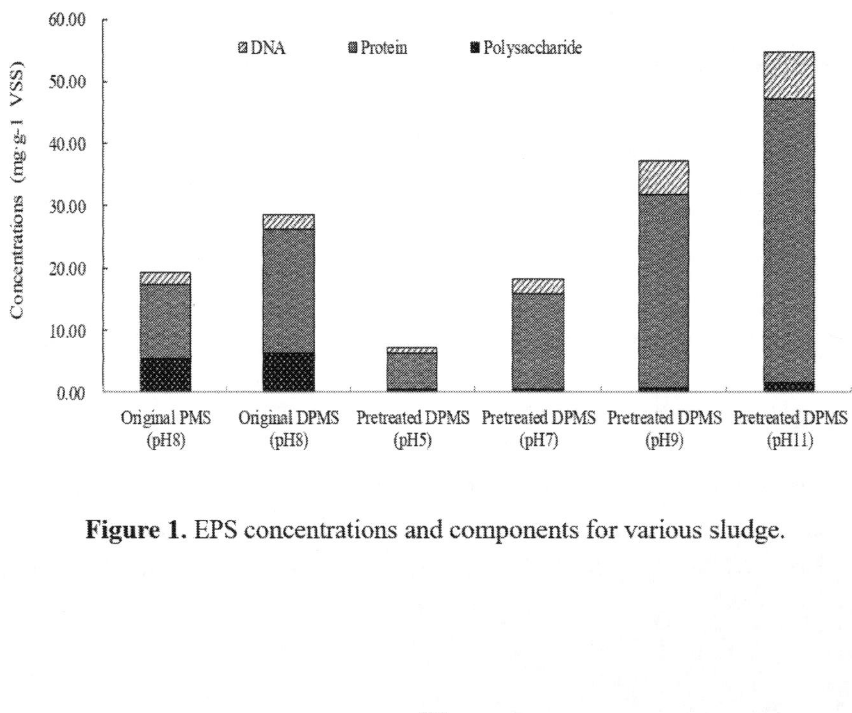

Figure 1. EPS concentrations and components for various sludge.

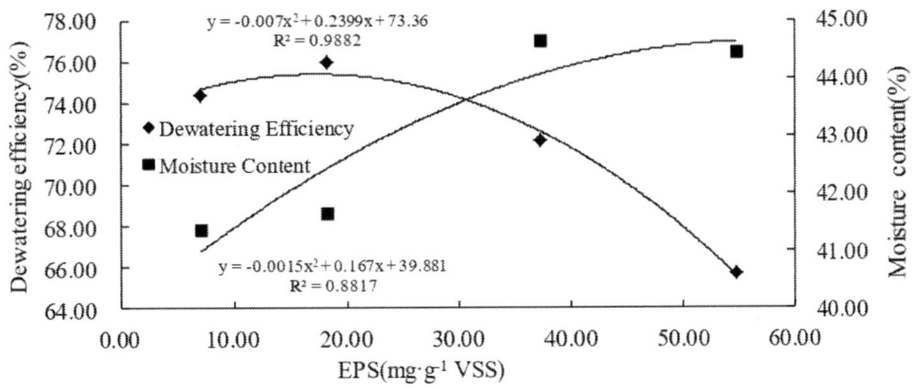

Figure 2. Correlation between EPS concentrations and dewaterabilities of pretreated DPMS.

3.3. Settling performance

As shown in figure 3, the settled sludge volume of the original PMS (pH 8) was lower than that of the original DPMS (pH 8), which indicated that the sludge dewaterability was reduced after anaerobic digestion. The settled sludge volumes of the pretreated DPMS (pH 5) and pretreated DPMS (pH 9) were significantly lower than those of the original DPMS (pH 8). This illustrated that acid or alkali pretreatment was helpful to enhance sludge settling performance.

The settling performances of different treatments ranked in the order of original PMS (pH 8) > pretreated DPMS (pH 5) > pretreated DPMS (pH 9) > original DPMS (pH 8) ≈ pretreated DPMS (pH 7) = pretreated DPMS (pH 11). Meanwhile, the sedimentation curves in Figure 3 also demonstrated that pretreated DPMS (pH 5) achieved the best settling performance among the four treatments with different pH values. This result is in accordance with the sludge dewaterability. Additionally, this result indicated that the settling performance of DPMS could be affected by pH values. Weakly acidic or alkaline conditions (pH 5 and pH 9) could enhance the sludge settling ability, while a strong alkaline environment (pH 11) would be negative for settling. Combined with EPS concentrations of pretreated DPMS (see section 3.2) and conclusions obtained by Liao et al. [12], the poor settling performance in strong alkaline conditions was possibly caused by the high concentration of EPS, which hindered further connections among the cells owing to the steric effect, and the increased diversity of density between sludge flocculation and water due to a close-grained hydrogel formation by EPS [9].

Regarding the settled volume of original PMS (pH 8), it was lower on average than that of the original DPMS. Since sludge dewaterability was consistent with its settling performance, and the settling performance was related to EPS concentrations [21, 22], taking the EPS concentrations and settling performances into consideration (figure 1 and figure 3), it can be inferred that anaerobic digestion reduced PMS dewaterability. This result was in line with the earlier results [2, 8].

By the way, the tiral with pretreated DPMS (pH3) had been designed in this study to test DPMS dewaterability in a strong acidic environment. Otherwise, an extremely intense reaction happened in the pretreatment reactor so that further tests could not continue. This phenomenon was caused by too much H+, which led to the failure of the reaction.

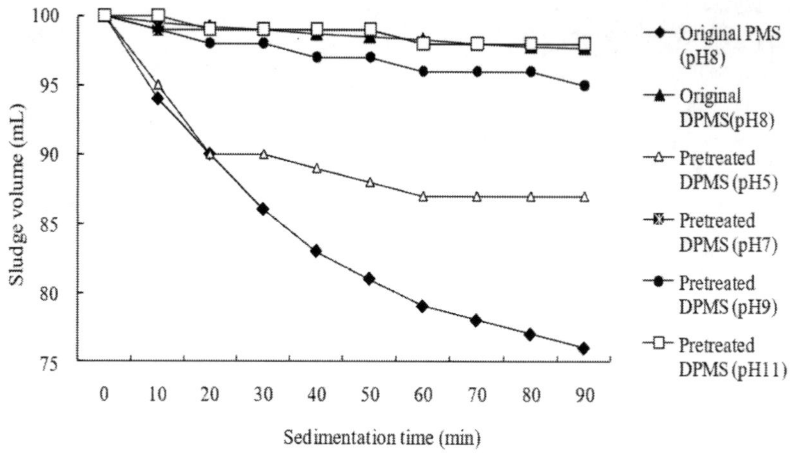

Figure 3. Settling performances of raw sludge and pretreated sludge.

3.4. Particle size distribution in sludge

Table 2 shows the observed differences in particle size distribution, and the differences are displayed by weight percentage of sludge particle size. Most of the particles presented had diameters of less than 180 μm (figure 4). However, for the particle size range above 1650 μm, the weight percentage in the original DPMS (pH8) was approximately 10% higher than those in the other acid/alkali pretreatments. Besides, it was noticed that when the particle size became small (<180 μm), the weight percentage of this size increased with the increase of the pH value. Moreover, the largest mean particle size of 121.82 μm was obtained in the original PMS (pH 8), and the smallest mean particle size of 99.15 μm was observed in the pretreated DPMS (pH 11).

With regard to the treatments after acid/alkaline pretreatment, it was observed that the pretreated DPMS (pH 11) with the smallest mean particle size of 99.15 μm had the worst dewaterability, as evaluated by a high moisture content and the lowest dewatering efficiency. In contrast, the pretreated DPMS (pH 5) with the largest mean particle size of 101.50 μm had the best dewaterability, as evaluated by the lowest moisture content and a high dewatering efficiency (tables 1-2, figures 1-3).

Therefore, a correlation between the dewaterability of pretreated DPMS and the mean particle size is presented in Figure 4. It indicates that the increased mean particle size in DPMS could obtain a better dewaterability, which confirms the earlier conclusion that sludge with a smaller mean particle size has worse dewaterability [10]. Thus, from the point of the mean particle size, it was proven again that AD would reduce the PMS dewaterability by decreasing the particle size in sludge.

Besides, Karr and Keinath [23] fractioned the sludge into various size ranges and found that its filterability reduced with the decrease of particle size. Meanwhile, some authors have also reported that sludge particle size could markedly influence sludge dewaterability [5, 24]. The sludge in an acidic environment was difficult to shear into smaller particles, while the sludge in an alkaline environment was gelatinous and could be cutted into small particles [23, 25]. As a result, the treatment of pH 5 with the largest mean particle size of 101.50 μm showed the best dewaterability (figures 1-4).

Therefore, these above results show that acid/alkali pretreatments under different pH values were able to change the particle size distribution in DPMS, which means that pH values had a considerable impact on DPMS dewaterability. Moreover, there was a significant inverse correlation between the dewaterability and the percentage of small particle size. In a word, acid pretreatment with pH 5 for DPMS was optimal to enhance its dewaterability.

Table 2. Particle size distribution in sludge measured by weight percentage (%).

Treatment	Particle size (μm)						Mean particle size
	>1650	**830~1650**	**380~830**	**250~380**	**180~250**	**<180**	**(μm)**
Original PMS[a)]	25.34	3.23	19.88	11.39	0.18	39.98	121.82
Original DPMS[b)] (pH8)	14.93	2.59	15.78	12.04	0.85	53.81	110.38
Pretreated DPMS (pH5)	6.13	3.3	10.24	10	0.96	69.37	101.50
Pretreated DPMS (pH7)	2.83	3.43	13.35	4.8	4.27	71.32	100.51
Pretreated DPMS (pH9)	3.64	5.84	10.02	3.95	3.38	73.17	99.71
Pretreated DPMS (pH11)	3.34	5.96	10.41	3.08	2.74	74.48	99.15

a) PMS: Paper mill sludge
b) DPMS: Digestd paper mill sludge

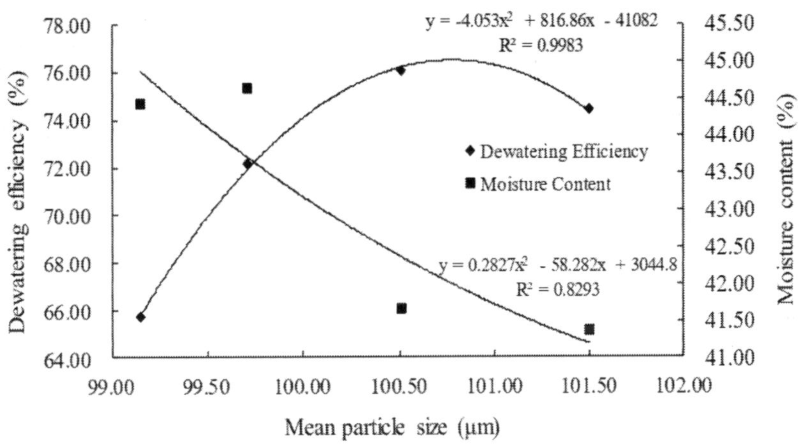

Figure 4. Correlation between mean particle sizes and dewatering efficiencies of pretreated DPMS.

3.5. Flocs structure

As shown in figure 5, flocs in the original DPMS (pH 8) had much more swelling than those in the original PMS (pH 8). Also, the fine particles in the original DPMS (pH 8) were more than those in the original PMS (pH 8). Likewise, the weight percentages of particles less than 180μm in the original PMS (pH 8) and original DPMS (pH 8) were 39.98% and 53.81%, respectively. This result demonstrated that the flocs structure of PMS was disrupted by anaerobic digestion, which resulted in a smaller particle size in DPMS and leading to a poor dewaterability [4]. Meanwhile, SEM images of the original and pretreated DPMS clarified that the flocs structure was further dispersed after pretreating by acid/alkali, which was in line with the previous report that pretreatment contributed to the rearranging/breaking of some chemical bonds [26]. Furthermore, considering the sludge flocs after acid/alkaline pretreatment, alkaline pretreatment was found to cause DPMS swelling and loosening, while acid pretreatment produced more particles with bigger size in DPMS. However, pretreated DPMS (pH 7) and pretreated DPMS (pH 9) had a similar flocs structure. Therefore, based on the flocs structure, the fact that pretreated DPMS (pH5) had the best dewaterability while pretreated DPMS (pH11) had the worst was reasonable.

Original PMS (pH 8) Original DPMS (pH 8) Pretreated DPMS (pH 5)

Pretreated DPMS (pH 7) Pretreated DPMS (pH 9) Pretreated DPMS (pH 11)

Figure 5. Microphotographs of sludge before and after pretreatment at 30 bar (1000 ×).

4. Conclusions

DPMS pretreated at pH 5 achieved the best dewaterability compared to other pretreated trials. It can be concluded that a weak acidic environment could enhance DPMS dewaterability, while a strong alkaline environment was detrimental to it. Compared to PMS, DPMS demonstrated poorer dewaterability because of the floc structure dispersion and finer particle formation in AD.

Acknowledgements

The authors would like to thank the Pear River Young Talents of Science and Technology in Guangzhou, China (No: 2012J2200082) and the Science and Technology Project of Guangdong Province under Contract No. 2015A010106012 and 2016A020210085 for financially supporting this research.

References

[1] Torsten M and Elizabeth A E 2014 Anaerobic digestion of pulp and paper mill wastewater and sludge *Water Research* **65** pp 321-49

[2] Ayol A 2005 Enzymatic treatment effects on dewaterability of anaerobically digested biosolids-I: performance evaluations *Process Biochemistry* **40**(6) pp 2427-34

[3] Lin Y Q, Wang D H and Xiao M Q 2011 Mesophilic batch anaerobic co-digestion of pulp & paper sludge and monosodium glutamate waste liquor for methane production in a bench-scale digester *Bioresource Technology* **102**(4) pp 3673-78

[4] Lin Y Q, Wang D H, Wu S Q and Wang C M 2009 Alkali pretreatment enhances biogas production in the anaerobic digestion of pulp and paper sludge *Journal of Hazardous Materials* **170**(1) pp 366-73

[5] Lawler D F, Chung Y J, Hwang S J and Hull B A 1986 Anaerobic digestion: effects on particle size and dewaterability *Journal of Water Pollutant Control Federal* **58**(12) pp 1107-17

[6] Bala S S, Yan S, Tyagi R D and Surampalli R Y 2010 Extracellular polymeric substances (EPS) producing bacterial strains of municipal wastewater sludge: Isolation, molecular identification, EPS characterization and performance for sludge settling and dewatering *Water Research* **44**(7) pp 2253-66

[7] Molla A H and Fakhru'-Razi A 2012 Mycoremediation-a prospective environmental friendly technique of bioseparation and dewatering of domestic wastewater sludge *Environmental Science and Pollutant Research* **19**(5) pp 1612-19

[8] Novak J T, Sadler M E and Murthy S N 2003 Mechanisms of floc destruction during anaerobic and aerobic digestion and the effect on conditioning and dewatering of biosolids *Water Research* **37**(13) pp 3136-44

[9] Neyens E and Baeyens J 2003 A review of thermal sludge pre-treatment processes to improve dewaterability *Journal of Fermentation Bioengineering* **98** pp 51-67

[10] Pei H Y, Hu W R, Li J and Chen L 2007 Dewaterability and particle size distribution of activated and digestion sludge *Environmental Science - China* **28**(10) pp 236-2241

[11] Costerton J W, Irvin R T and Cheng K J 1981 The bacterial glycocalyx in nature and disease *Annual Reviews of Microbiology* **35** pp 299-324

[12] Liao B Q, Allen D G, Droppo I G, Leppard G G and Liss S N 2001 Surface properties of sludge and their role in bioflocculation and settleability *Water Research* **35**(2) pp 339-50

[13] Liu H and Fang H H P 2002 Extraction of extracellular polymeric substances (EPS) of sludge. *Journal of Biotechnology* **95**(3) pp 249-56

[14] Frølund B, Palmgren R, Keiding K and Nielsen P H 1996 Extraction of extracellular polymers from activated sludge using a cation exchange resin *Water Research* **30**(8) pp 1749-58

[15] Chen Y G, Yang H Z and Gu G W 2001 Effect of acid and surfactant treatment on activated sludge dewatering and settling *Water Res.* **35**(11) pp 2615-20

[16] He W Y, Yang H Z and Gu G W 2006 Acid treatment of waste activated sludge for better dewaterability *Environmental Pollution & Control* **28**(9) pp 680-83

[17] APHA (American Public Health Association) 2005 *Standard Methods for the Examination of Water and Wastewater (21th ed.)* Washington, DC, USA

[18] Wang J T and Fang J 2010 Biochemistry experiments *Science and Technology Press in Huazhong University* Wuhan, China

[19] Yuan Y and Yang Z H 2003 Study on effect of surfactant and acid treatment on sludge dewaterability (in Chinese) *Sichuan Environment* **22**(5) pp 22, 1-8

[20] Shao L M, He P P, Yu G H and He P J 2009 Effect of proteinss, polysaccharidess, and particle sizes on sludge dewaterability *Journal of Environmental Science* **21** pp 83-88

[21] Forster C F 1971 Activated sludge surfaces in relation to the sludge volume index *Water Research* **5**(11) pp 861-870

[22] Urbain V, Block J C and Manem J 1993 Bioflocculation in activated sludge: An analytic approach *Water Research* **27**(5) pp 829-38

[23] Karr P R and Keinath T M 1978 Influence of particle size on sludge dewaterability *Journal of Water Pollutant Control Federal* **50** pp 1911-30

[24] Hougton J I and Stephenson T 2002 Effect of influent organic content on digested sludge extracellular polymer content and dewaterability *Water Research* **36**(14) pp 3620-28

[25] Novak J T, Goodman G L, Pariroo A and Huang J 1988 The blinding of sludge during filtration *Journal of Water Pollutant Control Federal* **60** pp 206-14

[26] Seehra M S, Akkineni L P and Yalamanchi M 2012 Structural characteristics of nanoparticles produced by hydrothermal pretreatment of cellulose and their applications for electrochemical hydrogen generation *International Journal of Hydrogen Energy* **37** pp 9514-23

IOP Conf. Series: Earth and Environmental Science **39** (2016) 012070 doi:10.1088/1755-1315/39/1/012070

Biological soil crust formation under artificial vegetation effect and its properties in the Mugetan sandy land, northeastern Qinghai-Tibet Plateau

Y F Li[1, 5], Z W Li[2], Y H Jia[3] and K Zhang[4]

[1]Nanjing Hydraulic Research Institute, Nanjing, 210029, China
[2] School of Hydraulic Engineering, Changsha University of Science &Technology, Changsha 410004, China
[3] East China Normal University, Shanghai 200052, China
[4] Key Laboratory of the Pearl River Estuarine Dynamics and Associated Process Regulation, Ministry of Water Resources, Guangzhou, 510611, China

Email: yfli@nhri.cn

Abstract. Mugetan sandy land is an inland desertification area of about 2,065 km^2 in the northeastern Qinghai-Tibet Plateau. In the ecological restoration region of the Mugetan sandy land, different crusts have formed under the action of vegetation in three types of sandy soil (i.e. semi-fixed sand dune, fixed sand dune and ancient fixed aeolian sandy soil). The surface sand particle distribution, mineral component and vegetation composition of moving sand dunes and three types of sandy soil were studied in 2010-2014 to analyze the biological crust formation properties in the Mugetan sandy land and the effects of artificial vegetation. Results from this study revealed that artificial vegetation increases the clay content and encourages the development of biological curst. The fine particles (i.e. clay and humus) of the surface layer of the sand dunes increased more than 15% ten years after the artificial vegetation planting, and further increased up to 20% after one hundred years. The interaction of clay, humus, and other fine particles formed the soil aggregate structure. Meanwhile, under the vegetation effect from the microbes, algae, and moss, the sand particles stuck together and a biological crust formed. The interconnection of the partial crusts caused the sand dunes to gradually be fixed as a whole. Maintaining the integrity of the biological crust plays a vital role in fixing the sand under the crust. The precipitation and temperature conditions in the Mugetan sandy land could satisfy the demand of biological crust formation and development. If rational vegetation measures are adopted in the region with moving sand dunes, the lichen-moss-algae biological curst will form after ten years, but it still takes more time for the sand dunes to reach the nutrient enrichment state. If the biological curst is partly broken due to human activities, reasonable closure and restoration measures can shorten the restoration time of the biological crust.

1. Introduction

Extensive arid area including desert, Gobi, and sandy land exists in the western China, where vegetation is scarce. However, sparse vegetation plays a crucial role on ecological environment on the edge of deserts [16,17,39,45]. Vegetation exerts good ecological functions of windbreak and sand fixation, moreover, it boosts formation of biological crust [4,9,40]. Biologic crust is the organic complex formed by interaction of soil particles, organic matter and its metabolites and microbes, algae, lichens, and mosses [1,23], which serves a significant ecological and environmental function in the desert ecosystem and is an indicator for the evaluation of the ecological state [26,34].Biological crust may partly change the local land desertification by transforming sand to soil to improve the ecological environment and land use [20]. Moss, as the dominant species in the biological crust, marks the maturity of the crust [19]. The development of the moss crust reaches the optimum when the sand dune has a strong capacity of water storage and fixed sand and the greatest resistance to mechanics disturbance [2,43,32]. Consequently, it is extremely important to repair damaged biological crust and accelerate the formation of biological crust in desertification control areas (e.g. in western China).

Biological crust develops slowly. Once the biological crust is damaged by human activities (e.g. fire and grazing), it takes a long time to fully restore. The recovery speed of biological crust depends on the crust type, disturbance intensity and frequency. Moreover, algae crust recovers faster than lichen and moss crusts [32]. In general, the full elementary recovery of the crust takes at least 1-5 years; the restoration to a thicker crust takes about 50 years; and the restoration of moss and lichen crusts takes about 250 years [37,29,14]. However, if the disturbance is effectively controlled, reasonable vegetation measures can result in the recovery of the biological crust in a relatively short time [46,30].

Mugetan sandy land is one of the most seriously desertified areas in the Gonghe Basin of the northeastern Qinghai-Tibet Plateau [3,27]. The Huangshatou zone is on the edge of the Mugetan sandy land, and has the strongest interaction between vegetation growth and sand storm activity. Since the late 1980s, the local government has implemented a series of control and restoration projects to fix moving sand dunes in the Huangshatou zone. Currently, moving sand dunes in the Huangshatou zone have been fixed with more than 30% coverage of shrubs and herbs, and even more than 60% of shrub-herb-moss structures in some areas. Moreover, biological crust has developed and reached a stable ecological sand-fixation state [49,18]. It is evident that the ancient sand dunes, most of which are several hundred years old [7], have lost their dune shape and covered by vegetation with 90% coverage. It is important to explore the formation and role of biological crust on the restoration process especially since the process of biological crust development in the Mugetan sandy land has not been documented. More importantly, the successful management experience of ecological restoration in the Huangshatou zone of the Mugetan sandy land is worth referencing and promoting. The objective of this study is to analyze the process and characteristics of biological crust in vegetation restoration zones base on the control practice and vegetation effect in the Mugetan sandy zone, the acceleration effect of artificial vegetation on biological crust, and the restoration time of the damaged biological crust.

2. Study area and methods

2.1. Study area

The Mugetan sandy land is a desertification area of about 2,065 km^2 near Yellow River source. It is located at 100°24 '~ 101°22' E, 35°27 '~ 35°47' N at an altitude of 3000~3400m, near east of LongyangxiaReservoir.It is a typical inland desert bottomland. The annual mean temperature in this area is 2.4°C, the annual mean precipitation isabout 400mm, and the annual mean evaporation is about 1,500mm. Dominant wind direction is south and southeast with an annual mean wind speed of 3m/s and the maximum annual mean wind speed of 14 m/s. The annual gale (wind speed equal to or higher than 7m/s) blowsfor12 days. Overall, this region has the characteristics of alpine, arid, and windy climate.

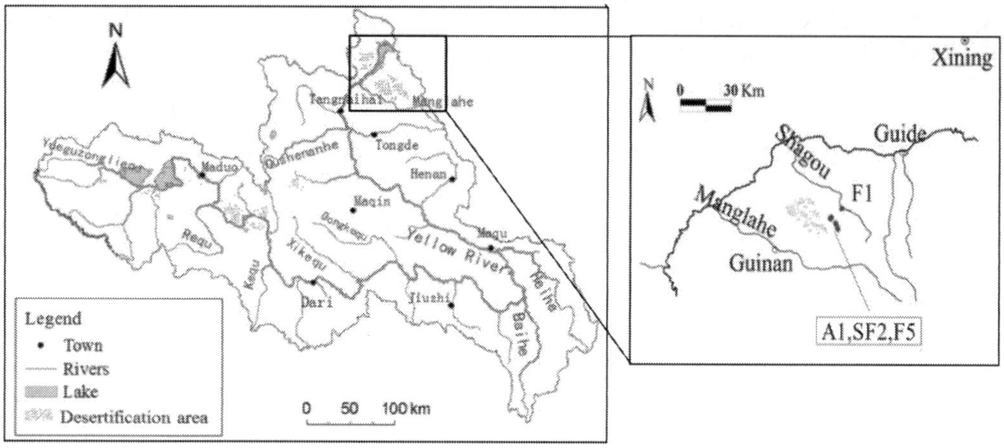

Figure 1. Location of the Mugetan sandy land near Yellow River source and four sampling points.

Figure 2 shows the topography and landscape of the Mugetan sandy land. Early lake sediments with the plateau uplift are exposed on the ground.The fine sand provides rich material sources for the aeolian sand [48]. The Mugetan sandy land is a transitionaltype between the northwest extremely arid desert zone and the semi-arid loess plateau zone.The desertification takes place in a periodic or non-periodic alternative process andindifferent degrees,and the desertification and its inverse processexist side by side [47].

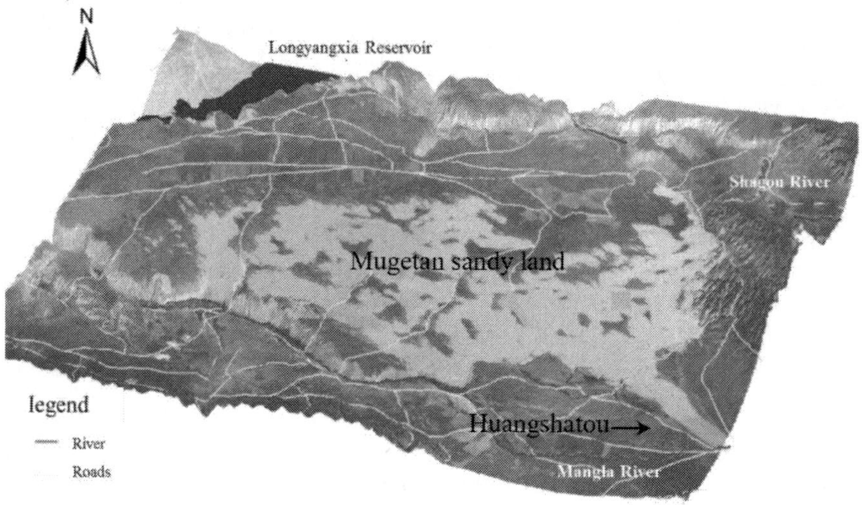

Figure 2. Topography and landscapeof the Mugetan sandy land.

Since 1970s, in order to reduce aeolian sand disaster, a series of control measures have been implemented, e.g. closing hillsides, afforestation, planting grass to fix sand dune and restore grassland, and forbidding overgrazing, deforestation, and unreasonable land use. Subsequently,vegetationstarted growing, the desertification area started reducing, and grassland was restored to some extent. In addition, on the front edge of the Mugetan sandy land, i.e. the Huangshatou zone, vegetationrestoration projects were implemented, such as afforestation and laying checkerboard barriers to fix moving sand dunes. Along with the reinforced sand control engineering and artificial vegetation influence, a belt of a few kilometres long with semi-fixed and fixed sand dunes appeared in the front edge of the Huangshatouzone as shown in figure 2.

The desertification area of Mugetan sandy areawas about 2,065km^2 in 2005.Compared to the measurements in 1990, the area of desertificationhad been gradually reducing, especially during 2000 to 2005, when the moving sand dune reducedby 41.76km^2 [3]. Obviously, implementing effective control measures and vegetation restoration can inhibit desertification expansion and encourage grassland recovery.

Figure 3. Afforestation zone of Huangshatou in the Mugetan sandy land (a) semi-fixed sand dune with20-30% vegetation coverage; and (b) fixed sand dune with 50-70% vegetation coverage.

2.2. Study method
In this study, the ecological restoration condition of the Mugetan sandy land is divided into four types: moving dune, semi-fixed dune, fixed sand dune, and ancient fixed aeolian sand soil. For each of these four representative types of sandy soil, the biological crust, fine sand beneath the crust, and vegetation composition were sampled and analyzed. In July and August of 2010-2014, field investigations were conducted in the vegetation restoration area of the Huangshatou, including sand sampling and vegetation measurement (A1, SF2, and F5), and a site of ancient fixed aeolian sand soil (F1) (shown in figure 1).

The meteorological data (i.e. rainfall, temperature, wind) in 1961-2011 were obtained from the meteorological station of Guinan County and relevant literatures [11,12,41]. Soil sampler drilling was used to obtain sand samples from typical dune surface and deep layer, with each sample weighing about 300 g. Sediment coarser than 0.075 mm was analyzed using the sieving method. For sediment finer than 0.075 mm, the LA-920 laser particle size analyzer was used to analyze the particle size gradation so as to obtain the particle size distribution. The total rock mineral and clay mineral composition of sandy soil samples were quantitatively analyzed by an X-ray diffraction instrument with 0.01% accuracy. The organic matter content was determined using the potassium dichromate oxidation plus heating method, with a detection standard of NY/T 1121.6-2006 and error less than 1 g/kg. The sodium pyrophosphate extraction-potassium dichromate oxidation method was used to determine the humus content, with a testing standard of LY/T 1238-1999 and error less than 1 g/kg. The uniformity of sediment particles composition was analyzed using the sorting coefficient , which is the square root of the ratio d84 to d16, i.e. , where d84 and d16 are the particle diameters for which 84% and 16%, respectively, of the sediment samples are finer.

For repeated sampling of the sand dunes within each time period, herbs of 1×1 m2 quadrat were taken (five repeats) and the shrubs of of 5×5 m2 quadrats were taken (two repeats). The number of vegetation species, number of trees, height, and coverage, and the species of the samples collected were identified. The vegetation coverage, richness (i.e. the number of species), and average height of the vegetation on the windward and leeward sides of the dunes in each sample were statistically analyzed. Based on the species composition of multiple samples and the numbers of the species, the importance parameter of each species were calculated, namely the total value of the relative density, relative coverage, and relative frequency of each species. The relative density, relative coverage, and relative frequency are the percentages of the density, coverage, and frequency of each species to the total density, coverage, and frequency of all species [8]. The summation of the important parameters of the same species is an important value of the vegetation components.

3. Results

3.1. Meteorological factor

Figure 4 (a) shows the average annual temperature recorded at the Guinan meteorological station from 1961 through 2010.Since 1961 the temperature has shown a rising trend with an annual average temperature of 2.4°C. , The annual temperature from 1998 through 2010was higher than the multi-year mean temperature from 1961 through 2010.Obviously, the rising temperature was conducive to afforestation, forest and vegetation development. Figure 4 (b) and (c) show the annual average precipitation and evaporation amount recorded at the Guinan meteorological station from 1961 through 2010.The average annual rainfall was about 400mm, and the annual average evaporation was 1500mm.The annual evaporation was much higherthan the annual rainfall.The annual rainfall from 2000 through 2010 was higher than themulti-year mean rainfall from 1961 through 2010, whereas, the annual evaporation from 2000 through 2010 was lower than the multi-year mean evaporation from 1961 through 2010.Since 2000,the precipitation condition has been conducive to the development of vegetation. Figure4 (d) showsthe annual maximum wind speed recorded atthe Guinan meteorological station from 1961 through 2010.The graph shows a downward trend since 1961.From1991 through 2010, the annual maximum wind speed became lower than the annual multi-year maximum wind speed from 1961 through 2010.In 2003, the maximum wind speed reduced to 10m/s. Decreasing wind speed reduces the wind-blown sand, which is advantageous to dune fixation. Therefore, in the past 20 years, the meteorologicalconditionshave beenin favor of desert vegetation restoration in the Mugetan sandy land.

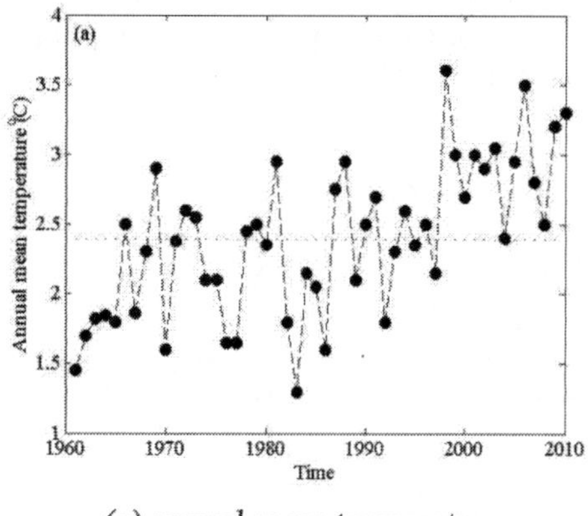

(a) annual mean temperature

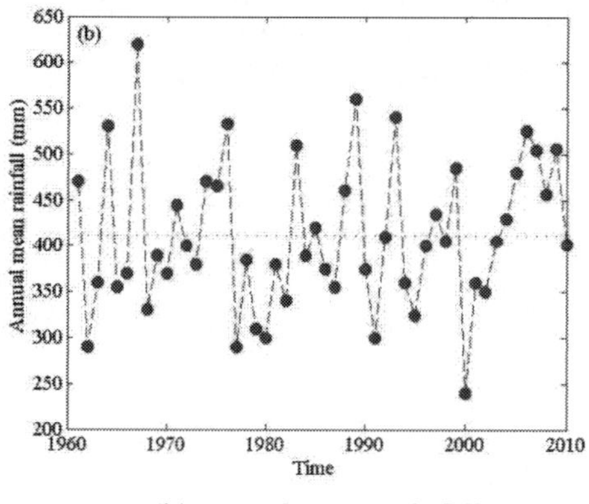

(b) annual mean rainfall,

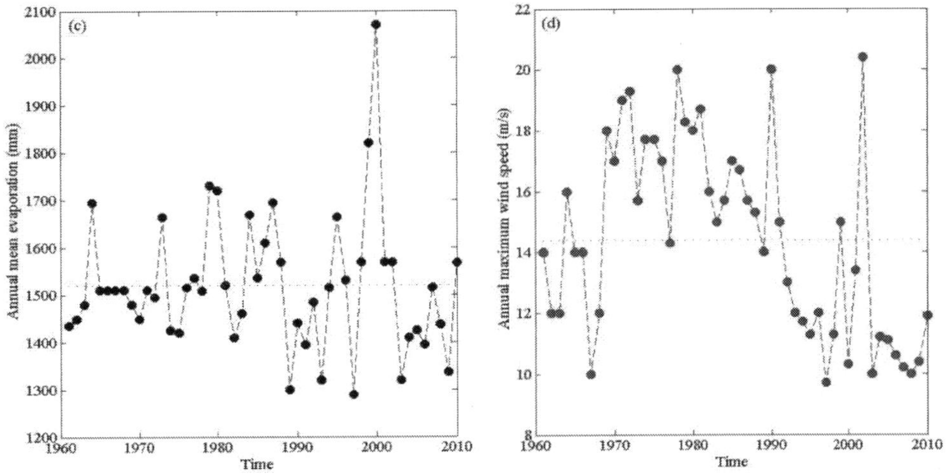

(c) annual mean evaporation (d) annual maximum wind speed

Figure 4. Change of four climate factors from 1961-2010 in the Mugetan sandy land.

3.2. Feature of sand dune

Figure 5(a) shows amoving sand dune (A1)with a crescent shape, a height of 7-10m, less than 5% surface vegetation cover, and the sand in the flow state. Figure 5(b) shows that a semi-fixed dune (SF2) which has vegetation coverage of 5-20%.A large area of the physical crust is coveredlocallywith weak algae crust after 14-15 years of artificial planting. However, these crusts are too fragile to effectively inhibit sand movement. If the crust is breached, sand below the crust will be activated again. Additionally, even though the semi-fixed sand dune is relatively flat, it still maintains a certain dune shape. Figure 5(c) shows fixed sand dune with 40% vegetation coverage and a large area of moss crust with 3-5 cm thickness after 30 years of artificial planting. Ancient fixed aeolian sandy soil is widely distributed in the Mugetan sandy land. One example (F1) is shown in Figure 5(d). After experiencing vegetation effect for hundreds of years, the ancient fixed aeolian sand soil has more than60% vegetation coverage and a thick humus layer (15-18 cm) including a surface biological crust (4-5 cm).The height of the ancient fixed aeolian sandy soil is less than 0.5m and the area is almost flat.

(a) moving sand dune (A1)

(b) biological crust of semi-fixed sand dune in ecological restoration zone (SF2)

(c) biological crust of fixed sand dune in ecological restoration zone (F5)

(d) humus layer with surface biological crust of ancient fixed aeolian sand soil (F1)

Figure 5. Surface coverage features of four types of sand dunes (distant viewon the left and close shot on the right).

3.3. Variation of particle composition

Figure 6 shows the grain distribution curves of the surface of sand dunes in different stages, and Table 1demonstratesthe particle compositioncharacteristicsof the surface of the sand dunes under the artificial vegetation effect. The composition of the moving sand dune is uniform with the median size d_{50}=0.15mm and the sorting coefficient σ=1.2. In particular, the grain distribution curve displays a steep slope between 0.2~0.1mm, and 95% of the particles have particle sizes of 0.1~0.2mm. After the implementation of vegetation restoration measures, under the effect of rainfall and vegetation, the surface particles of moving dunes became weathered,which resulted in asubstantial amount of fine particles (less than 0.05mm) including fine sand, silt, and clay. At the semi-fixed sand dune stage, d_{50}=0.13 mm, and σ=1.3. The 0.1-0.2mm particles counted for 80% of the total particles,and the fine particle content became 5%.At the fixed sand dune stage,d_{50}=0.11 mm, and σ=1.4. The 0.1-0.2mm particles counted for 70% of the total particles,and the fine particle content increased to 15%. The surface particle content of the ancient fixed aeolian sand soil indicates that after nearly one hundred yearsofvegetationrestoration, fine particles significantly increasedto 23.5%, the 0.1-0.2 mm particles dropped down to 50%, with d_{50}=0.09 mm, and σ=2.6. Consequently, after experiencing vegetation restoration for several years, the proportion of fine particles greatly increased, and the fine particle content increased to over 15% after several decades, and more than 20% after one hundred years.

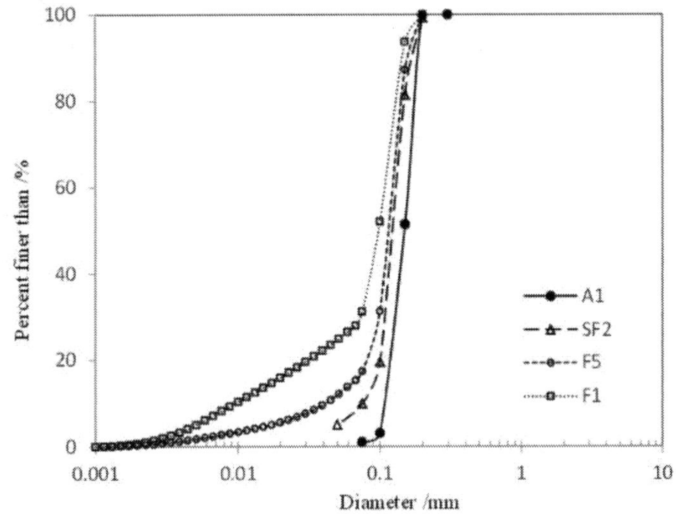

Figure 6. Grain size distributionsof biological crust samples of four sandy soils.

Table 1 shows the comparison of particle sizesat the surface and deep layers. According to Table 1, weathering was not obvious in the moving sand dune, i.e. no difference in the particle composition between the surface and deep layers. Weathering had occurred in the semi-fixed sand dune with locally forming crust and the fine particlecontent in the crust layer being 3-4% higher than that in the deep layer. For the fixed sand dune, due to long-term weathering and soil formation, the humus layer had formed and the fine particle content in the humus layer was 10% higher than that in the deep layer. For the ancient fixed aeolian sand soil,the fine particles inside the thick humus layer were 20% higher than that in the deep layer.

Table 1. Particle characteristics of four sandy soils under artificial vegetation restoration.

No.	Thickness of biological crust (cm)	Depth (cm)	Sorting coefficient σ	Median size(mm)	Fine particle content (%)	Grain size content (%)		
						0.25–0.1 (mm)	0.1–0.05 (mm)	<0.05 (mm)
A1	0	0	1.273	0.15	1.017	95.0	5.0	0.0
		10	1.225	0.13	1.244	95.4	3.0	1.3
SF2	1	0	1.316	0.13	5.019	80.0	15.0	5.0
		10	1.219	0.16	1.200	96.6	1.4	2.0
F5	2-4	0	1.449	0.11	15.730	70.0	18.0	12.0
		20	1.262	0.14	0.851	91.7	5.6	2.7

F1	4-5	0	2.613	0.09	23.460	50.0	25.0	25.0
	(humus layer of 15-18 cm)	50	1.348	0.14	2.429	91.8	5.4	2.8

3.4. Characteristics of mineral particle change

Figure 7 shows the mineral composition and content of the sand dunes at the four stages. The contents of the measured minerals were mainly quartz, feldspar,carbonatite, amphibole, and clay minerals which refer to layered silicate minerals (particle size less than 0.002mm), such as kaolinite, illite, smectite, chlorite, illite and smectite mixed layer, and green smectite interstratified minerals.

The content of clay minerals in the moving sand dune was about 9%, and the content of non-clay minerals was 91%. The content of clay minerals in the surface layer ofthe semi-fixed sand dune was 13.8%, while the content of the non-clay minerals was 86.2%. The content of clay minerals in the crust of the fixed sand dune was 16.2%, while the content of non-clay minerals was 83.8%. The content of clay minerals in the crust of the ancient fixed aeoliansand crust was 23.7%, while the content of non-clay minerals was 76.3%. These results show that the content of clay minerals increased to nearly twice as much after the sand dunewas fixed, and the clay content inthe ancient fixedaeolian sand crust was two times more than that in the moving sand dune after the long-term effect of vegetation. Moving sand dunes became fixed aeolian sandy soil after several decades with the clay minerals increasing greatly. During the following decades, the thickness of humus crust increased, and the clay particles inside the crust increased with time.

The humus content in the moving sand dune was 4.2g/kg, and that in the ancient fixed aeolian sandwas 16.67g/kg. Under the effect of vegetation, the increase of the clay mineral and humus content is the primary factor in the process of moving sand dune becoming fixed aeolian sand.

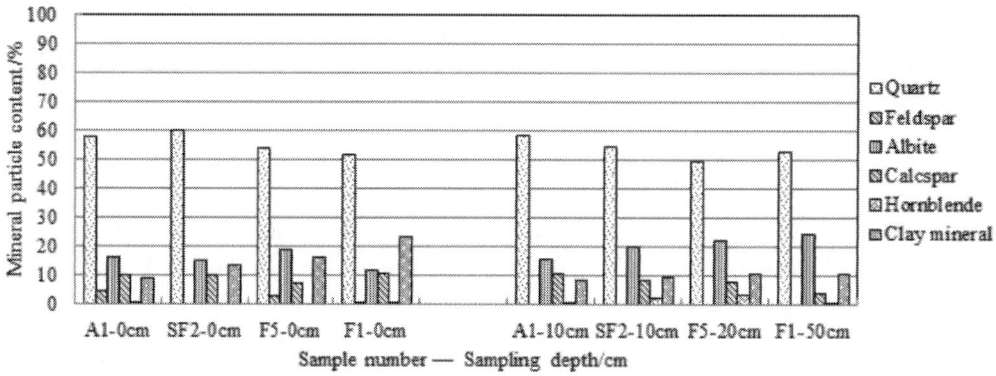

Figure 7. Mineral particle contents of four sandy soils.

3.5. Effect of vegetation on biological crust

The vegetation was planted in the Huangshatou ecological restoration area of the Mugetan sandy landasCathay poplar (*Populuscathayana*) since the late 1990s. Table 3 shows the composition of the herbaceous vegetationon the sand dunes in three stages under the planting vegetation effect including moving sand dune (A1), semi-fixed sand dune (SF2) and fixed sand dune (F5). The vegetation succession process and effect of vegetation on biological crust were analysed. And also the ancient fixed aeolian sandy soil (F1) of hundred years old was comparatively analysed.

During the stage of moving sand dune, the restoration measure was to plant poplar seedlings with tree diameters of 1-2cm at spacing of 2-3 m, and ground scattered sedge family (e.g. *Carexmoorcroftii*) and gramineousfamily (e.g.*Bromustectorum*).

During the semi-fixed sand dune stage, the artificialvegetation fixed the sand dune after 15 years [45], and herbaceous vegetation developed with invasion of microbes, algae, lichens in the surface particles layer. Meanwhile, under the climatic factors and vegetation effect, the surface particle composition and soil structure of the sand dune changed, i.e. hydrolysis or carbonation occurred to the feldspar and carbonate minerals andillite, kaolinite, chlorite, and illitesmectite mixed-layer clay minerals were generated. This process resulted in particles fining and gradual accumulation with time. Aggregation of plant residues and animal carcasses and faeces in the sand dune surface providedlarge amounts of organic matter that yielded the humus under the action of microorganisms.After 15 years of vegetation development and succession, the vegetation coverage reached 30%, and the herb layer was composedof 2 families and 2species, i.e. dominately*Carexmoorcroftii* and*Oxytropisfalcata Bunge*. At this time, the humus, clay particles, microorganism, and algae vegetation united with each other in localized areas of the semi-fixed sand dune and complex organic compound formed.This primary biological crust played an important role for fixing the sand dune. The biologicalcrust influences vegetation successionand turnsan early shrub-dominated system intoan herbaceous vegetation system.

During the stage of fixed sand dune, the fine sand, silt, and clay increasedgreatly, with the invasion of microbe, algae, lichens, and mosses, and bacteria and fungi growth. Meanwhile, vegetation interception and atmospheric dustfall (mainly clay) resulted in sediment deposition [6]. These effects promoted sand particles sticking and the formation ofsoil aggregate structure and stable organic complex, and finally a biological crust with a certain thickness developed. After the vegetation had experienced development and successionfor more than 30 years, the vegetation coverage reached to more than 60% with the height of*PopulusCathayana*reaching 3m and the tree trunk diametersreaching 8-12cm.The herb species composition included 8 families and 10 species, dominantly*Oxytropisfalcata Bunge* and*Carexmoorcroftii*, and accompanied by *Artemisiaannua L.*and *KalidiumfoliatumMoq*. After the sand fixing by planting wood Missy Populuscathayanaand vegetation restoration for ten years, plant communitieswith*Carexmoorcroftii* as the dominant species gradually developed in*OxytropisFalcata Bunge* in the ecological recovery area. Interaction betweenthe vegetation and sand particles with lichens and bryophytes encroachment accelerated the rate of mineral weathering and accumulated the clay and other particles for soil formation. In the ecological restoration area in the Mugetan sandy land, the fine particle content increased more than 15% after several decades and increased up to 20% after one hundred years.

In the ancient fixed aeolian sandy soil, due to vegetation development and succession after decades and even hundreds of years,the vegetationcoverage reached 95% with 10 species in 6 families, mainly including local perennial *Poatibetica*, *Elymusdahuricus*, and *Carexmoorcroftii*. These vegetation species in the Mugetan ecological restoration area are dominant species of the plant community. They have dense root, resistance to drought and sand burying, and play an importantrole on fixing sand dunes. Meanwhile, the biological crust changed the physical and biological properties of the subjacent sand, and transform sand to soil. To present, there formed a thick humus layer and the thickness has increased up to tens of centimetres, containing a large number of plant roots.

Compared to the ancient fixed aeolian sandy soil, after 30 years of the artificial planting, the vegetation coverage has gradually increasedto 45% (11 species), and the effect of herbaceous vegetation on sand fixation has been enhanced. The number of vegetation species in the artificial vegetation zone was the same as that in the ancient fixed aeolian sandy soil. It indicates that the artificial vegetation accelerates the vegetation succession process, which is consistent withconclusions by previous researchers[15,42]. In addition, after 15 years of artificial planting, the fine particles has increased by 15% and the clay mineral particles have reached 16%, equivalent to 67% and 68%, respectively, of that in the ancient fixed aeolian sandy soil. It indicates thatartificial vegetation promotes the increase of the clay content and the development of biological curst, but the clay and nutrient still need more time to reach the enrichment state.

Table 2. Herb characteristics of four sandy soils.

No.	Elevation (m)	Height of dune(m)	Slope type	Coverag e(%)	Richness	Species composition / Important value
A1	3363	9	Windward slope	2.5	1	-

			Leeward slope	3	3	Cyperus/0.523, Carexmoorcroftii/0.444, Bromustectorum/0.033
SF2	3369	4	Windward slope	3	1	Carexmoorcroftii
			Leeward slope	30	2	Carexmoorcroftii/0.800, Oxytropisfalcata Bunge/0.200
F5	3356	1	Windward and Leeward slope	45	11	Oxytropisfalcata Bunge/0.5188; Carexmoorcroftii/0.2814; Artemisia hedinii/0.0808; GentianaScabra Bunge/0.0747; Kalidiumfoliatum/0.0645; Pedicularischenocephala/0.0408 ; Pleurospermum/0.0408; Heteropappusaltaicusovopok/0. 02040; RosaceasePotentillaansrina/ 0.0204; Gentianamacrophylla Poll/0.02040
F1	3184	0.5	Windward and Leeward slope	95	11	Poatibetica/0.430; Elymusdahuricus/0.120; Carexmoorcroftii/0.046; Pedicularisalaschanica/0.091; Poapratensis/0.041; Pedicülariskansuensis/0.025; HerbaTaraxaci/0.161; Allium fistulosum/0.037,; Pedicularischenocephala/0.033; Stellerachamaejasme/0.015

4. Discussion

Many deserts and sandy lands in western China have developed localized mature biological crustunder the influence of artificial and natural vegetation. Previous studies by researchers have investigated biological crusts in many deserts and sandy lands in western China. Table 3 shows the characteristics of biological crustson sand dune surfaces under diverse vegetation effects in western China. Owing to the complexity of influencing biological crust growth, the thickness and development degree of the biological crusts in different sampling locations are totally different, even at the same time and inthe same area. For example, the Horqin sandy land in Inner Mongolia and the Mu Us sandy land in Yulin have better water and heat conditions (400 mm/yr rainfall and 6-8°C annual mean temperature) than many other sandy lands. Twenty-five yearsafter vegetation plantingand aerial seeding in the Horqin sandy land, a moss crust has developedwith a thickness of 15 mm and a fine particle content of 15% [13,46]. Thirty-one years after artificial vegetation planting in the Mu Us sandy land,a moss crust has developed with a thickness of 10.3 mm and a fine particle content of 12.6% [33]. In contrast, in the Shapotou area of the Tenggerdesert, the water and heat conditions are poor, with the annual mean rainfall of 180mm the evaporation of 3,000 mm, and the annual mean temperature is 9.7 °C. The biological crust only developed to a thickness of up to 11 mm and a fine particle content of 24.42%

[19,37]. The Gurbantunggut sandy land in Xinjiang Province also has low rainfall (150 mm) and high evaporation (2000 mm), the natural moss crust only developed to a thickness of 45 mm and a fine particle content of up to 25% [22].

The Mugetan sandy land is located in the plateau region at an elevation of 3000 m, and has a low annual mean temperature of 2.4 °Cwith the annual mean rainfall of 400 mm and the evaporation of1,500 mm. After 30 years of vegetation restoration in the desertification control area, a mature moss crust has developed on the sand dune surface with a thickness of 20-40 mm and a fine particle content of 12%. Therefore, theMugetan sandy land naturally has suitable water and heat conditions.Moisture and temperature play a crucial role on biological crust developmentbecausegood hydrothermal conditions benefitthe invasion and growth of microbial and plant, accelerate the micro-environment process, and encourage the formation of soil aggregate structure and biological crusts.As long as reasonable vegetation measures are adopted, biological crust can develop in a relatively short time. For the ancient aeolian sandy soil, a natural humus layer with a thickness of 10-20 mm formed under the action of biological crust, which has a thickness of 4-5mm anda fine particle content of 25%. It can be seen that mature biological crust can be developed as long as external disturbance is reduced or controlled, and change the subjacent sand to soil with humus layer.

The effect of vegetation exerts a significant role on crust formation. Vegetation restoration promotes composition change of the dune surface particles from loose particles to soil aggregate structure withbiological crust. The crust layer fixes dunes and promotesecological restoration, which is consistent with previous research see[13].The vegetation measuresto fix sand areto plantshrubs and trees, e.g. planting Populus in the Mugetan sandy land. The shrub intercepts atmospheric dustfall, and its own litter has important effect on the formation of biological crust. Silt, clay, organic matter, and humus significantly increaseunder the shrub canopy which reduces wind speed and improve the water and heat conditions and micro-environment in the ground, promoting the formation of biological crust [5,21,28,24]. Then the crusts under the canopy causethe surface sand and the surrounding sand to begradually fixed.Finally, theinterconnection of these crustsmakethe sand dunesbe fixed as an integral body [10,35]. Therefore, the sand fixing process is accelerated by implementing artificial vegetation measures, and the shrub-herb-biological crust community forms ina short time and good results of ecological restoration can be achieved.

Table 3. Biological crust characteristics of sand dunes under vegetationeffect in different deserts and sandy lands of western China.

Desert area	Elevation (m)	Mean annual precipitation|evaporation (mm)	Average Annual temp. /°C|wind speed (m/s)	Enclosing time (year)	Sand-fixing measures	Crust type	Thickness (cm)	Fine particle content (%)
Horqin Sandy Land ,Inner Mongolia	360	366|1935	6.4|3.2-4.1	15	Artifical planting((Populu)	Moss	1.2	22
				25	Artifical planting((Populu)	Moss	1.5	25
Mu Us sandyland, Yuling, Shannsi	1100	415|2500	7.9|2.1-3.3	31	Artificalplanting,seeding	Moss, algae	1.03	12.58

Shapotou, Tengeli desert	1300	180|3000	9.7|2.8	48	Artificial vegetation; Grass square	Moss,algae	11	24.42
				40			10.3	-
				23			2.2	-
Gurbantun-ggut Desert,Xingjiang	400	150|2000	6.0|1.4—2.7	-	Natural growth	Algae, lichen	0.5	15-25
							1.3	
						Moss	4.5	
Mugetan sandy land, Qinghai-Tibet Plateau	3000	400|1500	2.4|3	30	Artifical planting, Grass square	Moss,lichen	2-4	12
				-	Natural growth	-	4-5	25

A large area of biological crust can greatly improve the resistance to aeolian erosion. Wind tunnel tests have shown that under 25-30 m/s wind speed, sand dunes with biological crusts haveno erosion.Therefore,maintaining the integrity of the biological crust is extremely important in improving the resistance of aeolianerosion [36]. Biological crusts can well adaptto harsh environment. The biggest threat is grazing, firing, and mechanical disturbance [38]. Disturbance from grazing, firing, and mechanicalnot only affects the physical structure and chemical properties of soil, but also leads to fragmentation of the biological crust structure [25]. Once the biological crust is damaged, especially if the damage area exceeds the critical damage rate [31], even low wind speed can cause severe aeolian erosion on bare sand surfaces. According to the results from previous studies [31], the annual maximum wind speed is up to 14 m/s in the Mugetan sandy land, and the critical damage rateforthe moss crust and the alga crust are 30% and 10%, respectively. At present, moss-algae interactive biological crusts can control aeolian erosion in the Huangshatou restoration area under the measures of afforestation, straw checkerboard barriers, and vegetation fencing.

5. Conclusions

These studies reveal that the water and heat conditions satisfy the demand of the biological crust development.andartificial vegetation increases the clay content and promotes the development of biological crustin the Mugetan sandy land. The fine particle content of the sand dune in the Mugetan sandy land has increased by 15% in a short time scale (decades) and up to 20% in a long time scale (one hundred years). The content of clay minerals increased to nearly twice more than that in the movable sand dune.The clay and humus have formed the soil aggregate structure, promotes a stable formation of the biological crust and greatly enhances soil formation and the resistance to aeolian erosion. The lichen-moss-algae biologic crust will appear under reasonable measures after 30 years in moving sand dune regions in the Mugetan sandy land.If the biological crust suffers disturbance from human activities or extreme weather, reasonable closure and repairing measures can shorten the recovery time.

Acknowledgements
This study was supported by the National Natural Science Foundation of China(51309154, 51409114, 51309264) and China Postdoctoral Science Foundation(2014M550740).We deeply thank Han Lujiefor drawing the topography map, Wang Xuzhao, Zhu Haili, and Zhang Chendi for their field assistance in 2010-2014.

References
[1] Belnap J and Lange O L 2003a Biological soil crusts: Structure, function and management, *Springer-Verlag* Berlin Heidelberg
[2] Belnap J 2003b The world at your feet: desert biological soil crusts *Frontiers in Ecological Environments* **1**(5) 181-9
[3] Dong Z B, Hu G Y, Yan C Z, Lu J F and Wei Z H 2012 Aeolian desertification in the source regions of Yangtze River and Yellow River *Science Press*
[4] Duan Z H, Xiao H L, Li X R and Dong Z B 2004Evolution of soil properties on stabilized sands in theTengger Desert China *Geomorphology* **59** 237-46
[5] Duan Z H, Liu X L and Qu J J 1996 Study on formation mechanism of soil crust in the shapotou area *Arid zone Research* **13**(2) 31-7
[6] Fang H Y, Qu J J and Zu R P 2005 Research on effect of sand prevention and control engineering on formation of physical crust *Journal of Soil and Water Conservation* **19**(2) 17-20
[7] Fang X M, Li J J, Zhou S Z and Kang S C 1998 Aeolian sand deposition in the source area of yellow river and its significance*Acta Sedimentologica Sinica* **16**(1) 40-44
[8] Fu B Q 2006 *Principle and methods of ecological experience* Beijing: Science Press
[9] Guo Y R, Zhao H L, Zuo X, Drake S and Zhao X Y 2008a Biological soil crust development and its topsoil propertiesin the process of dune stabilization, Inner Mongolia, China *Environmental Geology* **54**(3) 653-62
[10] Guo Y R, Zhao H L, Zuo X, Li Y X, Huang Y X and Wang S K 2008b Crust development and subsurface soil properties under dominant shrubs in the process of dune restoration Horqin sand land *Environmental Science* **29**(4) 1027-34
[11] Guo L Y, Zhong C, Ding S Y and Han H F 2009 Local climate changes and their impacts on grassland degradation in Gonghe basin of Cuinan country of Qinghai province in past half-century *Chinese Journal of Agromeleorology* **30**(2)147-52
[12] Guo S S, He L B and Xu Z F 2010 Variety trend analysis of sunshine hours in Guinan country in recent 50 years *Journal of Anhui Agricultural Science* **38**(16) 8530-32 and 38
[13] Guo Y R, Zhao H L, Zhao X Y, ZuoX and Luo Y Y 2007Crust development and its influences on soil physicochemical properties in artificial forest of Horqin Sand Land *Journal of Desert Research* **27**(6) 1000-06
[14] Kidron G J and Temina M 2008 The mycobiont role in crustose lichen expansion oncobbles in the Negev Desert *Geomicrobiology Journal* **25**(2) 95–100
[15] Kong L J and Shen J Q 2003 Analysis on species diversity of plant communities in the southeastern Tenggel desert *Journal of Ningxia Agricultural College* (4) 25-28
[16] Li X R, Wang X P, Li T and Zhang J G 2002. Microbiotic soil crust and its effect on vegetation and habitaton artificially stabilized desert dunes in Tengger Desert, North China *Biology Fertil Soils*(35) 147-54
[17] Li X R, Xiao H L,Zhang J G and Wang X P 2004. Long-term ecosystem effects of sand-binding vegetation in the Tengger desert, Northern China *Restoration Ecology* **12**(3) 376-90
[18] Li X J, Li X R, Song W M, Gao Y P, Zheng J G and Jia R L 2008a Effects of crust and shrub patches on runoff, sedimentation, and related nutrient (C, N) redistribution in the desertified steppe zone of the Tengger Desert, Northern China *Geomorphology* **96**(1-2) 221-32
[19] Li S Z, Zhang H Z, Li S L and Shen B C 2008b Development Characteristics of biotic crusts on Shapotou vegetated sand dunes*Chiese Journal of ecology* **27**(10) 1675-79

[20] Li X R, Tian F, Jia R L, Zhang Z S and Liu L C 2010 Do biological soil crusts determine vegetation changesin sandy deserts? Implications for managingartificialvegetation. *Hydrological Processes* (24) 3621-30
[21] Li S Z, Shao H L, and Luo F 2005a. Regulation effect of microbiotic crusts on soil hydrological process in Shapotou vegetated sand dunes *Journal of Desert Research* **25**(2) 228-33
[22] Li W H, Ren T R, Ren T R, Zhou Z B and Liu J Z 2005b Study on the soil physicochemical characteristics of biological crusts on sand-dune surface in Gurbantonggt desert, Xinjiang Region *Journal of Glaciology and Geocryology* **27**(4) 619-26
[23] Li X R, Zhang Y M and Zhao Y G 2009 A study of biological soil crusts: recent development, trend and prospect *Advances in earth science* **24**(1) 11-24
[24] Liu P, Zhang H Y and Wang J F 2009 Summary on influencing factors of soil surface crust or seal *Arid land geography* **32**(5) 662-69
[25] Liang S M, Wu N, Wang H L, Nie H L and Zhang Y M 2005 Study on the effects of disturbances on the physical and chemical properties of microbiotic crusts in arid areas *Arid land geography* **28**(6) 818-23
[26] Maestre F T, Bowker M A, Cantón Y, Castillo-Monroy A.P, Cortina J, Escolar C, Escudero A, Lázaro R. And Martínez I. 2011 Ecology and functional roles of biological soil crusts in semi-aridecosystems of Spain *Journal of Arid Environments* (75) 1282-91
[27] Qiang M R, Chen F H, Song L, Liu X X, Li M Z and Wang Q 2013 Late Quaternary aeolian activity in Gonghe Basin, northeastern Qinghai-Tibetan Plateau, China *Quaternary Research* **79**(3) 403-12
[28] Su Y Z, Zhao H L and Li Y L et al 2004 Influencing mechanisms of several shrubs on soil chemical properties in semiarid Horqin Sandy Land, China *Arid Land Research and Management* (18) 251-63
[29] Thomas A D and Dougill A J 2006 Distribution and characteristics of cyano bacterial soil crusts in the Molopo Basin, South Africa *Journal of Arid Environments* (64) 270-83
[30] Wang W B, Liu Y D, Li D H, Hu C X and Rao B Q 2009 Feasibility of cyanobacterial inoculation for biological soil crusts formation in desert area *Soil Biology and Biochemistry* **41**(5) 926–9
[31] Wang X Q, Zhang Y M, Zhang W M and Han Z W 2004 Wind tunnel experiment of biological crust effect on wind erodibility of sand surface in Gurbantnnggnt desert, Xinjiang *Journal of Giaciology and Geocryology* **26**(5) 632-38
[32] Wu N, Zhang Y M., Zhang J, Wang H L, Nie H L and Chen R Y 2007 Differentiation of soil variables in different succession stages of biological soil crusts in Gurbantunggut Desert, Northern Xinjiang, China *Journal of Desert Research* **27**(3) 397-404
[33] Wu Y S, Ha S, Li S S, Liu H Q and Jia Z J 2010 Development characteristics of biological soil crusts on sand dune in Southern Mu Us sandy lands *Journal of soil and water conservation* **24**(5) 258-61
[34] Wu Y H, Gao Q and Yu X H 2003 Distribution influencing factors and monitoring of biological soil crusts *Journal of Ecology* **22**(3) 38-42
[35] Yan D R 2008 Study on fertile islands features of biological soil crusts in Kubuqi desert *Inner Mongolia Agricultural University* Doctoral Dissertation
[36] Yan D R 2009 Physicochemical properties of artificial biological soil crusts in Kubuqi desert. *Journal of Inner Mongotia Forestry Science and Technology* **35**(1) 1-4
[37] Yan D R, Ji M and Xue Y Y 2006 Development characteristics of desert biological crust. *Chinese Journal of Soil Science* **37**(5) 990-3
[38] Yan D R, Xue Y Y, Han F J, Zhao C G and Hao K 2007 Present study on biological soil crusts in abroad deserts *Journal of Inner Mongolia Forestry Science and Technology* **33**(1) 39-42
[39] Zhang J, Zhao H, Zhang T, Zhao X and Drake S 2005a Community succession along a chronosequenceof vegetation restoration on sand dunesin Horqin Sandy Land *Journal of Arid Environments* (62) 555-66

[40] Zhang Y M 2005b The microstructure and formation of biological soil crusts in their early developmental stage *Chinese Science Bulletin* **50**(2) 117-21

[41] Zhang C L, Dong G R, Zhou X Y, Cheng H and Yang S 2005c Contributing ratios of several factors to the steppe desertification in Guinan, Qinghai province *Journal of Desert Research* **25**(4) 511-8

[42] Zhang Y M and Wang X Q 2010 Summary on formation and developmental characteristics of biological soil crusts in desert areas *Acta Ecologica Sinica* **30**(16) 4484-92

[43] Zhang Y M, Cao T, Pan B R 2002 A study on bryophyte associated with formation of soil crust in south fringe of Gurbantunggut desert in Xinjiang *Acta Botanica Boreali-Occidentalia Sinica* **22**(1) 18-23

[44] Zhang Y X 2009 Study on quick control technique of migratory dune in Gonghe basin of Qinghai *Journal of Qinghai University (Nature Science)* **27**(4) 56-64

[45] Zhao H L, Zhou R L, Su Y Z, Zhang H, Zhao L Y and Drake S 2007 Shrub facilitation of desert land restoration in the Horqin Sand Land of Inner Mongolia *Ecological Engineering* (31) 1-8

[46] Zhao H L, Guo Y R, Zhou R L and Drake S 2010 Biological soil crust and surface soil properties in different vegetation types of Horqin Sand Land, China *Catena*(82)70-76

[47] Zhao G R 1993 Land desertification and prevention ways in Gonghe basin of Qinghai province *Science Press*

[48] Zhao X F 1991 The cause and its environmental significance of the wind sand accumulation along Qqinghai-Tibet highway *Resources and environment in arid areas* **5**(4) 61-9

[49] Zhu Z D 1989 China desert and its control measures *Beijing: Science Press* 405-12

A new threshold selection method for peak over for non-stationary time series

C R Zhou[1], Y F Chen[1, 3], S H Gu[2], Q Huang[1], J C Yuan[1] and S N Yu[1]

[1]Hohai University, College of Hydrology and Water Resources, Nanjing 210000, China
[2]Shanghai Hydrological General Station, Shanghai 200000, China

E-mail: 19870056@hhu.edu.cn

Abstract. In the context of global climate change, human activities dramatically damage the consistency of hydrological time series. Peak Over Threshold (POT) series have become an alternative to the traditional Annual Maximum series, but it is still underutilized due to its complexity. Most literature about POT tended to employ only one threshold regardless of the non-stationarity of the whole series. Obviously, it is unwise to ignore the fact that our hydrological time series may no longer be a stationary stochastic process. Hence, in this paper, we take the daily runoff time series of the Yichang gauge station on the Yangtze River in China as an example, and try to shed light on the selection of the threshold provided non-stationarity of our time series. The Mann-Kendall test is applied to detect the change points; then, we gave different thresholds according to the change points to the sub-series. Comparing the goodness-of-fit of the series with one and several thresholds, it clearly investigates the series that employs different thresholds performs much better than that just fixes one threshold during the selection of the peak events.

1. Introduction

In hydrology, the term non-stationarity refers to hydrological properties that are different from similar properties in the past. It has drawn wide attention, and many studies have contributed to non-stationary hydrological analysis[1-2]. Non-stationarity in floods can result from a variety of anthropogenic processes, including urban land-use modifications, climate change, and modifications to water infrastructure[3]. The interactions among these processes make it difficult to determine the cause of the non-stationarity.

Usually, flood frequency analysis (FFA) has a local basin perspective and is based on the assumption that floods are the consequences of random processes that are independent and identically distributed[4]; however, considering the changing environment, this assumption is no longer valid[5]. Therefore, to obtain a reliable flood estimation, new statistical models are needed to dynamically capture the changes of probability density functions over time. Numerous approaches[6-7] have been developed for FFA by using non-stationary observed data, namely, backward and forward restorations[8], the spatial Bayesian hierarchical approach[9], the climate-informed approach[10], and the Time- Varing Moments method[11].

Additionally, efficiently planning, designing, and operating hydrotechnical works require an in-depth understanding of the probabilistic behavior of extreme events[12]. Thus, the prerequisit lies in making a sound selection of extreme events. Extreme event series vary from each other due to different sampling techniques. An annual maximum series (AMS) is extracted from a flow series. A POT series tries to retain all peak values that exceed a certain threshold value, so it is not confined to only one event per year.

The differences between the two methods have been frequently discussed, and it is generally agreed that the POT approach gives more robust results as long as certain pre-conditions are met, such as peak event independence and selecting appropriate threshold levels[13].

Compared with the AMS sampling method, POT enlarges the sample number; at the same time, it can reflect both the flood magnitude as well as the flood process. Moreover, POT modeling provides more flexibility in representing floods, and a more comprehensive description of the flood generation process [14].

The extreme events of discharge observations can be described by many candidate distributions. For POT series, the Generalized Pareto distribution (GPD) has been proved to be ideal for POT samples[15-16]. Thus, the GPD was used in this paper.

2. Materials and methods

2.1. *Generalized pareto distribution*

Picklands first introduced GPD to describe exceedances over high thresholds. The GPD cumulative distribution function (with the 3 parameters μ,σ,ζ) is given below:

$$G(\mu,\sigma,\zeta)=\begin{cases}1-(1+\zeta\dfrac{x-\mu}{\sigma})^{-\frac{1}{\zeta}},\zeta\neq 0\\ 1-\exp(-\dfrac{x-\mu}{\sigma}),\zeta=0\end{cases} \tag{1}$$

Where ζ is the shape parameter, σ is the scale parameter, and μ is the location parameter.

The distribution moments (order = 0, 1, 2) can be easily deduced as follows:

$$E(X)=\mu+\frac{\sigma}{1-\zeta} \tag{2}$$

$$Var(X)=\frac{\sigma^2}{(1-\zeta)^2(1-2\zeta)} \tag{3}$$

$$Skew(X)=\frac{2(1+\zeta)\sqrt{1-2\zeta}}{1-3\zeta} \tag{4}$$

From the Equation(4), we can find that if we use moment methods (regardless of the method of moments or the probability weighted moments), they are theoretically applicable only for $\zeta\neq\frac{1}{3}$. Besides, ζ cannot exceed 0.5, because when ζ tend to be 0.5, the variance will tend to be infinite. Some researchers have suggested using the moment estimators for $\zeta<\frac{1}{4}$[17].

The quantile function, which we pay more attention to, especially in FFA, is given by Equation (5):

$$x(G)=\begin{cases}\mu-\dfrac{\sigma}{\zeta}\left[1-(1-G)^{-\zeta}\right],\zeta\neq 0\\ \mu+\sigma\left[-\ln(1-G)\right],\zeta=0\end{cases} \tag{5}$$

2.2. *The peak-over-threshold modeling*

Because the POT approach focuses only on higher maximum values, it allows for a wider range of events that can be considered floods; therefore, it can capture more information concerning the whole flood phenomena than can its AMS counterpart. In terms of information about the flood process, the POT approach is more advisable in FFA.

Typically, a POT series is obtained by setting a threshold and selecting the highest values from a continuous hydrological time series. So, a POT sample can be described with the following statement:

If X is a random variable, we define X_s as the maximum value of X in an episode. An episode is defined as a function of a threshold level "s"; it begins when $X_{(t)}$ exceeds "s" and ends when $X_{(t)}$ falls below "s". All $X_{(t)}$ in each episode will form a POT series.

There is still no universal method for choosing an available threshold. Referring to the existing research, the POT series should meet two criteria: independence criteria and threshold selection rules. The two criteria are described below.

2.2.1. Independence criteria. Meeting the independence condition is a prerequisite of any statistical frequency analysis. Regarding the POT series, we assume that the mean occurrence number of POT events per year meets the Poisson process. Researchers raised several independence criteria, such as those proposed by Water Resources Council (1976) (see Equation(6)) and Cunnane (1979).

$$\theta < 5days + \log(A) or X_{\min} > 0.75 * \min[Q_1, Q_2] \tag{6}$$

Where θ is the time interval between two peak events, $days$; A is the basin area, $mile^2$; $X_{\min}$ is the minimum runoff, m^3/s; Q_1, Q_2 are the runoff of two peak events, m^3/s.

2.2.2. Threshold selection. Selecting the threshold should ensure that the peak values meet independence condition, but a considerably low threshold will lead to the selection of minor peaks that are not always independent. Raising the threshold will obviously increase the series independence. However, if the threshold is too high, fewer peak events will be retained.

Some rules are imposed on threshold selection, such as on the basis of a given return period. For instance, researchers [18] in China concluded that if the mean number of POT per year varied within a range of 2-3, the POT series rivers in China can be described by a Poisson process; Begueria insisted that the number should be within 1.2-5.0[19]; Cunnane suggested that if we choose the POT approach rather than the AMS method, the average number of POT samples must exceed 1.65[20].

2.3. A new method of threshold selection for non-stationary time series

Previous studies tended to employ only one threshold to the time series regardless of whether the series is stationary or not. To our knowledge, the properties of hydrological time series have changed significantly, so it is not advisable to use the same threshold as the non-stationary series.

In this paper, we suggest using different thresholds when the series is no longer stationary, and it is especially important to the change points in the time series. Through this approach, we can obtain a POT series with different thresholds due to a thorough consideration of the changes of characteristics in the series. The steps are listed below.

- Step 1: Use a Mann-Kendall test to detect whether or not the hydrological time series is stationary. If the time series is stationary, use the same threshold to the series. If the results indicate that the points have changed, proceed to the next step.
- Step 2: Assuming λ change points are detected, divide the series into $(\lambda+1)$ parts ($(\lambda+1)$ sub-series).
- Step 3: Give different thresholds to these sub-series according to the rule that ensures the average occurrence number of peak events per year in each sub-series within the range of 2-3 (referring to the results of Chinese scholars). Thresholds that are relatively lower are preferable, as they obtain more peak samples.

Additionally, independence criteria proposed by Water Resources Council was used in this paper. For the sake of comparison, we use two POT series for FFA. The first POT series was selected by employing the same threshold to the whole series, and the second series was given special attention to considering the non-stationarity in the series due to the existence of change points.

After the POT series was selected, we used the retained peak events to estimate the parameters in GPD and compare the goodness-of-fit of two POT series. All the parameters were estimated by the

IOP Conf. Series: Earth and Environmental Science **39** (2016) 012071 doi:10.1088/1755-1315/39/1/012071

Probability Weighted Moments (PWM) method. Mean relative errors were used to estimate the adequacy of the two theoretical curves.

3. Case study

3.1. The introduction to the study area

The Yangtze River is China's longest river. The Yichang gauge station is a hydrological station on the main stream of the Yangtze River. The Gezhouba Water Control Project and the Three Georges Dam are all on the upper stream of this river. They are 6 and 44 kilometers away from the Yichang gauged station, respectively (see figure 1). Due to inconsistent recorded data prior to the year 1946, this study only examined runoff data collected from the year 1946-2012.

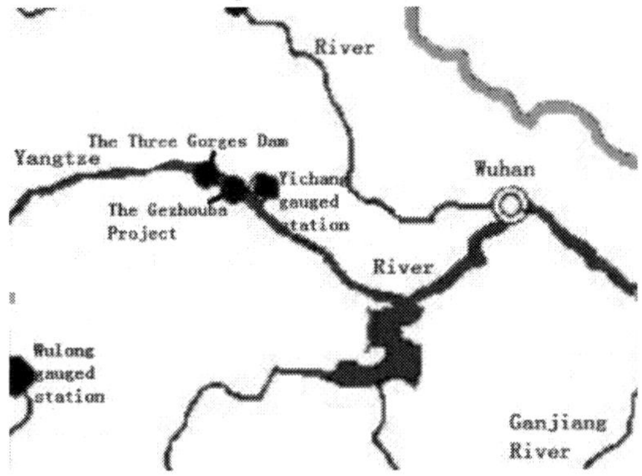

Figure 1. Location of Yichang gauged station.

It is an undeniable fact that human activities have significantly affected hydrological properties. Because of the construction of the Three Georges Dam, the hydrologic cycle in the local watershed has been changed dramatically, especially downstream of the Yangtze River [21]. We applied the new method detailed in 2.3 to the peak runoff of Yichang gauge station.

3.2. Results

3.2.1. Results: peak over threshold series selection and change points detection. If we ignore the existence of the possible change points in the time series, we can obtain the first POT series. From figure 2, we can see that when the threshold equals 35000 , the average number of peak events per year is close to 3, so 35000 is the minimum threshold. Thus, we get 182 peak events from the year from 1946 to 2012.

Additionally, figure 2 also reveals the absolute value of the average relative errors (ARE) between the empirical points and the theoretical curve (the parameters are estimated by PWM) of the series corresponding to different thresholds (35000, 36000, 37000, 38000, 39000, 40000). When the threshold is 37000 , the absolute value of ARE reaches the minimum. Hence, 37000 is the perfect threshold with the best goodness-of -fit to the whole series.

m^3/s

m^3/s

m^3/s

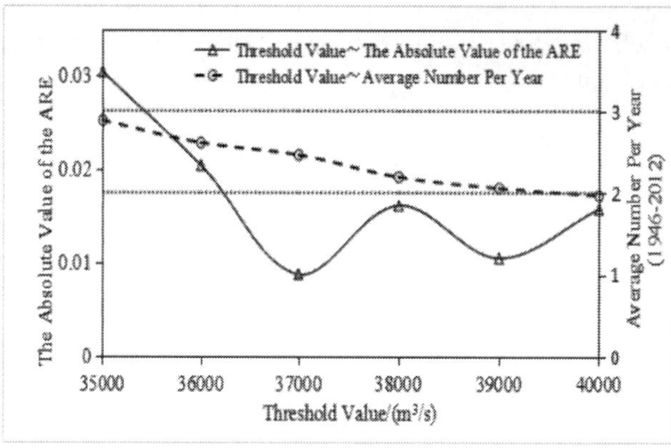

Figure 2. Relative curves between thresholds and
(a) the absolute value of ARE; (b) average number
(b) of peak events per year.

However, given human activities, some properties of our hydrological time series may have changed, so estimating hydrology information becomes a priority. The Mann-Kendall test was applied to test the mutability (or variant properties) of the POT series.

Researchers in China revealed through repeated trials that when the average occurrence per year was within 2~3[19], the theoretical curve fit with the observed data. Our hope is that the average occurrence per year is the largest one within the range of 2-3. Thus, we set 35000 m^3/s as the threshold (nearly 2.98 peaks per year) because 35000 m^3/s is the minimum threshold that can obtain as many peak events as possible.

Figure 3 below illustrates that two change points, in 1954 and 1984 respectively, separate the time series into three parts (1946-1954, 1955-1983, and 1984-2012). In each part, we can test its stationarity and the possible trend. The results show that each part of the series can be regarded as stationary.

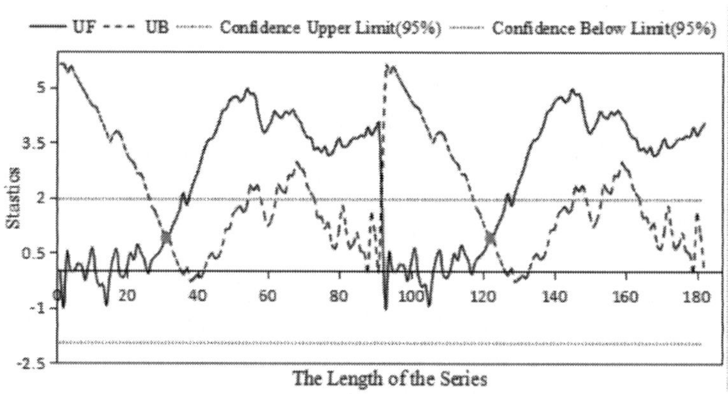

Figure 3. The results of the Mann-Kendall test (with a 95% significance level).

We take each part as a new population and study its threshold and peak event occurrence per year. Figure 4 illustrates how the mean numbers of over-threshold events per year change with the thresholds in the three sub-series. From figure 4, we can recognize the average occurrence per year of the first sub-series fell from 3.98 to 2.73, while the threshold increased from 35000 m^3/s to 40000 m^3/s, so we selected 38000 as the threshold (the average number of peak events per year was 2.889, the largest number within 2 to 3). The second sub-series showed a similar decrease as the first one; thus, we selected 36000 as the threshold to this sub-series (the average number of peak events per year is 2.9); 35000 m^3/s was selected as the threshold in the third sub-series. In conclusion, we gave three different thresholds (38000 m^3/s, 36000 m^3/s, and 35000 m^3/s) to three sub-series (1946-1953, 1954-1983 and 1984-2012, respectively).

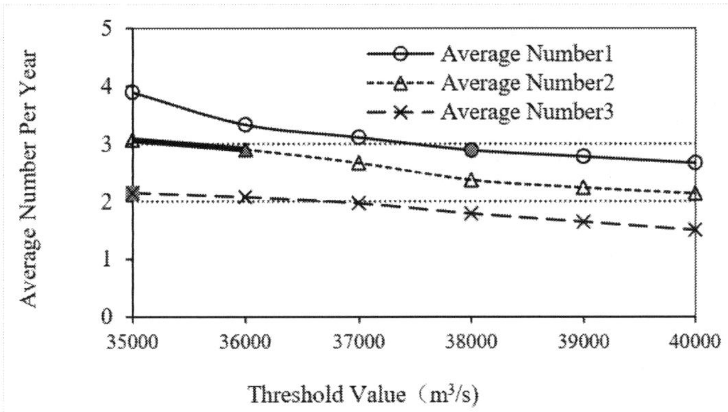

Figure 4. Selecting the threshold to the three sub-series according to how the average number of peak events per year changes with the thresholds.

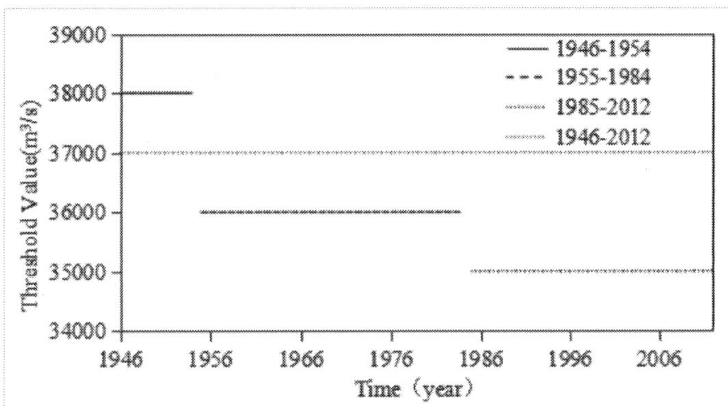

Figure 5. Comparison between the series threshold (daily runoff time series of 1946-2012, which ignores the non-stationarity) and the three sub-series (daily runoff time series of 1946-1953, 1954-1983 and 1984-2012, which takes change points into account).

3.2.2. Results: comparison of the goodness of fitting with the empirical points. According to the above threshold selection, we obtained two different POT series. Two POT series were used to FFA.

Samples of two series were applied to estimate the parameters in the GPD, and the PWM method is employed here [16].

In terms of frequency calculation, we should notice that the second POT series comprises data from three different sub-populations, so this data is less likely to be identically-distributed. For the sake of comparing the goodness-of-fit, the same frequency calculation method should be used. Consequently, the traditional frequency formula is applied here.

$$P(x=i)=\frac{i}{n+1}, i=0,1,...,n \tag{7}$$

Where n is the number of peak events.

Figure 6(a) and 6(b), respectively, show how the first and the second POT series fit with their theoretical curves. In figure 6(a), the theoretical curve is slightly above the empirical points. In figure 6(b), the theoretical curve fits much better than that in figure 6(a); most of the theoretical curve fits with the empirical points.

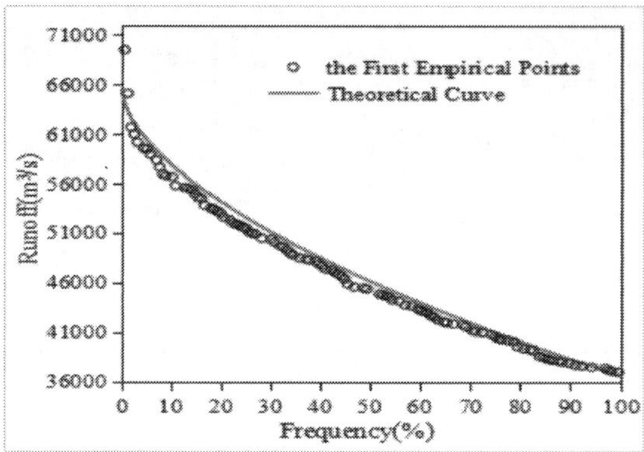

(a) The POT samples when the threshold equals 37000

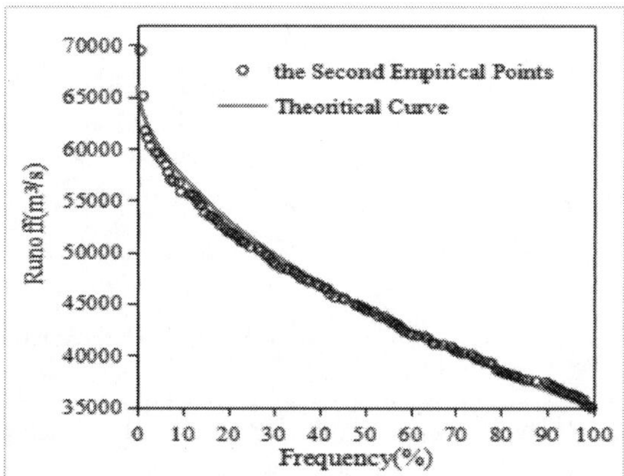

(b) The POT samples with different thresholds.

Figure 6. The comparison between the empirical curves and corresponding theoretical curves.

The average relative error (ARE) is used to compare the goodness-of-fit of two POT series, and the expression is given below (take μ as an example).

$$\delta = \frac{\mu_{obs} - \mu_{sim}}{\mu_{obs}} \tag{8}$$

Where μ_{obs} is the observed data of μ, and μ_{sim} is the theoretical value of μ.

Table 1 below shows the results. It is clear that regardless of the flood frequency, the theoretical curve of the POT series with full consideration to the change points shows a better fit with the empirical points. The theoretical curve obtained by the second POT series fits with the observed data especially well with the high frequency floods.

Table 1. The average relative errors between empirical points and the theoretical curves.

δ	The first POT series	The second POT series
Total	0.017253538	0.005606272
Low frequency flood	0.018960411	0.013587264
High frequency flood	0.015115555	-0.002906787

*Low frequency flood indicates a flood frequency higher than 50% ($P > 50\%$); if $P \leq 50\%$, this indicates high flood frequency.

4. Conclusions

The above study sheds light on a new method to select a threshold when the POT approach is used to FFA. The conclusions are given below.

- Due to climate change and human activities, the characteristics of hydrological data may encounter dramatic changes. Estimating hydrology information enables us to have a deeper insight into the change.
- Quantile estimation by other POT series, which gives different thresholds to sub-series according to the change points, performs better compared to the POT series without considering the change points detection. Besides, the theoretical curve corresponding to the peak data extracted on the basis of different thresholds shows superior goodness-of-fit to the POT points, especially in the upper tail.

If a POT series is non-stationary, it is more advisable to distribute different thresholds according to the results of the Mann-Kendall test, rather than to employ the same one.

References

[1] Tan X and Gan T Y 2015 Nonstationary analysis of annual maximum streamflow of Canada *Journal of Climate* **28**(5) 1788-805

[2] Vasiliades L, Galiatsatou P and Loukas A 2015 Nonstationary frequency analysis of annual maximum rainfall using climate covariates *Water Resources Management* **29**(2) 339-58

[3] Richard M V, Chad Y and Meghan W 2011 Nonstationarity: Flood magnification and recurrence reduction factors in the United States *Journal of the America Water Resources Association* **47**(3) 464-74

[4] Lima C H, Lall U, Troy T J and Devineni N A 2015 Climate informed model for nonstationary flood risk prediction: Application to Negro River at Manaus, Amazonia *Journal of Hydrology* **522** 594-602

[5] Tamer A G and Van Thanh Van N 2016 An at-site flood estimation method in the context of nonstationarity 1. A simulation study *Journal of Hydrology* **535** 710-21

[6] Ye C Q, Chen X H, Zhang L J, *et al.* 2013 Changing ruled influence of threshold for peak-over-threshold flood series under the Changing environment in Wujiang *Advances in Water Science* **24**(3) 392-401

[7] Xiong L H, Jiang C, Du T, *et al.* 2015 Review on Nonstationary Hydrological frequency analysis under Changing environments *Journal of Water Resources Research* **4**(4) 310-19

[8] Liang Z M, Hu Y M and Wang J 2011 Advances in hydrological frequency analysis of non-stationary time series *Advances in Water Science* **22**(6) 864-71

[9] Yan H and Moradkhani H 2015 A regional Bayesian hierarchical model for flood frequency analysis *Stochastic Environmental Research and Risk Assessment* **29**(3) 1019-36

[10] Yan H and Moradkhani H 2016 Toward more robust extreme flood prediction by Bayesian hierarchical and multimodeling *Natural Hazards* **81**(1) 203-25

[11] Du T, Xiong L H and Jiang C 2014 Nonstationary frequency analysis of rainfall time series in Weihe River Basin *Arid Land Geography* **37**(3) 468-79

[12] Lang M, Ouarda T B M J and Bobée B 1999 Towards operational guidelines for over- threshold modeling *Journal of Hydrology* **225**(3-4) 103-17

[13] Ngongondo C, Li L, Gong L, *et al.* 2013 Flood frequency under Changing climate in the upper Kafue River Basin, Southern Africa: A large scale hydrological model *Stochastic Environmental Risk Assessment* **27**(8) 1883-98

[14] Zhang L J, Chen X H, Ye C Q, *et al.* 2013 Pot flood frequency analysis with historical floods in Wujiang River *Shuili Xue Bao* **3**(44) 268-75

[15] Wang J F and Song S B 2010 Application of generalized pareto distribution in flood series

frequency analysis *Journal of Northwest A&F University(Nat&Sci Ed)* **38**(2) 191-6

[16] Bermudez P Z and Samuel K 2010 Parameter estimation of the generalized pareto distribution part-II *Journal of Statistical Planning and Inference* **140**(6) 1374-88

[17] Hosking J R M and Wallis J R 1987 Parameter and quantile estimation for the generalized Pareto distribution *Technometrics* **29**(3) 339-49

[18] Fang B, Guo S L, Chai X L, *et al.* 2005 Flood frequency analysis based on FPOT method *Water Power* **31**(2) 9-12

[19] Begueria S 2005 Uncertainties in partial duration series modeling of extreme related to the threshold value *Journal of Hydrology* **303**(1-4) 215-30

[20] Cunnane C 1973 A particular comparison of annual maxima and partial duration series methods of flood frequency prediction *Journal of Hydrology* **18**(3-4) 257-71

[21] Chen L, Xu W S, He X H, *et al.* 2006 Lower water level descending trend of Yichang section after water storing of Three Gorges and its solution *Yangtze River* (5) 59-64, 73

IOP Conf. Series: Earth and Environmental Science **39** (2016) 012072 doi:10.1088/1755-1315/39/1/012072

Research on environmental flow in Huai River Basin, China

H C Liu[1], Y Shui[1], L H Li[1], N Cao[2, 3], G Y Yu[1] and J Liu[1]

[1]Scientific Research Department of Water Resource Preservation on Huai He, Bengbu, China
[2]Ministry of Environmental Protection, Beijing, China

E-mail: 40656020@qq.com

Abstract. The estimation methods for environmental flow were discussed in this paper. Based on the comparative analysis of calculation methods of environmental flow, a "Huaihe Method" was proposed, which consists of the hydrological estimation of the minimum environmental flow with validation by the requirment space for fish. The minimum environmental flows of the important control sections in main stream and tributaries in the Huai River basin are estimated. The recommened minimum environmental flow has been adopted in the water resources planning, which provided reference for rational allocation of environmental flow, water resources usage, and sustainable development for the local society.

1. Introduction

With the area of 270000 km^2, Huai River basin is situated at China's north and south climate intermediate belt. North of the basin has been facing inadequate water issue for the temporal and spatial uneven water resources distribution. At present, water ecological system in most parts of the Huai River basin has been affected by extremely water resources exploitoration for economic development. A larger amount of dams have been constructed on the main stream and tributaries in Huai River for domestic, industrial, and agricultural utilization. There are various problems in water ecosystem. Research on environmental flow has become the key wayto maintain ecosystem health of the Huai River.

Combined with the development of the study on environmental flow and based on the comparative analysis about calculation methods of environmental flow, the minimum environmental flow of the important control sections in the Huai River basin is introduced. Finally, the prospective research on the ecological water requirement was proposed, so as to provide reference for rational allocation of environmental flow in future.

IOP Conf. Series: Earth and Environmental Science **39** (2016) 012072 doi:10.1088/1755-1315/39/1/012072

2. Development of research on environmental flow

Environmental flow is the amount of water needed in order to maximize the improvement of natural ecological system and maintain ecosystem integrity, and ensure the biodiversity of natural habitat. The concept was first proposed by Gleick[1] in 1996,but the research on environmental flow began in the 1940s. More than 70 years of study can be divided into three stages.

- Establishment of relevant concepts: In the 1940s, the U.S. Fish and Wildlife Conservation Association studied the relationship between fish growth and reproduction and river flows, and first proposed the concept of "In - stream Flow Requirement"-minimum river environmental flow in order to avoid the river ecosystem degradation. During this period,the concept of environmental flow was not clear, but some rules about flow compensation began to be developed and implemented.
- Quantitative study of environmental flow: In the 1950s and 1960s, the quantitative study on river environmental flow and research based on process started. Scholars focused on the energy flow, the carbon flux, and large invertebrate life history[2]. In the 1970s and 1980s, the relationship of fish growth to river flow was studied systematically in view of river ecosystems in the United States, Australia, South Africa, France, Canada, and other countries. There were some calculation and evaluation methods developed in this period. Preliminary research results have been achieved [3].
- The overall analysis of the ecosystem: Since 1990s, through study on correlation between water resources and ecological environment, environmental flow became one of the focuses of the world. The concept of environmental flow for a river was clearly put forward. Research objects extended from species (such as fish and invertebrates, etc.), and river physical form to maintenance of river flow considering the integrity of river ecosystem. At this point, some new research methods have appeared, such as BBM (Building Block Methodology) method, the overall analysis method and model simulation method based on the relationship between river flow and aquatic habitat.

Research on environmental flow in China began in the 1970s, which mainly concentrated on the determination of river environmental flow.

In the 1980s and 1990s, study on environmental flow aiming at ecological environment deterioration in the northwest inland areas began firstly in the northwest arid and semi-arid areas[4].Then, due to the discontinuous flow, siltation, river pollution, and other issues, scholars conducted research on environmental flow for rivers and lakes in Huang-Huai-Hai plain[5].

Since the 21st century, the study on environmental flow matured unceasingly in China. The analysis system of ecological water standard base was built in the national science and technology research project "Research on standard of regional ecological water in China". Characteristic value of environmental flow and ecological water control indexes of different development stages of four large river basin in the northern semi-humid and semi-arid areas were put forward[6]. Ecological water needed by ecosystem restoration in the Huang-Huai-Hai plain was studied by Ministry of Water Resources. Environmental flow have been known as content in the balance between supply and demand that must be considered in the projects such as water resources configuration of South-to-North water transfer, coordinated development between water conservancy and national economy, and the new planning of national water resources. The importance of ecological operation has been more emphasized during construction and operation of Xiaolangdi Reservoir [7]. At the same time, some scholars have carried on thorough studies on modeling environmental flow. China's first model of aquatic habitats fitness based on long series of field data has been established by Fengqing Li[8].

3. The calculation method of environmental flow for rivers

At present, there are as many as 200 calculation methods about environmental flow. These methods in general can be classified into four categories: hydrological methods, hydraulic methods, habitat simulation (or evaluation) methods, and entirety analysis methods.

- Hydrological methods: This method for environmental flow is based on historical flow, which has more advantages on the data collection cost than other methods. This method is a basic means in the research of environmental flow. The common methods are Tennant method, 7Q10 method, basic flow method, monthly frequency calculation method, etc. Among them, Tennant method is the most commonly used in China

Tennant method takes a percentage of annual average discharge as base flow, which is conformed to the environmental flow estimation demand for seasonal rivers. This method has the macroscopic qualitative guiding significance [9]. The method was implemented in rivers at Virginia region of the United States. 10% of the annual average flow is degraded or barren habitat conditions, 20% of the annual average flow provides appropriate standards to protect aquatic habitats, and 30% of annual average flow is close to the best standard of habitats in the small river[10].

- Hydraulic methods: Assuming that river physical form remains the same, river flow at downstream can meet the requirements of river ecological function when a section of the river satisfies a certain flow. The commonly used methods are wetted perimeter method, R2 - Cross and CASIMIR methods.
- Habitat methods: This method determines that the recommended flow is suitable for aquatic organisms survival based on the ecological hydraulics. Main methods are effective width method, the river flow increase method, etc.
- Entirety analysis methods: Starting from the whole ecological environment in the study area, the method determines the recommended value of flow through the comprehensive research on the relationship among river flow, sediment transport, bed configuration, and the riparian zone. The recommended value can simultaneously satisfy the overall ecological functions including biological conservation, habitat maintenance, erosion and siltation, pollution control and landscape maintenance, etc. Methods mainly include BBM (Building Block Methodology) method and the overall evaluation method [11, 12].

The advantages, disadvantages and scope of application of various calculation methods of environmental flow are summarized in table 1[13].

Table 1. Comparative aof calculation methods of environmental flow.

Method	Advantages	Disadvantages	Scope of application
Hydrological method	Simple; need least data; don not need extensive field work.	Lack of biological basis; Not considering seasonal variation and plentiful and withered circulation of flow; Environmental flow is often expressed in minimum; Historical flow data for years are needed.	River macro strategy management; The areas less controversial and no high priority.
Hydraulic method	Data acquisition by measurements or Manning formula; Measurement is relatively simple.	Not considering velocity changes; Not considering needs of specific species and life stage; Not considering channel instability with time; Selected section cannot represent the whole channel characteristics.	Small and medium-sized rivers (with top width less than 30 m wide) where pollution is not obvious and sediment content is low.
Habitat method	Considering the change of the habitat by species and its different stages of life.	Only consider the individual species, not considering the entire ecosystems; Need more biological data; Take a long time; Capital amount is big; Need a lot of field work.	Small and medium-sized habitat or the region of clear management objectives
Entirety analysis method	Considering the river ecosystem integrity; Satisfy variety of functions including biological conservation, habitat maintain and pollution control, etc. The most reasonable method at present.	Many subjects involved; Need expertise in ecology, geography, hydrology, etc; Consume a large amount of human resources and time cost.	Widely used in South Africa Currently; need a lot of correction when applied to other areas.

4. The determination of environmental flow of the important control sections in Huai River.

Through the relevant research results of comprehensive analysis in Huai River basin and summary of the environmental method, "Huaihe Method" is put forward as the calculation method of minimum environmental flow on the basis of Tennant method for the data availability. The river stages are raised at comparable higher level by the dense distributed dams along the main streams. The results of the survey about Tennant in Nebraska, U.S., have confirmed that 10% of the average annual flow provides the smallest habitat conditions for river ecosystem. For large rivers, 5%～10% of the annual average flow can maintain certain number of river flow, water depth and velocity. The flow can meet general requirements of the fish migration, fish survival, tourism and landscape. It is instantaneous minimum flow to keep the vast number of aquatic life for a short period of time. Thus, 5%～10% of the annual average flow was adopted as minimum environmental flow to keep the survival of most of aquatic organisms in a certain time. At the same time, the minimum water requirement space of fish was selected to correct the estimated environmental flows in the main stream and tributaries of the Huai River.

The main stream of Huaihe, Honghe and Yinghe have relatively stable and larger flows, while meso flows exist in Guohe according to the historical records at hydrological stations. For rivers with flow rates less than 80 m3/s, 10% of annual mean discharge is allocated for environmental flow. The 5% of annual mean discharge is assigned for environmental flow when the flow rates are larger than 80 m3/s. At the same time, minimum water requirement space of fish is calculated with the discharge and geometry of cross-section for validation of estimated environmental flow. The validated environmental flows at selected cross-sections can maintain minimum space for growth and reproduction of zoobenthos and plankton. Based on results of analyzing monthly average flows, the environmental flows are expressed for the dry season (October to March), period before flood season (April to May), and flood season (June to September), respectively. Then minimum environmental flows of main river section during the year are determined consequentially.

Huaihe method not only retains the simple and quick method of hydrologic index method, but also has a solid foundation of ecology through the hydraulic calculation based on ecological analysis. The hybrid method which combined by different methods can compensate for each other. Huaihe method is more efficiency and reliability. It can be used for the study of ecological water use of watershed planning scale and also can be used as the basis for further research.

Results of this method were adopted in Comprehensive Planning of Huai River Basin (2012-2030). Minimum environmental flows at seven sections in Huaihe, Yinghe, Guohe, Yihe, and Shuhe were determined with proposed methods. The selected seven sections include Wangjiaba, Bengbu, Xiaoliuxiang, Jieshou, Bozhou, Linyi, Daguanzhuang. The final control indicators of minimum environmental flow are listed in table 2.

Table 2. Minimum environmental flow results of important river control section.

The name of the river	Section	Minimum environmental flow(m^3/s)			
		October to March	April to May	June to September	The annual average

Huaihe	Wangjiaba	10.9	18.5	35.0	20.2
Huaihe	Bengbu	24.8	47.5	96.2	52.5
Huaihe	Xiaoliuxiang	28.5	54.6	111	60.3
Yinghe	Jieshou	5.5	5.8	16.7	9.29
Guohe	Bozhou	2.3	2.4	6.9	3.85
Yihe	Linyi	2.33	3.13	19.81	8.31
Shuhe	Daguanzhuang	1.14	1.76	10.52	4.38

5. Perspective of environmental flow in Huai River basin.

The application of the results in the *Comprehensive Planning of Huai River Basin (2012-2030)* indicates that the Huaihe River Basin will be officially putting ecological management into the management of the river basin.

Based on the recent studies of ecological water demand and the current situation of environmental flows in Huai River basin, we summarize the recent and future research focus, mainly in the following aspects:

5.1. Dynamics and coupling of environmental flow

River environmental flow is a dynamic value, which varies with the change of the seasons, times, and sections. River environmental flow is closely related to environment protection. For the moment, researchers focus on individual research on ecological water demand. However, for the total water demand in the area, the environmental flows cannot be achieved by simple addition or subtraction because of the overlapping water problems. Therefore, in the future studies, we should consider the watershed ecosystem from holistic point of view. By studying the status and mechanism of water cycle in each ecosystem, the ecological water demand would be solved by establishing a coupling model. With such a model, we can more scientifically determine the ecological water demand. Furthermore, research should be clearly carried out at different temporal and spatial scales. Finally, we should also strengthen coupling study about the water quantity and water quality.

5.2. Multi-methods and application of 3S (RS, GIS and GPS) and other technologies of ecological water demand.

Existing individual quantitative methods about ecological water demand are flawed, which are still non-unified and sophisticated for China's ecological environment construction. Therefore, various methods can be combined to overcome their own weakness. Such as hydrological-hydraulic method proposed by the researchers. In addition, the determination of environmental flows with strengthening GIS and remote sensing technology on the measurement and calculation of relevant data could be expected. A wildly acceptable quantitative method for estimating the environmental flow satisfying multiple objectives at various temporal scales will be accomplished with multiple approaches.

5.3. Establish evaluation system of ecological water demand.

Currently, due to the lack of reliable evaluation criteria, we cannot verify the correctness of the results of the ecological water demand. Therefore, the needs for a large number of field measurements are to establish evaluation criteria that can adapt to the sustainable development of Ecological Water Demand for Huai River Basin. The establishment of the ecological water demand system that associated with the biodiversity, habitat, and ecosystem health should be implemented.

Acknowledgments

This work was supported by the Major Science and Technology Program for Water Pollution Control and Treatment of the Key State Science and Technology Project under Grant No. 2014ZX07204-006-05.

Reference

[1] Gleick P H 1993 Water in crisis:paths to sustainable water use *Ecological Applications* **3**(8) 574-79

[2] Resh V H 1985 Periodical citations in aquatic entomology and freshwater benthic biology *Freshwater Biology* (15) 757-66

[3] Petts G E 1996 Water allocation to protect river ecosystems *Regulate River:Research &Management* (12) 353-65

[4] Liu C M 1999 Water supply and demand analysis in China in the 21st century: Study on ecological water conservancy *China Water Kesources* (10)18-20

[5] Cui B S and Yang Z F 2003 The Classification and case study on eco-environmental water requirement of wetlands *Resources Science* (01) 21-28

[6] Zhang H 2014 Review of theoretical researches on river ecological water demand and relevant progress *Tianjin Science & Technology* (08) 86-88

[7] Gao G M, Li X J and Ma Y P 2014 Research on ecological scheduling of Xiaolangdi Reservoir *Yellow River* (09) 76-79

[8] Li F Q, Cai Q H, Fu X C and Liu J K 2008 Establishment of habitat suitability model for the stream macroinvertebrate and researchment of river environmental flow - A case study of Xiangxi River in the Three Gorges Reservoir Area *Progress in Natural Science* **18**(12) 1417-24

[9] Tennant D L 1976 Instream flow regimens for fish,wildlife,recreation,and related environmental resources *Proceedings of Symposium and Specility Conference on Instream Flow Needs II, Maryland,. American Fisheries Society*

[10] Orth D J and Leonardp M 1988 Comparison of discharge methods and habitat optimaization for recommending instream flows to protect fish habitat *Regulated River* **5** 129-38

[11] Feng B P, et al. 2004 Review of calculation method for water demand of ecological environment *Advances In Science and Technology of Water Resources* **24**(6) 59-62

[12] King J M, Tharne R E and Villiers M S 2000 Environmental flow assessments for river:manual for the Building Block Methodology *Pretoria: Water Research Commission Report*

[13] Cui Y, Zhang Q and Chen X H 2010 Advances in the theories and calculation methods of ecological water requirement *Journal of Lake Sciences* (04) 465-80

Simulation of water temperature in two reservoirs with Delft3d

J Y Yang [1] and L Y Zhou [2, 3]

[1]Anhui Bengbu No.2 high school, Bengbu 233000, China

[2] China water Huaihe planning, design and research co., Ltd, Hefei 230000, China

E-mail: zy97@sohu.com

Abstract. The proposeled Guanjingkou and Fengdou reservoir will be constructed at Chongqing city and Muling city in China respectively. The water temperature in the reservoir, in the downstream, and the aquatic ecosystem would be altered by the construction of the reservoirs. This paper simulates the water temperature in the two reservoirs by using the Delft3d z-layer model, which uses the fixed elevation for layers. According to the simulation results, the temperature profile in the reservoirs can be divided into three layers: the upmost epilimnion layer, the beneathed thermocline layer, and the constant tepmerature layer at bottom. The temperature effects can be reduced by measurements of stoplogs gates and mutiple gates, respectively. Based on the simulation results in the wet, nomal, and dry year, the temperature of water released from the stoplogs gates at Guanjingkou reservior can be respectively increased by 5.7℃, 6.8℃, 9.6℃, and 5.5℃ in the irrigation season from May to August. The temperature of water released from the mutiple gates at Fengdou reservior can be respectively increased by 7.7℃, 1.9℃, 9.5℃, and 10.1℃ from May to August. The negative impacts from the water with lower temperature on the related ecosystem can be significently alleviated.

1. Introduction

The construction of reservoirs would form the vertical stratification of water temperature profile in front of the dam. The changes of water temperature have impacts on the growing, spawning, and propagation of fishes [1-2]. The biome in the river [3], aquatic ecosystem [4-5], and the crop yield [6] can be altered by the water with lower temperature. Hence, the downstream ecosystem can be improved by studying water temperature profile in the constructed reservoir and the corresponding measures.

The researches of reservoir water temperature have been conducted since1970s in China. The estimation of reservoir water temperature with statistically summarized formulas started in 1980s.

These experimental formulas include Zhang's formula [7], Zhu's formula [8], and statistical formula [9]. The disadvantages of the experimental estimation for the reservoir water temperature are large estimation variations in individual reservoir, ignoring the meteorological conditions and the effect from reservoir operations.

The methods adopted in this paper use numerical method with Delft3d software package to simulate the reservoir water temperature in Guanjingkou and Fengdou reservoir, which takes the reservoir operation, heat exchange, and the meteorological conditions into consideration. The simulation results provide the water temperature profile in the reservoirs, the temperature of water released from proposed gates, and the count measures effects on the temperature of released water. The negative impacts of the water with lower temperature will be alleviated.

2. Material and method

2.1. Method

The Delft3d software package (version 4.10, Deltares, the Netherlands) was selected for simulation the water temperature in the two reservoirs. The governing equations are:

$$\frac{\partial u}{\partial t}+\frac{u}{\sqrt{G_{\xi\xi}}}\frac{\partial u}{\partial \xi}+\frac{v}{\sqrt{G_{\eta\eta}}}\frac{\partial u}{\partial \eta}+\frac{w}{d+\zeta}\frac{\partial u}{\partial \sigma}-\frac{v^2}{\sqrt{G_{\xi\xi}}\sqrt{G_{\eta\eta}}}\frac{\partial\sqrt{G_{\eta\eta}}}{\partial \xi}+\frac{uv}{\sqrt{G_{\xi\xi}}\sqrt{G_{\eta\eta}}}\frac{\partial\sqrt{G_{\xi\xi}}}{\partial \eta}-fu$$
$$=-\frac{1}{\rho_0\sqrt{G_{\xi\xi}}}P_\xi+F_\xi+\frac{1}{(d+\zeta)^2}\frac{\partial}{\partial\sigma}\left(v_V\frac{\partial u}{\partial\sigma}\right)+M_\xi \tag{1}$$

$$\frac{\partial v}{\partial t}+\frac{u}{\sqrt{G_{\xi\xi}}}\frac{\partial u}{\partial \xi}+\frac{v}{\sqrt{G_{\eta\eta}}}\frac{\partial v}{\partial \eta}+\frac{w}{d+\zeta}\frac{\partial v}{\partial \sigma}+\frac{uv}{\sqrt{G_{\xi\xi}}\sqrt{G_{\eta\eta}}}\frac{\partial\sqrt{G_{\eta\eta}}}{\partial \xi}-\frac{u^2}{\sqrt{G_{\xi\xi}}\sqrt{G_{\eta\eta}}}\frac{\partial\sqrt{G_{\xi\xi}}}{\partial \eta}+fu$$
$$=-\frac{1}{\rho_0\sqrt{G_{\eta\eta}}}P_\eta+F_\eta+\frac{1}{(d+\zeta)^2}\frac{\partial}{\partial\sigma}\left(v_V\frac{\partial v}{\partial\sigma}\right)+M_\eta \tag{2}$$

The transport equation is :

$$\frac{\partial(\mathrm{d}+\zeta)c}{\partial t}+\frac{1}{\sqrt{G_{\xi\xi}}\sqrt{G_{\eta\eta}}}\left\{\frac{\partial\left[\sqrt{G_{\eta\eta}}(\mathrm{d}+\zeta)uc\right]}{\partial\xi}+\frac{\partial\left[\sqrt{G_{\xi\xi}}(\mathrm{d}+\zeta)vc\right]}{\partial\eta}\right\}+\frac{\partial\omega c}{\partial\sigma}=\frac{\mathrm{d}+\zeta}{\sqrt{G_{\xi\xi}}\sqrt{G_{\eta\eta}}}\bullet$$
$$\left\{\frac{\partial}{\partial\xi}\left(D_H\frac{\sqrt{G_{\eta\eta}}}{\sqrt{G_{\xi\xi}}}\frac{\partial c}{\partial\xi}\right)+\frac{\partial}{\partial\eta}\left(D_H\frac{\sqrt{G_{\xi\xi}}}{\sqrt{G_{\eta\eta}}}\frac{\partial c}{\partial\xi}\right)\right\}+\frac{1}{d+\zeta}\frac{\partial}{\partial\sigma}\left(D_V\frac{\partial c}{\partial\sigma}\right)-\lambda_d(d+\zeta)c+S \tag{3}$$

In which the ξ、η、ω are coordinates, ζ is the water level, d is water depth, $\sqrt{G_{\xi\xi}}$ is transfer coefficient in ξ direction, $\sqrt{G_{\eta\eta}}$ is transfer coefficient in η direction, u is velocity in ξ direction, v is velocity in η direction, w is velocity in ω direction, ρ is water density, P_ξ is static pressure in

ξ direction, P_η is static pressure in η direction, f is K's coefficient, F_ξ is turbulence flux in ξ direction, F_η is turbulence flux in η direction, M_ξ is energy transfer in ξ direction, M_η is energy transfer in η direction, v_V is vertical eddy viscosity.

The z-layer model was selected for simulating the water temperature profiles in reservoirs, which uses the constant layer thickness to provide the suitable layers for stoplogs gates better than the that provided by σ -layer model. The heat transfer package in the Delft3d consists of five input model, which can be decided with the data availability.

2.2. Study area

The proposed Guanjingkou reservoir (E106°52, N29°25') is located in the Wubu River, which is the first tributary of Yangze River in the Banan district of Chongqing city. The total area of the watershed is 439 km^2. The meteorological conditions in the watershed are: the average air temperature is 18.3℃, the annual average sunshine time is 1134 hours, the relative humidity is 81%, average evaporation is 702 mm/a, the average precipitation is 1104 mm/a. The total water capacity of the proposed Guanjingkou reservoir is 152 million cubic meter with useful storage of 119 million cubic meter. The characteristic water levels are normal water level of 281 meters, the minimum water level of 254 meters, and the average water level of 276 meters above sea level.

The proposed Fengdou reservoir (E130°16, N44°28') is located in Muling city, Heilongjiang province. The watershed covers 1740 km^2. The meteorological conditions in the watershed are: the average air temperature of4.2℃, the annual average sunshine time of 2613 hours, the relative humidity of 75%, the average precipitation of 500 mm/a. The average annual discharge in Muling river is 263 million cubic meter. The total water capacity of the proposed Muling reservoir is 165 million cubic meter with useful storage of 153 million cubic meter. The characteristic water levels are normal water level of 382 meters above sea level and the minimum water level of 362 meters.

2.3. Model setup

The simulation models were set up based on the proposed reservoir, the data availability and the research objects. The Digital Elevation Model with resolution of 30×30m from the International Scientific & Technical Data Mirror Site, Computer Network Information Center, Chinese Academy of Sciences (http://www.gscloud.cn) was selected for grid generation in Rgrid package. The total 164×8 grids for Guanjingkou reservoir and 40×8 grids for Fengdou reservoir were applied, respectively (figure 1).

The inflows of the model are the discharges at the upstream boundary of the two models while the outflows of the model were set at the downstream of the model domain. The 19 layers for Guanjingkou and 12 layers for Fengdou reservoir were selected based on the proposed project and the measures, respectively. Since the historical meteorological data are rare in the region, the monthly average air temperature could be used as excess temperature in the heat exchange package. The inflow and outflow discharges in normal, wet, and dry year are collected from the hydrological stations at 20%, 50%, and 75% frequency. In addition, one year date was also collected for special dry year at 95%

frequency for water temperature prediction at Fengdou reservoir. The initial temperatures for the two reservoirs were 10℃ and 2.3℃ for Guanjingkou and Fengdou reservoir from the observation in the nearby reservoirs, respectively. The roughness for Guanjingkou and Fengdou reservoir were 0.18 and 0.2 by trial and error with the water stage and capacity, respectively. The calibration of the model for two reservoirs was achieved by comparing the temperature observations at different depth in the nearby reservoirs with similar conditions and the simulated results.

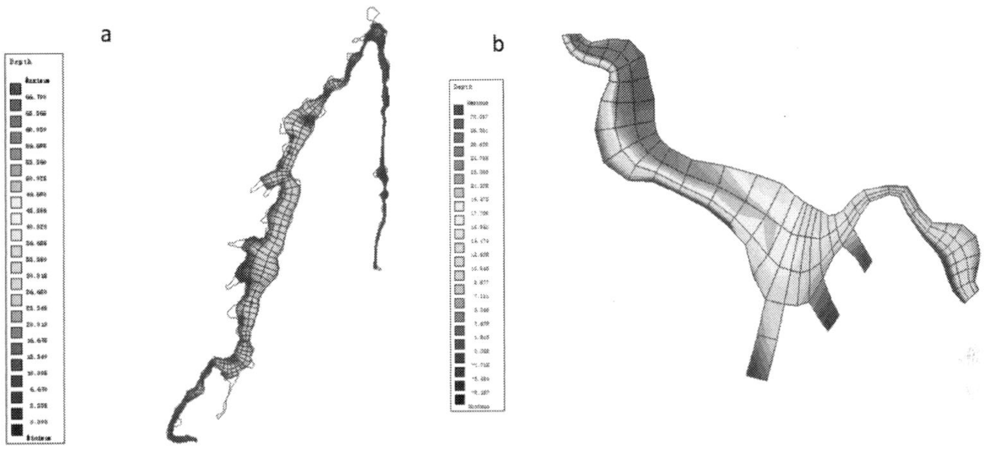

Figure 1. The computation grid for Guanjingkou (a) and Fengdou (b) reservoir.

2.4. Simulation scenarios

The original proposed water diversion scheme is single gate at two reservoirs, which were simulated at the bottom of the reservoir in the model. The proposed measures for water temperature are the stoplogs gates for the Guanjingkou and the three multiple gates for Fengdou reservoirs, respectively. Thus, the effects of stoplogs gates and the three multiple gates are simulated as that the water is released from the available upmost gate according to the predicted water level in the reservoir. The scenarios can be finalized as the water released from single gate and multiple gates during the normal, wet, and dry (special dry) year.

3. Results and discussion

The simulated water temperature profiles in front of the dam at the two reservoirs under normal, wet, and dry conditions are illustrated in figure 2. The comparison of water temperature with non-measures (single gate) and measures (multiple gates) at two reservoirs is shown in figure 3.

According to the simulation results, the water temperature profile in front of the Guanjingkou reservoir can be separated into three layers. The maximum depth of upmost epilimnion layer is 20 meter. The depth of thermocline layer is between 20 and 35 meter. The constant tepmerature layer is below 35 meter. The minimum water temperatures released from single gate are 9.7℃, 9.7℃, and 9.5℃ in January under wet, normal, and dry year. The water temperature can be increased by releasing the water from upmost gate in stoplogs gates system. During the main irrigation season from May to August, the water temperature could be increased 5.7℃, 6.8℃, 9.6℃, and 5.5℃, respectively.

Since the proposed Fengdou reservoir is located in north China, where the water is frozen during winter. So only the water temperatures from spring to autumn were selected for analysis. Based on the simulation results, the water temperature profile in front of the Fengdou reservoir can be separated into three layers. The maximum depth of upmost epilimnion layer is 15 meter. The depth of thermocline layer is between 15 and 25 meter. The constant temperature layer is below 25 meter. The minimum water temperatures released from single gate are 2.5℃, 2.3℃, and 2.3℃ in April in wet, normal, and dry year. The water temperature can be increased by releasing the water from upmost gate in three gates system. During the main irrigation seasons from May to August, the water temperature could be increased by 7.7℃, 1.9℃, 9.5℃, and 10.1℃, respectively.

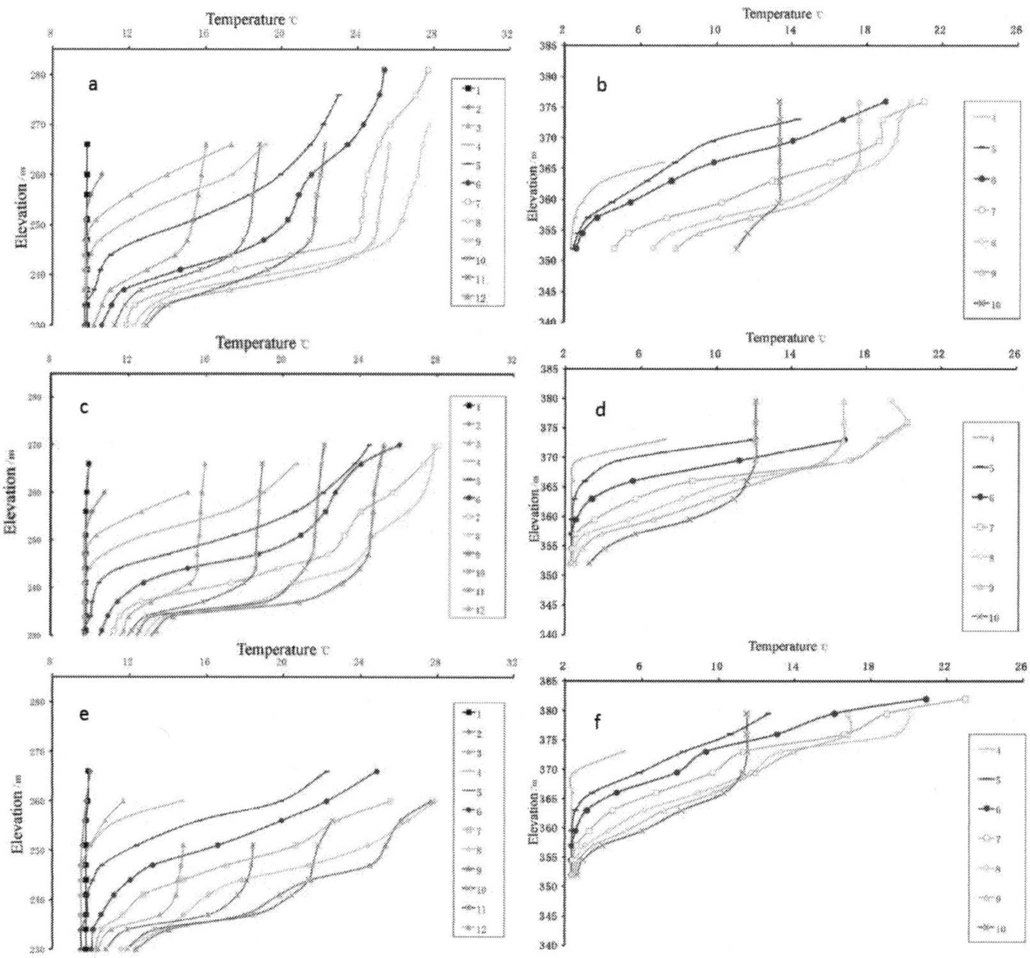

Figure 2. The simulated water temperature profiles in front of dam at Guanjingkou reservoir under normal (a), wet(c), and dry (e) condition and in front of dam at Fengdou reservoir under normal (b), wet (d), and dry (f) condition. The numbers are the different month.

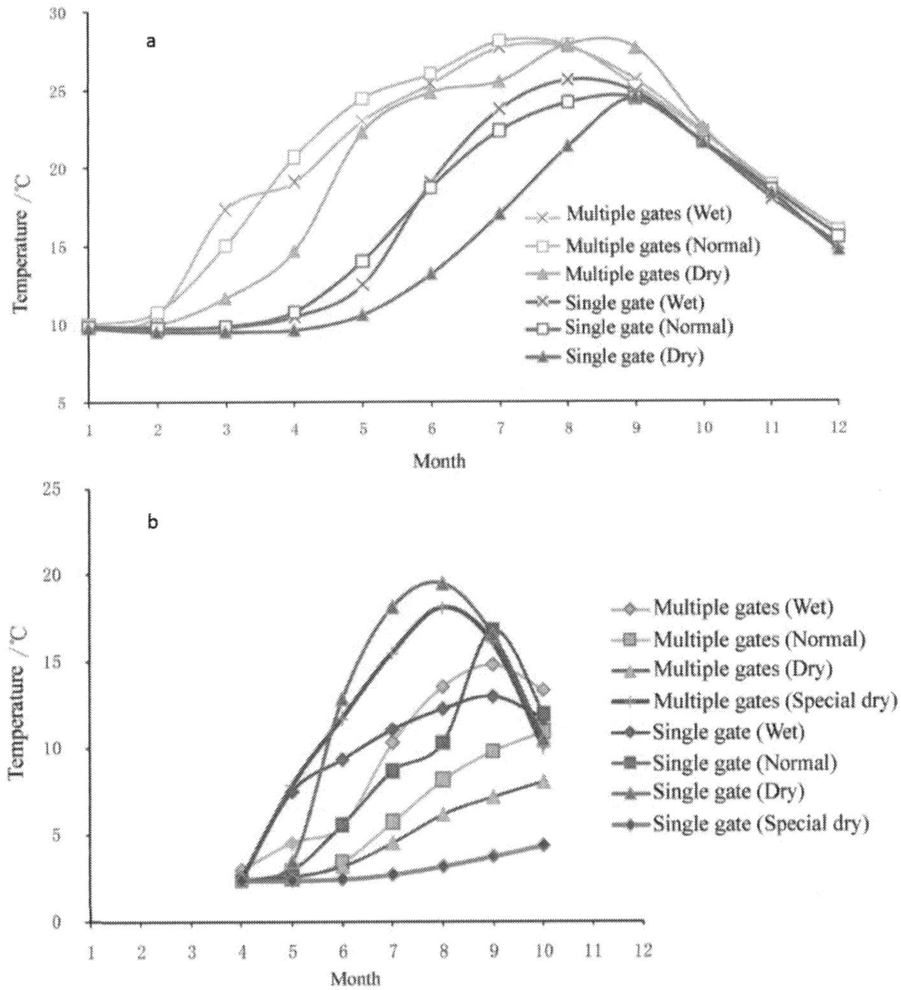

Figure 3. The comparison of simulated water temperature released from single gate and multiple gates from Guanjingkou reservoir (a) and Fengdou reservoir (b).

4. Conclusion

The numerical solution with Delft3d package can be the optimum or advanced method for reservoir water temperature prediction for the capability in simulating reservoir operation. Compared to those statistical experimental formula and the 2 dimensional simulations, one of the most suitable reasons for three dimensional simulations is that the variation of water levels can be included. The other advantage is that the horizontal convection can be simulated, which is crucial for the calculation of water temperature for the reservoir operation.

The downstream water temperature would be significantly decreased by the construction of the reservoir especially in the single gate scheme at the two reservoirs. The downstream aquatic ecosystem would eventually be affected by the original design of the two reservoirs. The revised solution of stoplogs gates and multiple gates at two reservoirs can improve the water temperature released from the two reservoirs. Thus, the negative effects of the water with lower temperature on the downstream aquatic ecosystem can be alleviated.

Acknowledge
This work was supported by the Major Science and Technology Program for Water Pollution Control and Treatment of the Key State Science and Technology Project under Grant No. 2014ZX07204-006-05. The editor and two reviewers are sincerely acknowledged for their critical comments on improving the paper.

References

[1] Zhang L and Sun D 2009 Impacts and count measures of water temperature released from high dam *Design of Hydropower* **25**(1) 76-8
[2] Ferguson R G 1958 The preferred temperature of fish and their midsummer distribution in temperate lakes and streams *Journal of the Fisheries Research Board of Canada* **15**(4) 607-24
[3] Jonsson N 1991 Influence of water flow, water temperature and light on fish migration in rivers *Nordic Journal of Freshwater Research*
[4] Huang L and Cheng H 2004 Analysis of water temperature changes impacts on aquatic ecosystem at Guanyintang reservoir *Journal of Hunan Environment Biological Polytechnic* **10**(4) 316-9
[5] Lou Y, Yang Z and Shui Y 2011 Analysis of water temperature structure and the impacts in Bailianya reservoir in Anhui province *Harnessing the Huaihe River* (12) 101
[6] Bao Q and Qiao G 2011 Impacts of reservoir water temperature stratification on agricultural irrigation *South-to-North Water Diversion and Water Science & Technology* **9**(2) 69-72
[7] Zhang D 1984 Reservoir Water Temperature analysis and estimation *Hydrology* (1) 19-27
[8] Zhu B 1985 Estimation of reservoir water temperature *Journal of Hydraulic Engineering* (2) 12-21
[9] Yue Y and Zhao Z 1997 Statistical analysis of water temperature in front of reservour dam *Water Resources and Hydropower Engeering* (3) 2-7